SpringerWienNewYork

Acta Neurochirurgica
Supplements

Editor: H.-J. Steiger

Intracranial Pressure
and Brain Monitoring XII

Edited by
W.S. Poon, C.J.J. Avezaat, M.T.V. Chan, M. Czosnyka,
K.Y.C. Goh, P.J.A. Hutchinson, Y. Katayama, J.M.K. Lam,
A. Marmarou, S.C.P. Ng, and J.D. Pickard (eds.)

Acta Neurochirurgica
Supplement 95

SpringerWienNewYork

Wai S. Poon
Matthew T. V. Chan
Keith Y. C. Goh
Joseph M. K. Lam
Stephanie C. P. Ng
The Chinese University of Hong Kong, Prince of Wales Hospital, Hong Kong, China

Anthony Marmarou
Medical College of Virginia Commonwealth University, Virginia, USA

Cees J. J. Avezaat
Erasmus University Rotterdam, Academisch Ziekenhuis Dijkzigt, Rotterdam, The Netherlands

John D. Pickard
Marek Czosnyka
Peter J. A. Hutchinson
University of Cambridge, Addenbrooke's Hospital, Cambridge, UK

Yoichi Katayama
Nihon University, School of Medicine, Tokyo, Japan

Printed in Austria
SpringerWienNewYork is a part of Springer Science+Business Media
springeronline.com

Printed on acid-free and chlorine-free bleached paper

With partly coloured Figures

ISSN 0065-1419

ISBN-13 978-3-211-99878-6 e-ISBN-13 978-3-211-32318-2

Preface

It is our greatest pleasure in editing this traditional monograph following the successful 12th International Symposium on Intracranial Pressure (ICP) and Brain Monitoring (16th–21st August, 2004), a 30 year-tradition since the 1972 Hannover Symposium. Despite the one year delay by the SARS epidemic in 2003, this Hong Kong International Symposium had attracted 240 delegates from 30 countries: in addition to the original neurosurgeon, clinical physicist and intensivist supporters from Europe, USA and Japan, we began to see quality original research presentations from Asia, notably Mainland China, Taiwan, Malaysia and Singapore. This monograph on ICP and Brain Monitoring is consisted of 88 short papers selected from the 115 oral and 88 oral-poster presentations after a peer-review process by the International Advisory Board members. I am grateful to all the editors (Cees Avezaat, Matthew Chan, Marek Czosnyka, Keith Goh, Peter Hutchinson, Yoichi Katayama, Joseph Lam, Anthony Marmarou, Stephanie Ng and John Pickard) for their untiring efforts in reviewing and proof-reading these manuscripts for timely publication in 2005.

The theme of this Hong Kong Symposium is to bring research findings into clinical practice. Anthony Marmarou's keynote lecture at the conclusion of the symposium underlined the importance of translational research of brain injury. A significant number of quality presentations have illustrated their scientific value and clinical relevance. The organization of these papers into nine sections in the sequence of their original presentation is kept as follows: ICP management in head injury, neurochemical monitoring and intracranial hypertension, neuroimaging, hydrocephalus, clinical trials, brain compliance, biophysics and experimental aspects. We have introduced cash prizes for three best oral and one best oral-poster presentations: Noam Alperin (on the importance of extracranial venous flow in idiopathic intracranial hypertension), Guohua Xi (on brain oedema and neurological deficits induced by thrombin), Mark O'Connell (on glucose metabolism in traumatic brain injury, a study combining microdialysis and PET) and Peter Smielewski (on ICP pattern after infusion study in hydrocephalic patients). Seven review papers of the Satellite Symposium on Neurochemical Monitoring occupy the final section of this monograph.

I hope the timely publication of this monograph will expedite dissemination of research findings, ask more questions and engage in more basic and clinical research. We look forward to the 13th International Symposium on ICP and Brain Monitoring in San Francisco in the year 2007.

Wai S. Poon
On behalf of the editors

Contents

Neuroimaging

Clinical trials

Brain compliance

Biophysics

Experimental aspects

Satellite symposium on neurochemical monitoring

List in Current Contents

Keynote lecture

Acta Neurochir (2005) [Suppl] 95: 3–5

The importance of translational research in brain injury

A. Marmarou

Division of Neurosurgery, Medical College of Virginia Commonwealth University, Richmond, Virginia, USA

Summary

Progress in the understanding of the pathophysiologic process and management of brain injury provides a unique challenge to the young investigator. Currently, there are teams both in the laboratory and clinical settings which may or may not interact as a result of the depth and breadth of the problems before them. For example, laboratory teams may be focused on cellular or molecular mechanisms of injury while the clinical investigator is attempting to deal with refractory ICP and it's treatment. It is clear that many aspects of biological research must indeed be approached using molecular techniques of neuroscience and much has been learned. However, the translational aspect of this effort may be elusive as it is unclear at present how these molecular techniques translate into the intensive care unit and the patient. For many, the research paths taken in the laboratory and ICU will always remain independent. For those who desire a more direct pathway to treatment, the "bench to bedside" approach is one which is more likely to yield results. Several examples will be given to strengthen the notion that we must not abandon the translational aspect of research but to embrace the process if we are to further improve outcome from brain injury.

Keywords: Brain injury; translational research; bench to bedside studies.

Introduction

I have been asked by Dr. Poon to lecture on the topic of the importance of translational research. This certainly is a broad assignment and it is hardly necessary to emphasize this topic to the investigators who attend the ICP symposiums. Most if not all are clinical scientists who merge data from laboratory and clinical settings in order to better understand the pathophysiology and develop better diagnostic and management techniques. In fact, this is perhaps one of the most important aspects of the ICP symposiums in that the "Bench to Bedside Investigator" is welcome and clinical data can be presented to an audience who understands both laboratory and clinical aspects of the problem in question. I thought that the best way to illustrate this interaction is to provide an example from own work which required both clinical and laboratory interaction in order to arrive at a sound scientific conclusion. It concerns the problem of brain swelling and how we utilized data from the laboratory to design experiments which we carried out in the clinic.

Methods

Many workers in the field of intracranial pressure were focused on the issue of brain swelling and it was the dictum at the time that brain swelling and subsequent rise in ICP was due to vascular engorgement. This notion was developed and promoted by several investigators among them Thomas Langfitt of the University of Pennsylvania and Professor Ishii of Japan [1]. Marshall, a neurosurgical resident in Langfitt' staff also contributed to the laboratory findings. These workers studied so called "vascular engorgement" in the laboratory and concluded this was the cause of pressure rise [2–4]. The engorgement was thought to be caused by a "vasoparalysis" and loss of autoregulation associated with the mechanical injury. So strong was the concept of vasoparalysis and vascular engorgement that all the ICP workers readily accepted this notion for three decades. At the same time, the newly developed cryogenic lesion model simulating a cortical contusion with progressive edema migration was the model of choice for studying not only traumatic brain edema but also the accompanying rise in ICP.

Cold injury studies in the laboratory setting

Workers, including our laboratory, utilized this cryogenic lesion model and attempted to understand the interaction between ICP and the increase in brain tissue water. We found the following.

a. The model did produce sizeable amounts of edema and ICP rise
b. The most significant finding was the dramatic compromise of the blood brain barrier which persisted for several days.

We concluded from these observations that the brain could not resolve the increase in tissue water, vessels were compressed and ICP increased. Our efforts were shifted to better understand how the brain absorbs the excess fluid. To simplify the experimental model, an infusion model of edema was developed which deposited a known amount of fluid within the brain parenchyma at a known rate via a needle positioned in the white matter of the internal capsule [5]. Using this model for several years we noted that

a. There appeared to be no increased resistance to the resolution of edema in injured animals
b. The edema resolved spherically and was not restricted to ventricular passage.
c. The resolution of edema was not different between injured animals and controls.

Given that the brain did not seem to be affected the increased extracellular volume and that the resolution proceeded equally in injured and controls, we now questioned wether or not the increased edema volume secondary to barrier compromise was truly responsible for the ICP rise in patients. We continued with our laboratory experiments and developed a model where we could measure both water and blood volume in the rat [6]. We realized that we had to be able to measure both edema and blood volume in injured animals, first to confirm the engorgement concept and secondly to observe if indeed the edema was responsible for the ICP rise observed in animals. We confirmed that

a. Blood volume was increased but was reduced following trauma
b. The edema was responsible for the observed brain swelling and ICP rise.

This completely altered our thinking and although autoregulation could have been effected in these experiments it was clear that edema was the culprit and responsible for brain swelling. But how was this possible, particularly in diffuse injury?

Studies in the intensive care unit

It was clear that we had to measure blood volume and edema content in injured Man with both diffuse and focal injury to determine if indeed blood volume was reduced and edema was increased and secondly, we had to determine if the BBB was altered. In the clinical setting, we decided to utilize stable Xenon for measurement of blood volume by integrating the product of CBF and transit time. We also developed our laboratory technique of gravimetric water content measurement and utilized this method in developing a means of measuring brain water in absolute terms, non-invasively utilizing magnetic resonance techniques and confirmed that the values corresponded to values obrained from gravimetric samples of edematous injured brain tissue harvested from patients [7]. Utilizing the combination of these methods, blood volume and water measurement, we studied head injured patients and found that blood volume was reduced following traumatic brain injury and brain edema was responsible for brain swelling [8].

Back to the laboratory

At a subsequent ICP congress, I recall vividly a comment made by Graham Teasdale who noted that in studies of patients, there was not an obvious barrier leakage as indicated by Gadolinium. Professor Teasdale did comment that only in contusion was there a slight "halo" surrounding the area of necrosis. This suggested to us that the BBB must close quickly following TBI. In the laboratory, utilizing the MR methodology, we studied the dynamics of the BBB permeability following TBI [9]. We observed that the BBB in injured animals was compromised soon after injury but was closed within 1 hour following trauma. Similarly, in studies of patients, we could not demonstrate any leakage to contrast in studies of head injured patients thereby confirming the observations by Professor Teasdale.

In summary

At this point we knew that the swelling was due to edema and somehow the brain continued to swell despite an early closure to the blood brain barrier. Could it be that the edema was not vasogenic but cellular in origin? We then embarked on studies utizing MR diffusion weighted imaging to measure the apparent diffusion coefficient for typing the type of edema that was present following experimental TBI [10]. We learned from these studies that the initial form of edema developing soon after injury was vasogenic but within twelve hours the predominant form of edema was cellular. Since that time we have continued to study edema type in Man utilizing the diffusion weighted imaging and and the results are similar. Current efforts are now focused on determining the cause of the cellular swelling and the pathways by which water enters the brain.

Conclusion

This short report is one example of "translational research" which has led to a better understanding of the pathophysioloy of brain swelling. Similar studies by other investigators are now taking place and the "bench to bedside" approach is firmly embedded in to those participants attending the ICP congress. It should continue.

Acknowledgments

A note of gratitude to all of the fellows from all parts of the globe who were responsible for these results.

References

1. Ishii S *et al* (1968) Experimental study on cerebral swelling: changes in the DNP-activated ATPase of mitochondria. Trans Am Neurol Assoc 93: 43–46
2. Langfitt TW, Weinstein JD, Kassell NF (1964) Cerebral vasomotor paralysis as a cause of brain swelling. Trans Am Neurol Assoc 89: 214–215
3. Langfitt TW, Tannanbaum HM, Kassell NF (1966) The etiology of acute brain swelling following experimental head injury. J Neurosurg 24(1): 47–56

4. Marshall WJ, Weinstein JD, Langfitt TW (1968) The pathophysiology of brain swelling produced by mechanical trauma and hypertension. Surg Forum 19: 431–432
5. Takagi H *et al* (1983) [Study of brain edema by an infusion edema Model – the method and characteristics of the model]. No Shinkei Geka 11(9): 957–964
6. Kita H, Marmarou A (1994) The cause of acute brain swelling after the closed head injury in rats. Acta Neurochir [Suppl] 60: 452–455
7. Fatouros PP, Marmarou A (1999) Use of magnetic resonance imaging for in vivo measurements of water content in human brain: method and normal values. J Neurosurg 90(1): 109–115
8. Marmarou A *et al* (2000) Contribution of edema and cerebral blood volume to traumatic brain swelling in head-injured patients. J Neurosurg 93(2): 183–193
9. Barzo P *et al* (1997) Acute blood-brain barrier changes in experimental closed head injury as measured by MRI and Gd-DTPA. Acta Neurochir [Suppl] 70: 243–246
10. Barzo P *et al* (1997) Contribution of vasogenic and cellular edema to traumatic brain swelling measured by diffusion-weighted imaging. J Neurosurg 87(6): 900–907

Correspondence: Anthony Marmarou, Department of Neurosurgery, Virginia Commonwealth University Medical Center, 1001 East Broad Street, Suite 235, Richmond, VA, USA 23219. e-mail: amarmaro@vcu.edu

ICP management in head injury

Acta Neurochir (2005) [Suppl] 95: 9–11

Cerebral blood flow (CBF)-directed management of ventilated head-injured patients

W. S. Poon[1], **S. C. P. Ng**[1], **M. T. V. Chan**[2], **J. M. K. Lam**[1], and **W. W. M. Lam**[3]

[1] Division of Neurosurgery, Department of Surgery, Prince of Wales Hospital, The Chinese University of Hong Kong, Hong Kong
[2] Prince of Wales Hospital, The Chinese University of Hong Kong, Hong Kong
[3] Department of Diagnostic Radiology & Organ Imaging, Prince of Wales Hospital, The Chinese University of Hong Kong, Hong Kong

Summary

Objective. Ischaemic brain damage has been shown to be an important contributing factor causing head injury fatality. Maintenance of an adequate cerebral perfusion pressure is difficult in patients with elevated intracranial pressure (ICP) and deranged cerebral vasoreactivity. Thirty-five cases of ventilated moderate-to-severe head-injured patients were prospectively studied, correlating their cerebral haemodynamic abnormalities, neurochemical disturbances (using microdialysis methodology) and clinical outcome.

Methods. Cerebral haemodynamic abnormalities were defined and classified by transcranial Doppler ultrasonography (TCD) and stable xenon-CT cerebral blood flow measurements (XeCT) into their status of CO_2 reactivity, pressure autoregulation, hyperaemia or non-hyperaemia. Two-hour episodes of these abnormalities with and without haemodynamic intervention were followed in their changes in ICP, CPP, intracerebral metabolites and finally their clinical outcome.

Results. Loss of CO_2 reactivity was associated with a significantly higher ICP, increasing intracerebral metabolites (lactate, glutamate and glycerol) and invariably a fatal outcome. Impaired pressure autoregulation was also associated with an elevated ICP, but no significant difference in intracerebral metabolites and incidence of favourable clinical outcome. Patients with intact CO_2 reactivity and impaired pressure autoregulation were treated with an elevated CPP in 32 episodes, resulting in a significant reduction in ICP, intracerebral glutamate and glycerol and non-survival. In patients with intact CO_2 reactivity and impaired pressure autoregulation, eleven episodes of hyperaemia were identified by XeCT. A modest 20% blood pressure reduction resulted in a trend towards a reduction of ICP, intracerebral glutamate and glycerol and non-survival.

Conclusions. The need for haemodynamic intervention in this group of ventilated patients with moderate-to-severe head inury can be made logical when these abnormalities are identified daily. The success of management was reflected by a stable or improved ICP, CPP, intracerebral metabolic deranagement and survival.

Keywords: Head injury; intracranial pressure; cerebral perfusion pressure; microdialysis.

Introduction

Ischaemic brain damage is consistently found in the majority of postmortem series of head injuries [2, 3]. Maintenance of adequate cerebral perfusion pressure (CPP) has been strongly recommended as "targeted therapy of CPP $>$ 70 mmHg" and is the key to achieving favourable clinical outcome [9, 10]. However, in patients with intractable intracranial hypertension and deranged cerebrovascular reactivity, an adequate CPP is impossible to achieve, resulting in death.[7] A recent well-conducted randomised trial involving 189 patients did not show an outcome benefit in CPP-targeted therapy of $>$70 mmHg [1, 8].

Traumatic brain injury is a heterogeneous group of patients and therefore both the ICP and CPP-targeted therapy have their limitations. In this prospective observational study, we have focused on daily assessment of cerebral haemodynamic status and its appropriate intervention. We hypothesise that early intervention of cerebral haemodynamic abnormalities may be associated with a favourable clinical outcome.

Patients and methods

Ethics Committee approval from the Chinese University of Hong Kong and informed consent from the family of these ventilated head-injured patients were obtained. Daily CO_2 reactivity (cerebral vasoreactivity, CVR) and pressure autoregulation (PAR) were documented,[5] using transcranial Doppler ultrasonography for the measurement of blood flow velocity (BFV) in each middle cerebral artery. A preservation of CO_2 reactivity or CVR is defined as $>$1% change in BFV per unit change (mmHg) in end-tidal CO_2. A preservation of PAR is defined as a $<$1.5% change in BFV per unit change (mmHg) in mean arterial blood pressure. Induced hypertension of $>$20% was achieved by intravenous infusion of phenylephrine.

Parameters that were continuous monitored included heart rate, arterial blood pressure of the radial artery, ICP by an intraventricular catheter, arterial oxygen saturation by a pulse oximeter, end-tidal CO_2 concentration by capnography, jugular bulb venous oxygen saturation by Oximetirx 3 System, and hourly monitoring of intracerebral extracellular metabolites (glucose, lactate, glycerol, glutamate and pyruvate) from a intracerebral microdialysis catheter (CMA 70, CMA, Stockholm, Sweden) placed in frontal lobe 3 cm anterior to the precoronal burr hole [7]. Cerebral blood flow (CBF) measurements [4, 6] by stable xenon-enhanced computed tomography (XeCT) (Xe/CT System 2, Diversified Diagnostic Products, Houston, USA) were carried out on day 1, 3 and 5 following the head injury. Patient management followed a standard protocol: (1) fully sedated for normal ventilation to a pCO_2 of 35–40 mmHg; muscle relaxant will be administered for patients with ICP > 20 mmHg; (2) ICP > 20 mmHg will be treated by intermittent CSF drainage ± bolus mannitol infusion (0.5 G/kg at 4–6 hour intervals); (3) CPP ≥ 70 mmHg unless hyperemia (CBF > 60 ml/100 g/min) is documented; (4) thiopentone infusion is considered when all medical and surgical options have exhausted.

Clinical outcome was assessed at six month after injury using Glasgow Outcome Scale. Data were analyzed by analysis of variance and t-test using SPSS. All tests used a 0.05 level of significance.

Results

This group of ventilated head-injured patients had a management mortality of 40%. The majority of them were young males (30/35, mean age 39 ± 19 years) with an admission Glasgow Coma Scale (GCS) score of 3–12 (median 8) and a focal to diffuse brain injury ratio of 25 to 10.

Patients with ICP > 20 mmHg carried a 50% mortality whereas patients with ICP ≤ 20 mmHg had a zero mortality. Patients with impaired CO_2 reactivity were associated with an elevated ICP, intracerebral lactate, glutamate and glycerol. Patients with bilateral loss of CO_2 reacitvity, intractable intracranial hypertension and inadequate CPP did not benefit from induced hypertension and carried a 100% mortality. Patients with intact CO_2 reactivity and impaired PAR had a significantly elevated ICP compared with those with intact PAR (20 ± 3 *versus* 9 ± 4 mmHg, $p = 0.05$). This group of patients benefited from induced hypertension, resulting in an improved CPP and a significant reduction in ICP, intracerebral glutamate and glycerol. For patients with intractable intracranial hypertension (ICP > 20 mmHg) and hyperaemia (as defined by XeCT CBF > 60 ml/100 g/min) with intact CO_2 reactivity, a therapeutic reduction in blood pressure resulted in a trend towards a reduction in ICP, intracerebral glutamate and glycerol.

Discussion and conclusions

This prospective observational study attempted to classify moderate-to-severe head injury according to their cerebral haemodynamic abnormalities, correlating intracerebral metabolic derangements and six-month clinical outcome. The widely held belief that CPP ≥ 70 mmHg did not pass a well conducted randomised trial [8], comparing CPP ≥ 70 mmHg with CPP of 50 mmHg. In our group of 35 patients, seven were shown to have an ICP ≤ 20 mmHg and intact vaso-reactivity: these patients did not require any treatment other than normal blood pressure and oxygenation to achieve favourable outcome in the majority (5 out of 7). The other extreme were another seven patients with intractable intracranial hypertension and bilateral loss of vaso-reactivity: a CPP of 70 mmHg could not be achieved despite maximum ICP therapies (induced hypertension and thiopentone infusion) and these patients were invariably fatal. In this small prospective study of 35 ventilated head-injured patients, we have identified two groups of patients with elevated ICP (>20 mmHg), intact CO_2 reactivity and impaired pressure autoregulation (PAP): the non-hyperaemic group benefits from induced hypertension to a CPP > 70 mmHg, whereas the hyperaemic group (as defined by XeCT CBF > 60 ml/100 g/min) may benefit from a lower level of CPP. The non-hyperemic group of patients that benefited from induced hypertension reacted to an elevated CPP of >70 mmHg by a significant reduction of ICP: this suggested that at this elevated level of CPP, the pressure autoregulation was preserved. A significant reduction of intracerebral glutamate and glycerol (the two-hour average after CPP treatment) suggested that appropriate treatment had been delivered. For the hyperemic group of patients, a reduction of CPP to <70 mmHg resulted in stabilisation of ICP and a trend towards an improvement of intracerebral glycerol.

Acknowledgment

This study was supported by a grant earmarked by the Research Grant Council, Hong Kong.

References

1. Goodman JC, Valadka AB, Gopinath SP, Uzura M and Robertson CS (1999) Extracellular lactate and glucose alterations in the brain after head injury measured by microdialysis. Crit Care Med 27: 1965–1973
2. Graham DI, Adams JH, Doyle D (1978) Ischaemic brain damage in fatal non-missile head inuries. J Neurol Sci 39: 213–234
3. Graham DI, Ford I, Adams JH, Doyle D, Teasdale GM, Lawrence AF, McLellan DR (1989) Ischaemic brain damage is still common in fatal non-missile head injury. J Neurol Neruosurg Psychiatry 52: 346–350
4. Marion DW, Bouma GJ (1991) The use of stable xenon-enhanced computed tomographic studies of cerebral blood flow

to define changes in cerebral carbon dioxide vasoresponsivity caused by a severe head injury. Neurosurgery 29: 869–873
5. Ng SC, Poon WS, Chan MT, Lam JM, Lam W, Metreweli C (2000) Transcranial Doppler ultrasonography (TCD) in ventilated head-injured patients: correlation with stable xenon-enhanced CT. Acta Neurochir [Suppl] 76: 479–482
6. Ng SC, Poon WS, Chan MT, Lam JM, Lam WW (2002) Is Transcranial Doppler ultrasonography (TCD) good enough in determining CO_2 reactivity and pressure autoregulation in head-injured patients? Acta Neurochir [Suppl] 81: 125–127
7. Poon WS, Ng SC, Chan MT, Leung CH, Lam JM (2002) Neurochemical changes in ventilated head-injured patients with cerebral perfusion pressure treatment failure. Acta Neurochir [Suppl] 81: 335–338
8. Robertson CS, Valadka AB, Hanny HJ, Contant CF, Gopinath SP, Cormio M, Uzura M, Grossman RG (1999) Prevention of secondary ischemic insults after severe head injury. Crit Care Med 27: 2086–2095
9. Rosner MJ, Daughton S (1990) Cerebral perfusion pressure management in head injury. J Trauma 30: 933–941
10. Rosner MJ, Rosner SD, Johnson AH (1995) Cerebral perfusion pressure: mangement protocol and clinical results. J Neurosurg 83: 949–962

Correspondence: W. S. Poon, Division of Neurosurgery, Prince of Wales Hospital, Shatin, Hong Kong. e-mail: wpoon@surgery.cuhk.edu.hk

Acta Neurochir (2005) [Suppl] 95: 13–16

Relationship of cerebral perfusion pressure levels to outcome in traumatic brain injury

C. J. Kirkness[1], **R. L. Burr**[1], **K. C. Cain**[2], **D. W. Newell**[3], and **P. H. Mitchell**[1]

[1] Biobehavioral Nursing and Health Systems, University of Washington, Washington, USA
[2] Office for Nursing Research, University of Washington, Washington, USA
[3] Neurological Surgery, University of Washington, Washington, USA

Summary

This study examined the relationship of cumulative percent time that cerebral perfusion pressure (CPP) fell below set thresholds to outcome in individuals with traumatic brain injury (TBI). The sample included 157 patients (16 to 89 years of age, 79% male) admitted to an intensive care unit at an academic medical center who underwent invasive arterial blood pressure and intracranial pressure monitoring. CPP levels were recorded continuously during the first 96 hours of monitoring. Initial neurologic status was assessed using the post-resuscitation Glasgow Coma Scale. Outcome was evaluated at hospital discharge and at six months post-injury using the Extended Glasgow Outcome Scale (GOSE). The relationship of cumulative periods of low CPP to outcome was evaluated using hierarchical and binary logistic regression analysis, controlling for age, gender, and injury severity.

Patients experiencing less cumulative percent time below specific CPP thresholds were more likely to have better outcome at discharge (55 mm Hg, $p = .004$; 60 mm Hg, $p = .008$; 65 mm Hg, $p = .024$; 70 mm Hg, $p = .016$). Although differences in GOSE scores at six months were not significant, those with less time below CPP thresholds were more likely to survive. Accumulated episodes of low CPP had a stronger negative relationship with outcome in patients with more severe primary brain injury.

Keywords: Traumatic brain injury; cerebral perfusion pressure; outcome.

Introduction

Maintaining optimal cerebral perfusion in the acute period following traumatic brain injury (TBI) is a key goal of management directed at preventing secondary ischemic brain injury. Cerebral perfusion pressure (CPP), calculated as the mean arterial blood pressure (ABP) minus the mean intracranial pressure (ICP), is a clinically relevant parameter that reflects the driving force of blood and metabolic substrates to brain cells. Inadequate CPP is linked with unfavorable neurologic outcome but there has been ongoing debate regarding the optimal level at which to maintain CPP [6, 8–10]. Recent evidence has resulted in a revision of the Guidelines for the Management of Severe Traumatic Brain Injury in adults to recommend maintaining a minimum CPP of 60 mm Hg, down from 70 mm Hg in the previous Guidelines [4, 5].

The purpose of this study was to examine the relationship of the cumulative percent time that CPP fell below fixed threshold levels to outcome at hospital discharge and at 6 months post-injury in a TBI sample.

Materials and methods

The study was approved by the local Institutional Review Board. Data were obtained from all subjects enrolled in a randomized controlled clinical trial examining the impact of a highly visible display of CPP on outcome in patients with TBI. Inclusion criteria were patients 16 years of age or older with a TBI admitted to an intensive care unit at an academic medical center who underwent invasive ABP and ICP monitoring. Exclusion criteria included bilateral fixed pupils and impending death. Medical management of patients with TBI at the institution where the study was carried out includes aggressive treatment of increased ICP with sedation and mannitol, and use of fluid volume and pharmacologic agents to maintain CPP. Demographic, diagnostic, and treatment data were recorded. Initial neurologic and injury severity status were assessed using the post-resuscitation Glasgow Coma Scale (GCS-PR) and the Injury Severity Scale. CPP levels were recorded during the first 96 hours of monitoring. ABP and ICP analog signals from the bedside monitoring system were input to a study computer through an analog-to-digital converter and saved on the computer in five-second summaries. CPP was calculated as mean ABP minus mean ICP. The primary outcome measure was the Extended Glasgow Outcome Scale (GOSE), assessed at hospital discharge and at 6 months post-injury.

The relationship of cumulative periods of low CPP to outcome was evaluated using hierarchical regression and binary logistic regression analyses, controlling for age, gender, and primary injury severity.

Table 1. *Demographic, clinical, and physiologic data*

Age, years, mean (SD)	37.1 (18.1)
Sex, male/female, %	79/21
GCS, post-resuscitation, mean (SD)	7.3 (2.8)
Injury Severity Scale score, mean (SD)	29.1 (9.6)
GCS, discharge (survivors), mean (SD)	13.0 (2.0)
CPP over first four days, mm Hg, mean (SD)	79 (12)
ABP over first four days, mm Hg, mean (SD)	95 (11)
ICP over first four days, mm Hg, mean (SD)	16 (7)
Percent time CPP less than 55 mm Hg	5
Percent time CPP less than 60 mm Hg	10
Percent time CPP less than 65 mm Hg	18
Percent time CPP less than 70 mm Hg	29
Mortality at discharge, N (%)	21 (13.4%)
GOSE score at discharge, mean (SD)	2.7 (.8)
Percent GOSE > 4 at discharge	0%
Mortality at six months, N (%)	27 (17.2%)
GOSE score at six months, mean (SD)	4.3 (2.1)
Percent GOSE > 4 at six months	56.7%

Results

A total of 157 patients were enrolled. The sample was primarily younger males (Table 1). The most common cause of TBI in this sample was motor vehicle crashes (46%), followed by falls (21%), and motorcycle crashes (12%). Post-resuscitation GCS scores (Table 1) reflected moderate to severe TBI, with 73% having a post-resuscitation GCS score of eight or less. The mean (SD) Injury Severity Scale score was 29.1 (9.6). Initial computed tomography findings included contusion (68%), subdural hematoma (45%), epidural hematoma (26%), traumatic subarachnoid hemorrhage (61%), intraventricular hemorrhage (28%), and shear injury (33%). Hospital mortality for this sample was 13%. Almost one half of hospital survivors were discharged to an acute inpatient rehabilitation unit, one quarter went to skilled nursing facilities, and 10% were discharged home. The mean GOSE score at discharge was 2.7 and increased to 4.3 at six months post-injury (Table 1). The mean levels of CPP, ABP, and ICP over the first four days of monitoring are presented in Table 1. On average, CPP levels were kept well above the currently recommended minimum of 60 mm Hg. The percent time that CPP was below the threshold levels over the first four days of monitoring ranged from 5% for the 55 mm Hg threshold to 29% for the 70 mm Hg threshold (Table 1).

The impact on GOSE scores of the percent time that CPP was below the set threshold levels was modeled using multivariate analysis, controlling for age, gender, and initial injury severity (Table 2). Patients experiencing less cumulative percent time below fixed CPP thresholds ranging from 55 mm Hg up to and including 70 mm Hg were significantly more likely to have better outcomes at discharge as reflected by higher GOSE scores. Differences in outcome based on percent time below any of the CPP thresholds, however, were not significant at six months post-injury when considering the full GOSE score as the outcome variable.

To assess the effect of time below the set CPP threshold levels on survival and on GOSE, dichotomized to greater than 4 versus 4 or less, binary logistic regression analyses were carried out for each CPP threshold level, again controlling for age, gender, and initial injury severity (Tables 3 and 4). In these models, for all CPP thresholds, greater percent time below the threshold was significantly associated with higher mortality, both at hospital discharge and at six months post-

Table 2. *Effect of percent time CPP below thresholds on extended Glasgow Outcome Scale Score*

CPP threshold	GOSE: discharge		GOSE: six months	
	Beta	p value	Beta	p value
55 mm Hg	−.185	.004	−.072	.228
60 mm Hg	−.170	.008	−.065	.277
65 mm Hg	−.145	.024	−.055	.359
70 mm Hg	−.156	.016	−.046	.447

Table 3. *Odds ratios for survival predicted by the percent time CPP below thresholds*

CPP threshold	Discharge		Six months	
	Odds ratio	p value	Odds ratio	p value
55 mm Hg	.080	.001	.097	.001
60 mm Hg	.122	.003	.122	.001
65 mm Hg	.168	.008	.191	.007
70 mm Hg	.143	.005	.158	.004

Table 4. *Odds ratios for GOSE > 4 predicted by the percent time CPP below thresholds*

CPP threshold	GOSE > 4: six months	
	Odds ratio	p value
55 mm Hg	.614	.426
60 mm Hg	.455	.118
65 mm Hg	.593	.268
70 mm Hg	.607	.317

injury. However, greater percent time below CPP thresholds was not significantly associated with recovery to a GOSE category of 5 or greater.

To determine if cumulative excursions of CPP below the specific thresholds had a differential effect on survival depending on TBI severity, binary logistic regression analyses were carried out with the sample divided into subgroups of GCS-PR scores of seven or greater, or less than seven. In the subgroup with lower GCS-PR scores, the odds ratio for survival was .200 ($p = .006$) for greater time CPP below 60 mm Hg and .222 ($p = .010$) for greater time below the 70 mm Hg CPP threshold. However, for those with GCS-PR scores of greater than 7, the odds ratio for survival was .600 ($p = .684$) for greater time below 60 mm Hg and .943 ($p = .983$) for 70 mm Hg.

Discussion

The critical role of ischemia in outcome following TBI and evidence of low global cerebral blood flow early following TBI and further focal reductions surrounding cerebral contusions and subdural hematomas provide a basis for elevating CPP in an attempt to prevent or minimize secondary ischemic brain injury [2, 7]. Although inadequate CPP is associated with poorer outcome following TBI, there is ongoing debate as to the minimum level at which CPP should be maintained to prevent or minimize secondary brain injury. The Guidelines for the Management of Severe Traumatic Brain Injury, published in 1996 and revised in 2000, recommended maintaining CPP at a minimum of 70 mm Hg [3, 4]. In 2003, an update to the CPP Guidelines reduced the recommended minimum CPP level from 70 mm Hg to 60 mm Hg and recommended against the aggressive use of fluid and vasopressors to maintain CPP above 70 mm Hg if cerebral ischemia is not present due to the risk of ARDS [5]. Nonetheless there is still not definitive evidence as to what the adequate minimum CPP threshold is following TBI.

This study examined the association with outcome of percent time CPP was below a range of thresholds from 55 mm Hg to 70 mm Hg during the first four days post-TBI. Greater percent time below all thresholds was associated with significantly poorer GOSE scores at hospital discharge, particularly for those who had a GCS-PR score of less than seven. This statistically significant difference, however, was not maintained at six months, nor was it present when GOSE was dichotomized to levels of greater than 4 versus 4 or less. In relation to survival, greater periods of CPP below each threshold from 55 mm Hg to 70 mm Hg were significantly associated with higher mortality at hospital discharge and this effect was maintained at six months post-injury. Thus the greatest impact of low CPP was in relation to hospital mortality and in those with more severe TBI.

This study and others demonstrate an association between CPP and outcome, most strongly between survival at discharge and 6 months post-injury. The association of CPP and quality of survival is not as clear. It is unlikely that there is one CPP threshold that ensures optimum perfusion for all injured brains, or even for all parts of one injured brain. Although research shows that CPP can be elevated without causing significant elevations in ICP [1], use of fluid volume and vasopressor drugs to elevate CPP is not without risk, particularly of ARDS [9]. The challenge is thus to identify the subgroups with TBI who require an elevation of CPP to prevent secondary ischemic brain injury and the corresponding optimum CPP level, and to individualize CPP management accordingly as part of their overall management. Further study designed to specifically examine the impact of CPP management at different levels on maintaining cerebral oxygenation to the injured brain and on long-term outcome is needed.

Acknowledgment

This study was supported by National Institutes of Health, National Institute of Nursing Research, R01 NR04901.

References

1. Bouma GJ, Muizelaar JP (1990) Relationship between cardiac output and cerebral blood flow in patients with intact and impaired autoregulation. J Neurosurg 73: 368–374
2. Bouma GJ, Muizelaar JP, Stringer WA, Choi SC, Fatouros P, Young HF (1992) Ultra-early evaluation of regional cerebral blood flow in severely head-injured patients using xenon-enhanced computerized tomography. J Neurosurg 77: 360–368
3. Bullock R, Chesnut RM, Clifton G, Ghajar J, Marion DW, Narayan RK, Newell DW, Pitts LH, Rosner MJ, Wilberger JE (1996) Guidelines for cerebral perfusion pressure. J Neurotrauma 13: 693–697
4. Bullock RM, Chesnut RM, Clifton GL, Ghajar J, Marion DW, Narayan RK, Newell, DW, Pitts LH, Rosner MJ, Walters BC, Wilberger JE (2000) Guidelines for cerebral perfusion pressure. J Neurotrauma 17: 507–511
5. Bullock RM, Chesnut RM, Clifton GL, Ghajar J, Marion

DW, Narayan RK, Newell DW, Pitts LH, Rosner MJ, Walters BC, Wilberger JE (2003) Update to guidelines for cerebral perfusion pressure. http://www.2.braintrauma.org/guidelines/downloads/btf_guidelines_cpp_u1.pdf?BrainTrauma_Session=e7f0e5d9dbe85b0c5c56b0abb0ede86c Last accessed 08/11/2004
6. Juul N, Morris GF, Marshall SB, The Executive Committee of the International Selfotel Trial, Marshal LF (2000) Intracranial hypertension and cerebral perfusion pressure: influence on neurological deterioration and outcome in severe head injury. J Neurosurg 92: 1–6
7. Marion DW, Darby J, Yonas H (1991) Acute regional cerebral blood flow changes caused by severe head injuries. J Neurosurg 74: 407–414
8. Nordstrom C-H, Reinstrup P, Xu W, Gargenfors A, Ungerstedt U (2003) Assessment of the lower limit for cerebral perfusion pressure in severe head injuries by bedside monitoring of regional energy metabolism. Anesthesiology 98: 809–814
9. Robertson CS, Valadka AB, Hannay HJ, Contant CF, Gopinath SP, Cormio M, Uzura M, Grossman RG (1999) Prevention of secondary ischemic insults after severe head injury. Crit Care Med 27: 2086–2095
10. Rosner MJ, Rosner SD, Johnson AH (1995) Cerebral perfusion pressure: management protocol and clinical results. J Neurosurg 83: 949–962

Correspondence: Catherine J. Kirkness, Biobehavioral Nursing and Health Systems, Box 357266, University of Washington, Seattle, WA 98195. e-mail: kirkness@u.washington.edu

Acta Neurochir (2005) [Suppl] 95: 17–20

Effects of moderate hyperventilation on cerebrovascular pressure-reactivity after head injury

L. A. Steiner[1,2], **M. Balestreri**[1], **A. J. Johnston**[2], **J. P. Coles**[2], **D. A. Chatfield**[2], **J. D. Pickard**[1], **D. K. Menon**[2], and **M. Czosnyka**[1]

[1] Academic Neurosurgery, Addenbrooke's Hospital, Cambridge, UK
[2] University Department of Anaesthesia, Addenbrooke's Hospital, Cambridge, UK

Summary

In volunteers, hyperventilation improves autoregulation. However, in head-injured patients, hyperventilation-induced deterioration and improvement of autoregulation have been reported. We have re-examined this question using an index of pressure reactivity. Thirty patients with severe or moderate head-injury were studied. Arterial blood pressure, cerebral perfusion pressure (CPP), and intracranial pressure (ICP) were recorded over 20 minute epochs separated by ten minutes of equilibration at baseline and during moderate (>3.5 kPa) hyperventilation. End-tidal CO_2 was constant during each phase of data acquisition. Pressure reactivity was assessed using an index 'PRx' based on the response of ICP to spontaneous blood pressure changes. Hyperventilation decreased $PaCO_2$ from 5.1 ± 0.4 to 4.4 ± 0.4 kPa ($p < 0.0001$). ICP decreased by 3.7 ± 2.2 mmHg ($p < 0.001$). CPP increased by 5.9 ± 8.2 mmHg ($p < 0.001$). Overall, PRx did not change significantly with hyperventilation. However, there was a significant negative correlation between baseline PRx and the change in PRx ($r = -0.71$, $p < 0.0001$). This suggests that patients with disturbed pressure-reactivity may improve, whereas patients with intact pressure reactivity remain largely unchanged. Our data suggest that the response of pressure reactivity to hyperventilation is heterogeneous. This could be due to hyperventilation-induced changes in cerebral metabolism, or the change in CPP.

Keywords: Head injury; autoregulation; hyperventilation.

Introduction

Intact cerebrovascular pressure autoregulation is associated with favourable outcome after head injury [5, 6, 8, 13]. The cerebral perfusion pressure (CPP) orientated approach that is used on many units aims at restoring autoregulation [12]. However, CPP augmentation is not without side effects [3], and it may be beneficial to optimise autoregulation by other means than CPP manipulation. In volunteers, autoregulation is improved by hyperventilation, as hypocapnia increases vascular tone, and therefore increases the range of CPP where pressure autoregulation is effective [10, 11]. However, after traumatic brain injury, the interaction between vascular tone and pressure autoregulation seems less straightforward, as both an improvement [7] and a deterioration [2] of autoregulation during hyperventilation have been reported. We used an index of pressure-reactivity in head-injured patients to re-examine the relationship between change in autoregulatory status and hyperventilation. Pressure-reactivity is a key component of autoregulation and disturbed pressure-reactivity implies autoregulatory failure.

Methods

Data were prospectively collected during routine CO_2-reactivity testing in adult head-injured patients (age ≥ 16 years) with severe ($GCS \leq 8$) or moderate ($GCS \leq 12$) traumatic brain injury with secondary neurological deterioration necessitating sedation and mechanical ventilation. Exclusion criteria included respiratory failure, and a baseline $PaCO_2 < 4.30$ kPa. Mean arterial blood pressure (MAP) was monitored from the radial artery using a standard kit (Edwards Lifesciences, Irvine, CA, USA). ICP monitoring was performed using an intraparenchymal probe (Codman MicroSensors ICP Transducer, Codman & Shurtleff, Raynham, MA, USA). Mainstream end-tidal CO_2 ($ETCO_2$) monitoring was used (Marquette Solar 8000M, GE Medical Systems, UK). For measurements of arterial partial pressure of CO_2 ($PaCO_2$), an AVL Omni blood gas analyser (AVL Omni, Graz, Austria) was used. All patients were treated according to a CPP orientated protocol aiming to keep CPP above 70 mmHg and ICP below 20–25 mmHg [9]. All patients were sedated with propofol (2–5 mg/kg/h) and fentanyl (1–2 μg/kg/h), artificially ventilated, and when necessary, paralysed with atracurium. Where appropriate, surgical removal of space occupying lesions, drainage of cerebrospinal fluid, osmotic agents, and moderate hypothermia (33–36 °C) were used. During the studies all physiological parameters were maintained within the limits specified in the treatment guidelines of our unit.

After recording baseline data at constant $ETCO_2$ for 20 minutes and obtaining a baseline value for $PaCO_2$, the minute volume of the ventilator was increased by 15 to 20%. If this intervention resulted in standard treatment guidelines being contravened, the protocol was abandoned. These limits included a $PaCO_2 \leq 3.5$ kPa and/or jugular bulb oxygen saturation (SJO_2) $\leq 55\%$. After a stabilisation period of 10 minutes $ETCO_2$ was kept stable for the next 20 minutes and data were acquired during this time. $PaCO_2$ was measured at the middle of this phase. Infusion rates of vasoactive drugs were not changed and body temperature was kept stable throughout the study period. Autoregulation was assessed continuously with methods developed in house that have been described in detail elsewhere [4]. An index of pressure-reactivity (PRx) was calculated every minute based on the responses of ICP to spontaneous slow waves of MAP during the preceding four minutes. PRx is zero or negative if autoregulation is intact, whereas disturbed autoregulation is reflected by positive values. This method has recently been validated against cerebral blood flow (CBF) measurements made with PET [15]. Analogue signals were sampled at 30 Hz and digitised. Using waveform time integration [16], averaged 6-second values for ICP, MAP and CPP were calculated and data were stored continuously on a computer for off-line analysis. For the final analysis all values were averaged for baseline and for the 20 minutes of the stable hyperventilation phase. All data are presented as mean $\pm$ standard deviation. Linear regression methods were used for assessment of associations. P-values < 0.05 were considered to represent statistical significance. Calculations were performed using SPSS 11.0 for Windows (SPSS Inc, Chicago, IL, USA).

Results

We investigated 30 head-injured patients (age: 38 ± 15 years, 5 women, 25 men, median admission GCS 6, range 3–12) for 3.6 ± 2.6 days after injury. Hyperventilation reduced $PaCO_2$ from 5.1 ± 0.36 to 4.4 ± 0.35 kPa ($p < 0.0001$). MAP did not change significantly due to hyperventilation (97 ± 9 vs. 100 ± 12 mmHg; $p = 0.13$). ICP decreased by 3.7 ± 2.2 mmHg ($p < 0.0001$), CPP increased by 5.9 ± 8.2 mmHg ($p < 0.001$). The ICP change per kPa (ICP-reactivity) was 6.4 ± 5.3 mmHg·kPa^{-1} (range 0.4–20.7 mmHg·kPa^{-1}). There was no significant difference between PRx at the two levels of $PaCO_2$. Responses of individual patients are shown in Fig. 1. It could be argued that an improvement in autoregulation can only be achieved when $PaCO_2$ reactivity is retained. However, there was no relationship between ICP-reactivity and the change in PRx. However, PRx during hypocapnia was significantly correlated with baseline PRx ($r = 0.56$, $p = 0.001$). Moreover, the change in PRx was inversely correlated to baseline PRx ($r = -0.71$, $p < 0.0001$) suggesting that patients with impaired pressure-reactivity improved autoregulatory efficiency whereas patients with intact pressure-reactivity remained largely unchanged or, in individual cases, showed worsening (Fig. 2).

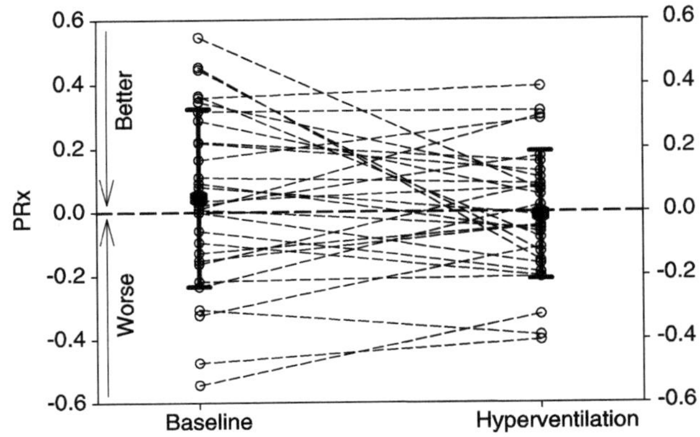

Fig. 1. Individual responses (hexagons and error bars represent mean $\pm$ standard deviation) of PRx to a reduction in $PaCO_2$. Overall no significant change was found and increases as well as decreases in PRx, i.e. improvement and deterioration of pressure reactivity, were found

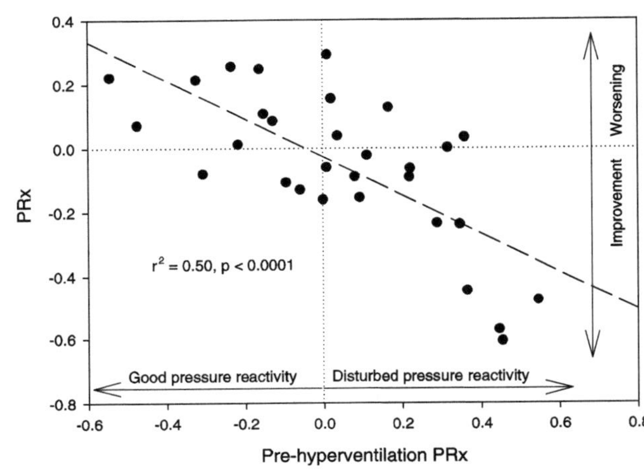

Fig. 2. This figure shows the change of PRx (*y-axis*) in response to a moderate reduction in $PaCO_2$ in relation to baseline PRx (*x-axis*). Patients with high PRx at baseline, i.e. disturbed pressure reactivity tended to improve (*PRx decreases*) whereas patients with low PRx at baseline, i.e. intact pressure reactivity tended to change little

Discussion

In contrast to earlier studies that showed either an improvement *or* a deterioration we have not found a uniform response of autoregulation to hyperventilation in this group of head-injured patients.

Ideally, the question of hyperventilation-induced changes in autoregulatory efficiency should be addressed by dynamic or static autoregulation measurements at identical CPP levels before and after a change in $PaCO_2$. However, this is difficult to achieve due to the hyperventilation-induced increase in CPP, and the use of vasoactive drugs to achieve identical CPP levels will possibly compound measurements. PRx is a global index of pressure-reactivity that gives a dynamic assessment of a key component of autoregulation. The strength of this index is that it reflects the status of the vascular system at a particular CPP and in contrast to the static rate of autoregulation no vasopressors are needed to perform the measurement. The method is however limited by the fact that if the two pressures at which measurements are made lie far apart, the comparison will only be valid if both measurements are made on the plateau of the autoregulatory curve. However, all our patients were managed at a comparable CPP and excluding patients with large increases in CPP (>10 mmHg) does not change our results. We therefore think it unlikely that the heterogeneity we observed is caused by the fact that we compare data obtained from different positions of a patient on his or her autoregulatory curve.

There are three hypothetical explanations to explain our results. First, as outlined above, the effect could be due to the change in CPP: patients who are shifted from below the lower threshold of autoregulation onto the autoregulatory plateau show an improvement, i.e. a decrease in PRx whereas those who are shifted from the plateau to above the upper threshold deteriorate. Those who move along the plateau remain unchanged. While we cannot exclude this mechanism in individual patients, based on subgroup analysis that did not show a systematic effect of high or low baseline CPP or large increases in CPP, we do not think that this is the prevailing mechanism responsible for the observed heterogeneity of responses. A second explanation arises from the possible independent effect of the oxygen extraction fraction (OEF) on PRx. We have shown previously that there is a significant relationship between global OEF and PRx [14], and the consistent increases in OEF produced by hyperventilation may be the cause of the observed changes in PRx. According to this model patients who become less hyperaemic due to hyperventilation would show an improvement in PRx whereas those who are shifted from the physiological range into the range of high OEF deteriorate. Those who are shifted within the physiological range of OEF remain unchanged. Anecdotal support for this hypothesis comes from an observation in a patient in whom the protocol had to be abandoned at a $PaCO_2$ of 4.2 kPa due to low SJO_2. In this patient we observed a rapid deterioration of PRx as SJO_2 approached 55% that was reversible upon normalisation of SJO_2. Finally, the fact that hyperventilation improves autoregulation in healthy volunteers is due to the fact that all segments of the vascular tree are equally affected. It is possible that this is not the case in head-injured patients. If the proximal increase in resistance is different from the distal response this could lead to a change of the autoregulatory reserve that is difficult to predict. For example, a constriction of proximal resistance vessels will result in a decrease in downstream perfusion pressure, and could trigger changes in resistance in distal vessels that result from the physiological consequences of hypocapnic vasoconstriction, rather than from the hypocapnia itself. This would potentially result in vasodilatation in the distal resistance vessels with a possible shift of the autoregulatory curve to the right. However, if hypocapnic vasoconstriction dominated this would result in constriction of the distal vessels, and could shift the autoregulatory curve to the left.

There are several differences between our study and earlier work on the same topic. Cold *et al.* [2] investigated eight patients on the first or second day after injury. They determined static autoregulation using angiotensin to control CPP and the ^{133}Xe method to measure CBF. The range of $PaCO_2$ they investigated was lower than ours. On average, $PaCO_2$ was reduced from 4.6 to 3.1 kPa. Responses at these lower $PaCO_2$ values are more likely to have been influenced by ischaemia resulting from intense hypocapnic vasoconstriction. The range over which autoregulation was tested was not comparable as during the higher level of $PaCO_2$ CPP increased from 60 to 85 mmHg whereas during hyperventilation CPP increased from 65 to 93 mmHg. The results reported by Newell *et al.* are more comparable to ours as far as $PaCO_2$ is concerned. They reduced $PaCO_2$ from 5.0 to 3.8 kPa in ten patients with severe head injury (GCS $\leq$ 8) and used the leg-cuff test [1] to determine dynamic autoregulation. Their patients had substantially lower values of autoregulatory efficiency at baseline than volunteers examined with the same methods, which perhaps explains the improvement they found. Measurements were performed at a CPP of 80 mmHg. Interestingly, hyperventilation did not lead to a significant increase in CPP in their group of patients despite an average normal CO_2-reactivity. This suggests that in this study the increase in CPP was not responsible for the improvement of autoregulation.

In summary, we report a non-uniform behaviour of pressure-reactivity in response to moderate hyperventilation. The expected improvement was only observed in patients with disturbed pressure-reactivity at baseline. Changes in $PaCO_2$ are not a reliable means of improving autoregulation in head-injured patients.

Acknowledgment

L. Steiner was supported by a Myron B. Laver Grant (Department of Anaesthesia, University of Basel, Switzerland), a grant from the Margarete und Walter Lichtenstein-Stiftung (Basel, Switzerland), by the Swiss National Science Foundation, and was recipient of an Overseas Research Student Award (Committee of Vice-Chancellors and Principals of the Universities of the United Kingdom). A. Johnston was recipient of a grant from Codman. J. Coles was funded by a Wellcome Research Training Fellowship and by a Beverley and Raymond Sackler Studentship Award. M. Czosnyka is on unpaid leave from the Warsaw University of Technology, Poland. This work was further supported by the MRC Acute Brain Injury Programme Grant No. G9439390 ID 56833.

References

1. Aaslid R, Lindegaard KF, Sorteberg W, Nornes H (1989) Cerebral autoregulation dynamics in humans. Stroke 1: 45–52
2. Cold GE, Christensen MS, Schmidt K (1981) Effect of two levels of induced hypocapnia on cerebral autoregulation in the acute phase of head injury coma. Acta Anaesthesiol Scand 5: 397–401
3. Contant CF, Valadka AB, Gopinath SP, Hannay HJ, Robertson CS (2001) Adult respiratory distress syndrome: a complication of induced hypertension after severe head injury. J Neurosurg 4: 560–568
4. Czosnyka M, Smielewski P, Kirkpatrick P, Menon DK, Pickard JD (1996) Monitoring of cerebral autoregulation in head-injured patients. Stroke 10: 1829–1834
5. Czosnyka M, Smielewski P, Piechnik S, Steiner LA, Pickard JD (2001) Cerebral autoregulation following head injury. J Neurosurg 5: 756–763
6. Lam JM, Hsiang JN, Poon WS (1997) Monitoring of autoregulation using laser Doppler flowmetry in patients with head injury. J Neurosurg 3: 438–445
7. Newell DW, Weber JP, Watson R, Aaslid R, Winn HR (1996) Effect of transient moderate hyperventilation on dynamic cerebral autoregulation after severe head injury. Neurosurgery 1: 35–43; discussion 43–34
8. Overgaard J, Tweed WA (1974) Cerebral circulation after head injury. 1. Cerebral blood flow and its regulation after closed head injury with emphasis on clinical correlations. J Neurosurg 5: 531–541
9. Patel HC, Menon DK, Tebbs S, Hawker R, Hutchinson PJ, Kirkpatrick PJ (2002) Specialist neurocritical care and outcome from head injury. Intensive Care Med 5: 547–553
10. Paulson OB, Strandgaard S, Edvinsson L (1990) Cerebral autoregulation. Cerebrovasc Brain Metab Rev 2: 161–192
11. Piechnik SK, Yang X, Czosnyka M, Smielewski P, Fletcher SH, Jones AL, Pickard JD (1999) The continuous assessment of cerebrovascular reactivity: a validation of the method in healthy volunteers. Anesth Analg 4: 944–949
12. Rosner MJ (1995) Introduction to cerebral perfusion pressure management. Neurosurg Clin N Am 4: 761–773
13. Steiner LA, Czosnyka M, Piechnik SK, Smielewski P, Chatfield D, Menon DK, Pickard JD (2002) Continuous monitoring of cerebrovascular pressure reactivity allows determination of optimal cerebral perfusion pressure in patients with traumatic brain injury. Crit Care Med 4: 733–738
14. Steiner LA, Coles JP, Czosnyka M, Minhas PS, Fryer TD, Aigbirhio FI, Clark JC, Smielewski P, Chatfield DA, Donovan T, Pickard JD, Menon DK (2003) Cerebrovascular pressure reactivity is related to global cerebral oxygen metabolism after head injury. J Neurol Neurosurg Psychiatry 765–770
15. Steiner LA, Coles JP, Johnston AJ, Chatfield DA, Smielewski P, Fryer TD, Aigbirhio FI, Clark JC, Pickard JD, Menon DK, Czosnyka M (2003) Assessment of cerebrovascular autoregulation in head injured patients – a validation study. Stroke 10: 2404–2409
16. Zablotny W, Czosnyka M, Smielewski P (1994) Portable software for intracranial pressure recording and waveform analysis. In: Nagai H, Kamiya K, Ishii S (eds) Intracranial pressure IX. Springer, Tokyo

Correspondence: Marek Czosnyka, Academic Neurosurgery, Box 167, Addenbrooke's Hospital, Cambridge CB2 2QQ, UK. e-mail: mc141@medschl.cam.ac.uk

Acta Neurochir (2005) [Suppl] 95: 21–23

Which paediatric head injured patients might benefit from decompression? Thresholds of ICP and CPP in the first six hours

I. R. Chambers[1], **P. A. Jones**[2], **R. A. Minns**[2], **L. Stobbart**[3], **A. D. Mendelow**[4], **R. C. Tasker**[5], and **F. Kirkham**[6]

[1] Regional Medical Physics Department, Newcastle General Hospital, Newcastle upon Tyne, UK
[2] Child Life and Health, University of Edinburgh, Edinburgh, UK
[3] School of Population and Health Sciences, University of Newcastle upon Tyne, Newcastle upon Tyne, UK
[4] Department of Neurosurgery, University of Newcastle upon Tyne, Newcastle upon Tyne, UK
[5] Addenbrookes Hospital, Cambridge, UK
[6] Southampton General Hospital, Southampton, UK

Summary

Severe head injury in childhood continues to be associated with considerable mortality and morbidity. Early surgical decompression may be beneficial and the objective of this study was to examine the relationship between age-related thresholds of mean intracranial pressure (ICP) and cerebral perfusion pressure (CPP) over the first 6 hours and age outcome in paediatric head injury patients.

A total of 209 head injured children admitted to five UK hospitals were studied. Patients aged 2 to 16 years were included if they had a minimum of six hours of invasive pressure monitoring. Mean values of ICP and CPP over this period were calculated and compared to those with independent (good recovery and moderate disability) and poor outcome (severe disability, and death) for different age groups.

There were 148 children with independent outcome (92 good recovery, 56 moderately disabled), and 61 with poor outcome (30 severely disabled, 31 deaths). There was a significant difference between those with independent compared to poor outcome in relation to ICP ($p < 0.001$) and CPP ($p < 0.001$). Patients were divided into three groups according to age. The sensitivity of ICP and CPP in predicting outcome was similar for all groups but the specificity differed between groups. At a CPP of 50 mmHg the specificity varied between the age groups (2 to 6 years: 0.47, 7 to 10 years: 0.28 and 11 to 16 years: 0.10) and similarly for an ICP of 25 mmHg (2 to 6 years: 0.53, 7 to 10 years: 0.44 and 11 to 16 years: 0.38).

Younger children may be able to tolerate lower perfusion pressures and still have an independent outcome. Our threshold values for young children are likely to be important in the identification of patients who might benefit from new treatments such as surgical decompression.

Keywords: Decompression; craniectomy; paediatric head injury; intracranial pressure; cerebral perfusion pressure; outcome.

Introduction

Severe head injury in childhood is associated with considerable mortality and morbidity. There is evidence that in head-injured adults, those with cerebral perfusion pressure (CPP) greater than 70 mmHg have better outcome and standardised protocols have been designed to maintain CPP above this level [1]. However, children have lower blood pressure when well and there are few data on optimal CPP for this age group [5, 4]. A randomized controlled trial (RCT) of early craniectomy after paediatric head trauma showed a clear benefit, with 7/13 independent after craniectomy compared with 2/14 randomised to conventional management [6]. This trial was stopped early, but has raised many questions, particularly concerning the intracranial pressure (ICP) and CPP levels at which randomization to craniectomy might be justified in different age groups. The objective of our study was to examine the relationship of ICP and of CPP to outcome related to age. This was achieved by examining the relative value of using mean ICP and mean CPP over the first 6 hours of monitoring to predict poor outcome (death, vegetative state or severe disability) in a large population of children with head injury.

This would provide grounds for a randomised controlled trial of craniectomy versus medical management.

Materials and methods

Monitoring data were available from 209 head-injured children admitted to five hospitals in the UK, with a median first recorded Glasgow Coma Score of 6 (range 3–15) and who required ICP monitoring. Patients were included if they were aged between 2 to 16 years of age and had a minimum of six hours of invasive monitoring (patients with non accidental injury were excluded). Mean values of mean arterial pressure (MAP), ICP and CPP over the first six hours of monitoring were calculated. After follow up of survivors for at least 6 months, no patients remained in a vegetative state. Mean ICP and CPP were compared between the independent (good recovery and moderate disability) and the poor (severe disability and death) outcome groups. Forward stepwise logistic regression was used to examine the prediction of poor outcome and the corresponding sensitivity and specificity of ICP and CPP in relation to independent or poor outcome calculated.

Results

There were 133 boys and 76 girls with a median age of 10 (range 2–16) years; 92 made a good recovery, 56 had moderate disability (148 independent outcome), 30 were severely disabled and 31 died (61 poor outcome). There was a significant difference between the independent and poor outcome groups in relation to ICP ($p < 0.001$) and CPP ($p < 0.001$) in the first 6 hours of monitoring. Patients were divided into three groups according to age, 2–6 years (Group A, $n = 49$), 7–10 years (Group B, $n = 68$) and 11–16 years (Group C, $n = 92$).

In predicting independent outcome, the sensitivity of ICP and CPP were similar for all age groups but in the younger children, 50% specificity for ICP was higher (Table 1) (Fig. 2). At a CPP of 50 mmHg the specificity varied between the groups ($A = 0.47$, $B = 0.28$, $C = 0.10$).

Discussion

Intracranial hypertension is common after head injury in childhood and mean CPP is an important

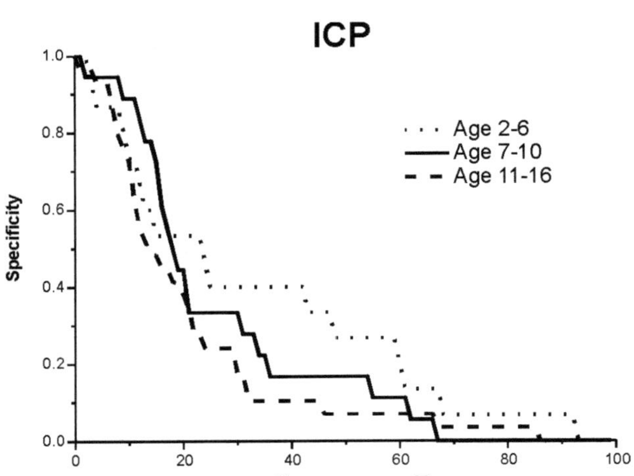

Fig. 1. Variation of specificity of ICP with pressure

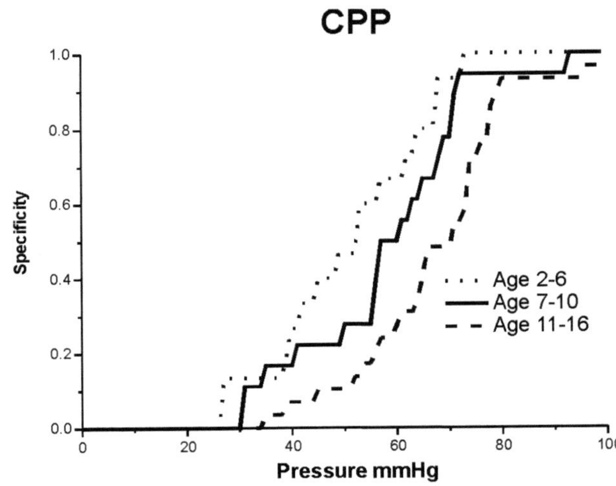

Fig. 2. Variation of specificity of CPP with pressure

Table 1. *Fifty percent values of specificity for ICP and CPP*

Age range	Number of cases	50% specificity	
		ICP	CPP
2–6	49	24 mmHg	53 mmHg
7–10	68	19 mmHg	58 mmHg
11–16	92	18 mmHg	67 mmHg

determinant of outcome [2, 4]. Our data provide some guidance on CPP thresholds above which to maintain CPP in different age groups. However, in adults, a controlled trial of treatment to maintain CPP > 70 mmHg showed no benefit in terms of outcome [3]. From adult studies, age has been frequently found to be one of the best predictors of good outcome, but this has not been confirmed for different paediatric age groups. A small trial of surgical decompression in children has suggested benefit, and therefore an RCT, stratified by age, should be considered. Our data may play a role in the identification of target levels for ICP and CPP in controlled treatment trials in severe paediatric head trauma.

References

1. Chambers IR, Treadwell L, Mendelow AD (2001) Determination of threshold levels of cerebral perfusion pressure and intracranial pressure in severe head injury by using ROC curves: an observational study in 291 patients. J Neurosurg 94: 412–416
2. Dark P, Woodford M, Vail A, Mackway-Jones K, Yates D, Lecky F (2002) Systolic hypertension and teh response to blunt trauma in infants and children. Resuscitation 54: 245–253

3. Jones PA, Andrews PJD, Easton VJ, Minns RA (2003) Traumatic brain injury in childhood: Intensive Care time series data and outcome. Brit J Neurosurg 17(1): 29–39
4. Robertson CS, Valadka AB, Hannay HJ, Contant CF, Gopinath SP, Cormio M, Uzura M, Grossman RG (1998) Prevention of secondary ischemic insults after severe head injury
5. Rosner MJ, Rosner SD, Johnson AH (1995) Cerebral perfusion pressure: management protocol and results. J Neurosurg 83: 949–962
6. Taylor A, Butt W, Rosenfield J, Shann F, Ditchfield M, Lewis E, Klug G, Wallace D, Henning R, Tibbalss J (2001) A randomised trial of very early decompressive craniectomy in children with traumatic brain injury and sustained intracranial hypertension. Childs Nerv Syst 17: 154–162

Correspondence: I. R. Chambers, Regional Medical Physics Department, Newcastle General Hospital, Newcastle upon Tyne, NE4 6BE, UK. e-mail: i.r.chambers@ncl.ac.uk

Acta Neurochir (2005) [Suppl] 95: 25–28

Association between outcome, cerebral pressure reactivity and slow ICP waves following head injury

M. Balestreri[1,2], **M. Czosnyka**[1], **L. A. Steiner**[1,3], **M. Hiler**[1], **E. A. Schmidt**[1,4], **B. Matta**[5], **D. Menon**[5], **P. Hutchinson**[1], and **J. D. Pickard**[1]

[1] Academic Neurosurgical Unit, Addenbrooke's Hospital, Cambridge, UK
[2] Department of Anaesthesia and Intensive Care (IInd) Policlinico San Matteo, Pavia, Italy
[3] Department of Anaesthesia, University Hospital Basel, Basel, Switzerland
[4] Department of Neurosurgery, Hopital Purpan, Toulouse, France
[5] Department of Anaesthesilogy, Addenbrooke's Hospital, Cambridge, UK

Summary

Objective. To investigate the relationships between slow vasogenic waves ('B waves') of intracranial pressure (ICP), pressure-reactivity and outcome after traumatic brain injury.

Material and method. 193 head-injured patients (age 34 ± 16.7 years; median GCS 6) were monitored from 1997 to 2002. ICP, arterial blood pressure (ABP) were continuously monitored. Pressure-reactivity index (PRx) and magnitude of ICP slow waves were evaluated using the bed-side computers.

Results. Distribution of PRx in different outcome groups indicated that pressure-reactivity was significantly worse in patients with fatal outcome. A magnitude of spontaneous slow waves of ICP was gradually decreasing in poorer outcome grades.

Mortality indicated threshold rise from 20% to 70% when averaged PRx increased above 0.3 ($p < 0.01$). There was no threshold for mortality observed along distribution of magnitude of ICP slow waves. Mortality gradually increased when the magnitude of slow waves decreased ($R = -0.26$; $p < 0.0001$).

Conclusion. Inadequate pressure-reactivity and low magnitude of slow vasogenic waves of ICP are associated with fatal outcome after head injury. Based on brain monitoring data, differentiation between favourable outcome and severe disability is more problematic than differentiation between survivors and non-survivors.

Keywords: Head injury; pressure reactivity; slow waves; outcome.

Introduction

Starting from September 1991, a bedside computer-supported system has been used [2] in University of Cambridge (UK) Neurosciences Critical Care Unit (NCCU). Its purpose is to monitor continuously physiological parameters such as ICP, ABP and CPP. The system provides one-minute averaged values for the monitored variables and calculates indices, which are potentially useful for the interpretation of cerebral haemodynamics after brain trauma.

While information about association between mean ICP, CPP and outcome is more readily available, secondary indices derived from ICP have not been sufficiently investigated. The spectral components of the ICP waveform, such as pulse amplitude, slow waves (e.g. Lundberg's B waves), plateau waves, etc., have been described in the past [1, 9], but not always with convincing proof about their value in the management and prognostication of head injury. In our previous study [1] we compared pressure reactivity and slow ICP waves in patients with intracranial hypertension who died and achieved a favourable outcome. In this study, based on a much larger population (193 versus 93), we investigated the relationship between slow waves, PRx and outcome following head injury.

Material and methods

This retrospective analysis was based on 193 head injured patients admitted to the NCCU between January 1997 and December 2001. Only patients with invasive monitoring of ICP and ABP over a period greater than 12 hours and connected to a bedside computerized system were included in the study. It is important to emphasize that the studied group was not representative for all admissions to this unit. Patients, who had been admitted and discharged promptly, or died soon after admission, admitted and discharged during or soon after the weekend or public holidays were not included in this analysis.

Patients were sedated, mechanically ventilated and paralyzed in order to maintain the ICP below 25 mmHg. Systemic hypotension was treated with fluids and vasoactive drugs. CPP was maintained

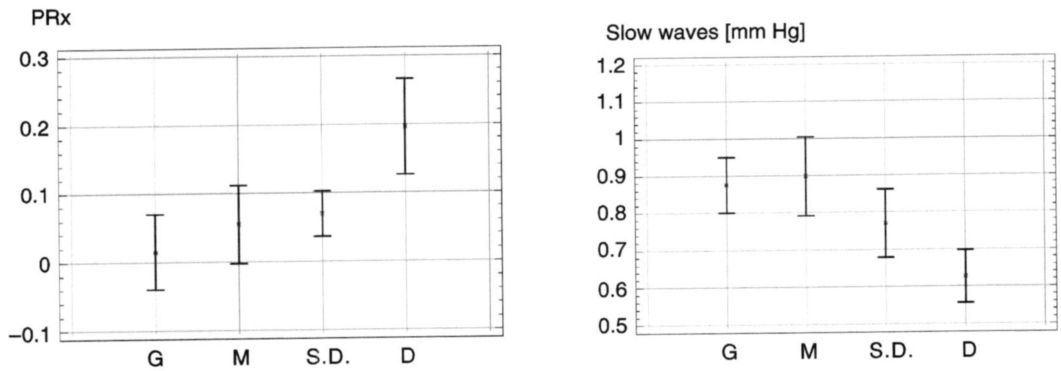

Fig. 1. Mean values and 95% confidence intervals (individual) for monitored parameters in different outcome groups (*G* – good, *M* – moderate, *S.D.* – severely disability, *D* – dead). Slow-slow waves of ICP and PRx-pressure reactivity index

above 70 mmHg to avoid secondary ischaemic insults. Episodes of intracranial hypertension were treated with mild hyperventilation ($PaCO_2 > 4.0$ kPa), moderate hypothermia, boluses of mannitol and thiopentone.

ICP was monitored by an intraparenchymal probe (Codman ICP MicroSensors). ABP was monitored invasively. Signals were sampled from the analogue output of the monitors at 30 Hz, digitized (12 bits analogue-to-digital converter) and subsequently analyzed as 8 seconds averages. Two derived variables, an index characterizing pressure reactivity and the amplitude of spontaneous slow waves of ICP were selectively studied. The pressure-reactivity index (PRx) was calculated as previously described [4]. Briefly, the correlation coefficient between 30–40 consecutive time-averaged samples of ABP and ICP (from four to little longer than five minute time-window); PRx > 0 is indicative of impaired pressure reactivity.

The amplitude of slow waves of ICP was calculated as the square root of the power of the signal in the frequency range between 0.05 and 0.0055 Hz (20 second to 3 minute epochs).

Data from every patient were summarized as mean values of the derived indexes (PRx and slow wave amplitude). Outcome was graded six months after injury according to the Glasgow Outcome Scale (GOS) [6]. Mean values were compared with t-tests when data were normally distributed; the Mann-Whitney U-test was used to compare medians of parameters that were not normally distributed.

Results

Demographic characteristics of this population were comparable to other studies: mean age was 36 (± 17 years) and median admission GCS was 7 (range 3 to 15; 25% of patients had initial GCS above 8; and the deteriorated later requiring full intensive care); 19.5% of these patients were female.

Overall fifty-two patients (27%) had a good outcome, 41 (21%) were moderately disabled, fifty (26%) severely disabled and 50 (26%) died. In patients who died compared to those who survived there pressure-reactivity was worse (0.2 ± 0.26 versus 0.05 ± 0.18, $p < 0.0004$). The magnitude of slow ICP waves was lower (0.64 ± 0.31 mmHg versus 0.855 ± 0.41, $p < 0.00007$) – see Fig. 1. There was a significant correlation between PRx, ICP, CPP and GOS, however, these parameters were powerful for differentiating between non-survivors and survivors and less so for discriminating between favourable and unfavourable outcome. The only exception is the magnitude of slow waves that was able to separate the group of severely disabled patients from the group with favourable outcome ($p < 0.03$).

Mortality rate was distributed unevenly along PRx and slow waves axis – see Fig. 2. PRx indicated the threshold value of 0.3 above which mortality was 69%, below only 19% (chi-square; $p < 0.0001$). Mortality rate was related to slow waves of ICP in rather non-threshold fashion. Correlation between slow waves and the mortality was significant ($R = -0.26$, $p < 0.0001$), however no threshold was found. Figure 3 examplified a patient who died on the second day after a severe head injury: PRx deteriorated shortly after admission at relatively adequate level of CPP; slow waves increased and then decreased secondarily at very low CPPs.

Discussion

In this retrospective analysis we have considered all the patients included in our multimodality neuromonitoring program over the past ten years, which represent 30–40% of the total number of the severely head injured patients admitted to the NCCU. Every patient admitted to the NCCU and receiving invasive monitoring of ICP and ABP was potentially eligible to be studied with our monitoring system. A 20% of patients had a GCS > 8 on arrival, but all of them developed a severe neurological deterioration at some stage of

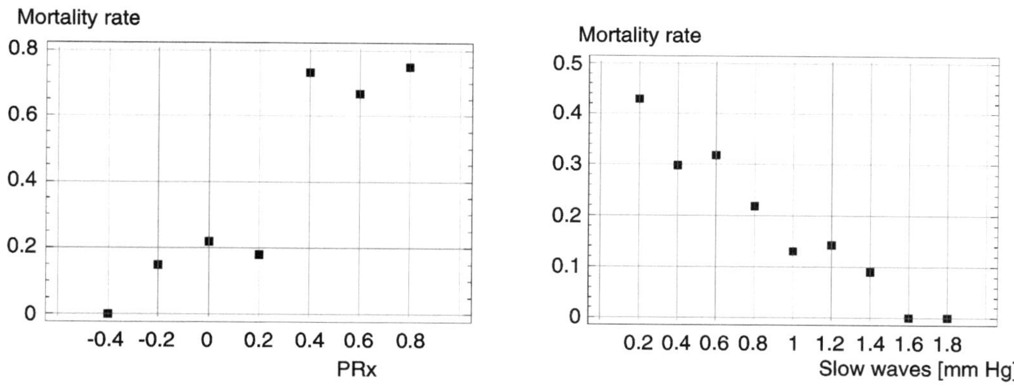

Fig. 2. Mortality rate expressed as a function of pressure-reactivity and magnitude of slow waves of ICP

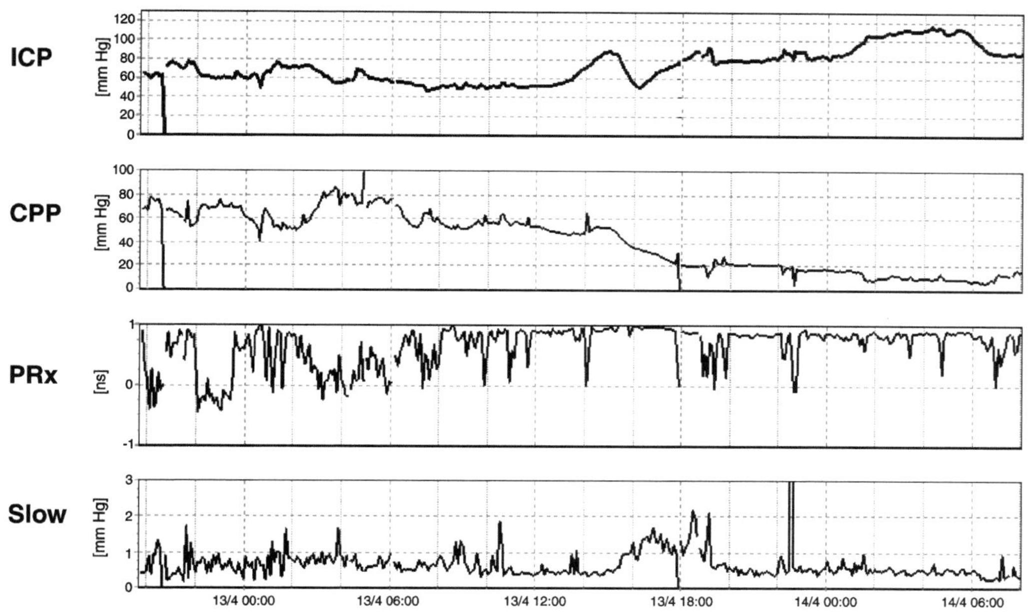

Fig. 3. Example of monitoring of ICP, CPP, PRx amd Slow waves in a patient who died on the second day after severe head injury. PRx deteriorated shortly after admission but at relatively good level of CPP. Slow waves increased and then decreased secondarily at very low CPPs

their admission requiring invasive monitoring and active medical treatment according to our guidelines. Patients whose monitoring period was short were excluded from the analysis.

The result of this analysis reconfirms PRx as one of the most important predictors of mortality after brain trauma [12, 13], but it also points out the limitation of this parameter as a tool to differentiate between patients who survive with favourable outcome and patients who survive with severe disability. In contrast, the only parameter that showed a clear trend in different outcome groups was the magnitude of slow waves of ICP that was able to differentiate the good recovery group from the severe disabled group.

Despite our improved knowledge of pathophysiology after head injury, untreatable intracranial hypertension remains one of the most important causes of mortality in the studied group of patients [14, 15]. Several protocols and therapies are in use [5, 7, 10, 11]. The higher values of PRx in patients who died, indicating an impaired cerebral autoregulation, were often reported throughout the monitoring period, suggesting a relationship between the degree of dysfunction of pressure reactivity and severity of the injury. The inability of these patients to maintain a constant CBF could be the trigger that initiates the secondary ICP rise.

In conclusion, impaired pressure-reactivity and low content of vasogenic ICP waves are associated with fatal outcome.

References

1. Balestreri M, Czosnyka M, Steiner LA, Schmidt E, Smielewski P, Matta B, Pickard JD (2004) Intracranial hypertension: what additional information can be derived from ICP waveform after head injury? Acta Neurochir (Wien) 146(2): 131–141
2. Czosnyka M, Whitehouse H, Smielewski P, Kirkpatrick P, Pickard JD (1994) Computer supported multimodal monitoring in neuro intensive care. Int J Clin Monitoring Computing 11: 223–232
3. Czosnyka M, Guazzo E, Whitehouse H, Smielewski P, Czosnyka Z, Kirkpatrick P, Piechnik S, Pickard JD (1996) Significance of intracranial pressure waveform analysis after head injury. Acta Neurochir (Wien) 138: 531–542
4. Czosnyka M, Smielewski P, Kirkpatrick P, Laing RJ, Menon D, Pickard JD (1997) Continuous assessment of the cerebral vasomotor reactivity in head injury. Neurosurgery 41: 11–19
5. Elf K, Nilsson P, Enblad P (2002) Outcome after traumatic brain injury improved by an organized secondary insult program and standardized neurointensive care. Crit Care Med 30(9): 2129–2134
6. Jennett B, Bond M (1975) Assessment of outcome after severe brain damage. Lancet 1(7905): 480–484
7. Meixensberger J, Jaeger M, Vath A, Dings J, Kunze E, Roosen K (2003) Brain tissue oxygen guided treatment supplementing ICP/CPP therapy after traumatic brain injury. J Neurol Neurosurg Psychiatry 74(6): 760–764
8. Menon DK (1999) Cerebral protection in severe brain injury: physiological determinants of outcome and their optimisation. Br Med Bull 55(1): 226–258
9. Naredi S, Olivecrona M, Lindgren C, Ostlund AL, Grande PO, Koskinen LO (2001) An outcome study of severe traumatic head injury using the "Lund therapy" with low-dose prostacyclin. Acta Anaesthesiol Scand 45(4): 402–406
10. Oertel M, Kelly DF, Lee JH, Glenn TC, Vespa PM, Martin NA (2002) Is CPP therapy beneficial for all patients with high ICP? Acta Neurochir [Suppl] 81: 67–68
11. Rosner MJ, Rosner SD, Johnson AH (1995) Cerebral perfusion pressure: management protocol and clinical results. J Neurosurg 83(6): 949–962
12. Steiner LA, Coles JP, Czosnyka M, Minhas PS, Fryer TD, Aigbirhio FI, Clark JC, Smielewski P, Chatfield DA, Donovan T, Pickard JD, Menon DK (2003) Cerebrovascular pressure reactivity is related to global cerebral oxygen metabolism after head injury. J Neurol Neurosurg Psychiatry 74(6): 765–770
13. Steiner LA, Czosnyka M, Piechnik SK, Smielewski P, Chatfield D, Menon DK, Pickard JD (2002) Continuous monitoring of cerebrovascular pressure reactivity allows determination of optimal cerebral perfusion pressure in patients with traumatic brain injury. Crit Care Med 30(4): 733–738
14. Unterberg A, Kiening K, Schmiedek P, Lanksch W (1993) Long-term observations of intracranial pressure after severe head injury. The phenomenon of secondary rise of intracranial pressure. Neurosurgery 32(1): 17–23; discussion 23–24
15. Whitfield PC, Patel H, Hutchinson PJA, Czosnyka M, Parry D, Menon D, Pickard JD, Kirkpatrick PJ (2001) Bifrontal decompressive craniectomy in the management of posttraumatic intracranial hypertension. Brit J Neurosurg 15(6): 500–507

Correspondence: Marek Czosnyka, Academic Neurosurgery, Box 167, Addenbrooke's Hospital, Cambridge CB22QQ, UK. e-mail: Mc141@medschl.cam.ac.uk

Acta Neurochir (2005) [Suppl] 95: 29–32

Quantification of secondary CPP insult severity in paediatric head injured patients using a pressure-time index

P. A. Jones[1], **I. R. Chambers**[2], **T. Y. M. Lo**[1], **P. J. D. Andrews**[3], **W. Chaudhry**[4], **A. Clark**[2], **J. Croft**[5], **R. Forsyth**[4], **B. Fulton**[6], **A. D. Mendelow**[5], **G. Wilson**[5], and **R. A. Minns**[1]

[1] Child Life and Health, University of Edinburgh, Edinburgh, UK
[2] Regional Medical Physics Department, Newcastle General Hospital, Newcastle upon Tyne, UK
[3] Department of Anaesthetics, University of Edinburgh, Edinburgh, UK
[4] Department of Paediatrics, Newcastle General Hospital, Newcastle upon Tyne, UK
[5] Department of Neurosurgery, Newcastle General Hospital, Newcastle upon Tyne, UK
[6] Department of Anaesthetics, Newcastle General Hospital, Newcastle upon Tyne, UK

Summary

This paper describes and validates a new *Cumulative Pressure-Time Index* (CPT) which takes into account both duration and degree of cerebral perfusion pressure (CPP) derangement and determines critical thresholds for CPP, in a paediatric head injury dataset.

Sixty-six head-injured children, with invasive minute-to-minute intracranial pressure (ICP) and blood pressure monitoring, had their pre-set CPP derangement episodes (outside the normal range) identified in three childhood age-bands (2–6, 7–10, and 11–16 years) and global outcome assessed at six months post injury.

The new cumulative pressure-time index more accurately predicted outcome than previously used summary measures and by varying the threshold CPP values, it was found that these physiological threshold values (≤ 48, ≤ 52 and ≤ 56 mmHg for 2–6, 7–10, and 11–16 years respectively) best predicted brain insult in terms of subsequent mortality and morbidity.

Keywords: Paediatric head injury; cumulative pressure-time index; critical CPP thresholds; outcome.

Introduction

Cerebral perfusion pressure (CPP) has been shown to be the best predictor of outcome in both adult and paediatric head injuries [1, 2, 4–6]. Previously a number of different summary measures of CPP, (e.g. means over hourly recordings or other time-frames, maximum or minimum values, percentage duration of derangement etc.), have been used as predictors of outcome. These are pragmatic and loose much of the detail by virtue of averaging the peaks and troughs of the perfusion pressure recording.

A measure is required that incorporates more than one dimension (duration and degree), to give a more accurate quantification of the total burden of secondary insult. It should be equally applicable to children and adult monitoring.

This paper describes a novel Cumulative Pressure-Time index (CPT) that has been applied to a paediatric head injury dataset, and has been used to help determine critical thresholds for CPP in children.

Materials and methods

Sixty-six head-injured children (aged 2–16 years) from two regional UK centres had minute-to-minute recordings of physiological parameters including intracranial pressure (ICP), systolic, diastolic and mean arterial blood pressure (MAP), with automatically calculated CPP. Data were downloaded from bedside monitors and analysed off-line. Entry to this study included i) a post-resuscitation, pre-intubation Glasgow Coma Score (GCS) of 12 or less, or an Injury Severity Score of ≥ 16 in association with a GCS of 13–15 after head injury; ii) monitoring within 24 hours from the time of injury until removal of the ICP monitor.

Demographic data including age, gender, cause of injury, data and time of injury were collected, and a modified Glasgow Outcome Scale score was assigned at 6 months post-injury from a postal questionnaire completed by parents/carers or General Practitioners. Outcome was dichotomised into independent (GOS 4 & 5) and poor (GOS 1, 2 & 3) outcome, and alive vs. dead.

Pre-set levels of CPP derangement (i.e. values that were outside the normal range) were used for each year of age [4]. The 66 children were then grouped into 3 practical age-bands (2–6, 7–10, 11–16) such that the mean CPP for each band was within a 5 mmHg span, giving mean CPP threshold values of ≤ 48 mmHg for the 2–6 year olds, ≤ 54 mmHg for the 7–10 year olds, and ≤ 58 mmHg for the oldest group (11–16 years of age).

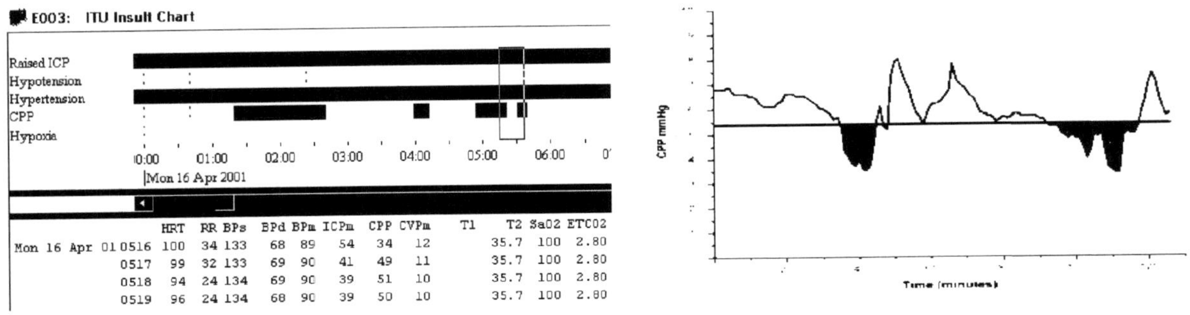

a) Browser screen identifying derangements

b) Enlarged view of section of CPP trace. CPT shown schematically, in grey

Fig. 1. Schematic representation of CPP and CPT: (a) The *Edinburgh Browser*© screen shows the recorded physiological values minute-by-minute below, and deranged physiology (i.e. >5 minutes) as horizontal black bars above. (b) A sample of the CPP trace from Fig. 1a is enlarged to illustrate the Cumulative Pressure-Time Index, shown as the area below the pre-selected threshold for CPP for age (horizontal line)

Method for calculating the CPT

Figure 1 outlines the steps in calculating the Cumulative Pressure-Time Index. Using the *Edinburgh Browser*© program [3] the minute-by-minute time series data were first validated to remove artefact and identify derangements lasting >5 minutes (Fig. 1a). Figure 1b shows a continuous trace of the resulting CPP data, with the selected threshold level. Each episode was identified and the summated areas between the CPP tracing and the threshold calculated.

This produced a single value representing both severity and duration of derangements. This can be shown mathematically by the following formula, where the greater the CPT value, the more CPP derangement was evident.

$$CPT = \sum (CPP_{\text{threshold}} - CPP) \times t_{\text{sample interval}}$$

The relationship between this new summary measure, and some alternative summary measures for CPP calculated from this dataset [i.e. the overall mean value for CPP (in mmHg); the accumulated total duration of all episodes of CPP derangement as specified by Browser as absolute time and as a percentage of the total CPP monitoring time] and outcome was explored.

Using the above, Receiver Operator Characteristic (ROC) curves were constructed. To test if there were significant differences between the ROC curves, 6 pairwise comparisons were made for each dichotomous outcome to establish which of the above summary measures were the best predictors of outcome.

The pre-set threshold levels for CPP for each age-band were then reduced by 10% and 20%, and the CPT recalculated for each, and related to outcome to identify which of the chosen thresholds were best discriminators of outcome.

Results

Of the 66 children, 47 suffered a severe head injury, 17 a moderate and 2 a mild head injury according to the entry criteria. There were 41 boys and 25 girls. Fifteen children were aged 2–6 years, 24 aged 7–10 years and 27 were in the oldest group (11–16 years). At 6 months outcome, 11 had died, none were vegetative, 6 were severely disabled (17 poor outcome), 19 were moderately disabled and 30 had made a good recovery (49 independent outcome).

CPT vs. outcome

The mean CPT values for those with poor and independent outcome were 31785 (mmHg × min) and 3391 respectively, and for the dead and alive were 48126 and 3220 respectively. Both were significantly predictive of outcome ($p < 0.001$). When the ROC curves were examined the areas under the curve for poor vs. independence and dead vs. alive were 0.839 (95% C.I. 0.731, 0.9470) and 0.957 (95% C.I. 0.901, 1.013) respectively.

CPT values by age vs. outcome

The CPT values for the 3 childhood age bands were also separately highly predictive for mortality ($p < 0.001$), and morbidity ($p = 0.027$, $p = 0.014$, and $p = 0.001$ for the 2–6, 7–10 and 11–16 year old groups respectively). The CPT derangement value was not significantly different (ANOVA) across the 3 childhood age bands for mortality and morbidity.

All summary measures related to outcome

A comparison of the other summary measures described above with the CPT index has been assessed utilising multiple ROC constructions (Fig. 2). The areas under the ROC curves for poor vs. independent outcome for the 4 measures were: – overall mean CPP

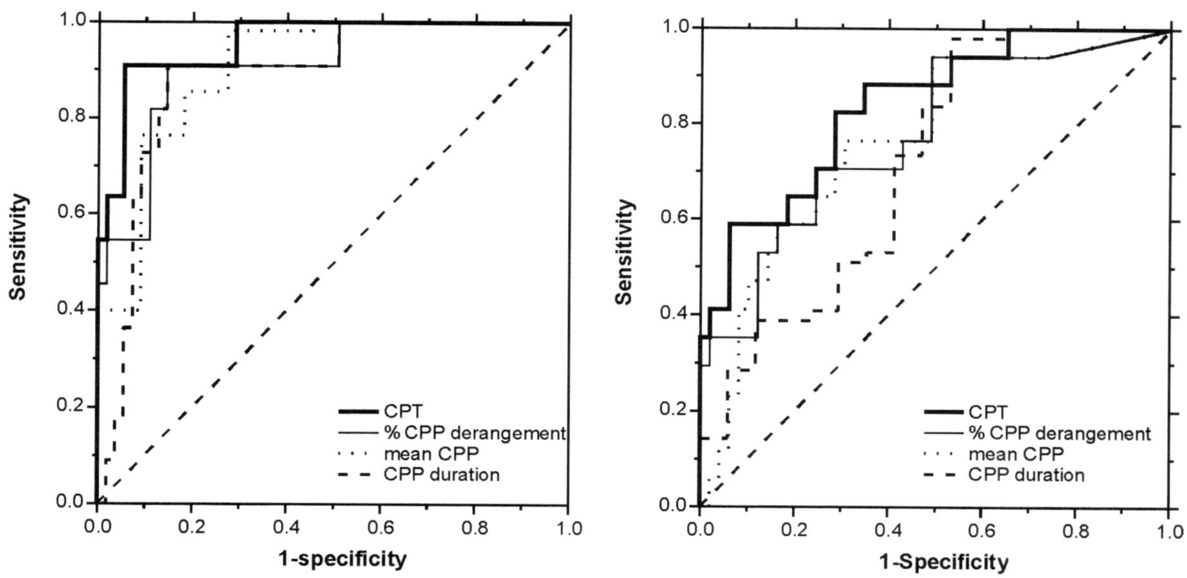

Fig. 2. ROCs for two dichotomised outcomes (death vs. survival on the left hand graph, and poor vs. independent outcome on the right hand graph) for CPP derangement when measured by i) overall Mean CPP in mmHg, ii) duration of CPP derangement; iii) derangement as a percentage of the acute monitoring time (% CPP derangement), and iv) using the Cumulative Pressure Time Index *(CPT)* for CPP

0.717; CPP insult duration 0.764; percentage CPP insult 0.776, and CPT at threshold 0.839.

Pairwise comparison of ROC curves for poor vs. independent outcome

The 6 possible pairwise comparisons were made of these ROC curves (MedCalc). This showed the CPT index was a significantly better measure than overall mean CPP ($p = 0.008$), CPP insult duration ($p = 0.02$), and percent duration of CPP ($p = 0.055$). None of the other possible paired comparisons were significant.

Pairwise comparison of ROC curves for dead vs. alive

Although all summary measures were highly predictive of mortality at 6 months, the area under the ROC curve for CPT index was 0.957, slightly greater than for the percentage insult duration (0.909), mean CPP (0.907) and CPP insult duration (0.886), but not statistically significant.

Threshold alteration and outcome

Although our chosen threshold levels above provide impressive ROC values, we considered whether a different CPP threshold might *even better* relate to outcome. Accordingly we carried out similar analyses of CPT and ROC curves when the CPP threshold value was reduced by 10% and 20% (Table 1). For independent and poor outcome the area under the ROC curve for our threshold value was 0.839, greater than the value 0.818 for threshold minus 10%, and the value 0.809 for threshold minus 20%, but not significantly different.

For non-survival the area under the ROC curve did not improve prediction by reducing the threshold values.

Table 1. *Values of the 'area under the ROC curves' for the cumulative pressure-time (CPT) at 'threshold CPP for age-band', 'threshold minus 10%', and 'threshold minus 20%' levels of CPP for mortality and morbidity*

	Area under the ROC curve poor vs. independent outcome	Area under the ROC curve death vs. survival
CPP at 'threshold level'	0.839	0.957
Threshold minus 10%	0.818	0.974
Threshold minus 20%	0.809	0.975

Discussion

The CPT index that we have described takes into account both the severity and the duration of the physiological derangement of CPP and this index is highly significantly related to outcome (classified as either independent vs. poor or alive vs. dead). This relationship to outcome also applies to children of different ages (2–6, 7–10 and 11–16 years of age). Our results show that there is a similar insult burden, measured by CPT, in young as in older children with head injury.

Previous studies on the relationship between CPP and outcome in head injury have normally used a unidimensional measure of the CPP insult, such as calculating the mean CPP value from the whole record, or the duration of reduced CPP, or the percentage time of the CPP reduction to the whole record, all of which we have similarly used. We have created a new index and demonstrated that the use of the CPT index at our threshold levels for children was a significantly better predictor of outcome.

Although CPT and the other indices were all significantly predictive of mortality, because of the absolute prediction of mortality with progressively lowered CPP values, there was no additional benefit from any one measure.

The threshold values chosen were inferred from physiological normative data [4] and it was not known at the outset whether those chosen values would prove to be definitive thresholds for sustaining brain "insult" (and be crucial to the determination of outcome). We therefore investigated alternative threshold levels of CPP, (of minus 10% and minus 20% below the physiological threshold), however these were not more predictive of outcome, indicating that the physiological threshold (after appropriate homeostatic compensation) best approximates the secondary brain insult threshold.

The brain insult may be acquired by either an exaggerated ICP or low MAP, and the CPT as a total CPP insult measure, will not recognise which is the major contributing cause. This is a common difficulty with any CPP measure.

Clearly the CPT index is a retrospective tool for use in future clinical trials and retrospective studies, although it could theoretically be applied after 24 hours or other epochs of patient monitoring. It is likely that the age-specific critical threshold CPP value will be the practical value which will guide the clinician in the bedside care of the head injured child.

The CPP threshold values described are particularly useful for discriminating brain insult, but "treatment" or "intervention" thresholds, which would be arbitrary, would probably require higher target levels of CPP for treatment intervention in any formal clinical trial.

Acknowledgments

This research was supported by the Chief Scientist Office, (Scottish Office), U.K., Grant K/MRS/50/C2784, and the Northern Brainwave Appeal.

References

1. Chambers IR, Treadwell L, Mendelow AD (2000) The cause and incidence of secondary insults in severely head-injured adults and children. Brit J Neurosurg 14: 424–431
2. Chambers IR, Treadwell L, Mendelow AD (2001) Determination of threshold levels of cerebral perfusion pressure and intracranial pressure in severe head injury by using receiver-operating characteristic curves: an observational study in 291 patients. J Neurosurg 94: 412–416
3. Howells TP (1994) *Edinburgh Monitor & Browser*© Computer Program
4. Jones PA, Andrews PJ, Easton VJ, Minns RA (2003) Traumatic brain injury in childhood: intensive care time series data and outcome. Brit J Neurosurg 17: 29–39
5. Jones PA, Andrews PJ, Midgley S, Anderson SI, Piper IR, Tocher JL, Housley AM, Corrie JA, Slattery J, Dearden NM (1994) Measuring the burden of secondary insults in head-injured patients during intensive care. J Neurosurg Anesthesiol 6: 4–14
6. Rosner MJ, Rosner SD, Johnson AH (1995) Cerebral perfusion pressure: Management protocol and clinical results. J Neurosurg 83: 949–962

Correspondence: Robert A. Minns, Child Life and Health, University of Edinburgh, 20 Sylvan Place, Edinburgh, EH9 1UW, Scotland, UK. e-mail: Robert.Minns@ed.ac.uk

Acta Neurochir (2005) [Suppl] 95: 33–37

The *BrainIT* Group: concept *and* current status 2004

P. Nilsson, I. Piper, G. Citerio, I. Chambers, C. Contant, P. Enblad, H. Fiddes, T. Howells, K. Kiening, and Y. H. Yau ***for the BrainIT Group***

Department of Clinical Physics, Institute of Neurological Sciences, Southern General Hospital, Glasgow, UK

Summary

Introduction. An open collaborative international network has been established which aims to improve inter-centre standards for collection of high-resolution, neurointensive care data on patients with traumatic brain injury. The group is also working towards the creation of an open access, detailed and validated database that will be useful for hypothesis generation. In Part A, we describe the underlying concept of the group and it's aims and in Part B we describe the current status of the groups development.

Methods. Four group meetings funded by the EEC have enabled definition of a "Core Dataset" to be collected from all centres regardless of specific project aim. A form based feasibility study was conducted and a prospective data collection exercise of core data using PC and hand held computer based methods is in progress.

Findings. A core-dataset was defined and can be downloaded from the *BrainIT* web-site (go to "*Core dataset*" link at: www.brainit.org). A form based feasibility study was conducted showing the overall feasibility for collection of the core data elements was high. Software tools for collection of the core dataset have been developed. Currently, 130 patient's data from 16 European centres have been recruited to the joint database as part of an EEC funded proof of concept study.

Interpretation. The *BrainIT* network provides a more standardised and higher resolution data collection mechanism for research groups, organisations and the device industry to conduct multi-centre trials of new health care technology in patients with traumatic brain injury.

Keywords: Head injury; multi-centre network; health technology assessment; neurointensive care; ICP monitoring; cooperative network.

Part A: *BrainIT* group concept

Head injury has devastating economic and social consequences both to the victim and to the society that supports the victim [1]. When assessing head injured patients' outcome from new therapies or the application of new monitoring devices, a large number of patients are required [4]. Recruiting patients from multiple centres will significantly reduce the time to assess new therapy and monitoring. However, despite the existence of guidelines for the management of severely head injured patients [2], this group of patients is subject to considerable variability in care [3]. As a first step towards improving management standards in this population, both the inter and intra-centre variability in the management and treatment of these patients needs to be assessed on a multi-national basis, and to do so requires a more standardised and higher resolution methodology for acquiring patient management and monitoring information.

Group formation

The idea for the Brain Monitoring with Information Technology (*BrainIT*) group came from discussions arising during the 10th International Symposium on Raised Intracranial Pressure and Neuromonitoring in Brain Injury held in Williamsburg, USA in May 1997. A few participants at this meeting, with a specific interest in neuro-monitoring, agreed that a more open and collaborative approach to the assessment of new monitoring technology would be a more efficient approach then continuing our current practice of conducting small scale, single centre studies. From those initial meetings in Williamsburg, a web site was setup (www.brainit.org) and from the interest generated, it became clear that *many* were interested in the concept of an open collaborative approach to developing standards in this area. Currently there are over 100 members from 25 countries who have registered interest in the group via the website. It is possible to summarise the interests of the group into three main aims:

The main aims of the group are:

1. To develop and disseminate improved standards for the collection, analysis and reporting of intensive care monitoring data collected from brain injured patients.
2. To provide an efficient multi-centre infra-structure for generating quality evidence on the utility of new forms of intensive care monitoring and methods for improving the care and outcome of brain-injured patients.
3. To develop and use a standardised database as a tool for hypothesis generation and the development, testing and validation of new data analysis methodologies.

The BrainIT *group approach – what are the differences?*

The Ethos of the *BrainIT* group is one of fostering open and free collaboration. The approach used, which we believe is novel in this field of medicine involves the following key elements:

1. *Only* high-resolution minute-by-minute monitoring and detailed management data is collected using computer based data collection tools. A basic set of data collection software tools are provided to all data contributing members free of charge. In addition to the free tools offered, the group is collaborating with industry on the development of more sophisticated data collection technology. A technical sub-group works towards developing tools and methods to assist with standardising data collection, analysis and database tools across centres.
2. A project-by-project based collection of data, where members voluntarily donate their time and effort towards collection of data for specific projects in which they are enrolled. The BrainIT group Internet based facilities (Web page and Forum) allow members either individually or in groups to form their own projects, enlist interest from other members, attract grant funding and manage their own project. Individual project PI's are responsible for project management, funding and publication of their results.
3. The data model used differs from previous collaborative groups working within the field of traumatic brain injury in that data collected as part of individual projects is also donated to a joint database. Data collection tools used in projects collect, as a minimum, a "Core Dataset" which once collected and anonymised is added to a common database. The common database will be openly accessible, through the Internet, to all "Data Contributing" centres. The database will be able to be queried over the Internet and datasets of interest can be downloaded to any member who has also contributed data to the database.
4. A steering group with overall responsibility for group management, does not dictate project selection but can help with project design if required. An important function of the steering group is to track data analyses being performed on the joint database to ensure a high level of analysis is maintained. Only officially registered and planned analyses conducted on "validated" data can lead to publication and presentation at meetings. The steering group will ensure that database access, analysis and publication criteria are adhered to.
5. An important element of the BrainIT group approach is to continuously work towards the development of improved "standards" for multi-centre collection and analysis of data in this patient population. We have achieved a key first step in this process by defining minimum data validation standards and have developed a mechanism for checking the validity of data against original documentation using regionally hired "data validation" staff. The BrainIT network provides an infrastructure supporting data quality control for trials of management or monitoring similar to that required by the Pharmaceutical industry in the conduct of trials of new drugs.

A detailed BrainIT Group "Operational Strategy" document can be viewed and downloaded from the group web-page (www.brainit.org → go to Operating Strategy Link). A sub-set of this document is summarised below. Database access and joint publication criteria can be viewed and downloaded from the group web-page (www.brainit.org → go to Operating Strategy Link) (Fig. 1).

Regional coordination

Each region (usually one but occasionally more than one country sharing a common language) with more than one neuro-intensive care centre contributing data to the *BrainIT* database has a *regional* coor-

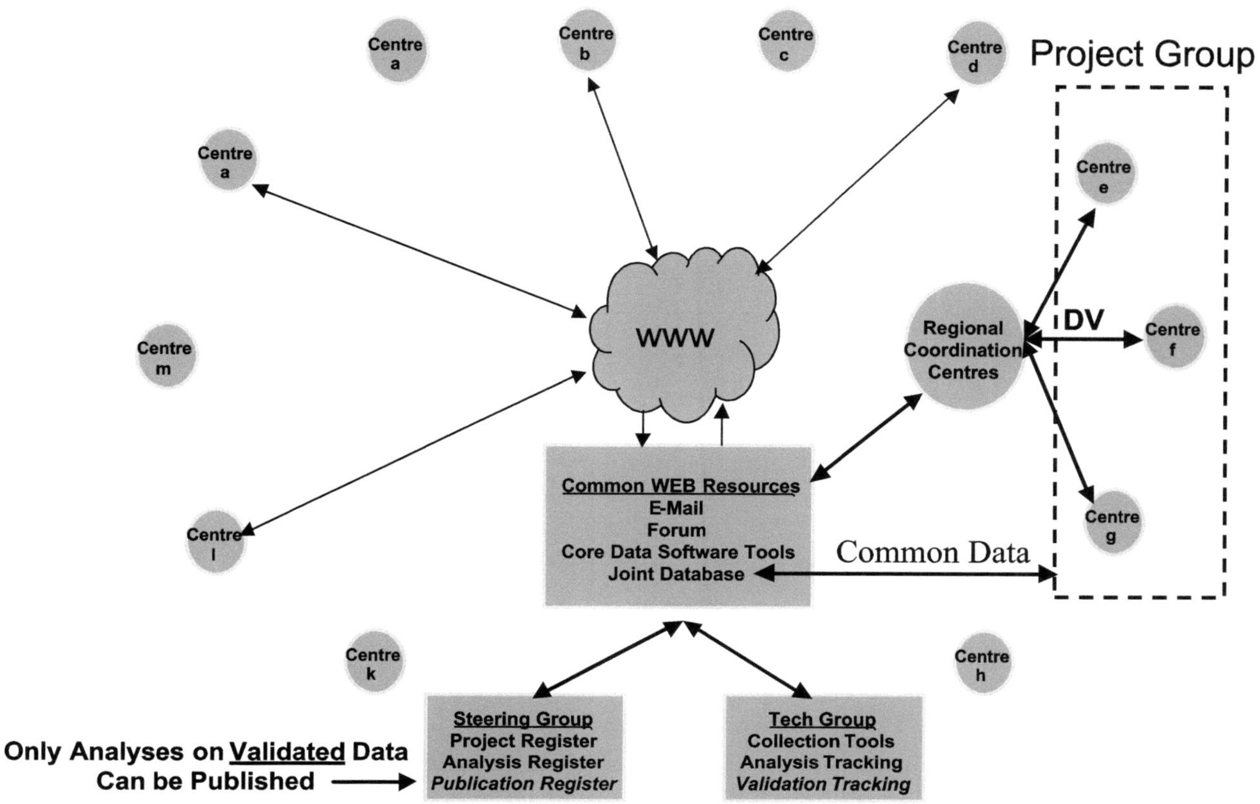

Fig. 1. *Graphical Representation of BrainIT Group Concept.* Using the Internet as a mechanism linking individual investigators, the *BrainIT* group provide web resources (mailbase forum, discussion forum and free access to common data collection tools) to foster formation of project groups. Project groups are responsible for managing, funding and publishing their own work. Collected data is anonymous and donated to a common database for the benefit of the entire network. Any data-contributing centre can access the entire common database useful for post-hoc hypothesis testing and generation. Only "Validated" data can lead to publication and the BrainIT group provides a region-by-region based mechanism for hiring and managing Data Validation (*DV*) staff to validate project group data. Validation costs will be generated from a range of resources, including a contribution from grant funding sourced from individual project group grants. Project and analysis duplication is prevented by a Steering group maintaining and managing a project and analysis register. A technical group helps develop data collection, analysis and database tools

dinator providing coordination support for centres within their region. They also hire, train and support "Data Validation" staff (funding dependant), used to travel to local centres to train local centre staff in the use of *BrainIT* data collection and analysis tools, as well as, conducting data validation exercises. *See the Operational Strategy Document for an overview of BrainIT Data Validation Approach.*

Group funding

The major resource cost of the BrainIT group is for the hiring and travel support of Data Validation Staff. These staff are currently grant funded. Grant funding will, for the most part, remain the predominate source of support, however, other sources of support are being considered including by industry and public donation. As the group expands and more project groups form and bring in their own funding, it may be possible to create a central DV staff resource fund based upon a fixed percentage of project funding.

Part B: current status 2004

Paper-based feasibility study

A paper form based feasibility study was conducted. Eighteen centres (82%) returned completed forms by the set deadline. Overall the feasibility for collection

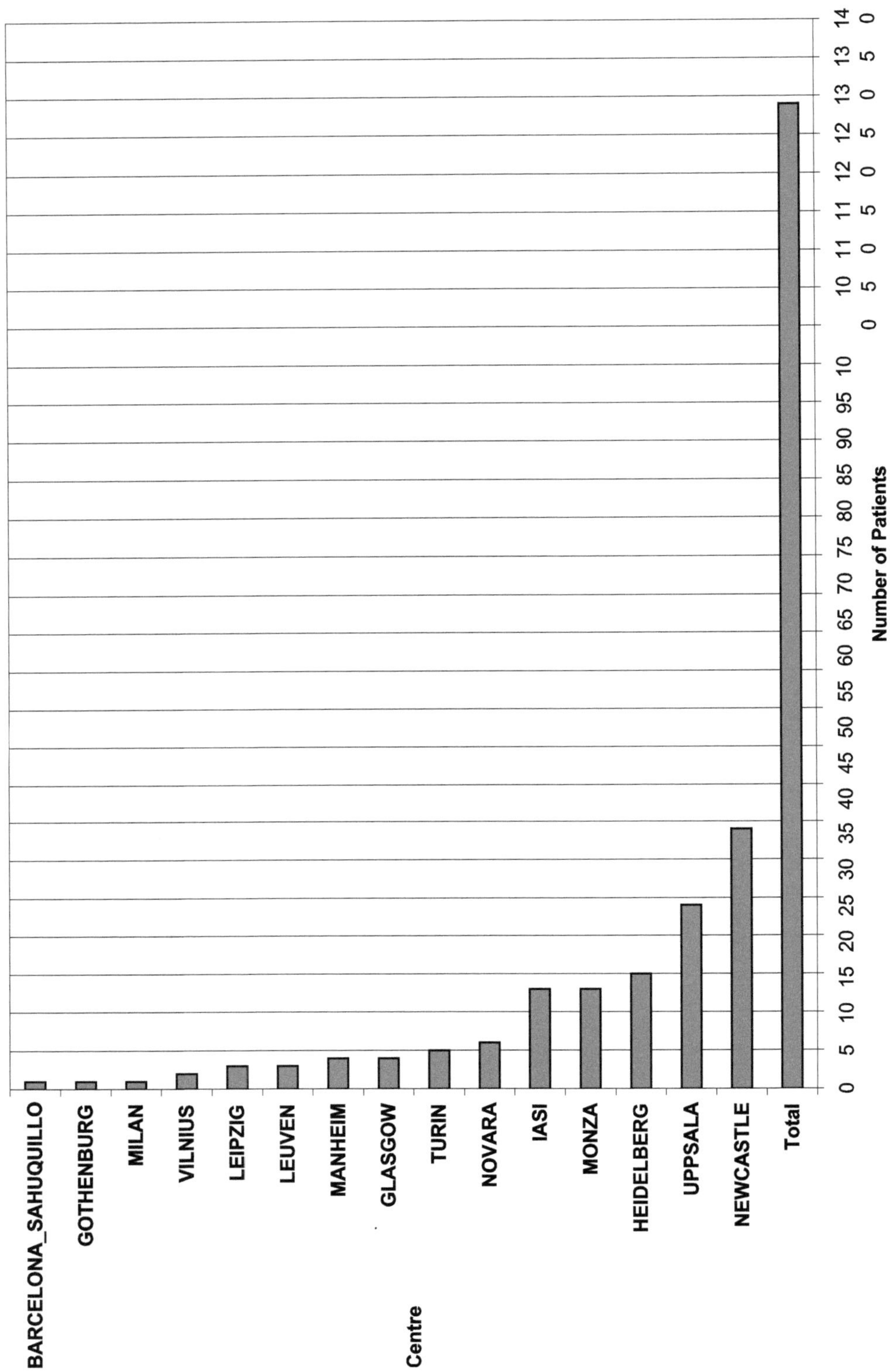

Fig. 2. Graph showing current recruitment by centre for the EEC funded proof of concept study

of the core data elements was high with only 10 of the 64 questions (16%) showing missing data. Of those 10 fields with missing data, the average number of centres not responding was 12% and the median 6%. An SQL database to hold the data has been designed and implemented Software tools for collection of the core dataset have been developed. Ethics approval has been granted for collection of multi-centre data as part of a pilot data collection study.

Proof of concept pilot data collection study

In October 2002, the *BrainIT* group received 3 year EEC research infrastructure support under the Quality of Life and Management of Living Resources Programme. This support enabled:

- Creation of a "Regional" Coordination Structure consisting of 6 country-specific coordinating centres via which data validation staff are hired and regional project and data acquisition is coordinated.
- Hire staff to Develop, Install and Train local centres in new IT tools to collect core dataset
- Prospectively collect (*and Validate using specially hired and trained staff*) at least 5 patients/year of core data from 30 centres across 13 EEC countries:
  - UK (7), Germany (4), Italy (4), Sweden (3), Spain (3), Lithuania (2), Belgium (1), Netherlands (1), France (1), Denmark (1), Switzerland (1), Czech (1), Romania (1)

Figure 2 above shows the current status of patient recruitment.

Discussion

This paper has outlined the concept underlying the BrainIT group approach to collaboration and describes the group's efforts to define and test the feasibility for collection of a core-dataset. The current status of a prospective proof of concept data collection study is described. Collectively working towards raising data collection and analysis standards is a critical aim of the BrainIT group and the current work towards defining minimum data validation standards and developing a mechanism for checking the validity of data against original documentation using regionally hired "data validation" staff will provide an infrastructure supporting data quality control for trials of management or monitoring similar to that required by the Pharmaceutical industry in the conduct of trials of new drugs.

References

1. Berkowitz M (1993) Assessing the socioeconomic impact of improved treatment of head and spinal cord injuries. J Emerg Med 11: 63–67
2. Ghajar *et al* (1995) Survey of critical care management of comatose head injured patients in the United States. Crit Care Med 23: 560–567
3. Maas AI, Dearden M, Teasdale GM *et al* (1997) EBIC guidelines for management of severe head injury in adults. European Brain Injury Consortium. Acta Neurochir (Wien) 139: 286–294
4. Signorini DF, Andrews PJD, Jones A *et al* (1999) Predicting survival using simple clinical variables: a case study in traumatic brain injury. J Neurol Neurosurg Psychiat 66: 20–31

Correspondence: Ian Piper, Department of Clinical Physics, Institute of Neurological Sciences, Southern General Hospital, 1345 Govan Road, Glasgow G51 4TF, UK. e-mail: ipiper@clinmed.gla.ac.uk

Acta Neurochir (2005) [Suppl] 95: 39–41

Accurate data collection for head injury monitoring studies: a data validation methodology

J. Barnes, I. Chambers, I. Piper, G. Citerio, C. Contant, P. Enblad, H. Fiddes, T. Howells, K. Kiening, P. Nilsson, and Y. H. Yau for the *BrainIT Group*

Regional Medical Physics Department, Newcastle General Hospital, Newcastle upon Tyne, UK

Summary

Background. BrainIT is a multi centre, European project, to collect high quality continuous data from severely head injured patients using a previously defined [6] core data set. This includes minute-by-minute physiological data and simultaneous treatment and management information. It is crucial that the data is correctly collected and validated.

Methods. Minute-by-minute physiological monitoring data is collected from the bedside monitors. Demographic and clinical information, intensive care management and secondary insult management data, are collected using a handheld computer. Data is transferred from the handheld device to a local computer where it is reviewed and anonymised before being sent electronically, with the physiological data, to the central database in Glasgow. Automated computer tools highlight missing or ambiguous data. A request is then sent to the contributing centre where the data is amended and returned to Glasgow. Of the required data elements 20% are randomly selected for validation against original documentation along with the actual number of specific episodic events during a known period. This will determine accuracy and the percentage of missing data for each record.

Conclusion. Advances in patient care require an improved evidence base. For accurate, consistent and repeatable data collection, robust mechanisms are required which should enhance the reliability of clinical trials, assessment of management protocols and equipment evaluations.

Keywords: Head injury; data collection; database; validation.

Introduction

The incidence of serious head injury is estimated at 1500 per 100,000 population per annum which equates to over one million head injuries per year in Europe. The long term care of these patients bears a great burden on society and has both social and economic implications for the injured person and carers involved [1]. The incidence and effects of secondary insults play a significant part in the outcome of patients. Prompt, and better medical management of these insults has improved the outcome in these patients [3]. The Traumatic Coma Databank used a common data collection protocol and provided information on patient recovery and outcome [5]. Recent pharmaceutical studies have endeavoured to protect the injured brain from further secondary insults therefore aiming to improve the outcome of a predominately young population. In neither case has high resolution monitoring data been recorded. Therefore the precise relationship between outcome and the deviation of physiological variables made. Clinical Trials of this sort have proved largely inconclusive, and a critical analysis of the possible reasons for this failure is given in an article by Maas [4]. With advances in monitoring and information technology interest has now shifted towards the accurate minute-by-minute monitoring of severely injured patients with a view to observing momentary secondary insults as they occur. Because of the relatively low number of head injuries seen in any one centre, and the number required to produce robust evidence based studies, it is necessary to pool data from several centres. This can only be achieved if the data from different centres can be combined in a standardised format.

The BrainIT Group http://www.brainit.org/ is a collaboration of Healthcare Professionals who came together during the 10th International Symposium on Raised Intracranial Pressure and Neuromonitoring in Brain Injury in Williamsburg, USA in May 1997. From this meeting and subsequent discussions, an EEC grant was obtained to fund the development of a network infrastructure that culminated in the formation of the BrainIT Group. The Group currently consists of 30 European Centres and led by a Steering

Group. The aim is to collect high quality continuous data from severely head injured patients and store it in a database that is available for analysis and review. The group strive to standardise data collection for this group of patients and also to establish and populate data for analysis. Each participating country has been assigned a data validator to help co-ordinate the setting up of centres each in their own country. A core data set has previously been defined [6] that includes minute-by-minute physiological data and simultaneous treatment and management information. Collected data is transferred via the BrainIT website to the Central Database in Glasgow where automated computer tools are used to highlight missing or ambiguous data. Of the required data elements 20% are randomly selected for validation against original documentation along with the actual number of specific episodic events during a known period. This will determine accuracy and the percentage of missing data for each record.

Methods

The defined core data set consists of four aspects of data collection: minute-by-minute physiological monitoring data, demographic and clinical information, intensive care management data and secondary insult management data. All these aspects of data collection are essential for the completion of each patient data file. Eligible patients for this study may be of any age with the inclusion criteria being insertion of an intracranial pressure monitoring device and arterial line. The minimum amount of time a patient may be monitored for is four hours. Data collection in the United Kingdom may begin as soon as the patient is admitted to intensive care as approved by the Multi Research Ethics Committee, Scotland, with written assent being obtained as soon as is practical after admission. In cases where assent is not given for any reason, then the previously collected data will not be retained.

Physiological data collection on such a large scale has never previously been attempted and various methods exist for collection of such data. Some centres use their own specifically devised data collection programmes, whilst others have enlisted commercial help to install and set up equipment to collect the required minute-by-minute data. Data from several areas (Intensive Care Unit, High Dependency Unit, Neurotheatre) may either be collected centrally via a networked system or at each bedside using a computer connected to the physiological monitor. Data from these systems is transferred to Glasgow via the BrainIT website. Collection of physiological data involves very little input from the ICU nursing and medical staff and requires one or two designated persons to be responsible for downloading and sending each patient's data to a dedicated research data server at the BrainIT Co-ordinating Centre in Glasgow.

Collection of demographic and clinical information, and intensive care treatment and management data is by use of handheld devices (PDAs). Specialist software developed by Kelvinconnect Limited enables the user to enter information relating to the patient's head injury and subsequent management whilst the patient is in intensive care. Secondary insult management data is also collected with a list of named target therapies and targets to which these therapies are aimed. All data is collected as long as the patient has arterial and intracranial pressure monitoring. Data is then transferred from the handheld device to a local computer, where it is stored in a database, therefore providing a list of recruited patients for each local centre. Prior to transfer to Glasgow the data collector will obtain a unique eight digit number from the BrainIT website and attach it to each patient's file.

Software on the local computer allows review of the data and all personal details are removed leaving the BrainIT number only as a means of patient identification. Data is then transferred to Glasgow to the central database. Data conversion tools convert data to the BrainIT standard format and it is then added to the central database. A web interface has been designed to allow members of the BrainIT Group access to the database and the information it provides.

Once the data has been received in Glasgow, automated computer tools convert data to a common file format, standardise units, parse the data and highlight missing or ambiguous data. A missing data request is sent to the local centre. Missing data, if available, is entered and the file is returned to Glasgow. This process continues until as much of the missing data as possible is found. Of the required data elements 20% are randomly selected for data validation against original documentation along with the actual number of specific episodic events during a known period. This part of the process requires assistance of the country data validator. A list of requested data for validation is sent to the data validator who will then visit the local centre from where the data was originally sent. Using the patient's notes and charts the data validator will check the list of data against the original documentation. A validation file will be created in the contributing centre using the BrainIT Core Data Collection tool which will then be exported and sent to Glasgow via the internet. Again missing or incorrectly entered data is identified and a new validation request is sent to the data validator. These steps are repeated until all of the missing validation data has been collected. It will be possible to search within the database for both validated and non-validated data. Only validated data should be used in analyses intended for publication or submission, whilst non-validated data may be accessed and used in analyses intended for hypothesis generation.

Data validation procedures are as follows. First of all fully anonymised raw data transferred to Glasgow is kept on the BrainIT group server which is separate from the common database. For all data entered to the common database one of four levels of data validation are progressively applied. Validation level one is a *"well-formed"* data check used to ensure that the data conversion stage functioned correctly, in particular the time-stamp format (YYYY-MM-DD HH:MM) is valid. Data validation level two checks all non-numeric categorical core dataset data for transcribing errors. Validation at this level differs depending on whether monitoring or non-monitoring data is being validated. Validation of monitoring data requires a start and end time for each particular monitoring channel, also that the difference between the range of monitored data values of the two methods (collected versus archived original source) compared with the average of the two methods does not show a significant bias and this will be achieved in a standard means by the application of a Bland and Altman form of statistical analysis [2]. Level three validation involves the conversion of units to BrainIT units if required whilst level four validation involves the data validators to check the accuracy of data collected against source documents. Of the four levels of validation, only the last (level four) requires human resources and so to cope flexibly with periods of low funding, the BrainIT group has also defined three "types" of level four validation: type one or *self validation* where the Principal Investigator validates his own data. Type two is *cross validation* and involves colleagues in nearby BrainIT centres validating each other's data whilst type three is *independent validation* and involves the data validator who has no link with the BrainIT centre. Each patient

Table 1. *Data collected from centres*

Data sent to Glasgow	
Number of centres contributing data	15
Total number of patients recruited	115
Number of patients on database	50
Requests for missing data	21
Average number of requests per patient	2

record in the database is sortable on both *level* and *type* of validation allowing individual investigators to prescribe their own validation security level. For example, for hypothesis generation studies levels 1–4 can be used, but for analysis intended for publication only level four data can be used but with a choice of validation *types* 1–3 (self validation → independent data validators).

Results

The project officially started in September 2002, although staff were not employed until January 2003. The first year of the project was spent developing hardware and software methods and in training staff. Currently 30 centres have agreed to participate and contribute data. Full data recruitment started in January 2004. A total of 115 patients from 15 centres currently have been recruited to the study of which all of the minute-by-minute physiological data, clinical management and demographic raw data have been transferred to Glasgow. To date, 50 patients' data has been converted to the common data format and entered into the project database (Table 1).

Twenty one requests for missing data were sent out from Glasgow to the relevant centre giving an average number of requests per patient of two. Some centres have still not recruited patients and there are several reasons for this. Out of range data will also be highlighted and requests for verification of such data will be sent to the contributing centre.

Discussion

The aim of this study is to standardise the collection of data from head-injured patients in a previously defined set format. This approach ensures that all those involved are collecting data in the same format. Using this approach, data from different centres can be combined together to provide a larger better standardised data set than would normally be available from a single centre. We have set up systems so data transfer can be done over the internet and tools for data validation produced and evaluated. The project is still developing and the techniques still require to be evolved into an automated process. However we have demonstrated that our data transfer process is successful and that missing data can be readily identified and requests made to local centres for such data. The production of a reliable fully populated database from head injured patients provides, for the very first time, a means to access data that can be used for analysis and hypothesis testing. Storage of validated data in this way ensures a firm evidence base which contributors may draw upon and use to ensure standardised practice for head injuries.

References

1. Berkowitz M (1993) Assessing the socioeconomic impact of improved treatment of head and spinal cord injuries. J Emerg Med 11: 63–67
2. Bland JM, Altman DG (1986) Statistical methods for assessing agreement between two methods of clinical measurement. Lancet 8(1): 307–310
3. Chesnut R (2004) Management of brain and spine injuries. Crit Care Clin 20(1): 25–55
4. Maas AIR, Steyerberg EW, Murray GD, Bullock R, Baethmann A, Marshall LF, Teasdale GM (1999) Why have recent trials of neuroprotective agents in head injury failed to show convincing efficacy? A pragmatic analysis and theoretical considerations. Neurosurgery 44(6): 1286–1298
5. Marshall LF, Becker DP, Bowers SA, Cayard C, Eisenberg H, Gross CR, Grossman RG, Jane JA, Kunitz SC, Rimel R, Tabbador K, Warren J (1983) The national traumatic coma data bank. Part 1: design, purpose, goals and results. J Neurosurg 59(2): 276–284
6. Piper IR, Citerio G, Chambers IR, Contant C, Enblad P, Fiddes H, Howells T, Kiening K, Nilsson P, Yau YH, for the BrainIT group (2003) The BrainIT group: concept and core dataset definition. Acta Neurochir (Wien) 145: 615–629

Correspondence: I. R. Chambers, Regional Medical Physics Department, Newcastle General Hospital, Newcastle upon Tyne, NE4 6BE, UK. e-mail: i.r.chambers@ncl.ac.uk

Acta Neurochir (2005) [Suppl] 95: 43–49

ICM+: software for on-line analysis of bedside monitoring data after severe head trauma

P. Smielewski[1,2], **M. Czosnyka**[1], **L. Steiner**[1], **M. Belestri**[1], **S. Piechnik**[1], and **J. D. Pickard**[1,2]

[1] Academic Department of Neurosurgery, Addenbrooke's Hospital, Cambridge, UK
[2] Wolfson Brain Imaging Centre, Addenbrooke's Hospital, Cambridge, UK

Summary

ICM software was developed in 1986 in Warsaw, Poland and has been in use at the University of Cambridge Neurocritical Care Unit for 10 years collecting data from bed-side monitors in nearly 600 severely head injured patients and calculating secondary indices describing cerebral autoregulation and pressure-volume compensation. The new software ICM+ includes a much extended calculation engine that allows easy configuration and on-line trending of complex parameters.

The program records raw signals, and calculates time trends of summary parameters. Configuration and analysis utilises arithmetic expressions of signal processing functions to calculate various statistical properties for each signal, frequency spectrums and derivatives, as well as correlations/cross-correlations between signals. The software allows configuration of several levels of analysis before calculating the final time trends. The final data are displayed in a variety of ways including simple time trends, as well as time window based histograms, cross histograms, correlations etc. All this allows complex information coming off the bed-side monitors to be summarized in a concise fashion and presented to medical and nursing staff in a simple way that alerts them to the development of various pathological processes.

The system provides a universal tool for clinical and academic purposes. Its flexibility and advanced signal processing is specialized for the needs of multidisciplinary brain monitoring.

Keywords: Brain monitoring; multimodal monitoring; on-line data analysis; cerebral autoregulation; head trauma; intensive care; neuro-monitoring.

Introduction

In an established environment of Clinical Neuroscience Department enormous quantities of data can be captured from bed-side monitors per each patient [5]. From that data information regarding cerebral autoregulation, cerebrospinal compensatory reserve, oxygenation, metabolite production and function can be obtained [1, 3, 6]. However, continuous assessment of changing cerebrovascular haemodynamics and oxygenation demands not only reliable monitoring techniques, but also sophisticated and time consuming signal analysis. This can be provided by dedicated computer support.

The intensive care multimodality monitoring system adopted in the Cambridge Neurosurgical Unit is based on software for a standard PC, equipped with a digital to analogue converter and RS232 serial interface, running Windows 2000/XP. First version of the software was introduced into clinical practice in Poland, Denmark and the UK in the middle 1980s and has subsequently been extended into a system for multimodal neuro-*i*ntensive *c*are *m*onitoring (ICM) and waveform analysis of intracranial pressure [1, 8] used in Cambridge, UK, and other centres in Europe (Goteborg, Toulouse) and United States (Detroit). Most data has been derived from head injured [3] and hydrocephalus patients [2]. However, the same or similar techniques are being increasingly applied to those suffering from severe stroke, subarachnoid haemorrhage, cerebral infections, encephalopathy, liver failure, benign intracranial hypertension, etc. In addition to monitoring of multiple variables describing dynamics of the studied pathology [5], some secondary indices have proved to be useful in clinical neurosciences [6]. The best known example is the cerebral perfusion pressure, calculated as a difference between mean arterial pressure and ICP. More sophisticated indices describing cerebrospinal compensatory reserve, pressure autoregulation and vascular reactivity were introduced to clinical practice recently and proved to be useful in head injury [3, 5] or poor grade subarachnoid haemorrhage. The aim of this paper was to present and discuss the concepts implemented in the new version of the monitoring software ICM+ and to show its applications in research of Neurosurgical diseases and head traumas.

Methods

The ICM+ software extends the ideas included in the original ICM program for DOS [1] as well as in the cerebrovascular reactivity test analyser CVRTest [8] by the same authors. In particular, the principle of multistage analysis of the bed-side monitors signals has been carried through (Fig. 1). It has however been extended heavily based on the authors experience over the years of using the software in the neuro-intensive care unit in the Addenbrookes Hospital, Cambridge.

The software is composed of several modules of which the core are:

1. Sampler module – collects data from variety of sources including analogue/digital converter and RS232 serial port. Configuration of the analogue input involves specifying sampling frequency, actual analogue channel to be used for particular signal as well as the voltage amplifier gain and the voltage-signal units conversion. Configuration of the serial input involves communication protocol, specifying record and data field separators as well as the position of the particular data field in the data record. All the signals are provided to sampler clients as one list with their respective sampling rates, hiding the actual signals sources. There is also an off-line version of the sampler that uses a data file (several formats are supported) instead of directly reading the data from bed-side monitors.
2. On-line Data Analysis module – collects signal samples from the Sampler and processes them according to requested analysis configuration (Fig. 2) producing time trends of calculated parameters.
3. Display module – allows viewing of the time trends using a selection of various charts, data browsing and analysis tools.
4. Diagnostic manoeuvres analysis module – provides a mechanism for plugging in optional analysis tools for various intervention tests like for example transient hyperaemic response test, CO_2 reactivity test or cerebrospinal fluid infusion study [2, 4, 9].

Results

The software has been used in a variety of applications covering various neurological pathologies ranging from hydrocephalus, stroke to severe head injury. For the last two years it has been used routinely in the neuro-intensive care unit in the Addenbrookes Hospital, Cambridge, UK. Total number of 78 patients has been monitored/diagnosed using this software. Case studies exemplifying the software application in different neurological condition are presented in Figure 3. More examples of research and clinical applications of the software can be found on the web site http://www.neurosurg.cam.ac.uk/icmplus.

Discussion

Data from various monitoring equipment used in an intensive care unit contains a wealth of information about the patient's state. Some of the signals coming from the monitors are more complex than others, i.e. include complex waveforms. Instantaneous values of those signals are often difficult to interpret. Trends of minute by minute time averages travel far to aide in interpretation of the monitoring data but they completely dispose of information carried by the waveforms. Also, it is often the strength and character of association between different signals that provides extra information rather than the signals themselves.

The first specialised computer-based systems for neuro-intensive care were introduced at the beginning of the 1970s. Initially these systems were oriented to the monitoring of ICP and ABP allowing calculation of CPP and a basic analysis of the pulse waveform. In contrast, contemporary systems, like the one presented here, are highly sophisticated multi-channel digital trend recorders with built-in options for complex signal processing [1, 8]. The considerable flexibility of such systems allows almost unlimited signal analysis, which can generate a state of data chaos. Thus the modern user is faced with the problem of which parameters should be considered, and how the data should be interpreted. The information should then be presented in a manner that is comprehensible to medical and nursing staff. Decision on which calculated parameters are clinically most relevant should be made by the clinical stuff in collaboration with their research colleagues. Those carefully selected parameters, and only those, should then be presented continuously on the computer screen. Flexible configuration of data views implemented in ICM+ software allows for designing a clear summary view page for the clinically relevant parameters and additional, readily available pages containing all the other calculated parameters for more advanced clinical users or research fellows.

Multiple data sources

In a research orientated intensive care unit there may be several other monitors available in addition to the standard bed-side monitors of ABP and ICP. Those devices, e.g. TCD, Neurotrend, NIRS etc, are often used only during certain time periods rather than for the whole time of patient's stay in the ICU. Rather than creating new file with each configuration change ICM+ can accommodate those changes in its native file format and presents the whole patient data across different configuration as one time strip.

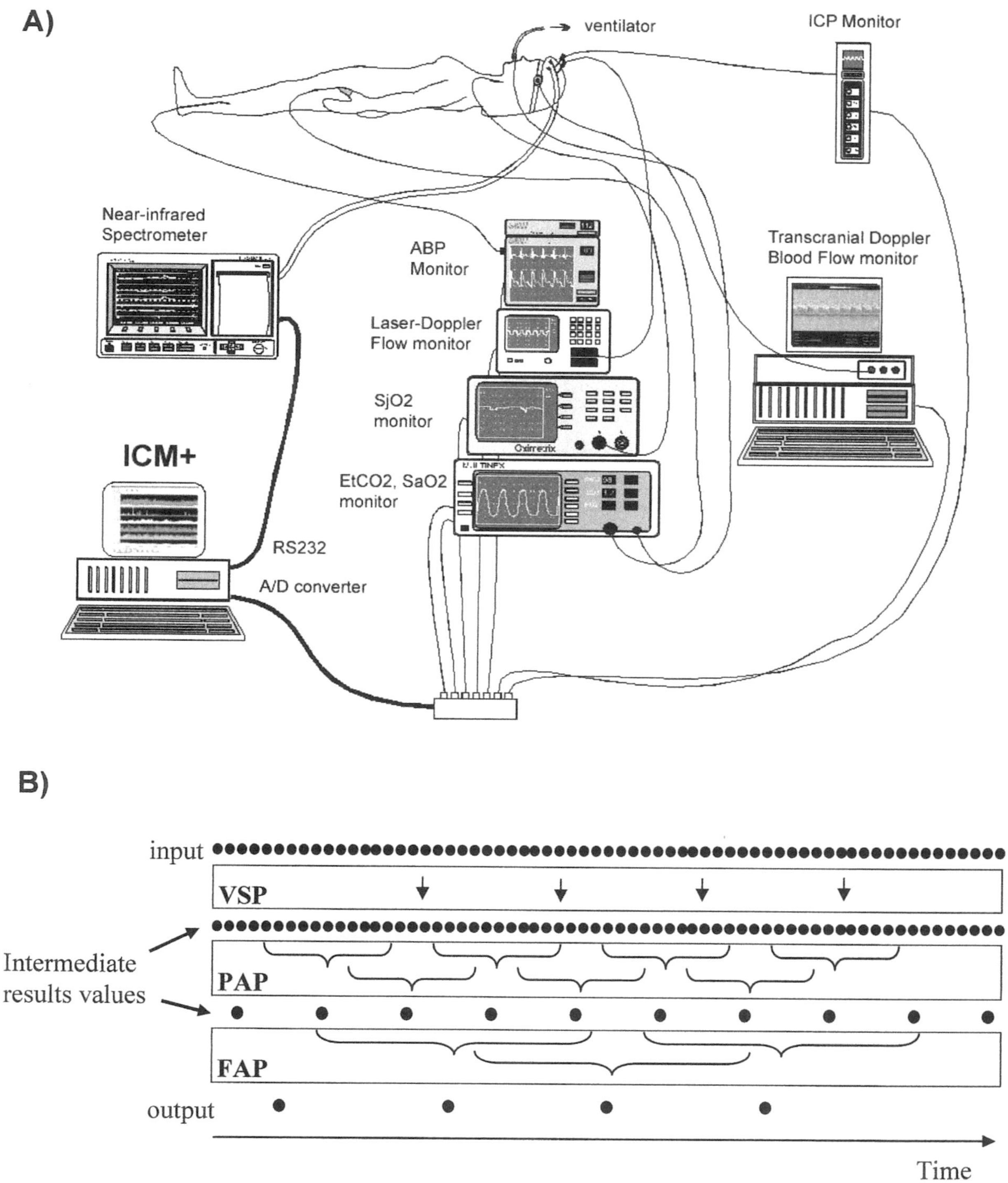

Fig. 1. (A) A symbolic diagram of a busy intensive care monitoring environment with various bed-side monitors hooked up to the data collection PC using either analogues or serial outputs. (B) A concept of multistage processing: raw data samples are first scaled using Virtual Signal Processor (*VSP*); the next stages (Primary Analysis Processor – *PAP* and Final Analysis Processor – *FAP*) operate on buffers of samples from the previous calculation stage – each buffer produces one intermediate result value at a time. The length of the buffer and how frequently the calculations are performed is fully configurable as are the calculation formulas

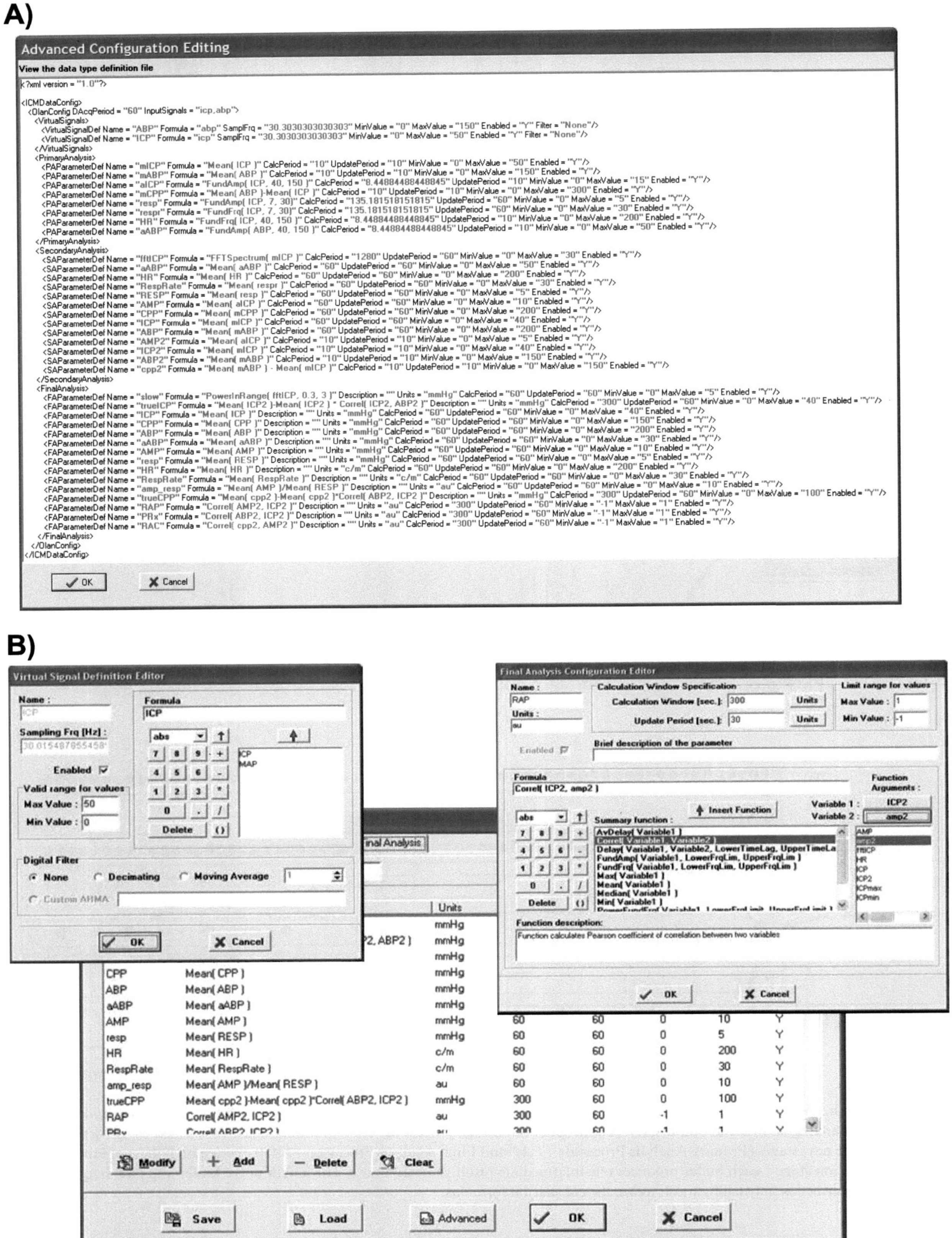

Fig. 2. Example of configuration of the on-line data analysis module. (A) The whole analysis prescription presented in its native XML format. (B) Snapshot of the graphical user interface for building up the analysis configuration

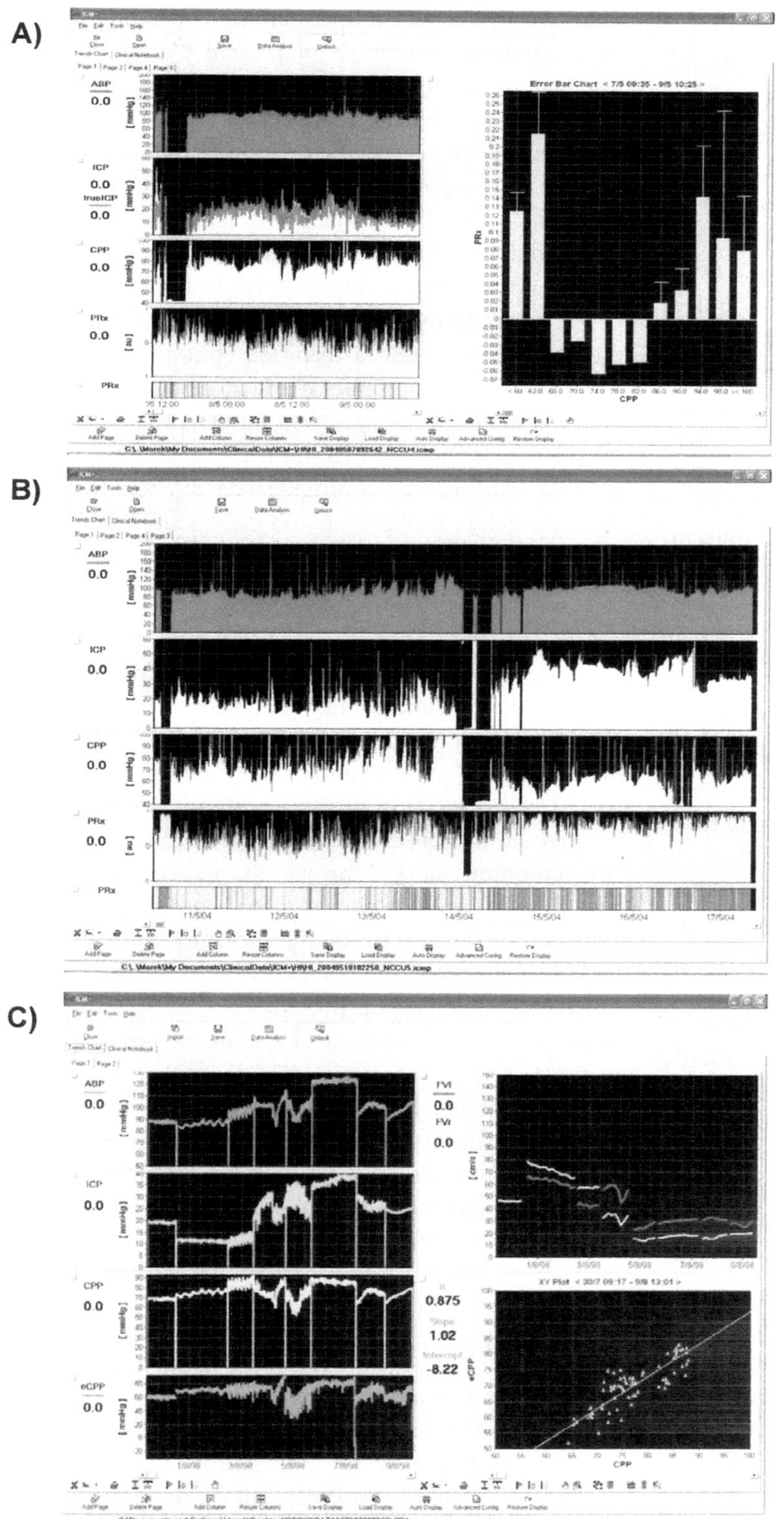

Fig. 3. Examples of ICP and ABP monitoring patients after head injury: (A) Patient admitted with GCS of 5. Initial ICP was around 40 mmHg and normalized after removal of large left subdural haematoma. Analysis of Pressure-reactivity (right panel) indicated 'optimal CPP' [10] for this patient of around 74 mm Hg; (B) Patient with multiple injuries. ICP was at first moderate but unstable, pressure-reactivity (*PRx*) was good. ICP deteriorated after abdominal surgery on the fourth day after accident. Pressure-reactivity index indicated permanent haemodynamic failure (see changing colour of bottom risk-chart: from green-light grey to red-dark grey). Patient died on day 6 after injury; (C) Day-by-day assessment of non-invasive CPP (*eCPP*) [7] after head injury. Correlation between non-invasive CPP and 'real' CPP is presented in right bottom corner

Clinical observations and comments

From the clinical research perspective the collected data is largely useless unless accompanied by the clinical picture of the patient. In order to decipher various episodes seen in the recorded time trends a scrupulous list of clinical observations should accompany the recording. Those observations can be recorded by the ICM+ software using three different methods: a notes tool, a system of user-definable events marking, and an annotation tool. The first two methods are used in an on-line mode with the event/note being stamped with the current time. Annotations are used off-line and get inserted at the point of time cursor, anywhere within the recording. These can be used to annotate the recording post-factum and can be modified/deleted at any time, unlike the notes and events which, once inserted, are read only.

Artefacts treatment

To ensure high quality of data analysis artefacts have to be dealt with. One cannot for example calculate mean value of ICP in the first 24 hours after injury without first taking out periods where ICP signal was lost or distorted. We feel strongly that rather than automatically substituting artefact periods with values predicted from the past and/or future history of the signal it is better to treat those as missing values. However in order to reduce influence of artefacts on the calculations they are divided into three categories depending on the where the fault occurred. Firstly there are global artefacts – i.e. reflected in all the signals. These are caused by events which involve temporary disconnection of the patient from all the monitoring devices and affect all the parameters. Then there are device related artefacts. These arise when a particular input device, like A/D converter, or a monitor using serial port for data transmission, gets temporarily disconnected, malfunctions or gets suspended. Here only parameters dependent on signals coming from the particular device will be affected. Other parameters will have valid values in those periods, unless of course affected by other overlapping source of artefact. And finally, there are artefacts that arise from temporary disturbance in measurement of individual signals. These can be for instance arterial line flushing, ICP calibration etc. As before, only parameters dependent on the particular signal are affected. The signal specific artefacts periods can be automatically detected and marked up during on-line analysis depending on the user configuration. Each of the calculation formula used in the analysis has a minimum and maximum value specified by the user. When the formula value falls outside of the specified region, signal(s) that contribute (directly or indirectly) to the calculations will have the period from which that value was calculated marked as an artefact. Such an automated processing will in no way cause data loss due to misclassification since it is only the artefact mark up that is being created. The data itself is recorded continuously and is not modified. The artefact mark up can later be modified manually, if necessary, in an off-line mode. Such an approach allows for cautious treatment of artefacts in the recording but yet provide a flexible tool for automated artefact recognition and treatment.

Optional diagnostic tools

In addition to continuous assessment provided by the calculated time trends of indices it is sometimes necessary to introduce external excitation to the measured system and quantify its response to it. This could be an increase of the ventilator rate to induce change in arterial CO2 content, brief compression of the common carotid artery to induce momentary drop in the cerebral perfusion pressure, or controlled infusion of saline into the cerebrospinal fluid space in order to challenge the compensatory reserve [2, 4, 9]. Such an intervention provides an opportunity for more accurate assessment of the queried system characteristics than the analysis of spontaneous fluctuations originated from it. Tools available on-line for assessment of these diagnostic tests help to gain additional insights into the developing pathology as well as allow for cross calibration of the continuous time trends.

Reuse of data

With the ongoing research on signal processing of data from the bed-side monitors it is important to build a data bank of raw signals for testing new hypothesis. The ICM+ software allows storing those in addition to time trends of calculated parameters. The raw signals can then be fed back into the software times and times again, with a different analysis configuration. This facilitates verification of new ideas for on-line data processing.

Data protection act

In order to comply with the Data Protection Act, all the data stored in the native ICM+ file format is strongly encrypted. In addition, the patient description data, i.e. information identifying the patient, is encrypted separately using a key individually selected by each research centre. That means that another centre will not be able to identify the patient without knowing the secret key but they will be able to browse the data and use it for the research purposes. That approach facilitates collaborative data exchange without worrying about violation of the Data Protection Act.

In conclusion, the system provides a universal tool for clinical and academic purposes. Its flexibility and advanced signal processing is specialized for the needs of multidisciplinary brain monitoring.

References

1. Czosnyka M, Whitehouse H, Smielewski P, Kirkpatrick P, Pickard JD (1994) Computer supported multimodal monitoring in neuro intensive care. Int J Clin Monitoring Computing 11: 223–232
2. Czosnyka M, Whitehouse H, Smielewski P, Simac S, Pickard JD (1996) Testing of cerebrospinal compensatory reserve in shunted and non-shunted patients: a guide to interpretation based on observational study. J Neurol Neurosurg Psych 60: 549–558
3. Czosnyka M, Smielewski P, Piechnik S, Steiner LA, Pickard JD (2001) Cerebral autoregulation following head injury. J Neurosurg Vol 95: 756–763
4. Gooskens I, Schmidt EA, Czosnyka M, Piechnik SK, Smielewski P, Kirkpatrick PJ, Pickard JD (2003) Pressure-autoregulation, CO2 reactivity and asymmetry of haemodynamic parameters in patients with carotid artery stenotic disease. Acta Neurochir (Wien) 145(7): 527–532
5. Kirkpatrick PJ, Czosnyka M, Pickard JD (1996) Multimodal monitoring in neuro-intensive care. J Neurol Neurosurg Psych 80: 131–139
6. Pickard JD, Czosnyka M (1993) Management of raised intracranial pressure. J Neurol Neurosurg Psych 56: 845–858
7. Schmidt EA, Czosnyka M, Gooskens I, Piechnik SK, Matta BF, Whitfield PC, Pickard JD (2001) Preliminary experience of the estimation of cerebral perfusion pressure using trascranial Doppler ultrasonography. J Neurol Neurosurg Psych 70: 198–204
8. Smielewski P, Czosnyka M, Zabolotny W, Kirkpatrick P, Richards HK, Pickard JD (1997) A computing system for the clinical and experimental investigation of cerebrovascular reactivity. Int J Clin Monitoring Computing 14: 185–198
9. Smielewski P, Czosnyka P, Kirkpatrick P, McEroy H, Rutkowska H, Pickard JD (1996) Assessment of cerebral autoregulation using carotid artery compression. Stroke 27: 2197–2203
10. Steiner LA, Czosnyka M, Piechnik SK, Smielewski P, Chatfield D, Menon DK, Pickard JD (2002) Continuous monitoring of cerebrovascular pressure reactivity allows determination of optimal cerebral perfusion pressure in patients with traumatic brain injury. Crit Care Med 30(4): 733–738

Correspondence: Peter Smielewski, Neurosurgery Unit, Level 4, A Block, Addenbrookes Hospital, Cambridge CB2 2QQ, UK. e-mail: ps10011@cam.ac.uk

Acta Neurochir (2005) [Suppl] 95: 51–53

Survey of traumatic brain injury management in European *Brain-IT* centres year 2001

P. Nilsson[1], P. Enblad[1], I. Chambers[2], G. Citerio[3], H. Fiddes[4], T. Howells[1], K. Kiening[5], A. Ragauskas[6], J. Sahuquillo[7], Y. H. Yau[8], C. Contant[9], and I. Piper[10] on behalf of the *Brain IT* Group

[1] Department of Neurosurgery, University Hospital, Uppsala, Sweden
[2] Regional Medical Physics, Newcastle General Hospital, Newcastle upon Tyne, UK
[3] Department of Anaesthesia, Ospedale San Gerardo, Monza, Italy
[4] Department of Neurosurgery, South Glasgow University Hospital NHS Trust, Glasgow, UK
[5] Department of Neurosurgery, Charité, Virchow Hospital, Berlin, Germany
[6] Department of BioEngeneering, Kaunas University of Technology, Kaunas, Lithuania
[7] Department of Neurosurgery, Vall d'Hebron University Hospital, Barcelona, Spain
[8] Clinical Neurosciences, Western General Hospital, Edinburgh, UK
[9] Department of Neurosurgery, Baylor College of Medicine, Houston, Texas
[10] Department of Clinical Physics, South Glasgow University Hospital NHS Trust, Glasgow, UK

Summary

Background. The aim of this study was to obtain basic knowledge about the current local conditions and neurointensive care of traumatic brain injury (TBI) in the new multi-centre collaborative *BrainIT* group.

Materials and methods. The survey comprised a background part on local policies (Part A), and a case study section (Part B). The information was gathered by questionnaire followed by telephone interviews. Twenty-three *BrainIT* centres participated in the survey and answers from two respondents were available from 18 of the sites.

Results. The average proportion of agreement between duplicate respondents was 0.778 (range 0.415–1.00). All *BrainIT* centres monitored ICP. The treatment protocols seem to have a pattern concerning escalation of treatment of intracranial hypertension: 1/ evacuation of mass lesions and head elevation; 2/ increased sedation and mannitol; 3/ hyperventilation; 4/ ventricular drainage; 5/ craniectomy and barbituates.

Conclusions. There seemed to be an agreement on neurointensive care policies within the *BrainIT* group. The suggested order of treatment was generally in accordance with published guidelines although the suggested order and combinations of different treatments varied. Variation of treatment within the range of prescribed standards provides optimal conditions for an interesting future analysis of treatment and monitoring data in reality using the *BrainIT* database.

Keywords: Head injury; survey; management.

Introduction

The *BrainIT* (The Brain monitoring with Information Technology) group is an open collaborative organisation that was created to enable evaluation of neurointensive care. With the advent of powerful computers in most units the group felt that some kind of system should be set up to use this hardware for: testing new monitoring devices, collecting minute-by-minute care data and creating a databas where it would be possible to pool this data [1]. The database will contain prospectively collected data from patients in the *BrainIT* centers.

The present study was undertaken as an initial step in analysing treatment of patients with traumatic brain injury (TBI) in *BrainIT* centres. The goal was to see if there was agreement on treatment within centres and if there were great differences between the different *BrainIT* centres.

Materials and methods

The survey was comprised of a background part on local policies and conditions (Table 1, Part A), and a case study section (Table 2, Part B) focused on discerning differences in treatment tradition for specific clinical situations in traumatic brain injury. The information was obtained by written questionnaire followed by telephone interviews a few days later. Twenty-four *BrainIT* centres participated in the survey and answers from two respondents were available from 18 of the sites. The complete questionnaire has been published [2].

Table 1. *Part A local policies and conditions*

Question	Predominating answer	Proportion with this answer
Prehospital care	"Scoop and run policy" i.e. basic life support	60%
Primary admission to NSU	<25% primary admission	55%
Reason for referral to NSU from primary hospital	TBI, unconscious and pathological CT	93%
Acute neurosurgery in primary hospital (without NS)	no neurosurgery in local hospitals	88%
What kind of ICU exist at the center	separate NICU	76%
Where are TBI patients admitted (BrainIT center)	NICU, irrespective of other injuries	44%
Are there written protocols for treatment	local or published guidelines are used	78%

NS Neurosurgery, *NSU* Neurosurgical Unit, *ICU* Intensive Care Unit, *NICU* Neurointensive Care Unit.

Results

The average proportion of agreement between duplicate respondents was 0.778 (range 0.415–1.00). All *Brain IT* centres monitored ICP and other cerebral monitoring techniques were available in most centers.

Part B case study

Questions based on a fictuous case admitted with traumatic brain injury including information on how the radiology, neurology and intracranial pressure developed:

The treatment routines seem to have a pattern concerning escalation of treatment of intracranial hypertension: evacuation of mass lesions and head elevation [1]; increased sedation and mannitol [2]; hyperventilation [3]; ventricular drainage [4]; craniectomy and barbituates [5].

Discussion

This survey of TBI treatment is different than those earlier carried out in the USA; UK and Ireland [3–7]. The earlier surveys have tried to establish what rou-

Table 2. *Part B case study*

Question	Predominating answer	Proportion with this answer
Initial information *Glasgow coma sum score 8 (motor score 5), spontaneous breathing and circulatory stable*		
ATLS principles applied	yes	95%
Intubate patient for neurological reasons	yes	83%
Neurological indication for intubation	GCSM 5 ≤	51%
Goal for ventilation	maintain ventilation ≥ 4 kPa	75%
New additional information *CT scan: Diffuse edema. Small kontusion right temporal lobe. Ventricles and basal cisterns slightly compressed*		
Indication for ICP monitoring	yes	90%
Type ICP monitor	intraparenchymal monitors	60%
Neurological indication for ICP monitoring	Glasgow motor score 5 ≤	51%
New additional information *Glasgow motor score decreases to 4. New CT scan: Acute subdural hematoma 5 mm thick right side. 7 mm midline shift*		
Is GCSM 4, ASDH and 7 mm shift clear indication for surgery	yes	80%
New additional information *Glasgow motor score still 4. New CT scan: Acute subdural removed, increase of right temporal contusion. Still 7 mm midline shift*		
Is GCSM 4, contusion and 7 mm shift a clear indication for surgery	no	60%
New additional Information *After 2 days. ICP increases to 25 mmHg and occasional plateau waves. New CT scan: No mass lesion, no midline shift. Ventricles and basal cisterns slightly compressed*		
Mannitol as part of treatment	yes	71%
CSF drainage as part of treatment	yes	56%
General policy questions on treatment of refractory high ICP		
Are barbituates used in centre	yes	86%
Are decompressive craniectomies sometime used	yes	88%
Is active hypothermia used	never	62%

tines were used in TBI care. This survey has been designed to gather basic information on organisation of care in the different BrainIT centres and on what the different centres believe that they are doing with the TBI patients [2].

There were 2 respondent from most centers and their answers concurred to a large extent on treatment strategies. There also seemed to be an agreement on indicatons for intubation, ventilation levels and ICP monitoring within the BrainIT group. The suggested order of treatment was generally in accordance with published guidelines [8, 9].

All centres in Brain IT are interested in NIC which means that the findings were not representative of all european centres. What is important is that the centres have been characterised as to what they think they are doing with their patients. This is an important basis for future analysis within the BrainIT network of what is actually going on with the patients. The BrainIT network should also give valuable information on the effects of the different treatment traditions that exist between centres.

References

1. Piper I, Citerio G, Chambers I, Contant C, Enblad P, Fiddes H, Howells T, Kienning K, Nilsson P, Yau YH for the Brain IT (2003) Core dataset definition. Acta Neurochir (Wien) 25; 145: 615–629
2. Enblad P, Nilsson P, Chambers I, Citerio G, Fiddes H, Howells T, Kiening K, Ragauskas A, Sahuquillo J, Yau YH, Contant C, Piper I; Brain IT group (2004) R-3 survey of traumatic brain injury management in European *Brain IT* centres year 2001. Intensive Care Med 30: 1058–1065
3. Ghajar J, Hariri RJ, Narayan RK, Iacono LA, Firlik K, Patterson RH (1995) Survey of critical care management of comatose, head-injured patients in the United States. Crit Care Med 23: 560–567
4. Jeevaratnam DR, Menon DK (1996) Survey of intensive care of severely head injured patients in the United Kingdom. BMJ 312: 944–947
5. Matta B, Menon D (1996) Severe head injury in the United Kingdom and Ireland: A survey of practice and implications for management. Crit Care Med 24: 1743–1748
6. Murray GD, Teasdale GM, Braakman R, Cohadon F, Dearden M, Iannotti F, Karimi A, Lapierre F, Maas A, Ohman J, Persson L, Servadei F, Stocchetti N, Trojanowski T, Unterberg A (1999) The European Brain Injury Consortium survey of head injuries. Acta Neurochir (Wien) 141: 223–236
7. Stochetti N, Penny KI, Dearden M, Braakman R, Cohadon F, Iannotti F, Lapierre F, Karimi A, Maas A, Murray GD, Ohman J, Persson L, Servadei F, Teasdale GM, Trojanowski T, Unterberg A, on behalf of the European Brain Injury Consortium (2001) Intensive care management of head-injured patients in Europe: a survey from the European Brain Injury Consortium. Intensive Care Med 27: 400–406
8. Bullock R, Chesnut RM, Clifton C *et al* (1996) Guidelines for the management of severe head injury. J Neurotrauma 13: 643–734
9. Maas AI, Dearden M, Teasdale GM, Braakman R, Cohadon F, Iannotti F, Karimi A, Lapierre F, Murray G, Ohman J, Persson L, Servadei F, Stocchetti N, Unterberg A (1997) EBIC-Guidelines for management of severe head injury in adults. Acta Neurochir (Wien) 139: 286–294

Correspondence: Pelle Nilsson, Department of Neurosurgery, University Hospital, 75185 Uppsala, Sweden. e-mail: pelle.nilsson@neurokir.uu.se

Acta Neurochir (2005) [Suppl] 95: 55–57

The importance of major extracranial injuries by the decompressive craniectomy in severe head injuries

U. Meier, A. Gräwe, and **A. König**

Department of Neurosurgery, Unfallkrankenhaus Berlin, Berlin, Germany

Summary

Neurosurgical therapy aims to minimise the secondary brain damage after a severe head injury. This includes the evacuation of an intracranial space occupying bleeding, the reduction of intracranial volumes, in hematocephalus an external ventricular drainage, and the conservative therapy in order to influence an increased intracranial pressure (ICP) and a decreased p(ti)0_2. When conservative treatment fails to act a decompressive craniectomy might be successful in lowering ICP. From September 1997 until July 2004 we operated on 737 patients with severe head injuries. 103 patients (14%) were treated by means of a decompressive craniectomy.

The prognosis after decompression depends on the clinical signs and symptoms on admission, the patients age and the existence of major extracranial injuries. Our guidelines for an indication for decompressive craniectomy after failure of conservative interventions and evacuation of space occupying hematomas include a patients age below 50 years without multiple trauma, a patients age below 30 years in the presence of major extracranial injuries, a severe brain swelling on CT scan, the exclusion of a primary brainstem lesion or injury and the intervention before irreversible brainstem damage and secondarily while monitoring ICP and p(ti)0_2 in an interval up to 48 hours after the accident before irreversible brainstem damage or generalised brain damage has occurred.

Keywords: Severe head injury; brain oedema; decompressive craniectomy; intracranial pressure.

Introduction

Bergmann described in 1880 a decompressive craniectomy and Cushing published in 1908 a case report about a subtemporal decompressive craniectomy for relief of intracranial pressure. There is still a controversy going on about the value of operative decompression after severe head injuries with traumatic brain oedema [1–4].

The aim of neurosurgical therapy after severe head injuries is the minimisation of the secondary brain damage. General principles of neurosurgical therapy are the evacuation of space occupying hematomas, the reduction of intracranial volume, respectively the drainage of hematocephalus, and the conservative therapy focussed on intracranial pressure (ICP), cerebral perfusion pressure (CPP) and brain tissue pO_2. In intractable intracranial hypertension which is refractory to conservative interventions a decompressive craniectomy is indicated in a few patients. Indications for decompressive craniectomy, course of disease, prognostic criterias are analysed and compared with the literature [8–10].

Patients and methods

All patients with a severe head injury at the Unfallkrankenhaus are treated by the neurosurgical service on the interdisciplinar ICU. The standard management includes an initial CAT-scan in the emergency room. Patients were included in our prospective study when we saw an indication for a neurosurgical operation. 737 patients with severe head injuries were operatively treated in the Department of Neurosurgery of the Unfallkrankenhaus Berlin between September 1997 to July 2004. The average age of the 534 male and 203 female patients was 39 years. 103 patients (14%) underwent decompressive craniectomy. In 68 patients craniectomy was performed additionally with removal of a space-occupying hematoma (subdural or epidural hematoma, contusion hemorrhage) because intraoperatively a generalised brain oedema was evident. The second group included 35 patients in who conservative treatment on the intensive care unit was therapy-resistant under neuromonitoring with ICP-, CPP-, MAP- and p(ti)0_2-measurement revealed an intractable brain oedema. The average age of the group of patients who were decompressed was 30 years and the male/female ratio was 3 : 1 (79 male, 24 female). In 93 patients we performed an unilateral craniectomy (right side 40, left side 53) and in 10 patients a bilateral craniectomy was necessary.

Results

In accordance with the results of other groups (16) the majority of head injuries with indication for craniectomy was due to motor vehicle accidents (57%),

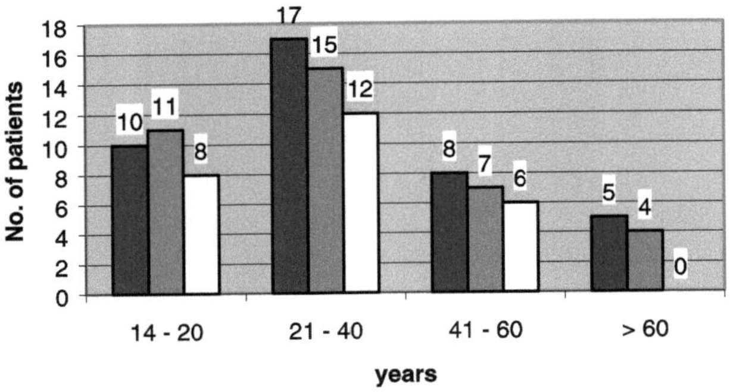

Fig. 1. Age of patients versus Glasgow Outcome Scale. ■ GOS 1; ■ GOS 2/3; □ GOS 4/5

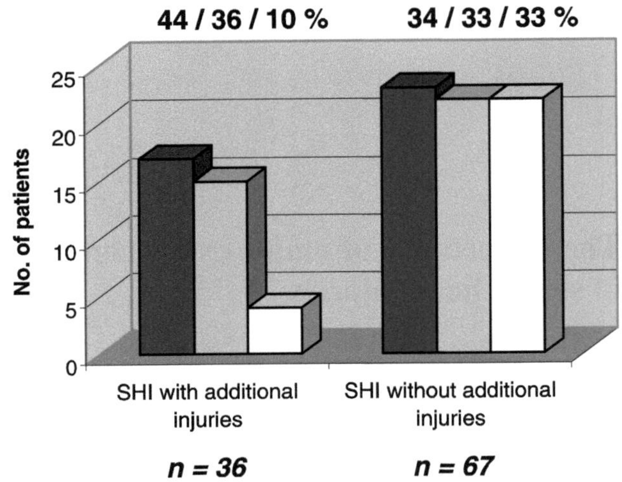

Fig. 2. SHI with/without additional Major Extracranial Injuries. ■ GOS 1; ■ GOS 2/3; □ GOS 4/5

followed by falls (27%), violences (7%), attempted suicide (6%) and horse riding accidents (3%). 55 percent had a craniectomy because of a diffuse brain injury, 25 percent after an acute subdural hematoma, 11 percent with an open head injury and 9 percent after an acute epidural hematoma. In patients with a diffuse brain injury the primary operative intervention was performed with the indication for evacuation of a space occupying contusion bleeding.

40 patients (39%) died despite decompression, 16 patients (16%) remained in a vegetative state, 21 patient (20%) had persistent severe neurological deficits, and 26 patients (25%) reached a GOS of 4 up to 5.

A comparison of the patients' age and the postoperative results is because of the small number of patients not sufficient (Fig. 1). Altogether, patients younger than 40 years have a better prognosis than the older ones. But also in the fourth and fifth decade satisfactory results are obtainable in 63 percent. According to our results a decompression is less favourable in patients over 60 years of age and of our nine patients in this group five died and four remained in a vegetative state.

There is evidence for an important influence on the outcome when major extracranial injuries are present in contrast to an isolated head injury (Fig. 2). Mortality was higher when there was a multiple trauma (47% versus 34%). Favourable outcome was associated with an isolated head injury in the absence of any extracranial injury (33% versus 10%). In our series we could not find a difference in the time span between accident and surgical decompression.

Discussion

The prognosis after severe head injuries depends on the clinical status on admission, the intracranial lesions, the patients age and the existence of accompanying injuries. Patients with epidural hematomas and open head injuries have a better prognosis than those with acute subdural hematomas. A poor prognosis is given in patients with brain contusions, diffuse brain injuries, traumatic injuries of the sinuses and/or less than 8 points on the Glasgow Coma Scale and an advanced age [8–10].

Contrast to our results referring to the comparison of patients with severe head injuries and those after craniectomy (Fig. 2) Polin *et al.* [11] found significantly better results after decompression. This finding might be a fact of a reduction on patients with diffuse brain injuries who were not operated. In our comparable group there are all operatively treated patients after severe head injuries summarised.

The influence of the patients age on the outcome after severe head injuries and craniectomy (Fig. 2) was also found by other authors [3, 6, 7, 11, 12]. We critically have to re-evaluate the indication for decompressive craniectomy in those nine patients over 60 years of age. Their postoperative results with a GOS score of 1 or 2 is one reason for our higher rate of patients which died or remained in a vegetative state and we postulate that there should be an age restriction. Like Karlen *et al.* [5] the prognosis was worse when there were major extracranial injuries present.

Under the influence of the results of Gaab *et al.* [3, 6, 12] we conclude the following guidelines for an indication for decompressive craniotomy after severe head injuries:

1. Patients age below 50 years without multiple trauma
2. Patients age below 30 years in the presence of major extracranial injuries
3. Severe brain swelling on CT scan
4. Exclusion of a primary brainstem lesion/injury
5. Intervention before irreversible brainstem damage
6. ICP increase up to 40 mm Hg and unsuccessful conservative therapy
7. Primarily for space-occupying hematomas with hemispheriel brain oedema
8. Secondarily while monitoring ICP and $p(ti)0_2$ in an interval up to 48 hours after the accident before irreversible brainstem damage or generalised brain damage has occurred.

References

1. Burkert W (1985) Zeitlicher Verlauf des posttraumatischen Hirnödems nach Ultraschalleinwirkung – Experimentelle Untersuchungen und Modellvorstellungen zum Zeitpunkt der operativen Dekompression und zur Größe der Trepanationsfläche. Dissertation B, Martin-Luther-Universität Halle
2. Burkert W, Plaumann H (1989) Der Wert der großen druckentlastenden Trepanation beim therapieresistenten Hirnödem. Tierexperimentelle Untersuchungen, erste klinische Ergebnisse. Zentralbl Neurochir 50: 106–108
3. Gaab MR, Rittierodt M, Lorenz M, Heissler HE (1990) Traumatic brain swelling and operative decompression, a prospective investigation. Acta Neurochir [Suppl] 51: 326–328
4. Gerl A, Tavan S (1980) Die bilaterale Kraniektomie zur Behandlung des schweren traumatischen Hirnödems. Zentralbl Neurochir 41: 125–138
5. Karlen J, Stula D (1987) Dekompressive Kraniotomie bei schwerem Schädelhirntrauma nach erfolgloser Behandlung mit Barbituraten. Neurochirurgia 30: 35–39
6. Kleist-Welch Guerra W, Gaab MR, Dietz H, Mueller JU, Piek J, Fritsch MJ (1999) Surgical decompression for traumatic brain swelling, indications and results. J Neurosurg 90: 187–196
7. Kunze E, Meixensberger J, Janka M, Sörensen N, Roosen K (1998) Decompressive craniectomy in patients with uncontollable intracranial hypertension. Acta Neurochir [Suppl] 71: 16–18
8. Meier U, Gartner F, Knopf W, Klotzer R, Wolf O (1992). The traumatic dural sinus injury-a clinical study. Acta Neurochir (Wien) 116: 91–93
9. Meier U, Heinitz A, Kintzel D (1994) Operationsergebnisse nach schwerem Schädel-Hirn-Trauma im Kindes- und Erwachsenenalter – Eine Vergleichsstudie. Unfallchirurgie 97: 406–409
10. Meier U, Zeilinger F, Kintzel D (1996) Prognose und Ergebnisse der operativen Versorgung von schweren Schädel-Hirn-Traumen. Akt Traumatol 26: 1–5
11. Polin RS, Shaffrey ME, Bogaev CA, Tisdale N, Germanson T, Bicchicchio B, Jane JA (1997) Decompressive bifrontal craniectomy in the treatment of severe refractory posttraumatic cerebral edema. Neurosurgery 41: 84–94
12. Rittierodt M, Gaab MR, Lorenz M (1991) Decompressive craniectomy after severe head injury. Useful therapy in pathophysiologically guided indications. In: Bock WJ *et al* (eds) Advances in neurosurgery, vol 19. Springer, Berlin, Heidelberg, New York, Tokyo, pps 265–273

Correspondence: Ullrich Meier, Department of Neurosurgery, Unfallkrankenhaus Berlin, Warener Straße 7, 12683 Berlin, Germany. e-mail: ullrich.meier@ukb.de

Acta Neurochir (2005) [Suppl] 95: 59–60

Beneficial effect of cerebrolysin on moderate and severe head injury patients: result of a cohort study

G. K. C. Wong, X. L. Zhu, and W. S. Poon

Division of Neurosurgery, Prince of Wales Hospital, Chinese University of Hong Kong, Hong Kong, China

Summary

Cerebrolysin is used as a neurotrophic agent for the treatment of ischemic stroke and Alzheimer's Disease. Exploratory studies in patients with post-acute traumatic brain injury have shown that this treatment might help improve recovery. Aim of this study was to investigate whether addition of Cerebrolysin to the initial treatment regimen of moderate and severe head injury patients would improve their outcome.

At 6 months, 67% of the patients (Cerebrolysin group) attained good outcome (GOS 3–5). The study group was compared with the historical cohort of patients from the hospital trauma data bank, with age, sex and admitting GCS matching. More patients tended to a good outcome in the Cerebrolysin group ($P = 0.065$). No significant side-effect requiring cessation of Cerebrolysin was noted.

It can be concluded that the use of Cerebrolysin as part of the initial management of moderate and severe head injury is safe and well tolerated. The results suggest that Cerebrolysin is beneficial in regard to the outcome in these patients, especially in elderly patients.

Keywords: Cerebrolysin; traumatic brain injury; clinical outcome.

Introduction

Traumatic brain injury is an important cause of morbidity and a costly disorder [5, 7, 12]. It also represents the most frequent reason for neurological illness after headache. Compared with other types of brain insult, traumatic brain injury produces more diffuse injury and more cognitive and neuropsychiatric disturbances [10]. Various drugs have been used in an attempt to improve the outcome of these patients.

Cerebrolysin is a peptide preparation produced by the biotechnologically standardized enzymatic breakdown of purified porcine brain proteins. It consists of approximately 15% peptides with a molecular weight not exceeding 10 kD and 85% amino acids, based on total nitrogen. The solution is free of proteins, lipids and antigenic properties.

The action mechanism is not clearly understood. It has been shown in animal studies to improve neuronal oxygen utilization, reduce cerebral lactic acid concentration and decreases free oxygen radical concentration. Cerebrolysin has produced significant improvement in cognitive function, non-cognitive psychiatric symptoms and daily activities in patients with Alzheimer's Disease and stroke [2, 8, 11].

In this study, we hypothesize that Cerebrolysin infusion during the acute phase of moderate and severe head injury improves functional outcome of patients [1], and second, our aim was to document its safety profile in these patients.

Patients and methods

Starting in March 2001, patients aged 18 years or above, with moderate or severe head injury within 48 hours prior to admission were recruited. Exclusion criteria included severe renal, liver, lung or cardiovascular disease, poor mental state due to drug or alcohol abuse, concomitant stroke, pregnancy or lactation, life-threatening multiple trauma, signs of brain stem failure, status epilepticus and grand mal fits. The study conformed to the Helsinki declaration. Written consent was obtained from suitable patients and their surrogate/guardian by one of the investigators.

All recruited patients received 50 ml Cerebrolysin through intravenous infusion over 15-minute-periods daily for 20 days. Data on age, sex, admission Glasgow Coma Score, hospital stay, 3-month Extended Glasgow Outcome Score and 6-month Extended Glasgow Outcome Score [3, 4] were prospectively collected.

Data were compared with a historical cohort of patients with moderate and severe head injury from the hospital trauma data bank. The historical control group was all matched by age, sex and admitting Glasgow Coma Score. Statistical analysis was carried out with SPSS for Window Release 10.

Results

Twenty one patients with moderate or severe head injury were recruited for this study between March

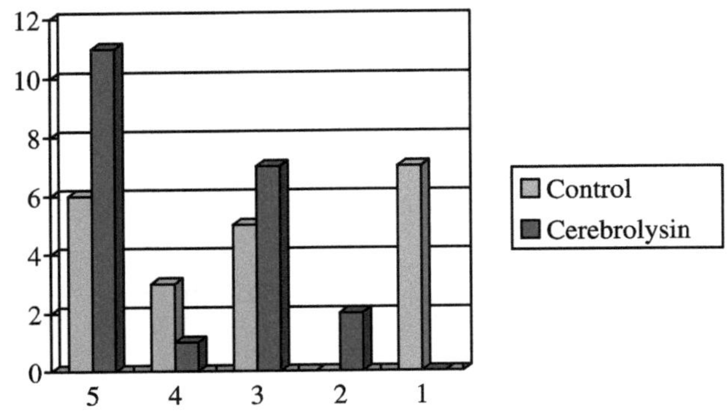

Fig. 1. Glasgow Outcome Score of the Crebrolysin group and cohort group. The X-axis represents the GOS and the Y-axis represents the number

2001 and March 2002. Mean age was 64 years, range from 28 to 82 years. The male to female ratio was 13:8. Results are charted in Fig. 1. At 6 months, 67% of patients had a good outcome as defined by Glasgow Outcome Score 3–5. They were compared with the historical cohort of patients from the trauma data bank of the same hospital. There was a trend of more patients with good outcome in the Cerebrolysin group ($P = 0.065$). No significant side-effects necessitating cessation of Cerebrolysin nor manifestation of allergy were noted.

Discussions

Previous studies have demonstrated that an important part of the recovery after traumatic brain injury occurs in the first six months after the injury [6, 7, 12]. A functional recovery is essential for the patient to return to society. Most of the moderate and severe head injury patients studied had cognitive and physical deficits, which hindered their eventual recovery [10].

The neurotrophic effects of Cerebrolysin are theoretically beneficial to traumatic brain injury patients, in particular when considering the cognitive function. With these data in mind, we performed this study to observe and record its clinical effect. The timing of Cerebrolysin treatment was designed to start within the first 48 hours and continue for 10 days to cover possible neuroprotective and neurotrophic effects. The results were encouraging with more patients attaining good outcome at six months in the Cerebrolysin group as compared to the historical control group.

No significant side effects were noted, as was also the case in the treatment of dementia patients [2, 8, 11]. The safety profile in head injury patients is confirmed with the current series.

The results suggest a beneficial effect of Cerebrolysin infusion in patients with moderate or severe head injury during the acute phase. It has also been shown in other studies to reduce EEG power ratio and enhance cognitive performance in tasks evaluating attention and memory functions [1]. This magnitude in terms of GOS improvement certainly is valuable to the patients and further studies to evaluate the role of Cerebrolysin in different neurological diseases are warranted.

References

1. Anton A, Sampedro C, Perez P *et al* (2003) Positive effects of Cerebrolysin on electroencephalogram slowing, cognition and clinical outcome in patients with postacute traumatic brain injury: an exploratory study. International Clinical Psychopharmacology 18: 271–278
2. Bae CY, Cho CY, Cho K *et al* (2000) A double-blind, placebo-controlled, multicenter study of cerebrolysin for Alzheimer's Disease. Journal of American Geriatric Society 48: 1566–1571
3. Jennett B, Bond M (1975) Assessment of outcome after severe brain damage, a practical scale. Lancet 1: 480–484
4. Jennett B, Teasdale G, Braakman R (1979) Prognosis of patients with severe brain injury. Neurosurgery 4: 283–289
5. Katz DI, Alexander MP (1994) Traumatic brain injury. Predicting course of recovery and outcome for patients admitted to rehabilitation. Arch Neurol 51(7): 661–670
6. Prigatano GP (1988) Rehabilitation intervention after traumatic brain injury. BNI Quarterly 4: 30–37
7. Rice-Oxley M, Turner-Stokes L (1999) Effectiveness of brain injury rehabilitation. Clin Rehab 13(S1): 7–24
8. Ruether E, Ritter R, Apecechea M, Freytag S, Gmeinbauer R, Windisch M (2000) Sustained improvements in patients with dementia of Alzheimer's type (DAT) 6 months after termination of cerebrolysin therapy. J Neural Transm 107(7): 815–829
9. Valouskova V, Francis-Turner L (1998) Can cerebrolysin influence chronic deterioration of spatial learning and memory? J Neural Transm [Suppl] 53: 343–349
10. Wilson JTL, Pettigrew LEL, Teasdale GM (2000) Emotional and cognitive consequences of head injury in relation to the Glasgow outcome scale. J Neurol Neurosurg Psychiatry 69: 204–209
11. Xiao S, Yan H, Yao P: The Cerebrolysin Study Group (2000) Efficacy of FPF 1070 (Cerebrolysin) in patients with Alzheimer's Disease. Clin Drug Invest 19(1): 43–53
12. Zhu XL, Poon WS, Chan CH, Chan SH (2001) Does intensive rehabilitation improve the functional outcome of patients with traumatic brain injury? Interim result of a randomized controlled trial. Brit J Neurosurg 15(6): 464–473

Correspondence: W. S. Poon, Department of Surgery, Prince of Wales Hospital, Shatin, New Territories, Hong Kong SAR, China. e-mail: wpoon@surgery.cuhk.edu.hk

Neurochemical monitoring and intracranial hypertension

Acta Neurochir (2005) [Suppl] 95: 63–66

Re-defining the ischemic threshold for jugular venous oxygen saturation – a microdialysis study in patients with severe head injury

M. T. V. Chan[1], **S. C. P. Ng**[2], **J. M. K. Lam**[2], **W. S. Poon**[2], and **T. Gin**[1]

[1] Department of Anaesthesia and Intensive Care, The Chinese University of Hong Kong, Prince of Wales Hospital, Hong Kong, China
[2] Division of Neurosurgery, Department of Surgery, The Chinese University of Hong Kong, Prince of Wales Hospital, Hong Kong, China

Summary

Neurological change is more likely to occur when jugular venous oxygen saturation ($SjvO_2$) is less than 50%. However, the value indicating cellular damage has not been clearly defined. We determined the critical $SjvO_2$ value below which intracerebral extracellular metabolic abnormalities occurred in 25 patients with severe head injury.

All patients received standard treatment with normoventilation and maintenance of intracranial pressure < 20 mmHg. $SjvO_2$ was measured from the dominant jugular bulb using a calibrated fibreoptic catheter. Intracerebral metabolic monitoring was performed by collecting perfusate from a microdialysis probe placed in the frontal lobe anterior to the intracranial catheter. Excitotoxin (glutamate) and other extracellular metabolites (lactate, glucose and glycerol) were measured frequently using enzymatic and colorimetric methods.

We observed biphasic relationships between $SjvO_2$ and all intracerebral metabolites. Analysis of variance showed that there were rapid increases in glutamate, glycerol and lactate when $SjvO_2$ dropped below 40, 43 and 45% respectively. Extracellular glucose decreased when $SjvO_2$ dropped below 42%.

Our findings suggested that the ischemic threshold for $SjvO_2$ in patients with severe head injury is 45%, below which secondary brain damage occurred.

Keywords: Cerebral ischemia; jugular venous oxygen saturation; head injury; intracerebral microdialysis.

Introduction

Ischemic brain damage is a common cause of morbidity and mortality after severe head injury [7]. Monitoring of jugular bulb venous oxygen saturation ($SjvO_2$) has been proposed as a simple method for detecting cerebral ischemia [14]. However, the utility of $SjvO_2$ monitoring requires the establishment of an ischemic threshold. Such threshold value is important because it is the lowest acceptable value for assured cerebral oxygenation, below which cerebral ischemia is deemed to occur. Therefore, therapy that maintains $SjvO_2$ above the ischemic threshold should produce more favorable outcome.

Previous studies have reported a wide range of threshold values. In volunteers breathing hypoxic mixture, electroencephalographic slowing was noted when $SjvO_2$ value was less than 35%, while loss of consciousness occurred at 24% [9, 12]. In another study on patients undergoing carotid endarterectomy with local anesthesia, neurologic deficits occurred at much higher $SjvO_2$ values (50%) [10]. There are few studies that correlate $SjvO_2$ with cerebral metabolism. In patients undergoing emergency craniotomies for evacuation of intracranial hematoma, Sapire *et al.* noted an increase in arterial to jugular venous lactate difference as $SjvO_2$ decreased below 50%, suggesting the start of anaerobic metabolism in the brain [13]. Similar results were reported in patients recovering from cardiopulmonary bypass [6]. However, jugular venous lactate concentration may be affected by systemic lactate production. Therefore, the actual point of ischemia may have been obscured. Recently, it is possible to measure changes of neurotransmitters and metabolites in the extracellular space using intracerebral microdialysis. The purpose of the present study was to define the ischemic threshold value for $SjvO_2$, below which cerebral energy metabolism fails. We correlate $SjvO_2$ value with extracellular metabolites of the brain in patients with severe head injury.

Materials and methods

The study was approved by the Clinical Research Ethics Committee and informed consents were obtained from patient's relatives. Patients were eligible for the study if they were 18 years or older suffering from severe traumatic head injury, defined as Glasgow coma scale score of 8 or less on admission. We excluded patients who are moribund with fixed dilated pupils, to whom active management was considered inappropriate. All patients received continuous intracranial pressure (ICP) monitoring, and were treated according to the protocols published by the Brain Trauma Foundation [1].

After initial resuscitation, a fiberoptic catheter (Opticath, Abbott Critical Care System, Park, IL) was placed in the dominant jugular bulb by retrograde cannulation of the internal jugular vein. The dominant jugular bulb was determined by sequential neck compression test [5]. This test based upon the principle that dominant jugular bulb receives the largest proportion of cerebral venous outflow. Therefore, manual compression of the dominant jugular vein results in greater rise in ICP. When the response was equivocal, the right jugular bulb was used. The tip of the catheter was advanced to the level of mastoid process as confirmed by lateral skull X-ray. $SjvO_2$ was measured continuously by reflection spectrophotomtery (Oximetrix-3 System, Abbott Laboratories, North Chicago, IL). Readings were calibrated twice daily against blood samples drawn from the same catheter. Venous oxygen saturation was measured by transmission co-oximetry (IL-482; Instrumentation Laboratories, Watertown, MA).

Intracerebral microdialysis was performed in the frontal cerebral cortex of maximal injury. In patients with diffuse head injury, a right frontal catheter was placed. During initial surgery, a microdialysis catheter (CMA70; shaft length 60 mm; membrane length 10 mm) was inserted in the frontal lobe, anterior to the ventricular drain via a single burr hole. The catheter was perfused with lactate free Ringer's solution (K^+ 4 mM, Na^+ 147 mM, Ca^{2+} 2 mM, Cl^- 155 mM) at 0.3 μL/min by a microinfusion pump (CMA 106, CMA Microdialysis, Stockholm, Sweden). Microdialysates were collected hourly for bedside analysis of extracellular concentrations of glucose, pyruvate, glutamate and lactate using an enzymatic colorimetric method (CMA 600 analyzer, CMA Microdialysis, Stockholm, Sweden).

Physiologic data including $SjvO_2$ readings were recorded on a personal computer using a purposely designed data acquisition program. Cerebral perfusion pressure (CPP), defined as the difference between mean arterial pressure and ICP, was also calculated. Data files were carefully examined, and artifacts were removed before analysis. Since microdialysis measured the average concentration of extracellular metabolites over the preceding hour, we calculated the hourly average $SjvO_2$ value by an interrupted time series technique [8]. $SjvO_2$ readings were initially averaged over an epoch of 60 seconds. The $SjvO_2$ value at the end of the hour was predicted according to the relative changes of the 60 epochs in this hour using the moving average function.

Changes in extracellular metabolites were correlated with $SjvO_2$ using a nonlinear regression model. The ischemic threshold when metabolites started to change over a range of $SjvO_2$ values was defined by analysis of variance (ANOVA) with repeated measures. The 95% confidence intervals (CI) of the ischemic threshold were calculated by bootstrap resampling technique with 3,000 replicates. We compare the differences in the ischemic thresholds among metabolites using ANOVA. $P < 0.05$ was considered statistically significant.

Results

Twenty five patients (15 males, 10 females) were recruited for the study. The median (range) age was 26 (18–67) years. The median (range) admission Glasgow coma scale score was 6 (4–8). At three months after injury, five patients died, eight were classified as severely disabled. The remaining 12 patients (48%) had a favorable outcome with good recovery or moderate disability.

A total of 3,490 patient-hour data were recorded. They contained epochs of paired $SjvO_2$ values and concentrations of extracellular metabolites. Nineteen percent of the $SjvO_2$ data were considered contaminated with artifacts, and were eliminated before analysis. They usually occurred during turning of the patient, tracheal suctioning and chest physiotherapy. We analyzed 2,827 hours of data. Figure 1 shows the relationships between cerebral metabolic changes and $SjvO_2$ during the course of treatment. Intracerebral concentrations of glutamate, glycerol and lactate increased rapidly as $SjvO_2$ fell below a breakpoint, corresponding to the ischemic threshold for metabolic failure and cellular damage. Accordingly, the ischemic threshold (95% CI) of $SjvO_2$ were 40 (31–49)%, 43 (32–54)% and 45 (34–56)% for glutamate, glycerol and lactate, respectively. Similarly, the ischemic threshold of $SjvO_2$ was 42 (35–50)%, below which intracerebral glucose concentration decreased precipitously. The corresponding critical CPP value, below which cerebral metabolic disturbances occurred, was 63 (58–66) mmHg.

Discussion

Using intracerebral microdialysis, we have defined the ischemic threshold for $SjvO_2$ values. In patients with severe head injury, we found that a $SjvO_2$ value less than 45% is associated with neurochemical changes and energy failure. This value is less than the commonly quoted critical value of 50% because metabolic abnormalities occurred late in the ischemic cascade [6, 9–11, 13, 14].

A number of factors may have influenced the results. We compared a global monitor of cerebral oxygenation ($SjvO_2$) with a regional measure of brain metabolism (intracerebral microdialysis). Therefore it is possible that the microdialysis would have missed critical ischemia in a region far away from the catheter. We inserted the microdialysis catheter in the frontal lobe of maximal injury. We believe this is an area most susceptible to secondary insult [7]. Therefore, local changes detected by the microdialysis should reflect, in large part, the global condition of the brain.

Another important concern of microdialysis monitoring is that the neurochemical changes associated

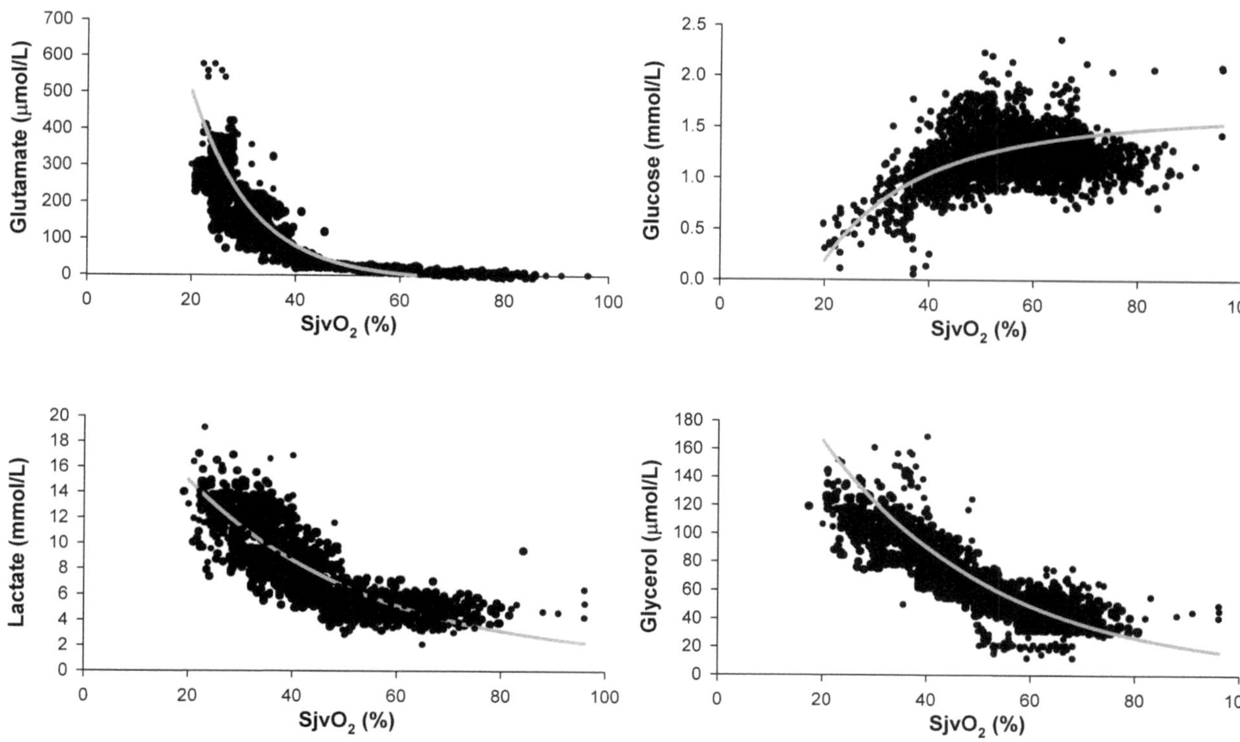

Fig. 1. Correlations of intracerebral extracellular concentrations of glutamate, lactate, glycerol, glucose and jugular venous oxygen saturation (*SjvO*$_2$). The line of best fit for nonlinear regression is shown in solid line

with catheter implantation can be confused with true secondary insult. In this regard, histological studies in rats suggested that catheter insertion is rarely associated with micro-hemorrhage and did not affect the integrity of blood brain barrier [2, 11]. In the present study, we allowed 60 min stabilization between collections of microdiasylates. This should maximize fluid recovery for online analysis and avoid spurious change in neurochemistry.

It should be emphasized that $SjvO_2$ is a global monitor [14]. In a group of patients with severe head injury, critical $SjvO_2$ values of 50% was only achieved when the ischemic brain volume, as measured by oxygen-15 positron emission tomography, exceeded 170 ml (10% of total brain volume) [4]. In another study on induced hyperventilation, $SjvO_2$ decreased by only 17% when there was over 130% increase in ischemic brain volume [3]. These data suggested that regional ischemia may be obscured with $SjvO_2$ monitoring. Our findings are consistent with these reports. Although the ischemic threshold is lower than 50%, significant metabolic derangements can be observed well above these breakpoints in a number of patients. Targeted therapy should be applied to keep $SjvO_2$ above these values in order to avoid metabolic changes in the brain.

Acknowledgments

The study was supported in part by a grant from the Research Grant Council of the Hong Kong Special Administrative Region, China (CUHK4223/98M).

References

1. The American Association of Neurological Surgeons. The Joint Section on Neurotrauma and Critical Care (2000) Management and prognosis of severe traumatic brain injury. The Brain Trauma Foundation, New York
2. Benveniste H, Diemer NH (1988) Early postischaemic 45Ca accumulation in rat dentate hilus. J Cereb Blood Flow Metab 8: 713–719
3. Coles JP, Minhas PS, Fryer TD, Smielewski P, Aigbirihio F, Donovan T, Downey SPMJ, Williams G, Chatfield D, Matthews JC, Gupta AK, Carpenter A, Clark JC, Pickard JD, Menon DK (2002) Effect of hyperventilation on cerebral blood flow in traumatic head injury: clinical relevance and monitoring correlates. Crit Care Med 30: 1950–1959
4. Coles JP, Fryer TD, Smielewski P, Rice K, Clark JC, Pickard

JD, Menon DK (2004) Defining ischemic burden after traumatic brain injury using ^{15}O PET imaging of cerebral physiology. J Cereb Blood Flow Metab 24: 202
5. Dearden NM (1991) Jugular bulb venous oxygen saturation in the management of severe head injury. Curr Opin Anesthesiol 4: 279–286
6. Gopinath SP, Cormio M, Ziegler J, Raty S, Valadka A, Robertson CS (1996) Intraoperative jugular desaturation during surgery for traumatic intracranial hematomas. Anesth Analg 83: 1014–1021
7. Graham DI, Ford I, Adams JH, Doyle D, Teasdale GM, Lawrence AE, McLellan DR (1989) Ischaemic brain damage is still common in fatal non-missile head injury. J Neurol Neurosurg Psychiatry 52: 346–350
8. Lam JM, Smielewski P, al-Rawi P, Griffiths P, Pickard JD, Kirkpatrick PJ (1997) Internal and external carotid contributions to near-infrared spectroscopy during carotid endarterectomy. Stroke 28: 906–911
9. Lennox WG, Gibbs FA, Gibbs EL (1935) Relationship of unconsciousness to cerebral blood flow and to anoxemia. Arch Neurol Psychiat 34: 1001–1013
10. Lyons C, Clark LC, McDowell H, McArthur K (1964) Cerebral venous oxygen content during carotid thrombintimectomy. Ann Surg 160: 561–567
11. Major O, Shdanova T, Duffek L, Nagy Z (1990) Continuous monitoring of blood-brain barrier opening to 51Cr-EDTA by microdialysis following probe injury. Acta Neurochir [Suppl] 51: 46–48
12. Meyer J, Gotoh F, Ebihara S, Tomita M (1965) Effects of anoxia on cerebral metabolism and electrolytes in man. Neurology 15: 892–901
13. Sapire KJ, Gopinath SP, Farhat G, Thakar DR, Gabrielli A, Jones JW, Robertson CS, Chance B (1997) Cerebral oxygenation during warming after cardiopulmonary bypass. Crit Care Med 25: 1655–1662
14. Schell RM, Cole DJ (2000) Cerebral monitoring: jugular venous oximetry. Anesth Analg 90: 559–566

Correspondence: Matthew Chan, FANZCA, Department of Anaesthesia and Intensive Care, The Chinese University of Hong Kong, Prince of Wales Hospital, Shatin, New Territories, Hong Kong. e-mail: mtvchan@cuhk.edu.hk

Acta Neurochir (2005) [Suppl] 95: 67–71

Estimated cerebral respiratory quotient and arteriovenous differences of CO_2 in the ultra early detection of global ischemia in severe head injury

A. Chieregato[1], **M. Marchi**[2], **C. Compagnone**[1], **V. Albarello**[1], **E. Fainardi**[3], **F. Tagliaferri**[1], and **L. Targa**[1]

[1] Neurorianimazione, Ospedale M. Bufalini, Cesena, Italy
[2] Servizio di Anestesia e Rianimazione, Arcispedale S. Anna, Ferrara, Italy
[3] Servizio di Anestesia e Neuroradiologia, Arcispedale S. Anna, Ferrara, Italy

Summary

The specificity of jugular bulb saturation (SjO_2) and arteriovenous oxygen difference ($AVDO_2$) to detect global cerebral ischemia remains controversial. An absolute increase in the arteriovenous difference of carbon dioxide tension ($AVDpCO_2$) and, more specifically, the estimated respiratory quotient ($eRQ = AVDpCO_2/AVDO_2$) may indicate anaerobic CO_2 production. We compared these variables with SjO_2 to predict global cerebral ischemia.

We selected 36 patients from a cohort of 69 consecutive patients suffering from severe traumatic brain injury. All patients had jugular bulb sampling within 6 hours after injury. Brain death at 48 hours was used as a surrogate index of irreversible ischemia to build a receiver operating characteristics (ROC) curve analysis.

The mean ($\pm$ standard deviation) eRQ in the 13 patients who died early (3.7 ± 3.2 mmHg/ml/dl) was higher than the survivors (1.78 ± 0.45 mmHg/ml/dl, $P = 0.03$). There was no differences in SjO_2 between groups. The area under the ROC curves for eRQ, but not that of $AVDpCO_2$, was greater ($P = 0.04$) than that of SjO_2.

The eRQ, more than $AVDpCO_2$, appears to be a potentially more informative index of global cerebral ischemia than SjO_2.

Keywords: Acidosis; carbon dioxide; head injury; intracranial pressure; ischemia; lactate; respiratory quotient; jugular bulb saturation.

Introduction

The early phase of head injury is influenced by secondary insults. Furthermore, there is an association between low cerebral blood flow (CBF) values and elevated intracranial pressure (ICP). While pathological signs of ischemia have been recognized in autopsy study [3], the in vivo diagnosis remains uncertain. Potentially, a decrease in SjO_2 and an increase in arteriovenous oxygen difference ($AVDO_2$) are capable to detect critical hypoperfusion. However, these values may also be associated with aerobic or anaerobic metabolism [1, 4]. An absolute increased arteriovenous lactate difference (AVDL) has also been detected early after trauma, but its measurement is dependent from a prompt availability of a lactate analyzer. Given that carbon dioxide (CO_2) is generated during abnormal lactate production, it seems reasonable to hypothesize that the difference in cerebral arterial and venous CO_2 tension ($AVDpCO_2$) may predict cerebral ischemia [2, 8]. However, an increase of jugular CO_2 may also be due to a decrease in venous blood flow (Fick's law), not associated with a concomitant ischemic CO_2 production. This can be compensated if we compare $AVDpCO_2$ with arteriovenous oxygen contents ($AVDO_2$) as an estimated respiratory quotient (eRQ) [$AVDpCO_2/AVDO_2$]. Thus, a disproportionate absolute increase in $AVDpCO_2$ with respect to a given $AVDO_2$ value is due to anaerobic CO_2 production [7]. We hypothesized that an increase in $AVDpCO_2$ and an increase in eRQ may indicate cerebral anaerobic metabolism and predict early brain death in patients with head injury. A secondary hypothesis was that eRQ would be more specific than $AVDpCO_2$ in predicting ischemia.

Materials and methods

Seventy-one patients admitted consecutively to the general intensive care unit of the S. Anna Hospital, Italy between June 1994 and August 1997, with a post-stabilization Glasgow Coma Scale score (GCS) ≤ 8, were recruited to this study. One patient was excluded because of brain death upon arrival in the emergency room. Another one was excluded because of associated carotid cavernous sinus fistula, which made monitoring unreliable. Among the remaining 69 patients, 36 underwent jugular bulb monitoring within 6 hours after injury.

A retrograde catheter was placed preferentially into the right superior jugular bulb. The position was confirmed by X-ray. During the first 6 hours after injury, paired arterial and jugular venous samples were collected according to clinical judgment and thereafter in association with pupillary reactivity or prolonged increase in ICP. Sampling was suspended in patients who have developed bilaterally non-reactive dilated pupils in relation to brain death. Lactate measurement and the accuracy of the assay (ACA Dupont, Dupont Company, Wilmington, USA) have been previously described [2]. Calculated jugular bulb variables were obtained as follow: $AVDpCO_2$ (mmHg) = arterial pCO_2 − jugular bulb pCO_2; eRQ (mmHg/ml/dl) = [$AVDpCO_2/AVDO_2$] lactate/oxygen index (LOI) = AVDL/($AVDO_2 \times 0.446$). The normal values of these variables were reported by Gibbs [5].

We defined hypoxia as arterial oxygen tension < 60 mmHg, and hypotension as systolic arterial pressure < 90 mmHg. GCS score and pupillary reactivity were evaluated at the scene of the accident, upon arrival in the emergency room, and after resuscitation. Pupillary patterns were summarized as: bilaterally reactive pupils, unilaterally dilated and non-reactive pupil, and bilaterally dilated non-reactive pupils. Only the worst and the best neurologic observations were included for analysis. After commencement of jugular monitoring, the GCS and pupillary status were recorded during each sampling. The first computed tomography (CT) scan was performed after arrival at the hospital and another one obtained during the first 48 hours. If multiple CT scans were performed during this period, the worst CT with maximal injury was identified. For data analysis, the Marshall CT classification was categorized as: diffuse injury (type I and II), swelling (type III and IV) and mass lesion (evolving and non-evolving mass lesions).

The outcome measurement was brain death (defined as the absence of brain stem reflexes and absent electrical cortical activity) due to intracranial causes within 48 hours after injury. Patients were classified as early brain death or early survivors.

Physiological variables, severity indexes were compared between groups using the pooled *t*-test, Fisher exact test or Chi-square test, as appropriate. *P* values below 0.05 were considered significant. For continuous variables, the extreme values commonly considered as the most representative for ischemia were analysed: they are the maximum eRQ, LOI and $AVDO_2$; the minimum $AVDpCO_2$, AVDL and SjO_2. A receiver operating characteristics (ROC) curve analysis was performed and the area under the curve (AUC) was calculated for each variable. The 95% confidence interval for AUC was used to test the hypothesis that the theoretical AUC was 0.5. If the confidence intervals were greater than 0.5, then there was evidence that the laboratory test had the ability to distinguish between groups. The AUC for each of the derived variables (eRQ, LOI, $AVDO_2$, $AVDpCO_2$ and AVDL) was compared with that of SjO_2 as well as to the best AUC found among these variables. An exploratory analysis of the threshold cut-off values was made. The value with the highest accuracy, minimal false negative and false positive results, was automatically searched by the statistical program (MedCalc 6.16, Frank Schoonijans, Belgium). The likelihood ratios for positive and negative tests were reported.

Results

The indexes of systemic severity were worse in the early brain death group (Table 1). Twenty-seven measurements were obtained in each group. The median (interquartile range, IQR) delay after injury for the

Table 1. *Patient characteristics*

	Early death	Early survival	*P* Value
Number of patients	13 (36)	23 (64)	
Age (years)*	39.1 (16)	38.2 (21.7)	0.33
Sex (male)	8 (33.3)	16 (66.7)	0.72
Multiple injuries	9 (69.2)	10 (43.5)	0.18
Hypotension	5 (38.5)	7 (30.4)	0.72
Hypoxia	12 (92.3)	17 (73.9)	0.38
Best initial GCS#	3 (2)	5 (1.75)	0.02
Worst initial pupillary response			0.02
– Bilaterally reactive	2 (15.4)	14 (60.9)	
– Unilaterally non-reactive	3 (23.1)	1 (4.3)	
– Bilaterally non-reactive	8 (61.5)	8 (34.8)	
Worst computed tomography scan			0.004
– Diffuse injury	0 (0)	12 (52.2)	
– Swelling	7 (53.8)	4 (17.4)	
– Mass lesions	6 (46.2)	7 (30.4)	
Minimum mean arterial pressure (mmHg)*	66 (27)	85.9 (15.2)	0.026
Minimum arterial saturation (%)*	96.5 (1.6)	96.5 (2.8)	0.36
Maximum arterial carbon dioxide tension (mmHg)*	4.89 (1.49)	4.56 (0.95)	0.49
Minimum hemoglobin (g/dL)*	10.2 (2.1)	11.6 (2.2)	0.06
Minimum temperature (C°)*	35.5 (1.1)	36.7 (0.5)	0.02
Maximum arterial lactate concentration (mmol/L)*	6.2 (4.1)	2.6 (1.8)	<0.001
Best 48 hrs GCS#	1 (1.25)	4 (2)	<0.001
Worst 48 hours pupillary response			<0.001
– Bilaterally reactive	1 (7.7)	17 (74)	
– Unilaterally non-reactive	5 (38.5)	3 (13)	
– Bilaterally non-reactive	7 (53.8)	3 (13)	

GCS Glasgow Coma Scale; Values are number (percent), or *mean (standard deviation), or median (interquantile range).

Table 2. *Univariate analysis and comparison of the receiver operating characteristics curve of derived jugular bulb variables. Precision of each point estimate was presented as mean (standard deviations) or 95% confidence intervals (95% CI). +LR is positive likelihood ratio and is −LR negative likelihood ratio*

Groups	Early death	Early survival	P values	AUC	Comparison to SjO_2 AUC	Comparison to LOI AUC	Cut-off	Sensitivity (95% CI)	Specificity (95% CI)	+LR	−LR
Number of patients	13	23	–	–	–						
Minimum SjO_2 (%)	50.9 (19.5)	51.5 (13.5)	0.76	0.523 (0.351–0.692)	–	0.02	44.7	38.5 (14.0–68.4)	82.6 (61.2–94.9)	2.21	0.74
Maximum eRQ (mmHg/ml/dl)	3.7 (3.2)	1.78 (0.45)	0.03	0.784 (0.616–0.903)	0.04	0.79	2.3	61.5 (31.6–86.0)	91.3 (71.9–98.7)	7.08	0.42
Minimum $AVDpCO_2$ (mmHg)	−18.6 (10.2)	−12.7 (4.3)	0.05	0.689 (0.513–0.832)	0.07	0.26	−21.4	38.5 (14.0–68.4)	95.7 (78.0–99.3)	8.85	0.64
Maximum $AVDO_2$ (ml/dL)	6.7 (2.2)	7.6 (2.6)	0.25	0.622 (0.445–0.778)	0.12	0.08	9	92.3 (63.9–98.7)	39.1 (19.7–61.4)	1.52	0.20
Minimum AVDL (mmol/L)	−1.0 (1.0)	−0.15 (0.23)	0.001	0.789 (0.621–0.907)	0.02	0.58	0.37	69.2 (38.6–90.7)	91.3 (71.9–98.7)	7.96	0.34
Maximum LOI	0.57 (0.96)	0.04 (0.08)	<0.001	0.808 (0.642–0.919)	0.02	–	0.14	61.5 (31.6–86.0)	95.7 (78.0–99.3)	14.15	0.40

SjO_2 Jugular bulb saturation; *$AVDO_2$* arteriovenous oxygen difference; *$AVDpCO_2$* arteriovenous difference of carbon dioxide tension; *eRQ* estimated respiratory quotient; *AVDL* arteriovenous lactate difference; *LOI* lactate/oxygen index; *AUC* Area under the curve.

first jugular bulb measurement was similar between groups (early brain death group, 4 (2.3) hours; early survival group 4 (2.8) hours, $P = 0.45$). ICP measurement was done within 6 hours after injury in only 6 patients (3 per group). Fifty-four percent of the measurements in the patients who died early were obtained in association to dilated non-reactive pupils as compared with 13% in the patients who survived.

There was no difference between groups in the minimum SjO_2 and maximum $AVDO_2$. The maximum eRQ in the 13 patients who died early was higher than the maximum eRQ in the early survival group. A nearly significant difference was found for minimum $AVDpCO_2$. Significant differences between groups were found for both minimal AVDL and maximal LOI (Table 2).

Analyses of the ROC curves are presented in Table 2. The maximal LOI and minimum value of AVDL had the best AUC, followed, in decreasing order, by maximal eRQ, minimum $AVDpCO_2$, maximal $AVDO_2$ and finally by minimum SjO_2. The confidence interval did not include the 0.5 value for LOI, AVDL, eRQ and $AVDpCO_2$. The AUC associated to the maximal eRQ, but not the one associated to the minimum $AVDpCO_2$, was greater than that of the minimal SjO_2 ($P = 0.04$) (Fig. 1). The AUC for eRQ and LOI were similar (Fig. 2).

The thresholds values automatically selected were associated to relevant likelihood ratio values only for LOI, AVDL, eRQ and $AVDpCO_2$.

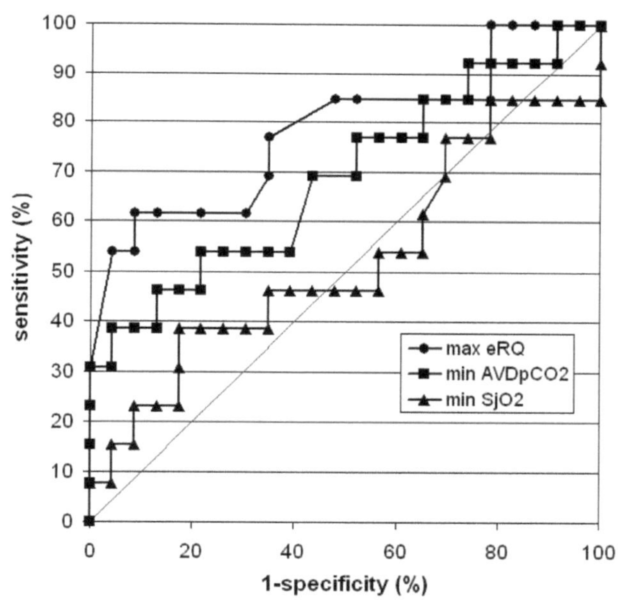

Fig. 1. Receiver operating characteristics curves for maximal estimated respiratory quotient (*eRQ*), minimum arteriovenous difference of carbon dioxide tension (*$AVDpCO_2$*), minimum jugular bulb saturation (*SjO_2*)

Discussion

The study aimed to analyse the association between the eRQ or $AVDpCO_2$ values and brain death within the first 48 hours after injury. Patients who died had worse clinical parameters, more severe abnormalities on CT scan, and more deranged physiologic variables.

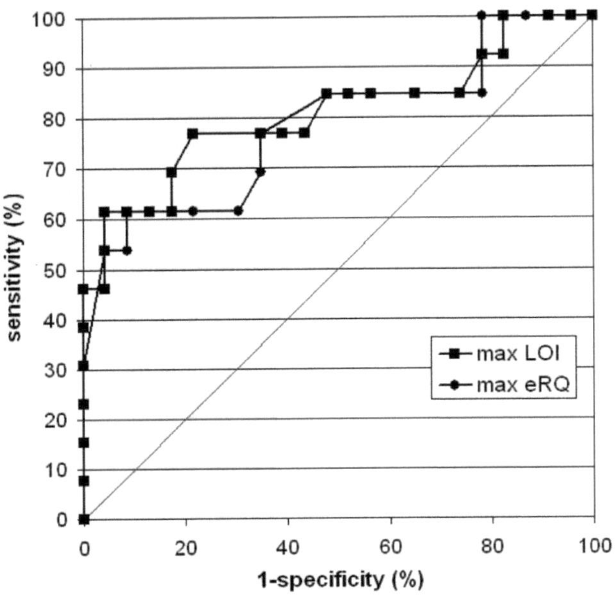

Fig. 2. Receiver operating characteristics curves for maximal estimated respiratory quotient (*eRQ*) and maximal lactate/oxygen index (*LOI*)

The present results confirm that in the patients who died early, global ischemia and anaerobic metabolism are the dominant events and suggest that it can be detected in vivo by an increase in eRQ.

The present study suggests that, an $AVDpCO_2$ of -21.4 mmHg indicates a threshold value for irreversible ischemia. However, the diagnostic superiority of $AVDpCO_2$ compared with SjO_2 has not been proven. This is probably because of its low specificity in detecting anaerobic metabolism. In fact, in an aerobic setting an increase in $AVDpCO_2$ is associated with an increase in $AVDO_2$, as a consequence of the Fick's equation. Assuming that CO_2 production and O_2 consumption remains constant, a reduction of CBF is phsyiologically associated with a compensatory increase in O_2 extraction (increase in $AVDO_2$ and decrease in SjO_2) and a decrease in CO_2 elimination in the venous capillary (increase of $AVDpCO_2$). On the other hand, during uncompensated ischemia anaerobic metabolism occurs. The reduction in CO_2 production is however less than oxygen consumption because CO_2 is actively produced to buffer the protons released during ischemia. The relative change, by comparing arteriovenous CO_2 ($AVDCO_2$) and oxygen contents ($AVDO_2$), give rise to the cerebral respiratory quotient ($AVDCO_2/AVDO_2$) [6]. Since measurement of blood CO_2 content is complicated [9], we used $AVDpCO_2$ instead $AVDCO_2$ to generate an estimated cerebral respiratory quotient, the eRQ. Physiologically, an uncompensated ischemia may be associated to an increase in eRQ, while a compensated ischemia may produce a stable eRQ, due to $AVDpCO_2$ and $AVDO_2$ coupling.

In the present study, we suggest that an increase in eRQ, above 2.4 mmHg/ml/dl, is associated to global irreversible cerebral ischemia. With respect to conventional SjO_2, eRQ add new information to jugular bulb monitoring. In fact SjO_2 was abnormally reduced in both patients who survive as well as those who died. In the setting of ultra-early fatal ischemia, oxygen extraction appeared still to be active, as indicated by jugular desaturation, even though anaerobic metabolism is already established. Conversely, in patients who survived, oxygen metabolism was maintained and jugular desaturation was associated with normal eRQ values.

In conclusion, eRQ seems to be a promising marker of global anaerobic cerebral metabolism. It is simple to calculate and may be useful for clinicians to interpret jugular bulb monitoring at the bedside, in the ultra early phases, by simply observing the full blood gases data, without the need of a lactate analyzer.

References

1. Bouma GJ, Muizelaar JP, Choi SC, Newlon PG, Young HF (1991) Cerebral circulation and metabolism after severe traumatic brain injury: the elusive role of ischemia. J Neurosurg 75: 685–693
2. Chieregato A, Zoppellari R, Targa L (1997) Cerebral arteriovenous PCO_2 difference and early global cerebral ischemia in a patient with acute severe head injury. J Neurosurg Anesthesiol 19: 256–262
3. Clifton GL, McCormick WF, Grossman RG (1981) Neuropathology of early and late deaths after head injury. Neurosurgery 8: 309–314
4. Cruz J (1996) Relationship between early patterns of cerebral extraction of oxygen and outcome from severe acute traumatic brain swelling: cerebral ischemia or cerebral viability? Crit Care Med 24: 953–956
5. Gibbs EL, Lennox WG, Nims LF, Gibbs FA (1942) Arterial and cerebral venous blood. Arterial-venous differences in man. J Biol Chem 144: 325–332
6. Maiolo AT, Negri VU, Della Porta P, Rossella E, Bianchi-Porro G, Faranda A (1965) Circolazione e metabolismo cerebrali nella crisi convulsiva e nel coma post-convulsivo dell'uomo sottoposto ad elettroshock in succinilcolina. Rass Studi Psichiatr 54: 723–738
7. Mekontso-Dessap A, Castelain V, Anguel N, Bahloul M, Schauvliege F, Richard C, Teboul JL (2002) Combination of

venoarterial PCO_2 difference with arteriovenous O_2 content difference to detect anaerobic metabolism in patients. Intensive Care Med 28: 272–277
8. Rossi S, Colombo A, Magnoni S, Roncati EZ, Conte V, Stocchetti N (2002) Cerebral veno-arterial pCO2 difference as an estimator of uncompensated cerebral hypoperfusion. Acta Neurochir [Suppl] 81: 201–204
9. Van Slyke DD, Neill JM (1924) The determination of gases in blood and other solutions by vacuum extraction and manometric measurement. I. J Biol Chem 61: 523–573

Correspondence: Arturo Chieregato, Servizio di Anestesia e Rianimazione, Ospedale M. Bufalini, Viale Ghirotti 286, 47023 Cesena, Italy. e-mail: achiere@ausl-cesena.emr.it

Acta Neurochir (2005) [Suppl] 95: 73–77

Linear correlation between stable intracranial pressure decrease and regional cerebral oxygenation improvement following mannitol administration in severe acute head injury patients

J. Gasco[1,2], **J. Sendra**[3], **J. Lim**[1], and **I. Ng**[1]

[1] Acute Brain Injury Research Laboratory, Department of Neurosurgery (Main Campus), National Neuroscience Institute, Singapore
[2] Neurosurgery Department, Hospital Universitario de Getafe, Madrid, Spain
[3] Neurosurgery Department, Hospital Universitario Ramon y Cajal, Madrid, Spain

Summary

Objectives. We investigated the relationship between stable decrease in intracranial pressure (ICP) following mannitol administration and variations in regional oxygenation ($PtiO_2$) in severe traumatic brain injured (STBI) patients.

Methods. Fourteen STBI patients (Glasgow Coma Scale score $\leq$ 8) admitted to the neurointensive care unit during a 12-month period were studied. Multiparameter data, including parenchymal tissue oxygen ($PtiO_2$) and carbon dioxide tension, cerebral perfusion pressure, mean arterial pressure; temperature, pH and pressure reactivity index were measured. Point values from 53 mannitol administrations were extracted every five seconds and divided into five consecutive 30-minute epochs. Mean values and trends were identified. Post-administration epoch with maximum decrease in ICP was selected along with the means of the corresponding variables. These data were compared with baseline to derive the delta values. Mean ΔICP was then compared with $\Delta PtiO_2$ in each patient using linear correlation.

Findings. In patients with $ICP_0 > 20$ mmHg (group A), $PtiO_2$ increased in 69.6% of samples, whereas in group B ($ICP_0 <$ 20 mmHg), $PtiO_2$ increased in only 50% of the samples. There was a significant negative correlation between mean ΔICP and $\Delta PtiO_2$ in both groups; Group A: $r = -0.79$ ($P = 0.01$); Group B: $r = -0.92$ ($P = 0.01$).

Conclusions. There is a strong negative correlation between stable decrease in ICP following mannitol administration and improvement in regional oxygenation around the peri-contusional area. The data suggest a potentially favourable interaction between mannitol therapy and cerebral internal milieu in STBI patients.

Keywords: Cerebral oxygenation; mannitol; head injury; neuromonitoring; multimodality monitoring.

Introduction

Multiparameter cerebral monitoring devices have brought the possibility of continuous surveillance of different variables (intracranial pressure, ICP; parenchymal tissue oxygen and carbon dioxide tensions, $PtiO_2$, $PtiCO_2$; pH and temperature) in severe head injured patients [1, 24]. The reliability of data from these probes, together with their physiological and ischemia threshold values have been investigated both in animal models and humans [7, 8, 10, 11, 17, 21, 35]. Data related to the severity of tissue damage induced by the probes suggest that brain tissue damage is minimal [4, 6, 24, 29, 32, 33]. Mannitol is widely accepted and used to decrease intracranial hypertension [5, 14–16, 18, 20, 22, 25–28, 31, 34].

While there have been descriptive articles on the effects of mannitol in global oxygenation [9, 26, 32], we have not found study evaluating regional (peri-contusional) tissue oxygenation. A described potential application of multiparameter cerebral monitoring probes is to detect critical hypoxic events in severe head injury patients [12, 13, 24, 29, 32]. In this setting, previous studies have reported no significant changes in global brain tissue oxygenation following mannitol administration [9, 32]. We therefore focused our study in the evaluation of the peri-contusional tissue aiming to determine the variations in regional oxygenation and its potential benefits in lowering ICP. Following Härtl's [9] conclusions, we considered the need to determine an approximation for cerebral perfusion pressure (CPP) thresholds in which therapeutic administration of mannitol could be beneficial. A secondary goal in this study was to observe the capability of a single mannitol administration to reverse hypoxic ($PtiO_2 < 10$ mmHg), hypoperfused (cerebral perfusion pressure, CPP < 70 mmHg; cerebral blood flow, CBF < 18 ml/min/100 g) and intracranial hypertensive (ICP > 20 mmHg) scenarios so as to provide further scientific background for the clinical use of mannitol.

Methods and materials

This study was developed between May 2002 and May 2003, during which 14 patients with STBI were consecutively admitted to the Neurointensive Care Unit of National Neuroscience Institute, Singapore. Glasgow Coma Scale (GCS) score at admission or deterioration to ≤8 points was a criterion for inclusion and insertion of multiparameter probe. Local ethics committee approval was granted and appropriate informed consent from designated next-of-kin was obtained. Intraparenchymal probes were inserted ipsilateral to the main insult in the frontal lobe. We used two types of multiparameter cerebral monitoring probes: Licox® brain tissue oxygenation probe (Integra NeuroSciences, Andover, Hampshire, England) in 7 patients, and Neurotrend® multiparameter sensor (Neurotrend, Diametrics Medical, Minneapolis, MN) in the remaining 7 patients. No major intracranial complications were associated with probe placement. Adaptation time of 1 hour was allowed. The parameters obtained were extracted on five-second intervals from the data provided by the probe monitors and subsequently extracted and displayed in a computerized database via an in-house developed software system. We used the Careview® software system for online monitoring of standard parameters and timely registrations of therapeutic interventions in the Neurointensive Care Unit. Pressure reactivity index (PRx) was calculated retrospectively. Indications to receive drug infusion were ICP > 20 mmHg or CPP < 70 mmHg. Daily surveillance of serum osmolarity values was performed to avoid values above 320 mOsm/kg. Therapy was limited to the acute phase of head injury. We studied a total of 53 mannitol administrations (20% mannitol in 100 ml solution infused over 30 minutes through a central vein). Previous literature thoroughly describes mannitol requirements and serum osmolarity dose responses [5, 14]. Observations started 30 minutes before infusion to determine baseline values and lasted for 2 hours. Observation period was then divided into five consecutive epochs. Mean values per time frame were determined along with the corresponding graphic trend of variation.

The time frame in which maximum ICP decrease after administration was selected (including running time) and the corresponding values of the remaining parameters were identified. These were further subtracted from the baseline values in order to obtain delta values of ΔICP, $\Delta PtiO_2$, ΔCPP, ΔMAP, Δtemperature, $\Delta PtiCO_2$, ΔpH and ΔPRx. To determine if the decrease in ICP after mannitol administration was followed by an improvement in regional oxygenation, we compared ΔICP with $\Delta PtiO_2$ using Pearson's correlation. Hypoxic, hypoperfused, and intracranial hypertensive situations within each pre-administration epoch were identified to determine if reversal was achieved during the 2-hour observation period. Two patients receiving a total of 10 mannitol administrations were discarded from analysis because of incomplete data captured as a result of failure to register the time stamp in the databases and the failure to complete continuous data extraction on 5-second intervals.

Results

We present the analysis of 43 mannitol administrations in 12 patients. The mean (± standard deviation) age was 44 ± 17 years, and the median GCS score after resuscitation was 10. Administrations were divided into two groups according to the baseline ICP values (Group A: $ICP_0 > 20$ mmHg, 23 administrations from 10 patients; Group B: $ICP_0 < 20$ mmHg, 20 administrations from 8 patients, 6 subsequently were transferred to this group due to a decrease in ICP below 20 mmHg). Mean follow-up period after probe insertion was 5 ± 1.8 days and the number of infusions per patient was 3.8 ± 3.1.

Maximum decrease in ICP following mannitol administration was observed to concentrate in the second and third intervals (68.6%): (0–30 min/running time): 2.9%; (30–60 min.): 40%; (1–1.5 h): 28.6%; (1.5–2 h): 28.5%. A comparison of baseline values for main parameters is listed in Table 1. Within group variations in ΔICP, $\Delta PtiO_2$ and corresponding CPP thresholds after mannitol administration is summarized in Table 2. The values were subdivided as to whether regional oxygenation was improved after drug infusion.

In patients with $ICP_0 > 20$ mmHg (Group A), improved regional oxygenation were observed in 16 out of 23 samples (69.6%) at the

Table 1. *Comparison of baseline values between Groups A ($ICP_0 > 20$ mmHg) and B ($ICP_0 < 20$ mmHg)*

Variable	Pre-administration Group A		Pre-administration Group B	
	Mean	SD	Mean	SD
ICP (mmHg)	35.8	19.4	17.5	2.5
$PtiO_2$ (mmHg)	31.4	23.8	24.6	14.8
CPP (mmHg)	64.8	16.3	77.2	11.2
MAP (mmHg)	101.2	16.5	86.9	20.7

ICP Intracranial pressure; *$PtiO_2$* parenchymal tissue oxygen tension; *CPP* cerebral perfusion pressure; *MAP* mean arterial pressure; *SD* standard deviation.

Table 2. *Within-group variations in samples with or without improvement in regional parenchymal tissue oxygen tension*

	Group A (ICP > 20 mmHg)				Group B (ICP < 20 mmHg)			
	Improved $PtiO_2$		Not improved $PtiO_2$		Improved $PtiO_2$		Not improved $PtiO_2$	
	Mean	SD	Mean	SD	Mean	SD	Mean	SD
No. of samples	16 (69.6%)		7 (30.4%)		10 (50%)		10 (50%)	
ΔMAP (mmHg)	−2.4	6.9	−1.4	8.5	18.6	35.2	−6.1	5.5
ΔICP (mmHg)	−6.2	6.3	−4.4	3	−3.3	2.8	−1.9	1.2
CPP (mmHg)	69	15.9	66.5	20.4	87.7	13.1	70.9	9.6
$\Delta PtiO_2$ (mmHg)	7.1	9.1	−3.6	3.9	2.4	4.7	−2.2	1.5

MAP Mean arterial pressure, *ICP* intracranial pressure; *CPP* cerebral perfusion pressure; *$PtiO_2$* parenchymal tissue oxygen tension.

lower CPP ranges, whereas only 50% of samples revealed $PtiO_2$ improvement in Group B (ICP < 20 mmHg). There was a negative correlation between mean ΔICP and mean $\Delta PtiO_2$ (Group A: $r = -0.79$; $P = 0.01$ (2-tailed), and Group B: $r = -0.92$; $P = 0.01$ (2-tailed), Fig. 1). The magnitude of improvement for ICP and $PtiO_2$ in group A, 6.2 ± 6.3 and 7.1 ± 9.1 mmHg, was higher than that in group B, 3.3 ± 2.8 and 2.4 ± 4.7 mmHg, respectively. Mean arterial pressure did not change and therefore was not responsible for the increase in $PtiO_2$.

A single administration of mannitol reversed 7 out of 33 (21%) episodes of intracranial hypertension, 4 out of 23 episodes of hypoperfusion and 2 out of 10 (20%) hypoxic episodes. In four episodes, all three scenarios were present at the same time prior to mannitol infusion. Intracranial hypertension was reversed in all four of them. Following drug administration, global reversal of all three parameters was observed in only one case. Mannitol did not reverse hypoxia in 2 episodes but did improve hypoperfusion. In the remaining sample the contrary occurred. With regards to autoregulation, the PRx index did not change following mannitol therapy.

Discussion

The use of multiparameter cerebral monitoring probes has many implications. Prognostic benefit obtained by the use of these probes is sometimes controversial [1, 12, 13, 24, 32, 29]. Physicians have to become familiarized with new hardware, software and technical skills. In surgical candidates, increased operative time and potentially harmful situations should be considered and thoroughly prevented [4]. Neurointensive care coordination for on-line monitoring with nursing personnel and physician orders is of utmost importance [1]. We recommend the implementation of Neurointensive Care Unit on-line software to monitor for continuous changes in patient status [1, 29]. Medico-legal issues and adequate permissions have to be obtained in scientific trials in order to comply with ethical rules.

The utility of peri-contusional probes to detect

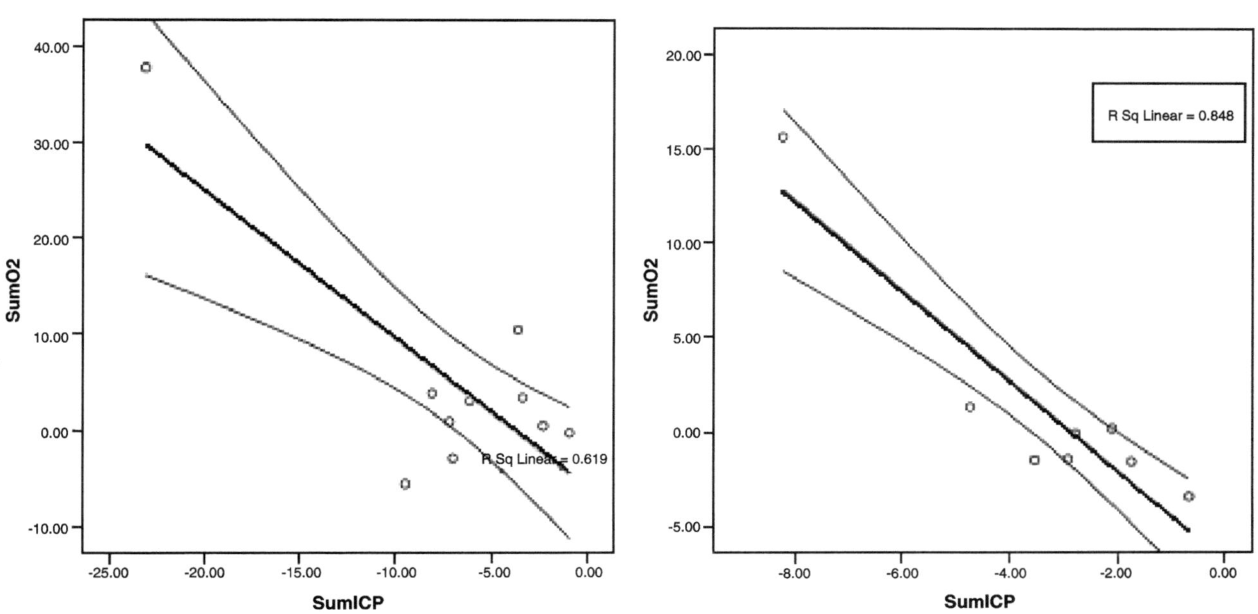

Fig. 1. Correlation between mean ΔICP and mean $\Delta PtiO_2$ following mannitol administration. Group A (left): Baseline ICP above 20 mmHg; $r = -0.79$; $P = 0.01$ (2-tailed). Group B (right): Baseline ICP below 20 mmHg; $r = -0.92$; $P = 0.01$ (2-tailed)

situations in which injured tissue is approaching irreversible damage is clear. To standardize methodology and materials used accordingly to the clinical goals should be a future aim. With regards to the results presented, stable decreasing trend of ICP following mannitol administration produced a strong negative linear correlation with improved regional oxygenation values. Interestingly the correlation was higher for samples without intracranial hypertension ($r = -0.92$ in Group B vs. $r = -0.79$ in Group A). The underlying mechanism remains unclear.

Rheological properties of mannitol may be the cause as why we observed an improved oxygenation in the peri-contusional area before a decrease in ICP was observed [2, 3, 18–20, 23, 26, 30]. We found a bigger benefit for brain oxygenation in patients with higher ICP (>20 mmHg) compared with patients without intracranial hypertension. CPP did not change significantly following mannitol administration.

Conclusion

Irrespective of baseline ICP values, a linear relationship between a decrease in ICP and an increase in pericontusional cerebral oxygenation has been established following mannitol infusion. Current data suggest that mannitol not only has the physiological effect of decreasing ICP but it has the potential of interacting with the internal milieu to improve regional cerebral oxygenation in patients with severe head injury.

References

1. Bardt TF, Unterberg AW, Kiening KL, Schneider GH, Lanksch WR (1998) Multimodal cerebral monitoring in comatose head-injured patients. Acta Neurochir (Wien) 140(4): 357–365
2. Burke AM, Quest DO, Chien S, Cerri C (1981). The effects of mannitol on blood viscosity. J Neurosurg 55: 550–553
3. Bruce DA, Langfitt TW, Miller JD, Schutz H, Vapalahti MP, Stanek A, Goldberg HI (1973) Regional cerebral blood flow, intracranial pressure, and brain metabolism in comatose patients. J Neurosurg 38: 131–144
4. Dings J, Meixensberger J, Jager A, Roosen K (1998) Clinical experience with 118 brain tissue oxygen partial pressure catheter probes. Neurosurgery 43: 1082–1095
5. Ferrer E, Vila F, Isamat F (1979) ICP response to mannitol in severe head injuries and other ICH pathologies: a histogram study [proceedings]. Acta Neurochir [Suppl] 28: 498
6. Gelabert-Gonzalez M, Villa JM (2003) Applications of the determination of brain tissue oxygen pressure ($PtiO_2$). Rev Neurol 36: 744–749
7. Gjedde A, Marrett S, Vafaee M (2002) Oxidative and nonoxidative metabolism of excited neurons and astrocytes. J Cereb Blood Flow Metab 22: 1–14
8. Harris RJ, Richards PG, Symon L, Habib AH, Rosenstein J (1987) pH, K^+, and PO_2 of the extracellular space during ischaemia of primate cerebral cortex. J Cereb Blood Flow Metab 7: 599–604
9. Härtl R, Bardt TF, Kiening KL, Sarrafzadeh AS, Schneider GH, Unterberg AW (1997) Mannitol decreases ICP but does not improve brain-tissue pO_2 in severely head-injured patients with intracranial hypertension. Acta Neurochir [Suppl] 70: 40–42
10. Hoffman WE (1999) Measurement of intracerebral oxygen pressure: practicalities and pitfalls. Curr Opin Anaesthesiol 12: 497–502
11. Hoffman WE, Charbel FT, Edelman G (1996) Brain tissue oxygen, carbon dioxide, and pH in neurosurgical patients at risk for ischemia. Anesth Analg 82: 582–586
12. Kiening KL, Unterberg AW, Bardt TF, Schneider GH, Lanksch WR (1996) Monitoring of cerebral oxygenation in patients with severe head injuries: brain tissue PO_2 versus jugular vein oxygen saturation. J Neurosurg 85: 751–757
13. Marshall LF, Smith RW, Rauscher LA, Shapiro HM (1978) Mannitol dose requirements in brain-injured patients. J Neurosurg 48: 169–172
14. McGraw CP, Howard G (1983) Effect of mannitol on increased intracranial pressure. Neurosurgery 13: 269–271
15. Mendelow AD, Teasdale GM, Russell T, Flood J, Patterson J, Murray GD (1985) Effect of mannitol on cerebral blood flow and cerebral perfusion pressure in human head injury. J Neurosurg 63: 43–48
16. Menzel M, Rieger A, Roth S *et al* (1998) Simultaneous continuous measurement of pO_2, pCO_2, pH and temperature in brain tissue and sagittal sinus in a porcine model. Acta Neurochir [Suppl] 71: 183–185
17. Muizelaar JP, Lutz HA 3rd, Becker DP (1984) Effect of mannitol on ICP and CBF and correlation with pressure autoregulation in severely head-injured patients. J Neurosurg 61: 700–706
18. Muizelaar JP, Wei EP, Kontos HA (1983) Mannitol causes compensatory cerebral vasoconstriction and vasodilation in response to blood viscosity changes. J Neurosurg 58: 822–828
19. Nath F, Galbraith S (1986) The effect of mannitol on cerebral white matter content. J Neurosurg 65: 41–43
20. Reinstrup P, Stahl N, Mellergard P, Uski T, Ungerstedt U, Nordstrom CH (2003) Intracerebral microdialysis in clinical practice: baseline values for chemical markers during wakefulness, anesthesia, and neurosurgery. Neurosurgery 47: 701–710
21. Roberts I, Schierhout G, Wakai A (2003) Mannitol for acute traumatic brain injury. Cochrane Database Syst Rev
22. Sahuquillo J, Biestro A, Mena MP, Amoros S, Lung M, Poca MA, De Nadal M, Baguena M, Panzardo H, Mira JM, Garnacho A, Lobato RD (2002) First tier therapeutic measures in the management of high intracranial pressure after severe head injury. Rationale and proposal for a protocol. Neurocirugia 13: 78–100
23. Sarrafzadeh AS, Kiening KL, Bardt TF, Schneider GH, Unterberg AW, Lanksch WR (1998) Cerebral oxygenation in contusioned vs nonlesioned brain tissue: monitoring of $PtiO_2$ with Licox and Paratrend. Acta Neurochir [Suppl] 71: 186–189
24. Sayre M, Daily S, Stern S, Storer DL, van Loveren HR, Hurst JM (1996) Out-of-hospital administration of mannitol does not change systolic blood pressure. Acad Emerg Med 3: 840–848
25. Schrot RJ, Muizelaar JP (2002) Mannitol in acute traumatic brain injury. Lancet 359: 1633–1634
26. Schwartz M, Tator C, Rowed D (1984) The University of Toronto Head Injury Treatment Study: a prospective randomised comparison of pento-barbital and mannitol. Can J Neurol Sci 11: 434–440
27. Smith HP, Kelly DL Jr, McWhorter JM, Armstrong D, John-

son R, Transou C, Howard G (1986) Comparison of mannitol regimens in patients with severe head injury undergoing intracranial monitoring. J Neurosurg 65: 820–824
28. Soehle M, Jaeger M, Meixensberger J (2003) Online assessment of brain tissue oxygen autoregulation in traumatic brain injury and subarachnoid hemorrhage. Neurol Res 25: 411–417
29. Takagi H, Saitoh T, Kitahara T *et al* (1983) The mechanism of ICP reducing effect of mannitol. In: Ishii S, Nagai H, Brock M (eds) Intracranial pressure V. Springer Berlin Heidelberg New York Tokyo, pp 729–733
30. The Brain Trauma Foundation. The American Association of Neurological Surgeons. The Joint Section on Neurotrauma and Critical Care (2000) Use of mannitol. J Neurotrauma 17: 521–525
31. Unterberg AW, Kiening KL, Hartl R, Bardt T, Sarrafzadeh AS, Lanksch WR (1997) Multimodal monitoring in patients with head injury: evaluation of the effects of treatment on cerebral oxygenation. J Trauma 42: 32–37
32. Van den Brink WA, Haitsma IK, Avezaat CJ, Houtsmuller AB, Kros JM, Maas AI (1998) The brain parenchyma/pO_2 catheter interface: a histopathological study in the rat. J Neurotrauma 15: 813–824
33. Von Berenberg P, Unterberg A, Schneider GH, Lanksch WR (1994) Treatment of traumatic brain edema by multiple doses of mannitol. Acta Neurochir [Suppl] 60: 531–533
34. Zauner A, Daugherty WP, Bullock MR, Warner DS (2002) Brain oxygenation and energy metabolism: part I-biological function and pathophysiology. Neurosurgery 51: 289–302
35. Zornow MH, Oh YS, Scheller MS (1990) A comparison of the cerebral and haemodynamic effects of mannitol and hypertonic saline in an animal model of brain injury. Acta Neurochir [Suppl] 51: 324–325

Correspondence: Ivan Ng, 11, Jalan Tan Tock Seng, National Neuroscience Institute, Level 3, Clinical Staff Office, Singapore 308433. e-mail: ivan_ng@nni.com.sg

Acta Neurochir (2005) [Suppl] 95: 79–81

Brain tissue oxygen ($P_{ti}O_2$): a clinical comparison of two monitoring devices

M. Jaeger[1], **M. Soehle**[2], and **J. Meixensberger**[1]

[1] Department of Neurosurgery, University of Leipzig, Leipzig, Germany
[2] Department of Anaesthesiology and Intensive Care Medicine, Rheinische Friedrich-Wilhelms-University, Bonn, Germany

Summary

Background. We investigated the difference between two commercially available sensors for continuous monitoring of brain tissue oxygen ($P_{ti}O_2$). One is a single parameter probe for $P_{ti}O_2$ monitoring (Licox®), the other is a multiparamter sensor (Neurotrend™) further including measurement of brain temperature, pH, and partial pressure of tissue carbon dioxide.

Methods. In seven patients after subarachnoid hemorrhage or traumatic brain injury continuous monitoring of $P_{ti}O_2$ was performed simultaneously using Licox® and Neurotrend™.

Findings. Mean $P_{ti}O_2$ was generally lower when assessed by the Neurotrend™, as compared with the Licox® (Licox® 27.7 mmHg vs. Neurotrend™ 20.9 mmHg; $P = 0.028$). The amplitude of $P_{ti}O_2$ elevations during ventilation with 100% oxygen was higher with the Licox®, but this did not reach statistical significance (Licox® 55.2 mmHg vs. Neurotrend™ 50.2 mmHg, $P = 0.082$). Regarding clinical stability of the sensors, only one Neurotrend™ sensor provided valid function over the desired monitoring period. Five Neurotrend™ sensors dislocated or broke and one sensor did not show any function after insertion. No malfunction occurred with the Licox® sensors.

Conclusions. Our results suggest that $P_{ti}O_2$ might be lower when assessed by the Neurotrend™ sensor. The clinical stability of the Neurotrend™ sensor was of concern and allowed monitoring in one of seven patients over the desired monitoring period of several days only.

Keywords: Brain tissue oxygen; neuromonitoring; cerebral hypoxia; Neurotrend™; Licox®.

Introduction

Continuous monitoring of the partial pressure of brain tissue oxygen ($P_{ti}O_2$) is being increasingly used in the treatment of patients suffering from severe acute brain disease such as traumatic brain injury (TBI) and aneurysmal subarachnoid haemorrhage (SAH) [1, 3, 8–10]. $P_{ti}O_2$ has been shown as an independent predictor of outcome after TBI, and a $P_{ti}O_2$ guided CPP management is successful in reducing cerebral hypoxic episodes after TBI which might improve outcome in this group of severely injured patients [5].

Currently, two different devices for continuous monitoring of $P_{ti}O_2$ are commercially available and have been used in the previously published literature: The Licox® and the Neurotrend™ system. Because until today both systems are considered similar in their ability to sufficiently continuously monitor $P_{ti}O_2$ and almost no data exists comparing both devices, with this study we sought to investigate differences between the two devices in the clinical setting.

Methods

Ethics committee approval to perform advanced neuromonitoring was given and informed consent from the patients' next relative was obtained. Seven patients with severe traumatic brain injury ($n = 3$) and severe aneurysmal subarachnoid haemorrhage ($n = 4$) were included in this study. All patients were sedated with midazolam and fentanyl. All patients received artificial ventilation throughout the study period. Ventilation aimed at keeping arterial oxygen at 100–120 mmHg and arterial carbon dioxide at 35–40 mmHg.

A Licox® catheter (Licox®, Integra NeuroSciences, Plainsboro, NJ, USA) and a Neurotrend™ catheter (Neurotrend™, Codman and Shurtlef, Raynham, MA, USA) were inserted in parallel in the frontal white matter at a depth of 25 mm subdurally with a distance between catheters of approximately 10 mm. Correct positioning of the catheters in normal tissue was confirmed by routine computed tomograhic scan.

The Licox® system contains a single parameter probe, that consists of a Clark type probe in the catheter tip with a oxygen sensitive area of 13 mm^2. The Neurotrend™ system is a multiparameter sensor including monitoring of the partial pressure of brain tissue carbon dioxide, pH, and temperature. This catheter uses an optode for $P_{ti}O_2$ monitoring with an oxygen sensitive area of approximately 2 mm^2. In this study, we meticulously aimed at inserting both O_2 sensitive areas at the same depth below dura.

Data of $P_{ti}O_2$ obtained from both catheters were stored on a computer at the patient's bedside with a rate of 2 min^{-1} and analysed retrospectively. Data recording started 2 hours after insertion of both catheters to allow sufficient "run in time". Statistical analysis was made using Wilcoxon-signed-rank test (SPSS 11.0, SPSS Inc., Chicago, Il, USA).

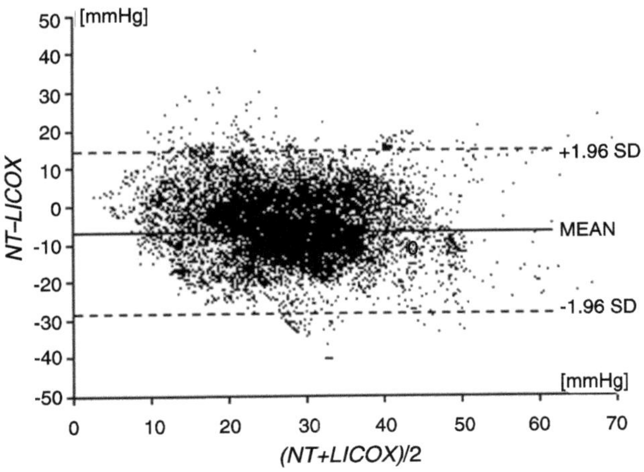

Fig. 1. Bland and Altman plot for agreement of $P_{ti}O_2$ data obtained with the Licox® and the Neurotrend™ (*NT*) catheter in six patients after SAH and TBI. Due to the different number of data pairs contributed from each patient the mean difference and the limits of agreement were calculated according to Bland *et al.* [2]

Results

The average of the individual mean ($\pm$ standard deviation) $P_{ti}O_2$ values was 27.7 ± 5.3 mmHg when assessed with the Licox®, as compared to 20.9 ± 6.1 mmHg when assessed with the Neurotrend™ ($P = 0.03$). Furthermore, the $P_{ti}O_2$ measurements obtained with the Licox® were significantly higher in each individual patient. Phases of higher Neurotrend™ measurements occurred, however, this accounted for a mean of 19% of the entire monitoring time, whereas Licox® measurements of $P_{ti}O_2$ were higher at an average of 81% of the monitoring time. Figure 1 shows the Bland and Altman plot for agreement of the Licox® and the Neurotrend™ system to measure $P_{ti}O_2$.

The mean amplitude of $P_{ti}O_2$ elevations during 31 O_2 tests, i.e. ventilation with 100% oxygen was 55.5 ± 44.8 mmHg with the Licox®, whereas it was 50.2 ± 49.5 mmHg when assessed by the Neurotrend™. This difference did not show statistical significance ($P = 0.09$).

While all Licox® catheters provided valid function over the desired monitoring period of an average of 6.3 days, only one Neurotrend™ catheter worked continuously without malfunction. One Neurotrend™ did not show any function after insertion, thus the data from only six patients could be included for analysis. Five Neurotrend™ catheters dislocated or broke during routine nursing interventions on our neurosurgical intensive care unit. Overall, monitoring with the multiparamter sensor was possible during a mean of 72% of the desired monitoring time only (151.9 vs. 123.2 h).

Discussion

These results of our study comparing the two commercially available systems for monitoring of $P_{ti}O_2$ (Licox® and Neurotrend™) by simultaneous measurement suggest that the two systems did not produce identical $P_{ti}O_2$ values in the clinical setting. The mean $P_{ti}O_2$ values measured by the Licox® system were significantly higher than those by the Neurotrend™. The results from the Bland and Altman plot [2] confirm this finding and we consider the mean $P_{ti}O_2$ difference of 6.25 mmHg is of importance in the clinical interpretation of the data. Furthermore, there was a trend towards higher $P_{ti}O_2$ amplitudes during O_2 tests with the Licox®. We believe that these findings should always be kept in mind when comparing results from studies using these two different devices.

It is unclear whether the differences in $P_{ti}O_2$ measurements are a result of the different technical properties of the devices or reflect "true" physiological O_2 differences in the measured tissue volumes. We did not perform tests on measurement accuracy of the probes after their removal. However, while both sensors were attempted to be inserted in the same depth in the frontal white matter, there may still be difference in the depths of insertion. This may cause different $P_{ti}O_2$ measurements, as $P_{ti}O_2$ values increase towards the cerebral cortex [4]. The remarkably smaller oxygen sensitive area of the Neurotrend™ sensor might also explain lower $P_{ti}O_2$ readings. Because it is assumed that O_2 is heterogeneously distributed in the cerebral tissue, the Licox® sensor with its larger oxygen sensitive area and larger sample volume may compensate better for local areas of different O_2 concentrations, probably providing a more accurate average estimation of $P_{ti}O_2$.

To our knowledge, two previous studies in the literature have used both systems and attempted a comparison of the data. While Sarrafzadeh *et al.* [7] did not find relevant differences using the sensors in parallel in non lesioned tissue in vivo, Valadka *et al.* [8] reported that the multiparamter sensor overestimated $P_{ti}O_2$ at very low levels in vitro. Notably, both studies used the Paratrend™ system, which contained a Clark type probe for $P_{ti}O_2$ monitoring. The Paratrend™ system was originally developed for continuous blood gas monitoring and then modified for intracranial use, however, since 1998 both the Paratrend™ and the Neurotrend™ consist of an optode for O_2 measurement. Thus, the results from Sarrafzadeh *et al.* and Valadka *et al.* might be difficult to compare with our results due to the different technical properties of the multiparameter sensors used.

We also consider the clinical stability and usability of the Neurotrend™ to be of special concern. While only one of seven sensors worked without malfunction over the desired monitoring period, five sensors dislocated or broke and one sensor did not show any function after insertion. Similar problems with the clinical usability of the catheter have been previously reported by others [6] and are clearly unsatisfactory for the physician using the system at the patients bedside. Because of these reported drawbacks it seems necessary to improve the clinical stability of the Neurotrend™ system.

References

1. Bardt TF, Unterberg AW, Hartl R, Kiening KL, Schneider GH, Lanksch WR (1998) Monitoring of brain tissue PO_2 in traumatic brain injury: effect of cerebral hypoxia on outcome. Acta Neurochir [Suppl] 71: 156
2. Bland JM, Altman DG (1999) Measuring agreement in method comparison studies. Stat Methods Med Res 8: 135–160
3. Dings J, Jäger A, Meixensberger J, Roosen K (1998) Brain tissue pO_2 and outcome after severe head injury. Neurol Res [Suppl] 20: 71–75
4. Dings J, Meixensberger J, Jäger A, Roosen K (1998) Clinical experience with 118 brain tissue oxygen partial pressure catheter probes. Neurosurgery 43: 1082–1095
5. Meixensberger J, Jaeger M, Väth A, Dings J, Kunze E, Roosen K (2003) Brain tissue oxygen guided treatment supplementing ICP/CPP therapy after traumatic brain injury. J Neurol Neurosurg Psychiatry 74: 760–764
6. Pennings FA, Bouma GJ, Kedaria M, Jansen GF, Bosch DA (2003) Intraoperative monitoring of brain tissue tissue oxygen and carbon dioxide pressures reveals low oxygenation in peritumoral brain edema. J Neurosurg Anesthesiol 15: 1–5
7. Sarrafzadeh AS, Kiening KL, Bardt TF, Schneider GH, Unterberg AW, Lanksch WR (1998) Cerebral oxygenation in contusioned vs. nonlesioned brain tissue: monitoring of $PtiO_2$ with Licox and Paratrend. Acta Neurochir [Suppl] 71: 186–189
8. Valadka AB, Gopinath SP, Contant CF, Uzura M, Robertson CS (1998) Relationship of brain tissue PO_2 to outcome after severe head injury. Crit Care Med 26: 1576–1581
9. van den Brink WA, van Santbrink H, Steyerberg EW, Avezaat CJJ, Suazo JAC, Hogesteeger C, Jansen WJ, Kloos LMH, Vermeulen J, Maas AIR (2000) Brain Oxygen Tension in Severe Head Injury. Neurosurgery 46: 868–878
10. Väth A, Kunze E, Roosen K, Meixensberger J (2002) Therapeutic aspects of brain tissue pO_2 monitoring after subarachnoid hemorrhage. Acta Neurochir [Suppl] 81: 307–309

Correspondence: Matthias Jaeger, Klinik und Poliklinik für Neurochirurgie, Universitätsklinikum Leipzig, Liebigstr. 20, 04103 Leipzig, Germany. e-mail: jaem@medizin.uni-leipzig.de

Acta Neurochir (2005) [Suppl] 95: 83–88

Extracellular amino acid changes in patients during reversible cerebral ischaemia

R. Kett-White[1], M. T. O'Connell[1], P. J. A. Hutchinson[1], P. G. Al-Rawi[1], A. K. Gupta[2], J. D. Pickard[1], and P. J. Kirkpatrick[1]

[1] University Department of Neurosurgery and the Wolfson Brain Imaging Centre, Addenbrooke's Hospital and University of Cambridge, Cambridge, UK
[2] University Department of Neuroanaesthesia and the Wolfson Brain Imaging Centre, Addenbrooke's Hospital and University of Cambridge, Cambridge, UK

Summary

This study investigated the changes in extracellular chemistry during reversible human cerebral ischaemia. Delayed analysis was performed on samples taken from a subgroup of patients during aneurysm surgery previously reported [12]. Frozen microdialysis samples from 14 patients who had all undergone temporary clipping of the ipsilateral internal carotid artery (ICA) were analysed for another 15 amino acids with HPLC and for glycerol with CMA-600. Changes were characterised according to whether cerebral tissue oxygen pressure (P_BO_2) decreases were brief or prolonged.

Brief ICA clipping (maximum duration of 16 minutes) in 11 patients was not associated with changes in amino acids or glycerol. Cerebral ischaemia, defined by a P_BO_2 decrease below 1.1 kPa for at least 30 minutes during ICA occlusion, occurred in 3 patients. None of whom developed an infarct in the monitored region. This prolonged reversible ischaemia was associated with transient delayed increases in γ-amino butyric acid (GABA) as well as glutamate and glycerol, each by two-to-three folds. This study demonstrates detectable transient increases in human extracellular glutamate, GABA and glycerol during identified periods of reversible cerebral ischaemia, maximal 30–60 minutes after onset of ischaemia, but not in other amino acids detected by HPLC.

Keywords: Cerebral aneurysm surgery; cerebral oximetry; cerebral monitoring.

Introduction

Ethical investigation of cerebral ischaemia in humans can only be conducted with serendipity. Deliberate temporary arterial clipping necessary during complex aneurysm surgery [11] offers such an opportunity. By microdialysis, cerebral extracellular chemical changes have been shown in reversible and irreversible cerebral ischaemia in experimental models, and during irreversible ischaemia clinically. This paper is novel in that it explores the effects of reversible cerebral ischaemia on an extended panel of amino acids in a series of patients. Clinical studies after head injury [3, 5, 20, 21] and subarachnoid haemorrhage [16–18, 20] constitute evidence that raised extracellular lactate [18, 21], lactate/pyruvate ratio (L/P) [18] and lowered glucose [18, 21] are associated with poor outcome, as are increased concentrations of the excitatory neurotransmitters glutamate and aspartate [14, 18]. These metabolic changes are similar to those originally demonstrated in an experimental permanent cerebral artery occlusion model [6] and in 4 patients during frontal lobe resection [7].

In the current paper, we report a broader analysis of extracellular amino acids in a subgroup of 14 patients, all of whom underwent temporary clipping of the ICA, including patients with giant or difficult aneurysms, often of poor clinical grade, where the chances of observing ischaemic complications were maximal. Changes in chemistry were compared against the degree of hypoxia that occurred, as indicated by a P_BO_2 decrease, as well as in the changes of other energy substrates. A P_BO_2 of 1.1 kPa has been proposed as a possible threshold for ischaemia after head injury, corresponding to a hypoxic threshold of 50% for jugular venous oxygen saturation [13].

Material and methods

A subset of patients, all those requiring ICA occlusion, was selected from a larger study of cerebral monitoring during aneurysm surgery [12] for subsequent HPLC. Table 1 lists the clinical and operative details of these patients, as well as the World Federation of Neurological Surgeons (WFNS) grades [19]. Clinical management and monitoring techniques are reported in the main study [12]. In short, cerebral oxygen was monitored using Neurotrend sensors

Table 1. *Clinical details of 14 patients who required ICA temporary clipping. No infarcts occurred in the regions monitored by the probes. On pre-operative Angiography, patients L and M had good calibre ACoAs and ipsilateral PCoAs. Indeed in M, during a trial balloon occlusion of the ipsilateral ICA, there was cross-flow from the contralateral ICA to the ipsilateral MCA, but with significant perfusion delay suggesting collateral inadequacy. In N, the angiogram showed a small ipsilateral PCoA, but both anterior cerebral arteries were of good calibre, although the ACoA was not demonstrated under normal flow conditions*

Case	Age (yrs)	WFNS grade*	Aneurysm	Duration temporary clips (min)	Total time (min)	Intra-op rupture	P_BO_2 < 1.1 kPa for >30 min	Infarct due to surgery	Outcome 6-months †
A	56	U	giant ICA	4,6	10	+	–	–	GR
B	63	2	PCoA.	3,4	7	–	–	–	GR
C	52	U	coiled ICA	5	5	–	–	PCA	MD
D	45	2	ICA	3,6,2	11	–	–	–	MD
E	65	4	PCoA	4,2,1,1,1	10	+	–	watershed	Dead
F	74	4	PCoA	7	7	–	–	–	MD
G	45	1	PCoA	3	3	–	–	–	MD
H	58	U	PCoA	9	9	+	–	–	MD
I	32	4	ICA	2,2,2	6	–	–	–	GR
J	55	5	MCA	2,1	3	–	–	–	GR
K	44	1	giant PCoA	4,1,7,4	16	+	–	–	GR
L	59	4	giant ICA	4,8,7,7,9,1	36	–	+	–	SD
M	39	U	giant ICA	1,4,20,20,3,3,2	53	+	+	partial MCA	MD
N	60	1	MCA	7,7,1,2,2,1,2,1, 5,4,1,5	38	+	+	–	GR

ICA Internal carotid artery; *PCoA* posterior communicating artery; *MCA* middle cerebral artery; *PCA* posterior cerebral artery; + present and – absent.
* World Federation of Neurosurgical Societies grade on presentation (1 good – 5 poor) *U* unruptured (19).
† Glasgow Outcome Score (*GR* good recovery, *MD* moderate disability, *SD* severe disability).

(Codman, Bracknell, UK), data was stored by an on-line personal computer and a contemporaneous operative log was kept. Samples of extracellular fluid were collected in vials changed every 10 minutes, using 10-mm microdialysis catheters (CMA-70, CMA, Solna, Sweden) perfused with Ringer's solution at 1 µl/min. Samples were immediately analysed for glucose, lactate, pyruvate and glutamate with a mobile analyser (CMA 600). Vials were stored at −70 °C for later glycerol analysis using a CMA-600 and HPLC for other amino acids [10]. Both monitors were introduced and secured through a triple-lumen access device (Technicam, Newton Abbott, UK) separate but close to the craniotomy, the Neurotrend to a depth of 40 mm and the CMA-70 to 10 mm below dura. The monitors were aimed to the vascular territory most likely to be at risk during surgery. Temporary clips were all placed on the ICA, proximal to the posterior communicating artery, except in case L where the size of the aneurysm necessitated control of the extra-cranial ICA.

HPLC technique

Amino acids in microdialysate were analysed by precolumn derivatisation, binary gradient liquid chromatography separation and fluorescence detection using an ortho-phthaldialdehyde (OPA)/3-mercaptopropionic acid (MPA) reaction mixture (Agilent Technologies, Waldbronn, Germany). Amino acids derivatisation products were identified and quantified by reference to standard solutions prepared from reference material (Amino acid standard mixture, Agilent Technologies, Waldbronn, Germany). Calibration curves were constructed to cover the range 0.1–100 µmol/l of each amino acid except glutamine (5–500 µmol/l) and GABA (0.002–2 µmol/l).

Statistical analysis

Data was amalgamated by synchronisation to the onset of the most significant period of ischaemia in each patient. In patients with only brief temporary clipping and therefore minimal P_BO_2 changes, (group 1), data were synchronised to the start of the first episode of temporary clipping. In 3 patients who required extended temporary clipping and in whom P_BO_2 fell below 1.1 kPa (8 mm Hg) for 30 consecutive minutes (group 2), synchronisation was to the time when P_BO_2 first decreased below 1.1 kPa during temporary clipping. The mean and standard deviation for variables during various time intervals were calculated.

Results

After probe insertion and mostly in the first hour, levels of glutamate, aspartate and glycerol decreased in an exponential fashion. The energy substrates and the other amino acids were less prone to this phenomenon (Tables 2 and 3). HPLC analysis was not possible in one patient (case G) because the vials were damaged in storage.

Pattern of changes with temporary clipping

In the majority of patients, brief temporary clipping produced only small, brief decreases in P_BO_2 with few microdialysis changes. However, there were considerable changes in 2 patients: Case L undergoing revision surgery after recurrent haemorrhage from a giant aneurysm previously clipped elsewhere, and Case M

Table 2. *Brief internal carotid artery temporary clipping (maximum total 16 minutes) caused little change in extracellular chemistry in 11 patients*

	Minutes before start		Minutes after start of temporary clipping				IF*
	−120 to −60	−60 to 0	1 to 15	16 to 30	31 to 60	61 to 90	
Glucose †	0.7 ± 0.4	0.8 ± 0.4	0.8 ± 0.3	0.8 ± 0.3	0.8 ± 0.3	0.7 ± 0.4	0.9
Lactate †	0.8 ± 0.4	1 ± 0.8	0.8 ± 0.3	1 ± 0.7	0.7 ± 0.3	0.6 ± 0.3	1
Pyruvate	48 ± 18	48 ± 20	42 ± 14	49 ± 18	41 ± 12	40 ± 11	1
L/P ‡	18 ± 6	20 ± 10	21 ± 8	19 ± 7	18 ± 6	16 ± 6	0.9
G/L §	0.9 ± 0.4	1.1 ± 0.6	1 ± 0.3	0.9 ± 0.3	1.2 ± 0.4	1.2 ± 0.4	0.9
Glycerol	101 ± 50	106 ± 106	74 ± 46	102 ± 124	66 ± 38	48 ± 30	1
Glutamate	34 ± 30	23 ± 18	18 ± 13	15 ± 10	11 ± 9	10 ± 7	0.7
Alanine	12 ± 5	17 ± 8	14 ± 9	17 ± 7	17 ± 13	15 ± 8	1
Arginine	6.8 ± 3.9	8.2 ± 4.3	6 ± 4.3	7.6 ± 3.2	6.6 ± 3.1	5.8 ± 4	0.9
Asparagine	5.9 ± 6.6	5.6 ± 5	3.7 ± 4.1	3 ± 1.6	3.1 ± 2.2	2.8 ± 1.8	0.5
Aspartate	9.5 ± 11	6.9 ± 6.5	4.7 ± 5.3	4.1 ± 4.2	3 ± 2.7	2 ± 2	0.6
Citrulline	1.5 ± 1	1.8 ± 1	1.3 ± 0.6	1.6 ± 0.9	1.4 ± 0.9	1.3 ± 0.7	0.9
GABA	0.21 ± 0.16	0.11 ± 0.05	0.06 ± 0.03	0.08 ± 0.02	0.08 ± 0.04	0.05 ± 0.04	0.7
Glutamine	392 ± 326	484 ± 302	427 ± 306	442 ± 173	430 ± 306	273 ± 97	0.9
Glycine	6.4 ± 2.8	9.5 ± 7.8	6.3 ± 4.4	7.6 ± 4.6	10.5 ± 12.2	7.9 ± 5.2	0.8
Phenylalanine	5.3 ± 3.6	6.3 ± 3.5	4.5 ± 2.5	5.7 ± 2.6	4.8 ± 2.5	4.3 ± 2.9	0.9
Serine	15 ± 8	22 ± 23	17 ± 5	18 ± 6	21 ± 14	16 ± 6	0.8
Taurine	7.4 ± 4.8	8 ± 6.3	5 ± 3.3	5.4 ± 3.9	5.4 ± 4.1	6.2 ± 5	0.7
Threonine	8.6 ± 3.2	11.5 ± 5.8	9.4 ± 4.3	10.9 ± 5.4	10.8 ± 5	10.7 ± 4.2	0.9
Tryptophan	1.8 ± 0.9	1.7 ± 1	1.3 ± 0.7	1.2 ± 0.9	1.8 ± 1.6	1.4 ± 1.1	0.7
Tyrosine	3.5 ± 2.1	4.7 ± 3.3	3.1 ± 2.4	3.5 ± 2.2	3.8 ± 2.3	3.7 ± 2.3	0.7
Valine	13 ± 10	18 ± 13	12 ± 8	17 ± 11	15 ± 9	14 ± 12	0.9

Values all mean ± SD μmol/l except: † mmol/l; ‡ lactate/pyruvate ratio; § glucose/lactate ratio.
* *IF* Increase factor (ratio between 16 to 30 min after and −60 to 0 min before the start of temporary clipping).

Table 3. *Extracellular chemistry changes during prolonged reversible cerebral ischaemia in 3 patients. Note increases in glutamate, γ-amino butyric acid and glycerol*

	Minutes before		Minutes after P_BO_2 decreased < 1.1 kPa for >30 min				IF*
	−120 to −60	−60 to 0	1 to 30	31 to 60	61 to 90	91 to 120	
Glucose †	1.1 ± 0.7	0.9 ± 0.5	0.5 ± 0.3	0.7 ± 0.4	0.7 ± 0.4	0.8 ± 0.4	0.8
Lactate †	0.9 ± 0.5	1 ± 0.4	1.5 ± 0.7	1.4 ± 1	1.9 ± 1.1	1.1 ± 0.8	1.4
Pyruvate	48 ± 14	54 ± 23	51 ± 34	58 ± 29	60 ± 17	72 ± 27	1.1
L/P ‡	20 ± 13	22 ± 16	49 ± 50	25 ± 25	29 ± 16	15 ± 12	1.2
G/L §	1.7 ± 1.3	1.2 ± 0.7	0.4 ± 0.4	1.5 ± 1.7	1.1 ± 1.7	1.5 ± 1.5	1.3
Glycerol	79 ± 26	49 ± 11	67 ± 19	114 ± 43	121 ± 34	97 ± 25	2.3
Glutamate	8 ± 5	6 ± 1	7 ± 4	14 ± 12	12 ± 9	9 ± 9	2.5
Alanine	48 ± 45	32 ± 28	20 ± 15	25 ± 14	31 ± 13	27 ± 12	0.8
Arginine	16 ± 9	12 ± 6	10 ± 5	10 ± 3	14 ± 7	14 ± 10	0.8
Asparagine	6.5 ± 3.7	4.4 ± 1.4	3.3 ± 0.4	3 ± 0.3	4 ± 0.8	3.4 ± 1.5	0.7
Aspartate	5.7 ± 4.2	5.5 ± 4.4	5.1 ± 4.3	6.2 ± 4.7	6.3 ± 5.5	5.1 ± 5.3	1.1
Citrulline	3.2 ± 2.2	2.4 ± 1.7	1.6 ± 0.5	1.7 ± 0.3	2.4 ± 0.5	2.5 ± 1.3	0.7
GABA	0.2 ± 0.1	0.1 ± 0.1	0.1 ± 0.1	0.3 ± 0.3	0.2 ± 0.2	0.4 ± 0.3	3
Glutamine	309 ± 179	247 ± 119	173 ± 69	203 ± 162	206 ± 112	215 ± 30	0.8
Glycine	36 ± 32	26 ± 22	17 ± 10	18 ± 10	25 ± 9	23 ± 11	0.7
Phenyl-alanine	10.3 ± 5.1	7.2 ± 3.5	5.3 ± 2.5	6 ± 2.1	8.5 ± 2.6	8.6 ± 3	0.8
Serine	24 ± 16	19 ± 14	17 ± 14	14 ± 10	20 ± 17	17 ± 19	0.7
Taurine	27 ± 29	12 ± 15	5 ± 6	6 ± 4	6 ± 3	6 ± 2	0.5
Threonine	27 ± 23	27 ± 25	13 ± 9	20 ± 14	18 ± 7	17 ± 8	0.7
Tryptophan	3.9 ± 2.3	4.2 ± 3.7	3.4 ± 2.3	2.4 ± 1.8	4.4 ± 3.4	4.4 ± 5.1	0.6
Tyrosine	9.9 ± 6.2	7.3 ± 3.6	5.3 ± 1.9	5.4 ± 1.3	7.4 ± 1.4	7.6 ± 2.8	0.7
Valine	33 ± 29	23 ± 17	18 ± 10	22 ± 10	32 ± 10	33 ± 13	1

Values all mean ± SD μmol/l except: † mmol/l; ‡ lactate/pyruvate ratio; § glucose/lactate ratio.
* *IF* Increase factor (ratio between 31 to 60 min after P_BO_2 decreased below 1.1 kPa and −60 to 0 min before).

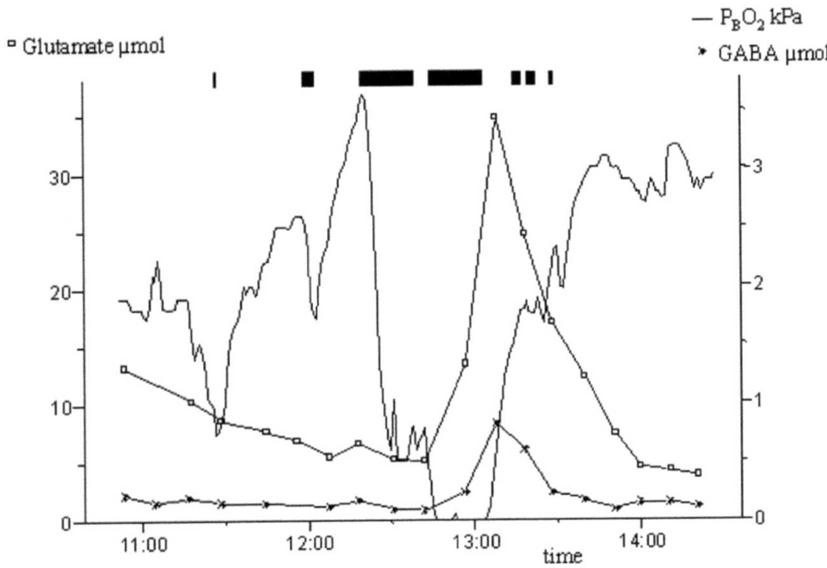

Fig. 1. In case M, prolonged temporary ICA clipping (blocks) caused brain oxygen (P_BO_2) to decrease below 1.1 kPa for 42 minutes with transient increases in glutamate and gamma-amino butyric acid (*GABA*), as well as glycerol to 108 µmol/l (not shown)

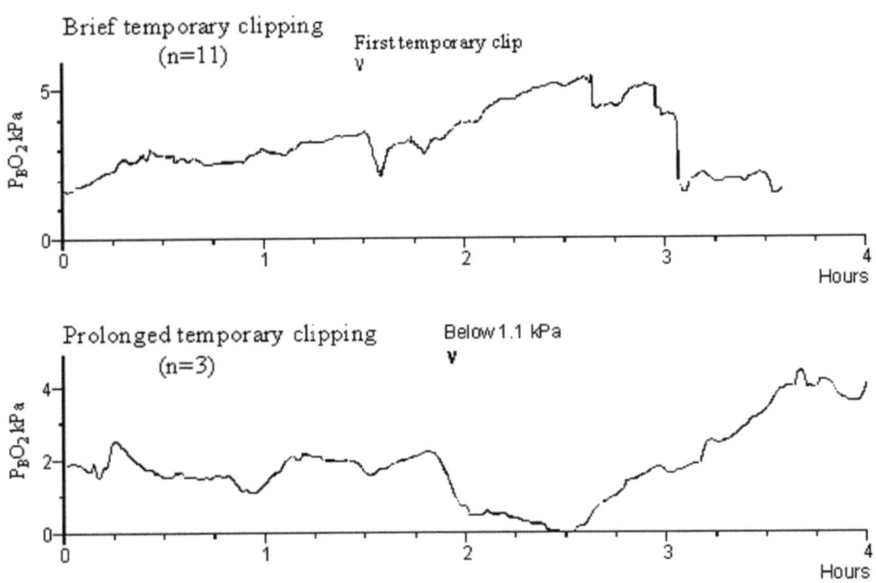

Fig. 2. The graph shows the mean value of brain oxygen (P_BO_2) for the two groups. Top: Group 1 (brief, small decreases in P_BO_2; n = 11) synchronised to the time of the first temporary clip. Bottom: Group 2 (n = 3) synchronised to the time when P_BO_2 decreased below 1.1 kPa for more than 30 minutes during temporary clipping

where an intra-operative rupture at the neck was difficult to control (Fig. 1). In both of these, P_BO_2 dropped below 1.1 kPa for over 30 minutes. Case N also met this criterion and showed small changes in chemistry. Patients were therefore divided into two groups according to these P_BO_2 criteria (Fig. 2): group 1 (cases A–K) and group 2 (cases L, M and N). Three patients developed infarcts as a result of surgery but none were in monitored regions (Table 1).

There were unremarkable microdialysis changes following only small decreases in P_BO_2 (Group 1, Table 2). In contrast, after prolonged low P_BO_2 (Group 2, Table 3) there were increases in Lactate/Pyruvate (L/P) ratio, by a factor of 2.3 within 30 minutes, and then, over the next 30 minutes, increases in GABA, glutamate and glycerol, each by twofold to threefold. There were no obvious changes in either group of the other quantifiable amino acids.

Discussion

In this study, we showed that a transient increase in GABA accompanies the previously reported increase in glutamate and L/P ratio during prolonged temporary clipping [12]. GABA has tended to receive less interest than glutamate, perhaps due to difficulty in detection at its normally low concentrations. The simultaneous extracellular increase in this inhibitory transmitter may be indicative of synaptic events, questioning the concept of excitotoxicity due to unopposed glutamate release [9].

The changes in extracellular chemistry were delayed, compared to changes in P_BO_2, an important appreciation if microdialysis is being used as a clinical monitoring tool. In addition, alterations in extracellular chemistry were generally uncommon; only 2 patients (Cases L and M) of the 46 patients monitored during surgery [12] had marked microdialysis changes, and both had long periods of tissue hypoxia during prolonged temporary clipping. The absence of chemical changes with brief hypoxia is consistent with an early microdialysis study in 9 patients, which failed to detect any changes after a 2-minute period of temporary clipping [1]. Similarly, there were no glutamate changes after 18 brief episodes of intracranial hypertension (ICP > 40 mm Hg) following trauma [3]. It could be inferred that the degree of insult necessary to produce glutamate changes appears to be considerable.

Changes due to permanent ischaemia in patients

Frontal lobe resection in 4 patients first demonstrated increased glutamate and GABA, aspartate, lactate, as well as taurine and glycine [11]. Changes greater than ours have been reported in brain death, where increases in most amino acids [5], lactate and glycerol are large [18]. Likewise, during infarction increasing glutamate has been demonstrated, after 40 minutes of temporary ACA clipping [15] or inadvertent arterial occlusion [18]. Glycerol increase in microdialysate is thought to be due to disintegration of cell membranes [8].

Experimental models of ischaemia

Transient complete cerebral ischaemia for a 10-minute period in a rat model was accompanied by an immediate eight folds increase in hippocampal extracellular glutamate and aspartate by a factor of 3, with GABA too low for measurement [2]. The absence of glutamate increases with ischaemia of equivalent duration in our study may reflect that our occlusions were often interspersed by periods of reperfusion and there was likely collateral blood flow through the communicating arteries (Table 1).

In a permanent occlusion experimental model, there were increases by a factor of 3.6 in lactate, 30 in glutamate, 12 in aspartate, 23 in GABA, and of 24 in taurine within 90 minutes, without changes in other amino acids [6]. The inference was that the extracellular fluid is flooded with both potentially harmful and protective chemicals during the acute stage of cerebral ischaemia.

Conclusion

The combination of these two cerebral monitoring techniques offers an opportunity to study the process of development of ischaemia in patients and, of surgical interest, to detect when temporary clipping is no longer being tolerated. The main findings of this study are that microdialysis detects transient increases in extracellular GABA, glutamate and glycerol without infarction, but only after prolonged tissue hypoxia, as defined by a decrease in P_BO_2 below 1.1 kPa for at least 30 minutes. Increases in the excitatory amino acid glutamate were matched by synchronous and proportional increases in inhibitory GABA.

Acknowledgments

This work was generously sponsored by The Stroke Association, Stroke House, Whitecross Street, London EC1Y 8JJ (Grant 13/98, received by Mr PJ Hutchinson). We wish to acknowledge the assistance of the nursing staff, Dr Stefan Piechnik and Dr Marek Czosnyka for help with data collection, Dr Hugh Richards for help with statistics, Dr Nicholas Higgins FRCR with radiological interpretation, Dr David Hutchinson for the initial design of the intracranial access device and Mrs Lynn Maskell for technical support.

References

1. Bachli H, Langemann H, Mendelowitsch A, Alessandri B, Landolt H, Gratzl O (1996) Microdialytic monitoring during cerebrovascular surgery. Neurol Res 18: 370–376
2. Benveniste H, Drejer J, Schousboe A, Diemer NH (1984) Elevation of the extracellular concentrations of glutamate and aspar-

tate in rat hippocampus during transient cerebral ischaemia monitored by intracerebral microdialysis. J Neurochem 43: 1369–1374

3. Bullock R, Zauner A, Tsuji O, Woodward JJ, Marmarou AT, Young HF (1995b) Patterns of excitatory amino acid release and ionic flux after severe human head trauma. In: Tsubokawa T, Marmarou A, Robertson C, Teasdale G (eds) Neurochemical monitoring in the intensive care unit. Springer Tokyo, pp 64–71
4. Doppenberg EM, Watson JC, Broaddus WC, Holloway KL, Young HF, Bullock R (1997) Intraoperative monitoring of substrate delivery during aneurysm and hematoma surgery: initial experience in 16 patients. J Neurosurg 87: 809–816
5. Goodman JC, Gopinath SP, Valadka AB, Narayan RK, Grossman RG, Simpson RK Jr, Robertson CS (1996) Lactic acid and amino acid fluctuations measured using microdialysis reflect physiological derangements in head injury. Acta Neurochir [Suppl] 67: 37–39
6. Hillered L, Hallstrom A, Segersvard S, Persson L, Ungerstedt U (1989) Dynamics of extracellular metabolites in the striatum after middle cerebral artery occlusion on the rat monitored by intracerebral microdialysis. J Cereb Blood Flow Metab 9: 607–616
7. Hillered L, Persson L, Ponten U, Ungerstedt U (1990) Neurometabolic monitoring of the ischaemic human brain using microdialysis. Acta Neurochir (Wien) 102: 91–97
8. Hillered L, Valtysson, Enblad P, Persson L (1998) Interstitial glycerol as a marker for membrane phospholipid degradation in the acutely injured human brain. J Neurol Neurosurg Psychiatry 64: 486–491
9. Hutchinson PJ, O'Connell MT, Al-Rawi PG, Kett-White R, Gupta AK, Maskell LB, Pickard JD, Kirkpatrick PJ (2000b) Changes in extracellular amino acid concentrations during cerebral ischaemia [abstract]. Cerebrovasc Dis [Suppl] 10: 99
10. Hutchinson PJ, O'Connell MT, Al-Rawi PG, Maskell LB, Kett-White R, Gupta AK, Richards HK, Hutchinson DB, Kirkpatrick PJ, Pickard JD (2000c) Clinical cerebral microdialysis – a methodological study. J Neurosurg 93: 37–43
11. Jabre A, Symon L (1987) Temporary vascular occlusion during aneurysm surgery. Surg Neurol 27: 47–63
12. Kett-White R, Hutchinson PJ, Al-Rawi PG, Czosnyka M, Gupta AK, Pickard JD, Kirkpatrick PJ (2002) Cerebral oxygen and microdialysis monitoring during aneurysm surgery: effects of blood pressure, cerebrospinal fluid drainage, and temporary clipping on infarction. J Neurosurg 96: 1013–1019
13. Kiening KL, Unterberg AW, Bardt TF, Schneider GH, Lanksch WR (1996) Monitoring of cerebral oxygenation in patients with severe head injuries: brain tissue pO_2 versus jugular vein oxygen saturation. J Neurosurg 85: 751–757
14. Koura SS, Doppenberg EM, Marmarou A, Choi SC, Young HF, Bullock R (1998) Relationship between excitatory amino acid release and outcome after severe head injury. Acta Neurochir [Suppl] 71: 244–246
15. Persson L, Hillered L (1992) Chemical monitoring of neurosurgical intensive care patients using intracerebral microdialysis. J Neurosurg 76: 72–80
16. Persson L, Valtysson J, Enblad P, Warme PE, Cesarini K, Lewen A, Hillered L (1996) Neurochemical monitoring using intracerebral microdialysis in patients with subarachnoid hemorrhage. J Neurosurg 84: 606–616
17. Saveland H, Nilsson OG, Boris-Moller F, Wieloch T, Brandt L (1996) Intracerebral microdialysis of glutamate and aspartate in two vascular territories after aneurysmal subarachnoid hemorrhage. Neurosurgery 38: 12–19
18. Schultz MK, Wang LP, Tange M, Bjerre P (2000) Cerebral microdialysis monitoring: determination of normal and ischaemic cerebral metabolisms in patients with subarachnoid haemorrhage. J Neurosurg 93: 808–814
19. Teasdale GM, Drake CG, Hunt W, Kassell N, Sano K, Pertuiset B, *et al* (1988) A universal subarachnoid haemorrhage scale: report of a committee of the World Federation of Neurosurgical Societies. J Neurol Neurosurg Psychiatry 51: 1457
20. Ungerstedt U, Nilsson OG, Nordstrom CH (1999) Microdialysis monitoring of human brain chemistry during neuro-intensive care. Int J Int Care 6: 102–106
21. Zauner A, Doppenberg EM, Woodward JJ, Choi SC, Young HF, Bullock R (1997b) Continuous monitoring of cerebral substrate delivery and clearance: initial experience in 24 patients with severe acute brain injuries. Neurosurgery 41: 1082–1093

Correspondence: Rupert Kett-White, Morriston Hospital, Swansea SA6 6NL, UK. e-mail: rupert.kett-white@swansea-tr.wales.nhs.uk

Acta Neurochir (2005) [Suppl] 95: 89–92

Cerebral metabolism and intracranial hypertension in high grade aneurysmal subarachnoid haemorrhage patients

A. S. Sarrafzadeh[1], **U.-W. Thomale**[1], **D. Haux**[1], and **A. W. Unterberg**[2]

[1] Department of Neurosurgery, Charité Campus Virchow Medical Center, Humboldt University of Berlin, Berlin, Germany
[2] Department of Neurosurgery, University of Heidelberg, Heidelberg, Germany

Summary

We evaluated the effect of intracranial hypertension on cerebral metabolism in patients with high grade aneurysmal subarachnoid hemorrhage (SAH) using bedside cerebral microdialysis (MD). Thirty-six patients with SAH were studied and classified into two groups (intracranial pressure, ICP > 20 mmHg, $n = 25$) and (ICP < 20 mmHg, $n = 11$). ICP was monitored hourly using an intraventricular drainage ($n = 36$). The MD catheter was placed after aneurysm clipping into the vascular territory of interest and was perfused with Ringer's solution (0.3 μl/min). The MD samples were collected hourly for measurements of glucose, lactate, and glutamate (CMA 600, Sweden). Lactate/pyruvate ratio was also calculated. To calculate group specific differences, the 24 hours median values of the first 7 days after SAH were compared. Differences were considered statistically significant at $P < 0.05$.

Patient groups were comparable for age, severity of SAH, Fisher's grade and duration of MD sampling. In patients with ICP > 20 mmHg from day 1 to 7 after SAH, extracellular concentrations of glucose were significantly lower, while the lactate/pyruvate ratio was higher compared to SAH patients with normal ICP values. The differences between groups in glutamate levels was only significant on day 1 after SAH due to high inter-individual differences. We concluded that intracranial hypertension in associated with an anaerobic cerebral metabolism indicated cerebral ischemia in high grade SAH patients.

Keywords: Cerebral metabolism; microdialysis; intracranial hypertension; subarachnoid hemorrhage.

Introduction

The outcome is poor for majority of patients with high grade aneurysmal subarachnoid hemorrhage (SAH). Many factors, such as intracranial hypertension, increase in cerebrospinal fluid outflow resistance, intracranial hemorrhage, cerebral edema, or a combination of these factors contribute to the poor outcome. With increasing diagnostic and monitoring activities, better overall outcome could be achieved [1]. Cerebral microdialysis, a recently available advanced neurointensive care monitoring technique, provides additional information on the metabolic state of the injured brain. Ischemia is associated with metabolic alterations of brain energy metabolism. This is characterized by an increase in anaerobic glycolysis with an increase in extracellular concentrations of lactate, lactate/pyruvate ratio and a decrease in glucose concentrations [8]. In trauma patients, it was shown that excitotoxic amino acids, such as glutamate and aspartate, are released into the extracellular fluid (ECF) in relation to ischemia and intracranial hypertension. These substances may have toxic effects leading to cell damage and membrane degradation with deliberation of glycerol [5]. Increased intracranial pressure (ICP) is well known to adversely affect patients with head injury. In contrast, the variables associated with intracranial hypertension following SAH and their impact on outcome has been less intensely studied. Heuer and collegues studied 433 SAH patients and concluded that an increase in ICP adversely affected outcome [4]. An increase in maximal ICP was associated with several admission factors such as a worse Hunt and Hess clinical grade, a lower Glasgow Coma Scale (GSC) motor score, a worse SAH grade based on results of computerized tomography (CT) scans, intracerebral hemorrhage and the severity of intraventricular hemorrhage. Since the subgroup of high grade SAH patients has the worst neurological outcome, an improved analysis of cerebral metabolic changes may lead to therapeutic options and a better outcome in these patiens. In our previous study, we identified the lactate/pyruvate ratio as the best metabolic independent prognostic factor for outcome at 12 months [7]. Microdialysis may help to select poor grade SAH patients at high risk for ischemia, in whom classical strategies for treatment of intracranial hypertension, such as hyperventilation therapy, might not be beneficial.

The objective of this study was to evaluate if regional cerebral metabolism measured by microdialysis is deranged in high grade SAH patients with intracranial hypertension (ICP > 20 mmHg) compared to those patients with normal ICP values.

Clinical material and methods

The study was approved by the Ethics Committee for conduct of Human Research at the Charite Campus Virchow Mediacl Center, Humboldt-University Berlin.

Patient characteristics and management

We studied 36 patients with high-grade aneurysmal SAH, defined as World Federation of Neurological Surgeons, WFNS grade IV and V [2], who were admitted to the Neurointensive Care Unit at the Charite Campus Virchow Medical Center of Berlin. Only patients with complete data sets of ICP and microdialysis monitoring for 7 days after SAH were included. Further inclusion criteria were SAH confirmed by cranial CT; cerebral angiogram demonstrating intracranial aneurysm(s); and patients underwent surgical therapy. Aneurysm location was evaluated using four vessels angiography on the day of admission. The distribution and pattern of the hemorrhage was graded as proposed by Fisher *et al.* [3].

For comparison or metabolic parameters, patients were classified into two groups, with ICP > 20 mmHg ($n = 11$) and those with normal ICP values ($n = 25$) during the first 7 days after initial bleeding. Demographic data are summarized in Table 1. All patients considered surgical candidates were managed according to an uniform protocol. After preoperative resuscitation, patients were treated with early ventricular drainage if hydrocephalus was present clinically or proven on CT scan. Aneurysm surgery was performed if patients responded with flexion to stimulation after sedation and paralysis were ceased for 24 hours. Patients with radiographic evidence of irreversible brain destruction were excluded from the study. Intracranial hypertension (ICP > 20 mmHg) was managed according to the AANS guidelines for the management of severe head injury. This included cerebrospinal fluid drainage, mannitol infusion (0.5 g/kg body weight over 20 minutes) and moderate hyperventilation ($PaCO_2$ 30–35 mmHg). Barbiturate coma was induced in otherwise uncontrollable intracranial hypertension and guided by a "burst-suppression" on EEG. Cerebral perfusion pressure was maintained above 60 mmHg using colloids and crystalloids, blood products and catecholamines, if necessary. ICP was monitored by means of a ventricular drainage and an additional intraparenchymal device in 10 cases (Camino, San Diego, CA; Codman, Johnson & Johnson). Decompressive surgery was performed when treatment was not sufficient to achieve an adequate ICP and CPP.

Table 1. *Demographic characteristics of patients with subarachnoid hemorrhage*

	ICP < 20 mmHg $n = 25$	ICP > 20 mmHg $n = 11$	*P* value
Age (mean ± SD)	55.2 ± 2.4	50.0 ± 3.6	0.15
WFNS grade	4.48 ± 0.1	4.73 ± 0.1	0.15
Time SAH – surgery (hrs)	44.0 ± 12.7	30.7 ± 7.8	0.13
Fisher-score	3.72 ± 0.1	3.8 ± 0.1	0.23

Data are expressed as mean ± standard deviation or numbers and percentage. *WFNS* World Federation of Neurological Surgeons Grading of SAH [8]

Bedside microdialysis

A microdialysis catheter (CMA 70, Sweden, length 10 mm, molecular weight limit of 20.000) was inserted immediately after clipping of the aneurysm into brain parenchyma of the respective vascular territory of the aneurysm. Care was taken to avoid insertion of the catheter into macroscopic abnormal brain tissue or into an intracerebral clot. Catheters were perfused with sterile Ringer's solution at a flow rate of 0.3 μl/min. On the outlet tube, perfusates were collected in microvials, exchanged hourly and immediately analyzed at bedside in a mobile photometric, enzyme kinetic analyzer (CMA 600). The estimated recovery for the system is 0.65–0.72 [6]. MD data are presented as microdialysate concentrations.

Statistical analysis

All summary data are expressed as mean ± SEM, if normally distributed or median (interquatile range) if the distribution was skewed. Between groups differences in each of the microdialysis variables and for ICP values was tested by Mann-Whitney U test. Differences were considered statistically significant at $P < 0.05$.

Results

Patient characteristics

The study population consisted of 35 patients, 38 to 76 years of age with SAH with high-grade aneurysmal SAH. Demographic data are listed in Table 1. Patient groups were comparable for age, WFNS grade, Fisher's grade and start of microdialysis monitoring after SAH. Four patients underwent decompressive surgery due to uncontrollable intracranial pressure.

Cerebral metabolism and intracranial pressure

Intracranial pressure was significantly higher from day 1 to 7 after SAH in patients classified in the intracranial hypertension group ($P < 0.001$, Fig. 1). These patients had significantly lower extracellular glucose concentrations and a higher lactate/pyruvarte ratio. This indicates an anaerobic metabolic condition in the monitored region (Fig. 2). Extracellular glutamate concentrations were higher in patients with intracranial hypertension, though the difference was only sig-

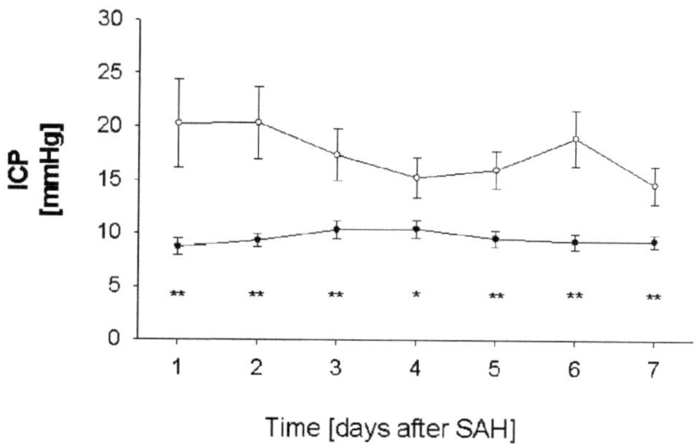

Fig. 1. Time course of intracranial pressure (*ICP*) in aneurysmal subarachnoid hemorrhage patients with intracranial hypertension (ICP > 20 mmHg) and normal ICP values. -•- ICP > 20 mmHg, $n = 11$; -o- ICP < 20 mmHg, $n = 25$

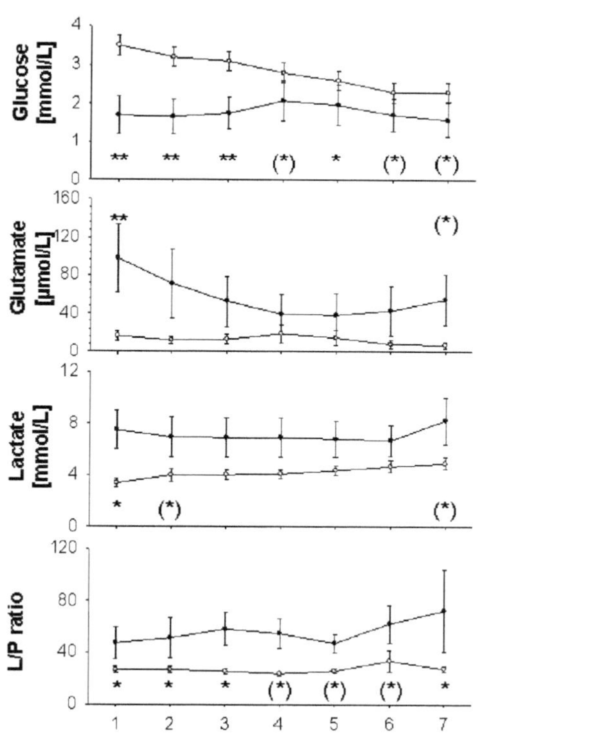

Fig. 2. Extracellular concentrations of lactate, glutamate, glucose and lactate/pyruvate (*L*/*P*) ratio in high grade subarachnoid hemorrhage (*SAH*) patients with intracranial hypertension ($n = 11$) and normal ICP values ($n = 25$), 1–7 days after surgery. In high grade SAH patients with intracranial hypertension, anaerobic metabolism (increased lactate and L/P ratio, low glucose) and high glutamate occurred compared to SAH patients with normal ICP values. (*) indicates a trend $P < 0.08$; $^{*}P < 0.05$; $^{**}P < 0.01$. Data are expressed as mean with standard deviation. -•- ICP > 20 mmHg, $n = 11$; -o- ICP < 20 mmHg, $n = 25$

nificant on day 1 after initial hemorrhage due to a high inter-individual variance.

Discussion

This study used prospective data collection in 36 patients following high grade aneurysmal SAH. In all patients cerebral metabolism was monitored by bedside microdialysis. We have evaluated the association of intracranial hypertension (ICP > 20 mmHg) and cerebral metabolism in these high risk patients. Our major findings were that the presence of intracranial hypertension was reflected in a deranged cerebral metabolism with lower extracellular concentrations of glucose and higher lactate/pyruvate ratios indicating anaerobic metabolism and regional ischemia. Glutamate was not significantly different between groups except on day 1 after SAH because of high inter-individual variances.

Bedside microdialysis is a well established technique for on-line detection of regional changes in neurochemistry [7]. Previous studies confirmed the usefulness of the microdialysis technique in clinical settings [6–8]. So far, no adverse effect has been observed in patients and because of its small size, the dialysis catheter is expected to causes only minimal damage. In this study there was no bleeding or infection associated with the microdialysis catheter.

Common risk factors for cerebral ischemia are the presence of intracranial hypertension and cerebral hypoxia produced by hyperventilation therapy. Both conditions are known to correlate with metabolic changes. In severely head injured patients, secondary increases of extracellular concentrations of glutamate, lactate and the lactate/pyruvate ratio were found to correlate with intracranial hypertension and severe ischemia. In patients with high grade SAH, only few data about ICP and metabolic parameters are available. Hyperventilation has traditionally been used in the treatment of intracranial hypertension. However prophylactic prolonged hyperventilation may be deleterious. It is known that hyperventilation induced ischemia and is reflected in a decrease in $PtiO_2$. Also moderate hyperventilation may be deleterious in individual patient. Therefore, treatment of intracranial hypertension with hyperventilation should only be used with alertness to potentially adverse effects and treatment should continuously be adapted to cerebral oxygenation and possibly be avoided during impending hypoxia in this specific group of patient.

Conclusion

In summary, our data demonstrated that the presence of intracranial hypertension was reflected in a deranged cerebral metabolism. The changes indicated anaerobic metabolism and regional ischemia. Cerebral monitoring including metabolism and oxygenation is desirable in these high risk aneurysmal SAH patients to avoid secondary cerebral hypoxia and ischemia.

References

1. Cesarini KG, Hardemark HG, Persson L (1999) Improved survival after aneurysmal subarachnoid hemorrhage: review of case management during a 12-year period. J Neurosurg 90: 664–672
2. Drake CG (1988) Report of World Federation of Neurological Surgeons Committee on a Universal Subarachnoid Hemorrhage Grading Scale. J Neurosurg 68: 985–986
3. Fisher CM, Kistler JP, Davis JM (1980) Relation of cerebral vasospasm to subarachnoid hemorrhage visualized by computerized tomographic scanning. Neurosurgery 6: 1–9
4. Heuer GG, Smith MJ, Elliott JP, Winn HR, LeRoux PD (2004) Relationship between intracranial pressure and other clinical variables in patients with aneurysmal subarachnoid hemorrhage. J Neurosurg 101: 408–416
5. Hillered L, Persson L, Ponten U, Ungerstedt U (1990) Neurometabolic monitoring of the ischaemic human brain using microdialysis. Acta Neurochir (Wien) 102: 91–97
6. Hutchinson PJ, O'Connell MT, al Rawi PG, Maskell LB, Kett-White R, Gupta AK, Richards HK, Hutchinson DB, Kirkpatrick PJ, Pickard JD (2000) Clinical cerebral microdialysis: a methodological study. J Neurosurg 93: 37–43
7. Sarrafzadeh A, Haux D, Kuchler I, Lanksch WR, Unterberg AW (2004) Poor-grade aneurysmal subarachnoid hemorrhage: relationship of cerebral metabolism to outcome. J Neurosurg 100: 400–406
8. Ungerstedt U, Hallstrom A (1987) In vivo microdialysis – a new approach to the analysis of neurotransmitters in the brain. Life Sci 41: 861–864

Correspondence: Asita S. Sarrafzadeh, Department of Neurosurgery, Charité Campus Virchow Medical Center, Humboldt University of Berlin, Augustenburger Platz 1, 13353 Berlin, Germany. e-mail: asita.sarrafzadeh@charite.de

Acta Neurochir (2005) [Suppl] 95: 93–96

Effect of ischemic preconditioning on brain tissue gases and pH during temporary cerebral artery occlusion

M. T. V. Chan[1], **R. Boet**[2], **S. C. P. Ng**[2], **W. S. Poon**[2], and **T. Gin**[1]

[1] Department of Anaesthesia and Intensive Care, The Chinese University of Hong Kong, Prince of Wales Hospital, Hong Kong, China
[2] Division of Neurosurgery, Department of Surgery, The Chinese University of Hong Kong, Prince of Wales Hospital, Hong Kong, China

Summary

Previous studies have demonstrated that a brief period of ischemia protect against subsequent severe ischemic insults to the brain, i.e. preconditioning. We evaluated the effects of ischemic preconditioning, produced by 2 min proximal temporary artery occlusion, on brain tissue gases and acidity during clipping of cerebral aneurysm.

Twelve patients with aneurysmal subarachnoid hemorrhage were recruited. All patients received standard anesthetics. After craniotomy, a calibrated multiparameter catheter was inserted to measure oxygen (PtO_2) tension, carbon dioxide ($PtCO_2$) tension and pH (pHt) in tissue at risk of ischemia during temporary artery occlusion. In patients assigned to the preconditioning group, proximal artery was occluded initially for 2 min and was allowed to reperfuse for 30 min. All patients underwent cerebral artery occlusion for clipping of aneurysm. The rate of change in PtO_2, $PtCO_2$ and pHt after artery occlusion were compared between groups using unpaired *t* test.

Baseline brain tissue gases and pHt were similar between groups. Following artery occlusion, the decline in PtO_2 and pHt were significantly slower in the preconditioning group compared with the routine care group.

These results suggested that ischemic preconditioning attenuates tissue hypoxia during subsequent artery occlusion. Brief occlusion of the proximal artery may be a simple maneuver for brain protection during complex cerebrovascular surgery.

Keywords: Cerebral ischemia; ischemic preconditioning; cerebral oxygenation.

Introduction

Ischemic preconditioning refers to an endogenous mechanism for brain protection against ischemia. In this phenomenon, a transient episode of minor cerebral ischemia produces tolerance to subsequent prolonged ischemic injury [2, 3, 6]. Cerebral ischemic preconditioning has been well defined in a variety of animal models after focal or global ischemia [2, 3, 6, 10]. There are however, few data to support the existence of ischemic preconditioning in human. In a number of observational studies, outcomes after ischemic strokes are generally better among patients with preceding transient ischemic attacks (TIAs) in the same vascular territory [1, 8, 11–14]. These data suggested that prodromal TIA induces protective response to subsequent ischemia. However, more recent data refuted these observations [5].

Unfortunately, despite its important therapeutic implications, in vivo ischemic preconditioning cannot be confirmed in TIA model because of the ethical and practical difficulties. However, it is possible to study patients undergoing microsurgical clipping of cerebral aneurysm. During this operation, proximal artery is often occluded temporarily to facilitate surgical dissection of the aneurysm, thus render the distal brain tissue at risk of ischemic damage. We hypothesized that brief episode of ischemia followed by reperfusion confers tolerance to subsequent ischemic injury during temporary artery occlusion. We evaluated the effect of ischemic preconditioning on brain tissue gases and acidity in patients undergoing clipping of cerebral aneurysm.

Materials and methods

After approval from the Clinical Research Ethics Committee, we studied 12 patients undergoing microsurgical clipping of aneurysm following subarachnoid hemorrhage (SAH). Patients were excluded if they were younger than 18 years or if there was a clinical plan not to apply temporary artery occlusion during surgery. Written informed consents were obtained from the patients or their relatives.

All patients received nimodipine infusion 1–2 mg/h. Monitoring included invasive arterial and central venous pressure measurements, electrocardiogram, capnography, pulse oximetry and oesophageal temperature recordings. Anesthesia was induced and maintained with target controlled infusions of propofol (3–4 μg/ml) and remifentanil (6 ng/ml). Neuromuscular blockade was achieved with rocuronium infusion (0.5 mg/kg/h). The lungs were ventilated with an air/oxygen mixture to maintain normocarbia. The inspired oxygen concentration was set as 30%. A warming blanket (Bair Hugger 505, Augustine Medical, Eden Prairie, MN) was used to maintain normothermia throughout the procedure.

We measured brain tissue oxygen (PtO_2), carbon dioxide tensions ($PtCO_2$) and acidity (pHt) with a calibrated multiparameter catheter (Neurotrend, Diametrics Medical, Minneapolis, MN). This catheter, 0.5 mm in diameter, incorporated a miniaturized optode electrode, two fibreoptic hydrogen electrodes and a thermocouple for PtO_2, pHt, $PtCO_2$ and brain temperature recordings, respectively. After a standard pteronial craniotomy was performed, the catheter was inserted into the ipsilateral middle frontal gyrus. Thirty minutes of equilibration was allowed. Readings were downloaded on a personal computer using a purposely designed data acquisition program.

Patients were randomly assigned to receive either ischemic preconditioning or routine care. In patients allocated to the preconditioning group, proximal feeding artery was briefly clamped for 2 min using a specifically designed, low-force, vascular clip. This was followed by 30 min reperfusion. In patients allocated to the routine care group, artery was exposed but not clamped. All patients then underwent cerebral artery occlusion for clipping of aneurysm. Immediately prior to artery occlusion, a bolus dose of thiopentone 2–3 mg/kg was given IV to produce electroencephalographic burst suppression. Arterial pressure was maintained within 20% of baseline using phenylephrine infusion. All patients were reviewed again one week after surgery for any development of new neurologic deficit.

The primary outcome measure in this study was the rate of change in PtO_2, $PtCO_2$ and pHt. As the delivery of substrates was interrupted with artery occlusion, a change in tissue gases and acidity is a measure of tissue vulnerability to ischemia. Therefore, a steep change in tissue gases or acidity indicates rapid depletion of intracellular substrate and accelerated energy failure. We calculated the rate of change in PtO_2, $PtCO_2$ and pHt during the first four minutes after temporary artery occlusion using linear regression. Values were compared between groups using Mann-Whitney test. Results are presented as median (range). A P value of less than 0.05 was considered significant.

Results

All twelve patients completed the study. Patient characteristics are listed in Table 1. Demographic data in the preconditioned patients were similar to that receiving routine care. The median (range) time from SAH to surgery was 47 (28–81) h in the preconditioned group and 53 (30–77) h in the routine care group, $P = 0.54$. The median (range) duration of temporary artery occlusion in the preconditioned group, 7.5 (6.1–13.6) min was also similar to that in the routine care group, 6.8 (5.0–15.1) min, $P = 0.73$. There was no intraoperative complication.

Following catheter equilibration, baseline values in the preconditioned group, pHt 7.11 (7.05–7.30), PtO_2 17 (13–29) mmHg, $PtCO_2$ 49 (32–58) mmHg, were similar to that in the routine care group, pHt 7.15 (7.03–7.28), PtO_2 20 (12–26) mmHg, $PtCO_2$ 53 (35–60) mmHg. After temporary artery occlusion, PtO_2 and pHt decreased and $PtCO_2$ increased in all patients (Fig. 1). But the rate of change in PtO_2 and pHt in the preconditioned patients was significantly slower than the controls (Fig. 2). However, the difference in the rise of $PtCO_2$ was small and was not statistically significant. PtO_2 in all patients receiving routine care fell below 10 mmHg before release of temporary artery occlusion, whereas only two of the preconditioned patients had ischemic change in PtO_2. A week after surgery, only one patient the routine care group had neurologic deficit. Proximal artery was occluded for 9 min in this patient. There was no adverse effect associated with catheter insertion.

Table 1. *Patient characteristics and severity of subarachnoid hemorrhage*

	Routine care	Ischemic preconditioning	*P* Values
No. of patients	6	6	
Age (year)	58 (38–72)	53 (40–69)	0.76
Body weight (kg)	72 (44–81)	59 (41–85)	0.89
Gender (Male/Female)	3/3	4/2	0.18
WFNS grade ≤ 3	3	4	0.80
Fisher's grade ≤ 3	4	3	0.82
Medical history			
– Hypertension	5	6	1.00
– Diabetes mellitus	4	3	0.37
– Current smoker	3	4	0.37
Location of ruptured aneurysm			
– Anterior cerebral artery	4	3	0.37
– Middle cerebral artery	2	3	

Values are median (range) or number of patients. *WFNS* World Federation of Neurological Surgeons.

Discussion

This study confirms the existence of ischemic preconditioning in humans. The changes in brain tissue gases and acidity suggested that the preconditioned patient will tolerate a longer period of ischemia during temporary artery occlusion for clipping of cerebral aneurysm. We are however, unable to detect a better outcome after ischemic preconditioning, mainly due to the small sample size. A larger study recruiting over 380 patients will be required to demonstrate the efficacy and safety of ischemic preconditioning during aneurysm surgery.

Ischemic preconditioning has been extensively evaluated in a number of animal models [2, 3, 6, 10]. On

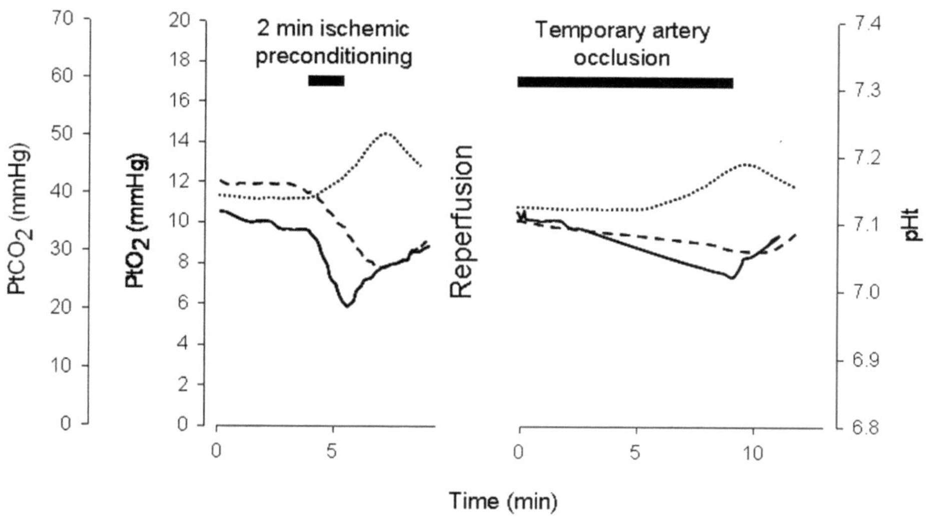

Fig. 1. Changes in tissue oxygen (*PtO*$_2$, *solid line*), carbon dioxide tension (*PtCO*$_2$, *dotted line*), and acidity (*pHt*, *dashed line*) following 2 min ischemic preconditiong, 30 min reperfusion and then temporary artery occlusion for clipping of cerebral aneurysm

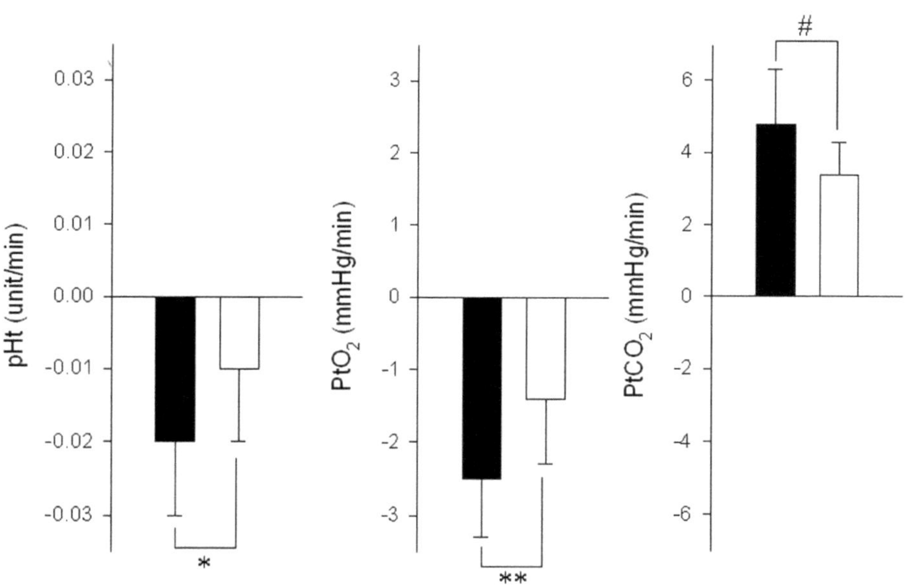

Fig. 2. Rate of change in tissue oxygen (*PtO*$_2$), carbon dioxide tension (*PtCO*$_2$) and acidity (pHt) after temporary artery occlusion in patients receiving prior ischemic preconditioning (black bar) or routine care (empty bar). $*P = 0.02$; $**P = 0.04$; $\#P = 0.07$

average, preconditioning with focal or global ischemia of 2–20 min duration, reduced the infarct size by 53% (range 20–91%) after subsequent prolonged ischemia [8]. While there is little difference among animal species, treatment efficacy depends largely on the duration of ischemia during induction of preconditioning and the interval to subsequent ischemia [2, 3, 6, 9, 10]. In the present study we induced preconditioning by 2 min proximal artery occlusion. This is followed by an inter-ischemic reperfusion for 30 min. This experimental paradigm is feasible during clinical practice and has been shown to produce the maximum protection against prolonged ischemia in rats [9].

The clinical evidence for ischemic preconditioning remains scanty in humans. The traditional TIA model suggested that prior occurrence of TIA decreases the severity of subsequent stroke [1, 8, 11–14]. However, it is not known whether the diverse etiologies of stroke may have biased the findings. A recent study of 180 patients with TIA and subsequent stroke within the next 90 days, showed no correlation between timing or duration of TIA and disability of subsequent stroke [5]. Nevertheless, the application of the TIA model is limited because neither the TIA itself nor the subsequent stroke can be staged. The use of ischemic preconditioning is more encouraging during proximal temporary artery occlusion for aneurysm surgery. Two observational studies suggested that repetitive brief occlusion of the proximal artery with intermittent reperfusion reduces the risk of stroke compared with uninterrupted ischemia of similar duration [4, 7]. Our data are in accord with these findings that brief occlusion of the proximal artery induces ischemic tolerance.

The underlying mechanism of ischemic preconditioning remains unclear. However, *N*-methy-D-asparate receptor activation would appear to be the primary event during induction of preconditioning [2, 3, 6, 10]. Subsequent delayed protection is best explained by genetic remodeling [2, 6, 10].

In conclusion, a brief (2 min) proximal artery occlusion induces tolerance to subsequent ischemia, 30 min apart. This is a simple and effective technique for brain protection when prolonged proximal artery occlusion is required during complex aneurysm surgery.

Acknowledgment

The study was supported in part by a Direct Grant for Research #2040874, The Chinese University of Hong Kong.

References

1. Castillo J, Moro MA, Blanco M, Leira R, Serena J, Lizasoain I, Davalos A (2003) The release of tumor necrosis factor-alpha is associated with ischemic tolerance in human stroke. Ann Neurol 54: 811–819
2. Davis DP, Patel PM (2003) Ischemic preconditioning in the brain. Curr Opin Anaesthesiol 16: 447–452
3. Dirnagl U, Simon RP, Hallenbeck JM (2003) Ischemic tolerance and endogenous neuroprotection. Trends Neurosci 26: 248–254
4. Ferch R, Pasqualin A, Pinna G, Chioffi F, Bricolo A (2002) Temporary arterial occlusion in the repair of ruptured intracranial aneurysms: an analysis of risk factors for stroke. J Neurosurg 97: 836–842
5. Johnston SC (2004) Ischemic preconditioning from transient ischemic attacks? Data from the Northern California TIA Study. Stroke 35[Suppl 1]: 2680–2682
6. Kirino T (2002) Ischemic tolerance. J Cereb Blood Flow Metab 22: 1283–1296
7. Lavine SD, Masri LS, Levy ML, Giannotta SL (1997) Temporary occlusion of the middle cerebral artery in intracranial aneurysm surgery: time limitation and advantage of brain protection. J Neurosurg 87: 817–824
8. Moncayo J, de Freitas GR, Bogousslavsky J, Altieri M, van Melle G (2000) Do transient ischemic attacks have a neuroprotective effect? Neurology 54: 2089–2094
9. Perez-Pinzon MA, Xu GP, Dietrich WD, Rosenthal M, Sick TJ (1997) Rapid preconditioning protects rats against ischemic neuronal damage after 3 but not 7 days of reperfusion following global cerebral ischemia. J Cereb Blood Flow Metab 17: 175–182
10. Schaller B, Graf R (2002) Cerebral ischemic preconditioning. An experimental phenomenon or a clinical important entity of stroke prevention? J Neurol 249: 1503–1511
11. Sitzer M, Foerch C, Neumann-Haefelin T, Steinmetz H, Misselwitz B, Kugler C, Back T (2004) Transient ischaemic attack preceding anterior circulation infarction is independently associated with favourable outcome. J Neurol Neurosurg Psychiatry 75: 659–660
12. Wegener S, Gottschalk B, Jovanovic V, Knab R, Fiebach JB, Schellinger PD, Kucinski T, Jungehulsing GJ, Brunecker P, Muller B, Banasik A, Amberger N, Wernecke KD, Siebler M, Rother J, Villringer A, Weih M (2004) Transient ischemic attacks before ischemic stroke: preconditioning the human brain? A multicenter magnetic resonance imaging study. Stroke 2004; 35: 616–621
13. Weih M, Kallenberg K, Bergk A, Dirnagl U, Harms L, Wernecke KD, Einhaupl KM (1999) Attenuated stroke severity after prodromal TIA: a role for ischemic tolerance in the brain? Stroke 30: 1851–1854
14. Yamamoto H, Bogousslavsky J, van Melle G (1998) Different predictors of neurological worsening in different causes of stroke. Arch Neurol 55: 481–486

Correspondence: Matthew Chan, FANZCA, Department of Anaesthesia and Intensive Care, The Chinese University of Hong Kong, Prince of Wales Hospital, Shatin, New Territories, Hong Kong, China. e-mail: mtvchan@cuhk.edu.hk

Acta Neurochir (2005) [Suppl] 95: 97–101

Changes in cerebral hemodynamics and cerebral oxygenation during surgical evacuation for hypertensive intracerebral putaminal hemorrhage

I. Ng, E. Yap, and **J. Lim**

Department of Neurosurgery, National Neuroscience Institute, Singapore

Summary

Objectives. The aim of this study was to evaluate the changes in cerebral hemodynamics, tissue oxygenation and blood flow before and after surgery for spontaneous intracerebral hematomas.

Methods. Eleven patients who underwent surgical decompression of spontaneous putaminal hematoma were studied. Intracranial pressure (ICP), cerebral perfusion pressure (CPP), brain tissue oxygen ($PtiO_2$), and carbon dioxide tensions ($PtiCO_2$), brain pH and regional cerebral blood flow (rCBF) were recorded prior to removing the bone flap and then on skin closure on completion of the operation.

Results. Following surgical decompression, mean ICP decreased significantly ($P < 0.05$); mean CPP, $PtiO_2$, brain pH and rCBF improved although the changes were not significant.

Conclusion. Surgical decompression for spontaneous intracerebral hematomas leads to significant reductions in ICP. This is accompanied by improvements in CPP, $PtiO_2$ and rCBF in the penumbra.

Keywords: Intracerebral hematoma; surgical decompression; intracranial pressure; brain tissue oxygen tension.

Introduction

Spontaneous intracerebral hemorrhage (ICH) accounts for 15 to 20% of all strokes in the western world [2]. The incidence is higher in Asian [20]. The 30-day mortality has been reported to be 45% [4]. As of today there is no effective medical treatment for ICH [16] and the value of surgical treatment is still controversial [8]. Experimental work in animals have shown that in addition to the initial destructive effects, there is evidence that the intracranial hematoma produces a "penumbra" of oedema and ischemia around the clot [1, 5, 14]. The additional damage that is due to hemorrhagic mass is therefore focal (from compressive ischemia) and global (from a reduction in the perfusion pressure associated with intracranial hypertension).

Surgical decompression in traumatic mass lesions has been noted to result in a clear improvement in substrate delivery [6]. However, surgery in spontaneous ICH has not been shown to improve outcome [3]. Nevertheless, there is little doubt that in selected patients, the role of surgery may produce a reasonable result by improving the neurometabolic milieu and preventing cell loss in the penumbra. Thus far, there has been little attention directed towards the changes that occur with surgical decompression for spontaneous intracerebral hemorrhage. In this study we investigated the effects of surgical decompression of hypertensive intracerebral hematoma on the brain tissue oxygenation ($PtiO_2$), intracranial pressure (ICP) and cerebral perfusion pressure (CPP) as well as regional blood flow (rCBF) using Laser Doppler Flowmetry (LDF) in the penumbral region.

Clinical materials and methods

This study was approved by the ethical committee at our institution and informed consent was obtained from families of the patients. We studied 11 patients with spontaneous intracerebral hemorrhage due to hypertension. All patients recruited into the study had surgery for intracerebral hemorrhage with its epicenter in the putamen and a clinical deterioration to a Glasgow Coma Score (GCS) of 8 or less together with radiological evidence of mass effect and midline shift (>5 mm). All patients were intubated, ventilated and empirical mannitol (20%; 100 ml) was given. $PaCO_2$ was controlled at 30 mmHg. Inspired oxygen fraction (FiO_2) was adjusted to achieve an arterial oxygen saturation $\geq 95\%$ throughout the operation. We employed 2 teams of surgeons as described by Verweij [22]. Briefly, this involved the senior author shaving and draping the head before inserting a multiparameter sensor (Neurotrend, Codman, Randolph, MA, USA), an intraparenchymal intracranial pressure monitor (Codman, Randolph, MA, USA) as well as an intraparenchymal laser Doppler probe (Laserflo BPM2; Vasamedics, St Paul, MN, USA) via a triple lumen bolt (Codman, Randolph, MA, USA) in the perceived penumbra region. In all cases, this was in the middle cerebral artery territory ipsilateral to the side of the hemorrhage. We inserted the catheters 3 cm from the pial surface along

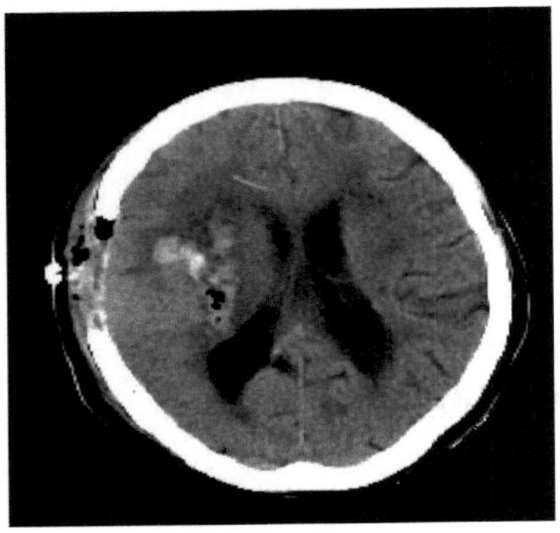

Fig. 1. Post-operative CT scan showing the position of the probes in the peri-hematoma region

the midpupillary line at the coronal suture, within the peri-hematoma region. The position of the catheters was verified by a post-operative computed tomographic (CT) scan (Fig. 1). The values were allowed to stabilize before the definitive surgical decompression was carried out. Data was then recorded at 5-second intervals. A formal craniotomy was made, and a duroplasty was then performed using lyophilised dura tacked loosely to the dural edges and the bone flap left out in all cases. A CT scan was done on the first post-operative day to assess the adequacy of evacuation. The ICP, CPP, $PtiO_2$, $PtiCO_2$, brain pH and regional CBF were recorded before the bone flap was lifted and after surgery when the skin was closed. Changes were analyzed using paired *t* test.

Results

Among the 11 patients enrolled in the study, the tissue oxygenation catheter failed to give meaningful readings in 3 patients. In 2 cases, this was due to an intrinsic failure of the catheter and the catheter was found to be in residual clot in another patient, hence giving erroneously low values. The data from these 3 patients were excluded from further study. Patient data of the remaining 8 patients are shown in Table 1. There were 3 females and 5 males. The mean ($\pm$ standard deviation) age was 62 ± 13 (range 46 to 81) years. The median GCS was 8 (range 4 to 11) and the mean volume of the hematoma was 68.4 ± 17.7 (range 45 to 98) ml, according to the method described by Kothari *et al.* [11]. Mean time from ictus to surgical decompression was 11.5 ± 7.8 (range 4 to 28) hours. Baseline values were compared with those after evacuation (Table 2).

Mean arterial pressure (MAP) remained relatively constant, preoperatively it was 95.3 ± 2.6 mmHg and postoperatively, it was 92.1 ± 11.3 mmHg ($P = 0.46$). Mean ICP was significantly reduced after surgery (preoperative mean ICP was 30.5 ± 11.8 mmHg; postoperative mean ICP was 12.3 ± 9.8 mmHg; $P = 0.01$). Of the 8 patients, 7 had intracranial pressures that were reduced following surgical evacuation. In all of these 7 patients, the maximal decrease in ICP appeared to be associated with opening the dura. There was a further graduated decrease in ICP with evacuation of the hematoma. The mean CPP improved from 68.5 ± 18.8 mmHg to 79.8 ± 15.4 mmHg ($P = 0.07$). In 6 of the 8 patients, CPP after decompression was less than 70 mmHg.

$PtiO_2$ was 18.1 ± 20.0 pre-operatively and was 29.8 ± 27.7 mmHg following evacuation of the hematoma ($P = 0.07$). In 3 patients (patients 1, 7 and 8), $PtiO_2$ values were above 20 mmHg before decompression and this improved further after surgery. In 2 patients, the values improved from less than 15 (ischemic threshold) to above 15 mmHg. Interestingly, the maximal increase in brain tissue oxygenation occurred after evacuation of the hematoma although in 2 patients, the values actually decreased to 0 mmHg after decompression (patients 3 and 5). $PtiCO_2$ tended

Table 1. *Patient demographics and clinical features*

Case	Age (years)	Gender	Glasgow coma scale score	Volume of putaminal hemorrhage (ml)	Time to operation (h)	Glasgow outcome score
1	46	male	4	64	6	3
2	72	female	10	45	6	4
3	66	male	8	98	16	1
4	65	male	11	54	28	4
5	81	female	9	60	4	1
6	46	male	7	90	10	3
7	67	female	8	66	13	2
8	49	male	8	71	9	2

Table 2. *Changes in mean arterial pressure (MAP), intracranial pressure (ICP), cerebral perfusion pressure (CPP), brain interstitial oxygen tension ($PtiO_2$), brain interstitial carbon dioxide tension ($PtiCO_2$), brain pH and regional cerebral blood flow (rCBF) using laser Doppler flowmetry with surgery*

Case	ICP (mmHg)		MAP (mmHg)		CPP (mmHg)		$PtiO_2$ (mmHg)		$PtiCO_2$ (mmHg)		Brain pH		LDF (ml/min/100 g)	
	Before surgery	After surgery	Before surgery	After surgery	Before surgery	After surgery	Before surgery	After surgery	Before surgery	After surgery	Before surgery	After surgery	Before surgery	After surgery
1	45	11	102	80	56	70	36	42	45	42	7.21	7.29	–	–
2	15	5	96	100	111	95	14	27	59	58	6.89	7.10	–	–
3	36	35	92	85	56	50	2	0	28	49	6.54	6.30	–	–
4	39	12	102	92	63	80	6	28	32	40	6.77	7.01	11	17.93
5	14	5	80	82	66	77	3	0	56	78	7.18	7.03	18	4.77
6	38	15	118	115	79	100	6	44	45	38	6.27	7.20	3	6.2
7	35	6	95	88	60	82	56	84	39	30	6.93	7.39	11	18.9
8	22	9	79	94	57	85	24	28	54	42	7.2	7.4	–	–

to decrease with decompression. In two fatal cases, $PtiCO_2$ increased. Mean $PtiCO_2$ 44.8 $\pm$ 11.3 to 45.9 $\pm$ 16.0 mmHg ($P = 0.13$).

Cerebral blood flow was measured by LDF in 4 out of the 8 patients. rCBF was uniformly low (<20 ml/min/100 g). The mean rCBF changed from 10.0 $\pm$ 6.3 pre-operatively to 12.0 $\pm$ 17.5 ml/min/100 g after operation ($P = 0.72$).

The rCBF was noted to increase immediately with lifting of the bone flap but the greatest increase was noted with evacuation of the hematoma. rCBF however, improved only modestly and did not reach beyond 20 ml/min/100 g in 3 of the 4 measured. In one patient (patient 5), there was actually a reduction after surgery despite improvements in ICP and CPP. In this patient, $PtiO_2$ corroborated by decreasing to 0 mmHg and $PtiCO_2$ increased.

Discussion

Animal models of SICH have shown that blood produces ischemia in the parenchyma by mechanical compression of surrounding microvasculature [13] resulting in an area of edema, oligemia and hemorrhagic necrosis. Surgical decompression may lead to amelioration of neurological damage [1, 5, 15].

While the goals of surgical intervention and evacuation in SICH are to reduce the mass effect, to block the release of neuropathic products from the hematoma and to prevent prolonged interaction between the hematoma and normal tissue which can initiate pathological processes leading to further cell death, its efficacy in improving outcome is controversial. A meta-analysis of 3 randomized controlled trials of SICH showed a higher rate of death and dependency for surgery compared to patients treated medically [10]. Despite a lack of proven benefit for surgery to improve outcome, it was estimated that 7000 such operations continue to be performed annually in the United States. There is to date paucity of data regarding specific changes in the peri-lesional tissue following intracerebral hemorrhage and how surgical decompression can affect the internal milieu in humans. There are practical and logistic difficulties of acquiring complex rCBF or neurochemical studies while other management issues are a prime concern. Previous similar clinical studies have been performed, although they have only analysed intracranial pressure and changes in cerebral perfusion [9].

Our study involved the insertion of all monitoring devices that measure local ICP, $PtiO_2$, brain pH and $PtiCO_2$ and rCBF to study changes in the peri-lesional tissue. A parenchymal intracranial pressure monitor was inserted in the peri-hematoma region to study peri-lesional tissue since there may be significant differences in regional ICP and corresponding CPP of the brain with ICH [17]. It was noteworthy that despite rather large volumes of hematoma with midline shift in our cohort of patients, two patients had ICP < 20 mmHg before decompression. Surgical decompression and evacuation led to a significant fall in ICP, improvement in CPP and this has been also previously demonstrated [9].

In our study, the poor rCBF before operation is in agreement with blood flow imaging data from the literature. Using PET, Zazulia *et al.* [23] found a lower CBF and cerebral metabolic rate ($CMRO_2$) in the penumbra region. Tanaka *et al.* [20] and Mayer *et al.* [12] showed an early hypoperfusion zone around the hematoma and this was resolved by 48 hours. Siddique *et al.* [19] showed that there was a zone of reversible ischemia. The improvement in mean blood flow in the surgical group using HMAO SPECT was about 3.9% ($n = 4$). Although the improvement was higher than those treated conservatively, it was at best modest. This improvement may be due to the continued effects of edema caused by red blood cell constituents leading to increased blood brain barrier permeability [23] and toxicity of hemoglobin breakdown products in the penumbra [18] limiting further increases in blood flow.

The very local effect of the measurement of $PtiO_2$ is ideal for monitoring in this specific situation as the catheters were placed in the peri-hematoma region to record changes of tissue oxygen in the microenvironment although brain shifts on raising the bone flap may change the location of the probe brain interface and lead to erroneous tissue oxygenation values. It was noteworthy that in the two patients who died, the values were below the critical threshold [7, 21]. Our data seems to support the fact that tissue oxygenation does increase following the procedure, although intrinsic changes in the penumbra may influence the internal milieu.

To a certain extent, our results do support the notion that there exists in human ICH the presence of a penumbra with low rCBF (as measured by LDF) and tissue ischemia (low $PtiO_2$ and brain pH) brought on by the physical effects of mechanical compression (high ICP, large putaminal hematoma with clinical and radiological evidence of mass effect). The clinical implication is that surgical evacuation of SICH may potentially lead to a fall in ICP, improvement of the CPP, $PtiO_2$ and rCBF in the peri-hematoma region. $PtiO_2$ and rCBF improvements however are at best modest following surgery suggesting that other pathophysiological processes are at play. This may explain why surgical decompression has got only limited success in selected patients.

Conclusions

Surgical decompression for SICH leads to a significant reduction in ICP and improvements in CPP. However, $PtiO_2$ are more variable and failure to improve may be predictive of mortality. Interestingly, regional blood flow is modest at best and this may explain the mixed results thus far of surgery.

Acknowledgments

This study was supported by a grant from the National Healthcare Group grant (NRN01/002).

References

1. Altumbabic M, Peeling J, Del Bigio MR (1998) Intracerebral hemorrhage in the rat: effects of hematoma aspiration. Stroke 29: 1917–1923
2. Bamford J, Sandercock P, Dennis M, Burn J, Warlow C (1990) A prospective study of cerebrovascular disease in the community: the Oxfordshire Community Stroke Project, 1981–1986, 2: incidence, case fatality rates and overall outcome at one year of cerebral infarction, primary intracerebral and subarachnoid hemorrhage. J Neurol Neurosurg Psychiatry 53: 16–22
3. Batjer HH, Reisch JS, Allen BC, Plaizier LJ, Su CJ (1990) Failure of surgery to improve outcome in hypertensive putaminal hemorrhage: a prospective randomised trial. Arch Neurol 47: 1103–1106
4. Broderick JP, Brott T, Tomsick T, Miller R, Huster G (1993) Intracerebral hemorrhage is more than twice as common as subarachnoid hemorrhage. J Neurosurg 78: 188–191
5. Deinsberger W, Vogel J, Fuchs C, Auer LM, Kuschinsky W, Boker DK (1999) Fibrinolysis and aspiration of experimental intracerebral hematoma reduces the volume of ischemic brain in rats. Neurol Res 21: 517–523
6. Doppenberg EMR, Watson JC, Broaddus WC, Holloway KL, Young HF, Bullock R (1997) Intra-operative monitoring of substrate delivery during aneurysm and hematoma surgery: initial experience in 16 patients. J Neurosurg 87: 809–816
7. Doppenburg EM, Zauner A, Watson JC, Bullock R (1998) Determination of the ischemic threshold for brain oxygen tension. Acta Neurochir [Suppl] 71: 166–169
8. Fernandes HM, Gregson B, Siddique S, Mendelow AD (2000) Surgery in intracerebral hemorrhage. The uncertainty continues. Stroke 31: 2511–2516
10. Hankey GJ, Hon C (1997) Surgery for primary intracerebral hemorrhage: is it safe and effective? A systematic review of case series and randomised trials. Stroke 82: 226–232
11. Kothari RU, Brott T, Broderick JP, Barsan WG, Sauerbeck LR, Zuccarello M, Khoury J (1996) The ABCs of measuring intracerebral hemorrhage. Stroke 27: 1304–1305
12. Mayer SA, Lignelli A, Fink ME, Kessler DB, Thomas CE, Swarup R, Van Heertum RL (1998) Perilesional blood flow and cerebral edema formation in acute intracerebral hemorrhage: a SPECT study. Stroke 29: 1791–1798
13. Nath FP, Kelly PT, Jenkins A, Mendelow AD, Graham DI, Teasdale GM (1987) Effects of experimental intracerebral hemorrhage on blood flow, capillary permeability and immunohistochemistry. J Neurosurg 66: 555–562
15. Qureshi AI, Tuhrim S, Broderick JP, Batjer HH, Hondo H, Hanley DF (2001) Spontaneous intracerebral hemorrhage. N Eng J Med 344: 1450–1460
16. Qureshi AI, Fareed KS, Ringer AJ, Guterman LR, Hopkins LN (2002) Regional intraparenchymal pressure differences in experimental intracerebral hemorrhage: Effect of hypertonic saline. Crit Care Med 30: 435–441

17. Regan RF, Guo Y (1998) Toxic effects of hemoglobin on spinal cord neurons in culture. J Neurotrauma 15: 645–653
19. Siddique MS, Fernandes HM, Wooldridge TD, Fenwick JD, Slomka P, Mendelow AD (2002) Reversible ischemia around intracerebral hemorrhage: a single photon emission computerized tomography study. J Neurosurg 1996: 736–741
20. Tanaka A, Yoshinaga S, Nakayama Y, Kimura M, Tomonaga M (1996) Cerebral blood flow and clinical outcome in patients with thalamic hemorrhage: a comparison with putaminal hemorrhage. J Neurol Sci 144: 191–197
21. Valadka AB, Gopinath SP, Contant CF, Uzura M, Robertson CS (1998) Relationship of brain tissue PO2 to outcome after severe head injury. Crit Care Med 26: 1576–1581
22. Verweij BH, Muizelaar JP, Vinas FC (2001) Hyperacute measurement of intracranial pressure, cerebral perfusion pressure, jugular venous oxygen saturation and laser Doppler flowmetry before, during and after removal of traumatic acute subdural hematoma. J Neurosurg 95: 569–572
23. Zazulia AR, Diringer MN, Videen TO, Adams RE, Yundt K, Aiyagari V, Grubb RL Jr, Powers WJ (2001) Hypoperfusion without ischemia surrounding acute intracerebral hemorrhage. J Cereb Blood Flow Metab 21: 804–810

Correspondence: Ivan Ng, Department of Neurosurgery, National Neuroscience Institute, 11 Jalan Tan Tock Seng, Singapore 308433. e-mail: ivan_ng@nni.com.sg

Acta Neurochir (2005) [Suppl] 95: 103–105

Open lung ventilation in neurosurgery: an update on brain tissue oxygenation

S. Wolf, D. V. Plev, H. A. Trost, and **C. B. Lumenta**

Department of Neurosurgery, Academic Hospital Munich-Bogenhausen, Technical University of Munich, Munich, Germany

Summary

Recently, we showed the feasibility of ventilating neurosurgical patients with acute intracranial pathology and concomitant acute respiratory distress syndrome (ARDS) according the *so-called* Open Lung approach. This technique consists of low tidal volume, elevated positive expiratory pressure (PEEP) level and initial recruitment maneuvers to open up collapsed alveoli. In this report, we focus on our experience to guide recruitment with brain tissue oxygenation ($p_{br}O_2$) probes.

We studied recruitment maneuvers in thirteen patients with ARDS and acute brain injury such as subarachnoid hemorrhage and traumatic brain injury. A $p_{br}O_2$ probe was implanted in brain tissue at risk for hypoxia. Recruitment maneuvers were performed at an inspired oxygen frcation (F_iO_2) of 1.0 and a PEEP level of 30–40 cmH_2O for 40 seconds.

The mean F_iO_2 necessary for normoxemia could be decreased from 0.85 ± 0.17 before recruitment to 0.55 ± 0.12 after 24 hours, while mean $p_{br}O_2$ (24.6 mmHg before recruitment) did not change. At a mean of 17 minutes after the first recruitment maneuver, $p_{br}O_2$ showed peak a value of 35.6 ± 16.6 mmHg, reflecting improvement in arterial oxygenation at an F_iO_2 of 1.0.

Brain tissue oxygenation monitoring provides a useful adjunct to estimate the effects of recruitment maneuvers and ventilator settings in neurosurgical patients with acute lung injury.

Keywords: ARDS; mechanical ventilation; Open Lung; ICP; brain tissue oxygenation; recruitment maneuver.

Introduction

Contemporary management of patients with acute respiratory distress syndrome (ARDS) consists of ventilating with low tidal volumes of 6 ml/kg body weight to avoid alveolar overdistention of the injured lung [1]. A more advanced approach advocates for the additional use of short periods of sustained high inflation pressure to open up collapsed alveoli. After this recruitment maneuvers, elevated levels of PEEP are used to maintain the recruited airspace, thus avoiding the repeated opening and closing of lung tissue, again to limit the damage of the lung induced by mechanical ventilation [12]. Proposed in theory by Lachmann more than a decade ago [8], this *Open Lung approach* showed improved survival in a small, randomised, controlled trial [2].

From the aforementioned as well as other ARDS studies, neurosurgical patients were excluded due to concerns of intracranial deterioration through the used elevated levels of PEEP. However, our group recently showed the feasibility of the Open Lung approach in patients with acute brain lesions [14]. While this previous study mainly addressed safety concerns and investigated the influence of Open Lung ventilation on ICP, the present report focuses on its impact on brain tissue oxygenation ($p_{br}O_2$). We hypothesized that $p_{br}O_2$ may provide a useful tool to estimate the efficacy of a recruitment maneuver and to guide the settings of mechanical ventilation.

Material and methods

We studied the clinical course of 13 patients with acute brain injury and concomitant ARDS, which was diagnosed according to the consensus conference criteria [3]. Main neurosurgical diagnosis was aneurysmal SAH (7 patients), traumatic brain injury (3 patients), intracranial hemorrhage (2 patients) and post surgery for brain metastasis (1 patient). In all patients, the trachea were intubated and the lungs were mechanically ventilated (Puritan Bennett, Tyco Healthcare, Mansfield, MA, USA). Patients were sedated with fentanyl, midazolam and/or propofol. Paralysis was used as required.

In all patients, $p_{br}O_2$ was measured with the Licox system (GMS, Kiel, Germany). The $p_{br}O_2$ probes were implanted in brain tissue estimated at highest risk of infarction, e.g. the vascular territory of the vessel harbouring a clipped or coiled aneurysm or the more affected hemisphere after traumatic brain injury. ICP monitoring was performed via parenchymal devices (CODMAN microsensor, Codman, Raynham, MA, USA and Spiegelberg III, Spiegelberg KG, Hamburg, Germany) or ventriculostomy. Monitoring data was collected with multimodal monitoring software (ICUpilot, CMA, Solna, Sweden) with a frequency of 1 per minute.

Respiratory management: Chest X-ray and bronchoscopy was performed as required and at least once before onset of Open Lung ventilation to rule out pneumothorax and lobar atelectasis. Recruitment maneuvers were performed on an inspired oxygen fraction (F_iO_2) of 1.0 with a CPAP/PEEP level of 30–40 cmH_2O for 40 seconds and optional ventilation with 20 mmHg above that PEEP for a few breathing cycles. After this recruiting maneuver, the PEEP level was set to 5 cmH_2O higher than before and at least 15 cmH_2O. All patients were ventilated with pressure controlled ventilation with an I:E ratio of 1:1 to 1.4:1. Inspiratory pressure was adjusted to achieve a tidal volume of 6 ml/kg body weight. Respiratory frequency was 15 to 30 per minute, depending on blood gas analysis to achieve normocapnia as feasible. Recruitment was considered sufficient if arterial p_aO_2 increased to more than 300 mmHg at a sustained F_iO_2 of 1.0.

After recruitment, we first tried to decrease F_iO_2 to at least 0.4, with regard to an p_aO_2 around 100 mmHg and then to decrease the PEEP level in steps of 1–2 cmH_2O. If derecruitment occurred after decreasing the PEEP level, after suctioning or disconnection of the ventilator system, another recruitment maneuver was performed and PEEP was set 2 cmH_2O higher than before.

ICP was treated if persistently increased above 25 mmHg. We aimed for a $p_{br}O_2$ above 15 mmHg. If desaturation was pending, we attempted to decrease ICP and to optimize mean arterial blood pressure, cardiac output, cardiac preload or p_aCO_2, thus providing an oxygenation oriented therapy [13].

For the cumulative analysis, the first recruitment maneuver of a patient was considered as start of Open Lung ventilation, even if this patient, as most were, was submitted to more than one. All statistic data are expressed as mean ± standard deviation. Analysis was performed with paired t-tests, as appropriate.

Results

In all but one patient, oxygenation improved after initiation of Open Lung ventilation. No patient had to be withdrawn from Open Lung ventilation due to a refractory increase in ICP or a desaturation episode of $p_{br}O_2$. No side effects of recruitment maneuvers like pneumothorax or lasting hemodynamic instability were noticed.

Mean F_iO_2 necessary for a p_aO_2 of 100 mmHg decreased gradually from 0.85 ± 0.17 before the first recruitment maneuver to 0.55 ± 0.12 after 24 hours (Table 1). The mean PEEP level used immediately after the first recruitment to prevent loss of alveolar recruitment was 18.2 ± 4.2 cmH_2O.

Mean ICP was 17.4 ± 6.2 mmHg before and 14.1 ± 8.2 mmHg after 24 hours. $p_{br}O_2$ was 24.6 ± 9.3 mmHg before and 24.2 ± 11.3 mmHg after 24 hours. After the first recruitment maneuver, a peak in $p_{br}O_2$ was seen after a mean of 17 minutes (Fig. 1). Mean $p_{br}O_2$ at this peak was 35.6 ± 16.6 mmHg. Afterwards, mean $p_{br}O_2$ increased further, reaching a plateau after 60 minutes with a value of 42.5 ± 19.6 mmHg.

Table 1. *Oxygenation, intracranial pressure (ICP) and Brain tissue oxygen tension ($p_{br}O_2$) before the first recruitment maneuver and 24 hours later*

	Before first recruitment	24 hours after first recruitment	*P* Values
F_iO_2 [mmHg]	0.85 ± 0.17	0.55 ± 0.12	<0.001
P_aO_2/F_iO_2	142 ± 42	257 ± 110	<0.001
P_aCO_2 [mmHg]	44.1 ± 6.6	40.3 ± 8.8	>0.05
ICP [mmHg]	17.4 ± 6.2	14.1 ± 8.2	>0.05
$p_{br}O_2$ [mmHg]	24.6 ± 9.3	24.2 ± 11.3	>0.05

PaO_2 Arterial oxygen tension; *$PaCO_2$* arterial carbon dioxide tension; *FiO_2* inspired fraction of oxygen. Data is expressed as mean ± standard deviation.

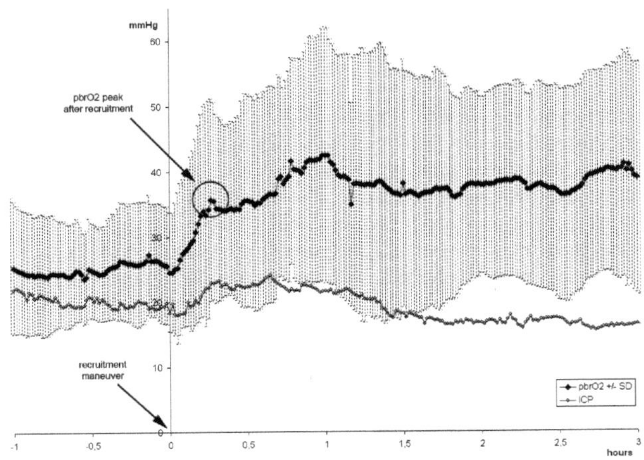

Fig. 1. Time course of brain tissue oxygen tension ($p_{br}O_2$) and intracranial pressure (*ICP*). At time point 0, the first recruitment maneuver was performed. The shaded area denotes standard deviation of $p_{br}O_2$, indicating heterogeneity in response to recruitment between different patients

Discussion

There is an inconsistency in the literature on the ways of performing recruitment as well as on the lasting efficacy of these recruitment maneuvers [10]. The largest study assessing recruitment, performed by the ARDS network [5], as well as a small observational study performed in a neurosurgical patient population were not able to demonstrate a lasting improvement in oxygenation [4]. However, the PEEP level to prevent loss of recruitment in these studies was not increased after recruitment, which may partly explain the lack of a beneficial effect of the recruitment maneuvers [9, 11]. In theory, the best method to exploit the recruitment capacity of an ARDS lung is a decremental PEEP trial as proposed by Hickling [7]. As this method is more aggressive than our recruitment scheme, we did not use it as a *first* approach of implementing

Open Lung ventilation in our patients with unknown stability of hemodynamic situation and ICP when exposed to a recruitment maneuver.

Our data shows that Open Lung ventilation with recruitment maneuvers had a lasting improvement on arterial oxygenation in neurosurgical patients with ARDS. In the 24 hours after the first recruitment maneuver we were able to reduce F_iO_2 keeping p_aO_2 above 100 mmHg and without detrimental effect on $p_{br}O_2$. Brain tissue oxygenation increased in the short term following an increased arterial oxygenation after pulmonary recruitment with unchanged F_iO_2. The peak of $p_{br}O_2$ at a mean of 17 minutes reflects the improved arterial oxygenation induced by this first recruitment maneuver. One patient failed to respond to multiple recruitment maneuvers, and thus did not show an increase in arterial oxygenation. This patient also lacked the peak in $p_{br}O_2$ observed in other patients. Presently we are unable to predict a certain response in an individual patient to recruitment and therefore unable to provide an recruitment algorithm suitable for all patients.

There are multiple reasons for a decrease in $p_{br}O_2$ after the observed peak. If the PEEP used after recruitment is not sufficiently high enough, the improvement in arterial oxygenation is not lasting in its full extent over time and partial derecruitment will occur again. Furthermore multiple parameters such as p_aO_2, p_aCO_2 and a mean arterial pressure might influence $p_{br}O_2$ [6]. These parameters as well as the change in PEEP level after recruitment may induce complex hemodynamic interactions which are not fully understood up to date and are difficult to monitor.

The presented data provides additional evidence that Open Lung ventilation is feasible in neurosurgical patients with ARDS. Albeit we did not experience adverse effects of recruitment maneuvers and Open Lung ventilation, we are aware that our data does not support the use of this ventilation strategy without concern. While careful hemodynamic and ICP monitoring is mandatory, brain tissue oxygenation monitoring provides a useful adjunct to visualize the effects of recruitment maneuvers and fine tuning ventilatory settings.

References

1. [The ARDS Network] (2000) Ventilation with lower tidal volumes as compared with traditional tidal volumes for acute lung injury and the acute respiratory distress syndrome. The Acute Respiratory Distress Syndrome Network. N Engl J Med 342: 1301–1308
2. Amato MB, Barbas CS, Medeiros DM, Magaldi RB, Schettino GP, Lorenzi-Filho G, Kairalla RA, Deheinzelin D, Munoz C, Oliveira R, Takagaki TY, Carvalho CR (1998) Effect of a protective-ventilation strategy on mortality in the acute respiratory distress syndrome. N Engl J Med 338: 347–354
3. Artigas A, Bernard GR, Carlet J, Dreyfuss D, Gattinoni L, Hudson L, Lamy M, Marini JJ, Matthay MA, Pinsky MR, Spragg R, Suter PM (1998) The American-European Consensus Conference on ARDS, part 2. Ventilatory, pharmacologic, supportive therapy, study design strategies and issues related to recovery and remodeling. Intensive Care Med 24: 378–398
4. Bein T, Kuhr LP, Bele S, Ploner F, Keyl C, Taeger K (2002) Lung recruitment maneuver in patients with cerebral injury: effects on intracranial pressure and cerebral metabolism. Intensive Care Med 28: 554–558
5. Brower RG, Morris A, MacIntyre N, Matthay MA, Hayden D, Thompson T, Clemmer T, Lanken PN, Schoenfeld D (2003) Effects of recruitment maneuvers in patients with acute lung injury and acute respiratory distress syndrome ventilated with high positive end-expiratory pressure. Crit Care Med 31: 2592–2597
6. Hemphill JC, III, Knudson MM, Derugin N, Morabito D, Manley GT (2001) Carbon dioxide reactivity and pressure autoregulation of brain tissue oxygen. Neurosurgery 48: 377–383
7. Hickling KG (2001) Best compliance during a decremental, but not incremental, positive end-expiratory pressure trial is related to open-lung positive end-expiratory pressure: a mathematical model of acute respiratory distress syndrome lungs. Am J Respir Crit Care Med 163: 69–78
8. Lachmann B (1992) Open up the lung and keep the lung open. Intensive Care Med 18: 319–321
9. Lapinsky SE, Aubin M, Mehta S, Boiteau P, Slutsky AS (1999) Safety and efficacy of a sustained inflation for alveolar recruitment in adults with respiratory failure. Intensive Care Med 25: 1297–1301
10. Lapinsky S, Mehta S (2004) Bench-to-bedside review: recruitment and recruiting maneuvers. Crit Care (in press)
11. Lim CM, Jung H, Koh Y, Lee JS, Shim TS, Lee SD, Kim WS, Kim DS, Kim WD (2003) Effect of alveolar recruitment maneuver in early acute respiratory distress syndrome according to antiderecruitment strategy, etiological category of diffuse lung injury, and body position of the patient. Crit Care Med 31: 411–418
12. Marini JJ, Gattinoni L (2004) Ventilatory management of acute respiratory distress syndrome: a consensus of two. Crit Care Med 32: 250–255
13. Meixensberger J, Jaeger M, Vath A, Dings J, Kunze E, Roosen K (2003) Brain tissue oxygen guided treatment supplementing ICP/CPP therapy after traumatic brain injury. J Neurol Neurosurg Psychiatry 74: 760–764
14. Wolf S, Schurer L, Trost HA, Lumenta CB (2002) The safety of the open lung approach in neurosurgical patients. Acta Neurochir [Suppl] 81: 99–101

Correspondence: S. Wolf, Department of Neurosurgery, Academic Hospital Munich-Bogenhausen, Technical University of Munich, Englschalkinger Ctraße 77, 81925 München, Munich, Germany. e-mail: Neurochirurgie.Bogenhausen@lrz.tum.de

Acta Neurochir (2005) [Suppl] 95: 107–111

Magnesium sulfate for brain protection during temporary cerebral artery occlusion

M. T. V. Chan[1], R. Boet[2], S. C. P. Ng[2], W. S. Poon[2], and T. Gin[1]

[1] Department of Anaesthesia and Intensive Care, The Chinese University of Hong Kong, Prince of Wales Hospital, Hong Kong, China
[2] Division of Neurosurgery, Department of Surgery, The Chinese University of Hong Kong, Prince of Wales Hospital, Hong Kong, China

Summary

We evaluated the effects of magnesium sulfate on brain tissue oxygen (PtO_2) tension, carbon dioxide ($PtCO_2$) tension and pH (pHt) in patients undergoing temporary artery occlusion for clipping of cerebral aneurysm.

We studied 18 patients with aneurysmal subarachnoid hemorrhage. All patients received standard anesthetics using target controlled infusion of propofol (3 µg/ml) and remifentanil (10 ng/ml). After craniotomy, a calibrated multiparameter sensor (Neurotrend, Diametrics Medical, Minneapolis, MN) was inserted to measure PtO_2, $PtCO_2$ and pHt in tissue at risk of ischemia during temporary artery occlusion. Patients were then randomly allocated to receive either intravenous saline or magnesium 20 mmol over 10 min followed by an infusion 4 mmol/h. Plasma magnesium concentration, brain tissue gases and pHt were determined at baseline, 30 min after study drug infusion and 4 min after temporary clipping. Data were analyzed by factorial ANOVA with repeated measures. Intergroup difference was compared with unpaired *t* test. *P* value < 0.05 was considered significant.

Patient characteristics, baseline brain tissue gases and pHt did not differ between groups. Magnesium infusion increased PtO_2 by 34%. Following temporary artery occlusion, PtO_2 and pHt decreased and $PtCO_2$ increased in both groups. However, tissue hypoxia was less severe and the rate of PtO_2 decline was slower in the magnesium group.

Our data suggested that magnesium enhances tissue oxygenation and attenuates hypoxia during temporary artery occlusion.

Keywords: Cerebral ischemia; magnesium; neuroprotection; cerebral oxygenation.

Introduction

Temporary artery occlusion is often required during aneurysm surgery. Although it facilitates surgical dissection and subsequent clip application, this technique imposes significant risk of cerebral ischemia [9]. This is particularly important when prolonged occlusion (>10 min) is required [9]. In this regard, a technique that enhances collateral circulation will improve cerebral tissue oxygenation and may reduce ischemic injury.

Magnesium is a cerebral vasodilator. In animal models, it blocks voltage-dependent calcium channels and reverses experimental vasospasm [14, 15]. Similarly, in ex-vivo segments of human middle cerebral artery, spasmogen-induced vasoconstriction can be readily inhibited by small doses of magnesium [1]. The clinical benefits of magnesium related cerebral vasodilation is further demonstrated in women with pre-eclampsia, a condition thought to be due to cerebral vasospasm and resultant ischemia. In these patients, magnesium infusion increases cerebral blood flow and reduces the incidence of seizure (eclampsia) [6, 11]. We therefore hypothesized that magnesium given before temporary artery occlusion enhances collateral perfusion, and hence attenuates ischemic insults. The purpose of the present study was to evaluate the effects of magnesium on brain tissue pH (pHt), oxygen (PtO_2) and carbon dioxide ($PtCO_2$) tensions in patients undergoing temporary artery occlusion for clipping of cerebral aneurysm.

Materials and methods

This study was approved by the Clinical Research Ethics Committee. Written informed consents were obtained from the patients or their next-of-kin. Eighteen patients, aged between 21 and 76 years, undergoing clipping of aneurysm following subarachnoid hemorrhage (SAH) entered the study. We excluded patients if they were pregnant, if they had significant renal impairment (plasma creatinine concentration > 200 µmol/L) or if there was a clinical indication or contraindication to magnesium infusion.

All patients were treated according to a standard protocol that included perioperative infusion of nimodipine 1–2 mg/h. Following placement of invasive arterial and central venous pressure monitoring, anesthesia was induced and maintained with target controlled infusions of propofol (3–4 μg/ml) and remifentanil (6 ng/ml). Neuromuscular blockade was achieved with rocuronium infusion (0.5 mg/kg/h). The lungs were mechanically ventilated with an air/oxygen mixture. Normocarbia was maintained throughout the procedure and the inspired oxygen concentration was initially set as 30%. Patients also received mild hypothermia using a forced air cooling unit (Polar Air, Augustine Medical, Eden Prairie, MN) to achieve a core temperature of 32.5–33.5 °C during dissection of the aneurysm and clip application.

A standard pteronial craniotomy was performed. After reflection of the dura, a calibrated multiparameter catheter (Neurotrend, Diametrics Medical, Minneapolis, MN) was inserted 3 cm into the middle frontal gyrus. This catheter, 0.5 mm in diameter, contained two optic fibers to measure brain tissue pHt and carbon dioxide tension ($PtCO_2$). It also incorporated a miniaturized optode for parenchymal oxygen tension (PtO_2) and a thermocouple for brain temperature measurements. Following 30 min of equilibration, patients were randomly assigned to receive either magnesium or placebo infusion. For patients receiving the active treatment, magnesium sulfate ($MgSO_4$) 20 mmol was administered over 10 min; this is followed by a continuous infusion of $MgSO_4$ 4 mmol/h. Equivalent volume of saline was infused in the placebo group. Plasma magnesium concentration was measured before surgery and 30 min after the start of infusion. Immediately prior to temporary artery occlusion, all patients received a bolus dose of thiopentone 2–3 mg/kg to produce electroencephalographic burst suppression. Arterial pressure was maintained within 20% of baseline using phenylephrine infusion. Patients were reviewed one week after surgery for any development of new neurologic deficit.

Physiologic parameters including those recorded from the Neurotrend catheter were downloaded on a personal computer using a purposely designed data acquisition program. Data at three time points during surgery were specifically sought for comparison. They are the baseline values recorded at the end of the 30 min equilibration period; data recoded at 30 min after infusion of study drug and those recorded at 4 min after temporary artery occlusion. Longitudinal data were compared between groups by factorial analysis of variance with repeated measures. Intergroup difference was compared between groups with unpaired t test. A P value of less than 0.05 was considered significant.

Table 1. *Patients characteristics and severity of subarachnoid hemorrhage*

	Magnesium	Saline	*P* values
No. of patients	9	9	
Age (year)	57 ± 15	55 ± 12	0.96
Body weight (kg)	68 ± 10	62 ± 11	0.82
Gender (Male/Female)	4/5	5/4	0.20
Medical history			
– Hypertension	8	7	0.40
– Diabetes mellitus	2	3	0.36
– Current smoker	5	4	0.33
– Others	1	0	1.00
WFNS grade			0.80
– 1/2 (good grade)	3	5	
– 3/4 (poor grade)	6	4	
Fisher's grade			0.82
– 1/2 (good grade)	3	2	
– 3/4 (poor grade)	6	7	
Location of ruptured aneurysm			
– Anterior cerebral artery	3	4	0.15
– Middle cerebral artery	6	5	
Time from SAH to surgery (h)	42 ± 15	45 ± 13	0.68

Values are mean ± SD. *WFNS* World Federation of Neurological Surgeons; *SAH* Subarachnoid hemorrhage.

Results

Demographic data were summarized in Table 1. Patient's age, weight and severity of SAH did not differ

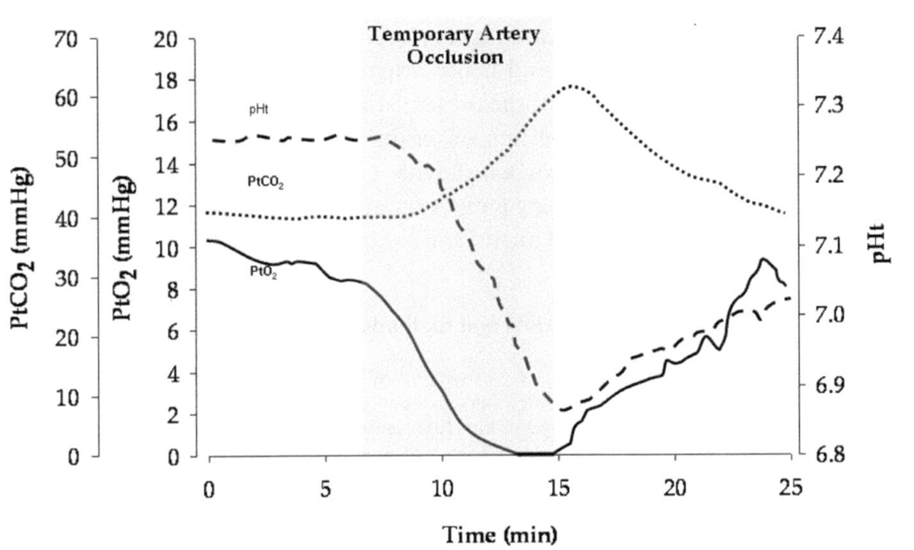

Fig. 1. Changes in tissue oxygen (*PtO_2, solid line*), carbon dioxide tension (*$PtCO_2$, dotted line*), and acidity (*pHt, dashed line*) following 8 min temporary artery occlusion for clipping of cerebral aneurysm in a patient receiving saline infusion. PtO_2 dropped below the detection limit at the end of the occlusion period

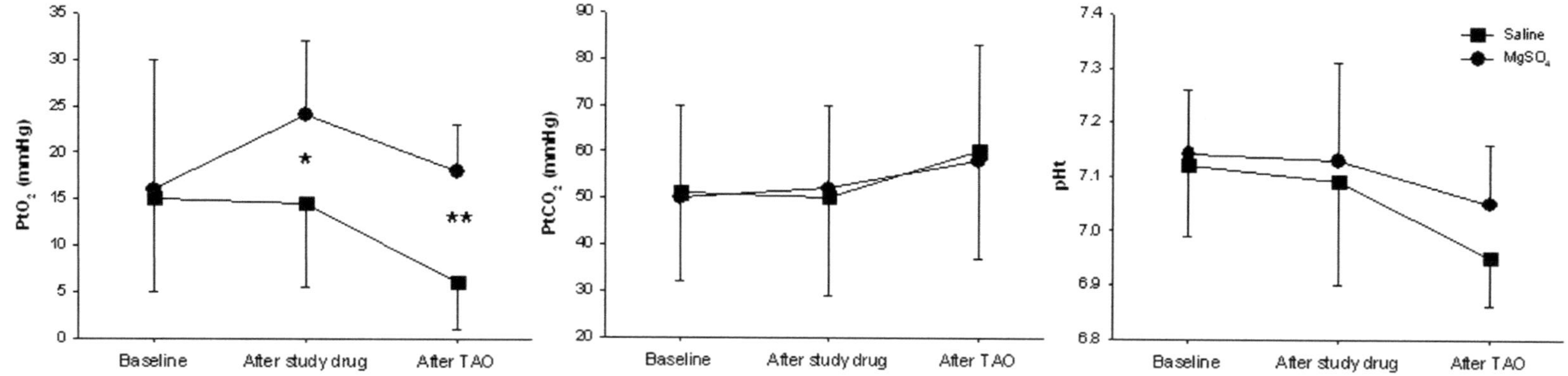

Fig. 2. Tissue oxygen (PtO_2), carbon dioxide tension ($PtCO_2$), and acidity (pHt) at baseline (30 min after catheter insertion), 30 min following study drug infusion and 4 min after temporary artery occlusion in patients receiving saline ($n = 9$) or magnesium sulfate ($n = 9$). $^{*}P = 0.04$; $^{**}P = 0.03$

between groups. Surgery was performed within 72 h after the onset of hemorrhage. The mean plasma magnesium concentration at baseline in the magnesium group, 0.76 ± 0.15 mmol/L was similar to that in the placebo group, 0.81 ± 0.41 mmol/L ($P = 0.76$). Following study drug infusion, plasma magnesium concentration in patients receiving the active treatment increased to 2.12 ± 0.52 mmol/L, whereas those having placebo was 0.79 ± 0.33 mmol/L ($P < 0.001$ magnesium *vs* placebo).

During baseline conditions, pHt, PtO_2 and $PtCO_2$ were similar between groups. Four patients in the magnesium group, and another three in the placebo group had severe tissue hypoxia ($PtO_2 < 10$ mmHg) or acidosis ($pHt < 7.0$). $MgSO_4$ infusion increased PtO_2 from 16 ± 14 to 24 ± 8 mmHg. However, pHt and $PtCO_2$ remained unchanged. There was no difference in brain tissue gases and pHt after placebo infusion.

Following temporary artery occlusion, PtO_2 and pHt decreased promptly in all patients (Fig. 1). However, the rise in $PtCO_2$ was more gradual. At four min after occlusion, PtO_2 was significantly higher in the magnesium group compared with the placebo group ($P = 0.04$). PtO_2 fell below the detection limit in three patients receiving placebo infusion. PtO_2 was set as zero for analysis (Figure 2). All these patients developed neurologic deficits on subsequent review. The differences in $PtCO_2$ and pHt are modest and did not reach statistical significance. The median (range) duration of temporary artery occlusion in the magnesium group, 7.2 (5.1–12.1) min was similar to that in the placebo group, 6.9 (4.9–14.3) min.

Discussion

Our data show that tissue ischemia is common after SAH. $MgSO_4$ infusion increased brain PtO_2 and attenuate tissue hypoxia during temporary proximal artery occlusion. The changes between groups in pHt and $PtCO_2$ were however small. These findings are consistent with an increase in oxygen delivery together with or without a decrease in oxygen consumption. Since magnesium has little effect on cerebral metabolism, we believed the enhanced tissue oxygenation is related to cerebral vasodilation [8].

Studies on the vasodilatory effect of $MgSO_4$, have produced conflicting results. Early reports in women with preeclampsia or eclampsia showed a decrease in middle cerebral artery flow velocity and pulsatility after $MgSO_4$ administration [2–6, 13]. But more recent studies in patients with SAH showed no changes on proximal cerebral artery flow velocity following acute high dose $MgSO_4$ infusion [7, 10]. The differential action of magnesium on different vascular beds may explain the discrepant findings. In this regard, the smaller resistance arteries are more sensitive to magnesium than the proximal conductance vessels [1, 14]. Since the small arteries are predominantly affected, it is expected that $MgSO_4$ will produce a more pronounced effect in preeclampsia. Further study will be required to define the effects of magnesium on tissue perfusion in SAH.

We found the changes in pHt and $PtCO_2$ were smaller than PtO_2 during baseline condition or after temporary artery occlusion. These data are consistent with the fact that carbon dioxide is more soluble than oxygen. Therefore, pHt and $PtCO_2$ can be maintained until very low level of tissue perfusion.

Magnesium may prevent brain damage by mechanisms other than enhanced tissue oxygenation. In animal models, magnesium inhibits presynaptic release of glutamate and blocks postsynaptic N-methy-D-asparate receptors. These changes prevent calcium influx and limit neuronal injury [12].

In conclusion, brain tissue hypoxia and acidosis are common among patients with SAH. $MgSO_4$ infusion dilates the small leptomeningeal arteries. This improves collateral blood flow and tissue oxygenation during brain artery clipping.

Acknowledgment

The study was supported in part by a grant from a Direct Grant for Research #2040874, The Chinese University of Hong Kong.

References

1. Alborch E, Salom JB, Perales AJ, Torregrosa G, Miranda FJ, Alabadi JA, Jover T (1992) Comparison of the anticonstrictor action of dihydropyridines (nimodipine and nicardipine) and Mg^{2+} in isolated human cerebral arteries. Eur J Pharmacol 229: 83–89
2. Belfort MA (1992) The effect of magnesium sulphate on blood flow velocity in the maternal retina in mild preeclampsia: a preliminary color flow Doppler study. Br J Obstet Gynaecol 99: 641–645
3. Belfort MA, Saade GR, Moise KJ (1992) The effect of magnesium sulfate on maternal retinal blood flow in preeclampsia: a randomized placebo-controlled study. Am J Obstet Gynecol 167: 1548–1553
4. Belfort MA, Moise KJ Jr (1992) Effect of magnesium sulfate on maternal brain blood flow in preeclampsia: a randomized, placebo-controlled study. Am J Obstet Gynecol 167: 661–666
5. Belfort MA, Saade GR, Moise KJ Jr (1993) The effect of mag-

nesium sulfate on maternal and fetal blood flow in pregnancy-induced hypertension. Acta Obstet Gynecol Scand 72: 526–530
6. Belfort MA, Anthony J, Saade GR, Allen JC Jr, Nimodipine Study Group (2003) A comparison of magnesium sulfate and nimodipine for the prevention of eclampsia. N Engl J Med 348: 304–311
7. Brewer RP, Parra A, Lynch J, Chilukuri V, Borel CO (2001) Cerebral blood flow velocity response to magnesium sulfate in patients after subarachnoid hemorrhage. J Neurosurg Anesthesiol 13: 202–206
8. Dubé L, Granry JC (2003) The therapeutic use of magnesium in anesthesiology, intensive care and emergency medicine: a review. Can J Anaesth 50: 732–746
9. Ferch R, Pasqualin A, Pinna G, Chioffi F, Bricolo A (2002) Temporary arterial occlusion in the repair of ruptured intracranial aneurysms: an analysis of risk factors for stroke. J Neurosurg 97: 836–842
10. Gin T, Chan MTV, Boet R, Poon WS (2002) Neuroprotection after intravenous magnesium sulphate is not related to vasodilatation in patients with cerebral vasospasm. Abstract presented at the Euroanaesthesia meeting, 6–9 April, 2002, Nice, France
11. Magpie Trial Collaboration Group (2002) Do women with pre-eclampsia, and their babies, benefit from magnesium sulphate? The Magpie Trial: a randomised placebo-controlled trial. Lancet 359: 1877–1890
12. Muir KW (1998) New Experimental and clinical data on the efficacy of pharmacological magnesium infusions in cerebral infarct. Magnes Res 11: 43–56
13. Naidu S, Payne AJ, Moodley J, Hoffmann M, Gouws E (1996) Randomised study assessing the effect of phenytoin and magnesium sulphate on maternal cerebral circulation in eclampsia using transcranial Doppler ultrasound. Br J Obstet Gynaecol 103: 111–116
14. Patel TR, Galbraith S, Graham DI, Hallak H, Doherty AM, McCulloch J (1996) Endothelin receptor antagonist increases cerebral perfusion and reduces ischaemic damage in feline focal cerebral ischaemia. J Cereb Blood Flow Metab 16: 950–958
15. Perales AJ, Torregrosa G, Salom JB, Miranda FJ, Alabadi JA, Monleon J, Alborch E (1991) In vivo and in vitro effects of magnesium sulfate in the cerebrovascular bed of the goat. Am J Obstet Gynecol 165: 1534–1538

Correspondence: Matthew Chan FANZCA, Department of Anaesthesia and Intensive Care, The Chinese University of Hong Kong, Prince of Wales Hospital, Shatin, New Territories, Hong Kong, China. e-mail: mtvchan@cuhk.edu.hk

Acta Neurochir (2005) [Suppl] 95: 113–116

Monitoring of autoregulation using intracerebral microdialysis in patients with severe head injury

M. T. V. Chan[1], **S. C. P. Ng**[2], **J. M. K. Lam**[2], **W. S. Poon**[2], and **T. Gin**[1]

[1] Department of Anaesthesia and Intensive Care, The Chinese University of Hong Kong, Prince of Wales Hospital, Hong Kong, China
[2] Division of Neurosurgery, Department of Surgery, The Chinese University of Hong Kong, Prince of Wales Hospital, Hong Kong, China

Summary

We evaluated the performance of continuous intracerebral microdialysis to indicate the autoregulatory reserve in 36 severely head-injured patients. All patients received standard treatment with intracranial pressure (ICP) monitoring. A microdialysis probe was placed in the frontal cortex anterior to the ICP catheter. Perfusate was collected frequently and extracellular concentration of glutamate was measured online using enzymatic method. Autoregulatory index was calculated by comparing glutamate concentration with CPP using Pearson's correlation. A correlation coefficient $(r) < -0.5$ is considered as loss of autoregulation, whereas r values approach 0 indicate preserved autoregulation. The change of autoregulatory status over time was correlated with outcome at 6 months.

Three patterns of autoregulatory profiles were identified. Patients with intact autoregulation had satisfactory outcome. Transient impairment of autoregulation may result in favorable outcome if patients responded to treatment. However, persistent loss of autoregulation was associated with poor outcome ($P < 0.001$).

The correlation between extracellular glutamate concentration (by microdialysis) and CPP is a useful index of autoregulation in head-injured patients. It predicts clinical outcome and may be used to guide therapy.

Keywords: Cerebral ischemia; jugular venous oxygen saturation; head injury; intracerebral microdialysis.

Introduction

The ability for the brain to maintain a constant cerebral blood flow (CBF) over a wide range of cerebral perfusion pressure (CPP) provides effective protection against ischemia. This mechanism, known as cerebral autoregulation, is frequently impaired in patients with severe head injury [3, 6, 7, 9]. Therefore, the head injured patient with deranged autoregulation is more susceptible to ischemic insult despite a small decrease in arterial pressure. In this regard, a loss of cerebral autoregulation was associated with unfavorable outcome [7, 9]. Furthermore, the choice of treatment depends largely on an accurate assessment of cerebral autoregulatory reserve. Thus, it is important to monitor cerebral autoregulation during the management of severe head injury.

A number of methods have been used to determine cerebral autoregulation. Common techniques include carbon dioxide reactivity [4], and pressure autoregulation with manipulations of systemic or regional arterial pressure [1, 5]. All these tests require a measure of dynamic CBF response. Glutamate, an excitatory neurotransmitter, is released during cerebral ischemia. The concentration is inversely proportional to cerebral perfusion and can be regarded as a surrogate marker of CBF [10]. Currently, extracellular glutamate can be measured by intracerebral microdialysis. We hypothesized that a change in glutamate concentration in response to spontaneous fluctuation of CPP is an index of autoregulatory reserve. We correlated this autoregulatory index with clinical outcome in patients with severe head injury.

Materials and methods

The present study involved a subgroup of patients from a large ongoing trial investigating the metabolic disturbances after head injury [11]. Patients were eligible for the study if they were 18 years or older, suffering from severe head injury. Patients were excluded if they had predominantly extracranial trauma, were pregnant or known to have a serious premorbid illness. We also excluded patients who are moribund on admission, to whom further active management was not considered. This study was approved by the Clinical Research Ethics Committee, and written informed consent was obtained from patients' relatives.

All patients were treated according to standard protocols targeting control of intracranial pressure (ICP). During episodes of intracranial hypertension, patients receive a treatment cascade that included sedation, muscle relaxation, mannitol administration, cerebrospinal fluid (CSF) drainage, moderate hyperventilation, mild-to-moderate hypothermia and finally barbiturate-induced coma. Mass lesions with pressure effects were removed promptly with surgery. Invasive arterial pressure was monitored continuously by arterial cannulation whereas ICP was monitored using ventricular catheter. CPP was defined as the difference between mean arterial pressure and ICP.

Intracerebral microdialysis was performed in the frontal cerebral cortex of maximal injury. In patients with diffuse injury, a right frontal catheter was placed. Technically, we used a single burr hole, where a ventricular drain was first inserted. This was followed by the placement of a microdialysis catheter (CMA70; shaft length 60 mm; membrane length 10 mm) anterior to the ventricular drain. The catheter was then perfused with lactate free Ringer's solution using a microinfusion pump (CMA 106, CMA Microdialysis, Stockholm, Sweden). Microdialysates were collected regularly. Extracellular glutamate concentration was measured by enzymatic colorimetry (CMA600 analyzer, CMA Microdialysis, Stockholm, Sweden).

Physiologic data including arterial pressure, ICP and CPP were captured on a personal computer using a purposely designed data acquisition program. Since microdialysis measured the average glutamate concentration over a specific period of time, we calculated the average CPP value over the same period using an interrupted time series technique [2, 8]. Data were divided into segments of 10 data points. An index of autoregulation was then calculated by comparing glutamate concentration with CPP in the epoch using Pearson linear regression. This index, derived from the correlation coefficient (r), varies from -1 to $+1$. An increase in glutamate concentration with decreasing CPP produces an index of high negative value that indicates loss of autoregulation. Conversely, preserved autoregulation maintains glutamate concentration with changes in CPP. This results in an index approaching zero. We defined a loss of autoregulation as the index less than -0.5 [10]. The changes of autoregulation over time was determined using a moving average technique, where new data point was added to the segment as old one was removed (Figure 1).

Patient outcome was evaluated independently at 6 months according to the Glassgow outcome scale. Autoregulatory status was correlated with outcome using Mantel-Haenzsel test. A P value < 0.05 was considered as statistically significant.

Results

We recruited thirty-six patients with severe head injury in the study. There were 22 males and 14 females. The median (range) age was 28 (18–67) years. The median (range) Glasgow coma scale score recorded on admission was 5 (4–8). At six months after injury, seven (19.4%) patients died, 15 (41.7%) were classified as severely disabled. The remaining 17 patients (47.2%) had a satisfactory outcome with good recovery ($n = 6$) or moderate disability ($n = 11$).

Intracerebral microdialysis was started within 12 hours after head injury. The median (range) duration of microdialysis monitoring was 5 (2–9) days. A total of 1,507 hours of data were recorded. After exclusion of artifacts, 1,428 segments of data were available for analysis.

Autoregulation, expressed as the correlation between glutamate concentrations and CPP, changed with time. Longitudinal analysis identified three patterns of autoregulatory profiles. 17 patients had no autoregulatory response despite treatment, seven (41.2%) of these patients died during the first month after injury. At six months follow up, six (35.3%) patients were classified as vegetative and another four (23.5%) patients were severely disabled. On the contrary, four patients demonstrated intact autoregulation throughout the course of treatment. All of these patients had favorable outcome. The remaining 15 patients had fluctuating course of autoregulatory response. 13 of these patients had responded to treatment (craniectomy $n = 3$; barbiturate coma, $n = 5$; others, $n = 5$) and resulted in favorable outcome. In the other two patients, autoregulation was lost for a considerable period of time with transient improvement after treatment. Both patients had poor outcome. The autoregulatory index correlated well with clinical outcome ($P < 0.001$, Table 1).

There was no complication associated with the placement and removal of micodialysis catheter.

Discussion

This study demonstrates that a change of intracerebral extracellular glutamate concentration in relation to CPP is a useful index of the state of cerebral pressure autoregulation. This autoregulatory index changed with time as the patients' conditions improved or deteriorated. Impairment of autoregulation may be caused by the primary brain injury or by secondary insults. In the present study, we have shown that a deterioration in the autoregulatory response is a good predicator of impending secondary insult. Therefore, timely resuscitation guided by the autoregulatory index should prevent further brain damage [7]. We found the overall progress of the autoregulatory responses correlated with the clinical outcome 6 months after injury.

The advantage of using intracerebral microdialysis as a surrogate measure of regional CBF is that the readings are unaffected by routine activities in the intensive care unit. This is in contrast to transcranial Doppler ultrasonography, jugular venous oximetry or laser Doppler flowmetry, for which recordings are often lost during chest physiotherapy, tracheal suction, and cerebrospinal fluid drainage [1, 3, 4, 6, 7]. Microdialysis is also resistant to brain shift and noises related to physiologic phenomena, including arterial pulsation and respiratory cycling. In the present study, only

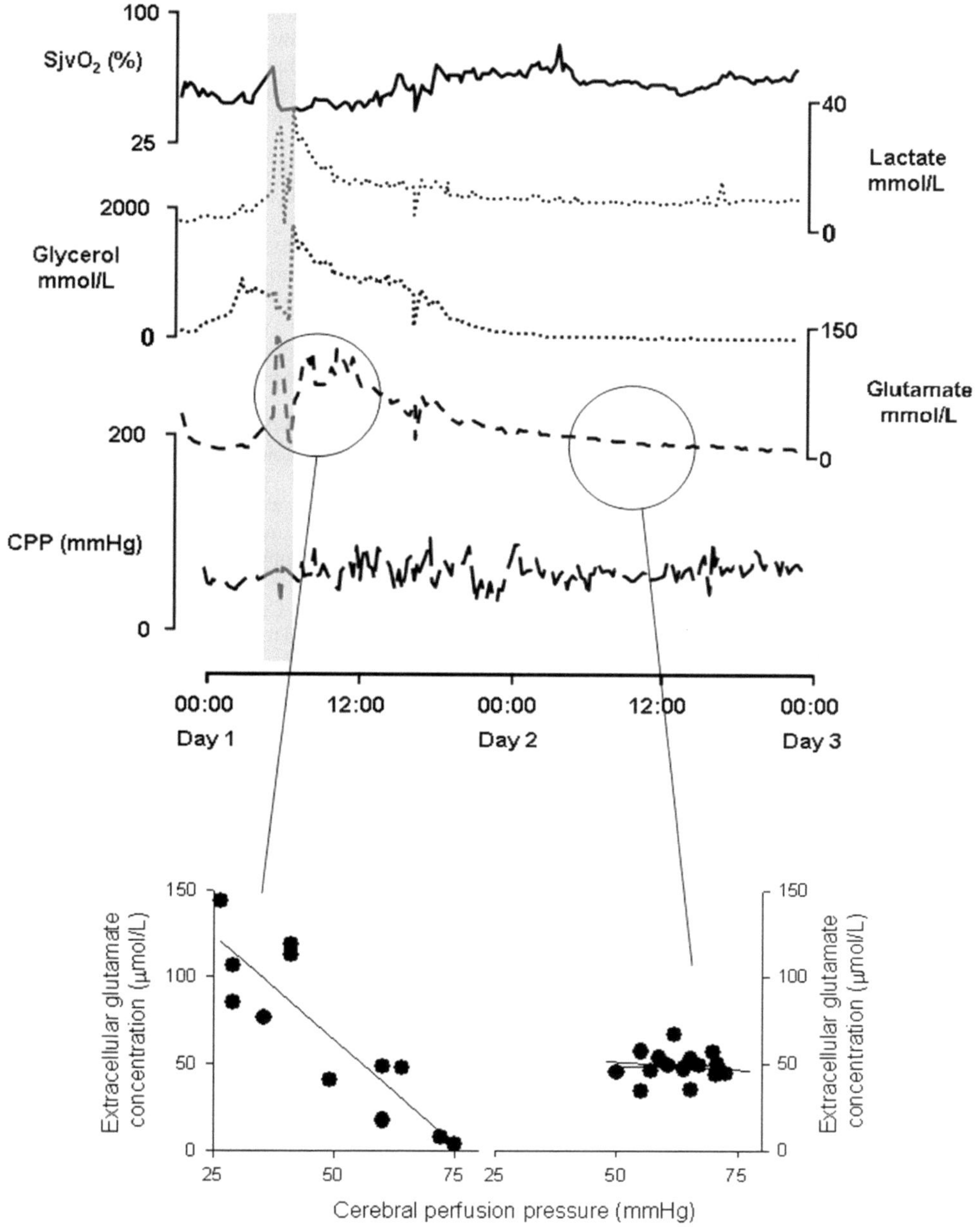

Fig. 1. Changes of intracerebral extracellular concentrations of glutamate, lactate and glycerol, jugular venous oxygen saturation ($SjvO_2$) and cerebral perfusion pressure (*CPP*) before and after evacuation of intracerebral hematoma in a 67 year-old male after road traffic accident (indicated by the gray rectangle). The correlation between extracellular concentrations of glutamate during absence of autoregulation is shown in the bottom left panel $r = 0.88$, whereas the period with restored autoregulation is shown in the bottom right panel, $r = 10$

5.2% of the data were excluded from analysis. This is due to a partially dislodged catheter and was readily identified as the volume of microdialysate became diminished.

The major problem of microdialysis is that it only measures regional metabolic changes. Therefore, regional ischemia at a distance from the catheter tip may have been missed [13]. We inserted the catheter in the hemisphere with maximal injury because this is the area that is most vulnerable to secondary insult. Sampling from multiple sites will improve the spatial resolution of microdialysis monitoring. However, this will increase the risk of brain hemorrhage and infection.

The safety of microdialysis monitoring has been a concern to clinicians. Animal report indicated a lack

Table 1. *Clinical outcome at 6 months after severe head injury in patients with transient or persistent loss of autoregulation*

Glasgow outcome score		Autoregulatory profile		
		Intact	Transient loss	Persistent loss
Favorable outcome	good recovery	3	3	0
	moderately disabled	1	10	0
Unfavorable outcome	severely disabled	0	1	4
	vegetative state	0	1	6
	death	0	0	7

of pathological changes around the insertion site following implantation of a microdialysis catheter for 7 days [12]. More recent human data suggested that long term monitoring is not associated with complications [13].

In summary, neurochemical monitoring with intracerebral microdialysis is a safe and feasible technique. Its correlation with CPP provides a useful index of cerebral autoregulation that predicts clinical outcome.

Acknowledgments

The study was supported in part by a Competitive Earmarked Research Grant, CUHK4223/98M, from the Research Grant Council, Hong Kong.

References

1. Aaslid R, Lindegaard KF, Sorteberg W, Nornes H (1989) Cerebral autoregulation dynamics in humans. Stroke 20: 45–52
2. Chan MTV, Ng SCP, Lam JMK, Poon WS, Gin T (2005) Re-defining the ischemic threshold for jugular venous oxygen saturation – a microdialysis study in patients with severe head injury. Acta Neurochir [Suppl] (in press)
3. Czosnyka M, Smielewski P, Kirkpatrick P, Menon DK, Pickard JD (1996) Monitoring of cerebral autoregulation in head-injured patients. Stroke 27: 1829–1834
4. Enevoldsen EM, Jensen FT (1978) Autoregulation and CO_2 responses of cerebral blood flow in patients with acute severe head injury. J Neurosurg 48: 689–703
5. Giller CA (1991) A bedside test for cerebral autoregulation using transcranial Doppler ultrasound. Acta Neurochir (Wien) 108: 7–14.
6. Hlatky R, Furuya Y, Valadka AB, Gonzalez J, Chacko A, Mizutani Y, Contant CF, Robertson CS (2002) Dynamic autoregulatory response after severe head injury. J Neurosurg 97: 1054–1061
7. Lam JMK, Hsiang JNK, Poon WS (1997) Monitoring of autoregulation using laser Doppler flowmetry in patients with head injury. J Neurosurg 86: 438–445
8. Lam JMK, Smielewski P, al-Rawi P, Griffiths P, Pickard JD, Kirkpatrick PJ (1997) Internal and external carotid contributions to near-infrared spectroscopy during carotid endarterectomy. Stroke 28: 906–911
9. Overgaard J, Tweed WA (1974) Cerebral circulation after head injury, 1: cerebral blood flow and its regulation after closed head injury with emphasis on clinical correlations. J Neurosurg 41: 531–541
10. Peerdeman SM, Girbes AR, Vandertop WP (2000) Cerebral microdialysis as a new tool for neurometabolic monitoring. Intensive Care Med 26: 662–669
11. Poon WS, Ng SC, Chan MT, Leung CH, Lam JM (2002) Neurochemical changes in ventilated head-injured patients with cerebral perfusion pressure treatment failure. Acta Neurochir [Suppl] 81: 335–338
12. Whittle IR, Glasby M, Lammie A, Bell H, Ungerstedt U (1998) Neuropathological findings after intracerebral implantation of microdialysis catheter. NeuroReport 9: 2821–2825
13. Siddiqui MM, Shuaib A (2001) Intracerebral microdialysis and its clinical application: a review. Methods 23: 83–94

Correspondence: Matthew Chan FANZCA, Department of Anaesthesia and Intensive Care, The Chinese University of Hong Kong, Prince of Wales Hospital, Shatin, New Territories, Hong Kong, China. e-mail: mtvchan@cuhk.edu.hk

Acta Neurochir (2005) [Suppl] 95: 117–118

Improvement of brain tissue oxygen and intracranial pressure during and after surgical decompression for diffuse brain oedema and space occupying infarction

M. Jaeger[1], M. Soehle[2], and J. Meixensberger[1]

[1] Department of Neurosurgery, University of Leipzig, Leipzig, Germany
[2] Department of Anaesthesiology and Intensive Care Medicine, Rheinische Friedrich-Wilhelms-University, Bonn, Germany

Summary

Background. We evaluated the perioperative and intraoperative changes of intracranial pressure (ICP) and partial pressure of brain tissue oxygen ($P_{ti}O_2$) after decompressive craniectomy in patients with diffuse brain oedema and space occupying infarction.

Methods. Ten patients suffering from medically intractable raised intracranial pressure (ICP) were included. The underlying diseases and causes for elevated ICP were diffuse brain oedema after subarachnoid haemorrhage (n = 3) and head injury (n = 3), or space occupying infarction of the middle cerebral artery territory due to vasospasm after SAH (n = 4). Continuous perioperative and intraoperative monitoring of $P_{ti}O_2$ and ICP was performed at the side of decompression.

Findings. ICP and $P_{ti}O_2$ improved significantly in a uniform pattern during bone flap removal and dura opening, irrespective of the underlying disease (mean ICP from 52 mmHg to 8 mmHg, mean $P_{ti}O_2$ from 9 mmHg to 25 mmHg). ICP, $P_{ti}O_2$, and cerebral perfusion pressure were further improved in the subsequent 12 hours after surgery, as compared to the preoperative 12 hours.

Conclusions. Decompressive craniectomy seems to be a successful option in the treatment of intractable intracranial hypertension with associated cerebral hypoxia. These positive effects may last for several hours after the procedure irrespective of the underlying disease.

Keywords: Decompressive craniectomy; intracranial pressure; brain tissue oxygen; treatment, cerebral hypoxia.

Introduction

Decompressive craniectomy is generally considered as a last step in the treatment of medically intractable intracranial hypertension. This procedure, where parts of the osseous skull are surgically removed is mainly performed in patients suffering from post-traumatic brain oedema or space occupying stroke. Despite the encouraging clinical studies on outcome in both diseases [2, 5], little knowledge exists about the pathophysiological changes. Therefore, with this study, we sought to investigate the perioperative changes of intracranial pressure (ICP) and partial pressure of brain tissue oxygen ($P_{ti}O_2$) in response to decompressive craniectomy.

Methods

Ten patients with medically intractable intracranial hypertension were included (age 47 ± 16 years; 8 female, 2 male). The underlying diseases and causes for elevated ICP were diffuse posttraumatic brain oedema ($n = 3$), diffuse swelling after severe aneurysmal subarachnoid haemorrhage (SAH, $n = 3$), and space occupying infarction of the middle cerebral artery territory due to cerebral vasospasm after SAH (n = 4). Patients were sedated with midazolam, fentanyl and artificially ventilated to keep the partial pressure of arterial oxygen at 100 mmHg and partial pressure of arterial carbon dioxide at 35 mmHg. Further therapy aimed at maintaining ICP below 25 mmHg using 30° head elevation, mannitol, hypertonic saline, and barbiturate coma. Cerebral perfusion pressure (CPP) was raised above 70 mmHg as individually required to elevate $P_{ti}O_2$ above 10 mmHg [3].

In all patients, routine neuromonitoring was started after admission to our neurosurgical intensive care unit (ICU), comprising continuous measurement of $P_{ti}O_2$ (Licox CC1.SB, GMS, Kiel, Germany), ICP (Codman & Shurtleff, Raynham, MA, USA), and CPP. The flexible microprobes for measurement of ICP and $P_{ti}O_2$ were inserted in CT-normal tissue of the frontal white matter via a double lumen skull bolt kit. Correct positioning was confirmed by CT scan. Neuromonitoring data were stored on a computer at bedside on the ICU and in the operating theatre at a rate of 2 min^{-1}. Due to technical incompatibility, no CPP data could be obtained in the operating theatre during the decompression, thus intraoperative values of ICP and $P_{ti}O_2$ are provided only.

We saw the indication for urgent surgical decompression after conservative measures failed to keep ICP below 25 mmHg. Craniectomy was carried out at the side of dominant brain swelling by removing a large unilateral fronto-temporo-parietal bone flap with duraplasty. The probes, all of which were implanted on the side chosen for decompression, were covered with a sterile dressing to avoid contamination of the surgical field. The procedure was performed at a mean of 5.2 days (range 2.0–10.4 days) after the trauma or SAH. Statistical analysis was made using Wilcoxon-signed-rank test.

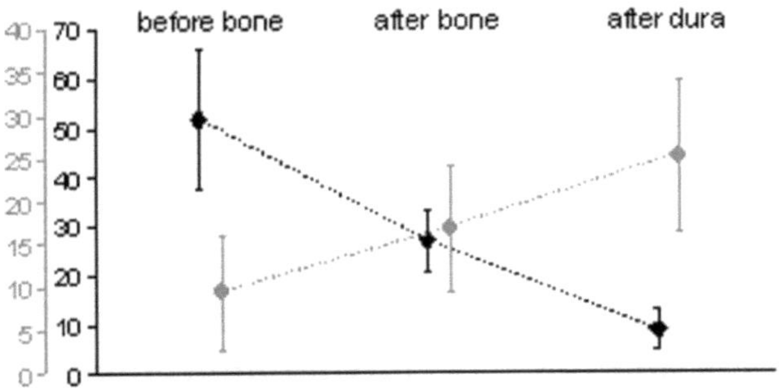

Fig. 1. Intracranial pressure (*ICP, black*) and brain tissue oxygen tension (*$P_{ti}O_2$, grey*) in mmHg obtained during the intraoperative steps of decompressive craniectomy. Differences between the intraoperative steps were statistically significant for both parameters ($P < 0.01$). Values are mean ± standard deviation

Results

In all ten patients, ICP and $P_{ti}O_2$ improved in a uniform two-step fashion relating to bone flap removal and opening of the dura. Mean values of ICP and $P_{ti}O_2$ before and after the intraoperative steps are shown in Fig. 1. This uniform pattern of a decrease in ICP and $P_{ti}O_2$ increases occurred irrespective of the underlying disease and cause for intracranial hypertension. Differences between the intraoperative steps revealed statistical significance ($P < 0.01$). Furthermore, ICP, CPP, and $P_{ti}O_2$ showed significant improvement in the 12 hours period after the operation, as compared to the preoperative 12 hours ($P < 0.05$, Table 1).

Discussion

The results of this study demonstrate the immediate reversal of pathologically elevated ICP during unilateral decompressive craniectomy. Furthermore, critically diminished $P_{ti}O_2$ improves in parallel to non-hypoxic values. These positive effects are not limited to the surgical procedure, but may last for several hours. The intraoperative course of ICP and $P_{ti}O_2$ from the three patients with diffuse oedema after SAH has been previously published [1], however, this study now adds further information on the uniform intracranial haemodynamic changes in response to decompressive craniectomy for posttraumatic oedema and space occupying stroke. The results are furthermore in agreement with a recent report on improved $P_{ti}O_2$ after decompression in seven patients with posttraumatic oedema. In contrast, however, in the study by Stiefel *et al.* $P_{ti}O_2$ probes were reinserted after the craniectomy and no intraoperative data were available [4]. In summary, decompressive craniectomy seems to be a successful option in the immediate treatment of elevated ICP and associated cerebral hypoxia. To avoid secondary brain damage, the study favours the early use of decompression, arguing that impending neuronal damage can be effectively treated. Still, the results of the currently ongoing prospective randomised trials for decompression after head injury (RESCUEicp trial) and stroke (Destiny trial) are awaited to apply this treatment according to the standards of evidence based medicine.

Table 1. *Intracranial pressure (ICP), brain tissue oxygen tension ($P_{ti}O_2$), and cerebral perfusion pressure (CPP) in the 12 hours period before and after decompressive craniectomy. Differences between intervals were statistically significant for all three parameters ($P < 0.05$). Values are mean ± standard deviation*

	12 h preoperative	12 h postoperative
ICP [mmHg]	27.4 ± 6.1	16.5 ± 8.4
$P_{ti}O_2$ [mmHg]	13.1 ± 8.0	21.4 ± 8.9
CPP [mmHg]	67.7 ± 10.4	77.4 ± 10.4

References

1. Jaeger M, Soehle M, Meixensberger J (2003) Effects of decompressive craniectomy on brain tissue oxygen in patients with intracranial hypertension. J Neurol Neurosurg Psychiatry 74: 513–515
2. Kleist-Welch Guerra W, Gaab MR, Dietz H, Mueller JU, Piek J, Fritsch MJ (1999) Surgical decompression for traumatic brain swelling: indications and results. J Neurosurg 90: 187–196
3. Meixensberger J, Jaeger M, Väth A, Dings J, Kunze E, Roosen K (2003) Brain tissue oxygen guided treatment supplementing ICP/CPP therapy after traumatic brain injury. J Neurol Neurosurg Psychiatry 74: 760–764
4. Stiefel MF, Heuer GG, Smith MJ, Bloom S, Maloney-Wilensky E, Gracias VH, Grady MS, LeRoux PD (2004) Cerebral oxygenation following decompressive hemicraniectomy for the treatment of refractory intracranial hypertension. J Neurosurg 101: 241–247
5. Uhl E, Kreth FW, Elias B, Goldammer A, Hempelmann RG, Liefner M, Nowak G, Oertel M, Schmieder K, Schneider GH (2004) Outcome and prognostic factors of hemicraniectomy for space occupying cerebral infarction. J Neurol Neurosurg Psychiatry 75: 270–274

Correspondence: Matthias Jaeger, Klinik und Poliklinik für Neurochirurgie, Universitätsklinikum Leipzig, Liebigstr. 20, 04103 Leipzig, Germany. e-mail: jaem@medizin.uni-leipzig.de

Acta Neurochir (2005) [Suppl] 95: 119–121

Clinic investigation and logistic analysis of risk factors of recurrent hemorrhage after operation in the earlier period of cerebral hemorrhage

S. C. Chen and **G. Feng**

Department of Neurosurgery, First Affiliated Hospital of Jinan University, Guangzhou, P. R. China

Summary

Objective of this study was to investigate the incidence, time, location, prevention, treatment and risk factors of recurrent hemorrhage in the earlier period of cerebral hemorrhage after operation. Three hundred and twenty two patients with operations in the earlier period of cerebral hemorrhage were analyzed retrospectively. The clinical data of hemorrhage and recurrent cerebral hemorrhage groups were compared and statistically analyzed. Logistic regression analysis was applied to evaluate the function of possible factors leading to recurrent hemorrhage after operation.

The incidence of recurrent hemorrhage was 21.4% in the earlier period of cerebral hemorrhage after operation. When the operation was performed after cerebral hemorrhage within 6 h, 6–12 h and 12–24 h, the incidence of recurrent hemorrhage after operation were 43.1%, 20.9%, 3.6% respectively. With regard to time of recurrent hemorrhage, the incidence was 3.1% within 12 h after operation, 15.5% between 12–24 h and just 2.8% after 24 h. Site of hemorrhage was in the basal ganglion in 92.6% of the cases. Mono-agent logistic analysis displayed that there is a significant correlation between high diastolic blood pressure, fluctuation of blood pressure after operation, taking anti-coagulant drugs for a long time, site of hemorrhage, difficult or not thorough hemostasis during operation and recurrent hemorrhage ($p < 0.01$). Multiple linear logistic regression analysis has shown that a remarkable diastolic blood pressure and fluctuation of blood pressure after operation are risk factors for recurrent hemorrhage. Their OR value were 10.32, 7.234. From this it is concluded that the incidence of recurrent cerebral hemorrhage after operation in the earlier period is 21.4%, which must never be ignored. The time period of 24 h after operation is a stage of high risk. Maintaining diastolic blood pressure below 85 mmHg and steadily controlling the pressure after operation are of great importance for prevention of recurrent cerebral hemorrhage.

Keywords: Cerebral hemorrhage; operation; recurrent hemorrhage; risk factor; logistics.

Introduction

With the advent of surgical treatment for hypertensive cerebral hemorrhage in the earlier period, the phenomenon of recurrent intracerebral hemorrhage after operation has been noted. To investigate the incidence, time, location, prevention, and risk factors of recurrent hemorrhage after operation for hypertensive cerebral hemorrhage in the earlier period, a total of 322 samples of cerebral hemorrhage treated surgically in the earlier period were obtained in our hospital between 1996 and 2002, of which 69 patients had recurrent intracerebral hemorrhage after operation. We compared and statistically analyzed the clinical data. Results were analyzed retrospectively.

Patients and methods

A total of 322 samples of hypertensive cerebral hemorrhage treated surgically in the earlier period were included in our study. There were 116 women and 206 men. Mean age of patients was 54.3 years (range 29–71 years). Hemorrhage location included thalamus (216 cases), thalamencephalon (29 cases), confined to ventricles of the brain (36 cases), cerebral hemisphere (24 cases), and cerebellum (17 cases). Operation time was within 6 h after cerebral hemorrhage (51 cases); 6–12 h (216 cases) after cerebral hemorrhage; 12–24 h (55 cases) after cerebral hemorrhage, respectively. According to duotian formula, the number of cerebral hemorrhage was 12–125 ml, mean 67.5 ml.

Among the 322 samples, a total of 69 cases were included in the group of recurrent hemorrhage after operation. 51 were men and 18 were women. Mean age of patients was 52 years, ranging from 30 to 70 years old. Recurrent hemorrhage time was within 12 h after operation (10 cases), within 12–24 h after operation (50 cases), within 48 h after operation (9 cases), respectively. The number and size of recurrent hemorrhage was 30–50 ml (28 cases), 50–70 ml (40 cases), >70 ml (1 case). The relationship of recurrent hemorrhage and time of first operation was as shown below: recurrent hemorrhage after operation within 6 hours after cerebral hemorrhage was seen in 22 patients, 6–12 hours in 45 patients, and 12–24 hours in one patient.

All cases were examined more than twice by CT scan of the brain to identify the location and number of cerebral hemorrhage. The following data were retained: age, gender, accompanying diseases, history of taking anti-coagulant, blood pressure, condition of hemostasis during operation, data of life symptom examined after operation. According to diastolic pressure, patients were divided into three groups: group A (<85 mmHg), group B (85–95 mmHg), and group

C (>95 mmHg). Chi-square test was performed to estimate the influence of diastolic pressure on recurrent hemorrhage after operation. Mono-agent logistic analysis was carried out to find out whether hemorrhage was a dependent variable or not, and the suspicious factors as mentioned above to be an independent variable. Thereafter a multiple linear logistic regression analysis was performed to determine the main risk factors. Data were statistically analyzed by SPSS 10.0.

Results

The incidence of recurrent intracerebral hemorrhage after operation in the earlier period of hypertensive cerebral hemorrhage was 21.4%. When the operation is performed within 6 h, 6–12 h and 12–24 h, respectively, after cerebral hemorrhage, the incidence of recurrent hemorrhage after operation were 43.1%, 20.9%, 3.6%, respectively. In 92.6% of patients location of recurrent hemorrhage was in the basal ganglia, and in 71% of them location of first cerebral hemorrhage and recurrent hemorrhage were both in the basal ganglia. A Chi-square test showed that there was a close correlation between diastolic pressure and occurrence of recurrent hemorrhage after operation. Mono-agent logistic analysis showed that there were 5 factors attributing to recurrent intracerebral hemorrhage: (1) high diastolic blood pressure; (2) fluctuation of blood pressure after operation; (3) hemorrhage site; (4) taking anti coagulant drug for a long time; (5) difficult or not thorough hemostasis during operation ($P < 0.01$). By multiple linear logistic regression analysis, a remarkable diastolic blood pressure and fluctuation of blood pressure after operation were the most significant risk factors for recurrent intracerebral hemorrhage.

Table 1. *Effects of diastolic blood pressure on recurrent hemorrhage after operation*

Results	<85 mmHg (A)	85–95 mmHg (B)	>95 mmHg (C)
Recurrent hemorrhage	1	18	50
No recurrent hemorrhage	32	190	30
Percent of recurrent hemorrhage	1.44%	26.08%	72.46%

Chi-square test showed, exception of group A and B, there was statistical difference among other groups. ($P < 0.01$).

Table 2. *Results of Mono-agent logistic analysis*

	b	SE (b)	P	OR	OR95%CI
High diastolic blood pressure	2.715	0.601	0.003	12.36	2.786~41.16
Fluctuation of blood pressure after operation	1.986	0.671	0.002	8.144	2.011~22.63
Hemorrhage site	1.091	0.568	0.029	2.861	1.124~10.32
Taking anti coagulant drug for a long time	0.987	0.472	0.041	1.724	1.091~5.167
Difficult or not thorough hemostasis during operation	0.653	0.328	0.031	1.928	1.011~4.896

Table 3. *Results of multiple linear logistic regression analysis*

	b	SE (b)	P	OR	OR95%CI
High diastolic blood pressure	2.532	0.597	0.004	10.32	2.231~28.17
Fluctuation of blood pressure after operation	1.654	0.603	0.005	7.234	1.987~21.02

Discussion

Operation in the earlier phase of intracerebral hemorrhage (within 24 h) could timely relieve the oppressive effect of hematoma on cerebral organization, reduce edema and necrosis around the hematoma, and precipitate the recovery of neural function to the greatest extent. But recurrent intracerebral hemorrhage can never be ignored.

There were few reports about recurrent intracerebral hemorrhage after operation in the earlier period of cerebral hemorrhage. According to the results of statistical analysis in our research, the incidence of recurrent intracerebral hemorrhage was 21.4%. When the operation was performed within 6 h, 6–12 h and 12–24 h, respectively, after cerebral hemorrhage, the incidence of recurrent intracerebral hemorrhage were 43.1%, 20.9%, 3.6% respectively. Our study indicates that the sooner an operation for cerebral hemorrhage is performed, the higher is the incidence of recurrent intracerebral hemorrhage. A possible explanation for this phenomenon according to our findings is that an early operation time and reduced edema around hematoma of the brain tissue could result in an incomplete obliteration of its small blood vessel, and slow accumulation of blood.

We analyzed the time of re-bleeding. The incidence of re-bleeding is 3.1% within 12 hours after surgery, then increases to 15.5% during 12 to 24 hours after surgery, while the incidence rate decreases to 2.8% 24 hours after surgery. Bea *et al.* are of the opinion that re-bleeding mainly occurs within 24 hours after surgery. Fujiison *et al.* concluded that the incidence rate of re-bleeding within 24 hours was 20% after analyzing the CT pictures of 107 patients. The incidence rate of re-bleeding in our group was 21.4%, which indicates that the period of highest risk for re-bleeding should be within 24 hours after surgery. Altogether 86.9% of re-bleeding occurred during this period.

The relative factors of re-bleeding: There was hypertension in all cases. The unstable blood pressure of hypertension-reduced cerebral hemorrhage patients after surgery has become the main factor for re-bleeding. The cerebral blood flow increases significantly when blood pressure increases suddenly after surgery, and fluctuation of blood pressure becomes the risk factor for re-bleeding. The logistic analysis of the data in our group indicated that the OR value of the unstable blood pressure after surgery was 7.234, and the 95% confidence interval was 1.987–21.02. This result coincides with the theory mentioned above. Gao'Xiaolan *et al.* emphasized that high diastolic pressure is the main risk factor for re-bleeding. Our statistical analysis indicated that the incidence of re-bleeding in patients categorized by different diastolic pressure were significantly different: the incidence rate in the group with diastolic pressure less than 85 mmHg was 1.44%, while the rate in the 85–95 mmHg diastolic pressure group was 26.08%, and the rate was 72.46% in the group with diastolic pressure above 95 mmHg. The OR value in the high-diastolic-pressure state was 10.32, and the 95% confidence interval was 2.231–40.17, which means that unstable pressure was the highest risk factor for re-bleeding. Hence, in order to decrease the incidence rate of re-bleeding, it is crucial to control the blood pressure after surgery, particularly to maintain diastolic pressure to no more than 85 mmHg. We think that high diastolic pressure and unstable blood pressure after surgery are two main risk factors of re-bleeding. The other risk factors indicated by clinical data include site of bleeding, method of operation, intra-operative hemostasis, use of anti-coagulants before operation, restlessness, choking and coughing. Concluding from our data, re-bleeding occurs most frequently in the basal ganglion zone – the incidence rate in this zone was 97.6%. There was no significant difference between the incidence rates in male and female patients. There were 71.07% of patients whose 1st and 2nd bleeding both occurred in the basal ganglion zone. This might be due to its location: the basal ganglion zone is close to the cerebral ventricles where sustaining strength is weak, and thus renders hemostasis difficult and allows hematoma to extend into the ventricles more easily. When analyzing according to methods of operations, we would find that the incidence rate of re-bleeding in the decompressive craniectomy group (28.8%) was significantly higher than that in the small bone windowed craniotomy (16.1%). This difference may due to initially larger hematoma from the previous method of operation, or due to the lower pressure in the hematoma after decompression, which makes it more difficult to stop bleeding. Furthermore, improper surgical techniques, such as too strong a suction, blind suction, too much tractions on the brain tissue, may cause re-bleeding as well.

It is critical to observe and closely monitor pupils, consciousness and blood pressure. A direct and effective method of monitoring is to detect the pressure through the bone window after the operation. If the pressure increases significantly within 48 hours after the operation, re-bleeding should be suspected. If there is a lot of pink drainage fluid from the hematoma cavity or from the subdural drainage tube after the operation, there should be no intracranial hypertension or bleeding.

Correspondence: S. C. Chen, Department of Neurosurgery, First Affiliated Hospital of Jinan University, Guangzhou, P. R. China. e-mail: drcsc@263.net

Acta Neurochir (2005) [Suppl] 95: 123–127

Cerebral blood flow augmentation in patients with severe subarachnoid haemorrhage

P. G. Al-Rawi[1], **D. Zygun**[2,3], **M. Y. Tseng**[1], **P. J. A. Hutchinson**[1], **B. F. Matta**[2], and **P. J. Kirkpatrick**[1]

[1] University Department of Neurosurgery, Addenbrooke's Hospital, Cambridge, UK
[2] University Department of Anaesthesia, Addenbrooke's Hospital, Cambridge, UK
[3] Department of Critical Care Medicine, University of Calgary, Calgary, Canada

Summary

Following aneurysmal subarachnoid haemorrhage (SAH), cerebral blood flow (CBF) may be reduced, resulting in poor outcome due to cerebral ischaemia and subsequent stroke. Hypertonic saline (HS) is known to be effective in reducing intracranial pressure (ICP) [16]. We have previously shown a 20–50% increase in CBF in ischaemic regions after intravenous infusion of HS [17]. This study aims to determine the effect of HS on CBF augmentation, substrate delivery and metabolism.

Continuous monitoring of arterial blood pressure (ABP), ICP, cerebral perfusion pressure (CPP), brain tissue oxygen (P_bO_2), middle cerebral artery flow velocity (FV), and microdialysis was performed in 14 poor grade SAH patients. Patients were given an infusion of 23.5% HS, and quantified xenon computerised tomography scanning (XeCT) was carried out before and after the infusion in 9 patients.

The results showed a significant increase in ABP, CPP, FV and P_bO_2, and a significant decrease in ICP ($p < 0.05$). Nine patients showed a decrease in lactate-pyruvate ratio at 60 minutes following HS infusion.

These results show that HS safely and effectively augments CBF in patients with poor grade SAH and significantly improves cerebral oxygenation. An improvement in cerebral metabolic status in terms of lactate-pyruvate ratio is also associated with HS infusion.

Keywords: Subarachnoid haemorrhage; hypertonic saline; brain tissue oxygen; microdialysis; xenon computerised tomography; cerebral blood flow.

Introduction

Death and disability following SAH is associated with cerebral ischaemia and subsequent stroke, which are often delayed by several days [9]. Combined mortality and morbidity approaches 50%, and long-term morbidity is substantial [1]. Processes are complex, combining to impair blood flow to several territories. Episodes of low CBF are common, particularly in patients who have a poor clinical grade or who deteriorate, and regional perfusion deficits are known to occur. Reduction in CBF below a critical threshold has been shown to result in a loss of function, which may be reversible or remains permanent depending upon the depth and duration of the ischaemic episode [18]. Episodes of low cerebral oxygenation and abnormal profiles of metabolic parameters, in particular a high lactate-pyruvate (L-P) ratio, are also associated with a poor outcome [13]. A threshold for low cerebral tissue oxygen has been identified, below which cerebral infarction occurs [4, 5].

It has been shown that in head injured patients, mannitol can effectively lower ICP whilst increasing CPP [6]. Hypertonic saline (HS) has been shown to have similar effects to mannitol but the duration of action is longer [14, 16]. Experimental studies have shown potential beneficial effects of HS on CBF [8]. However, studies on the use of osmotic agents, including HS, in patients with SAH are limited.

We have previously shown a 20–50% increase in CBF in ischaemic regions after intravenous infusion of hypertonic saline in patients with poor grade SAH, without compromising flow to other areas [17]. While the effects of augmenting CBF with HS appear impressive, a metabolic benefit to cerebral tissues has not been proven outside animal studies. Our hypothesis therefore is that low CBF following SAH can be reversed with HS, resulting in an improved oxygenation and metabolic profile. These effects may be greater in patients with poor grade SAH and who have poorer cerebral blood flow. By using bedside multimodal monitoring and XeCT, our aim was to determine the effect of HS on CBF augmentation, substrate delivery and metabolism.

Table 1. *Table of absolute values and % difference from baseline (% Diff.), at 30 and 60 minutes post infusion for mean arterial blood pressure (MAP), intracranial pressure (ICP), cerebral perfusion pressure (CPP) and MCA flow velocity (FV)*

	MAP		ICP		CPP		FV	
	mmHg	% Diff.	mmHg	% Diff.	mmHg	% Diff.	cm/sec	% Diff.
Baseline	102.2 (±10.2)	–	20.8 (±8.2)	–	81.5 (±11.0)	–	86.4 (±19.7)	–
30 Mins	114.0* (±9.5)	12.9* (±9.6)	5.9* (±2.6)	−66.0* (±15.5)	108.1* (±8.9)	36.6* (±15.6)	103.4* (±22.0)	50.5* (±24.6)
60 Mins	103.6 (±8.9)	2.4 (±7.9)	5.9* (±2.8)	−61.0* (±20.2)	88.5 (±17.1)	12.6 (±22.9)	97.6* (±14.9)	28.6* (±20.8)

Values are mean (±95% CI), * denotes $p < 0.05$.

Table 2. *Table of absolute values and % difference from baseline (% Diff.), at 30 and 60 minutes post infusion for brain tissue oxygen (P_bO_2), carbon dioxide (P_bCO_2) and pH (pH_b)*

	pH_b		P_bCO_2		P_bO_2	
	pH	% Diff.	kPa	% Diff.	kPa	% Diff.
Baseline	6.8 (±0.2)	–	6.5 (±0.65)	–	1.7 (±0.65)	–
30 Mins	6.9 (±0.2)	0.4 (±0.5)	6.4 (±0.65)	−1.2 (±3.3)	2.3* (±0.86)	45.9* (±29.4)
60 Mins	6.9 (±0.2)	0.4 (±0.6)	6.4 (±0.65)	−1.2 (±4.8)	2.1* (±0.86)	26.8* (±25.1)

Values are mean (±95% CI), * denotes $p < 0.05$.

Materials and methods

After Local Research Ethics Committee approval and following informed assent, 14 patients (7 male and 7 female) suffering from poor grade SAH underwent continuous multimodal monitoring at the bedside, including ABP, ICP, CPP and transcranial Doppler FV. Brain tissue oxygen, carbon dioxide and pH were monitored using a multiparameter sensor (Neurotrend™, Codman, Bracknell, UK) and brain tissue chemistry via a microdialysis catheter (CMA-70, CMA, Sweden), inserted through a triple lumen access device (Technicam, Newton Abbot, UK). This has been described in detail previously [4]. The mean age was 57 years (range 44–74). All patients had a Glasgow Coma Score (GCS) of 8 or less and required ventilation on the Neuro Critical Care Unit (NCCU).

The microdialysis catheter was perfused with CMA perfusion fluid (CMA, Sweden) at a rate of 0.3 μl/min. Vials were changed every 20 minutes. Microdialysis samples were analysed at the bedside for glucose, lactate, pyruvate and glutamate using a CMA 600 analyser (CMA, Sweden). Patients were given an infusion of 23.5% HS via a central venous catheter at a dose of 2 ml/kg. XeCT (Diversified Diagnostic Products, DDP[inc], Houston, Texas, USA) [18] was performed before and after HS administration in 9 of the 14 patients.

Data processing and analysis

CBF in a region of interest around the microdialysis and Neurotrend™ probes was calculated (ROI CBF). Data from all the monitored parameters was averaged at baseline and at 30 and 60 minutes post infusion. Absolute values and percentage difference from baseline (%Δ) was calculated. Comparison was made using Student's t-test. Values are given as mean ± 95% confidence intervals. Statistical significance was considered at $p < 0.05$.

Results

Pooling the data for all patients, baseline ICP was 20.8 mmHg (±3.8), with a baseline CPP of 81.5 mmHg (±5.1) (Table 1). At 30 minutes post HS infusion there is a significant increase in ABP and CPP, a significant drop in ICP and a significant increase in FV ($p < 0.05$) (Table 1). At 60 minutes, ICP & FV still showed significant changes, whilst the other parameters tended toward baseline levels. Average baseline ROI CBF was 30.4 ± 7 ml/100 g/min. In 2 patients, ROI CBF decreased after HS. In the remaining 7, a significant increase in ROI CBF following HS infusion was observed ($p < 0.05$).

Looking at the results from the Neurotrend™ sensor (Table 2), baseline brain tissue oxygen (P_bO_2) was low at 1.7 kPa (±0.3) and increased significantly at 30 and 60 minutes post infusion ($p < 0.05$) (Fig. 1). Very little change was seen in either brain tissue carbon dioxide (P_bCO_2) or pH (pH_b).

Turning to the microdialysis data, there was no significant change seen in any parameter when looking at the pooled data for all patients (Table 3). However, Fig. 2 demonstrates a graphical example of the bedside monitoring of L-P ratio seen in one patient. There is a

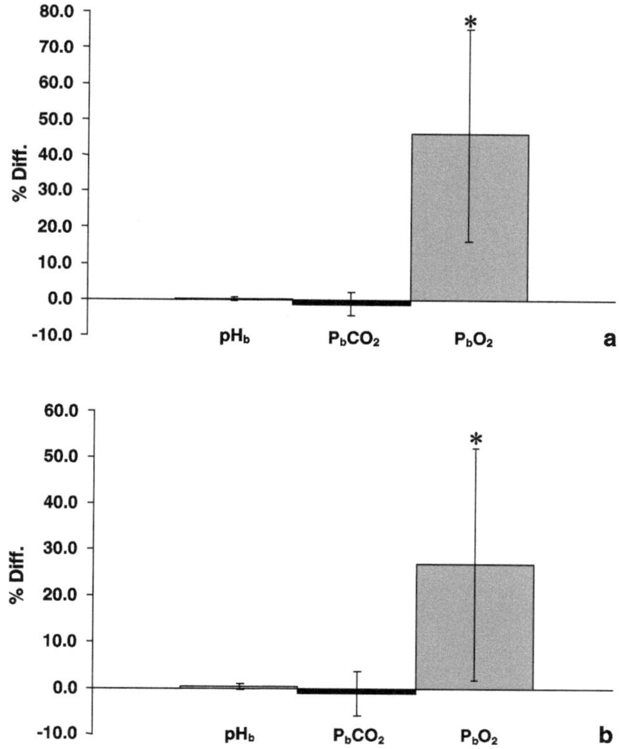

Fig. 1. Graphs of % difference from baseline (*%Δ*) for brain tissue oxygen (P_bO_2), carbon dioxide (P_bCO_2) and pH (*pHb*) at 30 (a) and 60 (b) minutes post hypertonic saline administration. Error bars represent 95% CI, * denotes $p < 0.05$

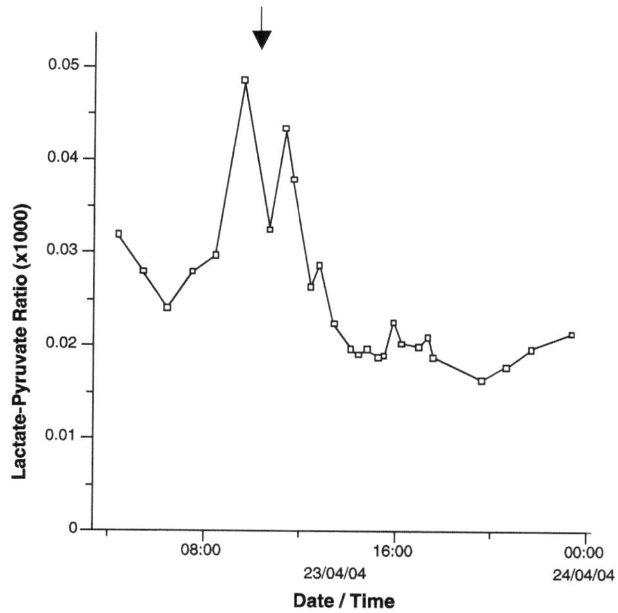

Fig. 2. Graph of lactate-pyruvate ratio monitored at the bedside for an individual patient. Time of administration of HS is marked (arrow)

distinct improvement (i.e. a decrease) in L-P ratio seen after administration of hypertonic saline. Figure 3 shows individual patients L-P ratio at baseline and at 30 and 60 minutes post infusion. At 60 minutes the L-P ratio can be seen to decrease in 9 of the 14 patients. However, one patient demonstrated a very high baseline L-P ratio (>40), which increased further after HS infusion. There were no complications throughout the duration of the study.

Discussion

In accordance with our previous study, this data has shown that 23.5% HS significantly decreases ICP whilst increasing CPP and FV. In addition, a significant improvement was seen in P_bO_2 with maximal effect at 30 minutes post infusion. A decrease in L-P ratio was seen in 9 patients although this did not reach significance for the group as a whole. Baseline glucose and glutamate are lower than previously reported, whilst baseline lactate and L-P ratio are higher. However, all the patients in this cohort were poor grade SAH patients. Previous studies have looked either exclusively at better grade SAH patients, or the data has been derived from a more heterogeneous group, in-

Table 3. *Table of absolute values and % difference from baseline (% Diff.), at 30 and 60 minutes post infusion for brain chemistry data*

	Glucose		Lactate		Pyruvate		Glutamate		L-P ratio	
	mg/ml	% Diff.	mM	% Diff.	μM	% Diff.	μM	% Diff.	Ratio	% Diff.
Baseline	0.16 (±0.06)	–	1.57 (±0.4)	–	65.2 (±18.6)	–	4.15 (±4.9)	–	25.2 (±8.0)	–
30 Mins	0.16 (±0.06)	−9.1 (±17.2)	1.67 (±0.6)	−0.7 (±13.5)	68.1 (±27)	−3.3 (±17.6)	4.01 (±3.7)	32.0 (±68.0)	35.1 (±27.4)	22.6 (±59.9)
60 Mins	0.19 (±0.09)	11.4 (±26.1)	1.89 (±0.9)	6.9 (±24.8)	78.9 (±32.4)	10.7 (±23.3)	3.93 (±2.6)	97.7 (±135.6)	29.5 (±19.9)	5.52 (±43.26)

Values are mean (±95% CI).

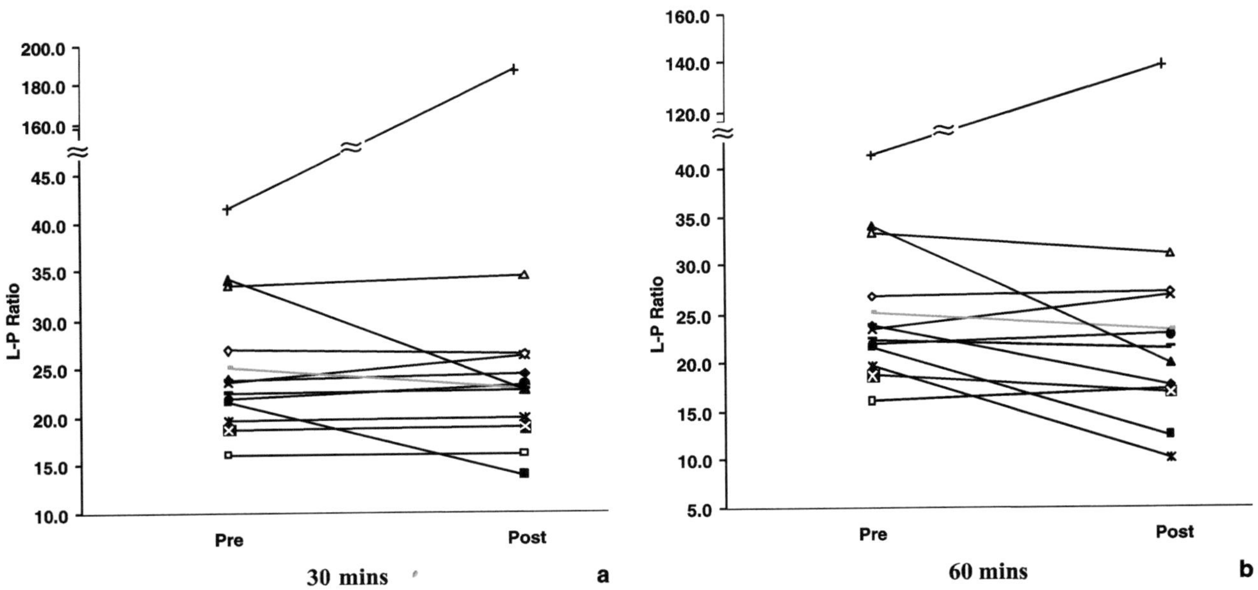

Fig. 3. Graphs showing the lactate-pyruvate (*L-P*) ratio for individual patients, pre and at 30 (a) and 60 (b) minutes post hypertonic saline infusion

cluding good grade SAH or unruptured aneurysms [2, 12]. Glutamate has been suggested to be a sensitive and early marker of cerebral hypoxia [10]. This is not our experience in this study, where considerable variation in glutamate was seen, as indicated by the large 95% CI (Table 3). Significant changes in L-P ratio have been seen in patients with SAH associated with cerebral infarction [11] and L-P ratio has been proposed as a more robust marker of acute ischaemia with a high sensitivity and specificity [4, 7].

Baseline pH_b was low, which may suggest metabolic disturbance as a result of injury severity or secondary insults. However these values may be typical of poor grade SAH patients. The fact that neither pH_b nor P_bCO_2 showed significant changes after HS compared to P_bO_2 may indicate that they are perhaps not such sensitive markers of improved metabolism. It has been suggested that pH_b and P_bCO_2 have a lower predictive value than P_bO_2 [3]. The mechanism of action of HS is probably threefold. It has a strong osmotic and haemodilution effect and improves haemorheology [15, 16]. Comparisons between HS and mannitol have suggested that HS is the more effective osmotic agent [14, 16]. Most importantly HS does not deplete intravascular volume. Whilst there are potential complications associated with administration of 23.5% HS, we did not find any short or long term side effects. Since compromised CBF is a precursor for stroke regardless of the underlying pathology, the findings in this study may have implications for other conditions. Prolonging neuroprotection may increase the window of opportunity for other therapeutic strategies, including thrombolysis, as well as reduce infarct size by protecting penumbral regions.

In conclusion, this initial data is encouraging and shows that HS safely and effectively augments CBF in patients with poor grade SAH and significantly improves cerebral oxygen. An improvement in cerebral metabolic status in terms of lactate-pyruvate ratio is also associated with HS infusion. The study is ongoing, and further numbers are needed to assess outcome and radiological stroke.

Acknowledgments

The authors would like to thank Dr Jurgens Nortje for help with the intracranial probe insertion and NCCU monitoring, Dr Mark O'Connell for advice on the microdialysis, and Dr HK Richards for assistance with statistical analysis. This study was supported by a grant from The Stroke Association, UK.

References

1. Hutchinson PJA, Seeley HM, Kirkpatrick PJ (1998) Factors implicated in deaths from subarachnoid haemorrhage: are they avoidable? Br J Neurosurg 12: 37–40
2. Kett-White R, Hutchinson PJ, Al-Rawi PG, Gupta AK,

O'Connell MT, Pickard JD, Kirkpatrick PJ (2002) Extracellular lactate/pyruvate and glutamate changes in patients during peroperative episodes of cerebral ischaemia. Acta Neurochir (Wien) 81: 363–365
3. Kett-White R, Hutchinson PJ, Al-Rawi PG, Gupta AK, Pickard JD, Kirkpatrick PJ (2002) Adverse cerebral events detected after subarachnoid haemorrhage using brain oxygen and microdialysis probes. Neurosurgery 50: 1213–1222
4. Kett-White R, Hutchinson PJA, Al-Rawi PG, Czosnyka M, Gupta AK, Pickard JD, Kirkpatrick PJ (2002) Cerebral oxygen and microdialysis monitoring during aneurysm surgery: effects of blood pressure, csf drainage and temporary clipping on infarction. J Neurosurg 96: 1013–1019
5. Kiening KL, Unterberg AW, Bardt TF, Schneider G-H, Lanksch WR (1996) Monitoring of cerebral oxygenation in patients with severe head injuries: brain tissue PO_2 versus jugular vein oxygen saturation. J Neurosurg 85: 7–13
6. Kirkpatrick PJ, Smielewski P, Piechnik S, Pickard JD, Czosnyka M (1996) Early effects of mannitol in patients with head injuries assessed using bedside multimodality monitoring. Neurosurgery 39: 714–721
7. Persson L, Valtysson J, Enblad P, Wärme P-E, Cesarini K, Lewén A, Hillered L (1996) Neurochemical monitoring using intracerebral microdialysis in patients with subarachnoid hemorrhage. J Neurosurg 74: 606–616
8. Qureshi AI, Wilson DA, Traystman RJ (2002) Treatment of transtentorial herniation unresponsive to hyperventilation using hypertonic saline in dogs: effect on cerebral blood flow and metabolism. J Neurosug Anaesthesiol 1: 22–30
9. Ross N, Hutchinson PJ, Seeley H, Kirkpatrick PJ (2002) Timing of surgery for supratentorial aneurysmal subarachnoid haemorrhage: report of a prospective study. J Neurol Neurosurg Psychiatry 72: 480–484
10. Sarrafzadeh AS, Kiening KL, Callsen T-A, Unterberg AW (2003) Metabolic changes during impending and manifest cerebral hypoxia in traumatic brain injury. Br J Neurosurg 17: 340–346
11. Sarrafzadeh AS, Sakowitz OW, Kiening KL, Benndorf G, Lanksch WR, Unterberg AW (2002) Bedside microdialysis: a tool to monitor cerebral metabolism in subarachnoid haemorrhage patients? Crit Care Med 30: 1070
12. Sarrafzadeh AS, Haux D, Sakowitz O, Benndorf G, Herzog H, Kuechler I, Unterberg A (2003) Acute focal neurological deficits in aneurysmal subarachnoid haemorrhage. Relation of clinical course, CT findings, and metabolite abnormalities monitored with bedside microdialysis. Stroke 34: 1382–1388
13. Schultz MK, Wang LP, Tange M, *et al* (2000) Cerebral microdialysis monitoring: determination of normal and ischaemic cerebral metabolisms in patients with subarachnoid haemorrhage. J Neurosurg 93: 808–814
14. Schwarz S, Georgiadis D, Aschoff A, Schwab S (2002) Effects of hypertonic (10%) saline in patients with raised intracranial pressure after stroke. Stroke 33: 136–140
15. Shackford SR, Zhuang J, Schmoker J (1992) Intravenous fluid tonicity: effect on intracranial pressure, cerebral blood flow, and cerebral oxygen delivery in focal brain injury. J Neurosurg 76: 91–98
16. Suarez JI, Qureshi AI, Bhardwaj A, Williams MA, Schnitzer MS, Mirski M, Hanley DF, Ulatowski JA (1998) Treatment of refractory intracranial hypertension with 23.4% saline. Crit Care Med 26: 1118–1122
17. Tseng MY, Al-Rawi PG, Pickard JD, Rasulo FA, Kirkpatrick PJ (2003) Effect of hypertonic saline on cerebral blood flow in poor grade patients with subarachnoid haemorrhage. Stroke 34: 1389–1397
18. Yonas H, Sekhar L, Johnson DW, Gur D (1989) Determination of irreversible ischemia by xenon-enhanced computed tomographic monitoring of cerebral blood flow in patients with symptomatic vasospasm. Neurosurgery 24: 368–372

Correspondence: P. G. Al-Rawi, University Department of Neurosurgery, Box 167, Level 4, A-Block, Addenbrooke's Hospital, Hills Road, Cambridge. CB2 2QQ UK. e-mail: pga20@medschl.cam.ac.uk

Acta Neurochir (2005) [Suppl] 95: 129–132

Evidence for the importance of extracranial venous flow in patients with idiopathic intracranial hypertension (IIH)

N. Alperin[1], **S. H. Lee**[1], **M. Mazda**[1], **S. G. Hushek**[2], **B. Roitberg**[3], **J. Goddwin**[4], and **T. Lichtor**[5]

[1] Department of Radiology, University of Illinois at Chicago, Chicago IL, USA
[2] iMRI Department, Norton Hospital in Louisville, Louisville, KY, USA
[3] Department of Neurosurgery, University of Illinois at Chicago, Chicago, IL, USA
[4] Eye and Ear Infirmary, University of Illinois at Chicago, Chicago, IL, USA
[5] Department of Neurosurgery, Rush-Presbyterian St. Luke's Medical Center, Cook County Hospital, Chicago, IL, USA

Summary

Idiopathic intracranial hypertension (IIH) is characterized by increased ICP without evidence for intracranial mass lesion. Although the pathogenesis remains unknown, some association was found with intracranial venous thrombosis. To our knowledge, the extracranial venous drainage was not systematically evaluated in these patients. This study compared extracranial cerebral venous outflow in eight IIH patients and eight control subjects using magnetic resonance (MR) Venography and flow measurements. In addition, the study identified extracranial factors that affect cerebral venous drainage.

In six of the IIH patients, either complete or partial functional obstruction of the internal jugular veins (IJVs) coupled with increased venous outflow through secondary venous channels was documented. On average, a four-fold increase in mean venous flow rate through the epidural and/or vertebral veins was measured in IIH patients compared with the healthy subjects.

In one of the healthy subjects, intracranial venous outflow was studied also during external compression of the IJVs. Over 40% of the venous outflow through the IJVs shifted to the epidural veins and intracranial pressure, measured noninvasively by MRI, increased from 7.5 to 13 mmHg. Findings from this study suggest that increased ICP in some IIH patients could be associated with increased extracranial resistance to cerebral venous outflow.

Keywords: Idiopathic intracranial hypertension (IIH); extracranial cerebral venous drainage; MR venography; vertebral venous plexus; effect of body posture on cerebral venous outflow.

Introduction

Idiopathic intracranial hypertension (IIH), also known as pseudotumor cerebri, is a disorder characterized by symptoms of increased intracranial pressure (ICP) without evidence of a mass lesion or ventricular obstruction. A lumbar puncture in IIH patients often documents a markedly elevated ICP with normal cerebrospinal fluid (CSF) chemistry and cell count. Although the pathogenesis remains unknown, associations were found with obesity in females, hypercoagulability states, thrombosis of the venous sinuses and other conditions. In one study, 26% of patients with IIH had radiographic evidence of dural sinus thrombosis [6]. However in another study, magnetic resonance venography (MRV) showed no evidence of venous sinus thrombosis in a group of overweight young female patients with typical pseudotumor cerebri [5]. It has also been reported that thrombosis of a transverse or sigmoid sinus secondary to a skull base or posterior fossa surgical procedure may result in IIH even if the sinus is non-dominant [4]. Other disorders postulated to be associated with IIH include various endocrine disorders such as Addison's disease or hypoparathyroidism, hypervitaminosis A, and systemic lupus erythematosus [8].

The treatment options for patients with IIH include carbonic anhydrase inhibitors known to reduce CSF production or loop diuretics. Although some patients with IIH present with venous sinus thrombosis, treatment directed to this particular problem is rarely employed; anticoagulation has been used on occasion. Periodic lumbar punctures to remove spinal fluid may also provide some relief of symptoms. Patients who develop progressive visual loss or experience persistent severe headache secondary to the elevated intracranial pressure despite maximal medical therapy may require surgical intervention. The most common procedures for this disorder include optic nerve sheath fenestration or placement of a lumbar-peritoneal shunt [8].

Venous pressure is known to influence intracranial

pressure in several ways. The absorption of CSF depends on the pressure difference between the subarachnoid space and the venous pressure in the superior sagittal sinus. Therefore, increased venous pressure in the sinus will increase ICP through reduced CSF absorption [3]. In addition, elevated venous pressure, either intra- or extracranial, would increase intracranial pressure through increased resistance to venous outflow, which causes venous congestion and increased ICP. Therefore, cerebral venous outflow may play an important role in pathophysiology of IHH. While obstruction to venous outflow inside the cranium has been postulated as one of the causes for IIH, to the best of our knowledge, the extracranial venous flow has not been systematically evaluated in these patients. This study aims to image and characterize the venous drainage through the extracranial pathways in IIH patients and in healthy control subjects. Extracranial cerebral venous outflow in patients and in the control subjects was studied under normal conditions in a supine posture. In addition, extracranial cerebral venous outflow was imaged in healthy subjects during external compression of the internal jugular veins (IJVs) and in supine and upright postures.

Material and methods

Eight IIH patients (mean age of 23 years, 6 females, and 2 males) and 8 healthy subjects (mean age of 24 years, 8 males) were scanned with either a 1.5T or 3T MRI scanner (GE Medical, Milwaukee, WI) to assess their extracranial venous drainage. Two subjects were scanned twice, one of the 8 patients was studied before and after treatment with carbonic anhydrase inhibitors, and one of the healthy subjects was studied at rest and during external compression of the IJVs. In addition, 4 of the healthy subjects were also scanned in upright and supine postures using a Signa SP/*i* 0.5T vertical gap open MRI scanner (GE Medical, Milwaukee) to evaluate the effect of posture on venous drainage. Imaging protocols were approved by the respective institutional review boards and an informed consent was obtained.

The extracranial venous vasculature was imaged using 2D time-of-flight (TOF) technique and 3D models were reconstructed using the maximum intensity projection (MIP) technique. The scanning parameters for the TOF technique were: TR = 23 ms, FA = 50 deg, FOV = 14–15 cm and slice thickness = 1.5–1.8 mm. Mean venous flow through the internal jugular veins and through the epidural, vertebral, and deep cerebral veins, when present, were quantified using velocity encoded MRI technique. Two retrospectively gated cine phase-contrast scans were used to measure venous outflow. A scan with high velocity encoding (VENC = 70 cm/s) was used to quantify flow in the IJVs. The scan parameters were TR = 18 ms, FA = 25 deg, FOV = 16 cm, NEX of 2, and slice thickness = 6 mm. A second scan with low velocity encoding (VENC = 7–9 cm/s) was used to quantify the venous flow in the vertebral plexus. Parameters of the lower VENC scan were similar except for FA = 25 degrees. Imaging planes were selected to be perpendicular to the direction of the flow. The high VENC scan was located above the carotid bifurcation and the low VENC scan was located at the level of C2. These two velocity encoded scans were also used to determine an MRI-derived ICP value (MR-ICP) using a previously published method [1]. This method utilizes the small changes in intracranial volume and pressure that occur with each cardiac cycle. The pressure change during the cardiac cycle was derived from the CSF velocities and the intracranial volume change was derived from the differences between arterial blood inflow, venous blood outflow, and CSF volumetric flow rates into and out from the cranial vault. A mean ICP value was then derived from the ratio of the measured pressure and volume changes based on a linear relationship that exists between intracranial elastance and ICP [7].

Results

Visual inspection of the 3D models of the extracranial venous vasculature revealed either bilateral functional occlusion or partial narrowing of the internal jugular veins with dense secondary extracranial venous networks (e.g., epidural, vertebral and deep cervical veins) in six of the eight IIH patients. One of the 2 IIH patients with normal IJV had a narrowing of the superior sagittal sinus seen on intracranial MRV, probably due to cerebral venous thrombosis. An extracranial MIP projection in the coronal plane of three MR venograms from a control subject (top) and from 2 different IIH patients (middle and bottom) are shown in Figure 1. The middle image is from an IIH patient with a narrowing in the dominant right jugular vein and the bottom image is from an IIH patient who had no flow through the IJV. Both internal jugular veins were functionally occluded. Increased flow in secondary venous channels such as epidural veins (arrow heads) and other vertebral plexus veins was clearly demonstrated. The mean venous flow through these secondary channels showed large inter-individual variability. However, on average, a four-fold increase in mean venous flow in these channels (60 mL/min vs. 15 mL/min) was measured in the IIH patients compared with the healthy control subjects.

The IIH patient with partial obstruction in the right IJV shown in Figure 1 (middle) was studied a second time a year later following treatment. In the follow up scan, the narrowing in the right IJV was less severe and a smaller MR-ICP value was measured compared to the initial study (24 mmHg vs. 29 mmHg). Results from the external manipulation in a healthy subject studied in the supine position are shown in Figure 2. Coronal MIP MRV images acquired during normal state (left) and during external compression of the IJV (right) are shown. Mean venous outflow rates through the IJV and the epidural veins during the normal state

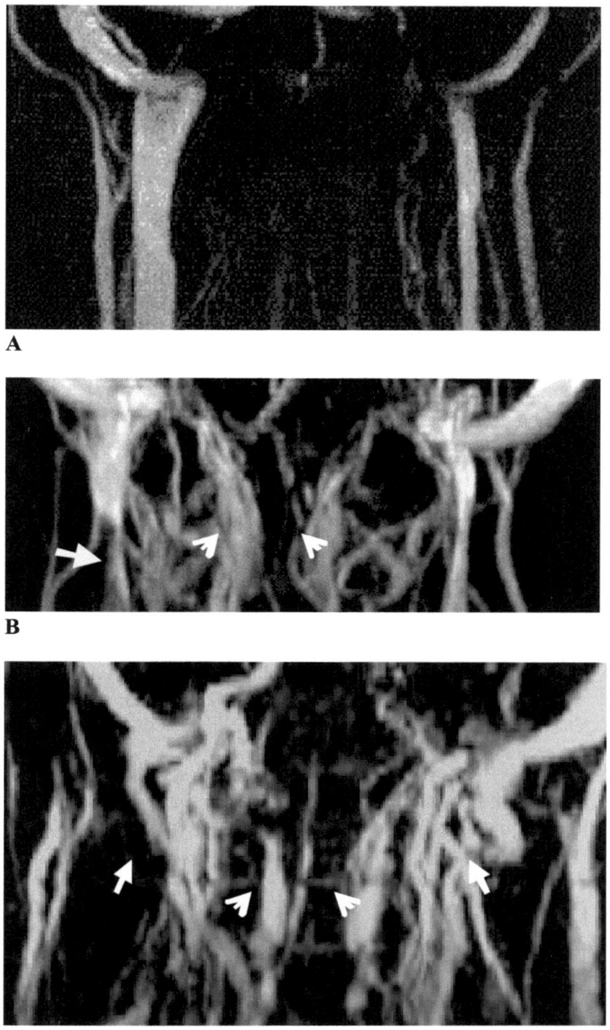

A

B

C

Fig. 1. Coronal MRI MIP images of the neck veins in a healthy subject (A), in an IIH patient with a partial obstruction of the flow in the right IJV (B), and in an IIH patient with no flow through the IJVs (C). The white arrows point to the location of the IJVs and the arrow heads identify the epidural veins

were 495 and 10 mL/min, respectively. The derived MR-ICP value was 7.5 mmHg. Mean venous outflow rates through the IJV and the epidural veins during the external compression state were 269 and 212 mL/min, respectively. The derived MR-ICP value was 13 mmHg. About 40% of the venous drainage through the IJV in the normal state shifted to the epidural veins during the external compression.

Results from imaging of the healthy control subjects demonstrated that in the supine posture, the dominant venous outflow is through the IJV, while in the upright posture, the dominant venous outflow occurs through secondary venous pathways with either complete (in three subjects) or partial (in one subject) collapse of the IJV. Coronal MIP MRV projections from one of the four subjects studied in the supine and upright postures are shown in Figure 3. The average MRI-derived ICP values for the four healthy subjects in the supine and the upright postures were 11.9 mmHg and 3.9 mmHg, respectively.

Discussion

In this report we compared cerebral venous outflow in the neck veins, imaged by MR venography, in a small group of IIH patients and in healthy subjects. We found that cerebral venous drainage is highly variable among individuals. Nevertheless, even in this small series of subjects clear differences are seen in the extracranial venous drainage pathways between patient and control groups. In 6 of the 8 patients studied, considerable amount of venous flow was seen through the venous plexus with either no flow or significantly reduced flow in IJV. In one of the two IIH patients with apparent normal extracranial venous drainage, narrowing was seen in the superior sagittal sinus. These findings further emphasize the already known

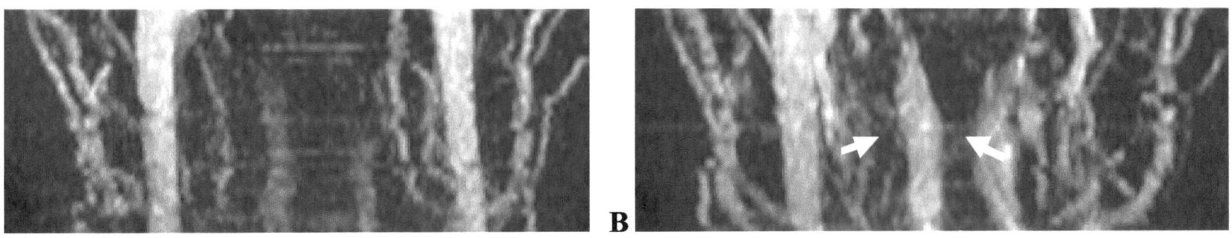

A B

Fig. 2. Coronal MRI MIP images of the neck veins in a healthy subject scanned in the supine posture. (A) Normal cerebral venous outflow. Venous outflow is mainly through the IJV. (B) Cerebral venous outflow during the application of external compression on the neck. Considerable amount of flow is now seen in secondary venous channels such as the epidural veins (arrows) and no flow is seen through the left IJV

A 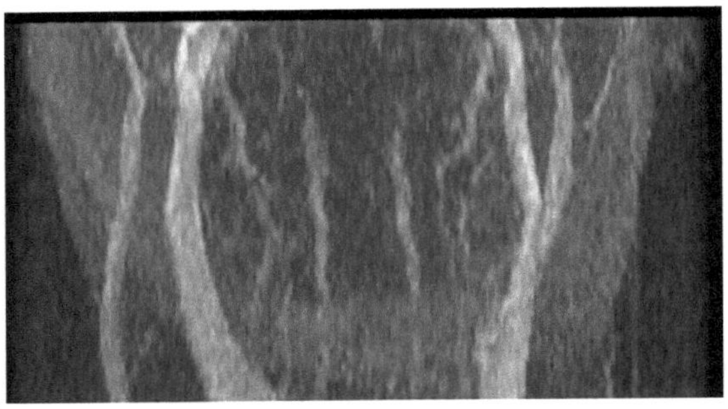 B 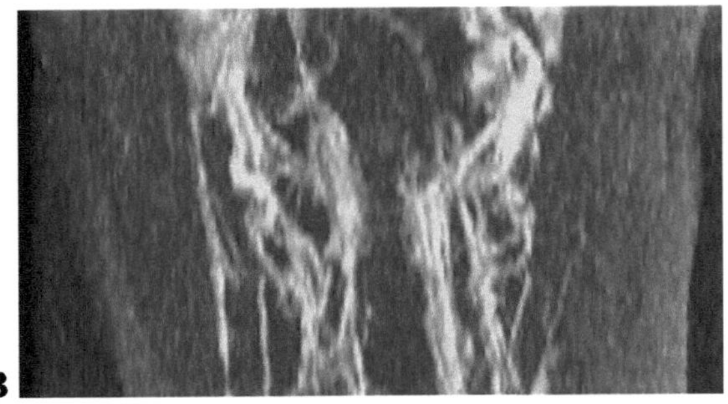

Fig. 3. Coronal MIP images of the neck veins in a healthy subject scanned in the supine posture (A) and in the upright posture (B). (A) In the supine posture, cerebral venous outflow is seen mainly through the IJVs. (B) In the upright posture, venous outflow is seen mainly through the vertebral plexus, i.e., epidural and vertebral veins

venous involvement in the pathogenesis of IIH. While previously, only intracranial involvement was reported, this study provides evidence for the involvement of extracranial vasculature.

The study further examines factors that influence extracranial venous flow. Shift in venous outflow from the IJV to the vertebral venous plexus was demonstrated in two very different circumstances. The study of a healthy subject in supine posture demonstrated that under external compression of the IJV, significant portion of the venous outflow through the IJV (over 40%) was shifted to the epidural veins. The restriction of venous outflow in the IJVs was associated with increased ICP, probably due to cerebral venous congestion. Further studies are needed to examine if this mechanism, i.e., increased resistance to cerebral venous outflow is the cause of elevated ICP in IIH.

Dominant venous outflow through the vertebral venous plexus can occur at a low ICP state as well, as seen in healthy subjects studied in the upright posture. In this case, the IJVs are fully or partially collapsed. Since there is also evidence that Positive Pressure Breathing (PPB) reverses the collapse of the IJV [2], the collapse of the IJV in the sitting position is probably associated with low venous pressure and not due to increased resistance to venous outflow. The functionally obstructed IJV in IIH patients occurs in the supine posture. Therefore, in some IIH patients, chronically elevated ICP is most likely a consequence of a prolonged increased extracranial resistance to venous outflow.

References

1. Alperin NJ, Lee SH, Loth F, Raksin PB, Lichtor T (2000) MR-Intracranial Pressure (ICP): a method to measure intracranial elastance and pressure noninvasively by means of MR Imaging: baboon and human study. Radiology 217: 877–885
2. Cirovic S, Walsh C, Fraser WD, Gulino A (2003) The effect of posture and positive pressure breathing on the hemodynamics of the internal jugular vein. Aviation Space Environ Med 74(2): 125–131
3. D'Avella D, Greenberg RP, Mingrino S, Scanarini M, Pardatscher K (1980) Alteration in ventricular size and intracranial pressure caused by sagittal sinus pathology in man. J Neurosurg 53: 656–661
4. Keiper GL, Sherman JD, Tomsick TA, Tew JM (1999) Dural sinus thrombosis and pseudotumor cerebri: unexpected complications of suboccipital craniotomy and translabyrinthine craniectomy. J Neurosurg 91: 192–197
5. Lee AG, Brazis PW (2000) Magnetic resonance venography in idiopathic pseudotumor cerebri. J Neuroophthalmol 20: 12–13
6. Leker RR, Steiner I (1999) Features of dural sinus thrombosis simulating pseudotumor cerebri. Eur J Neurol 6: 601–604
7. Marmarou A, Shulman K, La Morgese J (1975) Compartmental analysis of compliance and outflow resistance of the CSF system. J Neurosurgery 43: 523–534
8. Schutta HS, Corbett JJ (2004) Intracranilal Hypertension Syndrom. In: Joynt RJ, Griggs RC (eds) Baker's clinical neurology. Lippincott Williams and Wilkins, Philadelphia

Correspondence: Noam Alperin, Physiological Imaging and Modeling Lab, Department of Radiology (M/C 711), The University of Illinois at Chicago, 830 S. Wood St., Chicago, IL 60612, USA. e-mail: alperin@uic.edu

Acta Neurochir (2005) [Suppl] 95: 133–136

Subdural intracranial pressure, cerebral perfusion pressure, and degree of cerebral swelling in supra- and infratentorial space-occupying lesions in children

M. Stilling, E. Karatasi, M. Rasmussen, A. Tankisi, N. Juul, and **G. E. Cold**

Department of Neuroanaesthesia, Århus University Hospital, Århus, Denmark

Summary

To our knowledge comparative studies of intracranial pressure (ICP) and degree of cerebral swelling during craniotomy for supratentorial or infratentorial space occupying lesion in children are not available. In this prospective study subdural ICP, cerebral perfusion pressure (CPP), dural tension, and the degree of cerebral swelling were analysed in supine and prone positioned children subjected to craniotomy for space occupying lesions.

Material and method. 48 children with space occupying tumours were subjected to either isoflurane/nitrous oxide 50%/fentanyl ($n = 22$) or propofol/fentanyl/air/oxygen ($n = 26$). 25 children were operated supratentorially in supine position, while 23 patients were operated infratentorially in the prone position. Subdural ICP, mean arterial blood pressure (MABP), and CPP were measured just before opening of the dura. Dural tension was estimated before opening of dura, and the degree of cerebral swelling was estimated after opening of dura.

Results. The age and weight of children anaesthetised with isoflurane in the prone position were significantly lower than the propofol anaesthetised groups. No significant inter-group differences as regards tumour size, midline shift, rectal temperature, MABP or $PaCO_2$ were found. ICP in prone positioned children averaged 16.9 mm Hg against 9.0 mm Hg in supine positioned children ($p < 0.001$). In prone positioned children the dura was significantly tenser, and the degree of brain swelling after opening of dura was significantly more pronounced. No significant difference as regard ICP was disclosed when isoflurane/nitrous oxide/fentanyl and propofol/fentanyl anaesthetized children were compared, but MABP and CPP were significantly lower in isoflurane anaesthetised children.

Conclusion. In children with cerebral tumours ICP is higher, and the degree of cerebral swelling more pronounced in the prone-compared with supine positioned children.

Choice of anaesthesia did not influence ICP, but CPP was significantly lower during isoflurane anaesthesia.

Keywords: Intracranial pressure; cerebral perfusion pressure; children; craniotomy; brain tumour; supine position; prone position.

Introduction

Studies of intracranial pressure in adult supine positioned patients with supratentorial tumours [2], and prone positioned patients with infratentorial tumours [5] suggest a relationship between subdural intracranial pressure (ICP) on one hand and tension of dura and degree of cerebral swelling on the other hand. Moreover, the subdural ICP is higher in prone positioned patients compared with supine positioned patients. In adult propofol/fentanyl anaesthetised patients with cerebral tumours, ICP averages 21 mm Hg and 18.3 mm Hg in prone positioned with occipital and cerebellar tumours, respectively [13]. These values are considerably higher than ICP values found in adult propofol/fentanyl or isoflurane anaesthetized patients with supratentorial tumours operated in the supine position, where ICP averaged 7.5 mm Hg and 13.0 mm Hg, respectively [9]. To our knowledge, comparative studies of ICP and cerebral perfusion pressure (CPP) in supine positioned or prone positioned children are not available.

We hypothesize that ICP generally is higher and the degree of cerebral swelling is more pronounced in prone positioned children subjected to posterior fossa surgery, compared with supine positioned children subjected to craniotomy for space-occupying supratentorial lesions, and that ICP measured during craniotomy is lower in propofol/fentanyl anaesthetized compared with isoflurane/nitrous oxide anaesthetized patients. Therefore, the aims of the present study were as follows: 1) to compare ICP, CPP and degree of cerebral swelling in prone-positioned children subjected to posterior fossa surgery with supine positioned children subjected to supratentorial surgery, and 2) to compare ICP and CPP in children subjected to isoflurane/nitrous oxide with ICP and CPP in children subjected to supine positioned patients with supratentorial lesions.

Table 1. *Demographic data, neuroradiological data and data related to anesthesia*

	Isoflurane anaesthesia		Propofol anaesthesia	
	Supine	Prone	Supine	Prone
Demographic data	–	–	–	–
Number	14	8	11	15
Male/fem.	8/6	4/4	8/3	11/4
Age	6.1 ± 4.4	3.6 ± 2.1	9.3 ± 2.1*	8.3 ± 4.3*
Weight (kg)	24 ± 15	17 ± 4.5	34 ± 11*	29 ± 11
Steroid ($\pm$)	5/9	2/6	2/9	3/12
Neuroradiological data	–	–	–	–
Tumor area (cm^2)	10.4 ± 10	9.5 ± 4.7	10.4 ± 9.7	0.4 ± 8.5
Midline shift (mm)	0.7 ± 2.7	1.0 ± 1.9	1.4 ± 3.2	3.3 ± 6.4
Maintenance dose of anaesthesia	–	–	–	–
Propofol (mg/kg/h)	0	0	10.5 ± 3.8	12.7 ± 5.1
Fentanyl (µg/kg/h)	1.8 ± 1.3	1.2 ± 0.6	2.2 ± 1.1	2.3 ± 0.8
Isoflurane% (exp)	1.3 ± 0.5	1.2 ± 0.2	0	0

Mean $\pm$ SD are indicated, * difference from isoflurane, prone position ($P < 0.05$).

Material and methods

In this prospective study children scheduled for elective craniotomy for space-occupying cerebral lesions were included. Data were collected consecutively, and collected in a data bank. The data included relevant information about location and size of the cerebral lesion, demographic data, preoperative electrolytes and hemoglobin, and perioperative values of subdural ICP, CPP, and data obtained from arterial blood gas-analysis.

In our clinic we routinely use subdural ICP monitoring during craniotomy. The local ethic committee has approved this procedure.

Two groups of anesthesia were compared. Data in both groups were collected within the same time interval. Anaesthetic technique was chosen within the anaesthesiologist's discretion.

22 children with space occupying tumours were subjected to isoflurane/nitrous oxide 50%/fentanyl, 26 children were subjected to propofol/fentanyl anesthesia.

25 children were operated supratentorially in supine position, while 23 patients were operated infratentorially in the prone position.

All children were moderately hyperventilated, the aims being $PaCO_2$ 3.5–4.5 kPa, and $PaO_2 > 13$ kPa.

Subdural ICP, mean arterial pressure (MABP), and CPP were measured just before opening of the dura. The technique used for subdural ICP monitoring has been described recently [2]. The neurosurgeon estimated the tension of dura before opening of dura as normal tension, increased tension or pronounced increased tension. The degree of cerebral swelling was estimated after opening of dura as no swelling, swelling or pronounced swelling.

The data are reported as mean values and standard deviations. Inter-group differences were analysed with a one-way ANOVA model. Chi-square test was used for analyses of difference in distribution. A Sigma Stat version 2.0 was used for calculation. Post-hoc tests using Bonferroni corrections were performed. $P < 0.05$ was considered significant.

Results

The age and weight of children anaesthetised with isoflurane in the prone position were significantly lower than the propofol anaesthetised groups. No significant inter-group differences as regards tumour size, midline shift, histopathology, rectal temperature, MABP, PaO_2 and $PaCO_2$ were found (Table 1).

ICP in prone-positioned children averaged 16.9 mm Hg against 9.0 mm Hg in supine positioned children ($p < 0.001$). In prone-positioned children the tension of dura and the degree of cerebral swelling were significantly more pronounced (Table 2).

No significant difference as regard to ICP was disclosed when isoflurane and propofol/fentanyl anaesthetized children were compared, but MABP and CPP were significantly lower in isoflurane anaesthetised children. The tension of dura, and the degree of cerebral swelling were comparable in isoflurane- and propofol anaesthetized children (Table 3 and Table 4).

Discussion

In the present study we have demonstrated a significantly higher level of subdural ICP, and more pronounced cerebral swelling in prone positioned children compared with supine positioned children. The difference in subdural pressure was not related to the level of $PaCO_2$, MABP or tumour size.

The volume of the infratentorial compartment is considerably smaller than the supratentorial compartment. As a consequence space-occupying lesions in the posterior fossa should increase ICP more compared with supratentorial space-occupying lesions of the same size. To our knowledge only one study of ICP

Table 2. *Data related to cerebral circulation, tension of dura and degree of cerebral swelling*

Data related to cerebral circulation	Isoflurane	Propofol	Supine	Prone
$PaCO_2$ (kPa)	3.9 ± 0.6	4.2 ± 0.4	4.0 ± 0.5	4.1 ± 0.6
PaO_2 (kPa)	33 ± 12	36 ± 10	36 ± 14	36 ± 9
Temp. (°C)	36.8 ± 0.9	36.4 ± 0.7	36.3 ± 0	36.4 ± 0.7
ICP (mm Hg)	12.5 ± 9.1	13.0 ± 7.6	9.0 ± 6.9	16.9 ± 7.6*
MABP (mm Hg)	65 ± 9	72 ± 10*	67 ± 10	69 ± 11
CPP (mm Hg)	52 ± 10	59 ± 10*	59 ± 10	52 ± 11*
Tension of dura (number of patients are indicated)				
Normal	10	14	17	7
Increased	7	8	6	9
Pronounced increased	4	5	2	7
Degree of cerebral swelling (number of patients are indicated)				
No swelling	9	11	15	5
Swelling	6	10	6	10
Pronounced swelling	7	5	4	8

Anaesthesia and positioning are indicated. Mean ± SD are indicated, * indicates significant difference ($P < 0.05$).

Table 3. *Data related to cerebral circulation including $PaCO_2$, PaO_2, rectal temperature, intracranial pressure (ICP), mean arterial blood pressure (MABP) and cerebral perfusion pressure*

	Isoflurane anaesthesia		Propofol anaesthesia	
	Supine	Prone	Supine	Prone
$PaCO_2$ (kPa)	3.9 ± 0.6	3.9 ± 0.7	4.1 ± 0.4	4.2 ± 0.5
PaO_2 (kPa)	33 ± 12	33 ± 12	39 ± 17	39 ± 9
Temp. (°C)	36.7 ± 0.9	36.9 ± 1.1	36.1 ± 0.9	36.0 ± 0.6
ICP (mm Hg)	9.5 ± 6.8	17.6 ± 10.8*#	8.3 ± 7.3	16.5 ± 5.7*
MABP (mm Hg)	64 ± 10	66 ± 9	72 ± 10	71 ± 11
CPP (mm Hg)	55 ± 9	48 ± 10	63 ± 8	55 ± 11*

Positioning and anesthesia are indicated. The values are mean ± SD, * difference from propofol, supine position, # difference from isoflurane, supine position.

and CPP during infra- and supratentorial craniotomy is available in adults [13]. One reason being that supratentorial craniotomy is performed in the supine position, while infratentorial surgery is performed in the prone position. Thus, not only tumour localization but also position varies, making any conclusion concerning difference in ICP level difficult. In the prone positioned patient subjected to infratentorial surgery, the head is rotated downwards in order to provide acceptable surgical access. As a consequence the pressure in the cerebral venous system is increased, the reason being the hydrostatic difference between the central veins and the cerebral veins. Moreover, in adult patients the prone position results in an increase in abdominal pressure [4], and in experimental studies it has been demonstrated that intra-abdominal pressure is related to intracranial pressure during abdominal insufflations manoeuvre [3, 10]. A higher ICP in prone position compared with supine position has been demonstrated in patients with severe head injury [7].

In adult propofol/fentanyl anaesthetised patients with cerebral tumours, ICP averages 21 mm Hg and

Table 4. *Tension of dura and degree of cerebral swelling in children operated in prone or supine position with either propofol/fentanyl or Isoflurane/nitrous oxide/fentanyl*

	Isoflurane anaesthesia		Propofol anaesthesia	
	Supine	Prone	Supine	Prone
Tension of dura				
Normal	8	2	9	5
Increased	4	3	2	6
Pronounced increased	2	3	0	4
Degree of cerebral swelling				
No swelling	7	2	8	3
Swelling	4	2	2	7
Pronounced swelling	2	4	1	4

Number of patients are indicated.

18.3 mm Hg in prone positioned with occipital and cerebellar tumours, respectively [13]. These values are considerably higher than ICP values found in adult propofol/fentanyl or isoflurane anaesthetized adult patients with supratentorial tumours operated in the supine position, where ICP averaged 7.5 mm Hg and 13.0 mm Hg, respectively [9]. The average jugular bulb pressures in the two studies were 12.1 mm Hg and 14.3 mm Hg for prone positioned infratentorial and occipital tumours against 7.0 mm Hg in supine positioned patients. The differences in histopathology, tumour size, midline shift, $PaCO_2$ may contribute to the discrepancy in the obtained ICP values. Nevertheless, the differences in ICP and jugular bulb pressure in patients with occipital tumours operated in prone position compared with the values observed in supratentorial tumours operated in supine position are striking, suggesting that position more than tumour size elicit the higher ICP observed in prone positioned patients.

In adult patients with supratentorial cerebral tumours subdural ICP is significantly higher with isoflurane/fentanyl compared with propofol/fentanyl anaesthesia [9]. This finding is supported by studies in rabbits with brain tumour, indicating that CBF as well as CBV are significantly greater during isoflurane than with propofol [1]. To our knowledge comparative studies of ICP in isoflurane/fentanyl-compared with propofol/fentanyl anesthetized patients are not available in children. A dose related increase in ICP, however, has been demonstrated in children when isoflurane administration was changed from 0.5 to 1.0 MAC. In the same study ICP averaged 6 and 7 mm Hg with isoflurane 0.5 and 1.0 MAC, respectively [12]. In contrast, increasing propofol infusion rates stabilizes flow velocity on a low level in children [6]. Nitrous oxide also contributes to changes in ICP. Thus, addition of nitrous oxide to inhalation anaesthetics increases flow velocity in children [12]. Other differences include preserved CO_2 reactivity in children subjected to isoflurane anaesthesia [11], while the CO_2 reactivity is decreased during hypocapnic propofol anaesthesia [8]. In the present study we found that mean values of ICP was lower during propofol/fentanyl compared with isoflurane/nitrous oxide/fentanyl anaesthesia, but the difference was not significant. Taking the obtained values of difference in mean ICP (1 mm Hg) and the standard deviation (7 mm Hg) the sample size in a controlled trial should be around 800 to obtain a P value < 0.05 and a power of 0.8. We do not think such a study is realistic, partly because the dimension of such study is overwhelming, and partly because the importance of establishing evidence of such a little difference in ICP is without clinical relevance.

References

1. Cenic A, Craen RA, Lee TY, Gelb AW (2002) Cerebral blood volume and blood flow responses to hyperventilation in brain tumors during isoflurane or propofol anesthesia. Anesth Analg 94: 661–666
2. Cold GE, Tange M, Jensen TM, Ottesen S (1996) "Subdural" pressure measurement during craniotomy: correlation with tactile estimation of dural tension and brain herniation after opening of dura. Br J Neurosurg 10: 69–75
3. Halverson A, Buchanan R, Jacobs L, Shayani V, Hunt T, Riedel C, Sackier J (1998) Evaluation of mechanism of increased intracranial pressure with insufflation. Surg Endsc 12: 266–269
4. Hering R, Wrigge H, Vorwerk R, Brensing KA, Schöder S, Zinserling J, Hoeft A, Spiegel TV, Putnesen C (2001) The effect of prone positioning on intra abdominal pressure and cardiovascular and renal function in patients with acute lung injury. Anesth Analg 92: 1226–1231
5. Jørgensen HA, Bundgaard H, Cold GE (1999) Subdural pressure measurement during posterior fossa surgery. Correlation studies of brain swelling/herniation after dural incision with measurement of subdural pressure and tactile estimation of dural tension. Br J Neurosurg 13: 449–453
6. Karsli C, Luginbuehl I, Farrar M, Bissonnette B (2002) Propofol decreases cerebral blood flow velocity in anesthetized children. Can J Anaesth 49: 830–834
7. Lee ST (1989) Intracranial pressure changes during positioning of patients with severe head injury. Heart & Lung 18: 411–414
8. Leon JE, Bissonnette B (1991) Cerebrovascular responses to carbon dioxide in children anaesthetized with halothane and isoflurane. Can J Anaesth 38: 805–808
9. Petersen KD, Landsfeldt U, Cold GE, Petersen CB, Mau S, Hauerberg J, Holst P, Olsen KS (2003) Intracranial pressure and cerebral hemodynamic in patients with cerebral tumors. Anesthesiology 98: 329–336
10. Rosenthal RJ, Friedman RL, Chidambaram A, Khan AM, Martz J, Shi Q, Nussbaum M (1998) Effects of hyperventilation and hypocentilation on PaCO2 and intracranial pressure during acute elevation of intra abdominal pressure with CO2 pneumoperitoneum: large animal observations. J Am Coll Surg 187: 32–38
11. Rowney DA, Fairgrieve R, Bissonnette B (2004) The effect of nitrous oxide on cerebral blood velocity in children anaesthetized with sevoflurane. Anaesthesia 59: 10–14
12. Sponheim S, Skraastad O, Helseth E, Due-Tonnesen B, Aamodt GBH (2003) Effects of 0.5 and 1.0 MAC isoflurane, sevoflurane and desflurane on intracranial and cerebral perfusion pressures in children. Acta Anaesthesiol Scand 47: 932–938
13. Tankisi A, Larsen JR, Rasmussen M, Dahl B, Cold GE (2002) The effects of 10° reverse Trendelenburg position on ICP and CPP in prone positioned patients subjected to craniotomy for occipital or cerebellar tumours. Acta Neurochir (Wien) 144: 665–670

Correspondence: Georg Emil Cold, Department of Neuroanesthesia, Århus University Hospital, DK-8000 Århus C, Denmark. e-mail: gcold@akh.aaa.dk

Acta Neurochir (2005) [Suppl] 95: 137–140

The role of noninvasive monitoring of cerebral electrical impedance in stroke

L. X. Liu[1], W. W. Dong[1], J. Wang[2], Q. Wu[3], W. He[3], and Y. J. Jia[1]

[1] Department of Neurology, The First Affiliated Hospital, Chongqing University of Medical Sciences, Chongqing, China
[2] Department of Neurology, The Second Affiliated Hospital, Chongqing University of Medical Sciences, Chongqing, China
[3] College of Electrical Engineering, Chongqing University, Chongqing, China

Summary

Objective. To explore the change regularity of cerebral electrical impedance (CEI) in the healthy people and patients with intracerebral hemorrhage (ICH) and ischemic stroke.

Methods. CEI of 100 healthy volunteers, 52 patients with ICH and 33 patients with ischemic stroke was measured by noninvasive Brain-Edema Monitor. The results of perturbative index (PI) converted from CEI were compared with the volume of infarction, hematoma and surrounding edema, which calculated by image analyzing system according to MRI or CT.

Results. In the normal groups, PI in the left and right sides of cerebral hemispheres was respectively 7.76 ± 0.75 and 7.79 ± 0.58, and there was no significant difference between the two sides ($P > 0.05$). In the patients with ICH, PI in the hematoma side decreased and was lower than the other side, and then increased gradually, finally exceeded that of the other side. The average "cross" time was (16.25 ± 8.96) h. It showed that the volume of hematoma was no obvious change before and after the "cross" time [(31.25 ± 21.59) vs (37.59 ± 27.57)] ($P > 0.05$). However, the volume of peri-hematoma edema was significantly larger after the "cross" time than before the "cross" time [(26.35 ± 13.96) vs (14.68 ± 5.30)] ($P < 0.05$). There was a positive correlation between the PI of hematoma side and the volume of peri-hematoma edema ($r = 0.8811$, $P < 0.01$). In the patients with arterothrombotic cerebral infarction, PI in the infarct side had a positive correlation with the volume of infarction ($r = 0.8496$, $P < 0.01$).

Conclusions. CEI is a stable physical parameter reflecting the electrical character of human brain tissue. It is useful for monitoring edema and hematoma in stroke.

Keywords: Brain edema; electrical impedance; hematoma; monitoring; stroke.

Introduction

Ischemic stroke is the most common diseaswe affecting the neurological function. And intracerebral hemorrhage (ICH) is the most serious form of all acute strokes. The mortality in spontaneous ICH is high, and the mechanism of death is usually related to cerebral herniation, which results from a combination of the mass produced by the hemorrhage itself, and by the mass effect created by surrounding edema [1]. The mass effect of ischemic stroke is also associated with the volume and position of infarct. Therefore, monitoring brain edema or hematoma is important. As it is well-known, CT and MRI can accurately delineate the region of brain edema or hemorrhage, but it is difficult to repeat them frequently. Moreover, these could not be performed on the bedside. Could we find out a method which can monitor edema and hematoma continually, non-invasively and cost-effectively? In the previous study, we have found that edema and hemorrhage can lead to the change of cerebral electrical impedance (CEI). That is, edema can increase the CEI [5, 9, 10, 12, 13], while hemorrhage decrease it [5, 12, 13]. According to this principle, measurement of CEI is probably a new method in detecting and monitoring the evolution of cerebral edema and hemorrhage on the patients at bedside. Therefore, the purpose of our study is to explore the change of CEI in the normal people and patients with intracerebral hemorrhage (ICH) and ischemic stroke.

Materials and methods

Patients

In a prospective study from July 2003 to May 2004, we studied 52 patients with ICH and 33 patients with ischemic stroke on the wards of our department. All of the ICH patients were confirmed at the first day after onset by a CT scan which showed the volume of hematoma was between 7.69∼176.50 ml and the position of hematoma was respectively basal ganglia (39 cases), cerebral lobe (4 cases), cerebral ventricle (4 cases), brain stem (4 cases) and cerebella (1 case). The volume of infarction in 33 ischemic stroke was among 2.73∼210.30 ml according to MRI operated at the first or second day after admission. There were also 100 healthy volunteers allowed to the study in order to compare with the patients.

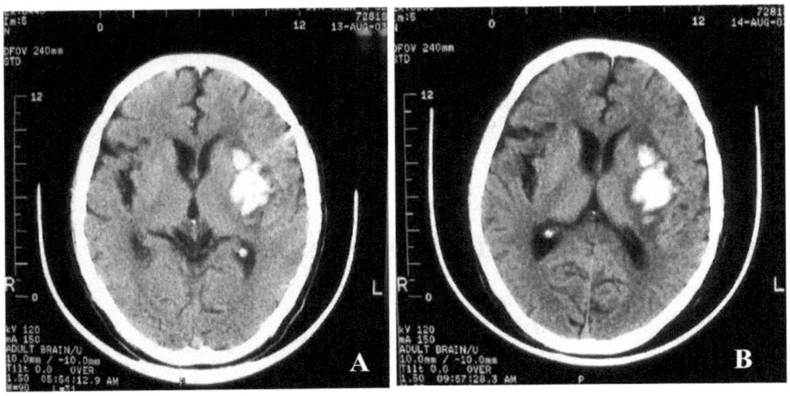

Fig. 1. Relationship between the PI and the volume of focal lesions in a patient with left ICH: (A) 4 h after onset, volume of hematoma was 18.26 ml, volume of peri-hematoma was 12.31 ml; PI in the left was 6.77, PI in the right was 7.07; (B) 30 h after onset, volume of hematoma was 20.80 ml, volume of peri-hematoma was 27.66 ml; PI in the left was 7.38, PI in the right was 7.09

Methods

We developed and used a kind of noninvasive Brain-Edema monitor (Born Science & Technology Corporation, China). A constant current was given into a person's brain, and the CEI of the two hemispheres was measured respectively. The measurement were then converted into perturbative index (PI). The position of three electrodes was: the left frontal, the right frontal, and the middle of occipital.

The volume of infarction, hematoma and surrounding edema was calculated automatically by image analyzing system according to MRI or CT scan.

Results

The normal group

In the normal group, PI in the left and right hemisphere was respectively 7.76 ± 0.75 and 7.79 ± 0.58, and there was no significant difference between the two sides ($P > 0.05$). Moreover, no person had any complaint of the device during 3~6 hours monitoring.

The ICH group

In the ICH group, the range of PI in all patients was 5.36~12.50. Thirty-three of 52 patients were monitored continuously within 24 h after onset. The PI of hematoma side decreased firstly, which was lower than the other side, and then increased gradually, finally exceeded that of the other side. The average "cross" time was (16.25 ± 8.96) hours after onset. We calculated the volume of hematoma and surrounding edema according to the results of CT scan. It showed that the volume of hematoma was no obvious change before and after the "cross" time [(31.25 ± 21.59) vs (37.59 ± 27.57)] ($n = 33$, $P > 0.05$). On the other hand, the volume of peri-hematoma edema became to increase and was significantly larger just after the "cross" time than before the "cross" time [(26.35 ± 13.96) vs (14.68 ± 5.30)] ($n = 33$, $P < 0.05$).

A typical example was interpretative in Fig. 1. It showed the change of PI in a patient with left ICH. PI of the left (PI = 6.77) was lower than that of the right (PI = 7.07) at 4 hours after onset, and the volume of peri-hematoma was 12.31 ml. It then increased and exceeded that of the right side at 16 hours after onset (PI in the left = 7.18, PI in the right = 7.12). At 30 hours after onset, PI in the left had increased to 7.38 and the volume of peri-hematoma had also increased to 27.66 ml. The situation persisted throughout the next 6 days. Above results suggested that it was the hematoma possessing predominant before the "cross" time, and the surrounding edema possessed predominant after the "cross" time.

However, there were 3 of 52 patients whose PI didn't increase but decreased during the continuous monitoring. We can see the situation in Fig. 2. It showed another patient with left ICH. Mannitol was injected immediately 2 h after onset, and repeated 2 h later. The patient's condition was deteriorating. CT scan showed one day later the volume of hematoma became much larger than before.

Moreover, there was a positive correlation between the PI of hematoma side and the volume of peri-hematoma edema in the patients who were all detected within 7 days after onset ($n = 52$, $r = 0.8811$, $P < 0.01$).

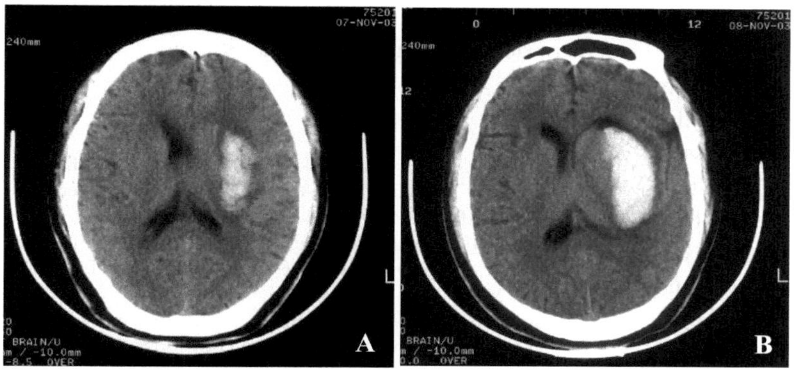

Fig. 2. Hematoma enlargement in the earlier period of ICH in a patient: (A) 1 h after onset, volume of hematoma was 33.72 ml, volume of peri-hematoma was 12.68 ml; PI in the left was 6.13, PI in the right was 6.31; (B) 28 h after onset, volume of hematoma was 55.53 ml, volume of peri-hematoma was 12.87 ml; PI in the left was 6.08, PI in the right was 6.38

Group of ischemic stroke

In the 33 patients with arterothrombotic cerebral infarction (CI), the range of PI in all patients was 6.43~12.96. Twenty-seven patients (81.8%) had a higher numeric data of PI in the infarct side than the other side, while the other 6 patients (18.2%) had no apparent difference. Two patients among all the 6 were multiple lacunar infarction, one patient's infarct position located in brain stem, and 3 patients' volume of infarction was less than 15 ml. The PI of infarct side in 27 patients increased to the highest level 3~5d after onset, and gradually returned to decrease in 7~14d. The larger of the volume of infarct, the higher of the PI, and there was a positive relationship between the two values ($r = 0.8496$, $P < 0.01$). (See Fig. 3.)

Discussion

Electrical impedance (EI) is a kind of physical parameter which can reflect the change of biologic tissue,

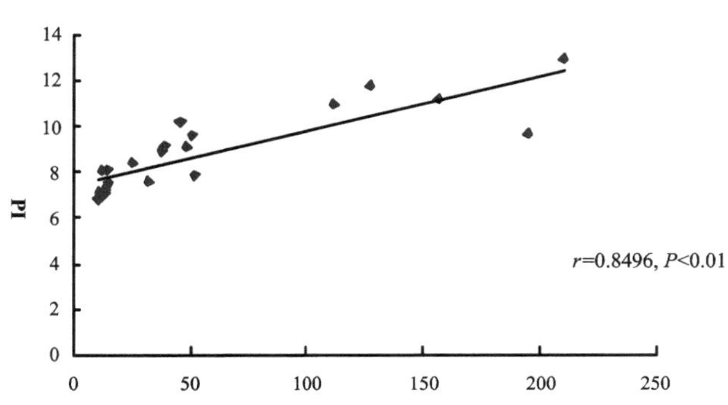

Fig. 3. Correlation between the PI and the volume of infarction in patients with CI

organ, and the whole organ. The principle of EI is mainly based on the low frequency bioelectrical impedance analysis [2, 3]. There are also some scholars who set up a disturbance method of electric current field [5, 13]. Its principle is that non-normal matter (e.g. hematoma or edema) which exists in the brain causes a disturbance to boundary potentials when a low frequency current flows through brain tissue. In terms of the measurement for the change of boundary potentials a reconstruction for the blood effusion or edema in the deep position of the brain can be done [5]. The method has begun to be applied to detecting the state of body activity and function of human [4, 11, 14], for it is noninvasive, safe, and can detect the EI on the bedside. Nowadays, however, not only has the EI research on brain tissue done mostly on the animal experiments, but also the experiments are invasive [6–10]. We have been studying the noninvasive CEI detecting system serving for animals since 1998 and have been successful. We then developed a new CEI detecting system serving for human being, and have found that the numeric values of CEI were symmetric in the two hemispheres of brain in different frequency (20 kHz, 50 kHz and 100 kHz). Age, sex and the time of monitoring have no obvious effect on the results of CEI [12]. It is again demonstrated in this study on the CEI monitoring of healthy volunteers and patients that the device has a high safety, for no person has any complaint of the device during 3~6 hours monitoring.

We can see that the PI of hematoma side in the ICH group was lower than the other side, and then increased and exceeded that of the other side. The average "cross" time was (16.25 ± 8.96) hours after onset. The volume of peri-hematoma edema was significantly larger just after the "cross" time than before the "cross" time. Above results suggests that it was the hematoma possessing predominant before the "cross" time, and the surrounding edema possessed predominant after the "cross" time. Therefore, when the PI of hematoma side is lower than that of the other side, it may be not necessary to treat with hyperosmotic solution (e.g. mannitol) unless there is obvious evidence of intracranial pressure. In this study, three ICH patients had larger hematoma compared with before in the earlier period of diseases, the main reason may be the inappropriate use of mannitol. Hence, the CEI monitoring is useful to direct the osmotherapy. In the CI group, there was a positive relationship between the PI and the volume of infarction. The larger of the volume of infarct, the higher of the PI, and the worse the outcome. CEI monitoring can reflect the change in the disease process and prognosis of the patients with cerebral infarct.

However, we have also paid attention to the problem that the CEI detecting system sometimes didn't detect the positive results in some patients who had been indeed suffered from ICH or CI confirmed by CT or MRI (false-negative). The probable reason is that thehe monitoring system is not sensitive if the focal lesions are located in the midline or near the midline (e.g. brain stem). It is also not sensitive if the volume of lesions is less than 15~20 ml. The results are often false-negative in patients with multiple lacunar infarction. Above are the limitations of the system, which needs to be improved in the future.

In summary, CEI is a stable physical parameter reflecting the electrical character of human brain tissue. It is useful for monitoring edema and hematoma in stroke. The higher the PI, the more severe of edema. Hyperosmotic solution (e.g. mannitol) should be used carefully when the PI of hematoma side is lower than that of the opposite side. The monitoring could be a good complement to CT and MRI scan of the brain. It may be useful for monitoring treatment on the bedside or determining neurological outcome.

References

1. Becker KJ, Tirschwell DL (2002) Intracerebral hemorrhage. In: Johnson RT, Griffin JW, McArthur JC (eds) Current therapy in neurologic disease, 6th edn. Mosby, St Louis p 209–214
2. Boone K (1997) Imaging with electricity: report of the European Concerted Action on Impedance Tomograph. J Med Eng Technol 21: 201–232
3. Chumlea WC, Guo SS (1994) Bioelectrical impedance and body composition: present status and future directions. Nutr Rev 52: 123–131
4. Cusick G, Holder DS, Birkett A (1994) A System for impedance imaging of epilepsy in ambulatory human subjects. Innov Tech Biol Med 15[Suppl] 1: 34–39
5. He W, Yao DG, Tian HY (2001) Disturbance method of electric current field and its application in impedance imaging. Chin J Med Phys 18: 20–22
6. Itkis ML, Roberts JK, Chajar JBG (1994) A square signals wave method for measurement of brain extra and intracellular water content. Acta Neurochir [Suppl] 60: 574–576
7. Lingwood BE, Dunster KR, Healy GN (2004) Effect of cooling and re-warming on cerebral and whole body electrical impedance. Physiol Meas 25: 413–420
8. Lingwood BE, Dunster KR, Healy GN (2003) Cerebral impedance and neurological outcome following a mild or severe hypoxic/ischemic episode in neonatal piglets. Brain Res 969: 160–167
9. Schuier FJ, Hossmann KA (1980) Experimental brain infarcts in cats. Ischemic Brain Edema Stroke 11: 563–601
10. Suga S, Mitani S, Shimamoto Y (1990) In vivo measurement of intra- and extracellular space of brain tissue by electrical impedance method. Acta Neurochir [Suppl] 51: 22–24
11. Tidswell AT, Gibson A, Bayford RH (2001) Electrical impedance tomography of human brain activity with a two-dimensional ring of scalp electrodes. Physiol Meas 22: 167–176
12. Wang J, Yang H, Wang P, Dong WW (2003) Noninvasive monitoring of hypertensive intracerebral hemorrhage and perihematomal tissue edema. Natl Med J China 83: 471–474
13. Wang P, He W (2003) A new technology of noninvasive monitoring for brain edema and its clinic usage. J Chongqing Univ 26: 83–85
14. Yasuda R, Takeuchi K, Funakoshi T (2003) Bioelectrical impedance analysis in the clinical management of preeclamptic women with edema. J Perinat Med 31: 275–280

Correspondence: Li-Xu Liu, Department of Neurology, The First Affiliated Hospital, Chongqing University of Medical Sciences, Chongqing, China. e-mail: chenlihua-llx@tom.com

Acta Neurochir (2005) [Suppl] 95: 141–145

The effect of subarachnoid sodium nitroprusside on the prevention of vasospasm in subarachnoid haemorrhage

J. Pachl[1], **P. Haninec**[2], **T. Tencer**[1], **P. Mizner**[1], **L. Houšt'ava**[2], **R. Tomáš**[2], and **P. Waldauf**[1]

[1] Department of Anaesthesiology and CCM, University Hospital Kralovske Vinohrady, 3rd School of Medicine, Charles's University, Prague, Czech Republic
[2] Department of Neurosurgery, University Hospital Kralovske Vinohrady, 3rd School of Medicine, Charles's University, Prague, Czech Republic

Summary

Objective. The aim of this study was to evaluate the effect of preventive and therapeutic use of subarachnoid sodium nitroprusside (SNP) administration in patients with non-traumatic subarachnoid haemorrhage (SAH).

Methods. All consecutive adult patients admitted in the period 2000–2003 with SAH, Hunt–Hess grade I–IV, indicated for neurosurgical intervention, were enrolled in the study. In the postoperative period they were treated with mechanical ventilation and triple H protocol with nimodipine. Subarachnoid preventive SNP was administred in initial dose of 1 mg by catheter inserted into basal cisterns during the neurosurgical procedure. The timing of following dosage was directed by the changes of respiratory parameters of brain tissue in the region of interest by multiparameter sensor (Codman Neurotrend™) and findings of blood flow velocity on the level of circle of Willis were measured by transcranial doppler ultrasonography (TCD).

Results. 17 patients were enrolled to study. All patients survived. No brain infarction developed. The increase blood flow velocity was found in three patients.

Conclusion. Preventive subarachnoid use of SNP in combination with multimodal monitoring might be a possible preventive strategy. Its efficacy has to be proved on a greater group of patients in the future. The therapeutical use of SNP requires an increase in application rate.

Keywords: Prevention of delayed vasospasm; subarachnoid haemorrhage; subarachnoid nitroprusside sodium administration; multimodal brain monitoring.

Introduction

Delayed vasospasm as a result of subarachnoid haemorrhage is seen in over half of patients and causes symptomatic ischemia in about one third of them. If left untreated, it leads to death or permanent deficits in over 20% of patients [3, 6]. Even though the pathogenesis of vasospasms is unclear, the results of many clinical studies support the hypothesis that vasospasms are caused by an imbalance of vasoconstrictive and vasodilating substances [9, 11, 13]. Presently, the outcomes of several experimental studies, which were designed to study the effects of nitric oxide donors for the treatment and prevention of this life-threatening condition, appear somewhat controversial [5, 7, 9, 12]. The strategy of therapy with sodium nitroprusside (SNP) in clinical trails was based on the treatment effect at the time of vasospasm development, which was proved by angiography and/or transcranial Doppler ultrasonography as well as by clinical neurological impairment. The purpose of this study was: 1) to specify the influence of prophylactic subarachnoid administration of SNP on the incidence of vasospasms and therapeutic efficiency of sodium nitroprusside application on already existed vasospasms 2) to determine the asset of multimodal monitoring of brain tissue in the region of interest for diagnosis and evaluation of the treatment effect.

Methods

Prospective interventional study took place in University Hospital Kralovske Vinohrady from the year 2000 till the year 2003 after approval of the project by the institutional ethical commitee. All patients rated according to the Hunt–Hess's scale I to IV with angiographically verified brain aneurysm and SAH in which neurosurgical intervention was indicated were enrolled. Asymptomatic patients or patients with altered neurologic status which were rated according to the Hunt–Hess's scale on the Vth grade were exluded. Other exlusion criteria were: unequivocal infarction by CT, patients with unsecured ruptured aneurysm, patients with significant hepatic and renal impairment and patients below 18 years of age. The data on patient's age, gender, neurologic state, localisation of aneurysm, duration of intraparenchymal monitoring, duration of SNP administration and dosage in milligrams, Fisher grade, occurrence of vasospasm, cerebral infarction and clinical outcome after 28 days since neurosurgical intervention were collected. Under general anesthesia a catheter was placed into the subarachnoid space of basal cisterns of each patient, through which sodium nitroprusside was administered and drainage of cerebrospinal fluid was done. For the monitoring of brain tissue oxygen pressure, carbon dioxide pressure, pH and temperature ($PtiO_2$, $PtiCO_2$, pHti, Tti) a multiparameter probe (Codman Neurotrend™ Multiparameter Sensor) was inserted in the brain tissue corresponding to the territory of operated artery at the end of the surgery. The longest period of sodium nitropruside administration (Nipruss Schwarzpharma AG) would not exceed 12 days. SNP solution (1 mg/ml) was delivered into the subarachnoid space by the way of catheter in basal cisterns with the minimal frequency of application of 1 ml/1 $\times$ 24 hours. This followed application of 1 ml SNP solution followed by 4 ml 0.9% NaCl to get the effective drug into region of interest. Preventive SNP application in intervals not exceeding 6 hours was indicated in the case of increasing $PtiO_2$ and/or the decreasing $PtiCO_2$ of over 10%. When the first application or some other applications of SNP led to perfusion imbalance, the development of decreasing $PtiO_2$ and the increasing $PtiCO_2$ of over 10%, the interval was prolonged to maximum of 24 hours. Postoperative treatment went according to the standard triple-H protocol with a continuous administration of nimodipine (Dilceren pro Infusione, Zentiva, a.s.) at doses of 15–30 ug/kg/hour. The neurosurgeon decided about an indwelling intraventricular catheter during surgery usually for patients that needed drainage of ventricle. The transcranial Doppler ultrasound (TCD Rimed Israel) was carried out at 12 hour intervals. The preventive SNP administration could be ended up if the altered level of consciousness was more than GCS 13, normal TCD findings and stable clinical status. Afterward we started to wean from the ventilator and the sedation. The indication for therapeticual use of SNP was: a) deterioration of the neurological status according to GCS more than 2 points and/or new focal neurological signs, in both cases with proving of TCD signs of vasospasm (tracing with mean velocity exceeding 120 cm/sec.) [1], b) decreasing level of $PbtiO_2$, below 10 Torr and increasing $PbtiCO_2$, above 60 Torr and the signs of vasospasms diagnosed by TCD. Descriptive statistical evaluation was undergone by using software of Microsoft Excel.

Results

The total number of 17 patients were included and evaluated. The Table 1 shows data of 17 patients (11M/6F), from who 16 patients underwent the protocol of preventive SNP administration and three out of seventeen patients underwent therapeutical management. The average neurologic status according to Hunt–Hess classification scale was III, average duration of multimodal monitoring in days was 5.9 (SD = 3.3) and average duration of SNP medication in days 5.5 (SD = 3.0). Vasospasm was observed in three cases, but only two patients had prophylactic using of SNP. In all of the patients with vasospasms, the diagnosis was supported by TCD. One out of three patients had a perfusion imbalance in microcirculation earlier, which was diagnosed by intraparenchymal multimodal monitoring before TCD findings of vasospasm. All patients survived. No brain infarction as a cause of vasospasm developed in studied group. Maximal dose of sodium nitroprusside, which was used in therapeutical application, reached 47 milligrams.

Discussion

Effect of protocols involved in the treatment of vasospasms is very difficult to compare with others. There are some proofs in the scientific literature, that prognosis of patiens after SAH depend on the localization of ruptured aneurysm, occurrence of intraparenchymal hematoma, entrance GCS, Fischer scale, concomitant diseases etc. The tested protocol of this study implemented preventive subarachnoid administration of SNP with the measurment blood perfusion changes by multimodal monitoring in the area of brain tissue close to the clipped artery and also by TCD on main arteries of circuit of Willii. Previously the influence of SNP on cerebral perfusion was studied in condition of systemic intravenous SNP administration in settings of artificial SAH on animal models [4].

In 16 cases we applied the preventive sodium nitropruside dose and in two cases the effect of subarachnoid application in vasospasm (proved by TCD) was tested. Six patients suffered from an intracerebral haemorrhage whose extent influenced the complete outcome of the treatment. Intracerebral haemorrhage, however, did not influence the main purpose of our study. We share the opinion that the administration of SNP into basal cisterns brings better results than its intraventricular administration. We used a lower dose of SNP in comparison to data presented in published studies [10]. A lower dose of SNP could be explicated by direct application of vasodilatator to main arteries. The effect of preventive subarachnoid administration was tested also by Pathak [8]. The proper indication for preventive administration in this study was based on the possibility of imminent clinical vasospasm, and so in time for neurological symptomatology impairment. Owing to specific characteristics of our application protocol, which was controlled by multimodal monitoring, it was set that the protocol considered the reaction of focal perfusion to SNP administration and minimalised the potential application risk of overdose. Some published data describe the increased vasospasm risk in pHti lower than 7,0–6,7 and $PtiCO_2$ upon 8 kPa [2]. However the $PtiO_2$ fluctuation changes should not be underestimated. It is an object of admi-

Table 1. *Basic clinical, monitoring a treatment profile of studied group*

Patient no.	Age	Gender	Localization of aneurysm other cerebral injuries	Fisher grade	H-H score	MMM (days)	SNP total (mg)	Localization of catheter	GOS 28 days
1.	48	F	ACM bilat.	2	IV	10	10	subarachnoid space in region of basal cistern	I
2.	55	M	ACI l.dx.	4	IV	8	8	subarachnoid space in region of basal cistern	II
3.	57	F	ACM l.sin + left frontal hematoma in SA space	3	IV	11	21	subarachnoid space in region of basal cistern	III
4.	46	M	ACoA + left frontobasal hematoma in SA space + hemocephalus	3	IV	10	15	subarachnoid space in region of basal cistern	I
5.	77	F	ACI l.dx.	3	II	3	3	subarachnoid space in region of basal cistern + in the third ventricle	I
6.	27	M	ACoA + left frontal intracerebral hematoma	4	III	6	44	subarachnoid space in region of basal cistern	I
7.	57	M	ACoA	3	II	3	3	subarachnoid space in region of basal cistern + in the third ventricle	I
8.	54	M	ACoA	2	I	3	1	subarachnoid space in region of basal cistern + in the third ventricle	I
9.	64	M	ACoA + frontal bilateral hematomas in SA space	3	I	8	25	subarachnoid space in region of basal cistern + in the third ventricle	III
10.	70	M	ACoA	3	III	4	17	subarachnoid space in region of basal cistern	II
11.	48	M	left ACM + expansive intracerebral hematoma	4	IV	10	12	subarachnoid space in region of basal cistern + in the third ventricle	IV
12.	57	F	right MCA	3	IV	7	8	subarachnoid space in region of basal cistern + in the third ventricle	I
13.	44	F	right MCA + left MCA l.sin + left ACoP + oedema	2	III	1	1	subarachnoid space in region of basal cistern + in the third ventricle	I
14.	51	M	ACoA	2	IV	3	16	subarachnoid space in region of basal cistern + in the third ventricle	III
15.	61	F	right MCA	3	II	7	5	subarachnoid space in region of basal cistern + in the third ventricle	III
16.	46	M	ACoA	3	II	2	15	subarachnoid space in region of basal cistern	I
17.	49	M	ACoA + 2 × left ACM	3	II	0	40	subarachnoid space in region of basal cistern	I

ACI Internal carotid artery, *ACM* middle cerebral artery, *AcoA* anterior communicating artery, *AcoP* posterior communicating artery, *MMM* multimodal monitoring (Neurotrend), *SAH* subarachnoid haemorrhage, *SNP* sodium nitroprusside.

H-H score (Hunt W. E., Hess R. M., 1968) Hunt–Hess score (I.–V.)

I. Asymptomatic or mild headache II. Moderate or severe headache, nuchal rigidity, can have oculomotor palsy III. Confused, drowsiness, or mild focal signs IV. Stupor or hemiparesis V. Coma, moribund, and/or extensor post

Glasgow Outcome Scale (Jennett B. 1975) – modified version

1 GOOD RECOVERY – able to return to work or school 2 MODERATE DISABILITY – able to live independently; unable to return to work or school 3 SEVERE DISABILITY – able to follow commands/unable to live independently 4 VEGETATIVE STATE – unable to interact with environment; unresponsive 5 DEAD

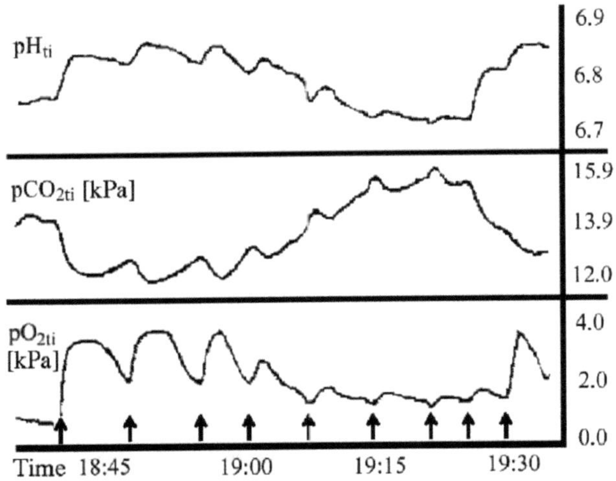

Fig. 1. Trend of pHti, pCO_2ti and pO_2ti in brain tissue in region of left arteria cerebri media during therapeutical sodium nitroprusside administration in 1 mg repeated boluses to subarachnoidal cysterns (arrows indicate the time of SNP application)

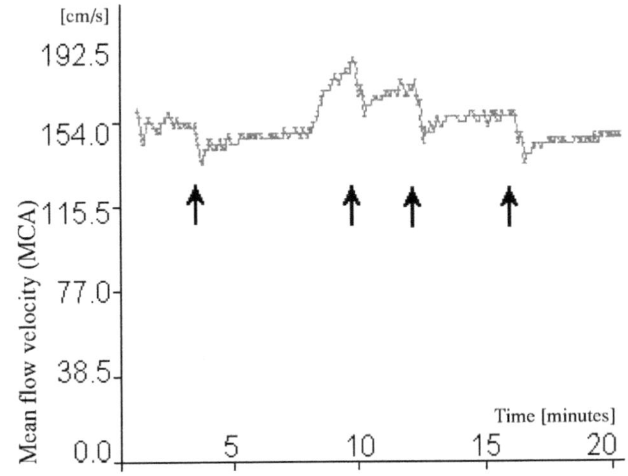

Fig. 2. Changes of flow rates measured by TCD during the vasospasm of the left arteria cerebri media during the SNP administration to subarachnoid space. The arrows indicate the time of SNP application

ration, that $PtiO_2$ could not be raised in some patients by an increase in FiO_2 of gas mixture in settings of mechanical ventilation. Developing vasospasm cannot be detected by multimodal monitoring in special cases, when its localisation is out of the territory of clipped artery. In these cases we emphasis the important role of angiography and TCD.

Vasospasm was observed in three of the seventeen patients. The first patient (in the Table 1 No. 6) developed vasospasm diagnosed by TCD and also by multiparameter sensor. Neurosurgical treatment was given to this patient late, six days after beginning of symptoms of SAH. The changes of the intraparenchymatous values measured by multiparameter sensor in connection with SNP administration are shown on Fig. 1.

Just curative subarachnoid SNP administration was used in one patient (No. 17 in the Table), because he didn't complete the criteria for preventive SNP administration. Signs of vasospasm diagnosed by TCD emerged simultaneously with impairment of consciousness at the level of soporous state on the 6th day after operation. The changes of flow velocity during SNP administration are documented on Fig. 2.

The adverse SNP reactions of subarachnoidal administration are migraine headaches associated with vegetative symptomatology including hypertension reaction. These problems were the main factors for excluding from the trial the patients with good postoperative recovery and short weaning from mechanical ventilation. According to our experiences this therapy is suitable for use in the setting of complex neurointensive care including the mechanical ventilation and deep analgosedation. It is necessary to finish this vasodilatative therapy latest in the time of weaning the patient from mechanical ventilation. Keeping the catheter in the subarachnoid space is suitable for the possible treatment of complications in the next period.

Conclusion

Preventive subarachnoid use of SNP in repeated bolus dose 1 mg into the subarachnoid region of basal cisterns might increase the effect of standard triple-H protocol with a calcium channel blocker. Multimodal brain tissue monitoring and TCD enables the low-dose preventive SNP therapy. The adverse effects of SNP administration (vertigo, systemic hypertension, headache, nausea) restrict the SNP usage only for a period of controlled mechanical ventilation, deep analgesia and sedation. A multicentre randomised controlled trial could be feasible in future.

References

1. Aasled R, Hubert P, Nornes H (1986) A transcranial Doppler method for the evaluation of cerebrovascular spasm. Neuroradiology 28: 11–16
2. Charbel FT, Du X, Hoffman WE, Ausman JI (2002) Brain tissue PO(2), PCO(2), and pH during cerebral vasospasm. J Neurosurg 97(6): 1302–1305
3. Dorsch N (2002) Therapeutic approaches to vasospasm in subarachnoid hemorrhage. Curr Opin Crit Care 8: 128–133

4. Fitch W, Pickard JD, Tamura A, Graham D (1988) Effects of hypotension induced with sodium nitroprusside on the cerebral circulation before, and one week after, the subarachnoidal injection of blood. J Neurol Neurosurg Psychiatry 51(1): 88–93
5. Gabikian P, Clatterbuck RE, Eberhart CG, Tyler BM, Tierney TS, Tamargo RJ (2002) Prevention of experimental cerebral vasospasm by intracranial delivery of a nitric oxide donor from a controlled-release polymer: toxicity and efficacy studies in rabbits and rats. Stroke 33(11): 2681–2686
6. Graves EJ (1992) Detailed diagnoses and procedures, National Hospital Discharge Survey. Vital Health Stat 113: 1–225
7. Macdonald RL, Zhang ZD, Curry D, Elas M, Aihara Y, Halpern H, Jahromi BS, Johns L (2002) Intracisternal sodium nitroprusside fails to prevent vasospasm in nonhuman primates. Neurosurgery 51(3): 761–768; discussion 768–770
8. Pathak A, Mathuriya SN, Khandelwal N, Verma K (2003) Intermittent low dose sodium nitroprusside therapy for treatment of symptomatic aneurysmal SAH-induced vasospasm. Br J Neurosurg 17(4): 306–310
9. Raabe A, Zimmermann N (2002) Effect of intraventricular SodiumNitroprusside on cerebral hemodynamics and oxygenation in poor – grade aneurysma patients with severe, medically refractory vasospasm. Neurosurgery 50(5): 1006–1013
10. Thomas JE, McGinnis G (2002) Safety of Intraventricular Sodium Nitroprusside and Thiosulfate for the treatment of Cerebral Vasospasm in the Intensive Care Unit. Stroke 33: 486–492
11. Thomas JE, Nemirovsky A, Zelman V, Giannotta SL (1997) Rapid reversal of Endothelin-1 induced cerebrál vasoconstriction by intrathecal administration of nitric oxid donors. Neurosurgery 40: 1245–1249
12. Thomas JE, Rosenwasser RH (1999) Reversal of severe cerebral vasospasm in three patients after aneurysmal subarachnoid hemorrhage: initial observations regarding the use of intraventricular sodium nitroprusside in humans. Neurosurgery 44: 48–57
13. Thomas JE, Rosenwasser RH, Armonda RA (1999) Safety of intrathecal sodium nitroprusside for the treatment and prevention of refractory cerebral vasospasm and ischemia in humans. Stroke 30: 1409–1416

Correspondence: Jan Pachl, University Hospital Kralovske Vinohrady, Srobarova 50, 10034 Prague 10, Czech Republic. e-mail: pachl@fnkv.cz

Neuroimaging

Acta Neurochir (2005) [Suppl] 95: 149–151

Mitochondrial injury measured by proton magnetic resonance spectroscopy in severe head trauma patients

A. Marmarou, S. Signoretti, P. Fatouros, G. A. Aygok, and **R. Bullock**

Department of Neurosurgery, Virginia Commonwealth University Medical Center, Richmond, VA, USA

Summary

The aim of this study was to evaluate the extent of mitochondrial injury by assessing N-Acetyl-Aspartate by MR spectroscopy in head injured patients and relating the extent of mitochondrial injury to outcome. The study population (n = 15) consisted of head injured patients (GCS < 8) in whom legal consent was obtained for MRS studies. Studies were performed on a 1.5 Tesla Vision/Siemens system. Size of Voxel equaled 8 cm^3 with location determined from T1 images. Voxels were positioned adjacent to the lesion and in the contralateral hemisphere for focal and bilateral for diffuse. Mitochondrial impairment was considered as percent reduction in NAA/Cr ratio compared to matched controls. Mitochondrial impairment gradually increases soon after injury reaching a nadir at 10 days. Subsequently, mitochondria recover in patients with favorable outcome, but remains impaired in patients with poor outcome. The prognostic value of NAA/Cr to assist in management and also to serve as a surrogate endpoint for clinical trials appears promising.

Keywords: NAA; N-acetyl-aspartate in patients; brain injury.

Introduction

It has been clearly established that following mechanical injury, in addition to the structural damage to the tissue, a series of complex neurotoxic mechanisms are initiated which extend over several hours for the development of irreversible axonal and neuronal damage [1, 2]. In addition, secondary insults of ischemia may contribute to the ongoing energy crisis, which limits the restorative ability of the brain to maintain ionic and cell volume homeostasis. These ionic shifts have been implicated as the major mechanisms for cellular and organelle swelling. The restoration of ionic and cell volume homeostasis implicates an energy crisis, which may occur in the presence of adequate blood flow and is characterized by a mitochondrial impairment and reduced ATP production [3]. The objective of this study was to assess mitochondrial function in the severely brain injured patient by measuring the N-Acetyl-Aspartate represented by the largest peak in a proton spectra (^{1}H MRS). A second aim was to establish the temporal course of NAA change following injury and correlate these changes to outcome.

Methods

After obtaining informed consent according to institutional procedures, patients of GCS 8 or less were transported from the Neuroscience ICU to the magnetic resonance (MR) suite. During transportation and the time of MRI-MRS, appropriate ventilation, sedation and continuous monitoring of blood pressure (arterial line), pulse oximetry, end-tidal CO2 and ICP was performed through MR compatible devices. The transport team consisted of a neurosurgical resident, respiratory technician, critical care nurse and research staff.

Measurement of proton spectroscopy

Semi-quantitative analysis of NAA, creatine containing compounds (Cr/PCr), and choline (Cho), were extracted from proton spectra using ^{1}HMRS in a 1.5 T (Vision/Siemens) system. In patients with focal lesions the location of an eight cm^3 single voxel was determined from T1-weighted images. For diffusely injured patients the voxel was placed bilaterally, in a standard region of the frontal lobes. Following localized shimming and water suppression a spectrum was obtained using the PRESS pulse sequence with 1500/135 TR/TE and 256 acquisitions. These results were compared with proton ^{1}HMRS studies performed in five normal, age-matched, control volunteers previously screened to exclude prior head injuries.

Analysis of single voxel NAA data

The NAA, Choline and Creatine were evaluated as ratios NAA/Cho, NAA/Cr respectively. In the voxel analysis measuring 20 mm^3, a single ratio of NAA/Cr or NAA/Cho was evaluated for each voxel. For diffuse injury, voxels were obtained at symmetrical sites so as to compare values within the brain hemispheres. In focal injury, voxels were placed: in the core of the lesion, peripheral to the lesion and at a distant site in the injured hemisphere. These were compared to symmetrical sites in the uninjured hemisphere.

Results

All patients entered into the study were transported from the neurointensive care unit (NCU) to the imaging suites and returned to the NCU without complication.

Table 1. *Single voxel proton spectroscopy: normal volunteers*

Volunteer	NAA	Cr	Cho	NAA/Cho	NAA/Cr
1	41.18	22.42	17.39	2.37	1.84
2	61.44	29.59	31.66	1.94	2.07
3	56.55	34.66	31.20	1.81	1.63
4	51.97	21.90	27.04	1.92	2.37
5	54.52	24.02	23.05	2.36	2.27
Av	53.13	26.52	26.07	2.08	2.04
S.D.	7.53	5.48	5.98	0.26	0.31

Single voxel MRS volunteer

After obtaining informed consent, single voxel MRS was obtained in 5 volunteers ranging in age from 26 to 34 years. The values obtained from spectra are shown in Table 1. The NAA integral values range from 41.18 to 61.44. The Integral values for Cr, Cho are also shown in Table 1. Based on these integrals, the ratios for NAA/Cho and NAA/Cr equal 2.08 ± 0.26 and 2.04 ± 0.31 respectively. There was no significant difference between ratios of NAA/Cho and NAA/Cr in normal volunteers.

Single voxel MRS in patients with diffuse injury

Of the 15 injured patients, 10 patients were classified as diffuse injury. The voxels were positioned in the white matter. (Figure 1) We observed that NAA/Cho

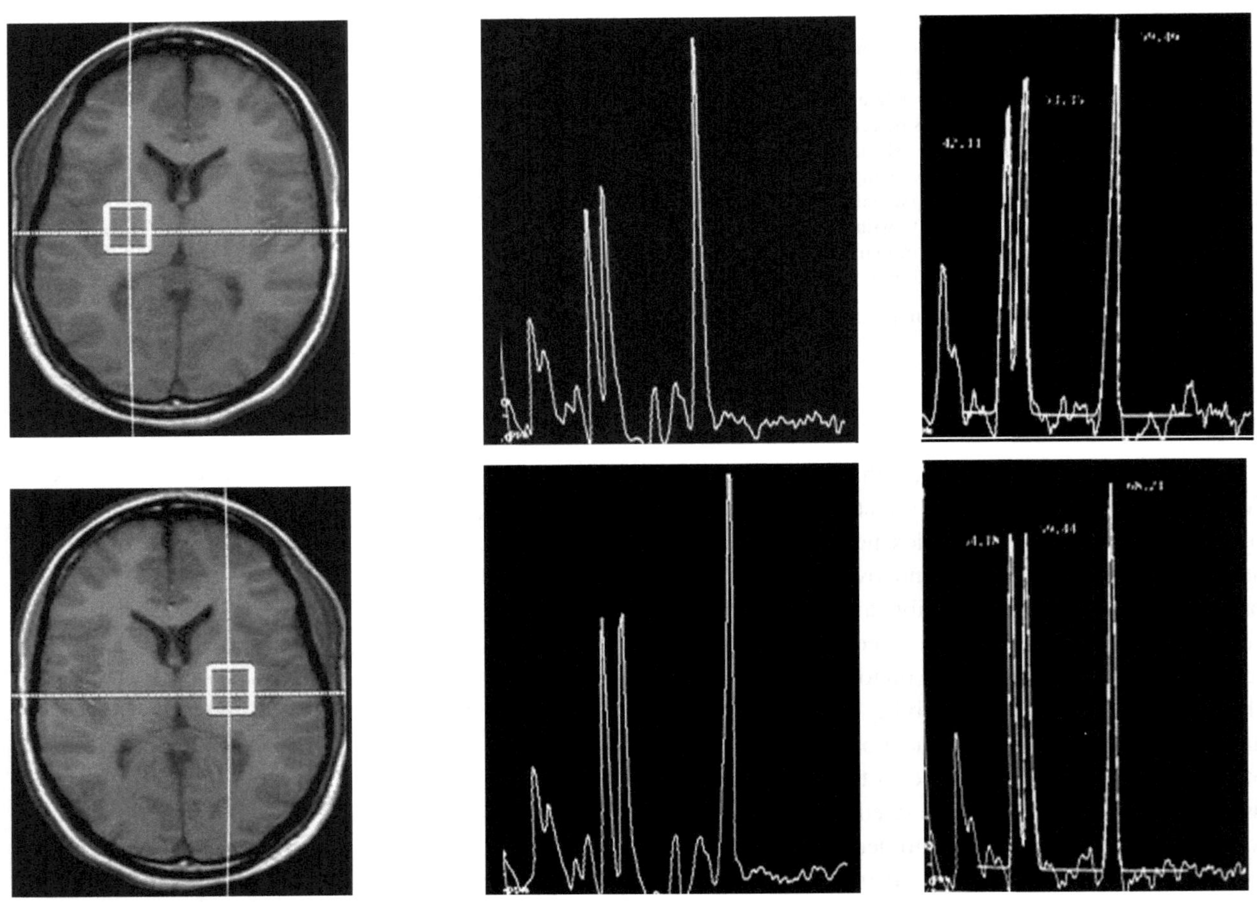

Voxel Placement   3 Days   10 days

Fig. 1. Proton spectroscopy diffuse brain injury – patient 18 years and NAA/Cr studies measured within the first 72 hours of injury were generally higher and gradually declined as time from injury increased. This is illustrated in Fig. 1 where the ratios are near normal levels in a patient with diffuse injury studied 72 hours post injury where NAA/Cho and NAA/Cr measured 1.82 and 1.55 respectively. However, as time progressed, NAA/Cho and NAA/Cr ratios declined to levels of 1.26 and 1.15 at 10 days post injury. This patient was graded severe disability at 6 months. Within the first 10 days the NAA/Cho and NAA/Cr for the entire group averaged 1.59 ± 0.46 and 1.44 ± 0.21 respectively. Beyond 10 days (range 10–75 days), the average NAA/Cho and NAA/Cr for diffuse injury fell to 1.05 ± 0.44 and 1.05 ± 0.30 respectively. The reduction in these ratios was statistically significant. ($p < 0.03$, $p < 0.015$)

Single voxel MRS in patients with focal injury

The NAA/Cho and NAA/Cr ratios of the patients with focal injury during the first 10 days post injury averaged 1.45 ± 0.19 and 1.37 ± 0.46. These ratios were close to the values obtained with diffuse injury during the first 10 days. Similar to diffuse injury, the ratios for focal injury declined as time progressed to values of 0.92 ± 0.26 and 1.07 ± 0.13. This reduction was statistically significant. (NAA/Cho, $p < 0.15$) Over time, the values reduced sharply up to 10 days and stabilized thereafter. We also compared NAA/Cho of the peri-lesion area and the contralateral site in patients with focal injury. We observed, as expected, the NAA/Cho ratio was reduced in the peri-lesion area measuring 1.14 ± 0.35 compared to the symmetrical site in the opposite hemisphere where NAA/Cho measured 1.74 ± 0.49. This difference was statistically significant. ($p < 0.006$)

Discussion

In many cases, it is not possible to ascertain the degree of functional brain damage in the severely injured patient soon after injury. It is particularly difficult to assess the prognosis in diffuse injury where the extent of frank structural damage observed on CT may be minimal. Our studies show that brain neurochemistry is altered significantly after brain injury. The reductions seen of NAA/Cr ratio after traumatic brain injury indicate that mitochondrial impairment has taken place and that in addition to the rampant axonal severe diffuse injury, the brain must overcome an energy crisis as signified by the reduction in ATP stores [3]. Of interest is the observation that the reductions of NAA/Cr and NAA/cho occur beyond 72 hours in diffuse injury. In fact, the reductions seen gradually reduce and do not plateau at a lower level until 10 days post injury. This would infer that the processes leading to neurochemical damage have been triggered but their effect is not realized for several days. It is not clear if these processes can be reversed but it is our sense that the window of opportunity for pharmacologic intervention is much wider than originally thought. The NAA/Cho and NAA/Cr ratios in patients with focal injury were not unlike those of diffuse injury. In general the same pattern of a slow decline was observed. However, if we compare peri-lesion area to the contralateral voxel, there was a clear difference soon after injury. With regard to prognosis, it is difficult to ascertain the direction of recovery or deterioration within the first 72 hours post injury. The NAA/Cr and NAA/Cho ratio reductions seen in traumatic brain injury are reversible. This would suggest that for purposes of prognosis, it is necessary to obtain MRS data at least 10 days after injury before an assessment of outcome is made. Finally, we express caution in the interpretation of NAA values, as it is entirely possible that the reduction might have occurred as a result of ischemic damage to the tissue. For this reason, current studies are focused on measuring both CBF and MRS parameters simultaneously. Nevertheless, the prognostic value of NAA/Cr to assist in management and also to serve as a surrogate endpoint for clinical trials appears promising.

References

1. Buki A, Siman R, Trojanowski JQ, Povlishock JT (1999) The role of calpain-mediated spectrin proteolysis in traumatically induced axonal injury. J Neuropathol Exp Neurol 58(4): 365–375
2. Povlishock JT (2000) Pathophysiology of neural injury: therapeutic opportunities and challenges. Clin Neurosurg 46: 113–126
3. Signoretti S, Marmarou A, Tavazzi B, Lazzarino G, Beaumont A, Vagnozzi R (2001) N-Acetylaspartate reduction as a measure of injury severity and mitochondrial dysfunction following diffuse traumatic brain injury. J Neurotrauma 18(10): 977–991

Correspondence: Anthony Marmarou, Department of Neurosurgery, Virginia Commonwealth University Medical Center, 1001 East Broad Street, Suite 235, Richmond, VA, USA 23219. e-mail: amarmaro@vcu.edu

Acta Neurochir (2005) [Suppl] 95: 153–158

Cerebral blood flow in mean cerebral artery low density areas is not always ischemic in patients with aneurysmal subarachnoid hemorrhage – relationship with neurological outcome

A. Chieregato[1], F. Tagliaferri[1], A. Tanfani[1], F. Cocciolo[1], W. Benedettini[1], C. Compagnone[1], M. Ravaldini[1], R. Pascarella[2], R. Battaglia[3], M. Frattarelli[3], L. Targa[1], and E. Fainardi[4]

[1] Neurorianimazione, Ospedale M. Bufalini, Cesena, Italy
[2] Neuroradiologia, Ospedale M. Bufalini, Cesena, Italy
[3] Neurosurgery, Ospedale M. Bufalini, Cesena, Italy
[4] Neuroradiologia, Arcispedale S. Anna, Ferrara, Italy

Summary

Aneurysmal subarachnoid hemorrhage (SAH) can be complicated by reduction of regional cerebral blood flow (rCBF) from large conductance vessels leading to focal edema appearing as an area of hypoattenuation on CT. In this study we included 29 patients with SAH due to aneurysmal rupture, having 36 CT low density areas within the middle cerebral artery territory in whom a total of 56 Xenon-CT (Xe-CT) studies were performed. Collectively, we evaluated 70 hypoattenuated areas. rCBF levels were measured in two different regions of interest drawn manually on the CT scan, one in the low density area and the other in a corresponding contralateral area of normal-appearing brain tissue. In the low density area (22.6 ± 22.7 ml/100 gr/min) rCBF levels were significantly lower than in the contralateral area (32.8 ± 17.1 7 ml/100 gr/min) ($p = 0.0007$). In the injured areas deep ischemia (CBF < 6 ml/100 g/min) was present in only 25.7% of Xe-CT studies, suggesting that hypodense areas are not always ischemic, whereas in 43.7% of the lesions/Xe-CT studies we found hyperemic values. Patients with a better outcome had hyperemic lesions, suggesting brain tissue recovery in injured areas.

Keywords: Aneurysm; cerebral blood flow; cerebrovascular circulation; cerebral circulation; ischemia; outcome; subarachnoid hemorrhage; tomography; xenon.

Introduction

Aneurysmal subarachnoid hemorrhage (SAH) can be complicated by cerebral ischemia due to inadvertent intraoperative vascular occlusion, thromboembolic events related to embolization, vasospasm or compression/distortion of arterial vessels following transient brain herniation. It is generally accepted that in animal models an acute reduction of cerebral blood flow (CBF) below 18 ml/100 gr/min in a major vascular territory leads to a focal post-ischemic cytotoxic edema formation [2]. In vivo, cytotoxic edema, representing irreversible ischemic brain damage, can be appreciated as an area of hypoattenuation on computed tomography (CT) [14]. However, a critical reduction of regional CBF (rCBF) may be transient because recanalization and reperfusion may occur, sometimes associated with a postischemic hyperperfusion due to "luxury-perfusion" [10]. In addition, CT scans are not able to provide any information on the extent and severity of perfusional deficits associated with ischemic lesions. Therefore, considering ischemia as a decrease of rCBF below accepted ischemic thresholds, it is conceivable that low density lesions corresponding to vascular territories observed after SAH may not always be actually ischemic.

The knowledge of rCBF values in low density areas related to SAH may be of potential interest in outcome prediction as well as in treatment planning. It is reasonable to believe that patients having low density areas associated with not-ischemic rCBF levels have more chances to recover.

The primary hypothesis of the study was that the low density areas complicating SAH are not always associated with ischemic rCBF levels. Secondly, we examined whether SAH patients having low density areas not associated with ischemic rCBF present a better clinical outcome. As regional ischemia affecting the mean cerebral artery (MCA) is highly relevant for the patient's prognosis, we measured rCBF levels by means of Xenon-CT (Xe-CT) in SAH patients having CT hypoattenuated lesions in the MCA territory.

Materials and methods

Patients

From January 2000 to December 2002, 191 patients with aneurysmal subarachnoid hemorrhage were admitted to the Intensive Care Unit at Bufalini Hospital, Cesena-Italy. Seventy patients underwent rCBF measurements by Xe-CT, 47 of them developed at least one hypoattenuated zone on the CT scan. We enrolled in a prospective study 29 patients having at least one hypodensity area greater than 1 cm^2 in the MCA territory. Patients with a hypodensity area on the initial CT scan and those with bilateral hypodensity zones were excluded. Parenchymal hypoattenuation on the CT scan was defined as a visually well-recognized cerebral region of abnormally increased radiolucency relative to other parts of the same structure or to its contralateral counterpart covering a vascular territory.

Patient's management

An intraventricular catheter to measure intracranial pressure (ICP) was placed in all the patients. Shortly thereafter, the patients were angiographically studied to detect aneurysms and to evaluate the best indicated stabilization procedure: embolization or surgery. After aneurysmal coiling or clipping, the patients were re-evaluated to plan the treatment. Cases with better Hunt & Hess grade and without impairment of vital functions were admitted to the ICU after surgical or angiographic procedure and evaluated for ventilatory weaning. In cases having a poor neurological status, poor control of airway reflexes, unstable or elevated ICP, a new low density area on CT or critical CBF areas on Xe-CT, sedation and analgesia were prolonged. A staircase treatment protocol was used to maintain an ICP of below 20 mmHg and a cerebral perfusion pressure (CPP) of above 70 mmHg. Patients with critical rCBF and/or suspected vasospasm were treated with a further elevation of CPP, roughly to 90 mmHg.

Specific management in patients with low density zones in MCA territory

Once a focal lesion was found and rCBF measured, the treatment applied was consistent with the ICP level, the extent of the lesion, the cardiovascular reserve status and the risk of medical complications. In patients with elevated ICP and large low density areas having either ischemic or hyperemic rCBF values, the treatment consisted of external decompression, CPP levels no higher than 70 mmHg and deep sedation. In patients having borderline values of rCBF in low density areas treatment aimed at maintaining elevated CPP (90 mmHg) was prolonged.

Outcome evaluation

Clinical and radiological severity on admission was graded according to Hunt & Hess [5] and Fisher [4] scores. Daily mean and maximum values of hourly ICP were recorded to calculate ICP summary values throughout the acute phase. Outcome was evaluated at one year after SAH. The following three categories of outcome were defined: good, equal to good recovery and moderate disability in the Glasgow Outcome Scale (GOS); poor, equivalent to severe disability, and finally dead and persistent vegetative state [6].

Planning of Xe-CT studies

Initial Xe-CT CBF studies were obtained shortly after stabilization of the aneurysm. Further serial Xe-CT studies were performed when a new low density area or an ICP elevation were found and/or an increased treatment level of ICP or persistent elevated CPP values had to be maintained to recover rCBF levels.

Study details

During the Xe-CT studies, all patients were artificially ventilated and sedated. Special care was taken to maintain the systemic parameters stable during transport from the intensive care unit to the Xe-CT room and also during the Xe-CT measurements. CBF measurements were performed using a CT scanner (Picker 5000) equipped for Xe-CT CBF imaging (Xe/TC system-2^{TM}, DDP, Inc., Houston, TX). The CT slices were placed to analyze the core region of low density area found in the corresponding standard CT. Repeated Xe-CT studies were carried out with the same positioning, angle of CT slices, scans and parameters as used in the previous studies.

Cerebral blood flow analysis

Two different regions of interest (ROIs) larger than 1 cm^2 were drawn freehand on the CT scan, one around the low density area and the other around a mirror region of the hypodensity area placed symmetrically on the corresponding normal-appearing brain tissue of the contralateral hemisphere. The rCBF mean values of each ROI were expressed in ml/100 gr/min. Different groups were defined according to rCBF levels detected in hypodense lesions: severe ischemia (CBF $<$ 6 ml/100 g/min) (Fig. 1), moderate ischemia (CBF $>$ 6 and $<$18 ml/100 g/min), reduced flow (CBF $>$ 18 and $<$33.9 ml/100 g/min), relative hyperemia (CBF $>$ 33.9 and $<$55.3 ml/100 g/min) and absolute hyperemia (CBF $>$ 55.3 ml/100 g/min) (Fig. 2) [8, 9, 11, 16].

Statistical analysis

Comparison of rCBF and areas between low density zones and mirror areas was performed by means of two tail paired t-test. The frequency of distribution of rCBF in different groups was analyzed by means of Chi-square t-test. Severity index variables were compared with parametric or non-parametric statistics. A value of $p < 0.05$ was considered significant.

Results

Patient characteristics

As reported in Table 1, more than half of the patients had a poor Hunt & Hess grade and Fisher score. We performed 56 Xe-CT studies in 29 patients who, collectively, developed 36 hypodensity areas on CT scans. Among the causes recognized, vasospasm and post-surgical or post-embolization impairment of vessel patency were predominant. As Xe-CT measurements were carried out at least twice in 16 of these patients, a total of 70 low density areas/Xe-CT studies were assessed. The median time elapsed between bleeding and aneurysmal surgery or embolization

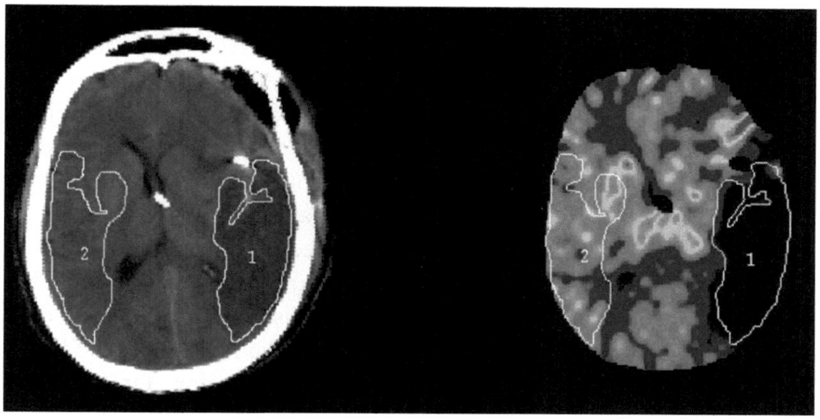

Fig. 1. A low density area associated with severe ischemia (1.5 ml/100 gr/min); rCBF in contralateral normal-appearing area was 41.8 ml/100 gr/min

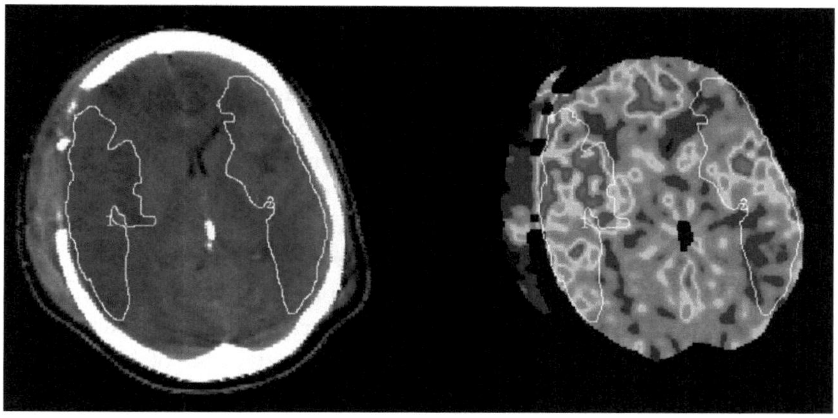

Fig. 2. A low density area associated with absolute hyperemia (rCBF 95.1 ml/100 gr/min); rCBF in contralateral normal appearing area was 45.3 ml/100 gr/min

and low density lesion/Xe-CT studies was respectively 194 hrs (IQR 220) and 121 hrs (IQR 223). The physiological parameters recorded during the Xe-CT studies suggest that they were performed in a stable condition: ICP was 22.9 ± 8 mmHg, CPP 75.1 ± 15.3 mmHg, PaCO2 38.7 ± 5.3 mmHg, Hb 10.2 ± 1.4 gr/dl, Ht $30 \pm 3.5\%$, SaO2 $96.7 \pm 1.8\%$, temperature °C 37.2 ± 1.0.

CBF values

As illustrated in Table 2, the size of CT hypoattenuated lesions was remarkably large (2357.9 ± 1699.6 mm^2). rCBF levels were significantly lower in the low density area (22.6 ± 22.7 ml/100 gr/min) than in the contralateral normal-appearing area (32.8 ± 17.1 7 ml/100 gr/min) ($p = 0.0007$). Only 25.7% of the low density areas examined had an rCBF under the threshold of severe ischemia, whereas 43.7% of these lesions had an rCBF above the hyperemic threshold.

Outcome

As indicated in Table 3, outcome categories were not related to differences in conventional predictors, except for higher maximum ICP levels in patients with poor outcome. On the other hand, the size of the low density area seems highly associated with poor outcome. Interestingly, a better outcome was found in patients having hyperemic low density lesions.

Discussion

The current study confirms the hypothesis that ischemic rCBF values are inconsistently found in low

Table 1. *Characteristics of 29 SAH patients having a focal low density lesion undergoing Xe-CT studies*

Patients, n		29
Age, median (IQR)	years	54 (20.7)
Sex, n (%)	female	19 (65.5)
Hunt & Hess scale, score, n (%)	I	6 (20.7)
	II	3 (10.4)
	III	6 (20.7)
	IV	9 (31)
	V	5 (17.2)
Location of bleeding aneurysm, n (%)	mean cerebral artery	13 (44.8)
	carotid artery	5 (17.2)
	anterior circulation	7 (24.2)
	posterior communicating artery	4 (13.8)
Fisher CT grade-1st CT, n (%)	grade I	2 (6.9)
	grade II	5 (17.2)
	grade III	14 (48.3)
	grade IV	8 (27.6)
Surgery, n (%)		20 (69)
Embolization, n (%)		9 (31)
Timing of procedure post bleeding, hrs		14 (43.5)
Causes of low density areas, n (%)	vascular distortion	4 (13.8)
	post-surgery or post-embolization	11 (37.9)
	vasospasm	14 (48.3)
Outcome at 12 months, n (%)	dead	10 (34.5)
	persistent vegetative state	1 (3.4)
	severe disability	8 (27.6)
	moderate disability	6 (20.7)
	good recovery	4 (13.8)

density areas after reduction of regional arterial flow in large conductance vessels. In fact, according to previous studies [7, 15], we found rCBF levels in low density areas lower than in contralateral normal-appearing tissue, indicating an ongoing focal hypoperfusion. However, rCBF levels suggestive of irreversible ischemic damage were present only in approximately one-fourth of the low density areas examined.

The pathogenesis of hypodense lesions included in this study was variable and, in decreasing order of frequency, they were: vasospasm, post-surgical-embolization impairment of vessel, patency impairment and, finally, distortion of vessels following reversible brain herniation. As a consequence, the pathogenetic mechanisms of initial injury as well as the time-evolution of rCBF changes in the affected areas could be heterogeneous, explaining why hypodense areas can be associated with a wide range of rCBF values. In fact, it seems reasonable to believe that while some vessels are permanently occluded leading to deep irreversible ischemia, others are only temporarily impaired due to the removal of the external compression and/or recanalization. On the other hand, a prolonged critical decrease of rCBF in feeding arteries may lead to a progressive distal microvascular ischemia, independent of the status of the main vessels.

This study may be of some interest for the clinical setting. Cerebral perfusion pressure management and triple H therapy, (hypervolemia, hemodilution hypertension), have been applied to prevent or revert focal ischemic delayed deficit in SAH [3]. The physiological background of these treatments suggests the possibility that an improvement of CBF values can be obtained both in large conductance vessels and in microcirculatory territories [3]. However, only in few cases have these physiological therapies been performed in association with a serial monitoring of rCBF values which could improve the cost/effectiveness of these treatments [1]. In fact, HHH therapy may be complicated by cardiovascular complications and, potentially, worsening of cerebral lesions due to vasogenic edema and hemorrhagic infarction [12].

Our data also suggest that, in the selected case-mix involving only patients with low density areas in MCA territory, a better outcome seems to be associated with hyperemic values. These data should be confirmed in a multivariate analysis since in the present univariate analysis the extension of low density area seems more relevant than the rCBF levels. Ischemic lesions with early hyperperfusional pattern have already

Table 2. *Characteristics of 36 low density areas in 70 Xe-Ct lesions studies*

		Low density area	Contralateral normal area	
Area	mm^2	2357.9 ± 1699.6	1859 ± 1223.8	$p \leq 0.0001$
rCBF	ml/100 gr/min	22.6 ± 22.7	32.8 ± 17.1	$p = 0.0007$
rCBF	<6 ml/100 gr/min, n/%	18 (25.7)	1 (1.6)	$p \leq 0.0001$
	>6 and ≤18 ml/100 g/min, n/%	23 (32.9)	13 (20.3)	
	>18 ml/100 gr/min <33.9 ml/100 gr/min, n/%	11 (15.7)	22 (34.4)	
	>33.9 and ≤55.3 ml/100 g/min, n/%	12 (17.1)	23 (35.9)	
	>55.3 ml/100 gr/min, n/%	6 (8.6)	5 (7.8)	

Table 3. *Univariate analysis of variables affecting the outcome*

	Dead or pvs	Severe disability	Good outcome	p
n (%)	11 (37.9)	8 (27.6)	10 (34.5)	
Age, Years	52.1 ± 11	56 ± 10.4	48.8 ± 15.2	0.4859
Sex, female	7 (63.6)	6 (75)	6 (60)	0.7904
Hunt-Hess grade IV and V, n/%	7 (63.6)	4 (50)	3 (30)	0.3032
Fisher grade III and IV, n/%	8 (72.7)	7 (87.5)	6 (60)	0.4310
Mean of mean daily ICP, mmHg	21.1 ± 7.1	16.5 ± 5.4	17.6 ± 4.8	0.2793
Mean of maximal daily ICP, mmHg	30.5 ± 8.6	23.5 ± 6.8	26.3 ± 6.6	0.1647
Maximal ICP, mmHg	55.4 ± 19.2*	31.9 ± 9.4*	39.7 ± 13.5	0.0112
				*p = 0.0134
Maximal low density area, mm^2	4513 ± 1425.2*^	2066.2 ± 1543.5^	2751.1 ± 1684*	0.0052
				*0.0495
				^0.0085
Min CBF in low density area, ml/100 gr/min	10.1 ± 14.8	15.3 ± 19.8	22.2 ± 24.2	0.3900
Max CBF in low density area, ml/100 gr/min	15.2 ± 16.2*	26.2 ± 21.6	42.6 ± 31.5*	0.0457
				p = 0.0465
Low density area min CBF $<$ 18 ml/100 gr/min, n/%	9 (81.8)	7 (87.5)	6 (60)	0.3364
Low density area max CBF $>$ 33.9 ml/100 gr/min, n/%	1 (9.1)	3 (37.5)	6 (60)	0.0485

been described in SAH patients with vasospasm [15]. In stroke patients, positron emission tomography studies have demonstrated that hyperperfusion is a beneficial phenomenon associated with spontaneous recanalization and good tissue outcome, when it occurs early in the acute ischemic phase, whereas it represents a harmful event more commonly related to poor tissue outcome if it arises later in the subacute ischemic stage [10]. The present results seem to indicate that late hyperperfusion, also, could produce a favourable clinical effect.

Last but not least, focal probes (ptiO2 and/or microdialysis) were placed in CT low density areas considering them to be ischemic and with the aim of monitoring and improving regional circulation and metabolism [13]. The study results suggest that a physiological interpretation may be incomplete if not associated with direct rCBF measurements.

In conclusion, rCBF studies in CT low density areas show that only in one-fourth of the cases CBF-ischemia is fully established. Conversely, hyperemia is frequent in these areas, suggesting that rCBF levels can be restored in the injured areas.

References

1. Darby JM, Yonas H, Marks EC, Durham S, Snyder RW, Nemoto EM (1994) Acute cerebral blood flow response to dopamine-induced hypertension after subarachnoid hemorrhage. J Neurosurg 80: 857–864
2. Del Zoppo GJ, von Kummer R, Hamann GF (1998) Ischaemic damage of brain microvessels: inherent risks for thrombolytic treatment in stroke. J Neurol Neurosurg Psychiatry 65: 1–9
3. Findlay JM (1997) Current management of aneurysmal subarachnoid hemorrhage. Guidelines from the Canadian Neurosurgical Society. Can J Neurol Sci 24: 161–170
4. Fisher CM, Kistler JP, Davis JM (1980) Relation of cerebral vasospasm to subarachnoid hemorrhage visualized by computerized tomographic scanning. Neurosurgery 6: 1–9
5. Hunt W, Hess R (1968) Surgical risk as related to time of intervention in the repair of intracranial aneurysms. J Neurosurg 28: 14–20
6. Jennett B, Bond M (1975) Assessment of outcome after severe brain damage. A practical scale. Lancet 1: 480–484
7. Jovin TG, Yonas H, Gebel JM, Kanal E, Chang YF, Grahovac SZ, Goldstein S, Wechsler LR (2003) The cortical ischemic core and not the consistently present penumbra is a determinant of clinical outcome in acute middle cerebral artery occlusion. Stroke 34: 2426–2435
8. Kaufmann AM, Firlik AD, Fukui MB, Wechsler LR, Jungries CA, Yonas H (1999) Ischemic core and penumbra in human stroke. Stroke 30: 93–99
9. Kelly DF, Kordestani RK, Martin NA, Nguyen T, Hovda DA, Bergsneider M, McArthur DL, Becker DP (1996) Hyperemia following traumatic brain injury: relationship to intracranial hypertension and outcome. J Neurosurg 85: 762–771
10. Marchal G, Young AR, Baron J-C (1999) Early postischemic hyperperfusion: pathophysiologic insights from positron emission tomography. J Cereb Blood Flow Metab 19: 467–482
11. Obrist WD, Langfitt TW, Jaggi JL, Cruz J, Gennarelli TA (1984) Cerebral blood flow and metabolism in comatose patients with acute injury. Relationship to intracranial hypertension. J Neurosurg 61: 241–253
12. Shimoda M, Oda S, Tsugane R, Sato O (1993) Intracranial complications of hypervolemic therapy in patients with a delayed ischemic deficit attributed to vasospasm. J Neurosurg 78: 423–429
13. Stocchetti N, Chieregato A, De Marchi M, Croci M, Benti R, Grimoldi N (1998) High cerebral perfusion pressure (CPP) improves low values of local brain tissue O2 pressure (PtiO2) in focal brain lesion. Acta Neurochir [Suppl] 71: 162–165

14. von Kummer R, Bourquain H, Bastianello S, Bozzao L, Manelfe C, Meier D, Hacke W for the European Cooperative Acute Stroke Study II Group (2001) Early prediction of irreversible brain damage after ischemic stroke at CT. Radiology 219: 95–100
15. Yonas H, Sekhar L, Johnson DW, Gur D (1989) Determination of irreversible ischemia by Xenon-enhanced computed tomography monitoring of cerebral blood flow in patients with symptomatic vasospasm. Neurosurgery 24: 368–372
16. Jones TH, Morawetz RB, Crowell RM, Marcoux FW, Fitz Gibbon SJ, DeGirolami U, Ojemann RG (1981) Thresholds of focal cerebral ischemia in awake monkeys. J Neurosurg 54: 773–782

Correspondence: Arturo Chieregato, Servizio di Anestesia e Rianimazione, Ospedale M. Bufalini, Viale Ghirotti 286, 47023 Cesena, Italy. e-mail: achiere@ausl-cesena.emr.it

Acta Neurochir (2005) [Suppl] 95: 159–164

Cerebral blood flow mapping in two different subtypes of intraparenchymal hemorrhagic traumatic lesions

A. Chieregato[1], C. Compagnone[1], A. Tanfani[1], M. Ravaldini[1], F. Tagliaferri[1], R. Pascarella[2], F. Servadei[3], L. Targa[1], and E. Fainardi[4]

[1] Neurorianimazione, Ospedale M. Bufalini, Cesena, Italy
[2] Neuroradiologia, Ospedale M. Bufalini, Cesena, Italy
[3] Neurotraumatologia, Ospedale M. Bufalini, Cesena, Italy
[4] Neuroradiologia, Arcispedale S. Anna, Ferrara, Italy

Summary

The pathogenesis and the viability of edematous tissue may be different in traumatic hematomas and traumatic contusions. We tested the hypothesis that mapping of regional Cerebral Blood Flow (rCBF) was different in these two subtypes of traumatic intraparenchymal lesions. We evaluated rCBF by means of Xenon-enhanced computerized tomography (Xe-CT) in 59 traumatic intracerebral lesions from 43 patients with severe head injury. One-hundred-nine intracerebral lesions/Xe-CT CBF measurements were obtained. The rCBF was measured in the hemorrhagic core, in the intralesional oedematous low density area and in a 1 cm rim of apparently normal perilesional parenchyma of both lesion subtypes. Not statistically significant lower rCBF levels were found in the edematous area of traumatic contusions. In traumatic hematomas rCBF levels were lower in the core than in the low density area, suggesting that rCBF in edematous area is marginally involved in the initial traumatic injury and that edema is probably influenced by the persistence of the hemorrhagic core. Conversely, in the traumatic contusions a difference in rCBF values was found between core, low density area and perilesional area, indicating that rCBF of the low density area is related to a concentrical distribution of the initial injury.

Keywords: Cerebral blood flow; cerebral circulation; head injury; ischemia; traumatic cerebral hemorrhage; traumatic contusions; traumatic hematomas; xenon-CT.

Introduction

Traumatic intraparenchymal lesions are consistently associated with edema, the extent of which is potentially relevant for the mass effect with elevation of intracranial pressure (ICP). The viability of the edematous tissue is still a matter of discussion because it is of interest for both medical and surgical management. As microvascular ischemia could be implicated in cytotoxic edema development, the measurement of regional cerebral blood flow (rCBF) has been applied to quantify regional ischemia and to predict tissue viability within the edematous area associated with cerebral contusions. However, the threshold of ischemia in comatose, sedated head-injured patients has not yet been defined. At the same time, different rCBF levels have been found in previous studies [3–5, 7, 8, 11, 14], probably due to the evaluation of different subtypes of contusions having potentially different mechanisms in edema formation. In addition, different forms of edema can be identified in various temporal phases which characterize the evolution of cerebral contusions. Vasogenic edema seems to be predominant in later days, roughly afther four days day post-injury, whereas osmotic and cytotoxic edema may occur in early phases.

It is well known that traumatic intraparenchymal lesions range from large traumatic hematomas, traumatic contusions, 'salt and peppers' contusions and low density contusions to petecchiae [9]. Thus, it seems reasonable to believe that in each type of lesion the pathogenesis and the form of edema could be different. In the clinical setting, the 'surgical' doubts concerning the viability of edematous areas are restricted to small traumatic hematomas and contusions. Concerning traumatic hematoma, the existence of a not-evacuated hemorrhagic core within the cerebral parenchyma is commonly associated with an enlargement of the perihematoma edema. This phenomenon has been described in intraparenchymal spontaneous hematomas [1] and growing evidence supports the opinion that

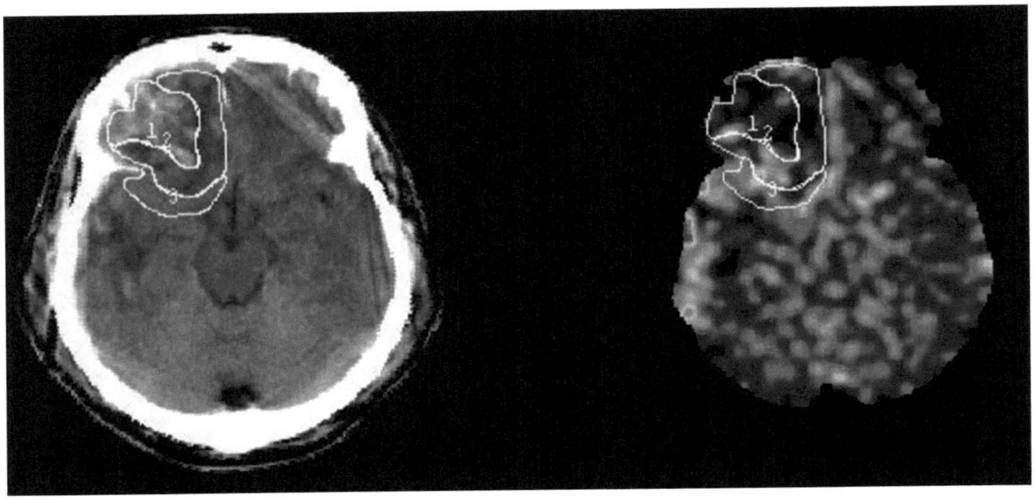

Fig. 1. Regional CBF in a traumatic contusion evaluated in three different ROIs: (*1*) within the hemorrhagic core; 10.5 ml/100 gr/min, (*2*) in the edematous low-density area, 23.2 ml/100 gr/min, (*3*) around a 1 cm. rim of normal-appearing brain tissue surrounding the perilesional low-density area, 32.4 ml/100 gr/min. The concentrical distribution of the ROIs is self-evident

the formation of vasogenic edema is mediated by the hemorrhagic core through the shedding of toxic molecules [15]. In contrast, in traumatic contusion the significance of the hemorrhagic component is, smaller. In these lesions, the scattered blood is embedded in necrotic edema [9], as a consequence of diapedesis and not of a primary bleeding, and the evolution of edema would in part be independent from the extent of the central hemorrhagic core, but could be related to cytotoxic phenomena due to a reduction of rCBF. Thus, edema surrounding the core in traumatic hematoma may be predominantly vasogenic and associated with a minor reduction of rCBF, whereas edema located around the clot in traumatic contusions could be cytotoxic and related to ischemic rCBF levels. To test this hypothesis we evaluated the absolute rCBF values in two different intraparenchymal lesions: traumatic hematomas, excluding large hematomas, and contusions, excluding purely low density traumatic contusions.

Material and methods

Patient's summary and case mix selection

From July 1999 to October 2001, 106 severely head-injured patients (GCS ≤ 8) with 148 traumatic intraparenchymal lesions were admitted to the Neuro Intensive Care Unit of our hospital and 63 of these patients underwent Xenon-CT. Among the patients consecutively submitted to Xenon-CT studies, at least one intraparenchymal lesion was present in 56 patients in whom, collectively, 84 intraparenchymal lesions were present. Among these lesions, large traumatic hematomas, 'salt and pepper' contusions, low density contusions and petecchiae were excluded. The 59 remaining intraparenchymal lesions, collected in 43 patients, were categorized by a single neuroradiologist (EF), unaware of the hypothesis of the study and not involved in statistical analysis in two subtypes: contusions in which the hemorrhagic core (larger than 1 cm in diameter) consisted of multiple confluent islands embedded in low density edematous tissue (Fig. 1) and traumatic hematomas, having a well-defined hemorrhagic core larger than 2 cm in diameter and, if any, a well distributed peri-core edema (Fig. 2).

Stable xenon-enhanced CT CBF measurement and CBF analysis

CBF studies were conducted using a CT scanner (Picker 5000) equipped for Xe-CT CBF imaging (Xe/TC system-2TM, Diversified Diagnostic Products, Inc., Houston, TX). All patients were artificially ventilated, sedated, and paralysed. Further details concerning the Xe-CT studies have been earlier described [4]. Lesional rCBF was calculated by dedicated software (Xe-CT System Version 1.0 w ©, 1998, Diversified Diagnostic Products, Inc, Houston, TX) and was expressed in ml/100 gr/min in three different ROIs larger than 1 cm^2 drawn freehand on the diagnostic CT as previously published [4]: 1) within the lesional hemorrhagic core (hemorrhagic core); 2) in the peri.-core low-density edematous area (intralesional low-density area); 3) within 1 cm of the normal-appearing brain tissue surrounding the low density area (perilesional normal-appearing area). An additional ROI was placed in a normal-appearing brain tissue area, with a shape that corresponded to the intraparenchymal brain lesion (hemorrhagic core plus intralesional low-density area), located symmetrically in the contralateral hemisphere (contralateral lesion "mirror" area). If the contralateral lesion "mirror" area displayed an altered coefficient of density at CT scan, this ROI was not analyzed. The differences between the values of the rCBF measured in the contiguous intraparenchymal lesion ROI were measured: delta core/edema (equal to the rCBF measured in the hemorrhagic core rCBF core minus the rCBF measured in the intralesional low-density area) and delta edema/normal (equal to the rCBF measured in the intralesional low-density area minus the rCBF measured in the normal appearing area), delta normal/contralateral (equal to the rCBF

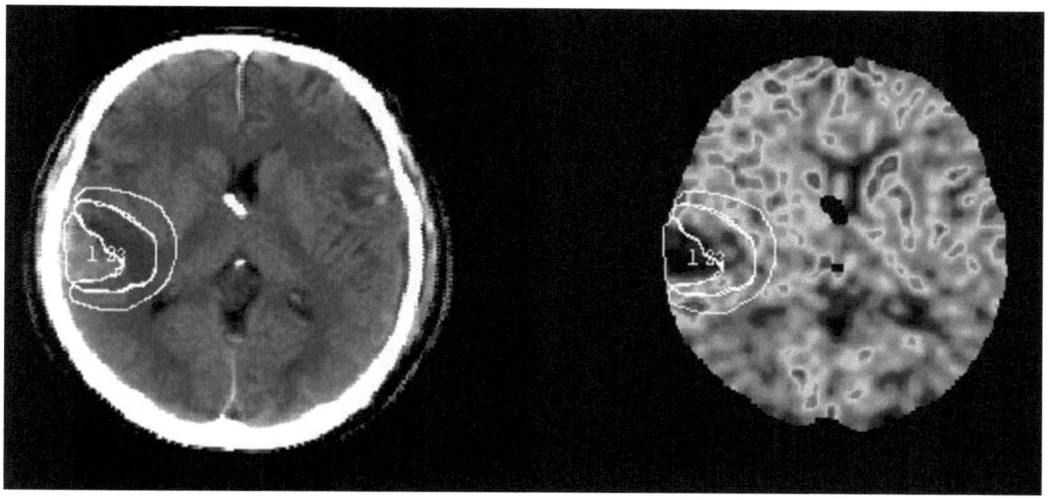

Fig. 2. Case example of rCBF in a traumatic hematoma evaluated in in three different ROIs: (*1*) within the hemorrhagic core; 9.9 ml/100 gr/min, (*2*) in the edematous low-density area, 44.5 ml/100 gr/min, (*3*) around a 1 cm. rim of normal-appearing brain tissue surrounding the perilesional low-density area, 63.5 ml/100 gr/min. Marked rCBF reduction in the core and minor derangement of rCBF in low density and perilesional normal appearing area can be observed

measured in the normal appearing area, minus the rCBF measured in the contralateral area). According to previous authors [7, 8, 11, 14], a value of CBF below 18 ml/100 gr/min was used to dichotomize the rCBF measurements.

Statistical analysis

The data were checked for normality. Mean values of CBF measured in the core, the intralesional low-density area and the perilesional normal-appearing area were compared by means of ANOVA. Comparison between the ROIs was made by applying the Scheffè test. To test the hypothesis that there were differences in the rCBF distribution in the three concentric ROIs, as well as in the rCBF delta between ROIs, the traumatic hematoma and traumatic contusion groups were compared by means of a pooled t-test. To test the hypothesis that the rate of ischemia was higher in the traumatic contusion group than in the traumatic hematoma group, we compared the frequency of ischemia by means of a chi-square test. A value of p below 0.05 was considered significant.

Results

The demographic and clinical parameters of our case material are reported in Table 1. Mean physiological values during the Xe-CT studies were consistent with a stabilized condition: ICP was 20.9 $\pm$ 10.5 mmHg, CPP 67.8 $\pm$ 13.7 mmHg, PaCO2 37.6 $\pm$ 5.1 mmHg, Hb 10.2 $\pm$ 1.6 gr/dl, SaO2 97.1 $\pm$ 2% and temperature 37.2 $\pm$ 0.9 °C. As some of the 59 contusions were repeatedly studied by 82 Xe-CT examinations during the acute phase, we analyzed a total of 109 contusion studies. The median time elapsed, respectively between injury and intraparenchymal lesion/Xe-CT studies, was 89 hours (IQR 106).

Table 1. *Characteristics of 42 severely head injured patients undergoing Xe-CT studies*

Patients, n		43	
Age, median (IQR)	years	35 (32)	
Sex, n (%)	male	31 (74.4)	
		best	worst
Initial GCS, median (IQR)		7(6)	4(4)
motor GCS, median (IQR)		4(3)	2(4)
Pupillary reactivity, n (%)	normal	24 (57.1)	22 (52.3)
	one pupil dilated unreactive	11 (26.2)	16 (38.9)
	bilaterally dilated unreactive pupils	0 (0)	4 (9.5)
Worst CT, n (%)	diffuse injury I	0 (0)	
	diffuse injury II	5 (15.7)	
	diffuse injury III	4 (12.5)	
	diffuse injury IV	0 (0)	
	evacuated mass lesions	21 (65.6)	
	not evacuated mass lesions	2 (6.2)	
Outcome at 12 months, n (%)	dead	7 (18.4)	
	persistent vegetative state	5 (13.2)	
	severe disability	7 (18.4)	
	moderate disability	5 (13.2)	
	good recovery	14 (36.8)	

The rCBF measured in the intralesional low-density area of contusions (29.7 $\pm$ 19.6 ml/100 gr/min) was lower than in the corresponding area of hematomas

Table 2. *Details concerning the mapping of rCBF measured during lesion/Xe-CT studies*

		Traumatic contusion	Traumatic hematoma	p
		52 (43 contra)	57 (47 contra)	
CBF (ml/100 gr/min)	lesional hemorrhagic core	19.6 ± 17.9*^	14.0 ± 11.8*^	0.055
	intralesional low density area	29.7 ± 19.6*$	35.1 ± 21.3*$	0.1795
	perilesional normal appearing area	41.4 ± 18.6^$&	41.0 ± 20.1^$&	0.9143
		Anova p ≤ 0.0001; *p = 0.028, $p = 0.0079, ^p ≤ 0.0001	Anova p ≤ 0.0001; *p ≤ 0.0001, $p = 0.24, ^=p ≤ 0.0001,	
	contralateral normal appearing area	39.0 ± 18.9&	36.8 ± 22.2&	0.6247
		&p = 0.5367	&p = 0.3211	
Area (mm2)	lesional hemorrhagic core	441.2 ± 337.2*	234.8 ± 230.6*	0.0003
	intralesional low density area	1088.5 ± 554.4*$	932.8 ± 625.5*$	0.1737
	perilesional normal appearing area	782.5 ± 283.0$	729.7 ± 346.2$	0.3620
		ANOVA p ≤ 0.0001; *p ≤ 0.0001, $p ≤ 0.001, ^p ≤ 0.0001	ANOVA p ≤ 0.0001; *p ≤ 0.0001, $p ≤ 0.001, ^p ≤ 0.0001	
	contralateral normal appearing area	1576.6 ± 533.6	1275.7 ± 715.3	0.0273
CBF values below 18 ml/100 r/min (%)	lesional hemorrhagic core	63.3	70.2	0.4507
	intralesional low density area	32.7	24.6	0.3473
	perilesional normal appearing area	9.6*	12.3*	0.6570
		p ≤ 0.0001	p ≤ 0.0001	
	contralateral normal appearing area	9.3*	12.8*	0.6015
		p = 0.9586	p = 0.9486	
CBF (ml/100 gr/min)	Delta hemorrhagic core – intralesional low density area	−10.2 ± 16.4*	−21.1 ± 14.4*	0.0004
	Delta intralesional low density area – perilesional normal appearing area	−11.6 ± 15* *t-test p = 0.6235	−5.8 ± 15.4*$ *t-test p ≤ 0.0001	0.0504
	Delta perinormal normal appearing – contralateral normal appearing	0.6 ± 14.1$ &t-test p = 0.0003	2.0 ± 14.9$ $t-test p = 0.0325	0.6336

(35.1 ± 21.3 ml/100 gr/min) but the difference did not reach statistical significance ($p = 0.1795$). Although not statistically relevant, rCBF levels below 18 ml/100 gr/min were more frequent in contusions (32.7%) than in hematoma (24.6%) intralesional low-density areas. The rCBF values of the hemorrhagic core was lower ($p = 0.055$) in hematomas than in contusions. No differences were found when the rCBF measured in the perilesional normal-appearing areas were compared between the two groups.

Differences in rCBF levels were observed when we cumulatively assessed the various ROIs in each of the two lesion subgroups ($p \leq 0.0001$). However in the traumatic contusion group (Fig. 1) we found a steady improvement of rCBF from the centre to the periphery of the lesion, whereas in traumatic hematomas (Fig. 2) a significant difference in rCBF was found only between core and intralesional low density area. In traumatic hematoma the reduced rCBF levels in the intralesional low density area with respect to the perilesional area did not reach statistical significance. The extent of rCBF delta core/edema and the delta edema/normal was comparable ($p = 0.6325$) in traumatic contusion, while consistent differences were found in traumatic hematoma ($p \leq 0.0001$). No differences were found between the perilesional normal-appearing area rCBF and the contralateral normal-appearing area rCBF in either of the lesion groups.

Discussion

The primary hypothesis that the edema surrounding the core was associated with a lesser reduction of rCBF in traumatic hematomas than in traumatic contusions was not confirmed statistically. Conversely, differences in flow mapping were found between the two groups. A constant degree of derangement of rCBF was detected from the periphery to the centre in traumatic contusions, whereas in traumatic hematomas we observed lower rCBF values in the core than in the edematous area. The rejection of the primary hypothesis may be due to an overlap in the case mix of the groups selected. This may also be one of the reasons explaining the wide range of rCBF values measured in traumatic intraparenchymal lesions in previous investigations [3–5, 7, 8, 11, 14]. Notwithstanding the potential limitations in the criteria of group selection, the study may have the advantage that it compares lesions in patients undergoing the same medical and surgical management and suggests that, although in absence of statistical differences in rCBF levels, mechanisms of edema formation could be divergent in traumatic contusions and hematomas. In fact, the pattern of rCBF in traumatic contusions is in agreement with the basic pathological knowledge suggesting a continuum in severity from the periphery to the centre of the contusion. The strongly reduced rCBF values found in the hemorrhagic core were consistent with the destruction of parenchyma at the centre of energy application associated with a central necrosis, vascular disruption, diapedesis of red blood cellsand osmotic edema. More peripherally the swelling of the astrocytes may generate a primary critical hypoperfusion, low rCBF and cytotoxic edema. The surrounding normal-appearing tissue and its rCBF levels may not be directly affected by the initial traumatic impact. A constant, average improvement of 10 ml/100 gr/min was found in the traumatic contusions moving from the core to the edematous area and thereafter to normal-appearing tissue, confirming von Oettingen study's results [14]. On the other hand, rCBF mapping in the traumatic hematoma category seems substantially different. In the hemorrhagic core, rCBF values were lower than in the surrounding edematous area. Even if the pathogenesis of traumatic hematomas is not fully know, it is probable that primary bleeding from small arterioles after traumatic injury represents the main factor. The hemorrhagic core may be detected at pathological examination as a clot, not as an hemorrhagic infarction or as a central necrosis. In contrast to the hemorrhagic core, similar rCBF values were recorded in perihematoma low density and normal-appearing areas suggesting that peri-core edema affects a previously normal tissue with mechanisms which only marginally pertubate cerebral perfusion. Considering that vasogenic edema does not affect rCBF [13] and seems preponderant in spontaneous hemorrhage [10], it is reasonable to presume that edema formation in traumatic hematomas may be primarily due to vasogenic mechanisms generated by the presence of the hemorrhagic core. This type of edema could be related to the volume of the hemorrhagic core [1], associated with a potentially reversible [12] non-ischemic reduction of rCBF [10], and coupled with a reduction of regional metabolism [16].

In conclusion, the results of the present study do not definitively confirm patterns in the pathogenesis of traumatic intraparenchymal lesions, but help to better define two possible mechanisms, previously hypothesized [3]. The first, probably more applicable to traumatic contusions, in which rCBF is directly affected by the initial destructive insult injury and involved in edema pathogenesis. The second, more applicable to traumatic hematomas, in which edema is the result of a diffusion of the pathogenetic process from the centre to the periphery and in which rCBF is marginally affected. The validation of this hypothesis may be of interest in order to improve the specificity and the timing of experimental medical therapies in traumatic hematoma [2] directed at reducing the progression of initial bleeding and edema as well as of surgery aimed at removing the central clot and preventing further edema [6]. Conversely, if low CBF values in edematous tissue surrounding traumatic contusions will be confirmed to be associated with a poor viability, there would be less doubt with regard to surgical evacuation, when an ICP elevation occurs or when an inappropriate elevation of the level of medical management is required.

References

1. Carhuapoma JR, Barker PB, Hanley DF, Wang P, Beauchamp NJ (2002) Human brain hemorrhage: quantification of perihematoma edema by use of diffusion-weighted MR imaging. Am J Neuroradiol 23: 1322–1326
2. Carr ME Jr, Martin EJ (2004) Recombinant Factor VIIa: clinical applications for an intravenous hemostatic agent with broad-spectrum potential. Expert Rev Cardiovasc Ther 2: 661–674
3. Chieregato A, Fainardi E, Tanfani A, Martino C, Pransani V, Cocciolo F, Targa L, Servadei F (2003) Mixed dishomogeneous hemorrhagic brain contusions: mapping of cerebral blood flow. Acta Neurochir [Suppl] 86: 333–337
4. Chieregato A, Fainardi E, Servadei F, Tanfani A, Pugliese G, Pascarella R, Targa L (2004) Centrifugal distribution of regional cerebral blood flow and its time course in traumatic intracerebral hematomas. J Neurotrauma 6: 655–666
5. Furuya Y, Hlatky R, Valadka AB, Diaz P, Robertson CS (2003) Comparison of cerebral blood flow in computed tomographic hypodense areas of the brain in head-injured patients. Neurosurgery 52: 340–346
6. Hankey GJ (2003) Evacuation of intracerebral hematoma is likely to be beneficial. Stroke 34: 1568
7. Hoelper BM, Reinert MM, Zauner A, Doppenberg E, Bullock R (2000) rCBF in hemorrhagic, non-hemorrhagic and mixed contusions after severe head injury and its effect on perilesional cerebral blood flow. Acta Neurochir [Suppl] 76: 21–25

8. McLaughlin MR, Marion DW (1996) Cerebral blood flow and vasoresponsivity within and around cerebral contusions. J Neurosurg 85: 871–876
9. Ribas GC, Jane JA (1992) Traumatic contusions and intracerebral hematomas. J Neurotrauma 9: 265–268
10. Schellinger PD, Fiebach JB, Hoffmann K, Becker K, Orakcioglu B, Kollmar R, Jüttler E, Schramm P, Schwab S, Sartor K, Hacke W (2003) Stroke MRI in intracerebral hemorrhage. Is there a perihemorrhagic penumbra? Stroke 34: 1674–1680
11. Schroder ML, Muizelaar JP, Bullock MR, Salvant JB, Povlishock JT (1995) Focal ischemia due to traumatic contusions documented by stable xenon-CT and ultrastructural studies. J Neurosurg 82: 966–971
12. Siddique MS, Fernandes HM, Wooldridge TD, Fenwick JD, Slomka P, Mendelow AD (2002) Reversible ischemia around intracerebral hemorrhage: a single-photon emission computerized tomography study J Neurosurg 96: 736–741
13. Sutton LN, Greenberg J, Welsh F (1990) Blood flow and metabolism in vasogenic oedema. Acta Neurochir [Suppl] 51: 397–400
14. von Oettingen G, Bergholt B, Gyldensted C, Astrup J (2002) Blood flow and ischemia within traumatic cerebral contusions. Neurosurgery 50: 781–790
15. Xi G, Wagner KR, Keep RF, Hua Y, de Courten-Myers GM, Broderick JP, Brott TG, Hoff JT (1998) Role of blood clot formation on early edema development after experimental intracerebral hemorrhage. Stroke 29: 2580–2586
16. Zazulia AR, Diringer MN, Videen TO, Adams RE, Yundt K, Aiyagari V, Grubb Jr, RL, Powers WJ (2001) Hypoperfusion without ischemia surrounding acute intracerebral hemorrhage. J Cereb Blood Flow Metab 21: 804–810

Correspondence: Arturo Chieregato, Servizio di Anestesia e Rianimazione, Ospedale M. Bufalini, Viale Ghirotti 286, 47023 Cesena, Italy. e-mail: achiere@ausl-cesena.emr.it

Acta Neurochir (2005) [Suppl] 95: 165–168

Glucose metabolism in traumatic brain injury: a combined microdialysis and [^{18}F]-2-fluoro-2-deoxy-D-glucose – positron emission tomography (FDG-PET) study

M. T. O'Connell[1,2], **A. Seal**[1], **J. Nortje**[3], **P. G. Al-Rawi**[1], **J. P. Coles**[3], **T. D. Fryer**[2], **D. K. Menon**[2,3], **J. D. Pickard**[1,2], and **P. J. Hutchinson**[1,2]

[1] University Department of Neurosurgery, Addenbrooke's Hospital and University of Cambridge, Cambridge, UK
[2] Wolfson Brain Imaging Centre, Addenbrooke's Hospital and University of Cambridge, Cambridge, UK
[3] University Department of Anaesthesia, Addenbrooke's Hospital and University of Cambridge, Cambridge, UK

Summary

Following traumatic brain injury, as a consequence of ionic disturbances and neurochemical cascades, glucose metabolism is affected. [^{18}F]-2-Fluoro-2-deoxy-D-glucose (FDG) – Positron Emission Tomography (FDG-PET) provides a measure of global and regional cerebral metabolic rate of glucose (rCMRglc), but only during the time of the scan. Microdialysis monitors energy metabolites over extended time periods, but only in a small focal volume of the brain.

Our objective in this study is to assess the association of parameters derived from these techniques when applied to patients with traumatic brain injury. Eleven sedated, ventilated patients receiving intracranial pressure monitoring and managed using Addenbrooke's Neurosciences Critical Care Unit protocols were monitored. Dialysate values for glucose, lactate, pyruvate, and glutamate, and the lactate to glucose (L/G), lactate to pyruvate (L/P) and pyruvate to glucose (P/G) ratios were determined and correlated with rCMRglc.

FDG-PET scans were performed within 24 hours (five patients), or between 1 and 4 days (two patients) or after 4 days (six patients). Two patients were rescanned 4 and 7 days after their initial scan. A 20 mm region of interest (ROI) was defined on co-registered CT scan on two contiguous slices around the microdialysis catheter. Mean ($\pm$sd) for rCMRglc was 19.1 $\pm$ 5.5 µmol/100 g/min, and the corresponding microdialysis values were: glucose 1.4 $\pm$ 1.4 mmol/L; lactate 5.3 $\pm$ 3.6 mmol/L; pyruvate 164.1 $\pm$ 142.3 µmol/L; glutamate 15.0 $\pm$ 14.7 µmol/L; L/G 11.0 $\pm$ 16.0; L/P 27.3 $\pm$ 7.9 and P/G 381 $\pm$ 660. There were significant relations between rCMRglc and dialysate lactate ($r = 0.58$, $P = 0.04$); pyruvate ($r = 0.57$, $P = 0.04$), L/G ($r = 0.55$, $P = 0.05$), and the P/G ($r = 0.56$, $P = 0.05$) but not between rCMRglc and dialysate glucose, L/P or glutamate in this data set.

The results suggest that increases in glucose utilization as assessed by FDG-PET in these patients albeit in mainly healthy tissue are associated with increases in dialysate lactate, pyruvate, L/G and the P/G ratio perhaps indicating a general rise in metabolism rather than a shift towards non-oxidative metabolism. Further observations are required with regions of interest (microdialysis catheters positioned) adjacent to mass lesions notably contusions.

Keywords: Brain glucose metabolism; FDG-PET; microdialysis; traumatic brain injury.

Introduction

Cerebral metabolism following traumatic brain injury can be quantified by imaging (Positron Emission Tomography) and catheter (microdialysis) techniques [1]. [^{18}F]-2-Fluoro-2-deoxy-D-glucose (FDG) – Positron Emission Tomography (FDG-PET) provides a measure of global and regional cerebral metabolic rate of glucose (rCMRglc), but only during the time of the scan. Microdialysis monitors energy substrates and metabolites (e.g. glucose, lactate, pyruvate) over extended time periods but only in a small focal volume of brain. We have previously employed these complementary techniques simultaneously in patients with traumatic brain injury, demonstrating a significant relation between the microdialysis-derived lactate/pyruvate ratio and the PET-derived oxygen extraction fraction [2].

In terms of glucose metabolism following traumatic brain injury there is now good evidence that hyperglycolysis (an increase in glucose utilisation) is present, particularly in and adjacent to focal mass lesions [3]. The objective of this study was to explore the relation between PET and microdialysis parameters of glucose metabolism, in particular to utilise microdialysis to determine the fate of glucose undergoing hyperglycolysis.

Material and methods

The study was approved by the Local Research and Ethics Committee, and UK Radiation Protection Committee. Informed assent from the patient's next of kin was obtained. Patients with traumatic brain injury older than 16 years, sedated, ventilated, and with intracranial pressure monitoring were eligible for the study. Patients were managed according to Addenbrooke's Neurosciences Critical Care Unit protocols [4, 5]. Eleven patients were studied using a combination of cerebral microdialysis and FDG-PET scans performed in the adjacent Wolfson Brain Imaging Centre. Two patients were re-scanned 4 and 7 days after their first scan. Comprehensive neuro-intensive care monitoring and treatment were maintained throughout the PET scans.

Cerebral microdialysis

Microdialysis catheters (CMA 70, 10-mm membrane, CMA, Stockholm, Sweden) were inserted into the frontal cerebral parenchyma, in conjunction with the intracranial pressure sensor (Codman, Raynham, MA, USA) using a triple-lumen cranial access device (Technicam, Newton Abbot, UK) [6]. The catheters were inserted to a depth of 30 mm monitoring predominantly white matter, and were perfused with Perfusion Fluid CNS for Microdialysis (NaCl 147 mmol/L, KCl 2.7 mmol/L, $CaCl_2$ 1.2 mmol/L, $MgCl_2$ 0.85 mmol/L, CMA Stockholm, Sweden) at a rate of 0.3 μl/min using the CMA106 pump (CMA, Stockholm, Sweden). Vials were changed at 20 minute intervals during FDG-PET scanning. Microdialysate was analysed for glucose, lactate, pyruvate and glutamate using a CMA600 bedside microdialysis analyser (CMA, Stockholm, Sweden). The delay from the microdialysis membrane to the collecting vial (the dead space) is 17 minutes, and this was accounted for in the study design. The catheters were inserted at least four hours before the PET scan to ensure baseline stability.

FDG-PET technique

All scans were performed in the Wolfson Brian Imaging Centre (WBIC) using standardized protocols on a GE Medical Systems Advance Whole-Body tomograph (General Electric Company, Milwaukee, WI, USA) with a spatial resolution of 5 mm × 5 mm × 6 mm. All FDG scans were acquired in 3D (septa out) mode. As standard procedure, a transmission scan was performed with external germanium-68 rod sources (10 minutes duration) to calculate the attenuation correction for each patient scan. Patients were injected intravenously with a 10 ml dose of circa 74 MBq [^{18}F] FDG by infusion over 60 seconds. The radiochemical purity of [^{18}F]-FDG exceeded 98%. Arterial blood samples were taken at 0.5, 1, 1.5, 2, 3, 4, 5, 7, 10, 15, 25, 35, 45, and 55 minutes post-infusion, resulting in 14 samples for the input function. For the first 5 minutes following FDG injection, the frame rate of the scanner was 1 frame per minute. Between 5 minutes and 55 minutes, the frame rate was reduced to 1 frame every 5 minutes, and in the final 20 minutes, reduced further to 1 frame every 10 minutes. Consequently, over 55 minutes 13 frames were recorded. The PET emission data was reconstructed using the PROMIS 3D filtered back-projection algorithm, with corrections applied for attenuation, scatter, random counts, and dead time.

After reslicing CT images to 5 mm, PET and CT images were co-registered and analyzed using custom-designed automated software (PETAN) [7] incorporating elements of several different software packages, including Statistical Parametric Mapping (SPM99, Wellcome Department of Cognitive Neurology, London, UK), Matlab 5.2 (MathWorks, Natick, MD, USA), Analyze 5 (AnalyzeDirect, Lenexa, KS, USA), and registration by multiresolution optimization of mutual information (mpr, Department of Radiologic Sciences, Guy's Hospital, London, UK) [8, 9].

Data analysis

A 20 mm region of interest (ROI) was defined on the co-registered CT scan around the sensor on two contiguous slices to represent the volume of brain measured with the microdialysis catheter. The size of the ROI has previously been determined [2, 10] aiming to keep the region as focal as possible whilst at the same time limiting parameter variance. By superimposing this ROI over the PET scans, ROI values for glucose utilization (rCMRglc) were obtained using the Huang operational equation [11].

Data obtained from FDG-PET and microdialysis were compared using least square linear regression and then analysis of variance (ANOVA).

Results

The age of patients ranged from 17 to 65 years with nine males and two females. Microdialysis monitoring started 24 ± 16 hours (range 8–62.5) post injury. FDG-PET scans were performed on average 85 ± 73 hours (range 12 hours to 7 days 21 hours) after injury. Five patients were scanned within 24 hours, two between 1 and 4 days and six patients were scanned more than 4 days after injury. Mean (±sd) for rCMRglc was 19.1 ± 5.5 μmol/100 g/min, and the corresponding microdialysis values were: glucose 1.4 ± 1.4 mmol/L; lactate 5.3 ± 3.6 mmol/L; pyruvate 164.1 ± 142.3 μmol/L; glutamate 15.0 ± 14.7 μmol/L; lactate to glucose ratio (L/G) 11.0 ± 16.0; lactate to pyruvate ratio (L/P) 27.3 ± 7.9 and the pyruvate to glucose ratio (P/G) 381 ± 660. In this data set there were significant associations between rCMRglc and dialysate lactate, pyruvate (see figures 1 a & b respectively), the L/G and P/G ratios but not between rCMRglc and dialysate glucose, L/P or glutamate (Table 1).

Discussion

In this study, we have shown that associations exist between measures of glucose utilisation, rCMRglc, derived from FDG-PET imaging and microdialysate parameters of energy metabolism, extracellular fluid lactate, pyruvate, L/G and P/G ratios. However, no significant correlation was observed between rCMRglc and dialysate glucose, glutamate or L/P ratio.

The cerebral microdialysis values reported here are similar to those of previous studies in head injury [2, 12–15]. A significant correlation between dialysate L/P ratio and PET derived oxygen extraction fraction in patients with head injury has previously been reported by our group [2]. Although this finding suggests an association between metabolic stress and tissue hypoxia, the OEF observed in these patients appeared to be within the normal range (i.e. not exceeding 0.7).

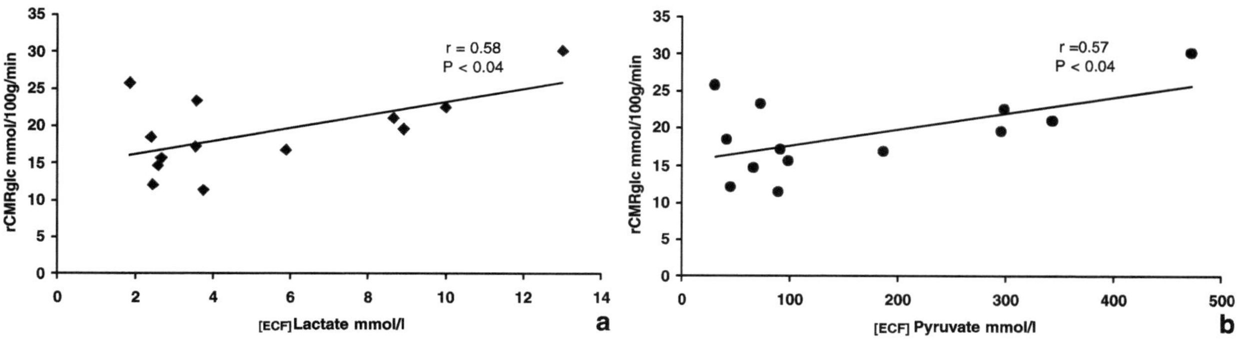

Fig. 1 a, b. Graphs showing dialysate levels for (a) lactate and (b) pyruvate for individual patients with head injury, plotted against FDG-PET derived regional cerebral metabolic rate of glucose utilisation, rCMRglc

Table 1. *Correlation between [18F]-2-fluoro-2-deoxy-d-glucose – positron emission tomography (FDG-PET) derived rCMRglc and microdialysis parameters of glucose utilisation and glucose extracellular space metabolites representing metabolism*

rCMRglc	Glucose	Lactate	Pyruvate	Glutamate	L/G	L/P	P/G
r	−0.31	0.58	0.57	0.414	0.55	0.19	0.56
P	n.s.	0.04	0.04	n.s.	0.05	n.s.	0.05

The rCMRglc is the region-of-interest cerebral metabolic rate of glucose utilisation. L/G, L/P and P/G are the dialysate lactate to glucose, lactate to pyruvate and pyruvate to glucose ratio, respectively. The r value is the regression coefficient. The value n.s. is non-significant to $P < 0.05$.

Another study examining the relationship between brain tissue oxygen, PbO_2, and tissue chemistry found that large increases in microdialysis parameters normally occurred only when tissue oxygen levels fell to very low levels [16]. Irrespective of the availability of oxygen, mitochondrial dysfunction may be a barrier to oxidative glucose use [17, 18].

Bergsneider *et al.* were first to report evidence of cerebral hyperglycolysis in severely head injured patients and presumed that over half the patients studied showed either global or regional hyperglycolysis within the first week following injury [3]. We have observed a case in which microdialysis was performed within a region of brain showing evidence of hyperglycolysis from FDG-PET and Oxygen-15-PET imaging. In this case, dialysate glucose values were very low (<0.25 mmol/L) whilst other parameters were high; lactate > 12 mmol/L; pyruvate > 450 mmol/L; L/P > 36; and glutamate > 50 μmol/L, consistent with hyperglycolysis.

Further work from our group on CPP augmentation (from ~70 mmHg to ~90 mmHg using norepinephrine to control CPP) showed significantly increased levels of PbO_2 and significantly reduced regional OEF [19]. However, this did not produce significant changes in regional chemistry as measured using microdialysis, albeit in a small patient group (eleven). The trends observed in association with CPP augmentation were average falls in lactate 30%, L/P ratio 20%, and glycerol 13% without changes in glucose and pyruvate levels. Furthermore, of three patients with baseline L/P ratio above 25 (25 being regarded as the upper limit of normal [20, 21], in one of these patients the L/P ratio fell to a normal level during CPP augmentation.

In patients with head injury the application of FDG-PET, in terms of the selection of operational equation, the use of population derived rate constants, and the general assumptions made regarding the relationship between the tracer, FDG, and endogenous glucose, requires further validation. In these patients we feel that correlation using a technique such as microdialysis may provide support for FDG-PET in the global and regional assessment of tissue glucose metabolism albeit as a snap-shot only. Whether, this approach can generate reliable estimates in terms of quantifiable values is at present a matter for debate.

On the other hand, microdialysis has the potential to measure metabolic rates but probably requires the use of labeled metabolite markers and use of calibration techniques. Nevertheless microdialysis does permit longer term monitoring of tissue chemistry although only within the immediate vicinity of the catheter. Consequently appropriate placement of the microdialysis catheter is critical. It should be remembered that microdialysis measures the extracellular fluid concentration of energy parameters, and not production, utilisation or elimination of substances per se. Consequent dialysate levels represent a summation of the processes described above and interpretation of results relies on an understanding of the tissue biochemistry and physiology involved. Despite these reservations we feel that this data suggests that higher levels of lactate, pyruvate, L/G and P/G but not L/P ratio are associated with PET-derived increased regional glucose utilisation. This indicates a general rise in metabolism rather than a shift towards non-oxidative metabolism. Further work is required to establish if this observation translates to patients in whom the region of interest is adjacent to contusions.

Acknowledgments

The authors wish to acknowledge the support of members of the Departments of Neurosurgery and Anaesthesia, Wolfson Brain Injury Centre and the Addenbrooke's Neuro Critical Care Unit, assistance of the nursing staff, Dr Hugh Richards for help with statistics, Dr Peter Smielewski for help with PET data analysis, and Dr David Hutchinson for the initial design of the intracranial access device. Mr Peter J Hutchinson is supported by an Academy of Medical Science/The Health Foundation Senior Surgical Scientist Fellowship.

References

1. Hutchinson PJ, O'Connell MT, Kirkpatrick PJ, Pickard JD (2002) How can we measure substrate, metabolite and neurotransmitter concentrations in the human brain? Physiol Meas 23: R75–R109
2. Hutchinson PJ, Gupta AK, Fryer TF, Al-Rawi PG, Chatfield DA, Coles JP, O'Connell MT, Kett-White R, Minhas PS, Aigbirhio F, Clark JC, Kirkpatrick PJ, Menon DK, Pickard JD (2002) Correlation between cerebral blood flow, substrate delivery, and metabolism in head injury: a combined microdialysis and triple oxygen positron emission tomography study. J Cereb Blood Flow Metab 22: 735–745
3. Bergsneider M, Hovda DA, Shalmon E, Kelly DF, Vespa PM, Martin NA, Phelps ME, McArthur DL, Caron MJ, Kraus JF, Becker DP (1997) Cerebral hyperglycolysis following severe traumatic brain injury in humans: a positron emission tomography study. J Neurosurg 86: 241–251
4. Menon DK (1999) Cerebral protection in severe brain injury: physiological determinants of outcome and their optimisation. Br Med Bull 55: 226–258
5. Patel HC, Menon DK, Tebbs S, Hawker R, Hutchinson PJ, Kirkpatrick PJ (2002) Specialist neurocritical care and outcome from head injury. Intensive Care Med 28: 547–553
6. Hutchinson PJ, Hutchinson DB, Barr RH, Burgess F, Kirkpatrick PJ, Pickard JD (2000) A new cranial access device for cerebral monitoring. Br J Neurosurg 14: 46–48
7. Smielewski P, Coles JP, Fryer TD, Minhas PS, Menon DK, Pickard JD (2002) Integrated image analysis solutions for PET datasets in damaged brain. J Clin Monit Comput 17: 427–440
8. Studholme C, Hill DL, Hawkes DJ (1996) Automated 3-D registration of MR and CT images of the head. Med Image Anal 1: 163–175
9. Studholme C, Hill DL, Hawkes DJ (1997) Automated three-dimensional registration of magnetic resonance and positron emission tomography brain images by multiresolution optimization of voxel similarity measures. Med Phys 24: 25–35
10. Gupta AK, Hutchinson PJ, Fryer T, Al-Rawi PG, Parry DA, Minhas PS, Kett-White R, Kirkpatrick PJ, Matthews JC, Downey S, Aighirhio F, Clark J, Pickard JD, Menon DK (2002) Measurement of brain tissue oxygenation performed using positron emission tomography scanning to validate a novel monitoring method. J Neurosurg 96: 263–268
11. Huang SC, Phelps ME, Hoffman EJ, Sideris K, Selin CJ, Kuhl DE (1980) Noninvasive determination of local cerebral metabolic rate of glucose in man. Am J Physiol 238: E69–E82
12. Persson L, Hillered L (1992) Chemical monitoring of neurosurgical intensive care patients using intracerebral microdialysis. J Neurosurg 76: 72–80
13. Zauner A, Doppenberg EM, Woodward JJ, Choi SC, Young HF, Bullock R (1997) Continuous monitoring of cerebral substrate delivery and clearance: initial experience in 24 patients with severe acute brain injuries. Neurosurgery 41: 1082–1093
14. Goodman JC, Valadka AB, Gopinath SP, Uzura M, Robertson CS (1999) Extracellular lactate and glucose alterations in the brain after head injury measured by microdialysis. Crit Care Med 27: 1965–1973
15. Hutchinson PJ, O'Connell MT, Al-Rawi PG, Maskell LB, Kett-White CR, Gupta AK, Richards HK, Hutchinson DB, Kirkpatrick PJ, Pickard JD (2000) Clinical cerebral microdialysis – a methodological study. J Neurosurg 93: 37–43
16. Valadka AB, Goodman JC, Gopinath SP, Uzura M, Robertson CS (1998) Comparison of brain tissue oxygen tension to microdialysis-based measures of cerebral ischemia in fatally head-injured humans. J Neurotrauma 15: 509–519
17. Verweij BH, Muizelaar JP, Vinas FC, Peterson PL, Xiong Y, Lee CP (2000) Impaired cerebral mitochondrial function after traumatic brain injury in humans. J Neurosurg 93: 815–820
18. Clausen T, Zauner A, Levasseur JE, Rice AC, Bullock R (2001) Induced mitochondrial failure in the feline brain: implications for understanding acute post-traumatic metabolic events. Brain Res 908: 35–48
19. Johnston AJ, Steiner LA, Coles JP, Chatfield DA, Fryer TD, Smielewski P, Hutchinson PJ, O'Connell MT, Al-Rawi PG, Aigbirihio FI, Clark JC, Pickard JD, Gupta AK, Menon DK (2004) Effect of cerebral perfusion pressure augmentation on regional oxygenation and metabolism after head injury (in press)
20. Persson L, Valtysson J, Enblad P, Warme PE, Cesarini K, Lewen A, Hillered L (1996) Neurochemical monitoring using intracerebral microdialysis in patients with subarachnoid hemorrhage. J Neurosurg 84: 606–616
21. Reinstrup P, Stahl N, Mellergard P, Uski T, Ungerstedt U, Nordstrom CH (2000) Intracerebral microdialysis in clinical practice: baseline values for chemical markers during wakefulness, anesthesia, and neurosurgery. Neurosurgery 47: 701–709; discussion 709–710

Correspondence: Mark T. O'Connell, Department of Neurosurgery, Box 167, Addenbrooke's Hospital, Hills Road, Cambridge CB2 2QQ, UK. e-mail: mt209@cam.ac.uk

Acta Neurochir (2005) [Suppl] 95: 169–172

Postoperative changes in SPECT-rCBF in hydrocephalus

S. K. Piechnik[1,2] and **L. Hultin**[1,3]

[1] Institute of Clinical Neurosciences, University of Goteborg, Goteborg, Sweden
[2] Institute of Electronic Systems, Warsaw University of Technology, Warsaw, Poland
[3] AstraZeneca, Goteborg, Sweden

Summary

Objectives. We investigated the effect of shunt surgery in patients with Normal Pressure Hydrocephalus (NPH) using Single Photon Emission Computerised Tomography (SPECT).

Materials & Methods. Thirteen patients diagnosed with NPH were assessed clinically and using (^{99m}Tc)-SPECT and MRI both pre- and post-operatively. Regions of interest were placed manually on T2 MRI and transferred to co-registered SPECT. Differences between pre- and post-operative cerebellum-normalised regional cerebral blood flow (rCBF) were calculated and analysed in relation to clinical findings represented by a new disability scale.

Results. The patients presented initially with $50 \pm 30\%$ disability score and improved following the surgery by $6 \pm 10\%$ ($p = 0.1$). We did not observe any significant rCBF changes in the whole group of patients (overall rCBF difference $= +0.3\%$, $p = 0.4$). Some improvement was in basal frontal lateral cortex, basal ganglia and thalamus (+5%, $p = 0.08$ to 0.2). Patients with <30% disability score initially ($N = 4$) had a reversed pattern of changes compared to those with more symptoms ($p < 0.05$).

Conclusions. The small patient sample failed to show significant changes in rCBF due to NPH or surgery. There is indication that in patients with good initial clinical presentation there is little space for relevant clinical improvement and increase in rCBF.

Keywords: Hydrocephalus; surgery outcome; disability score system; image processing; single Photon Emission Computerised Tomography.

Introduction

Ventricular enlargement and tissue distortion in hydrocephalic brain results in a range of functional disorders probably due to a disturbance of cerebral function. The latter can be estimated using a choice of imaging methods for cerebral blood flow (CBF) and metabolism offering a potential for objective evaluation of the disorder [5]. In this paper we contribute to the ongoing discussion by studying the relationship between the single photon emission tomography (SPECT) CBF and the main clinical symptoms of hydrocephalus combined into a novel disability assessment scale.

Material and methods

This study was performed on 13 patients with a clinical diagnosis of hydrocephalus (age: median 67, range = 19–79 years) of various aetiology: trauma ($N = 3$) cerebrovascular ($N = 2$) idiopathic ($N = 7$) other ($N = 1$) and a history of symptoms 1–7 years long (median 4 years). The patients were given surgical shunt treatment except of 4 who had aqueduct stenosis and were treated with 3rd ventriculostomy. All cases were assessed both before and 3 months after surgery using a battery of psychoneurological tests and imaging. From about 120 clinical variables we used only 7 factors to construct a composite scale of symptom prevalence representing a weighted proportion of functional disturbances with impact on the quality of life (Table 1).

Preoperative and postoperative cerebral blood flow imaging was performed using 1000-MBq of (^{99m}Tc)-d,l-HMPAO (Ceretec; Amersham) administered intravenously at rest with eyes closed. Scans were carried out on a three-headed SPECT gamma-camera system (General Electric, Neurocam 2). For anatomic reference we used contemporaneous T2 MRI scans performed on 1.5T camera (Magnetom Vision plus, Siemens). To obtain the outline of the

Table 1. *Scoring system for scalar disability scale in hydrocephalus. The final normalisation step limits the effect of missing values and substitution of specific criteria*

Categories and test conditions	Score range
Motor function	0–3
– Walking – any subjective problems	
– Walking slower than 10 m in 12 s	
– Romberg balance test < 1 minute	
Mental capacity	0–3
– MMSS < 27	
– Wakefulness – any problems	
– Sleep time > 8 h	
Incontinence	0–1
Total score = "none"–"all" symptoms	0–7/7 = 0%–100%

brain, the CBF volumes were thresholded at 50% of maximum intensity and corrected using morphological operations to close ventricle and white matter volume inside of the brain and exclude extracranial tissue. Similar procedure was performed on T2 MR volumes with a threshold in the first minimum of intensity histogram and appropriate procedure for exclusion of non-brain tissue, such as eyes. The segmentation was checked in all scans and manually adjusted if necessary. We matched the CBF volume to MR minimising the average distance between the outlines (top-hat method) and subsequently resampled it trilinearly to the MR coordinates. Thus, the regions of interest (ROIs) placed on MR could be directly referred to SPECT for pixel intensity analysis. Elliptical ROIs were placed bilaterally [6] in locations as named on Fig. 1. We analysed the maxima of pixel distributions in each ROI. After normalisation to the average of the maxima of the left and right cerebellum counts in each scan, the resulting maxima of relative CBF (rrCBF) they are assumed to carry the information about CBF distribution in grey matter [1]. As the control values we used 75 years old age group of controls from an earlier study.

Results

On clinical assessment 1 patient worsened, 7 remained the same, 5 improved after surgery; range −1 to +2 of 7 selected major NPH functional impairments. This corresponds to an improvement following the surgery by 5.5% (SD = 10%, p = 0.1) from an initial disability score of 50% (SD = 30%).

The regional blood flow values are presented in Fig. 1. In none of the ROIs, neither preoperatively nor postoperatively, there were no statistical differences from controls. None of the rCBF changes resulting from surgery were statistically significant either. Subdividing the patients did not yield any statistical differences depending on scale factors, outcome nor aetiology. Even without Bonferroni correction for multiple comparisons only a few changes were significant. E.g.: in response to surgery rrCBF increased in patients with noncomunicating hydrocephalus in thalamus (+0.06, p = 0.03), in continent patients increased in mesial temporal (+0.07, p = 0.02) and decreased in frontal medial regions (−0.05, p = 0.04). Avoiding such possibly spurious results, we pooled the data between the ROIs to relate the initial clinical presentation score to both the amount of clinical improvement and rrCBF change (Fig. 2).

Discussion

Despite the conceptually attractive link between blood flow and cerebral function, methods based on assessment of cerebral blood flow show a poor relation to clinical evidence as shown in a recent review of multimodal technology applied in normal pressure hydrocephalus [5]. Our negative findings of both a baseline difference from normal controls and the postoperative increase in the grey matter CBF are consistent with the conclusions of Owler and Pickard [5]. It is worth noting that the sample material in this study is small and considerably varied both in the patient's age, aetiology and clinical presentation. From the literature it appears that if there are changes in baseline CBF and corresponding compensation after shunt surgery, they are, without doubt, small in comparison with the reliability of any method currently available to estimate them. In contrast to our earlier study [7], in this one, due to SPECT camera upgrade we had to close material at 13 cases. It is possible that this number is not large enough to achieve statistical significance in the observed effects. Even though we can recognise a similar pattern of initial decrease and postoperative recovery in thalamus, mesial temporal and some frontal regions, the variability is too great to allow statistical significance (Fig. 1). The changes do not become more clear-cut with subdividing the material into groups based on initial presentation. Even though there was a perceptible trend suggesting that the patients who have had more clinical disability present more improvement both on clinical score and rrCBF, the reduction in number of cases after subdivisions makes the chance of non-spurious statistical significance even weaker. Only overall relation such as presented in Fig. 2 can be treated with moderate confidence.

Another potential factor in the analysis of our data may be the relatively low incidence of clinical symptoms in our group of patients. From figure 2 we can see that the initial minimal presentation with clinical symptoms is related to an overall decrease in rrCBF and does not prevent increasing the number of disabilities after surgery. Inclusion of such cases according to current trends [2] in a mixed group like this one, may result in averaging the effect to near zero, such as in our small study. It must be noted that the functional improvement depends most profoundly on the way, how the clinical symptoms are measured. Typically, each of the symptoms is assessed in its own arbitrary scoring scale or using physical measurement units obtained from an appropriate test [3, 4]. Unfortunately, even though the multiplicity of indexes enables detailed study of all clinical factors, the overall picture is often difficult to comprehend. To help with this, there are overall scales such as the NPH Scale for gait disturbance and dementia or the Rankin scale for handicap but they do not span across all clinical symptoms and

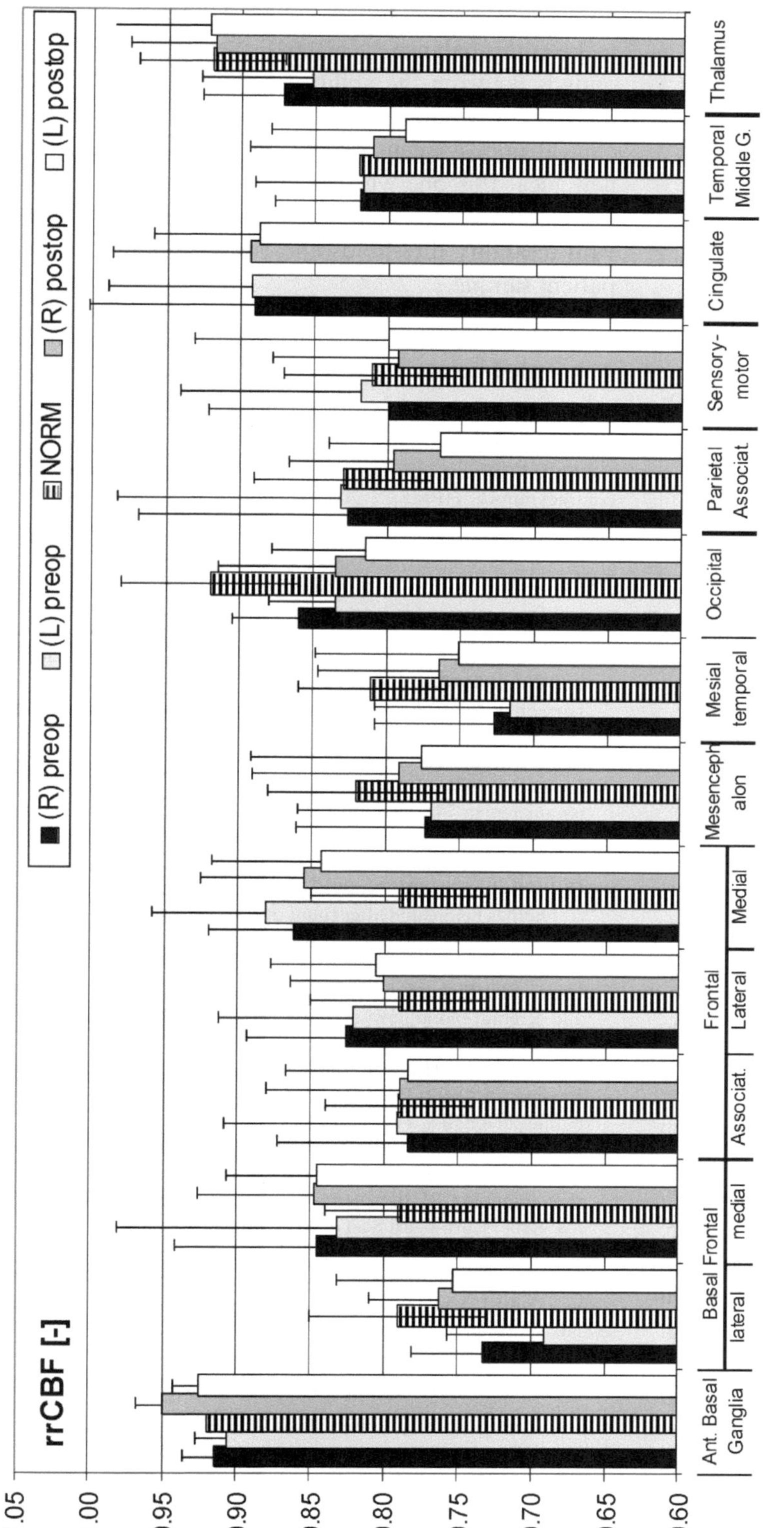

Fig. 1. Cerebellum normalised relative regional cerebral blood flow (rrCBF ± 1SD) in grey matter pre- and post-operatively compared to values obtained in normal controls

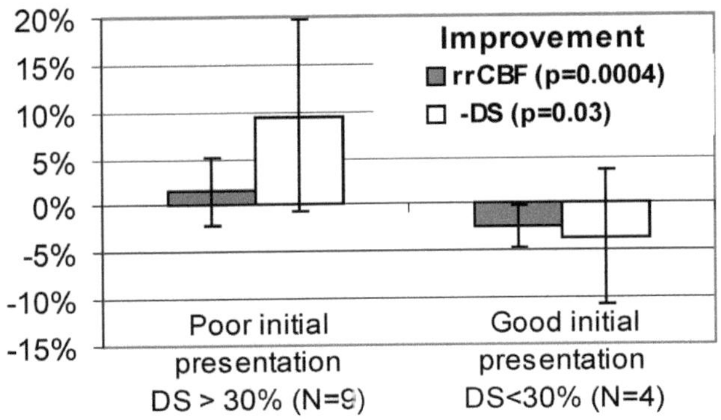

Fig. 2. Improvement in rrCBF across all ROIs and on clinical symptoms, depending on preoperative symptomatic disability score (*DS*). Whiskers denote ± 1SD range, p-value is for 2-tailed T Tests between the groups: paired for each ROI for rrCBF and unpaired for DS

give complex total scores. In this study we have constructed a blueprint of a symptomatic disability score, which makes it easier to compare the overall clinical presentation of a hydrocephalic patient using a percentage of the life affecting incapacities within three categories. We used 3 criteria in motor and mental sections compared to only one for incontinence reflecting their weight on daily life. The main advantage of the new scoring system is that overall normalised scalar scores can be compared even if some tests are replaced or skipped. The score carries similar information on the scale 0–100% of disability provided that allotment of tests within the sections is kept similar. Another potential advantage of the test is its close relation to the quality of life by concentrating on the differences between norm and disability rather than on parametric changes, which are often too small to have impact on daily activities. This last property is most specifically responsible for the lower outcome of shunt surgery than the outcome generally reported in the literature [3]. Even in this small sample, nearly all patients improved on at least one parameter out of the initial 120. In the group there is also statistically significant improvement from average 13 to 21 seconds on ability to keep the balance ($p = 0.04$). Nevertheless, 21 seconds is far from the ability to stand unattended and therefore it would not be recognised as a life affecting clinical improvement on the proposed scale. As any new idea, this one will benefit from further analysis, more systematic selection of clinical criteria and relevant disability thresholds and verification on a larger patient sample.

Acknowledgments

We thank Ms Ann-Christine Bergh for excellent technical assistance. S. K. Piechnik is supported by a Marie-Curie Fellowship of the Fifth European Community Framework Programme under the contract number HPMFCT-2002-01593.

References

1. Arlig A, Larsson A, Bergh AC, Jacobsson L, Wikkelso C (1994) A new method for the relative quantification of rCBF examined by 99Tcm-HMPAO SPECT. Nucl Med Commun 15: 814–823
2. Larsson A, Stephensen H, Wikkelso C (1999) Adult patients with "asymptomatic" and "compensated" hydrocephalus benefit from surgery. Acta Neurol Scand 99: 81–90
3. Larsson A, Wikkelso C, Bilting M, Stephensen H (1991) Clinical parameters in 74 consecutive patients shunt operated for normal pressure hydrocephalus. Acta Neurol Scand 84: 475–482
4. Lindqvist G, Andersson H, Bilting M, Blomstrand C, Malmgren H, Wikkelso C (1993) Normal pressure hydrocephalus: psychiatric findings before and after shunt operation classified in a new diagnostic system for organic psychiatry. Acta Psychiatr Scand Suppl 373: 18–32
5. Owler BK, Pickard JD (2001) Normal pressure hydrocephalus and cerebral blood flow: a review. Acta Neurol Scand 104: 325–342
6. Soderstrom H, Hultin L, Tullberg M, Wikkelso C, Ekholm S, Forsman A (2002) Reduced frontotemporal perfusion in psychopathic personality. Psychiatry Res 114: 81–94
7. Tullberg M, Hellstrom P, Piechnik SK, Starmark J-E, Wikkelso C (2004) Impaired wakefulness is associated with reduced anterior cingulate CBF in patients with normal pressure hydrocephalus. Acta Neurologica Scand (in press)

Correspondence: Stefan K. Piechnik, Institute of Clinical Neuroscience, Sahlgrenska University Hospital, 413 45 Goteborg, Sweden. e-mail: s.k.piechnik@iname.com

Acta Neurochir (2005) [Suppl] 95: 173–175

Dynamic susceptibility contrast-enhanced perfusion MR imaging in late radiation-induced injury of the brain

Y. L. Chan[1], **D. K. W. Yeung**[2], **S. F. Leung**[2], **S. F. Lee**[1], and **A. S. C. Ching**[1]

[1] Department of Diagnostic Radiology and Organ Imaging, The Chinese University of Hong Kong, Prince of Wales Hospital, Hong Kong, China
[2] Clinical Oncology, The Chinese University of Hong Kong, Prince of Wales Hospital, Hong Kong, China

Summary

The objective of the study was to evaluate radiation-induced cerebral injury on dynamic susceptibility contrast-enhanced (DSCE) perfusion MR imaging and study its relationship with morphological severity and disease progression.

Thirty-one patients with known radiation injury to the temporal lobes where studied. Gradient and spin-echo T2-weighted, gadolinium-enhanced T1-weighted and DSCE perfusion MR imaging were obtained in the coronal plane through the anterior temporal lobe. Regions of interest where selected in the anterior temporal lobes and the superior frontal lobe as control for analysis of perfusion parameters.

The mean transit time (MTT) was prolonged in both the High Dose Zone (HDZ) receiving from two-thirds to the total dose of 66–71. 2 Gy, and the Intermediary Dose Zone (IDZ) receiving up to 87% of the total dose. The HDZ but not the IDZ showed a low relative cerebral blood volume (rCBV) and relative cerebral blood flow index (rCBFi). The rCBV and rCBFi were significantly lower in both HDZ and LBZ in temporal lobes with severe lesions compared to the temporal lobes with mild lesions but there was no significant difference in bolus transit parameters. The rCBV and rCBFi were significantly lower in both HDZ and IDZ of the swollen temporal lobes compared to those without swelling.

It was concluded that DSCE perfusion MR imaging demonstrated a derangement in perfusion in radiation-induced cerebral injury in rCBV, rCBFi and MTT, which were related to the severity of the radiation-induced injury and the dose of irradiation delivered.

Keywords: Dynamic susceptibility enhanced MRT; radiation induced brain injury; cerebral blood volume; cerebral blood flow.

Introduction

Radiation-induced cerebral necrosis is one of the most serious complications of radiation therapy (RT) in cerebral and skull base tumors. Vascular injury as the primary event has been widely accepted [1–4]. Damage of the endothelium may be associated with thrombosis and narrowing of lumen from deranged proliferation. Dynamic susceptibility contrast-enhanced (DSCE) MR imaging has been applied clinically in the investigation of stroke, brain tumor, and dementia. The objectives of the current study were to evaluate quantitatively the various perfusion parameters of DSCE perfusion MR imaging in radiation-induced temporal lobe injury and to study their relationship with morphological severity and dose received.

Materials and methods

Thirty-one patients with known radiation injury to the temporal lobes after RT for nasopharyngeal carcinoma were studied. There were 26 men and five women, aged 36–75 years (mean, 52 years). The interval between RT and MR imaging ranged from 6 to 16 years (mean, 8.5 years). All patients were treated on standard RT plans using left, right and anterior radiation beams with a dose of 66–77 Gy covering the nasopharynx and adjacent at risk regions. All patients were recurrence free. The irradiated part of the temporal lobes could be segregated into an intermediate dose zone which was irradiated by the lateral radiation beams but not the anterior radiation beam (due to shielding by eyeshields in the anterior beam). This zone received 67% to 87% of the tumor dose, depending on the relative weighting of the lateral and anterior beams in different plans. The lower boundary of the intermediate dose zone corresponded to the plane of the floor of the pituitary fossa. The infero-medial portion of the anterior temporal lobes below the level of the floor of the pituitary fossa was irradiated by both the lateral beams and the anterior beam and thus subject to 100% tumor dose.

MRI was performed using a 1.5 T MR imager (Gyroscan ACS NT, Philips Medical System, Best, the Netherlands). Coronal 4 mm slices through the anterior temporal lobe were obtained on three pulse sequences:

(i) Gradient and spin-echo T2-weighted [Repetition time (TR) 4550 ms, echo time (TE) 90 ms];
(ii) DSCE perfusion MR imaging;
(iii) Gadolinium-enhanced T1-weighted (TR 500 ms, TE 15 ms).

DSCE perfusion MRI was performed with a gradient echo echo-planar technique. Acquisition was started 10 s after 20 ml intravenous gadodiamide (Omniscan, Nycomed) injection, in a concentration of 0.5 mmol/ml delivered by means of an automatic injector (Medrad, Pitsburgh, PA, USA) at a rate of 5 ml/s, with forty dynamic scans performed in a temporal resolution of 1.9 s (TR/TE 360/30 ms, flip angle 45°, EPI factor 11, FOV 25 cm, matrix 128 × 256, 1 NSA).

Four sections at the anterior temporal lobe were chosen for analysis. In each section, two regions of interest (ROIs) on each side were chosen:

(i) the lateral part of the upper half
(ii) the medial part of the lower half

ROI was selected with avoidance of contrast enhanced areas as shown on T1-weighted images. The lower ROI was targeted at a high dose zone (HDZ) > 60 Gy whilst the upper ROI was targeted at the intermediate dose zone (IDZ) < 60 Gy. ROI was selected with avoidance of contrast enhanced areas as shown on T1-weighted images.

An ROI in white matter of the superior frontal lobe on each side served as control.

Using the tracer kinetic method on the assumption that the T2* signal decrease is linearly and inversely related to the concentration of gadolinium [5], the data was used to generate signal intensity time curve using a gamma variate function based model with five relative perfusion parameters measured [6]:

TA (time of appearance): the time point the gamma-variate function starts;
TP (time to peak contrast concentration): the time point where the function reaches its maximum;
MTT (mean transit time): the first moment of the curve;
rCBV (regional cerebral blood volume): area under the curve;
Relative regional cerebral blood flow index (rCBFi = rCBV/MTT) is calculated by the central volume theorem [7].

Normalized perfusion parameters expressed as ratio were obtained by dividing the values in the ROIs of the anterior temporal lobes by the average of ROIs in the ipsilateral frontal lobe on the corresponding four sections.

Table 1. *Perfusion parameters in anterior temporal and frontal lobes in all patients*

Perfusion parameter	Frontal lobe	Intermediate dose zone in anterior temporal lobe	High dose zone in anterior temporal lobe
All Cases (62 lobes in 31 subjects)			
rCBV	1.90 ± 0.65	2.05 ± 1.17#	1.58 ± 0.97*
MTT	8.07 ± 1.10	9.09 ± 1.67**	9.04 ± 1.9**
rCBFi	0.24 ± 0.084	0.23 ± 0.14#	0.18 ± 0.11**
TA	14.7 ± 1.88	14.18 ± 2.21*#	15.3 ± 3.26
TP	20.2 ± 2.26	20.25 ± 2.59#	21.05 ± 2.47**

* $P < 0.01$, temporal lesion compared with frontal lobe, paired T-test, ** $P < 0.001$, temporal lesion compared with frontal lobe, paired T-test, # $P < 0.01$, superior temporal compared with inferior temporal lobe, paired T-test.

Table 2. *Normalized perfusion parameters with relation to severity*

Ratio of perfusion parameters in temporal lobe relative to ipsilateral frontal lobe	Necrotic group (n = 39)	Mild group (n = 23)	Unpaired T-test, 2 tail
Intermediate dose zone in anterior temporal lobe			
rCBV	0.92 ± 0.56	1.44 ± 0.47	P = 0.00037
MTT	1.15 ± 0.19	1.1 ± 0.12	NS
rCBFi	0.83 ± 0.55	1.32 ± 0.46	P = 0.0006
TA	0.97 ± 0.11	0.95 ± 0.07	NS
TP	1 ± 0.07	1 ± 0.03	NS
High dose zone in anterior temporal lobe			
rCBV	0.66 ± 0.41	1.18 ± 0.24	P < 0.0001
MTT	1.15 ± 0.26	1.08 ± 0.11	NS
rCBFi	0.57 ± 0.36	1.12 ± 0.22	P < 0.0001
TA	1.06 ± 0.17	1.0 ± 0.09	NS
TP	1.06 ± 0.09	1.02 ± 0.05	NS

Results

The perfusion parameters in the anterior temporal and frontal lobes of all patients are shown in Table 1. The normalized perfusion parameters with relation to morphological severity of temporal lobe lesions are given in Table 2.

Discussion

A mild but significantly increase in TP was seen in the HDZ of temporal lobes with necrosis but no significant delay in TP was observed in the IDZ in the same group of patients. This would suggest that the narrowing of the feeding vessel does not occur in the large supplying artery to the anterior temporal lobe (the middle cerebral artery and its major branch) which supplies both the HDZ and IDZ but in the more peripheral branches within the HDZ.

The elevation of MTT in the HDZ or IDZ indicates damage to the vascular unit which is associated with a rise in MTT with delayed emptying of contrast-material. This may occur as a result of narrowing of small arteries or arterioles, occlusion of small arterioles and capillaries, or telangiectatic change.

Significantly reduced rCBV was found in IDZ as well as HDZ of the temporal lobes with severe lesions. The changes may relate to direct RT damage to the capillary bed which may occur in both IDZ and HDZ. In severe lesions the change to necrotic cysts or microcysts is necessarily accompanied by loss of vascular spaces and reduction in rCBV.

In conclusion, DSCE perfusion MR imaging demonstrated a derangement in perfusion in radiation-induced cerebral injury in rCBV, rCBFi and MTT, which were related to the severity of the radiation-induced injury and the dose of irradiation delivered.

References

1. Tofilon PJ, Fike JR (2000) The radioresponse of the central nervous system: a dynamic process. Radiat Res 153: 357–370
2. Hopewell JW (1974) The late vascular effects of radiation. Br J Radiol 47: 157–158
3. Calvo B, Hopewell JW, Reinhold HS, Yeung TK (1988) Time- and dose-related changes in the white matter of the rat brain after single doses of X rays. Br J Radiol 61: 1043–1052
4. Yamaguchi N, Yamashima T, Yamashita J (1991) A histological and flow cytometric study of dog brain endothelial cell injuries in delayed radiation necrosis. J Neurosurg 74: 625–632
5. Fisel CR, Ackerman JL, Buxton RB *et al* (1991) MR contrast due to microscopically heterogeneous magnetic susceptibility: numerical simulations and applications to cerebral physiology. Magn Reson Med 17: 336–347
6. Petrella JR, Provenzale JM (2000) MR perfusion imaging of the brain: techniques and applications. AJR 19: 175–207
7. Klingensmith WC (1983) Regional blood flow with first circulation time-indicator curves: a simplified, physiologic method of interpretation. Radiology 149: 281–286

Correspondence: Yu-Leung Chan, Departments of Diagnostic Radiology and Organ Imaging, Prince of Wales Hospital, Chinese University of Hong Kong, Shatin, Hong Kong, China. e-mail: yl190chan@cuhk.edu.hk

Acta Neurochir (2005) [Suppl] 95: 177–181

MRI study of cerebral blood flow and CSF flow dynamics in an upright posture: the effect of posture on the intracranial compliance and pressure

N. Alperin[1], **S. G. Hushek**[2], **S. H. Lee**[1], **A. Sivaramakrishnan**[1], and **T. Lichtor**[3]

[1] Department of Radiology, University of Illinois at Chicago, Chicago, IL, USA
[2] iMRI Department, Norton Hospital in Louisville, Louisville, KY, USA
[3] Department of Neurosurgery, Rush-Presbyterian St. Luke's Medical Center, Cook County Hospital, Chicago, IL, USA

Summary

Postural related changes in cerebral hemodynamics and hydrodynamics were studied using Magnetic Resonance Imaging (MRI) measurements of cerebral blood flow and cerebrospinal fluid (CSF) flow dynamics. Ten healthy volunteers (mean age 29 ± 7) were studied in supine and upright (sitting) postures. A Cine phase-contrast MRI technique was used to image the pulsatile blood flow to the brain, the venous outflow through the internal jugular, epidural, and vertebral veins, and the bi-directional CSF flow between the cranium and the spinal canal. Previously published analyses were applied to calculate and compare total cerebral blood flow (TCBF), intracranial compliance and pressure in both postures. A lower (12%) mean TCBF was measured in the upright position compared to supine position. A considerable smaller amount of CSF flow between the cranium and the spinal canal (58%), a much larger intracranial compliance (a 2.8-fold increase), and a corresponding decrease in the MRI-derived ICP were also measured in the sitting position. These changes suggest that the increased cerebrovascular and intracranial compliances in the upright posture are primarily due to reduced amounts of blood and CSF residing in their respective intracranial compartments in the upright position. This work demonstrates the ability to quantify neurophysiologic parameters associated with regulation of cerebral hemodynamics and hydrodynamics from dynamic MR imaging of blood and CSF flows.

Keywords: CSF flow dynamics; cerebral blood flow; MR phase contrast; ICP; body posture; intracranial compliance.

Introduction

Cerebral hemodynamics and intracranial hydrodynamics are highly affected by body posture. Characterization of blood and CSF flow dynamics in upright and supine postures may provide insight into the coupling that exist between blood and CSF flow dynamics and may further explain complex neurophysiologic changes associated with different body postures. The ability to image and characterize posture related neurophysiologic changes may have practical implications for the diagnosis and treatment of neurological problems; it may enhance the use of sitting position for several cranial and cervical spine surgeries, the use of postural changes for management of increased intracranial pressure (ICP) in the intensive care, and may help optimize shunt therapy for hydrocephalic patients.

Since changes in cerebral hemodynamics and intracranial hydrodynamics are coupled, simultaneous measurements of hemodynamic and hydrodynamic parameters at different postures may contribute to our understanding of cerebral physiology. Studies in healthy subjects to date, however, were limited by the invasiveness of current methods for measurement of parameters such as intracranial compliance and pressure. Our lab has recently developed a system-based analysis of MRI measurements of transcranial blood and CSF flows for noninvasive quantitation of total cerebral blood flow (TCBF), intracranial compliance and ICP [1, 2]. Thus, the purpose of our study was to quantify the effect of posture on cerebral hemodynamics and hydrodynamics.

Materials and methods

Ten adult volunteers (3 males, 7 females, mean age 29 ± 7 years) with no known neurological problems were imaged in both upright and supine positions in a Signa SP/*i* 0.5T vertical gap MRI scanner (GE Medical Systems, Milwaukee). The imaging protocol was approved by the institutional review board and informed consent was obtained from all subjects. The vertical gap allows the subject to be positioned seated upright, with his or her upper cervical spine at the magnet isocenter as shown in Fig. 1.

Two retrospectively gated cine phase-contrast scans were used to

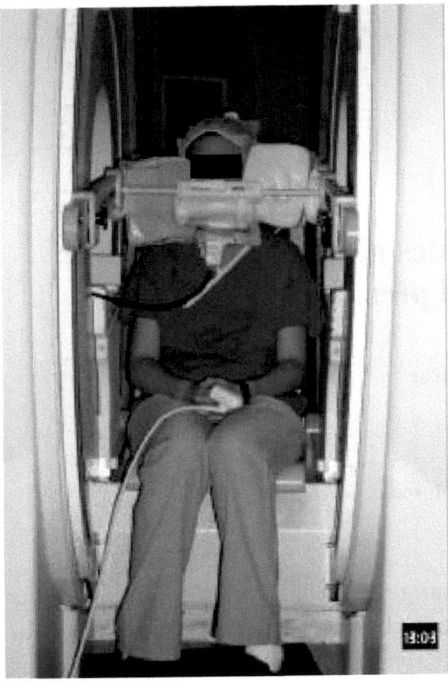

Fig. 1. A subject sitting in the vertical gap of the MRI scanner

measure arterial inflow, venous outflow, and CSF flow between the cranium and the spinal canal. A scan with high velocity encoding (VENC = 80 cm/s) located above the carotid bifurcation perpendicular to the direction of flow was used to quantify blood flow. The scan parameters were TR = 19 ms, FA = 25 deg, FOV = 16 cm, NEX of 2 or 4, and slice thickness = 8 mm. A second scan with low velocity encoding (VENC = 7–9 cm/s) located at the level of C2 perpendicular to the direction of the flow was used to quantify the slower CSF flow and venous flow in the vertebral plexus. The parameters of the low VENC scan were TR = 28 ms, FA = 30 deg, FOV = 16 cm, NEX of 2 or 4, and slice thickness = 8 mm.

Measurements of total cerebral blood flow and CSF flow

Each of the two cine phase-contrast MRI scans generates a set of 32 velocity-encoded images of the pulsatile flow during one cardiac cycle with pixel values proportional to velocity. Volumetric flow rate (VFR) through a blood vessel or the cervical spine CSF space was obtained by integrating the velocities through the lumen cross-sectional area. An automated lumen segmentation technique, pulsatility based segmentations (PUBS) [3], was used to delineate the boundary of the lumens for improved measurement accuracy and reproducibility. Mean VFR over the cardiac cycle is then calculated for each of the four vessels carrying blood to the brain, the right and left internal carotid and vertebral arteries. TCBF is obtained by summation of the mean VFR of the four vessels. Venous outflow was quantified in the following venous pathways, the internal jugular veins (IJV), vertebral veins, epidural veins, and the deep cerebral veins.

CSF flow during the cardiac cycle was measured from the low-velocity encoding phase-contrast image series. Since the CSF flow is bi-directional, outflow toward the spinal canal occurs in systole and reverse flow during diastole, the amount of CSF volume that flows back-and-forth (oscillatory CSF volume) is obtained by integrating the absolute values of the CSF flow waveform over the cardiac cycle and dividing the sum by two.

MRI-based derivation of intracranial and cerebral vascular compliance

Intracranial compliance (defined as the ratio of volume and pressure change, dV/dP) is measured invasively by injecting a known volume of fluid into the ventricles and measuring the resulting pressure change [4, 5]. The MRI-based method is analogous to the invasive method except for it utilizes the small change in intracranial volume and the corresponding pressure change that occur naturally with each cardiac cycle to derive the intracranial compliance and ICP, noninvasively [1]. The pressure change during the cardiac cycle is derived from the CSF pressure gradient waveform, which is calculated using the Naveir Stoke relationship between spatial and temporal derivatives of the CSF velocities and the pressure gradient. The intracranial volume change (ICVC) during the cardiac cycle is derived from the instantaneous difference between arterial blood inflow, venous blood outflow, and CSF volumetric flow rates into and out of the cranial vault. A mean ICP value is then derived based on a linear relationship that exists between intracranial elastance (the inverse of compliance) and ICP [5]. An intracranial compliance index and the derived ICP (MR-ICP) were calculated for each subject at the two postures.

Results

In the supine posture, the venous outflow is primarily through the IJV while in the upright posture the IJV's were either partially or fully collapsed and the main pathway for venous drainage was the cerebral venous plexus. An example of high-velocity encoding phase-contrast images from one of the subjects obtained in the supine and the upright postures is shown in Fig. 2. In these images, white pixels represent velocity in the cranio-caudal direction, i.e., venous flow, and black pixels are flow toward the brain, i.e., arterial flow. In the supine posture, venous flow velocities are seen mainly in the IJV while in the sitting posture, venous flow velocities are seen mainly in the vertebral veins.

A lower TCBF was measured in the sitting posture compared with the upright posture in each of the 10 subjects. The average and standard deviation (SD) TCBF in the supine and upright postures were 825 ± 166 and 724 ± 127 ml/min, respectively. The mean and SD percentage reduction in TCBF was $12 \pm 7\%$. Average and SD of mean venous flow through the IJV in the supine and upright postures were 614 ± 143 and 304 ± 261 ml/min, respectively.

The inter-subjects average and SD of the total CBF, the oscillatory CSF volume, intracranial volume change, intracranial compliance, and the derived MR-ICP are summarized in Table 1. Compared with the

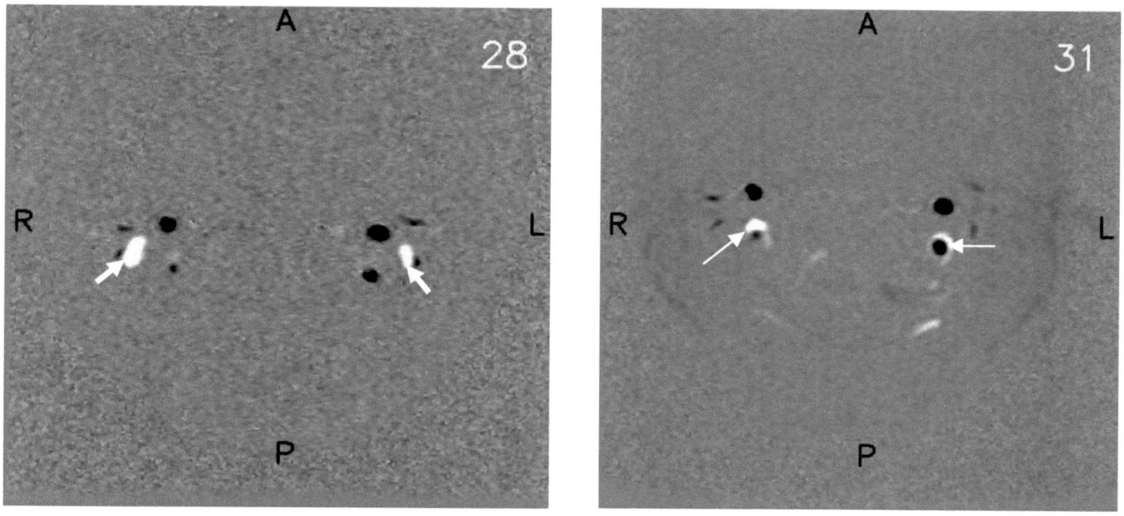

Fig. 2. One of the 32 high velocity-encoding axial phase-contrast Gradient Echo MR images obtained in the supine (A) and upright (B) postures. Black pixels indicate flow in the cranio-caudal direction (i.e., arteries) and white pixels indicate flow in the caudal-cranio direction (i.e., veins). For this subject, in the supine posture the dominant venous drainage is through the jugular veins (white arrows), and in the upright posture the dominant flow is through the vertebral veins (white arrows)

Table 1. *Mean and SD of main hemodynamic and hydrodynamic parameters measured in supine and upright postures*

	Supine (Mean ± SD)	Upright (Mean ± SD)
TCBF (ml/min)	825 ± 166	724 ± 127
Oscillatory CSF Volume (ml)	.55 ± .12	.23 ± .11
ICVC (ml)	.48 ± .15	.89 ± .44
Compliance Index	7.3 ± 2.6	20.2 ± 10.7
MR-ICP (mmHg)	10.6 ± 3.6	4.5 ± 1.8

* All differences are statically significant with a P value of 0.002 or smaller.

supine position, in the sitting position, on average, the oscillatory CSF volume is smaller by a factor of 2.4, the intracranial compliance is larger by a factor of 2.8, and the ICP is lower by a factor of 2.4. All these changes are statistically significant at a P value of 0.02 or smaller.

Plots of the measured CSF flow waveforms at the two postures are shown in Fig. 3. The CSF flow waveforms are plotted together with the net transcranial blood flow (arterial inflow-venous outflow) to demonstrate how the CSF flow is driven by the net blood flow. Note that the in the supine position, the CSF flow waveform follows the A-V flow waveform "more closely" – an indication of a lower intracranial compliance.

Discussion

A comprehensive characterization of the cerebral hemodynamics, venous drainage, and intracranial hydrodynamics in healthy subjects in the supine and sitting postures was performed using a vertical gap MRI scanner and a system based analysis of CSF and blood volumetric flow rates. Statistically significant differences between supine and sitting postures were documented and quantified. On average, approximately 50% of the venous outflow through the IJV in the supine position shifts in the sitting position to secondary venous channels, e.g., epidural, vertebral, and deep cervical veins. This is in agreement with previously reported findings of reduced flow in the IJV in the sitting posture [6, 7]. However, in this study, a smaller fraction of the flow through the IJV shifted to secondary veins compared with the approximately 90% reported recently by Valduaze *et al.* [7]. This difference can be explained by the reduced reproducibility and accuracy of ultrasound based VFR measurements. In addition, hydrodynamic parameters were also strongly affected by the body posture. Differences in the hydrodynamic parameters between the two postures included a 2.4-fold reduction in the CSF volume that moves in and out of the cranium and a large increase (2.8-fold) in the intracranial compliance with a corresponding decrease in mean ICP. All these changes are consistent with a more compliant intracranial compartment

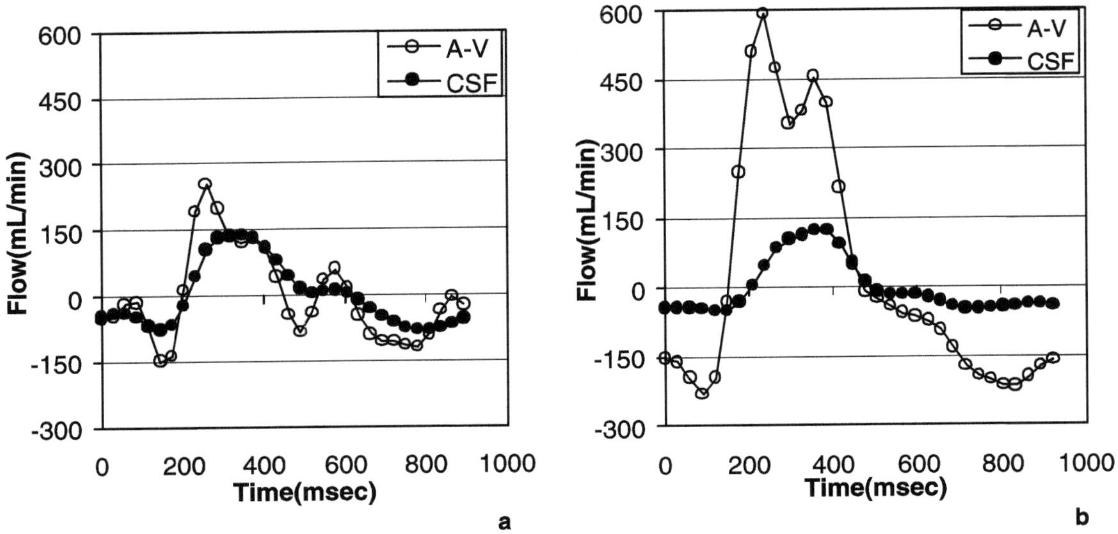

Fig. 3. Graphs depicting the A-V and CSF flow waveforms during one cardiac cycle in the supine (a) and upright (b) postures. Note that the CSF waveform follows the A-V less closely in the upright posture compared to the supine posture – an indication of a more compliance intracranial compartment

caused by smaller mean volumes of blood and CSF residing in the intracranial space in the sitting position.

The amplitude of the net transcranial blood flow is much larger in the sitting position than in the supine position (Fig. 3). A larger net transcranial blood flow results in a larger increase in blood volume during the systolic phase. However, because of reduced volumes of blood and CSF in the cranial vault in the sitting position, the blood entering the cranium can be easily accommodated even with less CSF being displaced into the spinal canal. The combination of smaller amount of CSF that leaves the cranium with each cardiac cycle, and larger net transcranial inflow explains the larger ICVC measured in the sitting position (Table 1).

The MRI-based hemodynamic and hydrodynamic measurements provide a clearer picture of the postural related physiological changes. An accepted explanation for a lower ICP in an upright posture argues that venous flow is increased in that position and this avoids compression of the jugular veins [8]. Since in steady state, mean venous outflow equals mean arterial inflow, the total venous outflow, as is TCBF, is actually reduced in the upright posture. The lower ICP is therefore, most likely, the result of the increased intracranial compliance due to a lower mean volume of the intracranial compartment caused by the reduced volume of CSF and blood. The shift of blood and CSF volume out of the cranium occurs during the transition from the supine to the upright posture. Further, the ability to quantify, noninvasively, the effect of posture on important neurophysiologic parameters offers a new diagnostic test for the evaluation of functions such as regulation of CBF and ICP, and the effect of pathologies on these functions.

Acknowledgments

This work is supported by a grant from the Dana Foundation's Clinical Hypotheses Program in Imaging.

References

1. Alperin N, Lee SH, Loth F, Raksin P, Lichtor T (2000) MR-intracranial pressure (ICP): a method for noninvasive measurement of intracranial pressure and elastance. Baboon and Human Study. Radiology 217(3): 877–885
2. Raksin P, Alperin N, Sivaramakrishnan A, Surapaneni S, Lichtor T (2003) Noninvasive intracranial compliance and pressure from dynamic MR imaging of blood and CSF flows: review of principles, implementation, and other noninvasive approaches. Neurosurg Focus 14(4): article 4
3. Alperin N, Lee SH (2003) PUBS: pulsatility based segmentation of lumens conducting nonsteady flow. Mag Resonance Med 49: 934–944
4. Ryder HW, Espey FF, Kimbel FD, Penka EJ, Rosenauer A, Evans JP (1953) The mechanism of change in cerebrospinal fluid pressure following an induced change in volume of the fluid space. J Lab Clin Med 41: 428–435
5. Marmarou A, Shulman K, La Mores J (1975) Compartmental analysis of compliance and outflow resistance of the CSF system. J Neurosurgery 43: 523–534
6. Cirovic S, Walsh C, Fraser WD, Gulino A (2003) The effect of posture and positive pressure breathing on the hemodynamics of

the internal jugular vein. Aviation Space Envir Med 74(2): 125–131
7. Valdueza JM, Munster TV, Hoffman O, Schreiber S, Einhaupl KM (2000) Postural dependency of the cerebral venous outflow. The Lancet 355: 200–201
8. Nathan BR (1999) Cerebrospinal fluid and intracranial pressure; treatment of increased intracranial pressure. In: Goetz CG, Pappert EJ (eds) Textbook of clinical neurology. WB Saunders Company, Philadelphia

Correspondence: Noam Alperin, Physiological Imaging and Modeling Lab, Department of Radiology (M/C 711), The University of Illinois at Chicago, 830 S. Wood St., Chicago, IL 60612, USA. e-mail: alperin@uic.edu

Acta Neurochir (2005) [Suppl] 95: 183–190

Quantitative evaluation of cerebrovascular reactivity in brain tissue by a refill kinetic method of transcranial ultrasonic perfusion imaging: a comparison with Doppler sonography

T. Shiogai[1], **A. Morisaka**[1], **N. Takayasu**[2], **K. Yoshikawa**[2], **T. Mizuno**[2], **M. Nakagawa**[2], and **H. Furuhata**[3]

[1] Department of Clinical Neurosciences, Kyoto Takeda Hospital, Kyoto, Japan
[2] Department of Neurology, Kyoto Prefectural University of Medicine, Kyoto, Japan
[3] Medical Engineering Laboratory, Tokyo Jikei University School of Medicine, Tokyo, Japan

Summary

To confirm the reliability of a refill kinetic method of ultrasonic harmonic perfusion imaging (HPI) capable of quantifying separate parameters of microvascular blood flow velocity and volume in brain tissue, we evaluated acetazolamide (ACZ) cerebrovascular reactivity by transcranial HPI in comparison with Doppler sonography (TCD).

Methods. HPI during continuous Levovist™ infusion with changing pulsing intervals (t) and TCD time-averaged maximum velocity (TAMX) in the middle and posterior cerebral arteries were evaluated before and after ACZ administration in 12 patients, 8 without and 4 with a temporal skull defect. Plateau value (A) and rise rate (β) of intensity (I) represented by HPI time-intensity curves of $I(t) = A(1 - e^{-\beta * t})$ were analyzed on the axial diencephalic plane.

Results. 1) A significantly decreased in proportion to the region of interest location depth only in the intact skull cases. 2) Despite inter- and intra-individual data scattering, in correspondence with TAMX increases after ACZ, significant β increases were more frequently identified than increases of A.

Conclusions. Cerebral vasoreactivity analysis utilizing refill kinetics of transcranial HPI can potentially provide separate quantification based on microvascular blood velocity and volume (capillary patency) with consideration of depth-dependant ultrasound attenuation. This should be suitable for bedside evaluation of neurointensive care patients.

Keywords: Transcranial ultrasonic perfusion imaging; second harmonic imaging; echo-contrast agents; transcranial Doppler sonography; acetazolamide cerebrovascular reactivity; refill kinetics; quantification; skull defects; intact skull.

Introduction

Transcranial ultrasonic harmonic imaging utilizing echo-contrast agents (ECA) has been introduced for repeatable non-invasive bedside measurements of brain tissue perfusion, and has been used in quantitative evaluation utilizing intravenous bolus tracer kinetics [10, 12, 17]. Parameters from time-intensity curve analysis after a bolus ECA injection have been evaluated [4, 6, 10, 12, 17] and correlated with dynamic CT [17] and perfusion MRI [8]. However, quantitative reliability of the transcranial harmonic perfusion imaging (HPI) has not yet been established, mainly due to skull- and depth-dependent attenuation of the ultrasound signals [10, 12, 17]. Furthermore, the dye-dilution principle [7] commonly utilized in neuroradiological perfusion imaging would not be applicable for the bolus kinetics of HPI due to the additional problems of bubble saturation [1] and shadowing effects [13].

To overcome problems of intravenous bolus kinetics, refill kinetics utilizing constant ECA intravenous infusion have been introduced in myocardial echocardiography [20] and have been tested in the quantitative evaluation of brain tissue perfusion in canine subjects with skull defects [11] and with an intact skull [14], and in normal subjects [15]. The refill kinetic method of HPI is able to quantify separate parameters of microvascular blood flow volume (A) and velocity (β) in brain tissue without knowledge of arterial input function [20]. Furthermore, calculated blood flow ($F = A \times \beta$) correlates neatly with cerebral blood flow measured by radiolabeled microsheres [11].

To overcome the problem related to depth-dependent attenuation, we have introduced acetazolamide (ACZ) vasoreactivity tests for quantitative evaluation at the same depth by HPI, and correlated these with transcranial Doppler sonography (TCD) and dynamic CT [18]. The objective of this study is to confirm

Table 1. *Clinical characteristics of patients with an intact skull (IS) and skull defects (SD)*

	IS	SD
n	8	4
Age mean ± SD	75 ± 10	64 ± 11
(range)	(58–85)	(55–79)
Primary diagnosis		
Cerebral infarction		
lacunar	2	0
atherothrombotic	1	0
embolic	3	0
Cerebral hemorrhage	1	2
Subarachnoid hemorrhage	0	2
Alzheimer dementia	1	0
*Major site of lesions**		
R	3	2
L	3	2
Diffuse or none	2	0
Examined side		
R	7	2
L	1	2

* CT, MRI, MRA, and/or Color Duplex Sonography.

the reliability of this method. Therefore, we evaluated ACZ cerebrovascular reactivity in neurological patients by transcranial HPI and compared this with TCD.

Materials and methods

The subjects were 12, mainly stroke patients (aged 55–85 years; mean, 72) with stable clinical conditions in the chronic stage: 8 patients had an intact skull (IS) and 4 patients had craniectomized skull defects (SD) (Table 1). Informed consent was obtained from patients and/or patients' family members.

Utilizing a SONOS 5500 ultrasound system with a S4 ultrabands (1.8/3.6-MHz) transducer (Philips) 124 images, taken by transient response harmonic B-mode imaging, were evaluated during continuous Levovist™ (400 mg/ml) infusion via the antecubital vein, with changing pulsing intervals (t: ms) of: 250, 500, 750, 1000, 1500, 2000, 3000, 4000 (Fig. 1a). The investigation depth was 12 cm with a focus on 6 cm. The mechanical index, system gain, and compression were 1.6, 75 (or 100) and 70, respectively. The images were recorded by T-INT mode and stored on an MO disk. The HPI was evaluated at resting state with 2 infusion rates (0.5 and 1 ml/min.), and 15 minutes (0.5 ml/min.) and 30 minutes (1 ml/min.) after Diamox® 500 mg intravenous injection (Figs. 1 and 2). Time-averaged maximum velocity (TAMX) in the middle and posterior cerebral arteries (MCA and PCA) was measured by TCD just before HPI. The relative changes (%Δ) of TAMX were also evaluated at rest and after ACZ administration (TAMX after ACZ – TAMX at rest/TAMX at rest × 100). Time-intensity curves of HPI were created for 3 regions of interest (ROI) on the axial plane involving the temporal lobe (TL), basal ganglia (BG), and thalamus (Th) via a temporal window (Fig. 1b). Quantification was performed by Acoustic Densitometry [2]. Plateau value (A) and rise rate (β) of intensity (I) represented by a curve of $I(t) = A(1 - e^{-\beta * t})$ and calculated F value ($= A \times \beta$) were analyzed. The curve-fittings for measured data were analyzed by Kyplot (version 2.0). Utilizing a Student t-test and one-way analysis of variance (ANOVA), statistical significance was set at p values less than 0.05.

Results

a) TCD (Table 2): TAMX both in the MCA and PCA after ACZ significantly increased (except for 30 min. in the MCA) in IS cases and tended to increase in SD cases. These increases (%Δ) in the MCA and PCA were 20–21% and 27–40% in IS cases and 24–30% and 25–37% in SD cases, respectively.

b) HPI (Table 3): 1) In IS cases only, A decreased in proportion to ROI location depth (Fig. 1b) in both infusion rates, and both at rest and after ACZ. The decreases of A were significant at rest during 1 ml/min. infusion. In contrast, in SD cases there was no such tendency of decreases in proportion to ROI location depth. However, A in BG tended to be higher than in TL and Th for both infusion rates and both at rest and after ACZ. Regarding infusion rates, there was no dependency in IS cases, but there was a tendency of increase in SD cases. 2) In terms of β, there was no tendency of decreases proportional to ROI location depth in either IS or SD cases. However, in comparison with A, inter- and intra-individual data scattering of β was so pronounced that it resulted in further data scattering of F. Furthermore, the data scattering was more apparent in SD cases than in IS cases. 3) Increases after ACZ were more frequently identified in β (Fig. 2b), thereby impacting on F, mainly in IS cases. Significant increases were observed only in IS cases of β and F in the Th and β in the BG during 0.5 ml/min. infusion and F in the BG during 1 ml/min. infusion. ACZ effects for A were not apparent, particularly in IS cases.

Discussion

In order to clarify the effects of temporal skull- and depth-dependant ultrasound attenuation, we evaluated parameters derived from the refill kinetics of 3 ROIs (TL, BG, and Th) both in IS and SD cases. As a result, A showed proportional decreases to ROI location depth, particularly in IS cases. The A is the plateau value of the echo-enhancement so that the intensity of echo-enhancement directly depended on the ultrasound attenuation. In contrast, we confirmed the result of a previous study [15] that β did not show attenuation in both IS and SD cases.

In the relationship between infusion rate increase

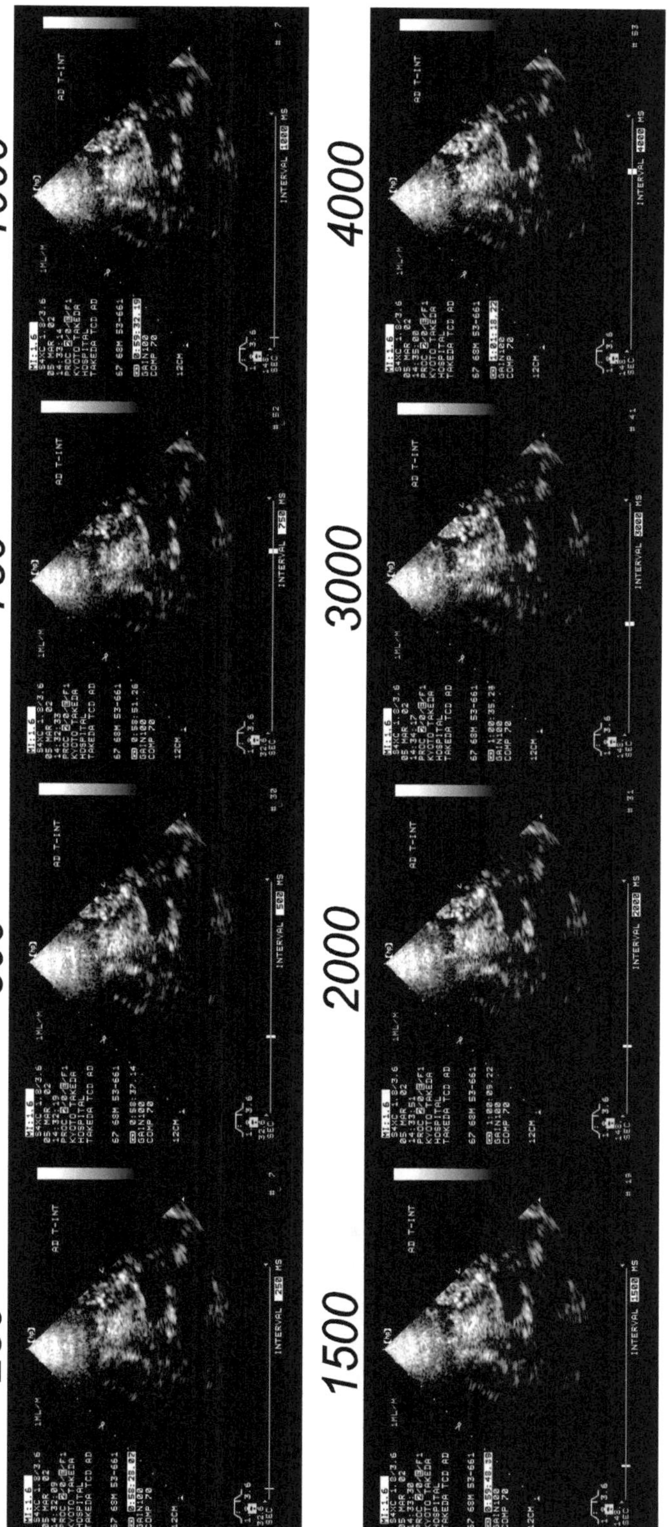

a

Fig. 1. A 68-year-old IS patient with lacunar infarctions in the Corona Radiata: (a) Serial harmonic perfusion images with increases of pulsing intervals (*t*) during continuous Levovist™ infusion (1 ml/min)

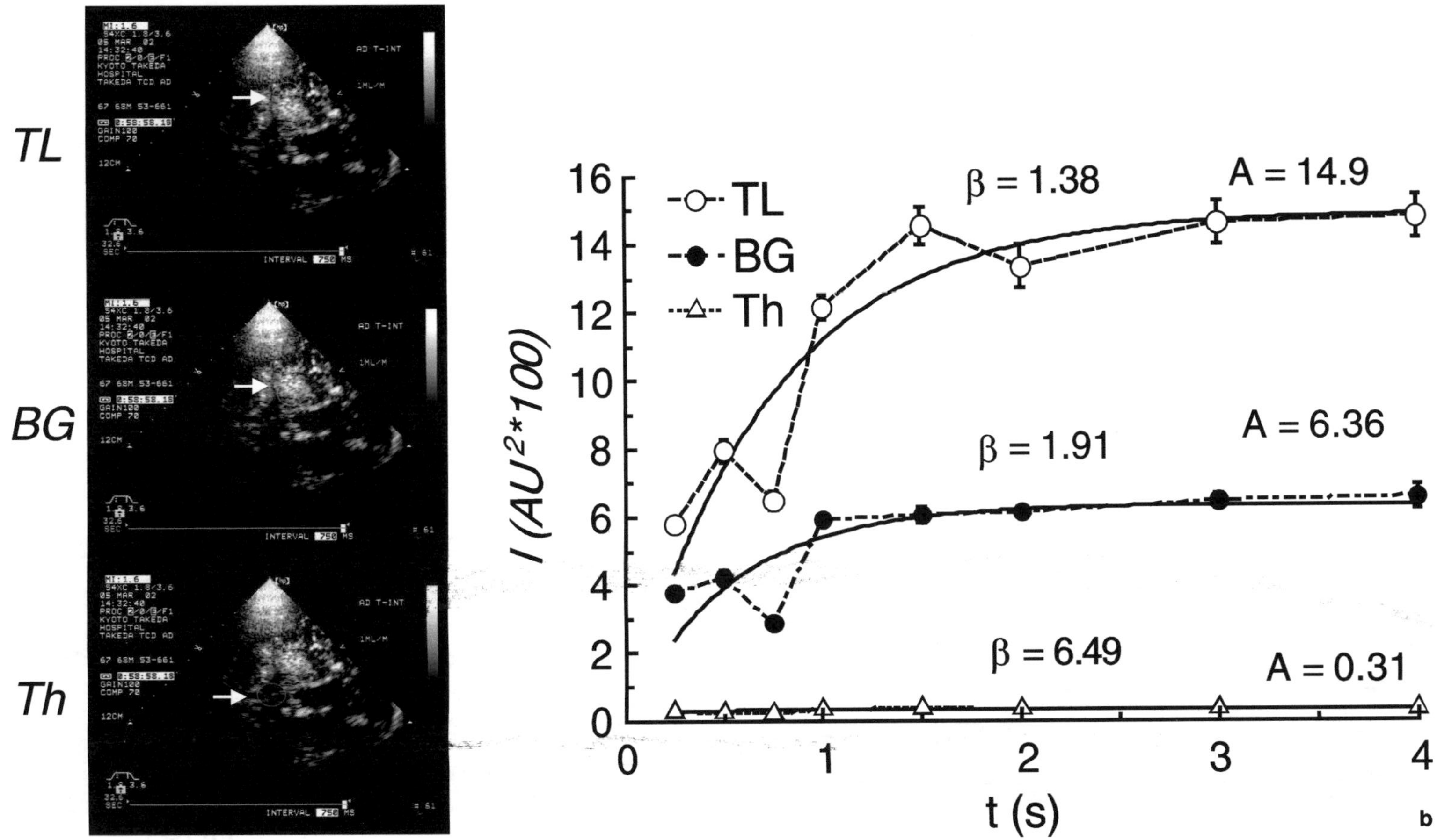

Fig. 1b (*continued*). Data and fitted curves in the 3 regions of interest (*ROI*) placed at the temporal lobe (*TL*), basal ganglia (*BG*), and thalamus (*Th*) (arrows) and are presented in open circles, closed circles, and open triangles, respectively. A and β values were calculated by the t-intensity curves in the 3 ROIs

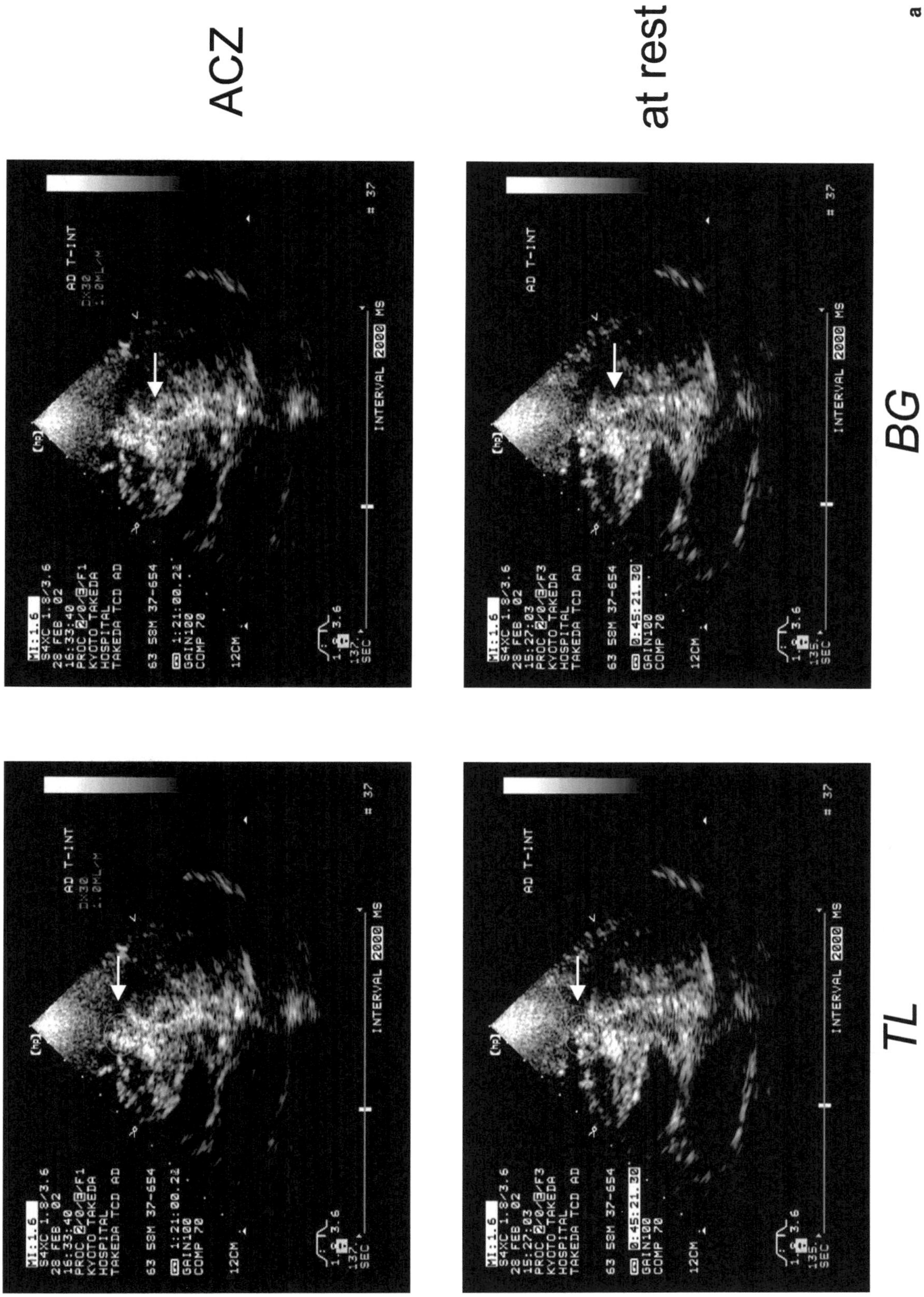

Fig. 2. A 58-year-old IS patient who has a large embolic infarction in the frontal-temporal lobes: (a) Harmonic perfusion images at rest (lower panels) and after ACZ (upper panels). The ROIs at the TL and BG (arrows) are shown in the left and right images, respectively.

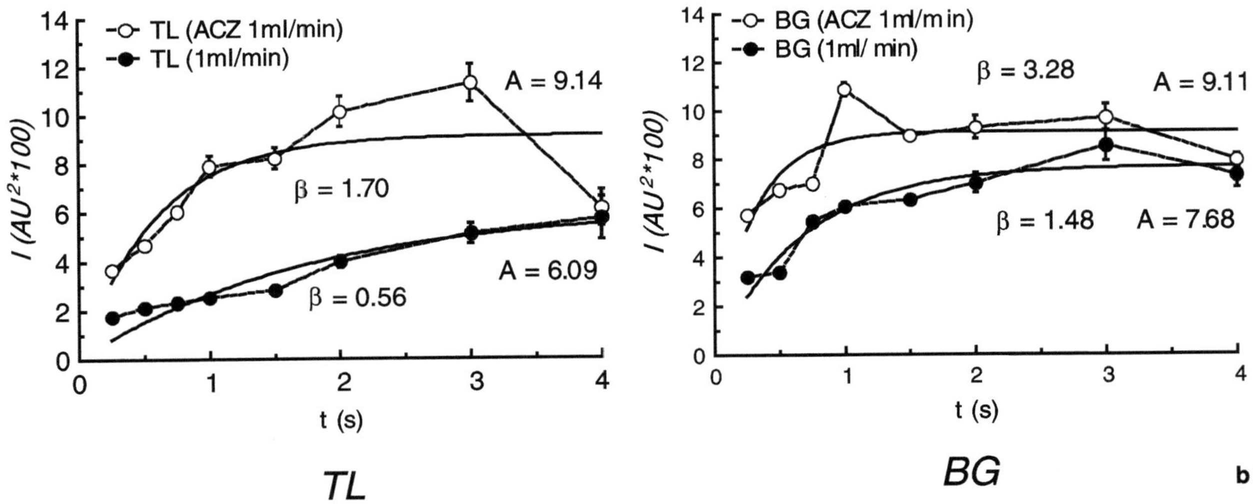

Fig. 2b (*continued*). A and β values were calculated by the t-intensity curves in the left panel of TL and in the right panel of BG. Data and fitted curves were presented at rest (closed circles) and after ACZ (open circles)

Table 2. *TCD Time-averaged Maximum Velocity (TAMX, cm/s) at rest and 15 minutes (ACZ15) and 30 minutes (ACZ30) after acetazolamide administration**

	IS	SD
MCA at rest	48 ± 17	55 ± 11
ACZ15 (%Δ)	59 ± 27† (21 ± 19)	70 ± 4 (30 ± 24)
ACZ30 (%Δ)	56 ± 23 (20 ± 26)	70 ± 7 (24 ± 26)
PCA at rest	28 ± 9	30 ± 5
ACZ15 (%Δ)	38 ± 10‡ (27 ± 16)	41 ± 8 (25 ± 14)
ACZ30 (%Δ)	38 ± 13‡ (40 ± 24)	43 ± 6 (37 ± 27)

* Mean ± SD, † $p < 0.05$, ‡ $p < 0.01$.

and refill kinetic parameters, dependent increases of A and/or F were identified in studies of a myocardial experiment [20] and in a cerebral experiment in normal subjects with an intact skull [14, 15]. However, we identified a tendency of A increases only in SD cases and no dependency in IS cases. This discrepancy in IS cases seems to be caused by skull-dependent ultrasound attenuation in relation to an ECA difference between the first generation of ECA Levovist™ used in our study and second-generation ECA Optison™ in the previous studies [14, 15]. Levovist™, which consists of air with galactose particles and palmitic acids, is necessary mainly to destroy bubbles when creating harmonic perfusion images and, therefore, is more suitable for SD cases [19] than IS cases. Optison™, which consists of a perfuluoropropane gas with albumin solution, is able to produce harmonic signals which come from resonant or oscillating phenomena produced by relatively low acoustic pressure and, therefore, is more suitable for images of microcircula-

Table 3. *HPI findings at rest and 15 minutes (ACZ15) and 30 minutes (ACZ30) after acetazolamide administration**

		IS				SD			
		0.5 ml/min.		1 ml/min.		0.5 ml/min.		1 ml/min.	
		at rest	ACZ 15	at rest	ACZ 30	at rest	ACZ 15	at rest	ACZ 30
A	TL	6.38 ± 3.76	7.58 ± 3.60	6.33 ± 1.80†	7.59 ± 3.83	4.07 ± 2.81	4.10 ± 1.29	7.96 ± 4.53	12.2 ± 5.67
	BG	3.08 ± 1.11	2.58 ± 0.96	3.18 ± 0.96†	3.23 ± 1.36	15.5 ± 6.29	35.0 ± 21.8	52.8 ± 38.2	39.9 ± 27.6
	Th	0.80 ± 0.37	1.13 ± 0.40	1.31 ± 0.70†	0.97 ± 0.48	8.05 ± 5.72	6.98 ± 3.17	21.6 ± 13.7	24.7 ± 12.1
β	TL	16.0 ± 7.12	20.1 ± 8.97	7.62 ± 6.22	18.8 ± 7.90	32.8 ± 10.6	26.9 ± 13.6	15.2 ± 12.4	21.5 ± 19.8
	BG	19.6 ± 7.14	37.8 ± 6.44§	3.23 ± 1.10	13.8 ± 7.15	29.1 ± 15.4	27.6 ± 11.9	16.8 ± 12.2	38.6 ± 34.5
	Th	15.6 ± 5.52	36.1 ± 5.92§	19.4 ± 6.90	14.1 ± 5.33	50.4 ± 7.53	17.8 ± 12.4	27.5 ± 15.7	25.3 ± 20.0
F	TL	36.5 ± 23.1	103 ± 79.5	52.6 ± 42.4	51.0 ± 29.1	188 ± 159	109 ± 52.2	86.7 ± 70.4	277 ± 272
	BG	45.9 ± 17.3	118 ± 52.0	10.5 ± 4.41	26.8 ± 10.8§	485 ± 321	289 ± 114	224 ± 76.3	2549 ± 2533
	Th	4.64 ± 1.39	49.0 ± 20.1§	16.0 ± 8.06	11.3 ± 5.77	517 ± 397	228 ± 201	445 ± 413	565 ± 552

* *TL* temporal lobe, *BG* basal ganglia, *Th* thalamus; mean ± SE, † $p < 0.05$ (ANOVA), § $p < 0.05$ (Student t-test).

tion [16], and would probably be advantageous in IS cases. As a previous study [14] indicated, we confirmed no dependence of infusion rates on β in either IS or SD cases.

In relation to TCD TAMX increases after ACZ, significant increases of β were more frequently identified than those of F in only IS cases. The ACZ effects for A were not always apparent in both IS and SD cases. An experimental study in IS cases showed higher increases of F compared to β after ACZ [14]. In another study involving craniectomized dogs during hyper- and hypo-ventilation, changes of A were more significant than β and F [11]. Several factors probably affect these discrepancies of vasoreactivity on the basis of A, β, and $F = A \times \beta$, e.g. pathological differences, ROI size involving gray and/or white matter, presence of temporal bone, temporal bone structure and thickness, $PaCO_2$, ACZ dose, and time after ACZ etc.

Ultimately, ACZ vasoreactivity utilizing refill kinetics is able to evaluate capillary patency by A increase, and flow velocity in the microcirculation by β increase [5], separately. Our results on the basis of ACZ vasoreactivity in mainly ischemic stroke patients indicated that cerebral capillary patency is more disturbed than microvascular velocity in the pathological brain tissue.

Inter- and intra-individual data scattering of parameters, particularly β impacting on F, were probably due to probe holdings during data acquisition (less than 3 minutes in this study), dynamic range between harmonic B-mode in our study and integrated backscatter in previous studies [14, 15], inhomogeneous insonation conditions in relation to inter-individual difference of temporal bone windows [14, 15] and skull defect size, and so on. Recently, to overcome the problems of refill kinetics, diminution [9] and depletion [3] kinetics have been introduced; however, further studies are essential for complete quantitative measurements of brain tissue perfusion. In conclusion, cerebral vasoreactivity analysis utilizing refill kinetics of transcranial HPI can potentially provide separate quantification based on microvascular flow velocity as β and volume (capillary patency) as A in the brain with consideration of depth-dependant ultrasound attenuation. In particular, β is suitable as a marker of vasoreactivity disturbance in ischemic brain tissue and advantageous for independence of depth and infusion rates. This would be suitable in bedside evaluations of neurointensive care patients with ischemic brain injuries.

References

1. Claassen L, Seidel G, Algermissen C (2001) Quantification of flow rates using harmonic grey-scale imaging and an ultrasound contrast agent: an in vitro and in vivo study. Ultrasound Med Biol 27: 83–88
2. D'Sa A (1999) Acoustic densitometry, Hewlett-Packard Company, USA
3. Eyding J, Wilkening W, Reckhardt M, Schmid G, Meves S, Ermert H, Przuntek H, Postert T (2003) Contrast burst depletion imaging (CODIM): a new imaging procedure and analysis method for semiquantitative ultrasonic perfusion imaging. Stroke 34: 77–83
4. Eyding J, Krogias C, Wilkening W, Meves S, Ermert H, Postert T (2003) Parameters of cerebral perfusion in phase-inversion harmonic imaging (PIHI) ultrasound examinations. Ultrasound Med Biol 29: 1379–1385
5. Frankel HM, Garcia E, Malik F, Weiss JK, Weiss HR (1992) Effect of acetazolamide on cerebral blood flow and capillary patency. J Appl Physiol 73: 1756–1761
6. Harrer JU, Klotzsch C (2002) Second harmonic imaging of the human brain: the practicability of coronal insonation planes and alternative perfusion parameters. Stroke 33: 1530–1535
7. Meier P, Zierler LK (1954) On the theory of the indicator-dilution method for measurement of blood flow and volume. J Appl Physiol 6: 731–744
8. Meves SH, Wilkening W, Thies T, Eyding J, Holscher T, Finger M, Schmid G, Ermert H, Postert T (2002) Comparison between echo contrast agent-specific imaging modes and perfusion-weighted magnetic resonance imaging for the assessment of brain perfusion. Stroke 33: 2433–2437
9. Meyer K, Seidel G (2002) Transcranial contrast diminution imaging of the human brain: a pilot study in healthy volunteers. Ultrasound Med Biol 28: 1433–1437
10. Postert T, Muhs A, Meves S, Federlein J, Przuntek H, Buttner T (1998) Transient response harmonic imaging: an ultrasound technique related to brain perfusion. Stroke 29: 1901–1907
11. Rim SJ, Leong-Poi H, Lindner JR, Couture D, Ellegala D, Mason H, Durieux M, Kassel NF, Kaul S (2001) Quantification of cerebral perfusion with "Real-Time" contrast-enhanced ultrasound. Circulation 104: 2582–2587
12. Seidel G, Algermissen C, Christoph A, Claassen L, Vidal-Langwasser M, Katzer T (2000) Harmonic imaging of the human brain. Visualization of brain perfusion with ultrasound. Stroke 31: 151–154
13. Seidel G, Meyer K, Algermissen C, Broillet A (2001) Harmonic imaging of the brain parenchyma using a perfluorobutane-containing ultrasound contrast agent. Ultrasound Med Biol 27: 915–918
14. Seidel G, Claassen L, Meyer K, Vidal-Langwasser M (2001) Evaluation of blood flow in the cerebral microcirculation: analysis of the refill kinetics during ultrasound contrast agent infusion. Ultrasound Med Biol 27: 1059–1064
15. Seidel G, Meyer K, Metzler V, Toth D, Vida-Langwasser M, Aach T (2002) Human cerebral perfusion analysis with ultrasound contrast agent constant infusion: a pilot study on healthy volunteers. Ultrasound Med Biol 28: 183–189
16. Seidel G, Meyer K (2002) Impact of ultrasound contrast agents in cerebrovascular diagnostics. Eur J Ultrasound 16: 81–90
17. Shiogai T, Tsunezuka C, Ohara T, Imai K, Makino M, Nakajima K, Furuhata (2001) Clinical significance of transcranial contrast-enhanced harmonic perfusion imaging (HPI) as a quantitative method of measuring cerebral blood flow. Neurotraumatology 24: 77–82 (In Japanese)

18. Shiogai T, Koshimura M, Murata Y, Nomura H, Doi A, Makino M, Mizuno T, Nakajima K, Furuhata H (2003) Acetazolamide vasoreactivity evaluated by transcranial harmonic perfusion imaging: relationship with transcranial Doppler sonography and dynamic CT. Acta Neurochir [Suppl] 86: 57–62
19. Shiogai T, Takayasu N, Mizuno T, Nakagawa M, Furuhata H (2004) Comparison of transcranial brain tissue perfusion images between ultraharmonic, second harmonic, and power harmonic imaging. Stroke 35: 687–693
20. Wei K, Jayaweera AR, Firoozan S, Linka A, Skyba DM, Kaul S (1998) Quantification of myocardial blood flow with ultrasound-induced destruction of microbubbles administered as a constant venous infusion. Circulation 97: 473–483

Correspondence: Toshiyuki Shiogai, Department of Clinical Neurosciences, Kyoto Takeda Hospital, Minamikinuta-cho 11, Nishinanajo, Shimogyo-ku, Kyoto 600-8884, Japan. e-mail: shiogait@pop11.odn.ne.jp

Acta Neurochir (2005) [Suppl] 95: 191–193

Relationship between total cerebral blood flow and ICP measured noninvasively with dynamic MRI technique in healthy subjects

N. Alperin[1], **S. H. Lee**[1], **A. Sivaramakrishnan**[1], and **T. Lichtor**[2]

[1] Department of Radiology, University of Illinois at Chicago, Chicago, IL, USA
[2] Department of Neurosurgery, Rush-Presbyterian St. Luke's Medical Center, Cook County Hospital, Chicago, IL, USA

Summary

Cerebral blood flow and ICP are important neurophysiologic parameters known to be affected by pathology and by trauma. Limited data on the relationship between these parameters following head trauma is inconsistent with regard to whether these parameters are correlated. Data on the relationship between these parameters in the healthy state is not readily available due to a lack of noninvasive means to measure these important parameters. A recently developed noninvasive MRI-based method for simultaneous measurement of total cerebral blood flow and intracranial pressure was applied to establish the relationship between ICP and TCBF values in healthy subjects. Seventy-one simultaneous measurements of CBF and ICP were obtained from 23 healthy young adults. These results demonstrated that CBF values span over a much narrower range compared with ICP. The relationship between the inter-individual CBF and ICP measurements suggest that in the healthy state and in rest these parameters are not correlated.

Keywords: ICP; cerebral blood flow; CSF flow dynamics; MRI; phase contrast MRI.

Introduction

Data on the relationship between cerebral blood flow (CBF) and ICP in the healthy state is not readily available due to a lack of noninvasive means to measure these important neurophysiological parameters. Limited information on the relationship between CBF and ICP following head injuries have been reported [1–3]. Uzzel *et al.* found significant occurrence between hyperemia and intracranial hypertension in young severely head-injured adults [1]. Muizelaar *et al.* who studied children with severe head injuries did not find occurrence between the course of ICP and hyperemia nor was there correlation between the level of CBF and ICP [2]. In contrast, Sharples *et al.* documented in pediatric patients with traumatic brain injuries an inverse relation between elevation in ICP and CBF [3]. Therefore, there is clearly a need to better characterize the relationship between ICP and CBF. This work aims to establish the distribution of total CBF (TCBF) and mean ICP measured simultaneously in young healthy adults using a noninvasive magnetic resonance imaging (MRI) based technique [4]. Total CBF is obtained by direct measurement of the volumetric blood flow through the four arteries supplying blood to the brain. ICP is derived from measurements of intracranial volume and pressure changes during the cardiac cycle. The study further tests whether the levels of CBF and ICP, measured at rest in the supine posture, are correlated.

Materials and methods

Twenty-three young adults (20 males, range 20 to 39, mean age 25 $\pm$ 5 years) with no known neurological problems participated in the study. The imaging protocol was approved by the institutional review board and informed consent was obtained from all subjects. CBF and ICP were measured in five subjects multiple times, both within the same MRI session and at different days in order to quantify the normal intra-individual variability of the MR-derived CBF and ICP measurements. All MRI scans were obtained when the subjects were at rest in the supine posture. A total of 71 simultaneous measurements of CBF and ICP were obtained.

TCBF and ICP were derived from MRI measurements of blood and cerebral spinal fluid (CSF) flows into and out from the cranial vault during the cardiac cycle [4]. Briefly, the method utilizes the small changes in intracranial volume and pressure that occur naturally with each cardiac cycle. The amplitude of the pulse pressure is derived from the CSF pressure gradient waveform, which is calculated from measurements of the CSF velocities. The intracranial volume change is derived from the instantaneous differences between volumetric arterial blood inflow, venous blood outflow, and CSF oscillatory flow. Elastance (the inverse of compliance) is then derived from the ratio of the measured pressure and volume changes. A mean ICP value is derived based on a linear relationship that exists between intracranial elastance and ICP [5]. The total CBF is calcu-

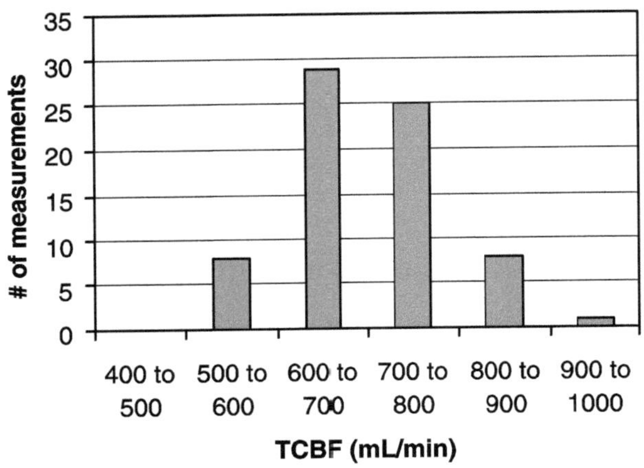

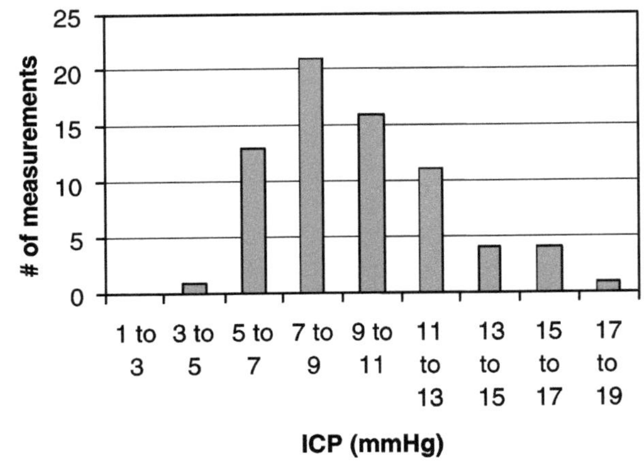

Fig. 1. The distribution of the total CBF (left) and the MRI-derived ICP (MR-ICP) measurements (right). The mean value and the relative standard deviation were 700 ml/min ± 13% and 9.6 ± 31%, respectively. Note that the ICP values are distributed over a wider range relative to the CBF values

lated from the sum of the mean volumetric flow rate through the four arteries supplying blood to the brain. The method has been previously validated in baboons, flow phantoms, with computer simulations and in measurements in healthy subjects and patients [6].

The MRI technique used to obtain the volumetric flow data is a retrospectively gated cine phase-contrast scan. A scan with high velocity encoding (VENC = 70 cm/s) located above the carotid bifurcation perpendicular to the direction of flow was used to quantify blood flow. The scan parameters were TR = 19 ms, FA = 25 deg, FOV = 14 cm, and slice thickness = 6 mm. A second scan with low velocity encoding (VENC = 7–9 cm/s) located at the level of C2 perpendicular to the direction of the flow was used to quantify the slower CSF flow and venous flow in the vertebral plexus. The parameters of the low VENC scan were TR = 19 ms, FA = 20 deg, FOV = 14 cm, and slice thickness = 8 mm.

Measurements of total cerebral blood flow

Each of the two cine phase-contrast MRI scans generates a set of 32 velocity-encoded images of the pulsatile flow during one cardiac cycle with pixel values proportional to velocity. Volumetric flow rates (VFR) through the blood vessels or the cervical spine CSF space were obtained by integrating the velocities through the lumen cross-sectional area. An automated lumen segmentation technique, pulsatility based segmentations (PUBS) [7], was used to delineate the boundary of the lumens for improved measurement accuracy and reproducibility. Mean VFR over the cardiac cycle is then calculated for the right and left internal carotid and vertebral arteries. Total CBF is obtained by summation of the mean VFR of the four vessels.

Data analysis

The distribution and frequency of the measured values, inter-individual mean and relatively standard deviations of the TCBF and MR-derived ICP values were calculated and compared. The inter- and intra-session measurement variability was also calculated as a reference. The values of TCBF were plotted against ICP values to assess the relationship between these parameters and a linear regression was applied to calculate the Pearson product moment correlation coefficient.

Results

The distribution of the TCBF and MR-ICP measurements are shown in Fig. 1. TCBF values range from 514 to 956 mL/min and ICP values range from 3.5 to 17.1 mmHg. The mean and the percentage standard deviation are 700 ml/min ± 13% and 9.6 mmHg ± 31%, respectively. The relationship between the inter-individual TCBF and ICP measurements is shown in Fig. 2. The trend line obtained with a linear regression analysis is shown superimposed on the data points. No significant correlation is found between TCBF and ICP. The correlation coefficient value was close to 0 (R = 0.042). The inter-session measurement variability ranged between 3% and 7% for the TCBF measurement and between 7% and 20%

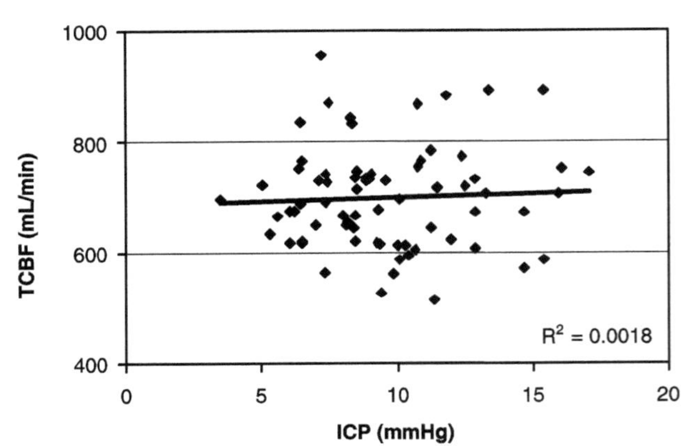

Fig. 2. Plot of total CBF measurements vs. the MR-ICP measurements. The trend line obtained by linear regression analysis demonstrates that CBF and ICP are not correlated

for the MR-ICP measurement. The respective inter-session measurement variability is smaller than the corresponding TCBF and ICP inter-individual measurement variability.

Discussion

Total CBF values measured in the current study are within the previously reported range measured in healthy young adults by either MRI [8] or by duplex ultrasound [9] achieved by summation of volumetric blood flow through the arteries leading blood to brain. The MRI derived ICP values ranged from 3.5 to 17.1 mmHg with the maximum number of measurements between 7 and 9 mmHg. Wide distribution of ICP values measured invasively in normal human subjects has been previously reported: from 2 to 16 mmHg in one study [10] and up to 18 mmHg in another [11]. Three of the 71 MRI measurements were between 16 mmHg and 17.1 mmHg – above the upper limit found in the 1937 study. However, it is not possible to determine whether this represents a normal variation of ICP, an MRI measurement error, or the true ICP value, as no invasive values were available for comparison. Based on the current view that ICP value of 20 mmHg is a critical threshold for elevated ICP [12], no false positives (elevated ICP) were measured with MRI.

It is interesting to note that the relative range of TCBF values spans a much narrower range than ICP. The relative SD of the ICP and TCBF measurements are 31% and 13% respectively. Since these values are larger than the inter-session measurement variability, it is reasonable to assume that the narrower distribution of TCBF reflects the autoregulation of CBF. Finally, it is apparent that there is no significant correlation between rest CBF and ICP values in the healthy state. The lack of correlation suggests that at rest these parameters can be considered independent of each other. The data presented in this study concerns the steady state rest values of these important neurophysiologic parameters. To answer whether changes in CBF and ICP following intervention are correlated, these measurements need to be repeated at rest as well as following intervention or a change in the steady state such as a different posture.

References

1. Uzzell BP, Obrist WD, Dolinskas CA, Langfitt TW (1986) Relationship of acute CBF and ICP findings to neuropsychological outcome in severe head injury. J Neurosurg 65: 630–635
2. Muizelaar JP, Marmarou A, DeSalles AA, Ward JD, Zimmerman RS, Li Z, Choi SC, Young HF (1989) Cerebral blood flow and metabolism in severely head-injured children. Part 1: Relationship with GCS score, outcome, ICP, and PVI. J Neurosurg 71: 63–71
3. Sharples PM, Stuart AG, Matthews DS, Aynslly-Green A, Eyr JA (2005) Cerebral Blood flow and metabolism in children with severe head injuries. J Neurol Neusurg Psychiatry 58: 145–152
4. Alperin N, Lee SH, Loth F, Raksin P, Lichtor T (2000) MR-Intracranial Pressure (ICP): A method for noninvasive measurement of intracranial pressure and elastance. Baboon and Human Study. Radiology 217(3): 877–885
5. Marmarou A, Shulman K, La Mores J (1975) Compartmental analysis of compliance and outflow resistance of the CSF system. J Neurosurg 43: 523–534
6. Alperin N (2004) MR-Intracranial Compliance and Pressure: A Method for Noninvasive Measurement of Important Neurophysiologic Parameters. In: Conn MP (ed). Methods in enzymology, vol 386. Imaging in biological research. Elsevier Academic Press
7. Alperin N, Lee SH (2003) PUBS: Pulsatility based segmentation of lumens conducting nonsteady flow. Magn Reson Med 49: 934–944
8. Enzmann D, Ross M, Marks M, Pelc N (1994) Blood flow in major cerebral arteries measured by phase contrast cine MR. AJNR 15: 123–129
9. Doepp F, Schreiber S, Brunecker P, Valdueza J (2003) Ultrasonographic assessment of global cerebral blood volume in healthy adults. J Cerebral Blood Flow Metabolism 23: 972–977
10. Ayer JB (1924) Cerebrospinal fluid pressure from the clinical point of view. Assn Res Nerv Mental Dis 4: 159–171
11. Merritt HH, Fremont-Smith F (1937) The cerebrospinal fluid. Saunders, Philadelphia
12. Marmarou A, Eisenberg HM, Foulkes MA (1991) Impact of ICP instability and hypertension on outcome in patients with severe head trauma. J Neurosurg 75: 59–66

Correspondence: Noam Alperin, Physiological Imaging and Modeling Lab, Department of Radiology (M/C 711), The University of Illinois at Chicago, 830 S. Wood St., Chicago, IL 60612, USA. e-mail: alperin@uic.edu

Hydrocephalus

Acta Neurochir (2005) [Suppl] 95: 197–199

Clinical comparison of tympanic membrane displacement with invasive ICP measurements

S. Shimbles[1], **C. Dodd**[1], **K. Banister**[1], **A. D. Mendelow**[2], and **I. R. Chambers**[1]

[1] Regional Medical Physics Department, Newcastle General Hospital, Newcastle upon Tyne, UK
[2] Department of Neurosurgery, Newcastle General Hospital, Newcastle upon Tyne, UK

Summary

Objective. Tympanic membrane displacement (TMD) measurements may be useful in the management of patients with hydrocephalus if they can be directly associated with measurements of ICP. We have compared TMD measurements using the Marchbanks Measurement System with invasive ICP monitoring.

Methods. Twenty-nine patients who were undergoing routine invasive monitoring using a Camino fibre optic ICP measurement system as part of their clinical management were studied. Simultaneous measurements of ICP and TMD in both sitting and supine positions were successfully made in thirteen patients.

Results. Thirty-nine pairs of readings were obtained. The invasive ICP readings varied from 1 to 36 mmHg in the supine position and from −12 to +35 mmHg sitting. Corresponding TMD values varied from −275 to +277 nL in the supine position and −133 to +466 nL sitting. Linear regression showed a significant negative relationship between the two measurements ($r = -0.57$, $p = 0.0013$).

Conclusions. There is a strong negative linear association between mean TMD and invasively measured ICP and this relationship is highly significant. Nevertheless, TMD is a poor surrogate for ICP in clinical terms because the predictive limits of the linear regression are too wide. However, serial intra-patient measurements may be useful to determine changes in ICP with time.

Keywords: Hydrocephalus; shunt; intracranial pressure; tympanic membrane displacement; non-invasive ICP.

Introduction

The failure rate of shunt valves implanted to treat hydrocephalus is high and often associated with abnormal ICP. Invasive monitoring normally involves either placing a transducer-tipped catheter into the subdural space or parenchyma of the brain, or measuring pressure directly via the ventricle. This invasive procedure can be poorly tolerated by patients, is not without risk, and requires significant hospital resources. A non-invasive estimate of ICP would help in the decision to introduce or replace a shunt system.

Stimulation of the acoustic reflex can induce a displacement in the tympanic membrane and this displacement may depend on ICP, according to the following rationale. The acoustic reflex is a protective mechanism whereby the stapedius muscle contracts in response to a loud noise. The resulting displacement of the tympanic membrane serves to protect the ear from potentially damaging noise levels. The kinematic chain involves the stapes, the innermost of the ossicles, which rests on the oval window of the cochlear. This is a thin, flexible membrane; consequently the pressure of the cochlear fluid determines the resting position of the stapes. This in turn influences the tympanic membrane displacement (TMD) resulting from the acoustic reflex [1].

If there is a patent communication between the intracranial and cochlear fluid spaces [2], TMD, as measured by the volume change in the outer ear, should reflect changes in ICP. Normal ICP results in a bidirectional displacement whereas in cases of high ICP the resting position of the stapes on the protuberant oval window causes an inward movement of the tympanic membrane. When the ICP is low, the stapes rests into the oval window, and contraction of the stapedius muscle causes an outward movement of the tympanic membrane. There is little published evidence on the relationship between TMD and ICP. This study tests the hypothesis that TMD and ICP have a linear relationship and studies a group of patients who were undergoing routine ICP monitoring.

Methods and materials

Twenty-nine patients who were undergoing invasive measurement of ICP as part of their clinical management were studied. The local

Ethics Committee approved the study and informed consent was obtained prior to patient recruitment to the study.

In each subject, TMD was measured as a response to an applied acoustic stimulus, using the Marchbanks Measurement System (MMS) analyser. Prior to each set of patient measurements, tympanometry was performed to ascertain normal middle ear function and pressure using a Kamplex KT20 hand-held UniTymp. The acoustic reflex threshold (ART) was determined for each ear. Measurement of TMD was recorded with the sound level set above the ART in both sitting and supine positions. In some cases normal middle ear function could only be determined in one ear and therefore only that ear was tested.

For each ear, cochlear aqueduct patency was tested using a simple postural test. In any individual, the ICP is naturally lower when sitting than when supine. If there was an incorrect or negligible change in TMD with posture using the Marchbanks system, it was assumed that the cochlear aqueduct was not patent and therefore the readings obtained were excluded from the analysis.

For all test procedures, the noise exposure was within specified safety limits – no sound level above 110 dB was used in any part of the test.

The non-invasive test was carried out by sealing the ear canal with a rubber ear tip of appropriate size. A 1 kHz tone was applied for 300 ms at intervals of one second, at a sound intensity 10 dB above the ART. The MMS analyser measured TMD over each stimulus response cycle, recording the average waveforms of ten reliable responses with a time resolution of 2 ms. Responses containing artefacts (heart beat or movement) were rejected automatically by the analyser; consistency of the responses to be averaged was confirmed by operator observation.

The mean displacement of the tympanic membrane from the point of maximum inward displacement to the end of the stimulus (Vm) was calculated (Fig. 1).

Invasive monitoring

In each patient the ICP was simultaneously measured directly using a Camino fibre optic transducer as part of the patient's routine clinical management. This involved drilling a burr hole through the skull under either local or general anaesthetic, screwing in a fixation bolt and introducing the transducer-tipped catheter so that its distal tip lay in the subdural space.

Results

Thirty-nine pairs of valid readings of ICP and TMD were obtained from 18 ears of 13 patients where a patent cochlear aqueduct was assessed. One patient had to be excluded from the analysis because their ART was too close to the safety limits to allow a sound level 10 dB higher to be used.

In eight patients, only one ear was tested. The TMD values obtained ranged from −275 to +212 nL in the supine position and from −133 to +434 nL in the sitting position. In the five patients with bilateral results the TMD values ranged from −55 to +457 nL in the supine position and from −197 to +529 nL in the sitting position. For these bilateral readings there was a left/right difference in TMD in the supine position of between 93 and 359 nL, and of between 125 and 353 nL in the sitting position. Since the right and left values are not independent, it was decided to use the mean values, in order to maintain statistical integrity. The combined range of TMD values was therefore −275 to +277 nL in the supine position and −133 to +466 nL sitting. ICP values, calculated in the same way, ranged from +1 to +36 mmHg supine and from −12 to +35 mmHg sitting.

Linear regression showed a significant negative relationship between displacement and ICP ($r = -0.569$, $p = 0.001$, Fig. 2). One patient attended for measurements on two occasions and manifested a change in

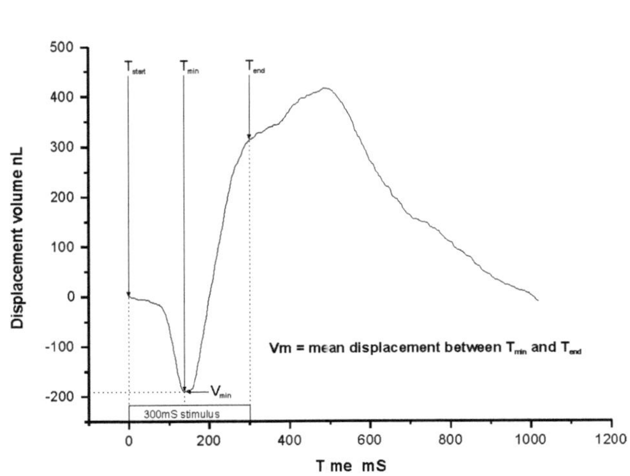

Fig. 1. Calculation of mean TMD (Vm) from displacement/time plot

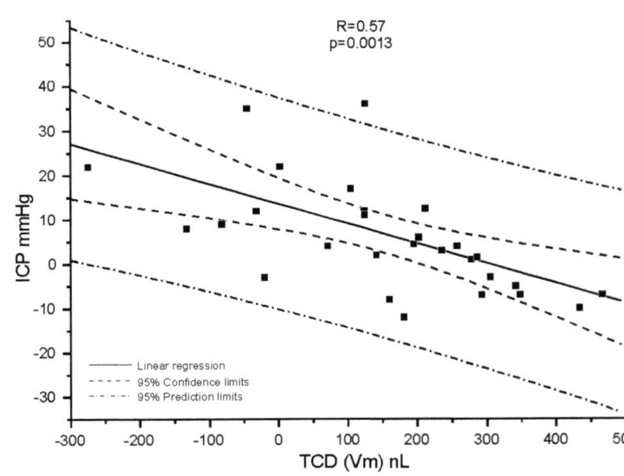

Fig. 2. Scatter plot showing linear regression of ICP vs TMD (Vm), including 95% confidence and prediction limits

TMD with ICP that was consistent with the trend of the complete data set.

Discussion

The aim of this study was to compare a non-invasive method of ICP measurement with invasive ICP monitoring in patients undergoing invasive monitoring as part of their routine clinical management. There is a strong negative linear association between mean TMD measured at ART+10 dB, and invasively measured ICP and this relationship is highly significant.

The relationship between supine and sitting values for the one patient who had two sets of measurements is consistent between the two sessions. Furthermore, the changes between supine and sitting closely follow the trend in the data as a whole. Nevertheless, TMD is a poor surrogate for ICP in clinical terms because the predictive limits of the linear regression are approximately ± 25 mmHg. This variability makes it difficult to make a clinically useful estimate of ICP. It has been suggested [3] that serial intra-patient measurements may be useful to determine changes in ICP over time, particularly if a baseline non-invasive measurement is recorded when the patient is asymptomatic. Our observations are consistent with this hypothesis and it may well be that such serial measurements can be used in the assessment of patients with hydrocephalus and in the determination of possible shunt malfunction. Further work is required on the repeatability and variability of these measurements to ascertain their exact clinical utility.

Acknowledgment

This study was supported by a grant from Action Medical Research, The Garfield Weston Foundation.

References

1. Marchbanks RJ (1989) Method and Apparatus for Measuring Intracranial Fluid Pressure. United States Patent no 4,841,986
2. Wagner N, Walsted A (2000) Postural-induced changes in intracranial pressure evaluated non-invasively using the MMS-10 tympanic displacement analyser in healthy volunteers. Acta Otolaryngol [Suppl] 543: 44–47
3. Madan S, Burge DM, Marchbanks RJ (1998) Tympanic membrane displacement testing in regular assessment of intracranial pressure in eight children with shunted hydrocephalus. J Neurosurg 88: 983–995

Correspondence: I. R. Chambers, Regional Medical Physics Department, Newcastle General Hospital, Newcastle upon Tyne, NE4 6BE, UK. e-mail: i.r.chambers@ncl.ac.uk

Acta Neurochir (2005) [Suppl] 95: 201–205

Gravity valves for idiopathic normal-pressure hydrocephalus: a prospective study with 60 patients

U. Meier

Department of Neurosurgery, Unfallkrankenhaus Berlin, Berlin, Germany

Summary

Objective. Are hydrostatic valves superior to conventional differential pressure shunts in patients with normal-pressure hydrocephalus, with regard to the postoperative results of treatment and possible complications?

Methods. From September 1997 to January 2002, 60 patients with idiopathic normal-pressure hydrocephalus were treated by surgical implantation of hydrostatic valve at the Unfallkrankenhaus Berlin. In a prospective study, the clinical examination and a CT examination were carried out preoperatively, postoperatively and 1 year after the operation.

Results. One year after the operation, the clinical picture was very good for 33% of the patients, good for 33% and satisfactory or poor for 17%, each, of the patients. 3 dislocations (5%) of ventricular or abdominal catheters and three valve infections were found as valve-independent. As valve-dependent complications, underdrainage was found in 4 patients (7%) and radiological signs of overdrainage in 2 patients (3%), while 1 patient (1.7%) showed symptomatic overdrainage.

Conclusion. According to our experience, hydrostatic Dual-switch valves Aesculap® are superior to conventional differential pressure shunts without an additional gravity unit especially with regard to the treatment of patients with idiopathic normal-pressure hydrocephalus, concerning both the postoperative results and the incidence of possible complications. A clinical improvement can be achieved for 83% of such patients.

Keywords: Normal-pressure hydrocephalus; idiopathic; outcome; complications; shunt operation; Dual-switch valve; hydrostatic valves.

Introduction

Despite the existence of modern, neuroradiological diagnostic methods, the accurate diagnosis and therapy of a normal-pressure hydrocephalus (NPH) still presents a challenge for the clinician. On the one hand, the clinical symptoms, e.g. the onset of gait disorders or the impairment of the short-term memory, often accompanied by vegetative symptoms like headaches and dizziness under physical strain, must be regarded as rather unspecific. On the other, the discrimination of the symptoms, through differential diagnostics, just cited from demential illnesses and degenerative genesis (Morbus Binswanger, Morbus Alzheimer) is often difficult [14].

For patients with idiopathic normal-pressure hydrocephalus (iNPH), conventional differential pressure shunts have the disadvantage that, when the patient stands up, they open abruptly, or stay open for too long and thereby exert suction on the cerebrospinal fluid (CSF) volume of the possibly already atrophic damaged brain. The principal question is if such suction phenomena, and the resulting complications due to overdraining, can be suppressed or prevented by using hydrostatic valves [11].

Materials and method

From September 1997 to January 2002, 60 patients with idiopathic normal-pressure hydrocephalus (iNPH) were surgically treated by implanting a hydrostatic valve (Dual-switch valve Aesculap®) at the Unfallkrankenhaus Berlin. The results of the clinical examination, the intrathecal infusion test and the cerebrospinal tap test served as a decision aid for the shunt operation [9]. The 41 male (68%) and 19 female patients (32%) were of an average age of 66 years (44–83 years). With all these patients, the clinical and CT examination was carried out preoperatively, postoperatively and 1 year after the operation in a prospective study.

Diagnostics

Following the clinical examination with findings of gait ataxia and additional symptoms [8] and the detection of a ventricle enlargement through neuroradiological imaging methods, the intrathecal infusion test was carried out (Fig. 1). The dynamic infusion test, which was performed on the patient in horizontal position, via lumbar puncture, in the computer-assisted constant flow technique at an infusion rate of 2 ml/min, was used for computing the ICP-dependent resistance to cerebrospinal outflow as a parameter of the CSF dynamics

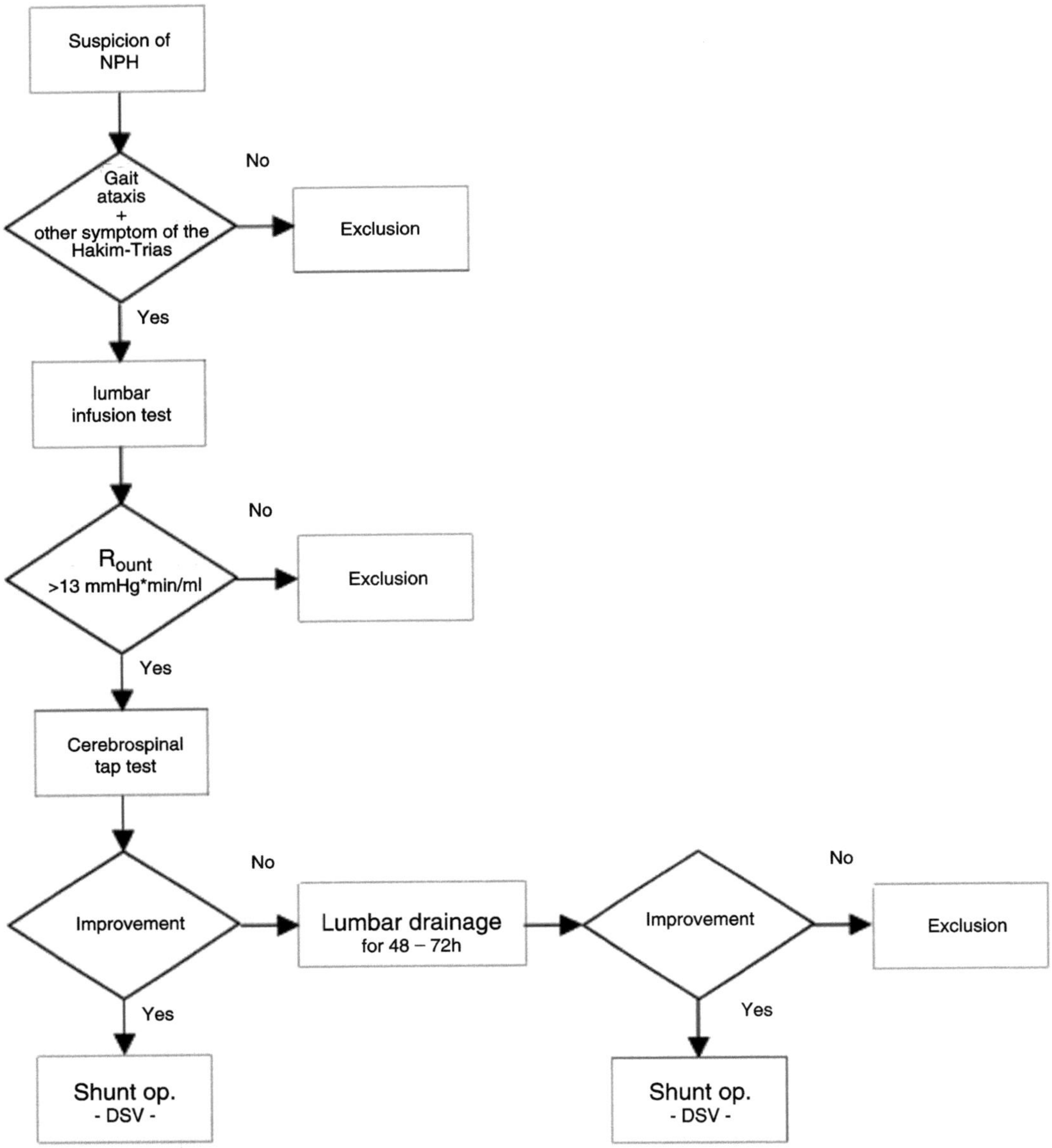

Fig. 1. Diagnostic pathway for normal-pressure hydrocephalus

[6, 7]. A resistance to outflow (R_{out}) exceeding 13 mmHg*min/ml was defined as pathological [9]. Immediately after the infusion test, a diagnostic CSF drainage of at least 60 ml was carried out (Cerebrospinal tap test). Improvement of the clinical symptoms within the next 2–3 days established the indication for a shunt operation. If the symptoms, especially the gait ataxia, did not improve, an external lumbar drainage was performed for 2–3 days (Fig. 1). If this resulted in an improvement of the symptoms, a DSV was implanted as a ventriculo-peritoneal shunt [11].

Working principle of the Dual-switch valve

The valve comprises two valve chambers operating independently, one for the lying patient and one for the standing or sitting patient. In this way, the intraventricular pressure (IVP) can be kept within the physiological range even when the patient is standing up, and accidental overdrainage is prevented systematically. The appropriate pressure chamber is activated by a tantalum ball, which blocks the CSF drainage at the low-pressure side of the valve when the patient sits or stands up. Only when a critical IVP is reached in the vertical position too, the high-pressure stage of the valve opens and thus prevents any further pressure rise (Fig. 2). The valve setting pressure ratings of the DSV for all of the 60 patients with iNPH are shown in Table 1.

Clinical grading

The results of the clinical examinations were assessed according to the Black grading scale for shunt assessment and the NPH recovery rate based on the clinical grading for normal-pressure hydrocepha-

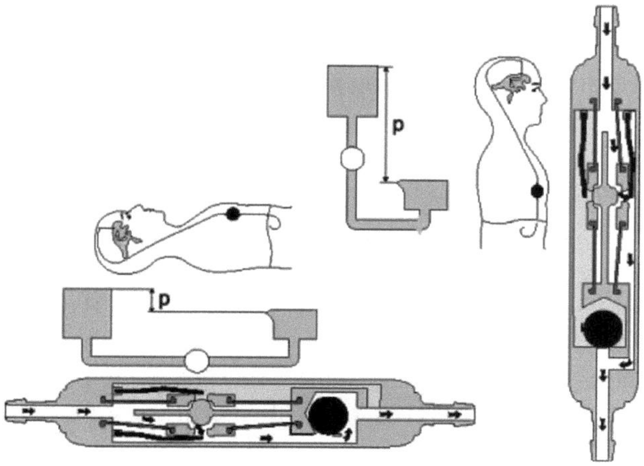

Fig. 2. Working principle of the Miethke dual-switch valve

Table 1. *Valve opening pressure of the DSV for all 60 patients with iNPH*

Valve opening pressure [mmH_2O]	300	400	500	Total
80	–	2	–	2
100	2	13	15	30
130	–	7	18	25
160	–	3	–	3

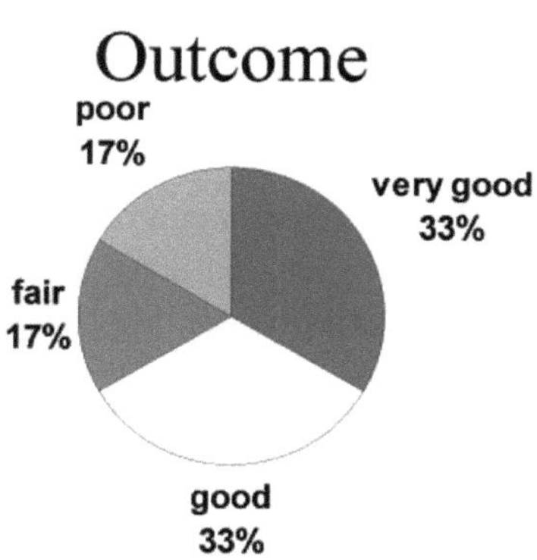

Fig. 3. Therapeutic results

lus, by Kiefer [5]. All graded examination results were classed into four groups. The group showing excellent results is characterized by a NPH recovery rate $\geq$ 7 points; the group showing a good development after the implantation scores an NPH recovery rate $\geq$ 5 points; the group showing a fair or satisfactory development contains patients with a NPH recovery rate $\geq$ 2 points, and the group showing poor results is characterized by a NPH recovery rate $<$ 2 points.

Results

A gait ataxia was diagnosed as the leading symptom of idiopathic normal-pressure hydrocephalus (iNPH) in all 60 patients. 26 patients (43%) showed a urinary incontinence as a late symptom. 32 patients (53%) showed mnestic disorders, mostly of the short-term memory and, in few cases, demential symptoms. The average resistance in the intrathecal infusion test was determined as $R_{out} = 21 \pm 7$ (13–47) mmHg*min/ml. For 53 patients (88%), the clinical symptoms improved, especially the gait ataxia after the cerebrospinal tap test. Because of a negative finding in the cerebrospinal tap test, 7 patients (12%) received a lumbar CSF drainage for 3 days. After that, the gait ataxia improved in these patients. All 60 patients were surgically treated with Dual-switch valves Aesculap® (DSV) as ventriculo-peritoneal shunt implants.

Outcome

The outcome for the 60 patients after implantation of a DSV indicated by iNPH are shown in table 3: 33% (20) of the patients showed a very good outcome, 33% (20) a good outcome, 17% (10) a satisfactory outcome and 17% (10) a poor outcome 1 year after the shunt was implanted. The average rating on the Kiefer scale, equivalent to the severity of the symptoms caused by the iNPH, dropped from 9 ± 3 points before the operation to 5 ± 3 points after the operation, or 4 ± 3 points at the follow-up examination, i.e. by more than half. The postoperative NPH recovery rate were compared to those obtained from the follow-up examination 1 year after the shunt implantation. This yielded a postoperative ratio of 5.1 ± 2.5, i.e. a result in the transition region between satisfactory and good outcomes, and a figure of 5.3 ± 3.5 one year after the operation, i.e. a good outcome, on average, or an average improvement of the clinical symptoms by 53%.

Complications

Regarding the valve-independent complications following the shunt implantation, the infection rate of 5% (3 patients) is the dominant factor. The implant was removed from all patients with shunt infections. After the remediation of the focus of infection, new valves were implanted in all patients. In 2 cases, the same low-pressure valve rating was chosen for the second implantation as for the initial one. In one case we decided for the lower rating for the low-pressure stage of the DSV. In two cases, a ventricular catheter had to be revised due to incorrect positioning, and in one case the abdominal drainage tube required revision because of its dislocation (dislocation rate: 5%). Thus, the valve-independent complications amounted to 10% (6 patients).

In 2 patients (3%) carrying an DSV, neuroradiological imaging procedures showed an obvious decrease of the ventricular size and narrow, subdural hygromas as a direct sign of overdrainage. One of those patients (1.7%) showed symptoms related to this neuroradiological finding, such as nausea, vomiting and headaches. In this patient a shunt revision was performed involving the implantation of higher-rate low-pressure stage of the DSV. In the other patient, the subdural hygromas was resorbed, without any clinical symptom, within 6 weeks. 4 patients (7%) suffered a worsening of their symptoms despite the shunt operation. In these patients, the ventricle size, as seen in the computer tomogram, increased slightly. A shunt revision involving the implantation of a lower-rate low-pressure stage of the DSV was performed on these 4 patients. The clinical development after that revision showed 2 patients with a significant marked improvement of their symptoms, while the clinical symptoms of the 2 other patients did not improve after the revision. Hence, the valve-dependent rate of complications amounts to 10% (6 patients).

The shunt-dependent morbidity thus amounted to 20 percent. Independent of the shunt implantation, 2 patients deceased within the period covered by this study, 5–7 month post operationem, one of a heart attack and one of a tumor illness, which had not been diagnosed prior to the operation. Another patient died of a lung embolism 6 days after the shunt operation, despite proper thrombotic prophylaxis. Thus, the operation-related lethality amounts to 1.7%, while the shunt-related lethality is 0 percent.

Ventricle width

The preoperative and postoperative Evans index and the catamnestic Evans index 1 year after the operation allowed drawing up a balance for the ventricle size. For 80 percent of the patients with iNPH, the evaluation did not yield any significant reduction of the ventricle width, despite an improvement in clinical symptoms and a good or very good development after the implantation of this valve type [10].

Discussion

All valves in use operate according to the differential-pressure principle. This means, the passage through the valve is opened as soon as the pressure difference between the inlet and the outlet of the valve exceeds its specific opening pressure.

Outcome

According to literature, the general improvement rates after a shunt operation for patients with normal-pressure hydrocephalus (NPH) vary between 31% and 96%, with an average of 53%. A meta-analysis by Vanneste [15] yielded improvement rates of 30–50 percent for idiopathic NPH and 50–70 percent for secondary NPH an. The meta-analysis by Hebb *et al.* [4] cites an improvement rate of 59% following a shunt operation on iNPH and a long-term improvement rate of 29%. The results of our study – 66% good or excellent outcomes and 17% satisfactory results, i.e. an overall improvement rate of 83 percent – clearly exceed these international findings. The necessity of diagnosis and therapy of normal-pressure hydrocephalus at an early stage cannot be emphasized enough, since, at a later stage, cerebral autoregulation disorders are already manifested and adversely affect the therapy outcome following a valve implantation [11].

Complications

Grumme *et al.* [3] cite a lethality of 0–6 percent and an incidence of overdrainage phenomena of 6–20 percent after shunt operations on patients with normal-pressure hydrocephalus. Hebb *et al.* [4] give a complication rate of 38%, a revision rate of 22% and an incidence of a combination of lasting neurological deficits and lethality of 6%. The lethality of 1.7% and the overdrainage rate of 3% found by this study are lower than those international figures [2, 12]. Our figures concerning peri- and postsurgical complications, at 20%, are at the lower edge limit of the 20–40 percent range quoted by Vanneste [46]. Especially the low overdrainage rate achieved with the DSV (3% according to radiological criteria, 1.7% symptomatically) compared to standard valves has to be emphasized at this point. The hydrostatic valves reduced the risk of overdrainage related to the design principle of the valve, but this also brought problems of underdrainage to the foreground. Four patients (7%) of our sample group had to be operated on again because of underdrainage, which was detected clinically and in intrathecal infusion tests. These 4 patients were treated by implanting an DSV of a lower pressure rating for the horizontal position.

In the British Shunt Registry [13], which covers more than 9000 cases, underdrainage is the dominant problem, at 52%. In that report, overdrainage as a cause for postoperative complications is quoted as very rare (3%). In contrast to this, Scandinavian groups assert [1] that 80% of all shunt complications are caused by overdrainage.

Conclusions

After 6 years of experience with the DSV, we arrived at a clearly positive assessment of the reliability of the design principle of this gravity-assisted hydrostatic valve. The technical principle of the parallel two-chamber system with large membranes allows minimizing the incidence of overdrainage. In this respect, it is clearly superior to conventional differential pressure valves as well as to adjustable valves and the majority of hydrostatic valves. Because of the low incidence of over- and underdrainage and the good postoperative results, we recommend implanting DSV as ventrico-peritoneal shunts for patients with iNPH.

References

1. Borgesen SE (1984) Conductance to outflow of CSF in normal pressure Hydrocephalus. Acta Neurochir (Wien) 71: 1–15
2. Decq P, Barat JL, Duplessis E, Lequerinel C, Gendrault P, Keravel Y (1995) Shunt failure in adult hydrocephalus: flow controlled shunt versus differential pressure shunts – a cooperative study in 289 patients. Surg Neurol 43: 333–339
3. Grumme T, Kolodziejczyk D (1995) Komplikationen in der Neurochirurgie, vol 2 Kraniale, zerebrale und neuropädiatrische Chirurgie. Blackwell, Berlin, pp 534–540
4. Hebb AO, Cusimano MD (2001) Idiopathic normal pressure hydrocephalus: a systematic review of diagnosis and outcome. Neurosurgery 49: 1166–1186
5. Kiefer M, Eymann R, Meier U (2002) Five years experience with gravitational shunts in chronic hydrocephalus of adults. Acta Neurochir (Wien) 144: 755–767
6. Meier U, Zeilinger FS, Kintzel D (1999) Diagnostic in normal pressure hydrocephalus: A mathematical model for determination of the ICP-dependent resistance and compliance. Acta Neurochir (Wien) 141: 941–948
7. Meier U, Kiefer M, Bartels P (2002) The ICP-dependency of resistance to cerebrospinal fluid outflow: A new mathematical method for CSF-parameter calculation in a model with H-Tx rats. J Clin Neuroscience 9: 58–63
8. Meier U, Zeilinger FS, Kintzel D (1999) Signs, symptoms and course of disease in normal pressure hydrocephalus in relation to cerebral atrophy. Acta Neurochir (Wien) 141: 1039–1048
9. Meier U, Bartels P (2001) The importance of the intrathecal infusion test in the diagnostic of normal pressure hydrocephalus. Europ Neurol 46: 178–186
10. Meier U, Mutze S (2004) Is decreased ventricular size a correlate of positiv clinical outcome following shunt placement in 80 cases of normal pressure hydrocephalus? J Neurosurg 100: 1036–1040
11. Meier U, Kiefer M, Sprung C (2004) Evaluation of the Dual-Switch-Valve in patients with normal pressure hydrocephalus. Surg Neurol 61: 119–128
12. Pollack IF, Albright AL, Adelson PD, the Hakim-Medos Invetigator Group (1999) A randomized, controlled study of a programmable shunt valve versus a conventionel valve for patients with hydrocephalus. Neurosurgery 45: 1399–1411
13. Richards HK, Seeley HM, Pickard JD (2000) Shunt revisions: Data from the UK shunt registry. Eur J Pediatr Surg 10 [Suppl] I: 59
14. Takeuchi T, Kasahara E, Iwasaki M, Mima T, Mori K (2000) Indications for shunting in patients with idiopathic normal pressure hydrocephalus presenting with dementia and brain atrophy (atypical idiopathic normal pressure hydrocephalus). Neurol Med Chir (Tokyo) 40: 38–47
15. Vanneste JAL (2000) Diagnosis and managment of normal-pressure hydrocephalus. J Neurol 247: 5–14

Correspondence: Ullrich Meier, Department of Neurosurgery, Unfallkrankenhaus Berlin, Warener Str. 7, 12683 Berlin, Germany. e-mail: ullrich.meier@ukb.de

Acta Neurochir (2005) [Suppl] 95: 207–211

The relationship between CSF circulation and cerebrovascular pressure-reactivity in normal pressure hydrocephalus

Z. Czosnyka[1], **F. van den Boogaard**[2], **M. Czosnyka**[1], **S. Momjian**[1], **L. Gelling**[1], and **J. D. Pickard**[1]

[1] Academic Department of Neurosurgery, Addenbrooke's Hospital, Cambridge, UK
[2] Faculty of Medicine, University of Utrecht, Utrecht, The Netherlands

Summary

Objective. Previously, we documented association between CSF circulation and transcranial-Doppler derived autoregulation in non-shunted patients suffering from hydrocephalus. In the present study we sought to investigate the relationship between the resistance to CSF outflow and pressure-reactivity both in shunted and non-shunted NPH patients.

Material and methods. Sixty-eight patients (47 non-shunted and 21 shunted) with NPH have been examined as a part of routine diagnostic procedure. Resistance to CSF outflow (Rcsf) was measured using a ventricular constant rate infusion test. Cerebrovascular pressure-reactivity was assessed as a moving correlation coefficient (PRx) between coherent 'slow waves' of ICP and arterial blood pressure (ABP). This variable has previously been demonstrated to correlate with the autoregulation of CBF in patients following head injury.

Results. In non-shunted patients cerebrovascular pressure-reactivity (PRx) was negatively correlated with Rcsf ($R = -0.5$; $p < 0.0005$). This relationship was inverted in shunted patients: a positive correlation between PRx and Rcsf was found ($R = 0.51$; $p < 0.03$).

Conclusion. Cerebrovascular pressure-reactivity is disturbed in patients with normal resistance to CSF outflow, suggesting underlying cerebrovascular disease. This result confirms our previous finding where transcranial Doppler autoregulation was investigated. After shunting the pressure-reactivity strongly depends on shunt functioning and deteriorates when the shunt is blocked.

Keywords: Hydrocephalus; autoregulation; CSF flow; shunt.

Introduction

The pathophysiology of hydrocephalus includes three major components: disturbed cerebrospinal fluid (CSF) circulation, poor pressure-volume compensation and the interference of abnormal CSF flux with cerebral blood flow (CBF).

The first component manifests as an impaired outflow or absorption of CSF usually seen in hydrocephalus [2], commonly expressed as an increased resistance to CSF outflow (Rcsf).

The second component, the pressure-volume compensatory reserve, can be regarded as a mechanism by which the cranial cavity adapts to a change in intracranial volume to maintain a stable intracranial pressure (ICP). It is probably predominantly expressed by a buffering capacity of low-pressure compartment of cranial venous volume.

The third component is responsible for a reduction of CBF and an impairment of mechanisms of CBF regulation. It appears that CBF is decreased in patients suffering from NPH, although it remains unclear whether this reduction is a cause or an effect of NPH [1]. It has recently been reported that cerebrovascular reactivity to changes in partial pressure of carbon dioxide in arterial blood and in reaction to acetozolomide are commonly depleted in NPH [3, 7]. In our previous study we demonstrated an association between pressure-autoregulation assessed with transcranial Doppler ultrasonography (TCD) and the resistance to CSF outflow [6]. Our present objective is to study the relationship between the resistance to CSF outflow and cerebrovascular pressure-reactivity in shunted and non-shunted NPH patients.

Material and methods

In this retrospective study we reviewed the data from 68 patients (40 men and 28 women), who presented with clinical and radiological symptoms of NPH (progressive dementia, gait disturbances with or without urinary incontinence, along with communicating hydrocephalus on brain CT/MRI with a bicaudate ratio > 0.25). The age range was 25–86 years (mean 58). 21 patients had a ventriculo-peritoneal shunt in-situ at the time of examination. All patients underwent a computerized CSF infusion test being a part of diagnostic procedure in the Addenbrooke's Hospital CSF Clinic. In non-shunted patients the aim was to determine their resistance to CSF

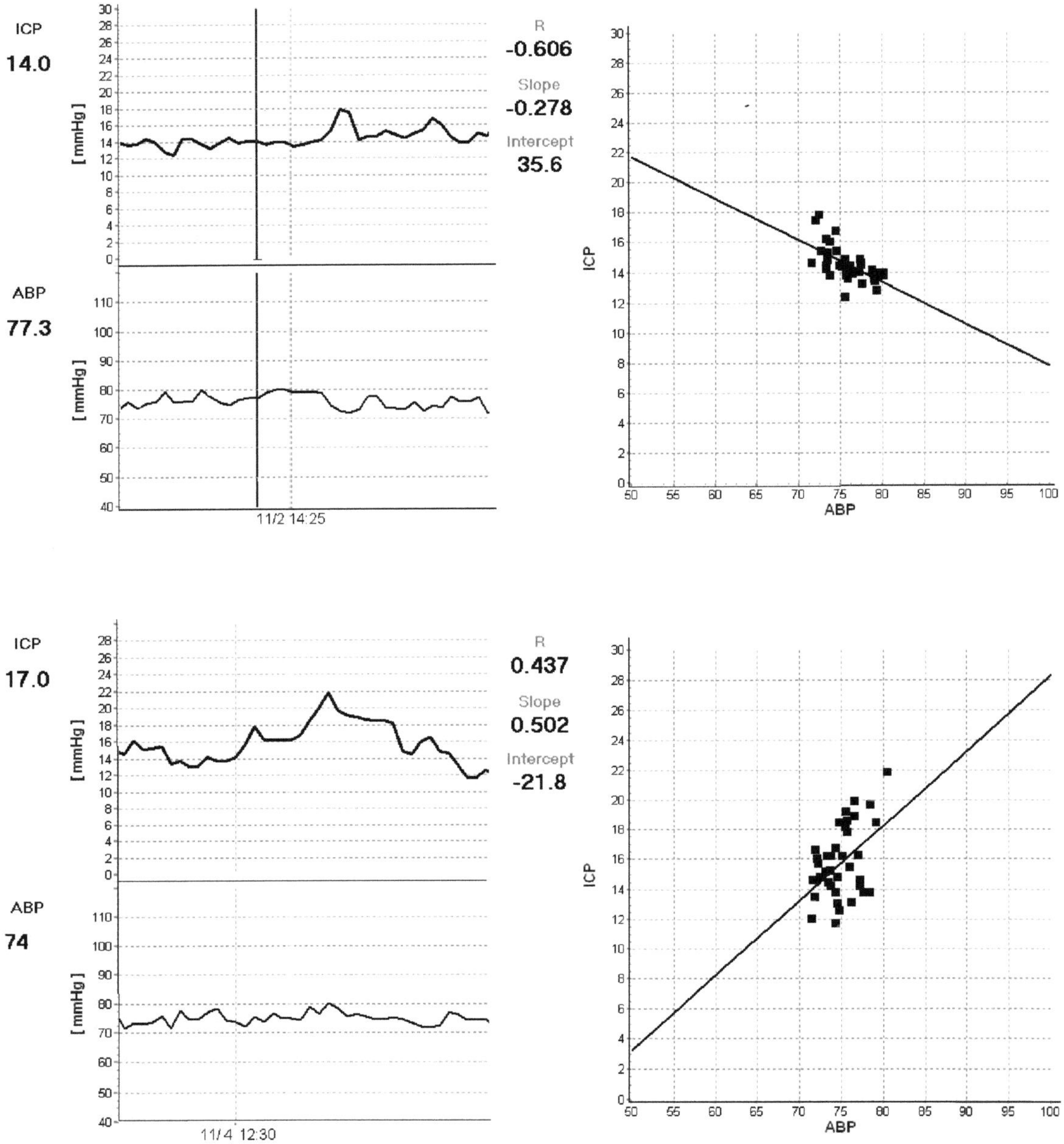

Fig. 1. Relationship between slow waves of arterial pressure (*ABP*) and intracranial pressure (*ICP*). Upper panel: *PRx negative* indicating good pressure-reactivity. Lower panel: *PRx positive* indicating poor pressure-reactivity

outflow, which is one of the predictors of response to shunting. In shunted patients the aim was to assess the shunt function in-vivo, as they had the persistence or recurrence of some clinical signs. This examination is a part of routine diagnostic procedure in our hospital. Data were analyzed retrospectively and anonymity and confidentiality were maintained throughout.

Two needles, which were used for the infusion and the pressure measurement, were placed either into a subcutaneous reservoir that was connected to the intraventricular catheter or into the shunt antechamber. Before the infusion was started, baseline ICP was recorded for 10 minutes. The infusion of a normal saline solution was then started at a rate of 1.5 ml/min or 1 ml/min. When a steady state of ICP plateau was reached, the infusion was stopped. The recording was continued until ICP decreased to a steady baseline level. During the whole period of recording a Finapress finger cuff measured the arterial pressure (ABP).

An IBM-compatible personal computer recorded and processed the data during the infusion test, to obtain mean CSF pressure, pulse wave amplitude of CSF pressure and to calculate the resistance to CSF outflow. ICP and ABP waveforms were processed during the infusion test to obtain an additional index describing cerebrovascular pressure-reactivity (PRx). The index is based on the concept of assessing vascular responses by observing repetitively the reaction of ICP to spontaneous fluctuations of ABP [4]. Using computational

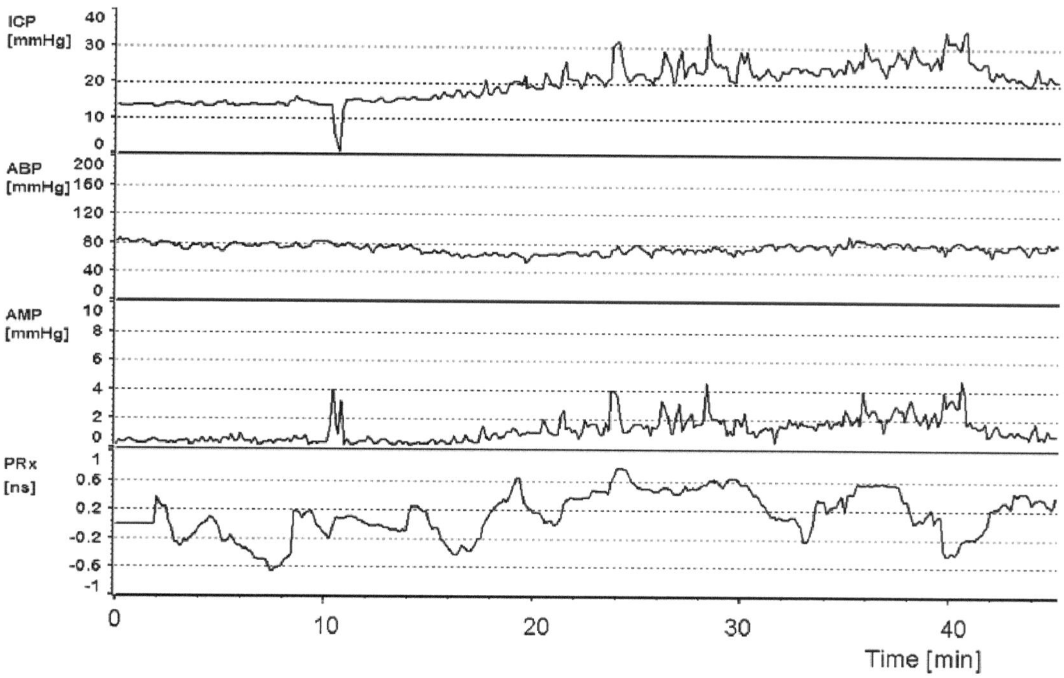

Fig. 2. Example of data recorded during the infusion test. Infusion of 1 ml/min started at 14 minutes and continued until the 40th minute. *ICP* Intracranial pressure, *ABP* arterial pressure, *AMP* pulse amplitude of ICP, *PRx* pressure reactivity index

methods PRx was determined by calculating the correlation coefficient between 40 consecutive, time-averaged (over 6 sec periods) data points of ICP and ABP. A positive PRx signifies a positive gradient of the regression line between the slow components of ABP and ICP, which has been shown to be associated with a passive behavior of a non-reactive vascular bed. A negative value of PRx reflects normal reactive cerebral vessels, as ABP waves provoke inversely correlated waves in ICP (Fig. 1). This index correlates well with indices of autoregulation based on transcranial Doppler ultrasonography. Furthermore, abnormal values of PRx indicative of poor autoregulation have been demonstrated to be predictive of a poor outcome following head injury [4].

Calculated variables were averaged over 10-min. periods at the baseline and during the plateau phase of infusion (Fig. 2) and a non-parametric paired test (signed-rank) was used for comparison of time-averaged values at the baseline and during the elevated ICP within the phase of constant rate infusion. Regression analysis was used, after checking the normal distribution of data, to compare compensatory and cerebrovascular-reactivity indices with the resistance to CSF outflow, separately for shunted and non-shunted patients.

Results

The measured parameters responded to the change in mean ICP between the baseline and the infusion. The mean values and standard deviations are given in Table 1 together with the significance levels for the differences (paired signed-rank test).

In the group of non-shunted patients 28 had a resistance to CSF outflow above 13 mmHg of which 12 above 18 mmHg/(ml/min). In the group of shunted patients 9 tests evaluated the shunt as functioning normally and 5 tests revealed a shunt blockage [5]. In 8 cases 'possible under drainage' was revealed (i.e. Rcsf was increased in comparison to the hydrodynamic resistance of the shunt, but lower than 10 mmHg/(ml/min)).

Of all the CSF compensatory parameters derived from the infusion test, the resistance to CSF outflow (Rcsf) and the baseline ICP demonstrated significant associations with cerebrovascular reactivity (PRx): In non-shunted patients, the regression between PRx and Rcsf indicated a negative linear relationship ($R = -0.5$; $p < 0.0005$) (Fig. 3a).

Table 1. *ICP, ABP and derived parameters before and during infusion*

	At the baseline	During infusion	P value
ICP [mmHg]	7.7 (5.7)	23.1 (10.1)	$2.47 * 10^{-12}$
ABP [mmHg]	85 (28)	92 (36)	0.000247
CPP [mmHg]	76.8 (28)	67.6 (34.3)	0.00019
PRx	0.11 (0.21)	0.18 (0.26)	0.047
Magnitude of B waves [mmHg]	0.71 (0.76)	2.26 (1.23)	0

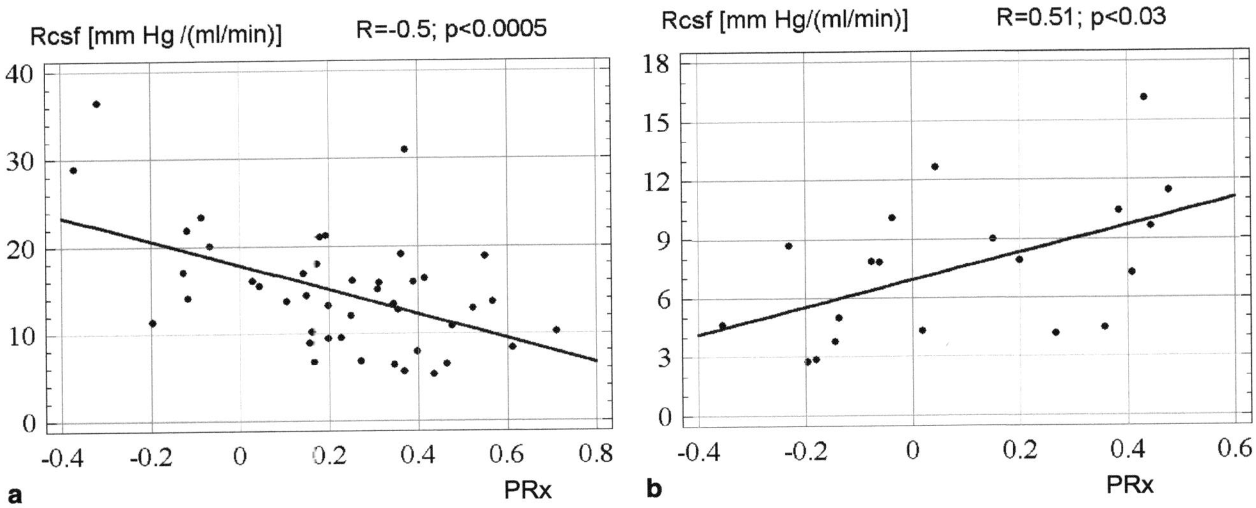

Fig. 3. (a) Relationship between PRx and resistance to CSF outflow in non-shunted patients (n = 44). (b) Relationship between PRx and resistance to CSF outflow in shunted patients (n = 20)

In shunted patients, we observed an opposite relationship between PRx and Rcsf: both parameters were positively associated ($R = 0.51$; $p < 0.03$) (Fig. 3b).

There was no correlation between PRx and ventricular size (bicaudate ratio), opening ICP, Rcsf or elastance coefficient of the intracranial space.

Following the infusion study 27 patients with increased Rcsf were shunted. 14 of them improved 6 did not show any signs of improvement (1 of them got worse). 8 patients were not available for the follow-up. When PRx was compared between improved and non-improved patients it appeared that it indicated nearly significantly ($p = 0.082$) worse pressure-reactivity in patients who did not improve ($PRx = 0.25 \pm 0.23$) in comparison to patients who demonstrated improvement ($PRx = 0.050 \pm 0.21$).

Discussion

The negative correlation between cerebrovascular pressure reactivity (PRx) and Rcsf for non-shunted patients was in agreement with our previous findings about CBF autoregulation assessed using transcranial Doppler ultrasonography. We found that patients with a higher Rcsf tended to have a better autoregulation than patients with a lower Rcsf. Similarly, in the present study, patients with higher Rcsf had better pressure-reactivity (lower PRx). This paradoxical finding was attributed to the possible higher rate of cerebrovascular disease in patients with ventricular dilatation and normal circulation of CSF. Consequently, such patients tend not to exhibit clinical improvement after shunting. The shunt cannot restore normal conditions of cerebral vascular reactivity (and hence improvement of CBF) when the problem lies only or predominantly within the cerebrovascular tree. In hydrocephalus, where there is a pathological link between disturbed CSF circulation (manifested by increased Rcsf) and reduced CBF, shunting can improve CSF distribution and hence CBF.

The relationship between PRx and Rcsf in shunted patients was found to be positive, i.e. inverse to the negative correlation between these two parameters in non-shunted patients. This indicates that the shunt may influence cerebrovascular haemodynamics.

The presented data suggest an association between the three pathophysiological components of hydrocephalus: CSF circulation and haemodynamic reactivity. We found a positive relationship between PRx and Rcsf in shunted patients, which is opposed to the negative relationship between these two parameters in non-shunted patients. This indicates an interference of the shunt with cerebrovascular haemodynamics.

References

1. Boon AJ, Tans JT, Delwel EJ, Egeler-Peerdeman SM, Hanlo PW, Wurzer HA, Hermans J (1999) Dutch normal-pressure hydrocephalus study: the role of cerebrovascular disease. J Neurosurg 90: 221–226
2. Boon AJ, Tans JT, Delwel EJ, Egeler-Peerdeman SM, Hanlo PW, Wurzer HA, Avezaat CJ, de Jong DA, Gooskens RH, Her-

mans J (1997) Dutch normal-pressure hydrocephalus study: prediction of outcome after shunting by resistance to outflow of cerebrospinal fluid. J Neurosurg 87: 687–693
3. Chang C, Kuwana N, Ito S, Ikegami T (2000) Impairment of cerebrovascular reactivity to acetazolamide in patients with normal pressure hydrocephalus. Nucl Med Commun 21(2): 139–141
4. Czosnyka M, Smielewski P, Kirkpatrick P, Laing RJ, Menon D, Pickard JD (1997) Continuous assessment of the cerebral vasomotor reactivity in head injury. Neurosurgery 41(1): 11–17
5. Czosnyka Z, Czosnyka M, Pickard JD (2002) Shunt testing in-vivo: a method based on the data from the UK shunt evaluation laboratory. Acta Neurochir [Suppl] 81: 27–30
6. Czosnyka Z, Czosnyka M, Whitfield PC, Donovan T, Pickard JD (2002) Cerebral autoregulation among patients with symptoms of hydrocephalus. Neurosurgery 50(3): 526–532
7. Klinge PM, Berding G, Brinker T, Knapp WH, Samu M (1999) A PET study of cerebrovascular reserve before and after shunt surgery in patients with idiopathic chronic hydrocephalus. J Neurosurg 91(4): 605–609
8. Momjian S, Owler BK, Czosnyka Z, Czosnyka M, Pena A, Pickard JD (2004) Pattern of white matter regional cerebral blood flow and autoregulation in normal pressure hydrocephalus. Brain 127(Pt 5): 965–972
9. Malm J, Kristensen B, Karlsson T, Fagerlund M, Elfverson J, Ekstedt J (1995) The predictive value of CSF dynamic tests in patients with the idiopathic hydrocephalus syndrome. Arch Neurol 52: 783–789

Correspondence: Zofia Czosnyka, Academic Neurosurgery, Box 167, Addenbrooke's Hospital, Cambridge CB2 2QQ, UK. e-mail: ZC200@medschl.cam.ac.uk

Acta Neurochir (2005) [Suppl] 95: 213–217

Evolution of intracranial pressure during the immediate postoperative period after endoscopic third ventriculostomy

D. Santamarta[1] and **J. Martin-Vallejo**[2]

[1] Department of Neurosurgery, University Hospital of Salamanca, Salamanca, Spain
[2] Department of Statistics, University Hospital of Salamanca, Salamanca, Spain

Summary

Objective. To establish a more accurate indication for endoscopic third ventriculostomy (ETV) in patients with noncommunicating hydrocephalus through the analysis of the evolution of postoperative mean intracranial pressure (ICP_M).

Method. Intracranial pressure (ICP) was recorded overnight during 8-hour periods with an intraventricular probe. A personal computer connected to the ICP monitor minutely recorded the values of ICP. Twenty-four patients were monitored from day 1 to day 3 after ETV. The evolution of ICP_M was analysed with an ANOVA test for repeated measures. The relevance of different factors (age, etiology, size of the lesion leading to hydrocephalus, clinical course and outcome) on the evolution of ICP_M was explored with a two-factor ANOVA.

Results. ICP_M progressively decreased from day 1 to day 3 after ETV ($p = 0.03$). ICP_M on the first postoperative day was 15.81 ± 2.04 mm Hg (mean $\pm$ standard error) and 13.43 ± 1.44 mm Hg on the third postoperative day. Different patterns in the evolution of ICP_M have been detected according to the age of the patient and the clinical course of hydrocephalus.

Conclusion. ICP_M progressively decreases after ETV. This pattern is not constant. It has been clearly detected in children and in acute forms of hydrocephalus.

Keywords: Noncommunicating hydrocephalus; endoscopic third ventriculostomy; intracranial pressure.

Introduction

Noncommunicating hydrocephalus is a protean clinical entity. The diagnosis is not always straightforward. Patients may present an acute intracranial hypertension syndrome lasting for a few days and, on the other hand, the diagnosis may be done in the presence of a more indolent clinical picture heralded by a gait disturbance, a memory disorder or chronic headache lasting for several months or even years. Endoscopic third ventriculostomy (ETV) has gained acceptance as the treatment of choice for this entity. If such different clinical scenarios should be treated with an ETV has not been convincingly answered [9].

The main hypothesis sustaining the design of this study is the existence of a transmantle pressure gradient in the early stage of development of hydrocephalus. Experimental models [2] and the introduction of biomechanics [4, 6, 7] to the study of hydrocephalus support this assumption. However, a pressure gradient with a higher pressure in the ventricles than in the subarachnoid space has not been clearly demonstrated in a clinical setting and in chronic forms of hydrocephalus there is no transmantle pressure gradient [8].

If ETV is an effective treatment for noncommunicating hydrocephalus, intracranial pressure (ICP) determined at the ventricular level should immediately or gradually decrease after surgery. Through the analysis of the evolution of the mean intracranial pressure (ICP_M) in the postoperative period, our goal has been to establish a more accurate indication for ETV in patients with noncommunicating hydrocephalus.

Methods

Overnight ICP recording was performed with an intraventricular probe (Probe 3, Spiegelberg GmbH & Co., Germany) during the first 3 postoperative days in 24 patients with noncommunicating hydrocephalus treated with ETV. ICP was monitored during 8-hour periods. In thirteen patients ICP monitoring was extended until the fifth postoperative day. The ICP monitor (Spiegelberg CPP Monitor) is connected to a personal computer. A software (Midas CPP Collection Program Version 3.0, developed by Ian Piper, Glasgow) minutely records the value of ICP. This yields a total of 480 lectures during an 8-hour period. The data are stored in an ASCII format file. These data are imported by common software to calculate the value of ICP_M every day.

The evolution of ICP_M in the immediate postoperative period was analysed with an ANOVA test for repeated measures. The level of

Table 1. *Baseline characteristics of the series analysed: 24 patients with noncommunicating hydrocephalus treated with endoscopic third ventriculostomy and postoperative ICP monitoring*

Characteristic	n	%
Sex		
– Male	9	*37.5*
– Female	15	*62.5*
Age		
– Adults (>15 yrs)	20	*83.3*
– Children (<15 yrs)	4	*16.7*
Etiology		
– Intracranial mass	16	*66.7*
– Primary aqueductal stenosis	8	*33.3*
Size		
– >30 mm	13	*54*
– <30 mm	11	*46*
Clinical onset		
– Acute intracranial hypertension	18	*75*
– Chronic hydrocephalus	6	*25*
Outcome		
– Satisfactory	21	*87.5*
– Shunt	3	*12.5*

significance chosen for this contrast was $p = 0.05$. A two-factor ANOVA with repeated measures on one factor (one within, one between factor) was used to analyse the influence of different factors of clinical relevance on the evolution of ICP_M. These factors include: age, etiology, size of the lesion leading to hydrocephalus, clinical course of the disease and outcome. The level of significance chosen for the interaction was $p = 0.10$. Informed consent was obtained from every patient. This study was approved by the Local Ethics Committee of Clinical Investigation.

This is an adult-based series (median age 51 yrs). Table 1 summarizes the baseline characteristics. Age was subdivided in two groups; children ($\leq$15 yrs) and adult patients. Etiology in primary aqueductal stenosis or intracranial mass. The size of the lesion leading to hydrocephalus was categorized as larger or smaller than 30 mm according to its maximal diameter. The clinical course of the disease was subdivided in acute and chronic forms of hydrocephalus. The former comprises patients with symptoms and signs of intracranial hypertension lasting less than 2 weeks. Chronic forms of hydrocephalus are characterized by a more indolent clinical picture lasting several months or even years usually heralded by gait disturbances, headache, memory disorders or a cognitive deficit detected by neuropsychological testing. The outcome has been classified as satisfactory, for those patients who improved and have remained clinically well up to their last follow-up visit, and failures or shunted patients in those cases in which clinical symptoms of hydrocephalus either returned or never resolved and the patients required additional surgery for treatment of hydrocephalus [5].

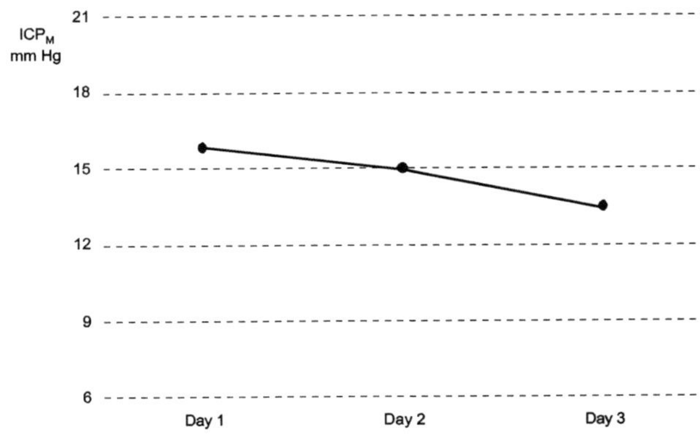

	n	Mean	SD	SE
Day 1	24	15.81	10.02	2.04
Day 2	24	14.95	8.66	1.76
Day 3	24	13.43	7.08	1.44

Fig. 1. Evolution of ICP_M during the first 3 postoperative days after endoscopic third ventriculostomy. *n* Number of patients; *SD* standard deviation; *SE* standard error

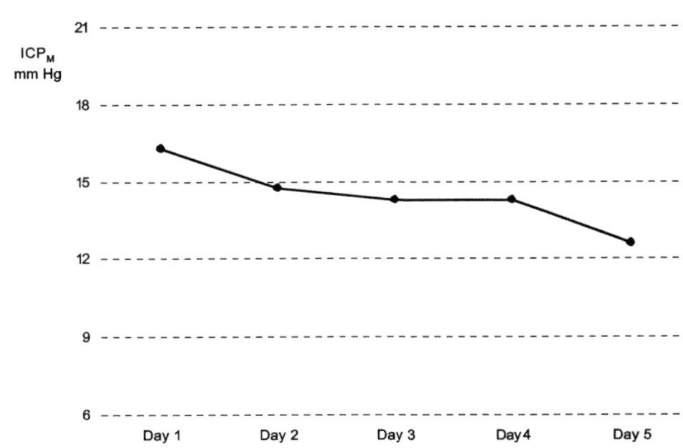

	n	Mean	SD	SE
Day 1	13	16.28	9.24	2.56
Day 2	13	14.77	8.09	2.24
Day 3	13	14.31	6.97	1.93
Day 4	13	14.31	6.60	1.83
Day 5	13	12.59	6.50	1.80

Fig. 2. Evolution of ICP_M during the first 5 postoperative days after endoscopic third ventriculostomy. *n* Number of patients; *SD* standard deviation; *SE* standard error

Results

ICP_M gradually decreased during the immediate postoperative period ($p = 0.03$). ICP_M on the first postoperative day was 15.81 ± 2.04 mm Hg (mean $\pm$ standard error) and 13.43 ± 1.44 mm Hg on the third postoperative day (Fig. 1). The gradient during this 48-hour period was 2.38 mm Hg. The same accounts for the subset of patients who were monitored during five days ($p = 0.01$). The gradient is larger when a more extended period of time is analysed (Fig. 2). ICP_M on the first postoperative day was 16.28 ± 2.56 mm Hg and 12.59 ± 1.80 mm Hg on the fifth

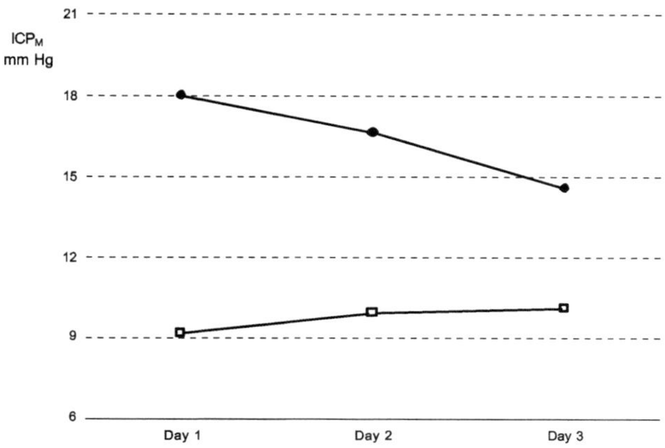

	ACUTE			CHRONIC		
	n	Mean	ES	n	Mean	ES
Day 1	18	18.02	2.42	6	9.18	2.32
Day 2	18	16.63	2.11	6	9.93	2.28
Day 3	18	14.55	1.69	6	10.06	2.45

Fig. 3. Interaction between the clinical course (acute or chronic hydrocephalus) and the evolution of ICP_M. -●- ACUTE; -□- CHRONIC

postoperative day, which represents a difference of 3.69 mm Hg.

Influence of the clinical course on the evolution of ICP_M: acute and chronic hydrocephalus

Patients with acute hydrocephalus displayed a trend towards higher values of ICP_M than patients with chronic hydrocephalus. The pattern of the evolution of ICP_M after ETV was different in both groups of patients ($p = 0.01$). ICP_M progressively decreased in acute hydrocephalus. By contrast, ICP_M in chronic forms of hydrocephalus remained stationary throughout the immediate postoperative period (Figs. 3 and 4).

Influence of age on the evolution of ICPM: children and adult patients

Children with noncommunicating hydrocephalus have higher ICP_M than adults. This subdivision of age determines different patterns of the evolution of ICP_M after ETV ($p = 0.002$). In the pediatric subset of patients there is a decrease of ICP_M from 31.81 ± 5.75 mm Hg on the first postoperative day to 22.73 ± 4.26 mm Hg on the third postoperative day. The magnitude of the descent is 9.08 mm Hg. In adults, the slope is much less pronounced (Fig. 5).

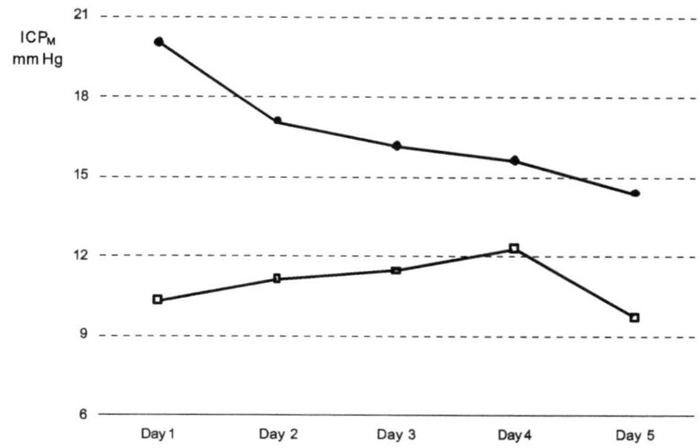

	ACUTE			CHRONIC		
	n	Mean	ES	n	Mean	ES
Day 1	8	20.01	3.30	5	10.31	2.48
Day 2	8	17.07	3.16	5	11.10	2.40
Day 3	8	16.12	2.64	5	11.43	2.50
Day 4	8	15.57	2.14	5	12.29	3.40
Day 5	8	14.39	2.39	5	9.71	2.44

Fig. 4. Interaction between the clinical course and the evolution of ICP_M throughout 5 postoperative days. -●- ACUTE; -□- CHRONIC

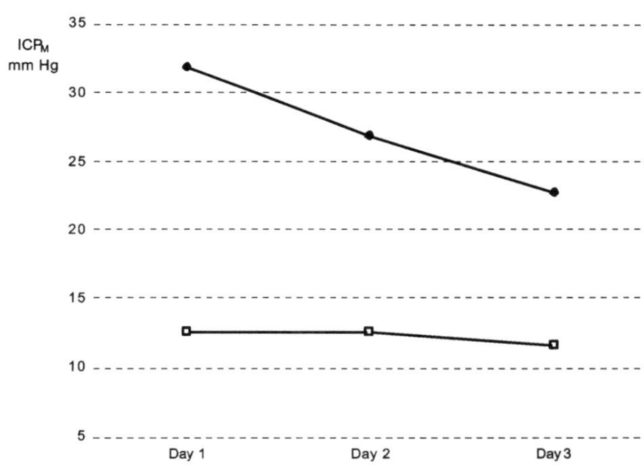

	CHILDREN			ADULTS		
	n	Mean	ES	n	Mean	ES
Day 1	4	31.81	5.75	20	12.61	1.34
Day 2	4	26.88	6.08	20	12.57	1.25
Day 3	4	22.73	4.26	20	11.57	1.17

Fig. 5. Interaction between the age (children or adults) and the evolution of ICP_M. -●- CHILDREN (>15 yrs); -□- ADULTS (>15 yrs)

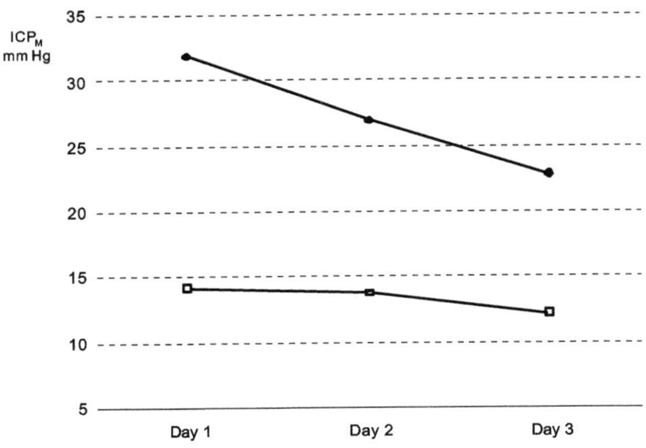

	CHILDREN with ACUTEHYD			ADULTS with ACUTEHYD		
	n	Mean	ES	n	Mean	ES
Day 1	4	31.81	5.75	14	14.08	1.52
Day 2	4	26.88	6.08	14	13.70	1.45
Day 3	4	22.73	4.26	14	12.22	1.33

Fig. 6. Interaction between the age and the evolution ICP$_M$ in the acute stage of hydrocephalus. -●- CHILDREN with acute hyd; -□- ADULTS with acute hyd

All pediatric patients had acute hydrocephalus. Adult patients with chronic hydrocephalus were deleted to avoid the confounding effect of the clinical course of the disease (Fig. 6). This yielded 18 patients with acute hydrocephalus monitored from day 1 to day 3 after ETV. The effect of age is maintained when only acute forms of hydrocephalus are considered ($p = 0.01$).

The etiology ($p = 0.68$), the size of the lesion ($p = 0.31$) and the outcome at the end of the follow-up period ($p = 0.34$) according to the above mentioned subdivision did not effect the postoperative evolution of ICP$_M$.

Discussion

ICP$_M$ gradually decreases during a 48-hour interval in the immediate postoperative period after ETV. The magnitude of the gradient is in the range of 2 to 3 mm Hg. Larger gradients are detected when the period of ICP monitoring is extended until the fifth postoperative day, confirming a progressive decline of ICP$_M$ throughout the immediate postoperative period after ETV.

This pattern, however, is not uniform. The age of the patient and the clinical course of the hydrocephalus seem to definitely effect the postoperative evolution of ICP$_M$. In the acute stage of development of hydrocephalus ICP$_M$ steeply decreases, while in chronic forms of hydrocephalus ICP$_M$ remains stationary. The subdivision of age in children and adults also effected a different response of the ICP$_M$ evolution of as a logical reflection of changes in stiffness and in elastic properties of the brain parenchyma related with age. The descent of ICP$_M$ in children is much more striking while the response in adults was less perceptible.

By contrast, other factors considered of clinical relevance did not seem to condition the postoperative evolution of ICP$_M$. Such factors include: subdivision of etiology in intracranial mass or primary aqueductal stenosis; the size of the lesion leading to hydrocephalus according to its largest diameter and the outcome at the end of the follow-up period.

Although ETV is a well-established indication for noncommunicating hydrocephalus, the scenario widely varies in terms of age at presentation, etiology, clinical manifestation and duration of symptoms. This variability may in part explain the rather consistent failure rates in the range of 30% reported in most unselected series of patients with noncommunicating hydrocephalus [1, 3]. Such results and the heterogeneity of the disease lead to define more precisely which patients with noncommunicating hydrocephalus are really candidates for ETV and which will be better treated with a shunt. The role of ETV in adult patients with chronic forms of noncommunicating hydrocephalus should be carefully considered. Both conditions are associated with a pattern of the evolution of ICP$_M$ in which changes are hardly detectable.

References

1. Buxton N, Ho KJ, Macarthur D, Vloeberghs M, Punt J, Robertson I (2001) Neuroendoscopic third ventriculostomy for hydrocephalus in adults: report of a single unit's experience with 63 cases. Surg Neurol 55: 74–78
2. Conner ES, Foley L, Black PMcL (1984) Experimental normal-pressure hydrocephalus is accompanied by increased transmantle pressure. J Neurosurg 61: 322–327
3. Hopf NJ, Grunert P, Fries G, Resch KDM, Perneczky A (1999) Endoscopic third ventriculostomy: outcome analysis of 100 consecutive procedures. Neurosurgery 44: 795–806
4. Hakim S: Biomechanics of hydrocephalus (1971) Acta Neurol Latinoam 17 [Suppl] 1: 169–194
5. Kulkarni AV, Drake JM, Armstrong DC, Dirks PB (2000) Imaging correlates of successful endoscopic third ventriculostomy. J Neurosurg 92: 915–919

6. Nagashima T, Tamaki N, Matsumoto S, Horwitz B, Seguchi Y (1987) Biomechanics of hydrocephalus: a new theoretical model. Neurosurgery 21: 898–904
7. Peña A, Bolton MD, Whitehouse H, Pickard JD (1999) Effects of brain ventricular shape on periventricular biomechanics: a finite-element analysis. Neurosurgery 45: 107–118
8. Stephensen H, Tissell M, Wikkelso C (2002) There is no transmantle pressure gradient in communicating or noncommunicating hydrocephalus. Neurosurgery 50: 763–773
9. Tisell M, Almström O, Stephensen H, Tullberg M, Wikkelsö C (2000) How effective is endoscopic third ventriculostomy in treating adult hydrocephalus caused by primary aqueductal stenosis? Neurosurgery 46: 104–111

Correspondence: David Santamarta, Servicio de Neurocirugía, Hospital Universitario de Salamanca "Virgen de la Vega", Paseo de San Vicente, 58-132, 37007 Salamanca, Spain. e-mail: dsantamarta@wanadoo.es

Acta Neurochir (2005) [Suppl] 95: 219–222

Clinical research on monitoring CSFP through lumbar epidural pressure

Z. Liu[1], **Y. Y. Dou**[2], **R. Chen**[1], and **X. Z. Zhang**[1]

[1] Neurosurgery Department, Fuzhou General Hospital, Fuzhou, Fujian, China
[2] Anaesthesia Department, Fuzhou General Hospital, Fuzhou, Fujian, China

Summary

Objective. Explore the relationship between lumbar Epidural Pressure (LEDP) and CSFP through clinical studies and to determine whether LEDP can represent CSFP.

Methods. We selected 150 cases of cerebral diseases at random for this study. A special mini transducer was implanted into the lumbar epidural space between the third and the 4th lumbar vertebra by means of a No. 18 Touhy needle for monitoring of LEDP. In the meantime, a traditional lumbar puncture was made between the 4th and the 5th lumbar vertebrae and CSFP, which was taken as the standard value, was measured using an ordinary catheter. The transducer was adjusted until the LEDP was equal to CSFP and was designated $LEDP_0$. The $LEDP_0$ was monitored continuously and was translated into a continuous curve. In each and every case, a lumbar puncture was repeated at 24, 48, 72 and 96 hours to obtain CSFP for comparison and to explore the relationship between $LEDP_0$ and CSFP. While monitoring $LEDP_0$, we observed the variation in pressure when the patient breathed, coughed, changed posture, and also if the clinical symptoms changed with intracranial hypertension.

Result. The values of the intracranial pressure measured by these two methods were identical.

Conclusion. LEDP responded to the changes of CSFP.

Keywords: Lumbar epidural pressure; intraventricular pressure.

Introduction

After considerable research on the lumbar epidural pressure (LEDP) measurement, many researchers came to the conclusion that LEDP could represent CSFP. However, in those studies, LEDP was not monitored continuously. By using a mini transducer we have conducted continuous monitoring with success using LEDP in 150 cases of cerebral disease, and compared the data with CSFP. These correlations is the subject of this report.

Material and method

Transducer & instruments

This system consisted of the following:

a. A catheter transducer: The special transducer, 0.9 mm in diameter was sufficiently small to pass through a No. 18 Touhy needle at its minimum and expanded to 8 mm $*$ 8 mm $*$ 10 mm at its maximum with the catheter graded in millimeters.
b. A high-precision physiological pressure display: showing the pressure (Unit: mmH_2O) from the transducer translated the pressure into a dynamic curve continuously;
c. A computer was used to record and save the dynamic pressure data for subsequent analysis.

Operational method

With the patient lying on his/her side horizontally, we adjusted the catheter transducer to zero based on the patient's spinous process line. A puncture was made between the third and the 4th lumbar vertebra with a No. 18 Touhy needle and the transducer was implanted into the lumbar epidural space by the graduation on the catheter. Once the probe was out of the Touhy needle, we withdrew the needle and connected the end of the catheter to the pressure display.

In the meantime, we, also made a lumbar puncture between the 4th and the 5th lumbar vertebra and measured CSFP with an ordinary catheter (mmH_2O). This reading was taken as the standard value for comparison. After obtaining the CSFP reading, we expanded the transducer slowly by means of a special device. The lumbar epidural pressure (LEDP) gradually increased. When it is was equal to CSFP, we designated this pressure as $LEDP_0$. We fixed the catheter with adhesive and withdrew the puncture needle and let the patient lie on his back while monitoring $LEDP_0$. The readings were translated into a curve and recorded on a computer.

These $LEDP_0$ readings were gathered for 4 to 5-days of continuous monitoring, during which several lumbar punctures are undertaken at between the 4th and the 5th lumbar vertebra to gather CSFP with an ordinary catheter. The two readings were compared. After the monitor period, we reduced the size of the transducer and withdrew it from the epidural space.

Clinical data

We included 150 cases of cerebral diseases taken at random: 101 cases male and 49 female ranging in age from 14 to 57 (29.8 on average). Among them were 110 cases of brain injury, 89 intracranial hematomas, 40 intracranial tumor treatment, and 37 surgical cases for tumor removal. No pathological changes such as a tumor in the vertebral canal was found in these clinical practices.

Research method

While monitoring $LEDP_0$ continuously in each of the 150 cases, we made three or four lumbar punctures at 24 hours (time1), 48 hours (time2), 72 hours (time3), and 96 hours (time4) respectively in order to gather CSFP for comparison. In time1 we designated P_{11} for $LEDP_0$, P_{12} for CSFP; for time2, P_{21} for $LEDP_0$ and P_{22} for CSFP; time3 coincides to P_{31} and P_{32}; and so on. We explored the correlation between $LEDP_0$ and CSFP by comparing these two readings. When monitoring $LEDP_0$ we observed if pressure changes corresponding to respiration, coughs, changes in posture, and if the clinical symptoms were altered with intracranial hypertension.

Results

In the process of continuous monitoring of $LEDP_0$ in these 150 cases, we measured these two pressures in four prescribed times. Hence the four groups of readings. The following results were obtained after making a statistical analysis using SPSS software: We observed the following.

1) The pressures measured by two different methods demonstrated the same distribution characteristic, i.e., the normal distribution. The result of the analysis is as shown in Table 1.
2) The pressures measured by two different methods displayed a similar curve. For example, at time4, the following chart (fshows their p-p normal probability distribution:
3) Table 2 comes into being after making interval estimate based on $\alpha = 0.1, 0.05$, and 0.10.

As shown, the interval difference at different periods is insignificant.

4) Again, based on $\alpha = 0.1, 0.05$ and 0.10, make T test and square deviation test on the means of these twin sample readings. The tests show that, given $\alpha = 0.1, 0.05$ and 0.10, the means and square deviations illustrate no significant difference in P_{11} and P_{12}, P_{21} and P_{22}, P_{31} and P_{32}, P_{41} and P_{42}.

Table 1. *Mean & square deviation*

	Mean	Standard deviation	Square deviation
P_{11}	265.8667	51.43454	2645.5123
P_{12}	266.2200	50.38463	2538.6157
P_{21}	274.2933	51.15072	2616.3966
P_{22}	274.1600	50.97625	2598.5782
P_{31}	266.4533	46.30285	2143.9541
P_{32}	267.2533	45.37187	2058.6065
P_{41}	263.9067	44.89650	2015.6959
P_{42}	263.5467	44.98360	2023.5246

Table 2. *Interval estimate*

	$\alpha = 0.1$	$\alpha = 0.05$	$\alpha = 0.10$
P_{11}	(254.9089,276.8244)	(257.5682,274.1652)	(258.9157,272.8176)
P_{12}	(255.4859,276.9541)	(258.0909,274.3491)	(259.4109,273.0291)
P_{21}	(263.3961,285.1906)	(266.0406,282.5460)	(267.3807,281.2059)
P_{22}	(263.2999,285.0201)	(265.9355,282.3845)	(267.2710,281.0490)
P_{31}	(256.5889,276.3178)	(258.9828,273.9239)	(260.1959,272.7108)
P_{32}	(257.5872,276.9195)	(259.9330,274.5737)	(261.1217,273.3850)
P_{41}	(254.3418,273.4715)	(256.6630,271.1503)	(257.8393,269.9741)
P_{42}	(253.9632,273.1301)	(256.2890,270.8044)	(257.4675,269.6258)

Table 3. *Frequency distribution (%)*

Intracranial pressure (mmH_2O)	P_{11}	P_{21}	P_{31}	P_{41}
80–180	1.3	0.7	0	4.6
180–270	54.0	48.6	60.7	59.4
270–421	44.7	50.7	39.3	36.0

Conclusion

The readings of the pressure measured through the two different ways are the same.

1. Classify CSFP as Lower than Normal (above 80 mmH_2O), Normal (80–180 mmH_2O), Slight Growth (180–270 mmH_2O), and Moderate Growth (270–570 mmH_2O). Among these 150 cases, the frequency distribution in P_{11}, P_{21}, P_{31} and P_{41} is as follows:
2. The $LEDP_0$ waveform is found through continuous monitor to fluctuate slightly with the patient's inhaling and exhaling. It's mainly constituted by the breathing wave with amplitude at 10–10 mmH_2O. The $LEDP_0$ goes up when the jugular vein or abdominal region is pressed upon and comes down when the pressure eases. But sharp increase or decrease is displayed when the patient coughs with the maximum amplitude of 410 mmH_2O. When the $LEDP_0$ rises and exceeds the normal CSFP, the clinical symptoms worsen as the patient feels headache and vomits. The $LEDP_0$ is observed to come

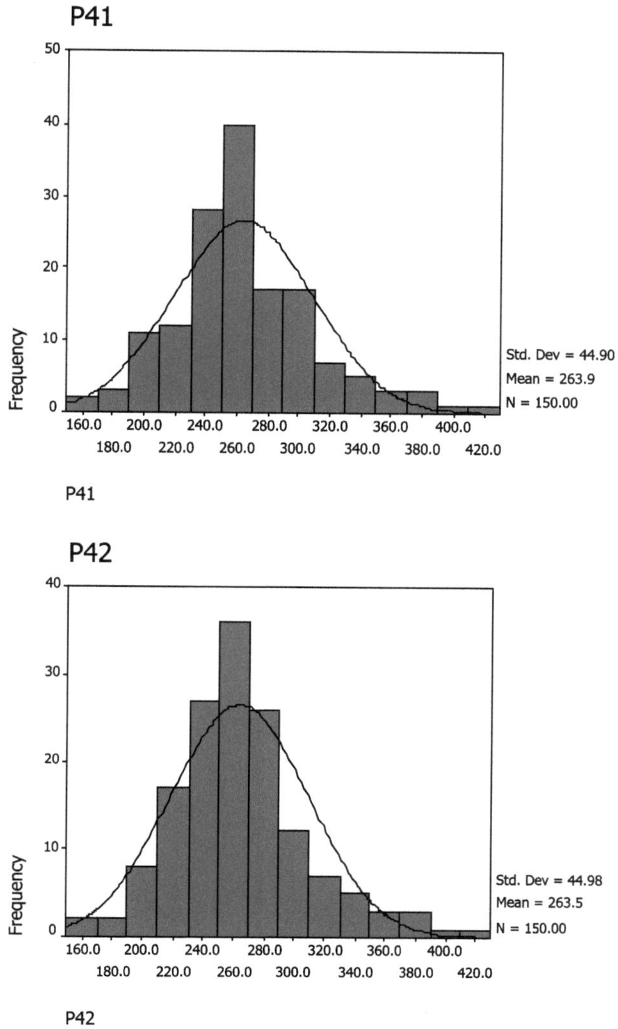

Fig. 1.

down as dehydrated medicine like mannitol is applied.

Discussion

The epidural space is a closed anatomic lacuna. As early as 1967 Usubia discovers that dural outer pressure is created when filling 10 to 20 ml physiological saline into the lumbar epidural space, and it displays a parallel relationship with CSFP. Shah also noted that LEDP could indirectly reflect CSFP. Unfortunately, this way of monitoring cannot last long because the created pressure eases as the saline is assimilated and diffuses. Huamadu tried to measure LEDP by placing Caelter transducer into dog's lumbar epidural space. However, due to the excessively small size of the transducer, the transducer is cut off from the contact with the dural as the CSFP gets down. Consequently, the communication is broken off. Therefore, it failed to continuously reflect CSFP.

As is known to all, cerebrospinal fluid is the essential volume compensatory substance when the intracranial pressure changes. It keeps flowing along the designated route. Dural is a flexible fiber membrane, keeping contact with or detached from the vertebral canal on and off. The surrounding hard vertebral lamina, yellow ligament and pyramid limit its expansion. In the case of a grown-up, from the third lumbar vertebra downward, the dural contains no vertebral cord but the CSF encircled in subarachnoid cavity. The CSF communicates its pressure to the latent lumbar epidural space, thus maintains the dural's pressure on the vertebral canal. This is a positive pressure equivalent to the CSF. There still exists valveless intervertebral venous plexus in the epidural space to receive blood from vertebral cord. The venous plexus transport the blood up into intervertebral vein through great occipital foramen and further on into innominate vein, and down through the lumbar veins in the intervertebral foramen into inferior vena cava. Therefore, the extension of the venous plexus will press upon the dural and make the CSFP change accordingly. So when the patient coughs or his jugular vein or abdominal region is under weight, the CSFP will go up. It is thus inferred that βCSFP = pure CSFP + epidural venous plexus' static pressure. Based on these anatomic features we invented a mini transducer whose size is able to expand as required. When it expands to be in contact with the vertebral canal wall and the dural and balances with the CSFP, the pressure measured through the transducer is exactly equivalent to the CSFP.

The result demonstrates that the $LEDP_0$ is basically the same with the CSFP when the patient's CSFP stays normal or increases. And long time continuous monitoring shows that the two readings are in parallel and conforms to the clinical symptoms. The $LEDP_0$'s translated waveform and amplitude are in conformity with the CSFP change, especially when the patient coughs and the jugular vein or the abdominal region is pressed upon. Therefore, we conclude that $LEDP_0$ can accurately correspond to the CSFP change. But it is restricted to apply in the case of cerebral hernia or vertebral canal obstruction.

References

1. Usubiaga JE, Moya F, Usubiaga LE (1967) Effect of thoracic and abdominal pressure change on the epidural space. Br J Anaesthesia 39: 612
2. Shah JI (1981) Influence of cerebrospinal fluid on epidural pressure. Anaesthesia 36: 627
3. Hamada J, Fujioka S, Ushio Y (1993) Experimental investigation of lumbar epidural pressure measurement. Neurosurgery 32: 820

Correspondence: Liu Zheng, Neurosurgical Department, Fuzhou General Hospital, Xi-er-huan Rd., Fuzhou, 350003, Fujian, PR China. e-mail: liuzheng@vip.cvn.com.cn

Acta Neurochir (2005) [Suppl] 95: 223–227

Evaluation of three new models of hydrocephalus shunts

Z. H. Czosnyka, M. Czosnyka, H. K. Richards, and **J. D. Pickard**

Academic Neurosurgical Unit, Addenbrooke's Hospital, Cambridge, UK

Summary

Objective. To assess the hydrodynamic properties of three new types of hydrocephalus valve.

Methods. Three new constructions have been recently tested in the UK shunt Evaluation Laboratory: the magnetically adjustable Strata Valve (Medtronic PS Medical), the gravitational Miethke Dual-Switch Valve (Aesculap) and the ventriculo-sinus SinuShunt (CSF Dynamics). Pressure-flow performance curves were assessed in a minimum of three samples of each valve to study their long-term variability, influence of temperature, negative outlet pressure, external pressure, presence of pressure pulsations, etc.

Results. The operating pressure of the Strata Valve can be adjusted magnetically in five steps. This Shunt prevents 'siphoning' but is sensitive to external pressure. The Dual Switch Miethke Valve is a system of two fixed-pressure ball-on-spring valves with a lower opening pressure operating in a horizontal body position and higher when vertical. This function is designed to cancel the effect of siphoning related to body posture. Both Strata and DSV valves have a low hydrodynamic resistance (less than 3 mm Hg/ml/min), and hence they cannot prevent overdrainage related to nocturnal vasomotor waves. The SinuShunt has a higher resistance (9 mm Hg/(ml/min)) and a lower opening pressure. The valve is intended to drain CSF from ventricles to the transverse sinus.

Conclusion. New shunt technology continues to evolve. Laboratory evaluation independent of the manufacturer forms an important link between R&D laboratories and clinical practice.

Keywords: Hydrocephalus; shunt; laboratory evaluation.

Shunting for the management of communicating hydrocephalus remains the mainstream strategy. To help the neurosurgeon choose from the many types of shunt available, information about each shunt's hydrodynamic properties should be available. The amount of technical information provided by the manufacturers varies. In the mid-1990s, the new ISO standard (ISO 7197) attempted to regulate the minimal requirements for the description of hydrodynamic properties of the shunt, but this has not been fully implemented by all manufacturers.

The UK Shunt Evaluation Laboratory has attempted to provide technical information, independent of the manufacturer and conforming to international standards [2]. The testing protocol has been developed and subsequently applied to at least 20 different models of hydrocephalus shunts currently in use or being introduced to clinical practice in the UK. This study describes the relevant hydrodynamic properties of three relatively novel shunt models.

Material and methods

Medtronic strata valve

The two primary attributes of this valve are opportunity to change the valve's operating pressure after implantation and its clinical ability to prevent siphoning. The Strata Valve is a differential-pressure hydrocephalus ball-on-spring valve (Fig 1a) incorporating a magnetically programmable mechanism to specify the operating pressure. The Strata valve is integrated with a Delta Chamber (also known from the literature as the Siphon Control Device [4]), which prevents overdrainage related to the body posture.

Dual switch miethke valve (DSV)

This valve belongs to the group of gravitational valves, which have been used for many years. It incorporates two different pathways for CSF for the horizontal and for the vertical body position – Fig 1b. The DSV is a fixed pressure valve. The valve is available in 9 combinations of performance levels in horizontal [10, 13, 16] and vertical position [30, 40, 50], coded xx/yy where xx and yy are opening pressure levels in cmH_2O relatively horizontally and vertically. Recently, xx = 5 cmH_2O has been introduced (not tested) to aid the treatment of Normal Pressure Hydrocephalus.

SinuShunt

The SinuShunt is a differential pressure valve designed to drain cerebrospinal fluid from brain ventricles to the cranial transverse venous sinus [1] – Fig 1c.

CSF is drained proportionally to the difference between intracranial pressure and transversal venous sinus pressure. The opening pressure of the SinuShunt is intended to be low (0 mm Hg) and its

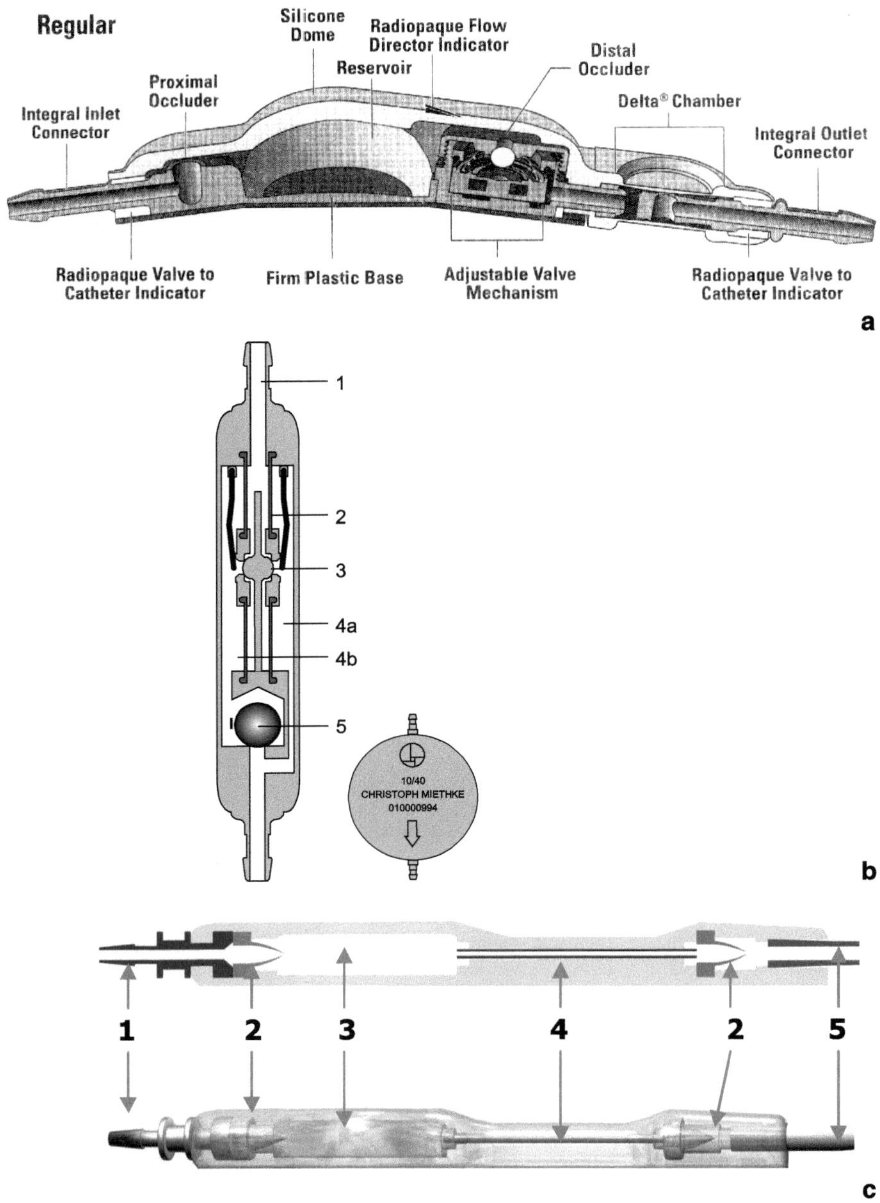

Fig. 1. Constructions of the three valves (provided by the manufacturers): (a) Strata Valve, (b) DualSwitch Valve. *1* Titanium casing; *2* Silicone diaphragm supported by flat spring. Left-low tension; Right-high tension spring; *3* Closing ball; *4* Outlet chambers. Right chamber (outlet from high-pressure valve) is always open. Left chamber (outlet from the low-pressure valve) is open only in horizontal position; *5* Heavy tantalum ball blocking low pressure valve in vertical position. (c) SinuShunt: *1* Connector; *2* One way valves; *3* Pre-chamber; *4* Resistance tube; *5* Distal catheter

hydrodynamic resistance-close to physiological resistance to CSF outflow (8 mm Hg/(ml/min)). With such parameters, the shunt is supposed to restore physiological conditions of CSF outflow, usually disturbed in hydrocephalus [1].

The testing rig has been previously described in detail [2], with its schematic diagram presented in Figure 2. Measurement is controlled by a standard IBM compatible personal computer with software designed in-house (M.C.) which precisely measures flow through the shunt and differential pressure. Three shunts of the same type are filled with deionised and deaerated water and mounted in three identical rigs.

Results

Numerical values of variables: opening pressures and hydrodynamic resistances with and without distal catheters of all valves are given in Table 1.

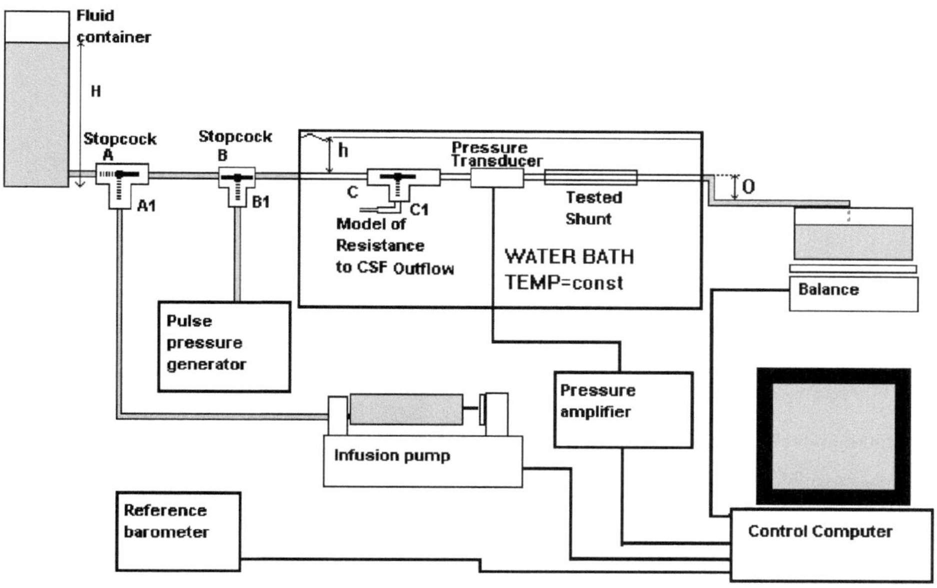

Fig. 2. Schematic diagram of testing rig

Table 1. *Basic hydrodynamic parameters of the tested shunts. Values are given as mean ± standard deviation*

Valve	Performance level	Opening Pressure [mm Hg]	Hydrodynamic resistance (no drain)	Hydrodynamic resistance (with drain)
Strata	0.5	1.5 ± 0.5	1.7 ± 0.6	2.8 ± 0.5
	1	4.2 ± 0.7	"	"
	1.5	6.1 ± 0.9	"	"
	2	8.4 ± 1.5	"	"
	2.5	10.7 ± 0.8	"	"
DSV	10 H	7 ± 0.6	2.2 ± 0.8	2.9 ± 1.1
	13 H	9.5 ± 0.4	"	"
	16 H	12 ± 1.1	"	"
	30 V	22 ± 2.1	"	"
	40 V	28 ± 1.9	"	"
	50 V	38 ± 2.2	"	"
SinuShunt	one level only	2.5 ± 1/2	9.6 ± 0.96	9.8 ± 0.8

DSV Dual Switch Valve. This valve consists of two parallel valves, coded by the values of their opening pressures in cm of water. H indicates that the valve works in horizontal and V – in vertical position. Opening pressure is given in mmHg and hydrodynamic resistance in mm Hg/(ml/min)

Strata

When tested without a distal catheter the valve had an almost a linear pressure-flow characteristic (Fig 3a). Measurement points showed a good convergence within and between tested samples. A negative outlet pressure (−17 cm H_2O, an equivalent to the value recently reported [5]) did not alter the valve's drainage rate. However, the valve's closing pressure changed when an external (environmental) pressure was applied.

Programming of the valve was checked and the pressure-flow curves found to be consistent with the nominal data provided by the manufacturer (see Table 1). The valve may be accidentally re-programmed by external non-homogeneous magnetic fields around 14 mT.

DualSwitch valve

The valve demonstrated an almost linear pressure-flow curve (Fig 3b) with a good convergence of the measurement points within and between tested samples. The DSV demonstrated an alteration in the

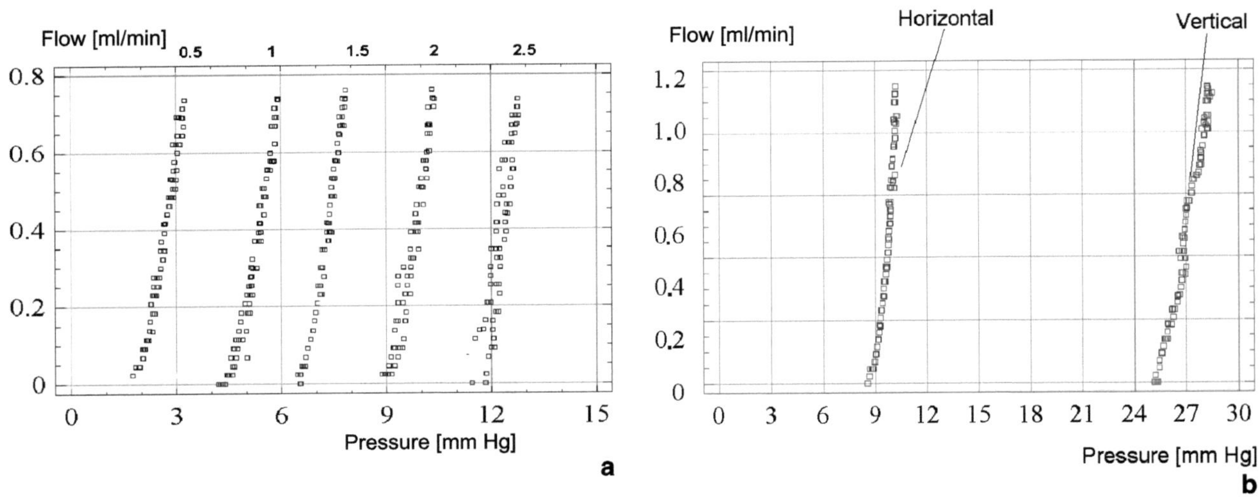

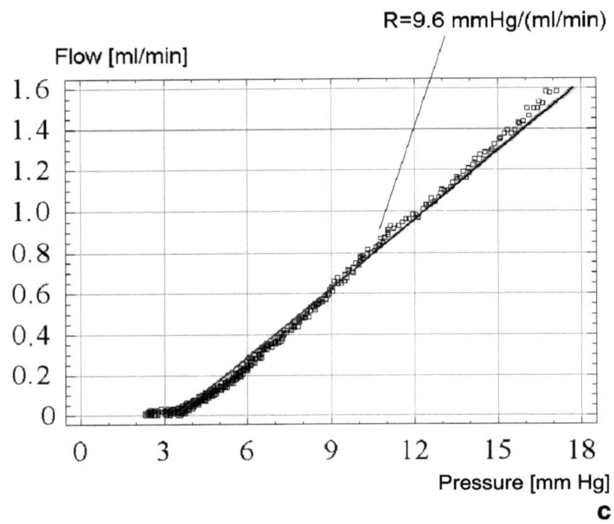

Fig. 3. Pressure flow performance curves of the tested shunts: (a) Strata Valve: five different performance levels, (b) DSV (level 10/30): two performance curves in horizontal and vertical position, (c) SinuShunt

drainage rate when a negative outlet pressure was applied. With the valve in horizontal position, the pressure–flow curve shifted toward low pressures. This is compensated by an increase in operating pressure due to the change in the body position to vertical by 20, 30 or 40 cm H_2O, depending on shunt operating pressure. There was no effect of external pressure. Pressure flow curve shifts to the right when the position of the valve switches from horizontal to vertical. There is no significant change of the slope of this curve (i.e. both low and high pressure valves have the same hydrodynamic resistances). Magnetic field 3T did not have any influence on the shunt's performance.

SinuShunt

The valve had almost linear characteristic. All the measurement points from the first 16 pressure-flow tests from three valves showed a good convergence within and between tested samples (see Fig 3c). The flow through the SinuShunt was affected by negative outlet pressure (siphoning) but this effect is eliminated in-vivo by positioning the inlet (ventricles) and outlet (transverse sinus) at the same hydrostatic level. None of the tested parameters changed when an external pressure was applied. 3 Tesla MRI did not alter the valve performance.

All three shunts shared some hydrodynamic properties

A pulse waveform applied to the input pressure with an amplitude changing from 1 to about 18 mm Hg produced a significant decrease in operating pressure of the range from 0.2 to 7.4 mm Hg. None of the parameters (opening, closing pressure and resistance) were altered by a temperature change from 30 °C to

40 °C. Opening and closing pressures displayed very limited (less than 1.5 mm Hg) variations during all the tests. The changes were not systematically time-related. Good agreement of the operating pressures with the nominal data has been recorded (differences less than 1 mm Hg). No significant differences in measured parameters were found between the three samples of each of the valves tested. The valves did not show any reflux when tested according to the ISO standard. Valves did not exhibit reversal of flow for an outlet-inlet differential pressure of up to 100 mm Hg.

Discussion

All of the assessed valves represent the latest progress in shunt technology. All of them resist the siphoning effect, each one in a different way. The Strata incorporates an anti-siphon device, the Miethke valve takes advantage of gravitational forces and the SinuShunt eliminates the problem by the way it can be implanted. All three valves thus prevent CSF overdrainage related to body position. However, in clinical practice, complications related to posture related overdrainage amount to only few percent of all reported shunt-related complications (unpublished data from UK Shunt Registry). Only the Strata Valve is adjustable. External adjustment may help decrease the number of revisions after shunting, but the programming mechanism is sensitive to external magnetic fields. The valve can be accidentally reprogrammed, which is a common feature of all currently used programmable valves (also Sophy and Codman-Hakim) [7].

The Strata Valve is sensitive to external pressure. The valve may be blocked by excessive subcutaneous scarring of the scalp. The Strata and DSV valves have low hydrodynamic resistance. This may make them prone to overdrainage of CSF in the presence of nocturnal ICP vasogenic waves [2]. No matter how well the opening/closing pressure mechanism is engineered, variations of ICP (via pulse or respiratory waves) may decrease a valve's performance levels. There are clinical reports in the literature about use of the presented shunts. The Miethke DSV valve has been enthusiastically described, particularly by German neurosurgeons [6] where it has been claimed to reduce the problem of overdrainage. The SinuShunt requires more complex surgery than other shunts. There has been a learning curve in the original centre – its implementation in other centres are awaited [1, 3]. The Strata Valve is becoming gradually more popular as an alternative to the Codman-Hakim Valve. It has only 5 programming levels as opposed to 18 in the Codman-Hakim model. However, this may be sufficient in the management of hydrocephalus. The Strata Valve, unlike Codman-Hakim, does not require X-ray to confirm the desired programming level. Patients with any of these three valves may be safely MRI-scanned. The Strata Valve should always be re-programmed after scanning.

Acknowledgment

R&D agreements between the Laboratory and the manufacturers were signed for the period of shunt testing and grants covering laboratory costs were awarded. The study was conducted independently and all presented findings and conclusions were not influenced by the manufacturers.

References

1. Borgesen SE, Gjerris F, Agerlin N (2002) Shunting to the sagittal sinus. Acta Neurochir [Suppl] 81: 11–14
2. Czosnyka Z, Czosnyka M, Richards HK, Pickard JD (2002) Laboratory testing of hydrocephalus shunts – conclusion of the UK Shunt evaluation programme. Acta Neurochir (Wien) 144(6): 525–538; discussion 538
3. Eklund A, Koskinen LO, Malm J (2004) Features of the Sinushunt and its influence on the cerebrospinal fluid system. J Neurol Neurosurg Psychiatry 75(8): 1156–1159
4. Horton D, Pollay M (1990) Fluid flow performance of a new siphon-control device for ventricular shunts. J Neurosurg 72: 926–932
5. Kajimoto Y, Ohta T, Miyake H, Matsukawa M, Ogawa D, Nagao K, Kuroiwa T (2000) Posture-related changes in the pressure environment of the ventriculoperitoneal shunt system. J Neurosurg 93(4): 614–617
6. Miethke C, Affeld K (1994) A new valve for the treatment of hydrocephalus. Biomed Tech (Berl) 39(7–8): 181–187
7. Schneider T, Knauff U, Nitsch J, Firsching R (2002) Electromagnetic field hazards involving adjustable shunt valves in hydrocephalus. J Neurosurg
8. Zeilinger FS, Reyer T, Meier U, Kintzel D (2000) Clinical experiences with the dual-switch valve in patients with normal pressure hydrocephalus. Acta Neurochir [Suppl] 76: 559–562

Correspondence: Z. H. Czosnyka, Academic Neurosurgical Unit, Box 167, Addenbrooke's Hospital, Cambridge CB2 2QQ, UK. e-mail: ZC200@medschl.cam.ac.uk

Acta Neurochir (2005) [Suppl] 95: 229–235

The enigma of underdrainage in shunting with hydrostatic valves and possible solutions

C. Sprung[1], **C. Miethke**[2], **H.-G. Schlosser**[1], and **M. Brock**[1]

[1] Neurosurgical Department, Charité, Humboldt-University, Berlin, Germany
[2] Miethke GmbH & Co. KG, Potsdam, Germany

Summary

Objective. Hydrostatic devices have considerable advantages compared to "conventional" differential-pressure-valves concerning overdrainage, but are thought to imply a tendency to underdrain or to clog. The aim of this study was to evaluate the ability of the hydrostatic gravitational Dual-Switch-Valve (DSV) to minimize overdrainage-related complications without increasing the danger of underdrainage.

Results. In a series of 202 adult patients with different etiologies treated with a ventriculo-peritoneal shunt including the hydrostatic Dual-Switch-valve (DSV), 21 cases were suspected of suffering from underdrainage. Using a new algorithm we were able to differentiate obstruction in 6 patients from functional underdrainage in 15 cases, thus we saw an indication to reimplant a DSV with a lower opening pressure in the latter.

Conclusion. The reasons for functional underdrainage were multifold in our series, especially the intraperitoneal pressure is still a "black box". Despite the ability of the DSV to avoid clogging and to minimize overdrainage by its high-pressure-chamber, it remains difficult to determine the optimal opening pressure of the low-pressure-chamber of the DSV for ideal clinical improvement. Therefore a new hydrostatic gravitational "programmable" valve (proGAV), entitled on avoiding the disadvantages of other adjustable devices, has been developed and implanted in 16 patients with promising results.

Keywords: Hydrocephalus; ventriculo-peritoneal shunt; underdrainage; hydrostatic valve; programmable shunt.

Introduction

The first reports about mechanical complications after shunting were dealing mainly with overdrainage-related problems. But especially since the introduction of hydrostatic valves, complications related to underdrainage gain increasing attention. However, an unequivocal evaluation of underdrainage is hindered by discrepancies in the definition of this entity, because in most series "functional" underdrainage is mixed with obstruction due to misplacement or kinking of catheters, disconnection of the shunt or blocking of the valve (Table 1). But if it is desired to evaluate the capability of a valve to avoid the two most important complications of shunting, over- and underdrainage, complications must be differentiated depending on the valve and those which are not directly related to the function of the device itself. For example, infection and surgical problems which include malpositioning of the ventricular or peritoneal catheter. As expected by the construction principle of a hydrostatic device like the DSV, the incidence of overdrainage-related complications was very low. However the question arose as to whether the decrease of overdrainage is counterbalanced by an increase of the rate of underdrainage.

By using our definition of functional underdrainage given in Table 1 we followed a new algorithm (Table 2) to differentiate between obstruction of different causes and functional underdrainage. Furthermore we think it is manditory to introduce a more objective grading system to determine the reduction of ventricular size for correlation with clinical outcome. We also wish to stress the importance of the intraperitoneal pressure for the function of a shunt and possible underdrainage. Therefore in every case of revision of the DSV implanted in the thoracic region we measured the intraabdominal pressure via the peritoneal catheter. Despite meticulous preoperative tests including: pressure measurements while performing a lumbar Tap-test before shunting; taking into account the etiology of hydrocephalus; radiological features and length of history, we could not avoid functional underdrainage due to the choice of a high pressure setting of the primarily implanted DSV. The incidence of functional

Table 1. *Discrepancies in definition and incidence of under- and overdrainage in different series in literature*

Author	Year	Patients and etiology	No. of cases	Type of valve	(functional) Underdrainage	Overdrainage	Under-drainage %	Over-drainage %	% related to
Saint-Rose [15, 16]	1991 1993	mainly children	1719 2062	DP-valves Orbis-S,	"underdrainage" different from obstr.	subdural eff. slit-ventricles isolated ventr.	69.7 (1.7)	12.3	all mechanic. complications
Di Rocco [3]	1994	mainly children	773	DP-valves + Orbis-S,	obstruction differ. from "insufficient drainage"	"excessive drainage"	70 (11)	2.3	all complications
Drake [4]	1996	children	344	DP- + Delta + Orbis-S,	assessed together with obstruction	subdural eff. slit-ventricles isolated ventr.	31.4	3.5	all patients
Boon [2]	1998	adults NPH	96	DP-valves	disconnect. displacem.	subdural effusions	19.8	53.1	all patients
Pollack [13]	1999	children and adults	377	adjust. Medos, DP-Delta-, Orbis-S, valves,	persistent ventriculomegaly	slit-like ventricles, extra-axial collections	7	6	all patients
Zemack [21]	2000	children and adults	583	adjust. Medos	based on symptoms or CT/MR	subdural eff., or signs and symptoms	39.1	35.4	adjustments of valves
UK-Shunt-Reg. [14]	2000	children and adults	10910	various valves	"shunt still functioning on removal"	subdural eff. craniostenosis slit-ventricles	52	3	all revisions
Hanlo *et al.* [5]	2003	children and adults	557	Orbis-sigma II	"valve hydrodynamics"	symptomatic overdrainage with slit-ventricles	22.6 (5)	1.8	all patients
Sprung	2004	adults different etiology	202	DSV	persistent ventr. size ass. with no clinical improving, despite functioning shunt	subdural eff., slit-ventricles clinical symptoms during orthostasis	11.9 (7.4)	5.0	all patients

Note the variations in patient material, different etiologies and valves and to what total number the incidence is related. The incidence of underdrainage in total is including obstruction and functional insufficiencies. The incidence of functional underdrainage in % is shown within brackets.

Table 2. *Algorithm to differentiate obstruction from functional underdrainage and to assess the necessity to revise the shunt/valve*

Diagnostic procedure	Result			Consequence
X-ray of Shunt: ↙	negative	/	Kinking Disconnection Misplacement PK →	Revision of Complication
CT-/MR-control: Shuntogram: ↙	negative	/	Increase ventricles Misplacement VC Shuntogram pos. →	Revision
ICP-measurement via reservoir recumbent + upright: ↙	negative	/	ICP significant higher than opening pressure ICP-drop in upright pos. inadequate →	Exchange of valve
Tap–test (50 ml): ↙ No operation	negative	/	Clinical improvement →	Exchange of valve

Table 3. *Clinical data of patients with DSV*

Etiology	n	Sex f	Sex m	Age	Opening pressure horiz./vertical [cmH$_2$O]	n	Replace-ments	Follow-up
Idiopathic NPH	59				13/40	117		
Typ. secondary NPH > 3 mos	44				10/40	72		
Acute/subacute NPH < 3 mos	72				8/40	5		
Hypertensive hydrocephalus	13	106	96	17–85 yrs ∅ 60.8 yrs	5/40	6	21	6–84 mos ∅ 33.2 mos
Pseudotumor cerebri	3							
Pat. not to wean from ext. drainage/narrow ventricles	11				16/50	2		

underdrainage in our DSV-series was one of the main reasons to look for a solution by adding the principle of adjustability to the advantages of gravitation-assisted hydrostatic devices. Therefore we saw an indication to construct a new adjustable valve (proGAV) avoiding the well known disadvantages of the "programmable" devices currently available.

Material and methods

From the beginning of 1995 to the end of 2002 we conducted a consecutive series of 202 adult patients, in whom we saw an indication for shunting, by implanting a frontal ventricular-peritoneal shunt including a burr-hole reservoir and a Dual-Switch-Valve. The collective comprised a majority of 76 cases with idiopathic and secondary Normal-Pressure-Hydrocephalus (NPH). 65 patients with development of subacute hydrocephalus (HC) within 3 months after the causative incident were differentiated, according to the Dutch NPH-Study [2], from typical secondary NPH (Table 3) with a time-gap of more than 3 months.

To avoid only subjective assessment of the ventricular regression postoperatively, we introduced an objective scale by grading the reduction of the Evans' Index (Fig. 1). The postoperative complications were differentiated in valve-related and those independent of valve-function.

A standarized assessment of clinical outcome was difficult due to the variety of etiologies in our series. We used different grading systems including the Stein and Langfitt-Scale and the Black-outcome-scale. The clinical outcome was determined not earlier than 6 months after shunting.

The proGAV comprises an adjustable unit in series with a gravitational hydrostatic unit, the Shunt-Assistant (Fig. 2). The adjustable unit consists of a differential-pressure-valve with a "brake" to avoid unintended changes of pressure-setting like in other adjustable devices. The brake working by friction of the rotor on the housing can be decoupled transcutaneously by an adjustment-pin to change the opening pressure of the valve. There is a wide pressure range from 0 to 20 cm H_2O. In the lying position the opening pressure is dependent only on the adjustable unit because the gravitational unit is not activated. If the patient moves into the upright position, the heavy tantalum-ball of the gravitational unit gradually closes the outlet of the valve. Thus in the upright position it remains closed until the intraventricular pressure plus the hydrostatic pressure is higher than the sum of the opening pressure of both units. By this mechanism the proGAV is counterbalancing the hydrostatic pressure in the vertical position of the shunted patient.

Since February 2004, 16 of the new adjustable proGAVs have been integrated in V-P-shunts of adult patients with different types of hydrocephalus. In this preliminary series we were primarily interested in the question whether the construction of the new valve can be implanted without surgical difficulties and able to serve as a draining-device at least as good as other gravitation-assisted valves. Our secondary aim was to evaluate the security of different new tools to adjust the valve transcutaneously and the safety to determine the adjusted opening pressure avoiding the burden of multiple X-ray-controls. Furthermore we were interested in the capability of the new mechanism to avoid unintended adjustments.

Results

The outcome in our series of 202 patients with DSV was characterized by excellent and good clinical results accompanied by an only minimal reduction of ventricular size in the majority of cases. A total of 16 patients (7.9%) suffered from infections and 6 cases from minor surgical complications requiring revisions. The latter included 3 misplacements plus 1 kinking of the ventricular catheter, and 2 occlusions or malpositioning of the peritoneal catheter. We included 3 patients, in whom we primarily suspected functional underdrainage in this category of complications (not depending on valve-function), because the patients did not improve clinically despite operative change to a valve with a lower opening pressure. Probably this non-responsiveness was due to the severe damage done to the brain by the causative incidents pre-shunting (2 cases of ICB and 1 SAH), which could not be compensated by drainage of the hydrocephalic state.

The 17 cases of functional underdrainage could be subdivided in 15 with a wrong (too high) choice of pressure level, 1 change of pressure setting in vivo 2 years following implantation and 1 patient with abnor-

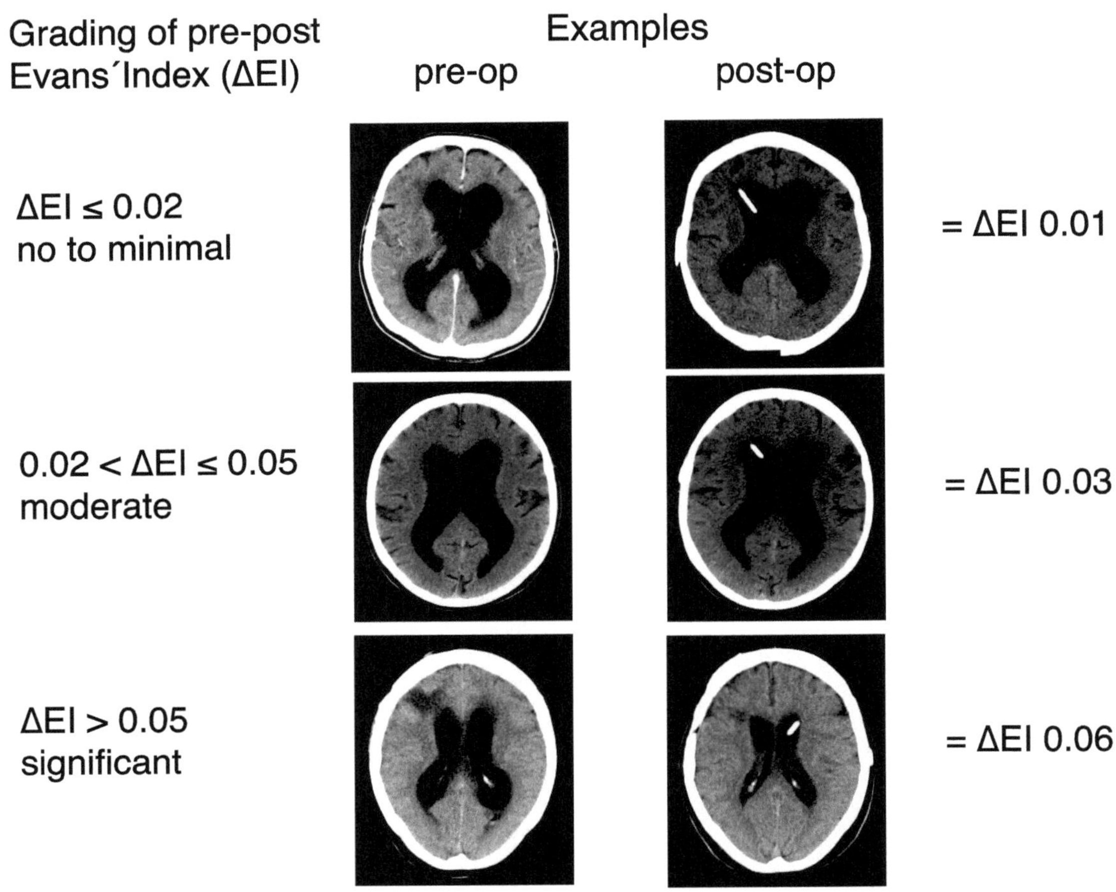

Fig. 1. Objective grading of the reduction of ventricular size by measuring the Evans-Index pre-/postoperatively

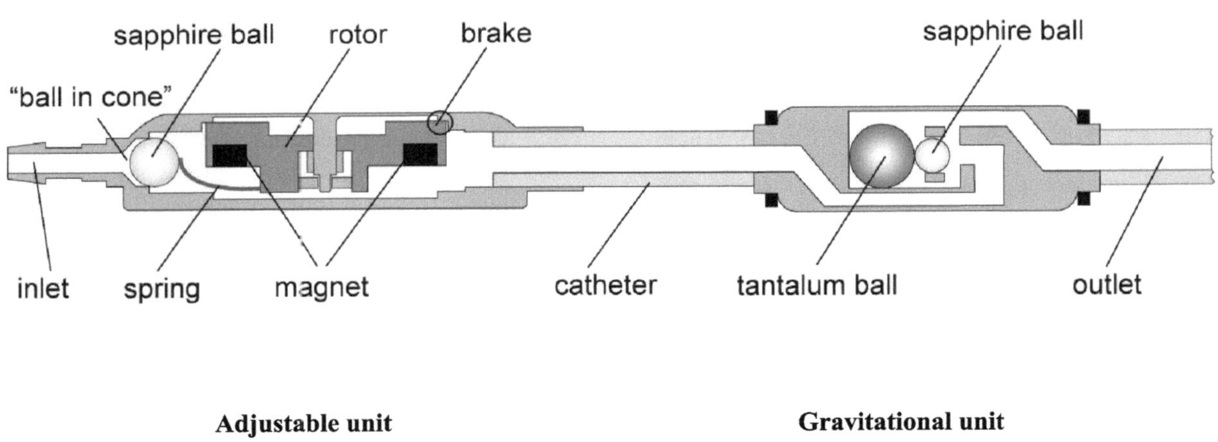

Fig. 2. Schematic depiction of proGAV

mal high intraperitoneal pressure. The 10 cases with overdrainage comprised 1 patient with slit-like ventricles five days postoperatively, 4 hygromas (3 transient and 1 persistent) and 5 subdural hematomas.

Of the patients with overdrainage related to the function of the DSV, all 10 presented clinical and radiological evidence of overdrainage. In only 7 cases, the overdrainage was persistent. Surprisingly one of

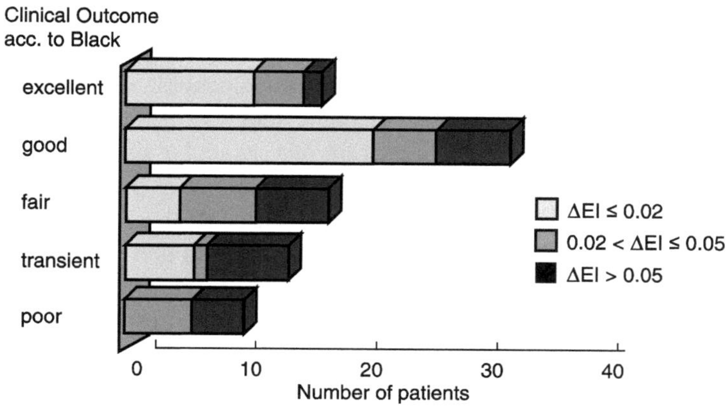

Fig. 3. Correlation of clinical outcome and reduction of ventricles

the cases with pseudotumor cerebri and narrow ventricles developed subdural hematoma. All other hydrocephalic cases with overdrainage – related complications necessitating revisions revealed predisposing factors like aqueductal stenosis due to isolated IV.ventricle, pre-shunt small subdural hematoma, too inclined implantation of the valve and one case with spontaneous fronto-basal fistula and pneumatocephalus. None of the patients with idiopathic or typical secondary NPH in our series developed persistent overdrainage-related complications. Thus, we believe the results of our collective concerning subdural effusions were superior to comparable series in the literature.

"Functional" underdrainage was suspected in 25 of our patients. Following the exclusion of obstruction due to other causes by the usual diagnostic procedures, all these cases underwent a continuous pressure-measurement in the recumbent as well as in the upright position via the Rickham-reservoir on a tilting-table, followed by a Tap-test. By this method we could clearly differentiate between obstruction and functional underdrainage in most cases (Table 2).

We did not find one obstruction of the valve itself neither by this pressure-measurement nor by re-examination of the explanted valves despite the fact that our series included several cases with high protein levels in the subgroup of cases with acute/subacute hydrocephalus. All explanted valves were re-evaluated by the manufacturer and with only one exception revealed the exact pressure-levels as stated before implantation.

After implanting a valve with a lower opening pressure for the recumbent position, 17 cases improved clinically but 3 did not. Thus in the latter cases the failure of shunting was judged as not depending on the valve-function.

A comparison of this complication to other series was not possible because other authors did not differentiate between obstruction and functional underdrainage (Table 1).

An only minimal or no reduction of ventricular size was documented on follow-up CTs in about 70% of our series with the hydrostatic DSV (Fig. 3). However, there were unequivocal differences concerning reduction of ventricles between the various forms of hydrocephalus in our patient group. Whereas hypertensive hydrocephalus revealed significant reduction of the ventricles in the majority of cases, the percentage of significant regressions was reduced in patients with sub-acute hydrocephalus and lowest in NPH-cases. Contrary to the radiological outcome a comparison of the clinical results depending on etiology did not reveal significant differences.

In agreement with the experience of the majority of authors reporting series with hydrostatic valves no correlation was found between the grade of regression of ventricles and overall clinical outcome (Fig. 3).

Despite numerous bench-tests in our 3-Tesla-MRT unit which provided evidence of the safety of pressure-level setting, the clinical series of 16 cases with proGAV is much too small and the follow-up is too short to assess one of the main theoretical advantages of the proGAV, the capability to avoid unintended adjustments by the "brake". There were no difficulties in releasing the brake for changing the opening pressure and we did not have to register an unintended adjustment in the clinical course of our 16 patients up to now.

Discussion

Since 1995 the hydrostatic gravitation-assisted Dual-switch-valve has been implanted for shunt-treatment of hydrocephalus [11, 17]. Since that time, the device has proven to be capable to decrease the rate of overdrainage in different forms of hydrocephalus [6, 8–10, 18, 19] without significantly increasing the complication of underdrainage. Nevertheless it remains difficult to define the ideal opening pressure of the valve for an optimal clinical result in the individual patient and in some cases it is unclear whether a better result could have been achieved with a lower opening pressure of the valve. The problem of underdrainage is multifold up to now and all efforts should be undertaken to solve the enigma for the successful shunting of hydrocephalic patients.

A prerequisite for a thoughtful comparison of shunt valves are corresponding definitions of the complications and similarity of the most important items of the series like patient inclusion, etiology of hydrocephalus and uniformity of implanted valves. However in the literature these criteria are not fulfilled in the vast majority of reports about shunt series. In some of the most important publications mostly children are included [3, 4, 15, 16], various groups comprise children and adults [13, 14, 21], whereas others exclusively adults with only one etiology [2] or with different types of hydrocephalus [17–19]. Whereas the definition of overdrainage is comparable in most of the series (Table 1), the complication of underdrainage is delineated differently by the various authors. While some reports differentiate clearly between obstruction and functional underdrainage [3, 16], others presume obstruction together with functional underdrainage [4], and do not mention the latter important complication at all [2, 13] or give equivocal definitions like "shunt still working on removal" [14] or "valve hydrodynamics" [5]. Another reason for the extremely divergent results concerning the incidence of these two most important complications after shunting is the different choice of the parameter under- and overdrainage is related to in these series (Table 1). In addition to the wrong choice of a too high pressure level by the surgeon, the intraperitoneal pressure remains a "black box" up to now necessitating the measurement of the patency of the peritoneal catheter and the intraperitoneal pressure during the revision. By this relatively simple manauver you can exclude or prove other possible reasons for non-responsiveness of shunts [20].

The grade of reduction of ventricular size with this type of valve depends on the etiology of hydrocephalus, but also on the pressure-setting. If there is a significant or moderate reduction, obstruction and functional underdrainage can be ruled out. On the other hand an only minimal or no reduction of ventricular size does not rule out good or even excellent clinical results. Apparently older brains of adults with more chronic forms of hydrocephalus and this type of hydrostatic valve react different from juvenile hydrocephalus and treatment with differential-pressure valves. Furthermore, in many cases brain tissue damage may be severe. In agreement with the majority of authors [2, 3, 6, 8, 9, 19, 22] but contrary to others [1, 5, 7, 12] no correlation was found between grade of regression of ventricles and overall clinical outcome (Fig. 3).

The outcome of our series also stresses the necessity to improve the preoperative diagnostic tools and the importance of the possibility to adjust the valve in order to avoid functional underdrainage. The clinical and radiological results of our small series of 16 patients with the adjustable proGAV provides evidence that this construction principle can help to minimize the danger of functional underdrainge. But the cohort of our pro-GAVs is too small and the follow-up is too short to compare our results with series comprising other adjustable devices.

Conclusions

The outcome of our series proves the capability of the DSV to minimize overdrainage – related problems also in patients suffering from idiopathic NPH. The discrepancies regarding the definition of underdrainage are the main reason for the differences regarding the incidence of this mechanical complication reported in the shunt-literature. The results also point out that the choice of the ideal opening-pressure of the valve to achieve the optimal clinical result after shunting remains difficult and a matter of controversy up to the present time. This study of the new adjustable pro-GAV provides evidence which minimizes the problem of functional underdrainage in shunts. However, a larger series and a longer follow-up is necessary before we can decide whether the enigma of underdrainage can be solved by this device and lead us toward a new era of shunting.

References

1. Bergsnider M, Peacock WJ, Mazziotta JK, Becker DP (1999) Beneficial effect of siphoning in adult hydrocephalus. Arch Neurol 56: 1224–1229
2. Boon AJ, Tans JT, Delwel EJ, Egeler-Peerdeman SM, Hanlo PW, Wurzer HA, Avezaat CJJ, de Jong DA, Gooskens RHJM, Hermans J (1998) Dutch Normal-Pressure Hydrocephalus Study: randomized comparison of low- and medium-pressure shunts. J Neurosurg 88: 490–495
3. Di Rocco C, Marchese E, Velardi F (1994) A survey of the first complication of newly implanted CSF shunt devices for the treatment of nontumoral hydrocephalus. Child's Nerv Syst 10: 321–327
4. Drake JM, Kestle JRW, Milner R, Cinalli G *et al* (1998) Randomized trial of cerebrospinal fluid shunt valve design in pediatric hydrocephalus. Neurosurgery 43: 294–305
5. Hanlo PW, Cinalli G, Vandertop WP, Faber JA, Bogeskov L, Borgesen SE, Boschert J, Chumas P, Eder H, Pople IK, Serlo W, Vitzthum E (2003) Treatment of hydrocephalus determined by the European Orbis Sigma Valve II survey: a multicenter prospective 5-year shunt survival study in children and adults in whom a flow-regulating shunt was used. J Neurosurg 99(1): 52–57
6. Kiefer M, Eymann R, Meier U (2002) Five Years Experience with Gravitational Shunts in Chronic Hydrocephalus of Adults. Acta Neurochir (Wien) 144: 755–767
7. Mc Connell KA, Zou KH, Chabrerie AV, Bailey NO, Black P McL (2004) Decreases in ventricular volume correlate with decreases in ventricular pressure in idiopathic normal hydrocephalus patients who experienced clinical improvement after implantation with adjustable valve shunts. Neurosurgery 55: 582–593
8. Meier U, Kintzel D (2002) Clinical experiences with different valve systems in patients with normal-pressure hydrocephalus: Evaluation of the Miethke dual-switch valve. Child's Nerv Syst 18: 288–294
9. Meier U, Paris S, Gräwe A, Stockheim D, Hajdukova A, Mutze S (2003) Is there a correlation between operative results and change in ventricular volume after shunt placement? A study of 60 cases of idiopathic normal-pressure hydrocephalus. Neuroradiology 45: 377–380
10. Meier U, Kiefer M, Sprung C (2004) Evaluation of the Miethke Dual-Switch-Valve in patients with normal pressure hydrocephalus. Surg Neurol 61: 119–128
11. Miethke C, Affeld K (1994) A new valve for the treatment of hydrocephalus. Biomedizinische Technik 39: 181–187
12. Pang D, Altschuler E (1994) Low-pressure hydrocephalic state and viscoelastic alterations in the brain. Neurosurgery 35: 643–655
13. Pollack IF, Albright AL, Adelson PD, the Medos-Hakim Investigator Group (1999) A randomized study of a programmable shunt valve versus a conventional valve for patients with hydrocephalus. Neurosurgery 45: 1399–1411
14. Richards HK, Seeley HM, Pickard JD (2001) Examining Received Wisdom for CSF Shunt Using the UK Shunt Registry. Eur J Pediatric Surg 11 [Suppl] I: 60
15. Sainte-Rose C, Piatt JH, Renier D, Pierre-Kahn A, Hirsch JF, Hoffman HJ, Humphreys RP, Hendrick EB (1991) Mechanical complications in shunts. Pediatr Neurosurg 17: 2–9
16. Saint-Rose C (1993) Shunt obstruction: a preventable complication? Pediatric Neurosurg 19: 156–164
17. Sprung C, Miethke C, Trost HA, Lanksch WR, Stolke D (1996) The dual-switch-valve. A new hydrostatic valve for the treatment of hydrocephalus. Child's Nerv Syst 12: 573–581
18. Sprung C, Miethke C, Shakeri K, Lanksch WR (1998) Pitfalls in shunting of hydrocephalus – clinical reality and improvement by the hydrostatic dual-switch valve. Eur J Pediatric Surg 8 [Suppl] I: 1–3
19. Sprung C, Miethke C, Shakeri K, Lanksch WR (1999) Importance of intraventricentricular pressure and ventricular size for outcome of shunting. Proc of 25th Annual Meeting of EANS, ed Monduzzi, 1680–1684
20. Williams MA, Razumovsky AY, Hanley DF (1998) Evaluation of shunt function in patients who are never better, or better than worse after shunt surgery for NPH. Acta Neurochir [Suppl] 71: 368–370
21. Zemack G, Romner B (1999) Seven years of clinical experience with the Codman Hakim programmable valve: a retrospective study of 583 patients. J Neurosurg 92: 941–948
22. Zemack G, Romner B (2002) Adjustable valves in normal-pressure hydrocephalus: a retrospective study of 218 patients. Neurosurgery 51(6): 1392–1400

Correspondence: Christian Sprung, Neurosurgical Department, Charité, Humboldt-University, Campus Virchow, Augustenburger Platz 1, 13353 Berlin, Germany. e-mail: christian.sprung@charite.de

Acta Neurochir (2005) [Suppl] 95: 237–240

Guidelines for management of idiopathic normal pressure hydrocephalus: progress to date

A. Marmarou[1], **P. Black**[2], **M. Bergsneider**[3], **P. Klinge**[4], **N. Relkin**[5], and **the International NPH Consultant Group**

[1] Department of Neurosurgery, Virginia Commonwealth University Medical Center, Richmond, VA, USA
[2] Department of Neurosurgery, Brigham & Women's Hospital, Boston, MA USA
[3] Division of Neurosurgery, University of California, Los Angeles, CA USA
[4] Department of Neurosurgery, Medical School, Hannover, Germany
[5] Burke Research Institute and Rehabilitation Hospital, White Plains, New York USA

Summary

The aim of this project was to develop evidenced based guidelines for the diagnosis and management of idiopathic normal pressure hydrocephalus (iNPH). An advisory panel consisting of the authors assisted by international experts met on several occasions and formulated preliminary guidelines for iNPH managemen. The authors developed evidentiary tables based on available literature from 1966 to the present. Additional meetings to refine the evidentiary tables and incorporate expert opinion when necessary resulted in the development the iNPH guidelines. Evidence based guidelines identifying the value of clinical examination, brain imaging, Tap Test, CSF drainage, ICP monitoring and Surgical Management in diagnosing and treating the iNPH patient were developed. These are the first international evidence based guidelines focused on iNPH. Class I data were scant and guidelines relied mostly on class II and III evidence. It became clear that more prospective randomized studies are needed to resolve some of the controversial issues such as iNPH classification and sensitivity of diagnostic tests for identifying shunt responsive iNPH.

Keywords: Guidelines for idiopathic normal pressure hydrocephalus; NPH; shunt responders.

Introduction

The symptoms of gait disturbance, incontinence and dementia associated with Normal Pressure Hydrocephalus (NPH), a syndrome introduced by Hakim and Adams [1, 2], seriously impacts upon the quality of life of senior citizens. However the diagnosis of NPH remains controversial, as there are no specific guidelines for diagnosis as well as management. In part, this controversy has evolved as a result of mixing idiopathic NPH from those cases of known cause resulting from trauma, stroke, subarachnoid hemorrhage. Another reason for controversy is the general acceptance by many clinicians that an accurate diagnosis of NPH depends upon the response to shunting. Taking these issues in concert, it is clear that an evidenced based set of guidelines for diagnosis and management of NPH is needed. This report provides an overview of the progress made in development of these guidelines and provides recommendations for future research.

Methods

The guidelines were compiled from a review of the MEDLINE literature in combination with references provided by an expert panel made up of clinical scientists from the U.S., Europe and Japan. Articles from 1965 to the present were considered and the evidence was classified as class I, II or III. Class I evidence was derived from prospective, randomized well-controlled clinical trials. Class II evidence was derived from a prospective data collection with retrospective analysis of clearly reliable data and class III data referred to retrospective data, chart reviews, databases or registries and expert opinion. When expert opinion was required, it was obtained from advisory panels made up of international experts.

Results

The absence of prospective randomized clinical trials of any aspect of NPH required that the guidelines be constructed primarily on class II and III data. A decision was made to classify NPH into two major categories, idiopathic (INPH) and secondary (SNPH) following the initial separation by Black [3, 4].

Clinical diagnosis of INPH

It was also decided that the diagnosis of INPH required convergent evidence from the clinical history, physical examination and brain imaging. The symptoms of INPH typically manifest during adult life as an insidiously progressive chronic disorder that lacks an antecedent cause. Gait and/or balance impairments are usual symptoms and findings may also include disturbances in cognition or control of urination. Documention of ventricular enlargement (Evans Index ≥ 0.30) by brain imaging is necessary but not sufficient in itself to establish an INPH diagnosis. Results of neuroimaging must be interpreted in conjunction with the clinical history and physical findings in order to accurately diagnose INPH and differentiate it from other disorders. It was recommended that it may be useful to classify INPH into Probable, Possible and Unlikely categories, operationally defined by the extent to which the expected elements of INPH are present and diagnostic cofounders can be excluded. Note that shunt responsiveness did not enter into this diagnostic formulation and was considered separately.

Supplementary tests for identifying shunt responders

Drainage of cerebrospinal fluid (CSF) via a lumbar tap can be of prognostic value if the response is significant. Lumbar puncture tap tests should withdraw 40–50 cc since lesser volumes (25 cc or less) have low sensitivity [5, 6]. However, the tap test cannot be used as an exclusionary test due to the inherent low sensitivity (25–61%). The prognostic value of this procedure for identifying patients who will benefit from shunt diversion of fluid increases as greater amounts of fluid are removed by external lumbar drainage. The highest sensitivity and specificity is associated with prolonged controlled lumbar drainage (500 cc/3 days). (50–100%) [7]. The utilization of methods to compute outflow resistance of cerebrospinal fluid (Ro) is also helpful in identifying shunt responders. The informationavailable for assessing the usefulness of Ro in INPH is limited and the two reports cited for INPH are the largest series of data currently available. Although data are scant, the reported accuracy of resistance measures may be higher than that of the CSF tap test [8]. Therefore, determination of CSF outflow resistance may be helpful in increasing prognostic accuracy for identifying SRINPH when tap test results are negative.

Identifying shunt responsive patients: summary

Step 1 – clinical evaluation

Based on the history, neurological exam, and basic neuroimaging (CT and/or MRI), the patient is categorized as Probable, Possible, or Unlikely INPH. In an otherwise healthy patient in whom the clinical diagnosis appears highly probable, it may not be unreasonable to proceed directly to treatment (the placement of a shunt) without supplemental tests keeping in mind the 50 to 61% degree of certainty.

Step 2 – supplemental testing

To avoid complication and improve the certainty of a positive shunt response beyond 50 to 61%, all Probable and Possible INPH patients should be considered for supplemental testing (CSF tap test, Infusion study, ELD).

Step 3 – tap test

Given its ease, it is reasonable to proceed initially with a CSF tap test. A positive response to a 40–50 cc tap test has a higher degree of certainty for a favorable response to shunt placement than that which can be obtained by clinical examination. However the tap test cannot be used as an exclusionary test due to its low sensitivity.

Step 4 – resistance testing

Determination of the CSF outflow resistance via an infusion test carries a higher sensitivity (57–100%) compared to the tap test but a similar positive predictive value of 75 to 92%.

Step 5 – external lumbar drainage

Prolonged external lumbar drainage in excess of 300 cc is associated with high sensitivity (50–100%) and high positive predictive value (80–100%). It is a most effective test for identifying SRINPH but requires hospital admission and carries a higher complication rate than CSF tap or Resistance studies. Of the three recommended supplemental tests, the prognostic value of the ELD is likely retained even with Possible and Unlikely INPH clinical designations.

Surgical considerations

There are no Class I studies that have addressed the question of comparing operative versus conservative management of INPH. The risk-benefit ratio must be individualized for each patient with the following issues in mind: 1) shunt-responsive INPH exists with reasonable certainty, 2) there are low surgical risks related to co-morbidities, and 3) the degree of INPH-related morbidity warrants the shunt-related risks. The two most commonly used configurations are the VP and VA shunts. The choice of valve type and setting should be based on empirical reasoning and a basic understanding of shunt hydrodynamics. The most conservative choice is a valve incorporating an anti-siphon device (ASD) with the understanding that under-drainage (despite a low opening pressure) may occur in a small percentage of patients due to the ASD. Based on the results of retrospective studies, the use of an adjustable valve may be beneficial in the management of INPH due to the ability to non-operatively manage both under- and over-drainage problems.

Studies of outcome following shunt surgery for INPH

To date, there is no standard for outcome assessment of shunt treatment in idiopathic NPH. The variable improvement rates reported are not only due to different criteria for selection of patients but also due to different postoperative assessment procedures and follow-up intervals. Studies that have established fixed protocols for follow-up have shown that short- and long-term periods after shunting are determined by many factors. While short-term results were more likely to be influenced by shunt-associated risks, long-term results were independent upon factors inherent to the shunt procedure and shunt complications, i.e. death and morbidity related to concomitant cerebrovascular and vascular diseases. Studies have shown that beyond one year post surgery these factors definitely influence the clinical effect of shunting, making the one year post shunt period a potential determinate of the shunt-outcome.

A firm description of shunt outcome can be based on the documentation of either the clinical impairment, improvement following treatment or both. Grading of either the functional status of the idiopathic NPH patient or grading the clinical criteria of gait, incontinence and dementia should be performed. Examples of reported scales are: Black, Stein and Langfitt, Boon, Mori and Krauss [9–13]. Besides gait, improvement in cognition is also correlated with the patient's daily function. Neuropsychological testing may be of value in evaluating subtle cognitive deficits or changes with treatment. The latter have the advantage of having established norms for age and education level, however, the contribution of the various neuropsychological tests in the assessment of clinical outcome of shunt treatment remains to be elucidated. Efforts should be made to investigate how and when clinical outcome from shunt treatment is best assessed with respect to short- (3, 6 months) and long-term (1 year or greater) prognosis. The long-term prognosis may be affected by life expectancy and co-morbid factors not related to the shunt procedure.

In addition, there is a need of standardized reporting of shunt related complications and their effect on both the clinical outcome and the benefit of shunt treatment in INPH.

Conclusion

Although much has been learned regarding the diagnosis and treatment of idiopathic NPH, prospective, randomized multi-center trials are needed to resolve many of the issues regarding selection of patients for shunt surgery, type of valve configuration, value of resistance testing, effectiveness of probable, possible and unlikely diagnostic categories. These studies are necessary to elevate the guidelines from class II and III data to class I.

References

1. Adams RD, Fisher CM, Hakim S, Ojemann RG, Sweet WH (1965) Symptomatic occult hydrocephalus with "normal" cerebrospinal-fluid pressure. A treatable syndrome. N Engl J Med 273: 117–126
2. Hakim S, Adams RD (1965) The special clinical problem of symptomatic hydrocephalus with normal cerebrospinal fluid pressure. Observations on cerebrospinal fluid hydrodynamics. J Neurol Sci 2(4): 307–327
3. Black PM (1980) Idiopathic normal-pressure hydrocephalus. Results of shunting in 62 patients. J Neurosurg 52(3): 371–377
4. Black PM (1982) Normal-pressure hydrocephalus: current understanding of diagnostic tests and shunting. Postgrad Med 71(2): 57–61, 65–67
5. Malm J, Kristensen V, Fagerlund M, Koskinen LO, Ekstedt J (1995) Cerebrospinal fluid shunt dynamics in patients with idiopathic adult hydrocephalus syndrome. J Neurol Neurosurg Psychiatry 58(6): 715–723
6. Walchenbach R, Geiger E, Thomeer RT, Vanneste JA (2002) The value of temporary external lumbar CSF drainage in predicting the outcome of shunting on normal pressure hydrocephalus. J Neurol Neurosurg Psychiatry 72(4): 503–506
7. Williams MA, Razumovsky AY, Hanley DF (1998) Comparison of Pcsf monitoring and controlled CSF drainage diagnose normal pressure hydrocephalus. Acta Neurochir [Suppl] 71: 328–330
8. Malm J, Kristensen B, Karlsson T, Fagerlund M, Elfverson J, Ekstedt J (1995) The predictive value of cerebrospinal fluid dynamic tests in patients with th idiopathic adult hydrocephalus syndrome. Arch Neuro 52(8): 783–789
9. Krauss JK, Droste DW, Vach W, Regel JP, Orszagh M, Borremans JJ, Tietz A, Seeger W (1996) Cerebrospinal fluid shunting

in idiopathic normal-pressure hydrocephalus of the elderly: effect of periventricular and deep white matter lesions. Neurosurgery 39(2): 292–299; discussion 299–300
10. Black PM, Ojemann RG, Tzouras A (1985) CSF shunts for dementia, incontinence, and gait disturbance. Clin Neurosurg 32: 632–651
11. Stein SC, Langfitt TW (1974) Normal-pressure hydrocephalus. Predicting the results of cerebrospinal fluid shunting. J Neurosurg 41(4): 463–470
12. Boon AJ, Tans JT, Delwel EJ, Egeler-Peerdeman SM, Hanlo PW, Wurzer HA, Avezaat CJ, de Jong DA, Gooskens RH, Hermans J (1997) Dutch normal-pressure hydrocephalusstudy: prediction of outcome after shunting by resistance to outflow of cerebrospinal fluid. J Neurosurg 87(5): 687–693
13. Takeuchi T, Kasahara E, Iwasaki M, Mima T, Mori K (2000) Indications for shunting in patients with idiopathic normal pressure hydrocephalus presenting with dementia and brain atrophy (atypical idiopathic normal pressure hydrocephalus). Neurol Med Chir Tokyo 40(1): 38–46; discussion 46–47

Correspondence: Anthony Marmarou, Department of Neurosurgery, Virginia Commonwealth University Medical Center, 1001 East Broad Street, Suite 235, Richmond, VA, USA 23219. e-mail: amarmaro@vcu.edu

Acta Neurochir (2005) [Suppl] 95: 241–245

Three-year outcome of shunted idiopathic NPH patients

G. Aygok, A. Marmarou, and **H. F. Young**

Department of Neurosurgery, Virginia Commonwealth University Medical Center, Richmond, VA, USA

Summary

The incidence of idiopathic normal pressure hydrocephalus (iNPH) has increased as a result of improved longevity. This report describes the 3-year outcome of shunted iNPH patients compared to three-month outcome after shunting. Patients (n = 50) (Age 70.4 ± 8.9) admitted to our service were diagnosed and treated according to a fixed protocol for management of iNPH and after shunting were followed at least three times per year in clinic. The outcome of 50 patients was graded according to the level of improvement in symptoms as Excellent/Good, Partial or None in each category of Gait, Incontinence and Dementia. If we lump favorable (excellent, good, partial recovery) vs poor recovery (none), we found from 3 months to 3 years, a moderate decline in gait performance (91% to 75%), a retention of memory improvement (80%–80%) and an improvement in incontinence occurred over time (70%–82.5%).

With proper diagnosis and management of iNPH, shunting of patients is associated with a favorable risk/benefit ratio that is reasonably long lasting.

Keywords: NPH; normal pressure hydrocephalus; long term outcome.

Introduction

A unique triad of clinical symptoms defining normal pressure hydrocephalus (NPH), which include gait disturbance, dementia and incontinence in association with enlarged ventricles and normal cerebrospinal fluid (CSF) pressure, was first introduced by Hakim in 1965 [5]. He reported that diversion of fluid using a shunt relieved the symptoms. Since that time, significant progress has been made in shunt technology and several authors have reported improvement following shunting in NPH patients [2, 6, 9]. However, studies of the long-term outcome of shunted patients are few, partly due to the inherent loss of follow up and also the confounding of outcome assessment by co-morbid factors. An additional problem is the lack of standardized outcome scales, which makes such reporting difficult. Overall improvement rates vary, ranging between 30% and 96 [4, 6, 11, 15]. These outcome results were graded based on clinical judgment by the physician as well as patient and caregiver opinions and were relatively subjective in nature [7, 8, 13]. The objective of this study was to document the three-year outcome of shunted patients diagnosed with idiopathic normal pressure hydrocephalus and utilizing a scale, which describes the outcome of the symptoms separately.

Patients and methods

Patients admitted to Medical College of Virginia were diagnosed and treated for INPH according to a fixed protocol. The diagnostic clinical criteria were based on gait disturbance, dementia and incontinence in combination with ventriculomegaly without antecedent cause. On the day of their clinical visit, a video was obtained of the patient rising from a chair and walking for a distance of 10 meters and the patient was asked to make a tandem walk if possible. The video then was analyzed for many characteristics of gait; shuffling, wide based, ability to tandem walk, rising from a chair, and number of steps per minute required to traverse 30 feet. When necessary, spinal films were requested and examined in order to exclude any reason for gait difficulty like spinal stenosis. Gait impairment was rated as a gait scale in four categories; Normal, Mild, Moderate and Severe (Table 1). Mental Disturbance was evaluated with the mini mental state examination (MMSE) and with a series of neuropsychological tests. Impairment in memory and bladder function were assessed similar to gait (Table 1). CT or MRI was examined for ventricular enlargement with a frontal horn index of equal to or greater than 0.40 and/or Evans Index ⇒ 0.3.

The infusion test

A lumbar pressure measurement coupled with CSF resistance study was obtained in all patients in an outpatient setting. The resistance to outflow (Ro) and pressure–volume index (PVI) was measured by the bolus technique [10]. via the lumbar space.

External lumbar drainage

Following the measurement of Ro, the patient was admitted to the hospital for external lumbar drainage (ELD) within 2 weeks. Before discharge, a cranial CT was obtained to insure that drainage did not

Table 1.

GAIT impairment
Severe: unable to walk even with assistance
Moderate: unable to walk without assistance
Mild: unsteady but can walk without assistance
Normal
MEMORY impairment
Severe: disoriented in time and place
Moderate: difficulty noted in recent memory by physician, support staff and family
Mild: testing required for demonstrating deficit
Normal
INCONTINENCE
Severe: frequent urgency or incontinence
Moderate: infrequent urgency or incontinence
Mild: urinary urgency
Normal

cause any complication. Patient and caregiver were instructed to complete a 10-day survey to assess clinical symptoms after discharge.

Selection criteria for shunt surgery

The videos for gait that were recorded on the day of the first clinical visit and after the 3 days lumbar drainage procedure were compared. A positive improvement in gait was considered when the patient's gait disturbance was improved to a less severe form of impairment in the gait scale. MMSE which was administered before and after drainage was also compared as well as the neuropsychological tests. Improvement in incontinence was obtained from the patient and caregiver. The patients were admitted for shunt surgery if they had responded positively to the ELD.

Postoperative assessment

The patients were routinely examined 3 times per year for 3 consecutive years.

The post operative investigation was made for each individual symptom including the gait video assessment and MMSE in addition to the patient's and family member's estimations for the results. The evaluation of post-operative improvement was based on a scale devised by Marmarou, which described the improvement in three categories for each INPH symptom as Excellent/Good, Partial and None (Table 2).

Statistics

For statistical analysis, chi-square testing and Student's t test were used. A level of 0.05 was considered significant.

Table 2.

Outcome assessment scale scored individually *for Gait, Memory, Incontinence*
Excellent/Good: Return back to normal/Towards normal with only mild residual symptoms
Partial: Moderate improvement with pronounced residual symptoms
None: No change from the pre-operative baseline symptoms

Results

Age and gender distribution

There were 50 patients in our study population, which included 29 men and 21 women. The mean age was 70.4 ± 8.8 years (range 61–85 years-old). Within the study group, 52% (26/50) of the patients manifested the clinical triad of gait disturbance, memory impairment and incontinence. Gait difficulty was present in 90% (44/50) of patients, 56% (28/50) with memory impairment and 46% (23/50) with incontinence. All patients but two had either mild or moderate impairments. In the clinical diagnosis of INPH, the ability to tandem walk was mostly impaired and shuffling gait with unsteadiness was more pronounced. Neuropsychological tests showed no significant difference before and after 3 days lumbar drainage procedure. MMSE showed a significant difference between the patients who improved with drainage and who did not. ($p = 0.003$)

Opening pressure and Ro

The average opening pressure was 8.6 ± 2.8 mmHg (range 4–14 mmHg). Average PVI was 25.51 ± 12.10 ml (range 9.84–56.65 ml). Of these 50 patients, 39 patients had Ro measurements. 31 patients presented with Ro equal or greater than 4.0 mmHg/ml/min., which is considered a prognostic threshold for shunt responders.

ELD outcome

The decision for shunt was based strictly on our ELD results. Therefore, in this patient population, 94% (47/50) of our patients had improvement after ELD.

3 month outcome

Three months after shunt placement, 41% (18/45) of the patients had excellent/good recovery in gait whereas 50% (23/45) had partial gait improvement. In general, overall gait improvement equaled 91%. For memory impairment, 40% (11/28) of the patients had excellent and good recovery as well as the other 40% (11/28) with partial recovery. For incontinence, 44%

Table 3.

3 month–3 year outcome		
Gait	3 Month	3 Year
– Excellent/Good	41%	50%
– Partial	50%	25%
– None	9%	25%
Incontinence		
– Excellent/Good	44%	65%
– Partial	26%	17.5%
– None	30%	17.5%
Dementia		
– Excellent/Good	40%	60%
– Partial	40%	20%
– None	20%	20%

(10/23) had improved dramatically and another 26% (6/23) had partial improvement (Table 3).

3 years outcome

At the end of three years, 92% (46/50) had survived. The remaining 4 patients died of causes unrelated to the shunt. At three years, the improvement in gait was preserved for 75% (32/43). Better results were seen in memory improvement in 3 years compared with 3 months. 80% (22/27) of the patients with memory impairment had sustained their recovery in 3 years and even 20% (6/27) had reached a better improvement category. More than 80% (17/22) of improvement was observed in incontinence in excellent/good/partial improvement categories. This was even higher from the rate of improvement that was achieved in 3 months (Table 3).

Complications and co-morbidities

Major and minor complications due to surgery totaled 6% and 14% respectively. Subdural hematoma (SDH) and hygroma was seen in 2 patients and in 1 patient respectively. 2 patients had hearing loss, 3 patients developed double vision and 3 patients developed headaches after the shunt implantation. All of these patients improved without surgical intervention. If a programmable valve was used, these complications were managed by adjusting the pressure. There was significant co-morbidity in this patient group 24% (12/50) not related to the surgery. There was a wide variety of co-morbidities which included stroke, cardiac problems, vertigo and psychiatric problems.

Discussion

This study shows that with proper identification of "shunt responders" using external lumbar drainage, the improvement in outcome seen immediately after shunting is reasonably sustained over a period of three years. At three years post shunt, the improvement in gait declined in 15% of patients. However the gains seen in dementia and incontinence were remarkable. Of 80% improvement in dementia seen postoperatively, the improvement in dementia was sustained in all these patients.

The importance of supplementary testing

In a retrospective study by Vanneste *et al.* [15], 91 INPH patients were selected for surgery according to their clinical and radiological criteria and they observed a 65% improvement after surgery. In another retrospective analysis for 26 INPH patients out of 74 NPH, Larsson *et al.* [8] concluded that there was no clinical parameter correlated with the outcome in the 2-year post-operative period. This would suggest that optional tests such as CSF resistance testing or external lumbar drainage, which more efficiently identify shunt responders should be utilized. Resistance testing identifies the opening pressure and absorption capacity of the patient and can be performed in an outpatient setting. In our study we had resistance test results for 39 (78%) patients. With regard to resistance testing, an Ro value greater than 4.0 mmHg/ml/min is likely to predict the ELD response. ($p = 0.0001$) In some cases, therefore, it may be more optimal to shunt on the basis of resistance testing alone when ELD may not be used (e.g. extensive spinal surgery).

The improvement in symptoms was seen immediately at the conclusion of the external drainage. It is our sense that the high success rate of shunting was due to the proper selection of patients using ELD. In a prospective study of 43 INPH patients, Walchenbach *et al.* [16] reported a positive predictive value of 87% with ELD and is close to the results reported here.

The need for standardized rating scales

Results of outcome after surgery in many reports are still unclear because there is no standardized follow-up protocol. Moreover, there is no consensus on the time period for short and long term. In a prospective study for 42 INPH patients, Malm *et al.* [9] reported an overall improvement of 64% at 3 months when compared with 26% in 3 years. At 3 months, 74% of his patients were independent, at 3 years the independent group was only 28%. Sustained improvement at 5 years was noted for 91% of INPH patients by Raftopoulos *et al.* [11]. However, 57% of his patients died before their 5-year follow-up.

We considered 3 months for our short-term follow-up and 3 years for long-term follow-up. As we discussed earlier, at 3 years follow-up, even though we had a slight decline, patients still retained a 75% improvement in gait. Moreover, 80% retained their improvement in memory and incontinence at 3 years post shunt.

However, to improve our objectivity, we need grading systems specified for INPH. Currently, there are no standardized outcome scales that are widely used, for example, like the Glasgow outcome scale (GOS) for the head injury management. Rating scales, which measure the overall improvement or change in activities of daily living, make it difficult to draw a clear picture. Patients may have a decline in these overall functional rating scales because of co-existing comorbidities. In early studies, Stein and Langfitt [14] introduced the assessment of outcome based on the activities of daily living and termed Stein and Langfitt Scale. Boon *et al.* [3] used Rankin Scale as a disability score and developed 'Dutch-NPH Scale'. They include neuropsychological tests and the assessment of 10 features of gait. There is no consensus on scales and therefore many authors decided to develop their own scales. Using Stein's scale, Black *et al.* [2] reported 33% improvement in 62 INPH patients. For the same patients, he also assessed the improvement, which was graded based on the pre-operative and post-operative changes in gait, dementia and incontinence. Using the Black Scale and with their classification, improvement rate was noted 50% in contrast to 33%. We used an outcome assessment developed by Marmarou, which was based on each symptom separately. With this method it is easy to focus on objective assessments using different tools. We compared pre and post-operative videos for many different features of gait assessment as mentioned earlier. We also use MMSE and our clinical judgment to report memory improvement. However, for incontinence, this was based on the patient's own assessment. When necessary, family members assessments were also used.

The effect of co-morbid factors on outcome assessment

There is no doubt that health conditions like coexisting Parkinson's Disease, hip or knee problems may affect the patient's gait as well as prostate problems that worsens the incontinence. It is also possible that memory impairment can be a part of Vascular Disease or Alzheimer's Disease. In summary, these conditions not related to shunt have influenced the patient's clinical outcome and morbidity [1, 11, 12].

Conclusion

It is important to assess the risk/benefit ratio before proceeding to surgery in this elderly population. Complications, co-morbidities and deaths are affecting the post-operative assessment but using scales that measure the improvement of the triad elements objectively for both short and long-term assessments will hopefully improve outcome assessment. Finally, the most important aspect of NPH management is correct diagnosis and the use of supplementary tests to identify shunt responders. This study reports sustained improvement in the greater majority of shunted patients that extends to three years and as such presents a favorable risk/benefit profile for the elderly population with NPH.

References

1. Benzel EC, Pelletier AL, Levy PG (1990) Communicating hydrocephalus in adults: prediction of outcome after ventricular shunting procedures. Neurosurgery 26(4): 655–660
2. Black PM (1980) Idiopathic normal-pressure hydrocephalus. Results of shunting in 62 patients. J Neurosurg 52(3): 371–377
3. Boon AJ, Tans JT, Delwel EJ, Egeler Egeler-Peerdeman SM, Hanlo PW, Wurzer HA, Avezaat C, deJong (1998) Dutch Normal-Pressure Hydrocephalus Study: randomized comparison of low- and medium-pressure shunts. J Neurosurg 88(3): 490–495
4. Greenberg JO, Shenkin HA, Adam R (1977) Idiopathic normal pressure hydrocephalus – a report of 73 patients. J Neurol Neurosurg Psychiatry 40(4): 336–341
5. Hakim S, Adams RD (1965) The special clinical problem of symptomatic hydrocephalus with normal cerebrospinal fluid pressure: observations on cerebrospinal fluid dynamics. J Neurol Sci 2: 307–327
6. Hughes CP, Siegel BA, Coxe WS, Gado MH, Grubb RL, Coleman RE, Berg L (1978) Adult idiopathic communicating hydrocephalus with and without shunting. J Neurol Neurosurg Psychiatry 41(11): 961–971
7. Krauss JK, Droste DW, Vach W, Regel JP, Orszagh M, Borremans JJ, Tietz A, Seeger W (1996) Cerebrospinal fluid shunting in idiopathic white matter lesions. Neurosurgery 39(2): 292–299

8. Larsson A, Wikkelso C, Bilting M, Stephensen H (1991) Clinical parameters in 74 consecutive patients shunt operated for normal pressure hydrocephalus. Acta Neurol Scand 84(6): 475–482
9. Malm J, Kristensen B, Stegmayr B, Fagerlund M, Koskinen LO (2000) Three-year survival and functional outcome of patients with idiopathic adult hydrocephalus syndrome. Neurology 55(4): 576–578
10. Marmarou A, Shulman K, Rosende RM (1978) A nonlinear analysis of the cerebrospinal fluid system and intracranial pressure dynamics. J Neurosurg 48(3): 332–344
11. Raftopoulos C, Massager N, Baleriaux D, Deleval J, Clarysse S, Brotchi J (1996) Prospective analysis by computed tomography and long-term outcome of 23 adult patients with chronic idiopathic hydrocephalus. Neurosurgery 38(1): 51–59
12. Savolainen S, Hurskainen H, Paljarvi L, Alafuzoff I, Vapalahti M (2002) Five-year outcome of normal pressure hydrocephalus with or without a shunt: predictive value of the clinical signs, neuropsychological evaluation and infusion test. Acta Neurochir (Wien) 144(6): 515–523
13. Spanu G, Sangiovanni G, Locatelli D (1986) Normal-pressure hydrocephalus: twelve years experience. Neurochirurgia (Stuttg) 29(1): 15–19
14. Stein SC, Langfitt TW (1974) Normal-pressure hydrocephalus. Predicting the results of cerebrospinal fluid shunting. J Neurosurg 41(4): 463–470
15. Vanneste J, Augustijn P, Dirven C, Tan WF, Goedhart ZD (1992) Shunting normal-pressure hydrocephalus: do the benefits outweigh the risks? A multicenter study and literature review. Neurology 42(1): 54–59
16. Walchenbach R, Geiger E, Thomeer RTWM, Vanneste JAL (2002) The value of temporary external lumbar CSF drainage in predicting the outcome of shunting on normal pressure hydrocephalus. J Neurol Neurosurg Psychiatry 72: 503–506

Correspondence: Anthony Marmarou, Department of Neurosurgery, Virginia Commonwealth University Medical Center, 1001 East Broad Street, Suite 235, Richmond, VA, USA 23219. e-mail: amarmaro@vcu.edu

Acta Neurochir (2005) [Suppl] 95: 247–251

Clinical testing of CSF circulation in hydrocephalus

Z. Czosnyka[1], **M. Czosnyka**[1], **B. Owler**[1,2], **S. Momjian**[1,3], **M. Kasprowicz**[1,4], **E. A. Schmidt**[1,5], **P. Smielewski**[1], and **J. D. Pickard**[1]

[1] Academic Neurosurgical Unit, Department of Neurosciences, Addenbrooke's Hospital, Cambridge, UK
[2] Department of Surgery, University of Sydney, Sydney, Australia
[3] Service de Neurochirurgie, Hôpitaux Universitaires de Genève, Geneva, Switzerland
[4] Division of Measurements and Measuring Systems, Faculty of Electronics, University of Technology, Wroclaw, Poland
[5] Service de Neurochirurgie, Hôpital Universitaire Purpan, Toulouse, France

Summary

Introduction. Recent 'NPH Dutch trial' has re-emphasised the importance of the resistance to cerebrospinal fluid (CSF) outflow (Rcsf) in the diagnosis of hydrocephalus. We re-evaluated the clinical utility of the physiological measurements revealing CSF dynamics. The results were summarized from our previous publications.

The Computerised Infusion Test was designed to perform quick and low-invasive assessment of CSF dynamics described by parameters as Rcsf, brain compliance, elasticity coefficient, estimated sagittal sinus pressure, CSF formation rate and other variables. Overnight ICP monitoring with quantitative analysis of CSF dynamics was used in those cases where infusion study was unreliable or producing results close to the borderline. We found that the threshold of normal and increased Rcsf should be age-matched because in patients older than 55 Rcsf increases 0.2 mm Hg/(ml/min) per year ($p < 0.04$; $N = 56$). Rcsf was positively correlated with cerebral autoregulation ($R = 0.41$; $p < 0.03$; $N = 36$) indicating that in patients with symptoms of NPH but normal Rcsf underlying cerebrovascular disease is more frequent.

Computerized infusion tests and overnight ICP monitoring are useful diagnostic technique alone or in conjunction with other forms of physiological measurement.

Keywords: Hydrocephalus; cerebrospinal fluid; intracranial pressure; shunt.

Introduction

Hydrocephalus manifests with excessive accumulation of fluid within the brain. Implantation of hydrocephalus shunt is a standard way of management in communicating hydrocephalus. As shunting is almost purely mechanistic treatment, which radically affects pressure-volume compensation, the hydrodynamics of patient's own compensation should be ideally examined before a shunt is implanted.

Testing of CSF dynamics, although invasive, may help with the decision about surgery. It also provides basic information for further management of shunted patient when complications, such as shunt blockage, under-, and over-drainage, arise. In such cases, physiological measurement may aid the decision regarding shunt revision.

This paper reviews our 10 years of experience with physiological measurements in patients suffering from hydrocephalus [3–5, 11, 13, 15].

Material and methods

1250 infusion studies and 256 overnight ICP monitoring were performed in 653 patients over a 10 year period. All patients presented with clinical symptoms of hydrocephalus (solely or overlapping other CNS diseases) and various degree of ventricular enlargement on brain CT/MRI scans. 63% of tests were performed to assess CSF dynamics before shunting and 37% to assess shunt function. In total, 64% of tests were performed to pre-implanted Ommaya Reservoirs connected with the ventricular space, 11% to shunt pre-chamber and 25% into lumbar space. There were only two serious complications related to the infusion study (one brain abscess-treated successfully and one internal CSF leak related to a rapid mobilization of a patient after lumbar puncture – also managed with good results. Minor adverse effects such as headaches or nausea reported during the study resolve quickly after infusion and affect less than 10% of patients.

The aim of CSF infusion study was

1. To measure the Rcsf along with other compensatory parameters and consider whether the cerebrospinal circulation is disturbed and can be improved by shunting
2. In patients with shunts, to assess quality of shunt functioning in-vivo.

Almost all authors agree that in hydrocephalus the drainage of CSF is disturbed. This may be expressed quantitatively by an elevated resistance to CSF outflow. The limit of resistance is reported

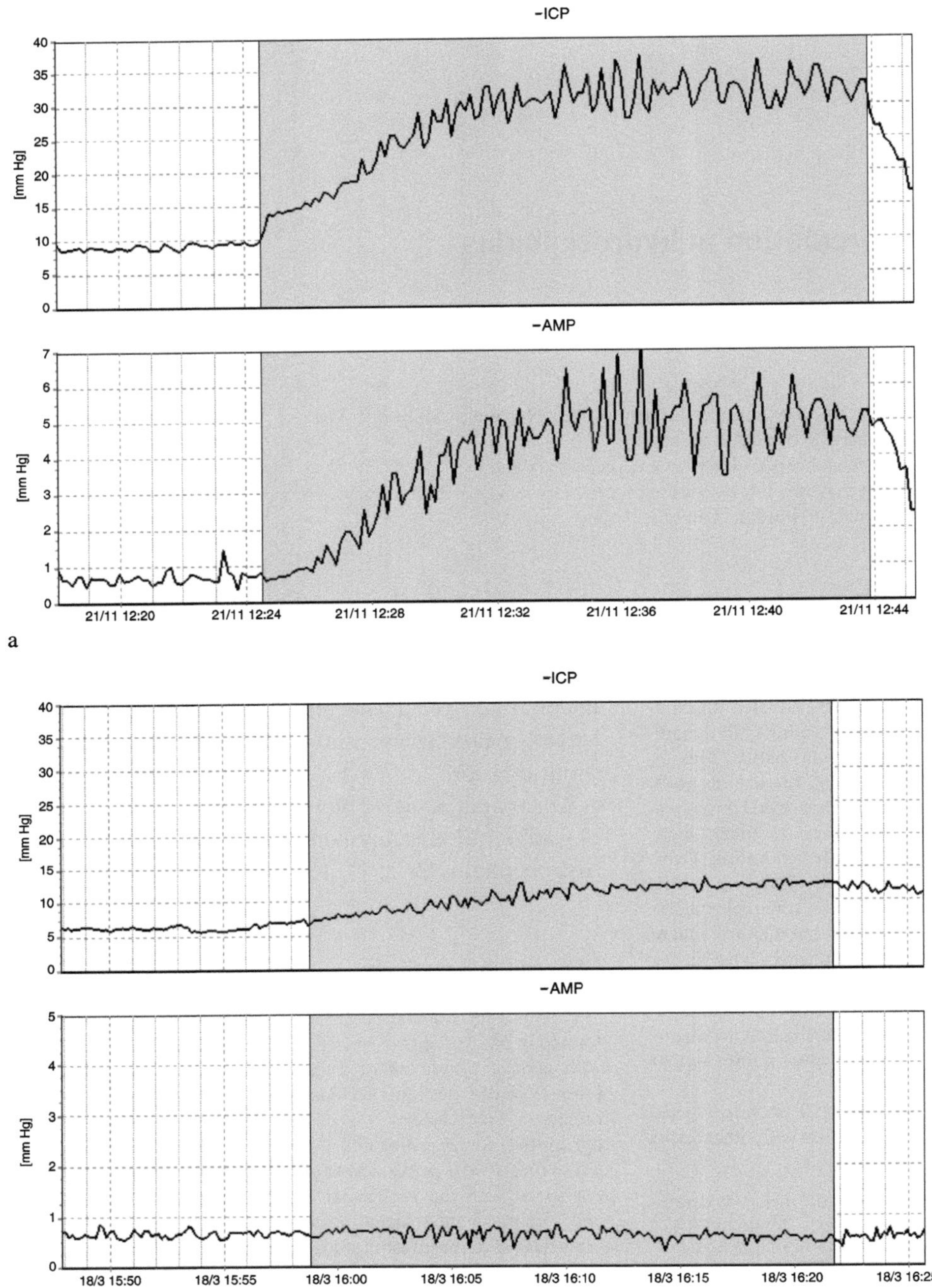

Fig. 1. Examples of constant rate infusion test. ICP – mean ICP (10 second averages) and AMP – pulse amplitude of ICP. Vertical lines mark beginning and the end of infusion of Hartman solution at a rate of 1.5 ml/min. (a) In patient suffering from normal pressure hydrocephalus (*NPH*). Baseline pressure is normal, resistance to CSF outflow increased, there are plenty of strong vasogenic waves, and changes in pulse amplitude are very well correlated with changes in mean ICP. (b) Cerebral atrophy. Baseline pressure is also low, resistance to CSF outflow is in norm, there are no vasogenic waves and pulse amplitude responds to changes in mean ICP very sluggish or (like in this particular case) there is no response

to range from 13 mm Hg/ml/min (in younger patients) [2] to 18 mm Hg/ml/min [1]. The computerized infusion test [3] is a modification of the traditional constant rate infusion as described by Katzman [10]. The method requires a fluid infusion to be made into any accessible CSF compartment. A lumbar infusion, even if it has understandable limitations, is less invasive than ventricular infusion. However, the most frequent approach in our centre is an intraventricular infusion into a subcutaneously positioned reservoir, con-

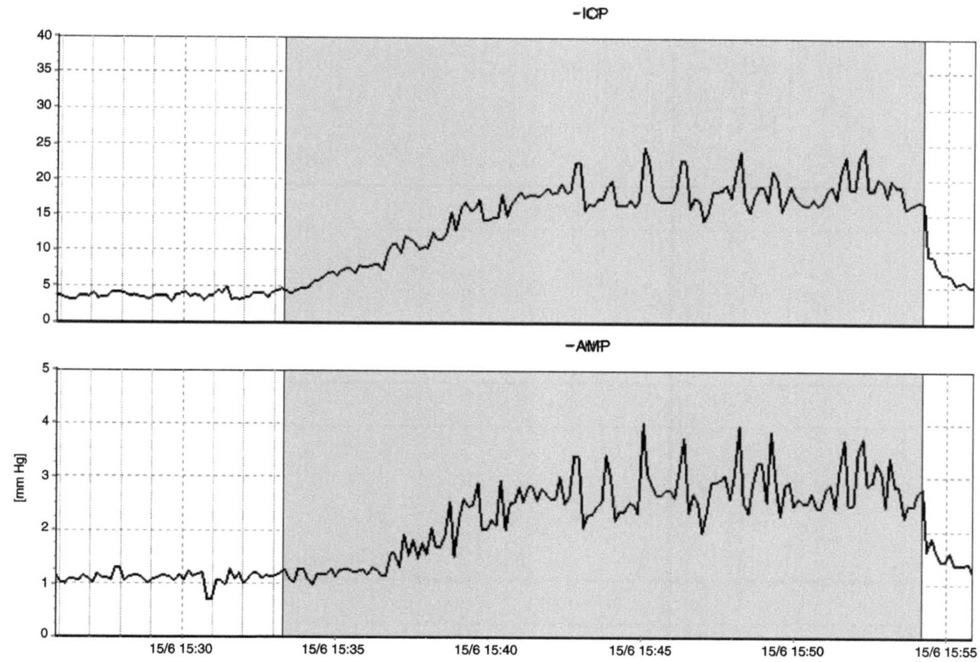

a

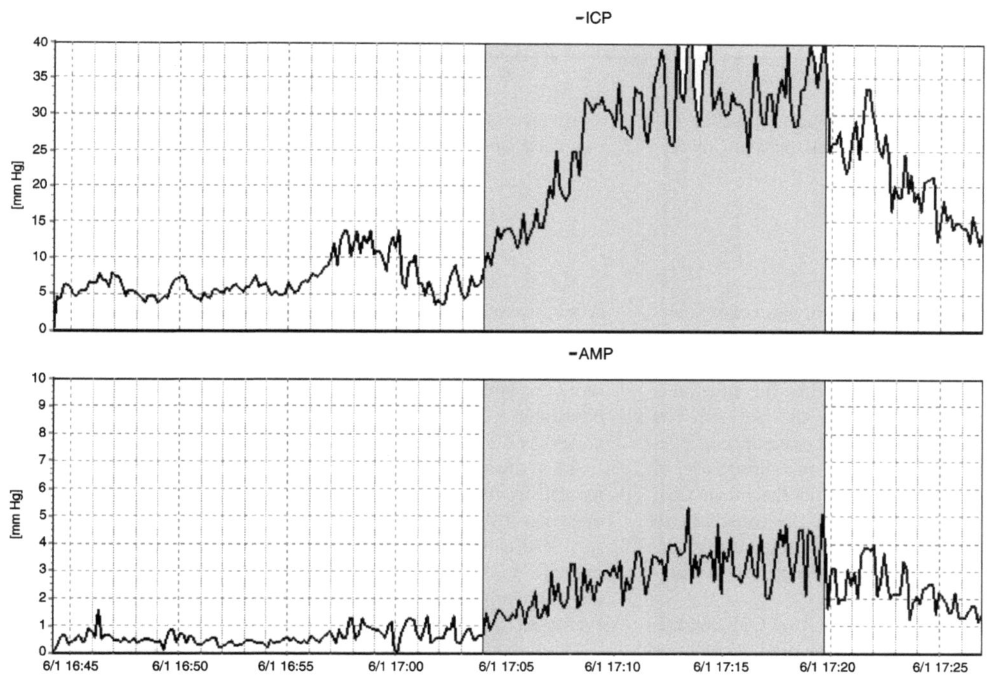

b

Fig. 2. Examples of two infusion tests performed to assess shunt functioning in vivo. In both cases ICP increased in the response to infusion. (a) Response was lower than the threshold level for the used valve (Strata, level 1.5). (b) Baseline pressure was low but the response of ICP during the test exceeded the critical threshold level. Valve is blocked. (Strata, level 1.5)

nected to an intraventircular catheter or shunt antechamber. In such cases two needles (gauge 25) are used: one for the pressure measurement and the second for the infusion.

During the infusion, the computer calculates and presents mean pressure and pulse amplitude (with time along the x axis – see Fig. 1). The resistance to CSF outflow can be estimated using simple arithmetic as the difference between the value of the plateau pressure during infusion and the resting pressure divided by the infusion rate.

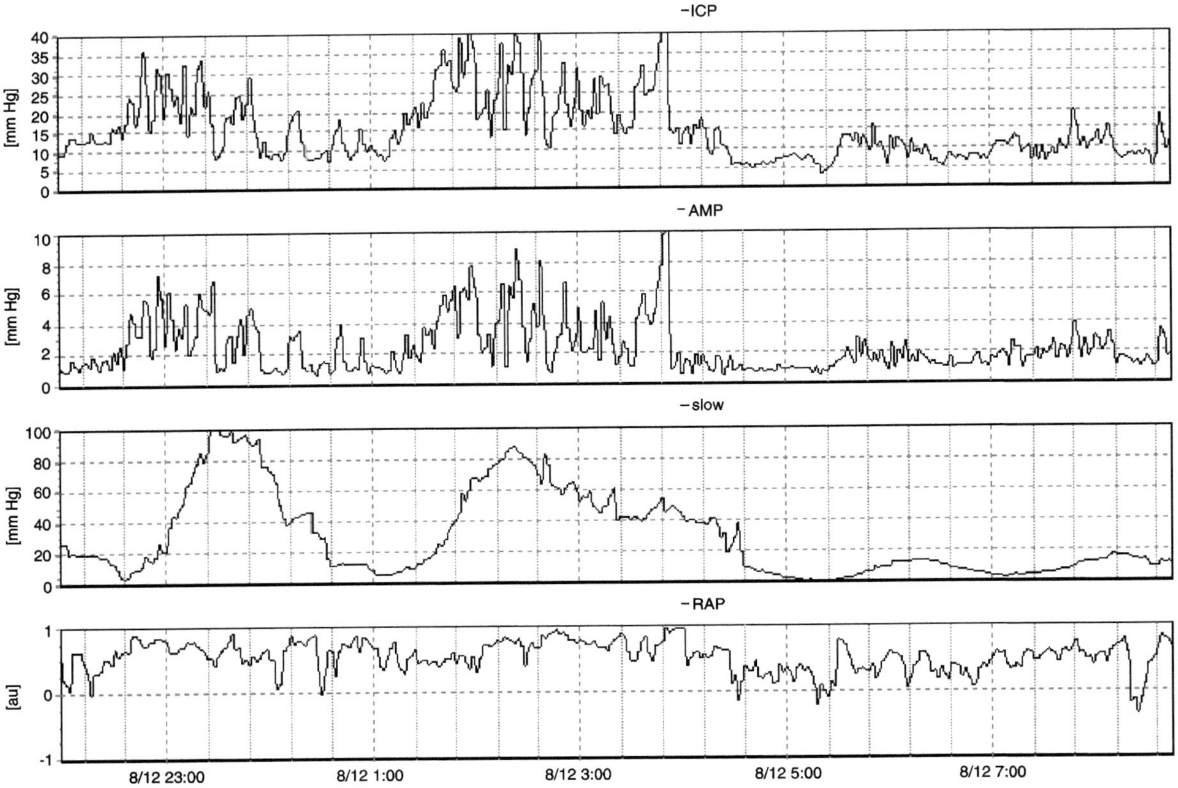

Fig. 3. Example of overnight monitoring of ICP (using Camino bolt) revealing increased CSF dynamics. Slow – magnitude of slow waves (B waves) is elevated. RAP – correlation coefficient between mean ICP and pulse amplitude (*AMP*) indicates poor compensatory reserve (RAP > 0.6 overnight)

However, in many cases strong vasogenic waves or an excessive elevation of the pressure above the safe limit of 40 mm Hg do not allow the precise measurement of the final pressure plateau. Computerized analysis, produces results even in difficult cases when the infusion is terminated prematurely (i.e., without reaching the end-plateau). The algorithm utilizes a time series analysis for volume-pressure curve retrieval, the least-mean-square model fitting and an examination of the relationship between the pulse amplitude and the mean CSF pressure. Apart from resting CSF pressure and the resistance to CSF outflow, the elastance coefficient or pressure-volume index, cerebrospinal compliance, CSF formation rate and the pulse wave amplitude of CSF pressure are estimated.

However, not all patients presenting with abnormal CSF circulation may improve after shunting [1, 11]. As positive predictive power of infusion study is usually reported as satisfactory, some patients with apparently normal profile of CSF circulation may still get better after surgery [1]. Therefore, the infusion test does not offer a definite indication for the management of hydrocephalus. It should be always interpreted in conjunction with other forms of investigations (neuropsychological, brain imaging, gait analysis, CSF tap test or diagnostic drainage [17], vascular reactivity [4], biochemical composition of CSF, etc.).

In hydrocephalus a shunt is used to drain excess CSF to elsewhere in the body according to a pressure difference between inlet (ventricles) and outlet (peritoneal or atrial) compartments. Ideally, the resistance of an open shunt taken together with the natural CSF outflow resistance (usually increased in hydrocephalus) should be close to the normal resistance to CSF outflow, i.e. 6 to 10 mm Hg/ml/min [7]. The flow through the shunt should not depend on the body posture or be affected by body temperature, external pressure (within the physiological range for subcutaneous pressure) or the pulsatile component of CSF pressure.

After shunting, the model of CSF space [12] should be supplemented by the branch representing the property of the shunt. The most sensitive indicator of the shunt partial blockage is the steady state level achieved during the test [5]. With known value of the shunt pressure-flow curve (opening pressure and its hydrodynamic resistance where the curve is quasi-linear) the critical threshold may be evaluated for each individual type of the shunt:

5 mm Hg + shunt opening pressure

+ infusion rate hydrodynamic resistance of shunt*

Examples of the tests revealing properly functioning shunt and the blocked shunt are presented in Fig. 2.

Parameters describing vascular effects and pressure-volume compensation can also be evaluated during the infusion study, using for example non-invasive transcranial Doppler ultrasonography. As the study usually starts with 10–15 minutes of period of baseline assessment, which can be easily extended. Vasogenic waves of ICP, analogical to these monitored during ICP overnight (see Fig. 3) recording, can be calculated.

Results

In patients with clinical symptoms of NPH ($N = 133$) resistance to CSF outflow is associated with severity of symptoms (modified Stein-Langfitt score: $R = 0.18$; $p < 0.05$).

In different subgroup [6] ($N = 56$) the resistance to CSF increased with age ($R = 0.57$; $p < 0.0001$) and estimated CSF formation rate decreased ($R = 0.49$; $p < 0.002$).

Vasogenic waves (pulse amplitude, slow waves and respiratory waves) of intracranial pressure are positively correlated with the resistance to CSF outflow [13], with the strongest association between respiratory waves and Rcsf ($R = 0.52$; $p < 0.003$; $N = 35$).

Baseline ICP measured after the test is usually greater than the value measured before the test. This phenomenon can be described as hysteresis of the pressure-volume curve [11]. The width of hysteresis is positively associated with width of ventricles ($R = 0.63$; $P < 0.03$; $N = 35$) and negatively with modified Stein-Langfitt score ($R = -0.61$; $p < 0.02$).

Cerebral autoregulation assessed using transcranial Doppler ultrasonography during the test proved to correlate with Rcsf [4] in a manner suggesting worse autoregulation in those having lower resistance to CSF outflow ($R = -0.41$; $p < 0.03$; $N = 36$).

Conclusion

Physiological monitoring can be useful in the management of hydrocephalus. It helps to exclude patients from unnecessary shunting, to evaluate shunt functioning in-vivo, and to detect vascular components of hydrocephalus.

References

1. Boon AJ, Tans JT, Delwel EJ, Egeler-Peerdeman SM, Hanlo PW, Wurzer HA, Avezaat CJ, de Jong DA, Gooskens RH, Hermans J (1997) Dutch normal-pressure hydrocephalus study: prediction of outcome after shunting by resistance to outflow of cerebrospinal fluid. J Neurosurg 87(5): 687–693
2. Borgesen SE, Gjerris F (1982) The predictive value of conductance to outflow of CSF in normal pressure hydrocephalus. Brain 105: 65–86
3. Borgesen SE, Albeck MJ, Gjerris F, Czosnyka M, Laniewski P (1992) Computerized infusion test compared to steady pressure constant infusion test in measurement of resistance to CSF outflow. Acta Neurochir (Wien) 119: 12–16
4. Czosnyka Z, Czosnyka M, Whitfield PC, Donovan T, Pickard JD (2002) Cerebral autoregulation among patients with symptoms of hydrocephalus. Neurosurgery 50(3): 526–532
5. Czosnyka Z, Czosnyka M, Pickard JD (2002) Shunt testing in-vivo: a method based on the data from the UK shunt evaluation laboratory. Acta Neurochir [Suppl] 81: 27–30
6. Czosnyka M, Czosnyka ZH, Whitfield PC, Donovan T, Pickard JD (2001) Age dependence of cerebrospinal pressure-volume compensation in patients with hydrocephalus. J Neurosurg 94(3): 482–486
7. Ekstedt J (1978) CSF hydrodynamic studies in man. Normal hydrodynamic variables related to CSF pressure and flow. J Neurolog Neurosyrg Psychiatry 41: 345–353
8. Hakim S, Adams RD (1965) The special clinical problem of symptomatic hydrocephalus with normal cerebrospinal fluid pressure. Observations on cerebrospinal fluid hydrodynamics. J Neurol Sci 2(4): 307–327
9. Kasprowicz M, Czosnyka Z, Czosnyka M, Momjian S, Juniewicz H, Pickard JD (2004) Slight elevation of baseline intracranial pressure after fluid infusion into CSF space in patients with hydrocephalus. Neurol Res 26(6): 628–631
10. Katzman R, Hussey F (1970) A simple constant infusion manometric test for measurement of CSF absorption. Neurology (Minneap) 20: 534–544
11. Malm J, Kristensen B, Karlsson T, Fagerlund M, Elfverson J, Ekstedt (1995) The predictive value of cerebrospinal fluid dynamic tests in patients with th idiopathic adult hydrocephalus syndrome. J Arch Neurol 52(8): 783–789
12. Marmarou A (1973) A theoretical model and experimental evaluation of the cerebrospinal fluid system. Thesis, Drexel University, Philadelphia, PA
13. Momjian S, Czosnyka Z, Czosnyka M, Pickard JD (2004) Link between vasogenic waves of intracranial pressure and cerebrospinal fluid outflow resistance in normal pressure hydrocephalus. Br J Neurosurg 18(1): 56–61
14. Silverberg GD, Mayo M, Saul T, Rubenstein E, McGuire D (2003) Alzheimer's disease, normal-pressure hydrocephalus, and senescent changes in CSF circulatory physiology: a hypothesis. Lancet Neurol 2(8): 506–511
15. Tisell M, Edsbagge M, Stephensen H, Czosnyka M, Wikkelso C (2002) Elastance correlates with outcome after endoscopic third ventriculostomy in adults with hydrocephalus caused by primary aqueductal stenosis. Neurosurgery 50: 70–76
16. Williams MA, Razumovsky AY, Hanley DF (1998) Evaluation of shunt function in patients who are never better, or better than worse after shunt surgery for NPH. Acta Neurochir [Suppl] 71: 368–370
17. Wikkelso C, Andersson H, Blomstrand C, Lindqvist G, Svendsen P (1986) Normal pressure hydrocephalus. Predictive value of the cerebrospinal fluid tap-test. Acta Neurol Scand 73(6): 566–573

Correspondence: Zofia Czosnyka, Academic Neurosurgery Unit, Box 167, Addenbrooke's Hospital, Cambridge CB2 2QQ, UK. e-mail: zc200@medschl.cam.ac.uk

Acta Neurochir (2005) [Suppl] 95: 253–256

Intracranial baroreflex yielding an early Cushing response in human

E. A. Schmidt[1,2], **Z. Czosnyka**[1], **S. Momjian**[1,3], **M. Czosnyka**[1], **R. A. Bech**[1,4], and **J. D. Pickard**[1]

[1] Academic Neurosurgery Unit, Addenbrooke's Hospital, Cambridge, UK
[2] Service de Neurochirurgie, Hôpital Universitaire Purpan, Toulouse, France
[3] Service de Neurochirurgie, Hôpitaux Universitaires de Genève, Switzerland
[4] Department of Neurosurgery, Rigshospitalet, Copenhagen, Denmark

Summary

The Cushing response is a pre-terminal sympatho-adrenal systemic response to very high ICP. Animal studies have demonstrated that a moderate rise of ICP yields a reversible pressure-mediated systemic response. Infusion studies are routine procedures to investigate, by infusing CSF space with saline, the cerebrospinal fluid (CSF) biophysics in patients suspected of hydrocephalus. Our study aims at assessing systemic and cerebral haemodynamic changes during moderate rise of ICP in human.

Infusion studies were performed in 34 patients. This is a routine test perform in patients presenting with symptoms of NPH during their pre-shunting assessment. Arterial blood pressure (ABP) and cerebral blood flow velocity (FV) were non-invasively monitored with photoplethysmography and transcranial Doppler.

The rise in ICP (8.2 ± 5.1 mmHg to 25 ± 8.3 mmHg) was followed by a significant rise in ABP (106.6 ± 29.7 mmHg to 115.2 ± 30.1 mmHg), drop in CPP (98.3 ± 29 mmHg to 90.2 ± 30.7 mmHg) and decrease in FV (55.6 ± 17 cm/s to 51.1 ± 16.3 cm/s). Increasing ICP did not alter heart rate (70.4 ± 10.4/min to 70.3 ± 9.1/min) but augmented the heart rate variance (0.046 ± 0.058 to 0.067 ± 0.075/min).

In a population suspected of hydrocephalus, our study demonstrated that a moderate rise of ICP yields a reversible pressure-mediated systemic response, demonstrating an early Cushing response in human and a putative intracranial baroreflex.

Keywords: Infusion studies; Cushing response; transcranial Doppler; cerebral haemodynamics; systemic haemodynamics; baroreflex; intracranial.

Introduction

The Cushing response is a well-known pathophysiological phenomenon: rising intracranial pressure (ICP) to high values produces an increase in arterial blood pressure (ABP) [1]. The Cushing response has been demonstrated to be a sympatho-adrenal systemic response to brainstem ischemia [5, 8, 9]. However, few animal studies [4, 6, 7] found that a moderate rise of ICP could also influence ABP, via a putative pressure-mediated response. If the Cushing response is known in clinical practice, the pressure-mediated response to moderate rise of ICP has never been described in human.

Infusion studies are clinical routine procedures to investigate the circulation of cerebrospinal fluid (CSF) in patients suspect of hydrocephalus. ICP is moderately augmented by mean of an infusion in the ventricular or sub arachnoid space with mock CSF; subsequently the biophysical properties of the CSF circulation are computed [2]. Our work aims at using infusion studies to assess the cerebral and systemic haemodynamics changes during moderated rise of ICP in human.

Material and method

34 patients suspected of normal pressure hydrocephalus (cognitive impairment, gait disturbance and/or urinary incontinence, enlarged ventricles on CT) were included in this study. There were 15 females and 19 males (mean age: 54 years; range 28–76 years). A computerized CSF infusion test with constant–rate infusion [2] was undergone *via* a surgically pre-inserted Ommaya Reservoir connected to the ventricular catheter. The patients stayed supine in a comfortable tilting armchair. The ventricular pressure was measured at baseline for ten minutes, and termed "baseline" ICP. Then the ventricular space was infused with normal saline solution (0.9%) at room temperature (20 °C) with a rate of 1.5 ml min^{-1}. Subsequently ICP rose until a steady-state plateau has been achieved, termed "plateau" ICP (*cf.* Fig. 1). This "plateau" ICP corresponds to the pressure equilibrium at which all the mock CSF infused is reabsorbed.

During the infusion study, arterial blood pressure (ABP) and cerebral blood flow velocity (FV) were continuously and non-invasively measured using photoplethysmography and transcranial Doppler (TCD). The signals of CSF pressure (*i.e.* ICP), ABP and FV were amplified, converted and stored on a computer running our in-house software [2].

Off line, we calculated: ABP mean, systolic, diastolic and pulse amplitude (respectively ABPm, ABPs, ABPd and ABPa), cerebral

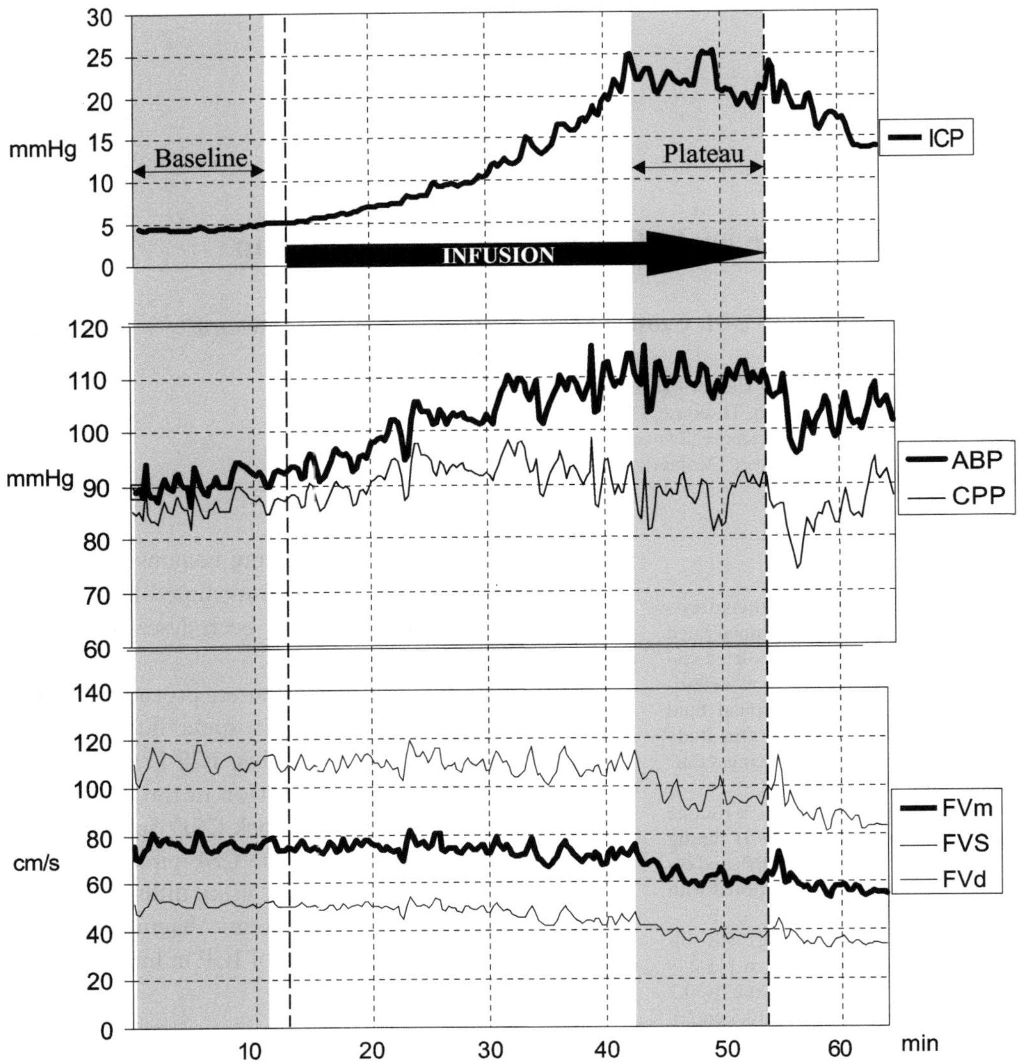

Fig. 1. An example of infusion test: ABPm, ABPs, ABPd and ABPa for ABP mean, systolic, diastolic and pulse amplitude (respectively), ICP and CPP for intracranial pressure and cerebral perfusion pressure; HRm, HRsd and HRcv for heart rate mean, standard deviation and coefficient of variance; FVm, FVs, FVd and PI for flow velocity mean, systolic, diastolic and Gosling pulsatility index

perfusion pressure (CPP = ABPm-ICP), heart rate mean, standard deviation and coefficient of variance (HRm, HRsd and HRcv), FV mean, systolic and diastolic (FVm, FVs and FVd), and the Gosling pulsatility index (PI = (FVs-FVd)/FVm). For every infusion test, we carefully selected ten minutes of recording during the "baseline" period, and ten minutes of ICP steady-state "plateau" (*cf.* Fig.1).

Results

The "baseline" and "plateau" values of the different variables are displayed on Table 1. The arithmetic difference between baseline and plateau, plus the significance level (paired t test) are also shown.

The rise in ICP yielded a significant increase in ABP mean, systolic, diastolic and in ABP pulse amplitude. The concurrent rise in ICP and in ABP had a net negative effect on the CPP, which significantly decreases. There was no change in the mean value of the heart rate, whereas its standard deviation and coefficient of variance significantly increased. The rise in ICP yielded also a significant drop in FV mean, systolic, and diastolic, but an increase in pulsatility index.

Conclusions

The Cushing response should be considered as a four-step process: i) rise in ICP approaching ABP, ii)

Table 1. *Changes in ICP and other haemodynamic parameters*

	Baseline	Plateau	Difference	
ICP (mmHg)	8.2 ± 5.1	25 ± 8.3	16.8 (+204.8%)	<0.0001
ABPm (mmHg)	106.6 ± 29.7	115.2 ± 30.1	8.6 (+8.1%)	<0.0001
ABPs (mmHg)	145.3 ± 36.4	156.7 ± 37.4	11.4 (+7.8%)	<0.0001
ABPd (mmHg)	76.5 ± 31.4	90.5 ± 26.5	14 (+18.3%)	<0.0001
ABPa (mmHg)	35.1 ± 19.3	38.0 ± 22	2.9 (+8.3%)	0.049
CPP (mmHg)	98.3 ± 29	90.2 ± 30.7	−8.2 (−8.3%)	<0.0001
HR (min^{-1})	70.4 ± 10.4	70.3 ± 9.1	−0.08 (−0.1%)	0.903
HRsd	3.11 ± 3.31	4.5 ± 4.5	1.4 (+44.8%)	<0.0001
HRcv	0.046 ± 0.058	0.067 ± 0.075	0.021 (+46%)	0.0005
FVm (cm s^{-1})	55.6 ± 17	51.1 ± 16.3	−4.5 (−8.2%)	<0.0001
FVs (cm s^{-1})	83.0 ± 22.9	80.7 ± 23.3	−2.3 (−2.8%)	0.011
FVd (cm s^{-1})	37.9 ± 12.8	33.4 ± 12.2	−4.5 (−12%)	<0.0001
PI	0.828 ± 0.209	0.946 ± 0.257	0.118 (+14.3%)	<0.0001

oligemia or ischemia of the medulla oblongata, iii) sympathetic and sympatho-adrenal activation, finally iv) systemic effects. In a way, the Cushing response ought to be considered as a pre-lethal phenomenon [3]. Our study demonstrates that, in awake subjects, a moderate rise of ICP produces i) a rise in ABP, ii) a drop in CPP, iii) a reduction in FV and iv) an increase in the heart rate variance.

In the population suspected of hydrocephalus, our data support the notion that a moderate rise of ICP yields a reversible pressure-mediated systemic response, demonstrating an early Cushing response in human.

The early Cushing response raises three issues: the concept of intracranial baroreflex, its influence on cerebral haemodynamics and on autonomic status.

1. The concept of intracranial baroreflex is supported by the notion that intracranial pressure can influence directly *via* a reflex loop systemic haemodynamics. As we demonstrated an intracranial baroreflex, therefore there should be an intracranial baroreceptor to trigger the reflex.
2. ICP bears direct and indirect influences on cerebral haemodynamics. The direct influence on cerebral perfusion is driven by CPP, altered directly by ICP. However, the changes in cerebral haemodynamics during changes in CPP are dampened by the mechanism of autoregulation. ICP has an indirect effect on cerebral haemodynamics, *via* the changes in systemic haemodynamic related to the putative early Cushing response. Therefore, when ICP rises, the secondary increase in ABP lessens the drop in CPP. The intracranial baroreflex could be considered as a protective mechanism to maintain cerebral haemodynamics. However, the significant reduction of flow velocity during the rise in ICP remains to be clarified, as one would expect flow velocity to be maintained.
3. The rise in ICP probably interferes with the sympathetic/parasympathetic balance of the cerebral vasculature, as suggested by the change in heart rate variance. The drop in flow velocity can be related to changes in autonomic status.

The early Cushing response demonstrated in our subgroup of patients, is a new physiological concept and might be of clinical importance.

References

1. Cushing H (1901) Concerning a definite regulatory mechanism of vasomotor centre which controls blood pressure during cerebral compression. Bull Johns Hopkins Hosp 12: 290–292
2. Czosnyka M, Whitehouse H, Smielewski P, Simac S, Pickard JD (1996) Testing of cerebrospinal compensatory reserve in shunted and non-shunted patients: a guide to interpretation based on an observational study. J Neurol Neurosurg Psychiatry 60(5): 549–558
3. Fitch W, McDowall DG, Keaney NP, Pickerodt VWA (1977) Systemic vascular responses to increased intracranial pressure. 2. The "Cushing" response in the presence of intracranial space-occupying lesions: systemic and cerebral haemodynamics studies in the dog and the baboon. J Neurol Neurosurg Psychiatry 40: 843–852
4. Fitch W, McDowall DG, Paterson GM, Hain WR (1977) Systemic vascular responses to increased intracranial pressure. 3. Effects of individual balloon inflations on intracranial pressure and the systemic effects. J Neurol Neurosurg Psychiatry 41: 340–344
5. Johnston IH, Rowan JO, Harper AM, Jennett WB (1972) Raised intracranial pressure and cerebral blood flow. I. Cisterna magna infusion in primates. J Neurol Neurosurg Psychiatry 35(3): 285–296
6. Hayreh SS, Edwards J (1971) Vascular response to acute intracranial hypertension. J Neurol Neurosurg Psychiatry 34: 587–601

7. Nakamura K, Osborn JW Jr, Cowley AW Jr (1987) Pressor response to small elevations of cerebroventricular pressure in conscious rats. Hypertension 10(6): 635–641
8. Rowan JO, Teasdale G (1977) Brain stem blood flow during raised intracranial pressure. Acta Neurol Scand [Suppl] 64: 520–521
9. Van Loon J, Shivalkar B, Plets C, Goffin J, Tjandra-Maga TB, Flameng W (1993) Catecholamine response to a gradual increase of intracranial pressure. J Neurosurg 79(5): 705–709

Correspondence: Eric A. Schmidt, Service de Neurochirurgie, Hôpital Purpan, TSA 40031 31059 Toulouse Cedex 9, France. e-mail: schmidt.e@chu-toulouse.fr

Acta Neurochir (2005) [Suppl] 95: 257–259

Does the ventricle size change after shunt operation of normal-pressure hydrocephalus?

U. Meier[1] and **S. Mutze**[2]

[1] Department of Neurosurgery, Unfallkrankenhaus Berlin, Berlin, Germany
[2] Department of Radiology, Unfallkrankenhaus Berlin, Berlin, Germany

Summary

Objectives. The aim of this study was to determine whether ventricular size correlates with a positive clinical outcome following shunt placement. Hydrostatic valves (Dual-Switch-Valves) were implanted in 80 patients with NPH at Unfallkrankenhaus Berlin between September 1997 and January 2002. One year postoperatively, these patients underwent computerized tomography scanning, and their ventricular size was ascertained using the Evans-Index. Among 80% of the patients who showed no postoperative change in ventricular volume, 59% nonetheless had good to excellent clinical improvements, 17% satisfactory clinical improvement, and 24% no improvement. Furthermore, a moderate reduction in ventricular size was observed in 14% of patients in this cohort. Among these, 36% experienced a good to excellent clinical improvement, 28% a satisfactory improvement, and 36% unsatisfactory improvement. A marked reduction in ventricular size was observed in 6% of the patients. Of this group, 60% demonstrated good to excellent outcomes, whereas 40% had unsatisfactory outcomes. Favourable outcomes following the implantation of a hydrostatic shunt in patients with NPH did not correlate with decreased ventricular volume 1 year after surgery. In fact, better clinical outcomes were observed in patients with little or no alteration in ventricular size, compared with those in patients with marked decrease in ventricular size. A postoperative change in ventricular volume should be assessed differently in patients with NPH compared with those suffering from hypertensive hydrocephalus.

Keywords: Normal-pressure hydrocephalus; ventricle; Evans Index; shunt; dual-switch valve; hydrostatic valve.

Introduction

There is a great deal of disagreement among neurosurgeons, neuroradiologists and neurologists as to whether a clearly positive correlation exists between postoperative reductions in ventricular volume and clinical outcomes following shunt placement in patients with NPH. Bear in mind that all studies of clinical outcome following implantation of differential pressure valves have yielded varying results. The advent of shunts with hydrostatic valves has only heightened this controversy, because the prevention of overdrainage may be associated with an increased risk of relative underdrainage and with the related phenomenon of a reduced rate of ventriculomegaly regression [1, 3].

Clinical material and method

Hydrostatic valves (dual-switch valves; Christoph Miethke GmbH & Co. KG, Berlin, Germany) were implanted in 80 normal pressure hydrocephalus patients at Unfallkrankenhaus Berlin between September 1997 and January 2002. The cohort consisted of 43 men and 37 women, with a mean age of 65 years at the time of the study. Diagnoses were established and the decisions to place shunts were made on the basis of the patients' histories as well as gait ataxia as a cardinal symptom, dementia and urinary incontinence as additional and late symptoms, and performance of an intrathecal infusion test [7]. In addition, resistance to cerebrospinal fluid outflow R_{out} exceeding 13 mmHg/ml/min was regarded as evidence of derangement of cerebrospinal fluid outflow in the craniospinal pathways [7]. After performing an intrathecal infusion test, no less than 60 ml of CSF was siphoned during a tap test, which was performed as an auxiliary diagnostic measure. Indications for shunt placement were as follows: gait ataxia, pathologically raised resistance to CSF outflow, and improvement in clinical symptoms (particularly gait ataxia) following CSF tap test. The hydrostatic Miethke dual-switch valve was implanted in all 80 patients [5]. CT studies were obtained in all 80 patients preoperatively and one year postoperatively. During all of these CT scanning sessions, ventricular volume in the frontal horn was ascertained using the Evans index, and any postoperative changes were measured in accordance with this index as well. Preoperative and postoperative results for all patients were evaluated using the Kiefer normal hydrocephalus grading scale [4] and the NPH-recovery rate.

Results

Both clinical and CT findings in 80 patients with NPH were evaluated following implantation of a hydrostatic valve. A reduction in the Evans index of at

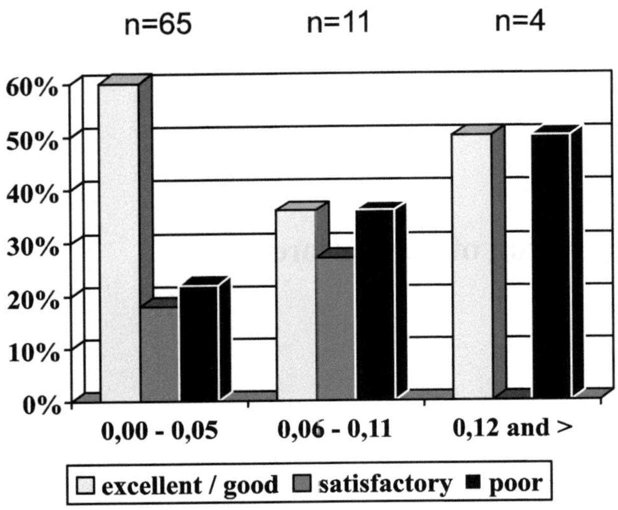

Fig. 1. Change of the ventricle size versus course of disease

least 0.12 was defined as a clear or a significant reduction in ventricular volume, whereas a reduction of at least 0.06 to 0.11 was classified as a moderate improvement. According to the Evan's-Index, a differential of 0.05 constitutes little or no reduction in ventricular volume.

The following changes in ventricular volume were observed one year postoperatively in patients who underwnet a shunt insertion: a slight reduction in 64 patients (80%), a moderate reduction in 11 patients (14%), and marked reduction in 5 (6%). A comparison of outcomes one year postoperatively with changes in ventricular volume during the same period (Fig. 1) revealed the following: 59% of patients with a minimal reduction in ventricular volume showed good to excellent clinical outcomes, 17% had satisfactory outcomes and 24% had unsatisfactory outcomes; among the patients with a moderate decrease in ventricular size 36% had good to excellent clinical results, 36% had unsatisfactory results, and the remaining 28% had satisfactory clinical outcomes; and among the patients with a marked reduction in ventricular volume 60% had good to excellent clinical outcomes and the remaining 40% had unsatisfactory outcomes. No statistical analysis of these results could be performed due to the relatively small number of patients who showed a marked reduction in ventricular volume. Generally speaking, treatment outcome was characterized as only a minor reduction in ventricular size, accompanied in the majority of cases, by good to excellent or at least satisfactory clinical improvement. The results of the CT studies are surprising given that in 80% of cases, little or no reduction in ventricular volume (according to our definition of this parameter) was observed, notwithstanding the fact that good to excellent improvement was observed in a majority (59%) of cases.

Discussion

Hydrostatic valves were the first hydrocephalus valves to compensate for the facts that patients with hydrocephalus spend a great deal of time sitting and standing, and having completly different needs with respect to CSF dynamics in the recumbent position versus the upright. Although hydrostatic valves have reduced the risk of overdrainage induced by valve design [8], problems of underdrainage have also become increasingly prevalent. Three patients (4%) in the present study underwent revision operations because of clinical underdrainage that had been confirmed on performing an intrathecal infusion test. Miethke dual-switch valves with a lower pressure setting for recumbent position were implanted in these patients. A review of the literature on underdrainage following shunt surgery reveals marked discrepancies and inconsistencies in the definitions and classifications of postoperative complications. Some authors calculate the overall number of complications, while others counted only mechanical complications and excluded infections. Thus, for example, Sainte-Rose [10] fails to define underdrainage at all and characterizes overdrainage as a subdural hygroma, slit ventricle or isolated ventricle. Drake *et al.* [3] classify underdrainage and occlusion as a combination of complications, defining overdrainage as a subdural hydroma and a slit ventricle syndrome. In our view, underdrainage can be said to exist only when the pressure level of the valve is unduly high; when, due to a manufacturing defect, the actual valve pressure level is higher than the recommended level; when shunt function changes in vivo; or when cerebrospinal fluid outflow is reduced because of raised intraperitoneal pressure caused by adiposis or insufficient cerebrospinal fluid resorption provoked by a pseudocyst. Analysis of data by the UK shunt registry [9], which is composed of more than 13,000 cases, shows underdrainage to be a significant complication following shunt placement (46.7 $\pm$ 1.2% of cases). Overdrainage provoked postoperative complications in only 3% of the patients in our study, which stands in stark contrast to the results of some Scandinavian researchers who reported that 80% of all postoperative shunt complications are attributable to overdrainage.

It is difficult to perform a comparison of underdrainage associated with other types of shunts for two reasons: first, only very few studies have reached any clear conclusions regarding this complication, and second, the researchers of the majority of potentially comparable studies in patients with NPH fail precisely to define or record clear data about this complication [2, 10]. The complications of underdrainage might be divided into three different types: enlargement of ventricle width after shunt placement together with a persistent or worsening clinical pathology, no change of ventricle and a clear worsening of clinical pathology, or late worsening clinical pathology after a clear postoperative recovery. In all cases of suspected underdrainage, the intrathecal infusion test was conducted to confirm the diagnosis and a shuntogramm routinely was obtained. A positive correlation between ventricular size reduction and clinical improvement following shunt placement has been reported in several studies.

In the absence of a valve that is both adjustable and authentically programmable – one that could continuously measure pressure and flow, and whose feedback mechanism would enable a regenerative coupling function – we must make the most of the technical capabilities offered by currently available valves as we strive to achieve optimal levels of ICP in individual patients. The valves available today lack the capacity to respond to ventricular volume, cerebral blood flow, or alterations in cerebral metabolism. Furthermore they cannot optimally regulate CSF circulation. These valves only respond to the differential pressure between the region of the proximal catheter and the environment at the site of distal diversion. We must therefore continue to make efforts – through the now widely accepted radiological examinations, measurement of blood flow, measurments of ICP and differentiated infusion tests – to develop criteria that allow us to reliably predict the optimal pressure ratios needed by individual patients postoperatively in both the recumbent and upright positions.

Conclusions

Our results clearly indicate the following: in the majority of cases, good to excellent clinical results are accompanied by only a slight reduction in ventricular volume following implantation of a hydrostatic valve (Miethke dual-switch valve); and a maximum reduction in ventricular volume is not a prerequisite for clinical improvement in patients with NPH. Five years of experience with the Miethke dual-switch valve have demonstrated that the design engineering of this gravitational, hydrostatic valve delivers dependable performance. The technical design principle consisting in a parallel two-chamber system containing large membranes [6] that minimize overdrainage, making the Miethke dual-switch valve greatly superior in this respect to diferential pressure valves and adjustable valves, as well as to the majority of other hydrostatic valves. The fact that the functionality of the Miethke dual-switch valve is determined neither by subcutaneous pressure nor by implantation height is an additional contributing factor to the valves excellent perfomance. In our view, the low overdrainage rate of 3% and the generally positive outcomes achieved with the Miethke dual-switch valve make it the therapy of choice for patients with normal-pressure hydrocephalus [5].

References

1. Bergsneider M, Peacock WJ, Mazziotta JC, Becker DP (1999) Beneficial effect of siphoning in treatment of adult hydrocephalus. Arch Neurol 56: 1224–1229
2. Di Rocco C, Marchese E, Velardi F (1994) A survey of the first complication of newly implanted CSF shunt devices for the treatment of nontumoral hydrocephalus. Cooperative survey of the 1991–1992 Education Committee of the ISPN. Childs Nerv Syst 10: 321–327
3. Drake JM, Kestle JR, Milner R, Cinalli G, Boop F, Piatt J Jr, Haines S, Schiff SJ, Cochrane DD, Steinbok P, MacNeil N (1998) Randomized trial of cerebrospinal fluid shunt valve design in pediatric hydrocephalus. Neurosurgery 43: 294–303
4. Kiefer M, Eymann R, Meier U (2002) Five years experience with gravitational shunts in chronic hydrocephalus of adults. Acta Neurochir (Wien) 144: 755–767
5. Meier U, Kiefer M, Sprung C (2002) The Miethke Dual-Switch valve in patients with normal pressure hydrocephalus. Neurosurgery Q 12: 114–121
6. Meier U, Kintzel D (2002) Clinical experiences with different valve systems in patients with normal-pressure hydrocephalus: evaluation of the Miethke dual-switch valve. Childs Nerv Syst 18: 288–294
7. Meier U, Zeilinger FS, Kintzel D (1999) Signs, symptoms and course of normal pressure hydrocephalus in comparison with cerebral atrophy. Acta Neurochir (Wien) 141: 1039–1048
8. Pudenz RH, Foltz EL (1991) Hydrocephalus: overdrainage by ventricular shunts. A review and recommendations. Surg Neurol 35: 200–212
9. Richards HK, Seely HM, Pickard JD (2000) Shunt revisions: data from the UK shunt registry. Eur J Pediatr Surg [Suppl] 10: 1–59
10. Sainte-Rose C, Hooven MD, Hirsch JF (1987) A new approach in the treatment of hydrocephalus. J Neurosurg 66: 213–226

Correspondence: Ullrich Meier, Department of Neurosurgery, Unfallkrankenhaus Berlin, Warener Straße 7, 12683 Berlin, Germany. e-mail: ullrich.meier@ukb.de

Clinical trials

Acta Neurochir (2005) [Suppl] 95: 263–264

Intravenous magnesium sulfate to improve outcome after aneurysmal subarachnoid hemorrhage: interim report from a pilot study

R. Boet[1], **M. T. V. Chan**[2], **W. S. Poon**[1], **G. K. C. Wong**[1], **H. T. Wong**[1], and **T. Gin**[2]

[1] Division of Neurosurgery, Department of Surgery, The Chinese University of Hong Kong, Prince of Wales Hospital, Hong Kong, China
[2] Department of Anaesthesia and Intensive Care, The Chinese University of Hong Kong, Prince of Wales Hospital, Hong Kong, China

Summary

Objectives. Magnesium sulfate ($MgSO_4$) may be useful in preventing neurological injury after subarachnoid haemorrhage (SAH). In this randomized, double-blind study we evaluated the safety and efficacy of $MgSO_4$ infusion to improve clinical outcome after aneurysmal SAH.

Methods. With ethics committee approval and informed consents, 45 patients with SAH were randomly allocated to receive either $MgSO_4$ 80 mmol/day or saline infusion for 14 days. All patients also received intravenous nimodipine. Episodes of vasospasm were treated with hypertensive, hypervolemic therapy. Neurological status was assessed 3 months after haemorrhage using Barthel index and Glasgow outcome scale (GOS). Incidences of cardiac and pulmonary complications were also recorded. Data were compared between groups using Mann-Whitney or Fisher exact tests as appropriate. $P < 0.05$ was considered significant.

Results. Patient characteristics, severity of SAH and surgical treatment did not differ between groups. Although the number of episodes was not reduced, $MgSO_4$ shortened the duration of vasospasm. Patients receiving $MgSO_4$ tended to have fewer neurological deficits, better functional recovery and an improved score in GOS. However, none of these outcome variables reached statistical significance. The incidence of cardiac and pulmonary complications in the $MgSO_4$ group (43%) was also similar to that in the saline group (59%), $P = 0.14$.

Conclusions. $MgSO_4$ infusion after aneurysmal SAH is well tolerated and may be useful in producing better outcome. A larger study is required to confirm the neuroprotective effect of $MgSO_4$.

Keywords: Intracranial aneurysm; subarachnoid haemorrhage (SAH); vasospasm; magnesium; randomized controlled trial.

Introduction

Patients who survive an aneurysmal subarachnoid haemorrhage (SAH) may still suffer from two delayed events: first, a second haemorrhage, the consequences of which may be fatal, and second, delayed cerebral ischemia (vasospasm) that may lead to brain infarction, focal deficits and a poor neurological outcome.

Contemporary management of patients suffering from ASAH has been streamlined over the last two to three decades. Early treatment of the aneurysm either by surgical clipping or endovascular occlusion has largely decreased the incidence of aneurysmal rebleeds. Despite the best medical treatment, such as nimodipine [3], triple-H therapy and angioplasty, more than 20% of patients still suffer from clinical vasospasm resulting in unfavourable outcome. We believe that magnesium sulfate may improve clinical outcome because of (1) its vasodilatation effect, and (2) the association of poor outcome with hypomagnesaemia [4]. We have carried out a prospective randomized double-blind placebo controlled trial (a pilot study) with the hypothesis that magnesium sulfate improves clinical outcome by decreasing the incidence and severity of vasospasm.

Patients and methods

Patients with aneurysmal subarachnoid haemorrhage (ASAH), proven by CT and cerebral angiography (CTA or DSA), were randomized within 48 hours to receive intravenous magnesium sulfate infusion or placebo (normal saline). Approval from the Clinical Ethics Committee of the Chinese University of Hong Kong and informed consent to participate in this study from either the patient or his/her family were obtained. Following randomization, a bolus infusion of 20 mmol of magnesium sulfate (or placebo) was administered in 30 minutes followed by a maintenance infusion of 80 mmol in 24 hours [1]. The intravenous infusion was maintained for 14 days. Clinical vasospasm (or delayed cerebral ischaemia) is defined as a Glasgow Coma Score (GCS) reduction of two or more, or a focal neurological deficits, for more than six hours, when rebleed, progressive hydrocephalus and electrolyte disturbances are excluded. The dichotomized Glasgow Outcome Scale into favourable (good and moderate disability) and unfavourable outcomes was employed as the primary outcome measure of the study.

Table 1. *Outcomes after $MgSO_4$ or saline infusion in patients with SAH*

	Magnesium ($n = 23$)	Saline ($n = 22$)	P Values
Plasma magnesium concentration (mmol/L)	2.2 (1.6–2.5)	0.9 (0.7–1.2)	<0.01
Incidence of vasospasm, *n* (%)	6 (26%)	12 (55%)	0.11
MCA flow velocity (cm/s)			
during vasospasm	145 (127–210)	139 (121–220)	0.82
during asymptomatic period	65 (32–125)	57 (28–130)	0.69
Barthel index	83 (45–100)	71 (15–100)	0.06
Glasgow outcome score			
Favorable outcome	15 (65%)	10 (45%)	0.10
Good recovery	10 (43%)	7 (32%)	
Moderately disabled	5 (22%)	3 (13%)	
Unfavorable outcome	8 (35%)	12 (55%)	
Severely disabled	5 (22%)	8 (36%)	
Vegetative state	0 (0%)	1 (5%)	
Death	3 (13%)	3 (14%)	

MCA middle cerebral artery.

Results

Forty five patients entred this pilot study: twenty three were randomized to receive magnesium sulfate and 22 placebo. The mean age (57 years), gender (female ratio 19/23 versus 18/22), clinical grading (WFNS grade 3–4: 19/23 versus 19/22; Fisher grade 3–4: 20/23 versu 21/22), aneurysm treatment modality (endovascular treatment 8/23 versus 8/22), median time from SAH to study drug infusion (27 hours versus 35 hours) and median time from SAH to aneurysm treatment (2 days) were comparable in both groups. While the plasma magnesium concentration in the treatment group was shown to be adequate (2.2 mmol/l versus 0.9 mmol/l), there was a trend towards a reduced incidence of clinical vasospasm (26% versus 55%, $p = 0.11$) and an improved outcome at three months (Barthel Index 83 versus 71, $p = 0.06$; favourable GOS 65% versus 45%, $p = 0.1$, table 1). In this pilot study, the median duration of vasospasm was significantly reduced (2.7 days [2–6] versus 4.2 days [2–8], $p = 0.02$).

Dicussion and conclusion

This pilot study has demonstrated a trend towards an improved neurological outcome in the dichotomized Glasgow Outcome Scale (incidence of favourable outcome in the magnesium treated group 65% versus 45%, $p = 0.1$, table 1) and Barthel Index. Interestingly, the duration of clinical vasospasm was significantly reduced from a median of 4.2 days to 2.7. Magnesium sulfate infusion has been shown to be simple and safe by a previous study [1] as well as the present study. A trend towards improved neurological recovery was also demonstrated by previous case control study [2] and a pilot randomized controlled trial [5]. A randomized, double-blind, placebo-controlled multicentre trial has been designed to recruit 340 patients to prove the trend towards better outcome in Venya's [5] and our study [www.surgery.cuhk.edu.hk/imash-trial].

Acknowledgment

This study was supported by a grant earmarked by the Research Grant Council, Hong Kong.

References

1. Boet R, Mee E (2000) Magnesium sulfate in the management of patients with Fisher Grade 3 subarachnoid hemorrhage: a pilot study. Neurosurgery 47: 602–606
2. Chia RY, Hughes RS, Morgan MK (2002) Magnesium: a useful adjunct in the prevention of cerebral vasospasm following aneurysmal subarachnoid haemorrhage. J Clin Neurosci 9(3): 279–281
3. Pickard JD, Murray GD, Illingworth R, Shaw MDM, Teasdale GM, Foy PM, Humphrey PRD, Lang DA, Nelson R, Richards P, Sinar J, Bailey S, Skene A (1989) Effect of oral nimodipine on cerebral infarction and outcome after aneurysmal subarachnoid haemorrhage: British aneurysm nimodipine trial. Br Med J 298: 636–642
4. Van Den Bergh WM, Algra A, Van Der Sprenkel JW, Tulleken CA, Rinkel GJ (2003) Hypomagnesemia after aneurysmal subarachnoid hemorrhage. Neurosurgery 52: 276–282
5. Veyna RS, Seyfried D, Burke DG, Zimmerman C, Mlynarek M, Nichols V, Marrocco A, Thomas AJ, Mitsias PD, Malik GM (2002) Magnesium sulfate therapy after aneurysmal subarachnoid hemorrhage. J Neurosurg 96: 510–514

Correspondence: W. S. Poon, Division of Neurosurgery, Prince of Wales Hospital, The Chinese University of Hong Kong, Shatin, Hong Kong, China. e-mail: wpoon@surgery.cuhk.edu.hk

Acta Neurochir (2005) [Suppl] 95: 265–267

Secondary insults and outcomes in patients with hypertensive basal ganglia hemorrhage

Z. Fei, X. Zhang, and **S. J. Song**

Department of Neurosurgery, Xijing Hospital, Fourth Military Medical University, Xi'an, P.R. China

Summary

This study was designed to monitor secondary insults and their impact on outcomes of patients with hypertensive basal ganglia hemorrhage (HBGH). One hundred and twelve patients with HBGH (male 73, female 39) of age 42 ± 8 years (range from 38 to 57 years) were studied. Operations included craniotomy or trephination drainage with urokinase thrombolysis. Conventional therapies were also given to the patients including the administration of mannitol, crystalloid and colloid solution. In the meantime, blood pressure (MAP), temperature (T) and SaO_2 and other parameters were recorded in the intensive care unit. The ICP values were recorded, and the early clinical outcome was assessed upon discharge according to Glasgow Outcome Scale. Cerebral Perfusion Pressure was calculated as CPP = MAP-MICP. Outcomes in the group without secondary insults were better than that in the group with secondary insults ($P < 0.01$). No unfavorable outcomes were found in the 59 cases managed by ultra-early surgery whereas 36.1% of the cases operated after 6 hours of onset had unfavorable outcomes. It is concluded that the high incident rate of secondary insults in HICH patients influences outcome. Ultra-early surgery may also contribute to improved quality of survival.

Keywords: Hypertensive intracerebral hemorrhage; secondary insults; ICP.

Introduction

Despite recent advances in understanding and application of new techniques to the treatment of hypertensive basal ganglia haemorrhage (HBGH), the outcome remains unacceptable. Many predictive factors should be taken into consideration, including haematoma volume, location, and patients' age, but secondary insults may also affect the outcome, including systemic hypotension, compromised cerebral perfusion pressure (CPP), raised intracranial pressure (ICP), elevated temperature, and hypoxemia, leading to initiation of destructive inflammatory and biochemical cascades. Detection and correction of secondary insults may improve outcome [1–3]. From August 2001 to April 2004 one hundred and twelve cases with HBGH were operated on in our department, and some of secondary insults were monitored and analyzed in relation to clinical outcome. The experimental protocol received prior approval by the hospital review body and informed consent was obtained from each patient or next of kin.

Materials and methods

Patients

One hundred and twelve cases with HBGH (male 73, female 39) of mean age 42 years (range from 38 to 57 years) were studied. They were operated on by craniotomy or trephination drainage with urokinase thrombolysis. All patients had no prior medical history of hypertension, diabetes or coronary heart disease, or other illness of important organs preoperatively, and all suffered from HBGH for the first time. Fifty one of 112 cases had a history of alcohol, tiredness, defaecation or overexcitement prior to the onset. All patients were diagnosed with HBGH by CT scan. Hemorrhage due to tumor, trauma, aneurysms, AVMs, cases without basal ganglia hemorrhages, and cases with organ failure were excluded. The patients' condition were estimated according to the Kanetani's grading scale, including 21 cases in grade III (19%), 53 cases in grade IVa (47%), and 38 cases in grade IVb (34%, Table 1). The volume of the hematoma was measured [4].

Methods

Surgical timing and treatment

The operations were all performed within 2 to 24 hours from the onset, and 59 (53%) of the cases were operated within 6 hours. The hematoma volumes ranged from 20∼120 ml in both groups. Conventional therapies were also given to the patients including the administration of mannitol, crystalloid and colloid solution.

General systemic observations

The blood pressure (MAP), temperature (T) and SaO2 and other parameters were recorded in the intensive care unit.

Table 1. *Parameters of secondary insults between two groups* ($\bar{\chi} \pm s$)

Groups	No.	MAP (mmHg)	BT (°C)	ICP (mmHg)	CPP (mmHg)	SaO_2 (%)
With insults	67	144 ± 10.7	38.7 ± 1.24	39.4 ± 8.8	45.6 ± 9.82	90.7 ± 3.6
Without insults	45	106 ± 12.4*	37.1 ± 1.02*	28.7 ± 6.3*	67.2 ± 11.86*	94.8 ± 2.6*

* Compared with that of group with secondary insults $P < 0.01$.

ICP, CPP monitoring and clinical outcome

For patients who underwent traditional craniotomy, following surgical evacuation of the hematoma under microscopic illumination and magnification, a catheter was placed into the hematoma cavity for drainage. In addition, another catheter was placed in the anterior horn of the contralateral ventricle to monitor the ICP. For patients who underwent trephination drainage with urokinase thrombolysis, a catheter was placed into the hematoma cavity for hematoma aspiration, and a bolus of 20000u of urokinase in 2 ml of normal saline was slowly injected through the catheter. Normally the catheter was shut for 2 hours and then released. After urokinase was injected, another catheter was placed in the anterior horn of the contralateral ventricle to monitor the ICP. The injection of urokinase was repeated once a day or 6 hours later as necessary after operation. After the operation, all the patients were in the intensive care unit. ICP was monitored continuously by Model V24 Multifunctional Physiological Monitoring Systems of Hewlett Packard Ltd. The ICP values were recorded after operation immediately and at 24, 72 hours and 1 week after operation respectively. The early clinical outcome was assessed upon discharge according to the Glasgow Outcome Score (GOS) including good recovery (G), moderate disability (M), severe disability (S), persistent vegetative state (PVS) and death (D). CPP was calculated as CPP = MAP-MICP.

Statistical analysis

The ICP values were expressed as mean $\pm$ SE. Statistical differences were determined by the t-test with SPSS 1.0 software and Chi-squared for the comparison of the outcomes of patients.

Results

Patients were divided into two groups: these with secondary insults and these without. There were 67 (60%) cases in the group with secondary insults showing elevated MAP, ICP, body temperature and decreased CPP, SaO2 compared with that in the group (45 cases) without secondary insults (P < 0.01, Table 1). As shown in Table 2, the outcomes in the group without secondary insults were better than that in the group with secondary insults (P < 0.01). No unfavorable outcomes were found in the 59 cases managed by ultra-early surgery whereas 36.1% of the cases operated after 6 hours of onset had unfavorable outcomes.

Table 2. *Glasgow outcome scale on discharge between two groups*

Groups	No.	GOS				
		G	M	S	P	D
With insults	67	18	29	13	4	3
Without insults	45	27*	15*	2*	1*	0

* Compared with that of group with secondary insults $P < 0.01$.

Discussion

Following HGBH, the mass effect of the hemorrhage is complicated by brain edema. HGBH is accompanied by different degrees of hydrocephalus, which may occur at 24 to 96 hours and peaks at 3 to 6 days after acute onset. These changes may restrict the compensatory ability of the patients' intracranial compliance and impair the self-adjustment function of cerebral vessels resulting in the acute increased ICP and the decrease of CPP. If HGBH patients were not treated in time, a cerebral hernia may occur [5]. Therefore HGBH must be treated as early as possible, and the ICP should be monitored, which not only guides reasonable therapeutic methods, but also provides direct clinical evidence with which to assess prognosis. In the current study, the drainage catheter was placed in lateral ventricle of HBGH patients to monitor ICP, which can also be used to drain cerebrospinal fluid [6].

The theory of secondary insults was suggested by Miller in 1978 and refers to the impact of secondary insults such as blood pressure, body temperature, ICP, cerebral blood flow and CPP on secondary brain injury. These insults will exacerbate the primary head impact and cerebral edema [1–3].

There is no relevant research on incidence of secondary insults in HGBH. Our data showed that there has been a high incidence rate of secondary insults in HGBH patients which may influence prognosis.

From the 1970s, many surgeons proposed ultra-early surgery. After 7 hours of HICH onset, hydrocephalus follows and necrosis will form around the hematoma. Brain function may recover poorly and the consequences will be serious if the operation is carried out after 7 hours of onset [7, 8]. In our study, 59 of 112 cases underwent ultra-early surgery and satisfactory results were achieved. This suggests that ultra-early surgery may contribute to releasing the pressure of the hematoma on the brain tissue as early as possible, thereby reducing the occurrences of secondary insults and increasing the quality of survivals.

Acknowledgments

This work is supported both by National Natural Science Foundation of China (30270534) and Foundation for University Key Teacher by the Ministry of Education.

References

1. Dearden NM (1998) Mechanisms and prevention of secondary brain damage during intensive care. Clin Neuropathol 17(4): 221–228
2. Fei Z, Zhang X, Lu PL (2001) The expression and changes of heat shock protein 70, MDA and hemorheology in rat cortex after diffuse axonal injury with secondary insults. J Clin Neurosci 8(3): 250–252
3. Fei Z, Zhang X, Lu PL (2001) Level of the neuronal free Ca2+, BBB permeability and ultrastructure in head injury with secondary insults. J Clin Neurosci 8(6): 557–563
4. Broderick J, Brott T, Duldner J, Tomsick T, Huster G (1993) Volume of intracerebral hemorrhage. A powerful and easy-to-use predictor of 30-day mortality. Stroke 24(7): 987–993
5. Jung-Il Lee, Do-Hyun Nam, Jong-Soo Kim, Seung-Chyul Hong, Hyung-Jin Shin, Kwan Park, Whan Eoh, Jong Hyun Kim (2003) Stereotactic aspiration of intracerebral haematoma: significance of surgical timing and haematoma volume reduction. J Clin Neurosci 10(4): 439–443
6. Chambers IR, Banister K, Mendelow AD (2001) Intracranial pressure within a developing intracerebral haemorrhage. Br J Neurosurg 15(2): 140–141
7. Zuccarello M, Andaluz N, Wagner KR (2002) Minimally invasive therapy for intracerebral hematomas. Neurosurg Clin N Am 13(3): 349–354
8. Zhang Y, Yan S, Chen D (1997) Ultra-early operative treatment for severe hypertensive intracerebral hemorrhage. Clin Neurol Neurosurg 99[Suppl] 1: S58

Correspondence: Z. Fei, Department of Neurosurgery, Xijing Hospital, Fourth Military Medical University, Xi'an, P.R. China. e-mail: feizhou@fmmu.edu.cn

Acta Neurochir (2005) [Suppl] 95: 269–272

Balance of risk of therapeutic hypothermia

S. Hayashi[1], **M. Takayasu**[2], **S. Inao**[1], and **J. Yoshida**[2],
Nagoya Therapeutic Hypothermia Study Group

[1] Department of Neurosurgery, Nagoya First Red Cross Hospital, Nagoya, Japan
[2] Department of Neurosurgery, Nagoya University Graduate School of Medicine, Nagoya, Japan

Summary

The complications of therapeutic hypothermia sometimes undermine its clinical effects. In this study we investigated the efficacy and safety of therapeutic hypothermia based on analysis of 20 severe head injury cases from 6 institutions treated with therapeutic hypothermia in 1999.

The twenty patients with severe head injury were enrolled prospectively based on the following indications; Glasgow Coma Scale of 7 or less on admission, age 60 or younger, and systric BP over 100 mmHg. A control group consisting of 21 patients with severe head injury met the same criteria but were treated without therapeutic hypothermia in other institutions. Clinical benefit were evaluated by a comparison of clinical result in the two groups defined according to the Glasgow Outcome Scale six months after injury. The hypothermia group was divided into two groups based on a target temperature [mild hypothermia group: 32∼34 °C (n = 10); very mild hypothermia group: 35∼36 °C (n = 10)]. The complication rate, clinical results and the duration of therapeutic hypothermia were analyzed between two groups.

In the hypothermia group, 12 patients obtained a favorable outcome (Good Recovery or Moderate Disabled in GOS) and the mortality rate was 35%. In the control group, however only 5 patients had a favorable outcome and the mortality rate was 57%. Comparison between mild hypothermia and very mild hypothermia groups revealed no difference in clinical outcome. In the hypothermia group, severe pneumonia was seen in three patients, all in the mild hypothermia group with a hypothermic duration of over 120 hours.

Mild hypothermia should be ended within 120 hours to avoid severe complication. When long-lasting therapeutic hypothermia of more than 120 hours is planned, very mild hypothermia is the treatment of choice.

Keywords: Hypothermia; complication; clinical outcome; severe head injury.

Introduction

The mechanism of cerebral damage by mechanical injury and by ischemia, have much in common [1]. Many experimental studies have reported that hypothermia could reduce cerebral damage by either ischemia or mechanical injury. However, in large clinical studies, therapeutic hypothermia reported improved the patient outcome after cardiac arrest [2, 3], and substantial clinical benefit was not proven for the patient with severe head injury treated by therapeutic hypothermia in a recent large clinical study [4]. Early induction and longer-term therapeutic hypothermia were expected to assure a more favorable outcome for patients with severe head injury treated with therapeutic hypothermia [5, 6]. On the other hand, therapeutic hypothermia of such long duration could mean a higher complication rate that will undermine clinical benefits [7]. In the present study, we investigated the appropriate temperature and the duration of therapeutic hypothermia from the standpoint of therapeutic effects and possible complications.

Clinical materials and methods

From Jan 1999 to Dec 1999, sixty five patients with severe head injury (Glasgow Coma Scale score of 8 or less) admitted to our six institutions. The hypothermia group consisted of twenty out of sixty five patients who were enrolled in this prospective study based on our indications. Another forty-five patients were excluded for following reasons: 1) GCS score of 8; 2) older than 60 years of age; or 3) systolic blood pressure (SBP) below 100 mmHg. In each case, informed consent to perform therapeutic hypothermia was obtained from the patient's family.

The target temperature and the duration of therapeutic hypothermia were decided individually according to the patient condition. The patients were all treated according to the guideline for the management of severe head injury [8]. In the emergency room, they were initially intubated under deep sedation. We immediately cooled patients to the target temperature using nasogastric lavage with iced water and the cooling blanket. Surgical procedure was performed for the patients with intracranial hematoma. Continuous infusions of a paralytic drug (vecuronium bromide) and intravenous sedation (midazolam) until rewarming were recommended and performed in most cases. In surface cooling, the target temperature was maintained for 72 to 384 hours (mean 141 h) depending on the patient condition. Patients were then slowly rewarmed (approximately 1 °C/day).

Twenty-one sequential patients who met our criteria and had been treated in other institutions where therapeutic hypothermia was not performed were enrolled as a control group, retrospectively. Clinical benefits of therapeutic hypothermia were then evaluated as follows by a comparison of clinical outcome in the hypothermia and control groups defined according to the Glasgow Outcome Scale (GOS) six months after injury; 1, death; 2, vegetative state; 3, severe disability – unable to live independently; 4, moderate disability – capable of living independently but unable to return to work or school; and 5, mild or no disability – able to return to work or school. Patients with a GOS score of 4 or 5 were classified as having a favorable outcome, and those with a GOS score of 1, 2, or 3 as having an unfavorable outcome.

The hypothermia group was then divided into two groups based on the range of temperature control [mild hypothermia group: 32∼34 °C (n = 10); very mild hypothermia group: 35∼36 °C (n = 10)]. In this study, uncontrollable hypotension; mean blood pressure (MBP) less than 90 mmHg, bradycardia; heart rate (HR) less than 60/min, severe pneumonia; PaO_2/FiO_2 less than 200, and occurred until the end of rewarming were evaluated as the complications associated with therapeutic hypothermia. Complication rate, clinical result (GOS) and the duration of therapeutic hypothermia were analyzed in the two groups. The duration of therapeutic hypothermia included both the target temperature period and also the rewarming period.

All values are expressed as the mean ± SD. Baseline characteristics, complications and outcome in the two groups were compared using chi-square tests, Fisher's exact tests or Mann-Whitney U-test, as appropriate. Significance was assigned when the probability value was less than 0.05.

Results

Clinical characteristics of patients in the hypothermia group and control group are summarized in Table 1. The two groups did not differ significantly in age, initial GCS score, or number of the patients with craniotomy. The hypothermia group consisted of more patients with midline shift over 5 mm or compressed basal cistern in the head CT scan taken on admission, but there was no statistically significant difference (p = 0.09).

Table 1. *Clinical characteristics of hypothermia and normothermia group*

Characteristic	HT	Cont
Age (yrs): mean ± SD	28.9 ± 15.5	33.6 ± 16.6
GCS: mean ± SD	5.0 ± 1.4	4.7 ± 1.3
Midline shift (%)	12/20 (60%)	7/21 (33%)
Surgery (%)	6/20 (30%)	6/21 (29%)

HT Hypothermia group, *Cont* Control group, *GCS* Glasgow Coma Scale on admission, *Midline shift* patients with midline shift over 5 mm or compressed basal cistern in the head CT scan taken on admission, *Surgery* patients with craniotomy.

Table 2. *Glasgow Outcome Scale scores 6 months after injury in two treatment groups*

GOS	HT (n = 20)	Cont (n = 21)
Good recovery	9	1
Moderate disability	2	4
Severe disability	2	3
Vegetative state	0	1
Death	7	12

HT Hypothermia group, *Cont* Control group.

Table 3. *Clinical characteristics of mild and very mild hypothermia group*

Characteristic	m HT (n = 10)	vm HT (n = 10)
Age (yrs): mean ± SD	27.4 ± 10.1	30.4 ± 13.7
GCS: mean ± SD	4.5 ± 1.4	5.5 ± 1.3
Midline shift (%)	6/10 (60%)	6/10 (60%)
Surgery (%)	4/10 (40%)	2/10 (20%)

m HT Mild hypothermia group (32∼34 °C), *vm HT* Very mild hypothermia group (35∼36 °C), *GCS* Glasgow Coma Scale on admission, *Midline shift* patients with midline shift over 5 mm or compressed basal cistern in the head CT scan taken on admission, *Surgery* patients with craniotomy.

Mean duration from injury to the achievement of the target temperature was 8.4 h. In 12 cases, we achieved target temperature within 6 hours after injury. The body temperature was equal to or less than 35 degrees on admission in 4 cases.

Six months after injury, in the hypothermia group, 55% of the patients showed a favorable clinical outcome (GCS of 4 or 5), while in the control group only 24% of the patients had favorable clinical outcome (Table 2). Clinical outcome of the hypothermia group was significantly better than that of the control group (p = 0.02).

The hypothermia group was then divided into two groups by a target treatment temperature [mild hypothermia group: 32∼34 °C (n = 10), very mild hypothermia group: 35∼36 °C (n = 10)] (Table 3). The two groups did not differ significantly in age, initial GCS score, or number of patients with craniotomy. The mild hypothermia group consisted of more serious cases than the very mild hypothermia group, but there was no statistically significant difference (p = 0.12). In the mild hypothermia group, 30% of the patients had a favorable clinical outcome (GOS of 4 or 5) and in the very mild hypothermia group, 80% of the patients showed favorable clinical outcome. Lower temperature seems to be selected for more serious cases.

Table 4. *Glasgow Outcome Scale scores 6 months after injury and complication rate in the four treatment groups*

Outcome and complication	120 h≧	120 h<
m HT: Favorable Outcome	3/6 (50%)	0/4 (0%)
Complication rate	0/6 (0%)	3/4 (75%)
vm HT: Favorable Outcome	2/2 (100%)	6/8 (75%)
Complication rate	0/2 (0%)	0/8 (0%)

m HT Mild hypothermia group (32∼34 °C), *vm HT* Very mild hypothermia group (35∼36 °C).

Each group was divided according to the duration of hypothermia (Table 4). Patients treated with mild hypothermia over 120 hours had an extremely high complication rate. Three patients suffered from severe pneumonia, and one from acute renal failure. No patient died of these severe complications, but no patients treated with mild hypothermia for over 120 hours had favorable outcome. Very mild hypothermia group had no severe complication in spite of long duration of hypothermia and achieved good clinical outcome that could not reach statistical significant difference ($p = 0.13$).

Discussion

A recent large-scale randomized controlled study conducted by Dr. Clifton *et al.* indicated that therapeutic hypothermia for severe head injury had no clinical effect compared to normothermia [4]. The mean intracranial pressure (ICP) was within the normal limit in both groups. Their target temperature was 33 °C and mean duration of hypothermia was 66 hours with rapid rewarming (approximately 4 °C/18 h). They reported 43% that of the patients in both the hypothermia and normo-thermia group had a favorable outcome.

Our control group may have poor clinical result because patients with GCS of 8 were excluded and may been given inappropriate temperature management.

In our hypothermia group, although ICP was not measured in all cases, 60% of patients had a midline shift over 5 mm or compressed basal cistern in the head CT scan on admission, and 30% of the patients had surgical lesions. These findings suggested our hypothermia group could include higher ICP patients [9]. Our patients were exposed for 141 hours, then slowly rewarmed (approximately 1 °C/24 h). Dr. Shiozaki *et al.* treated patients with low intracranial pressure with a target temperature of 34 °C for 120 hours, the same as in our study, and concluded that there was no therapeutic effect [7]. These findings suggest that the therapeutic hypothermia had a clinical effect only for the patient with increased ICP and even very mild hypothermia yielded sufficient clinical effect depending on the case.

If we included all cases of pneumonia or electrolytes, 50% of the patients given very mild hypothermia had complications and 70% of those treated with mild hypothermia, similar to findings in other studies [4, 7]. In this study, we evaluated only severe, uncontrollable complications because of their clinical importance.

It is well known that lower temperature and longer duration of therapeutic hypothermia should bring more neuro-protective effect for the patients. On the other hand, such lower temperature and longer duration result in complications that will undermine clinical benefits. Our data suggested that mild hypothermia should not be prolonged more than 120 hours when we treated the patient with severe head injury by therapeutic hypothermia.

Conclusions

Mild hypothermia should not be prolonged for more than 120 hours or the complication rate will increase and the patient outcome cannot be improved.

Appendix

Centers for Nagoya Therapeutic Hypothermia Study Group: Kainan Hospital, Kariya Hospital, Kasugai Municipal Hospital, Shizuoka Saiseikai Hospital, Tajimi Prefectural Hospital, Toyohashi Municipal Hospital.

References

1. Teasdale GM, Graham DI (1998) Craniocerebral trauma: protection and retrieval of the neuronal population after injury. Neurosurgery 43: 723–738
2. Holzer M (2002) Mild therapeutic hypothermia to improve the neurologic outcome after cardiac arrest. N Engl J Med 346: 549–556
3. Bernard SA, Gray TW (2002) Treatment of comatose survivors of out-of-hospital cardiac arrest with induced hypothermia. N Engl J Med 346: 557–563
4. Clifton GL, Miller ER (2001) Lack or effect of induction of hypothermia after acute brain injury. N Engl J Med 344: 556–563
5. Hayashi S, Inao S, Takayasu M (2002) Effect of early induction of hypothermia on severe head injury. Acta Neurochir (Wien) 81: 83–84
6. Jiang J, Yu M, Zhu C (2000) Effect of long-term mild hypothermia therapy in patients with severe traumatic brain injury: 1-year follow-up review of 87 cases. J Neurosurg 93: 546–549

7. Shiozaki T (2001) A multicenter prospective randomized controlled trial of the efficacy of mild hypothermia for severely head injured patients with low intracranial pressure. J Neurosurg 94: 50–54
8. Bullock R, Chesnut RM, Clifton G (1996) Guideline for the management of severe head injury. J Neurotrauma 13: 643–734
9. Eisenberg HM, Gary HE (1990) Initial CT findings in 753 patients with severe head injury. A report from the NIH Traumatic Coma Data Bank. J Neurosurg 73: 688–698

Correspondence: Hayashi Shigemasa, Department of Neurosurgery, Nagoya First Red Cross Hospital, 3-35 Michishita-cho Nakamura-ku 453-8511 Nagoya, Japan. e-mail: hayashi1216@yahoo.co.jp

Acta Neurochir (2005) [Suppl] 95: 273–275

The influence of mild hypothermia on ICP, CPP and outcome in patients with primary and secondary brain injury

M. Smrčka[1], **M. Vidlák**[1], **K. Máca**[1], **V. Smrčka**[1], and **R. Gál**[2]

[1] Neurosurgical Department, University Hospital Brno, Brno, Czech Republic
[2] Anaesthesiological Department, University Hospital Brno, Brno, Czech Republic

Summary

Aim of this study was to examine the hypothesis that only a subgroup of patients with lesser primary brain damage after severe head injury may benefit from therapeutic hypothermia.

We prospectively analysed 72 patients with severe head injury, randomized into groups with (n = 37) and without (n = 35) hypothermia of 34 °C maintained for 72 hours. The influence of hypothermia on ICP, CPP and neurological outcome was analysed in the context of the extent of primary brain damage.

Patients with normothermia and primary lesions (n = 17) – values: GCS on admission 5 (median), ICP 18.9 (mean), CPP 73 (mean), GOS 4 (median). Patients with normothermia and extracerebral hematomas (n = 20): GCS 4, ICP 16, CPP 71, GOS 3. Patients with hypothermia and primary lesions (n = 21): GCS 4,62, ICP 10,81, CPP 78,1, GOS 4. Patients with hypothermia and extracerebral hematomas (n = 14): GCS 5, ICP 13.2, CPP 78, GOS 5.

Hypothermia decreased ICP and increased CPP regardless of the type of brain injury. Hypothermia was not able to improve outcome in patients with primary brain lesions but this pilot study suggests that it significantly improves outcome in patients with extracerebral hematomas.

Keywords: Severe head injury; primary and secondary brain damage; mild hypothermia.

Introduction

Severe head injuries (Glasgow Coma Scale – GCS ≤ 8) generally have a very serious prognosis. About 30% of patients have a good result, but 25% are severely disabled, 5% end up in a vegetative state and about 40% of patients die [6]. Most important for the final result is the extent of primary brain damage and the development of secondary ischemic brain damage. Modern methods of the management of patients after severe head injury are therefore based on the principle of maintaining an adequate cerebral perfusion pressure (CPP) above 60 mmHg (see European and American guidelines) [3, 4]. Despite all effort, secondary ischemic changes may continue to evolve in many patients after severe head injury. It seems that neither the early evacuation of the hematoma, nor the correct management on the intensive care unit is able to protect many patients from secondary ischemic brain damage. One option to improve the results is the use of mild hypothermia. Hypothermia is not a new neuroprotective method, it has been in use already for several decades. Contemporary technology, however, has improved. Hypothermia now is used in accordance with modern knowledge of the pathophysiology of brain injury. In the clinical setting it is necessary to use hypothermia together with multimodal monitoring of the brain physiology, which was not available previously.

The laboratory results after application of mild hypothermia (30–34 °C) confirm less neuronal damage, decreased release of neurotransmitters and a prevention of blood-brain barrier breakdown as a consequence of ischemic insult, which is the most frequent cause of secondary brain damage [2].

Not all experimental and clinical studies, however, confirm a positive effect of hypothermia. Despite the fact that hypothermia influences physiological variables important for the patient s prognosis (intracranial pressure – ICP and cerebral perfusion pressure – CPP), the results of clinical studies are often controversial [1, 5].

This study was supported by a grant of the Internal Grant Agency of the Czech Ministry of Health No 6844-3 and 7671-3

Material and method

We studied 72 patients after severe head injury (GCS ≤ 8) treated in our hospital during the years 2001–2003. Their mean age was 41 years and there were 51 males and 21 females. Patients older 60 years of age and those with severe primary brain damage with no possibility to survive (bilateral mydriasis, no reflexes above C1) were excluded.

Patients with significant hematomas were urgently operated on. The intraparenchymal probe for ICP monitoring (Codman, Lelocle, Switzerland), jugular bulb oxymetry probe (Edwards Lifesciences LLC, Irvine, USA) and invasive blood pressure monitoring (a.radialis) were instituted. The patients were randomized into two groups: with hypothermia (n = 35) and without hypothermia (n = 37). There were no statistical age or sex differences between the groups. In those randomized for hypothermia, mild hypothermia (34 °C) was started as soon as possible for a period of 72 hours by means of surface cooling (Blanketrol II, Cincinnati, USA). Central body temperature was measured in the urinary bladder. After 72 hours the patients were passively warmed-up to reach normothermia. All patients were cooled down within 15 hours after injury and we were able to reach the desired temperature within 3 hours after initiation of cooling. Intensive care for these patients was otherwise performed according to the international standards for severe head injuries [2, 3].

The observer who had access only to the CT scans divided this group into those with dominant primary brain injury (diffuse injury, contusions) (n = 38) and with dominant extracerebral hematoma (21 patients with subdural and 13 patients with epidural hematoma) (n = 34). Finally we evaluated the influence of mild hypothermia on ICP, CPP and $SvjO_2$ (jugular bulb oxygen saturation) and Glasgow Outcome Scale (GOS) 6 months after trauma taking into account the dominant type of injury (primary lesions versus extracerebral hematomas). The results are expressed as mean and standard deviation. We tested the conformity of scatters by F-test and we used an unpaired t-test to compare the differences between the groups.

Results

The results are presented in Tables 1–4.

Discussion

Our study differed from many studies using controlled hypothermia in severe head injuries by the fact that we used a milder degree of hypothermia (34 °C) for a relatively longer time (72 hours). This temperature is more easily reached in the clinical setting and is accompanied by no or minimal side effects with a similar influence on ICP and CPP. With the use of multimodal monitoring, this method can also be safely used in patients with multiple trauma.

Jugular bulb oximetry monitoring, in particular, is advantageous during therapeutic hypothermia because it can give a warning regarding a global decrease of cerebral metabolism.

Hypothermia did not influence the systemic blood pressure. We observed only 2 episodes of bradycardia below 40/min which reacted well to atropine. The occurrence of pneumonia did not differ significantly between the groups.

Mild hypothermia of 34 °C was able to decrease ICP and improve CPP in hypothermia patients (although

Table 1. *Results of all patients after severe head injury (hypothermia of 34 °C for 72 hours versus normothermia)*

	Normothermia (n = 37)	Hypothermia (n = 35)	
MABP (mmHg)	123.11 ± 5	123.45 ± 4	p = 0.9013
CPP (mmHg)	72.16 ± 4	78.23 ± 6	p < 0.0001
ICP (mmHg)	17.65 ± 7	11.77 ± 5	p < 0.0001
$SvjO_2$ (%)	71.12 ± 4	69.05 ± 4	p = 0.2436
GCS	4.27 ± 1.24	4.57 ± 1.27	p = 0.3115
GOS	3.16 ± 1.69	4.09 ± 1.36	p = 0.0131

MABP Mean arterial blood pressure; *CPP* cerebral perfusion pressure; *ICP* intracranial pressure; *$SvjO_2$* jugular bulb oxygen saturation; *GCS* Glasgow Coma Scale; *GOS* Glasgow Outcome Score.

Table 2. *Results of patients with primary brain injury*

	Normothermia (n = 17)	Hypothermia (n = 21)	
MABP (mmHg)	129.83 ± 10	123.44 ± 14	p = 0.3692
CPP (mmHg)	73.71 ± 5	78.10 ± 6	p = 0.0240
ICP (mmHg)	18.88 ± 6	10.81 ± 5	p < 0.0001
GCS	4.59 ± 1.37	4.62 ± 1.24	p = 0.9426
GOS	3.47 ± 1.74	3.71 ± 1.62	p = 0.4470

Table 3. *Results of patients with brain compression through extracerebral hematoma*

	Normothermia (n = 20)	Hypothermia (n = 14)	
MABP (mmHg)	118.67 ± 11	123.33 ± 11	p = 0.4402
CPP (mmHg)	70.85 ± 2	78.43 ± 6	p < 0.0001
ICP (mmHg)	16.60 ± 7	13.21 ± 5	p = 0.1077
GCS	4.00 ± 1.08	4.50 ± 1.34	p = 0.2377
GOS	2.90 ± 1.65	4.64 ± 0.50	p = 0.0006

Table 4. *Morbidity and mortality of patients according to GOS (mean 6 months after injury)*

GOS	Normothermia (n = 37)	Hypothermia (n = 35)	Total (n = 72)
5 (good)	13 (35%)	18 (51%)	31 (43%)
4 (moderate disability)	5 (13.5%)	12 (34%)	17 (24%)
3 (severe disability)	5 (13.5%)	0	5 (7%)
2 (vegetative state)	3 (8%)	0	3 (4%)
1 (death)	11 (30%)	5 (15%)	16 (22%)

the decrease of ICP in the extracerebral hematoma group was not significant) which correlates with previous studies [1, 5]. The outcome (GOS) was significantly better in the hypothermic group of patients with extracerebral hematomas but did not differ in patients with primary brain injury.

Hypothermia has basically three important roles in influencing brain physiology. First it lowers ICP, second it improves CPP, and third it has a direct neuroprotective effect on cerebral neurons. We consider the use of hypothermia in patients with brain compression as very well-founded since these patients are threatened by the development of a secondary ischemic brain damage. Hypothermia may help these patients not only by its effect on ICP and CPP but also by its neuroprotective effect. This might be why there were better results in patients with extracerebral hematomas. The very good results in this group were the reason why the hypothermia group had better results even when evaluated regardless of the type of injury. In patients with primary brain injury, hypothermia may affect mainly ICP and CPP. Brain ischemia does not play such an important role in this type of injury (with some exceptions) and therefore the neuroprotective effect of hypothermia in primary injuries does not influence pathophysiology that much.

This does not mean, however, that hypothermia in primary head injury is useless and should be neglected. We think that in these patients, hypothermia should be used in cases with otherwise intractable intracranial hypertension.

References

1. Clifton GL, Miller ER, Choi SC *et al* (2001) Lack of effect of induction of hypothermia after acute brain injury. N Engl J Med 344(8): 556–563
2. Clark RS, Kochanek PM, Marion DW, Schiding JK *et al* (1996) Mild posttraumatic hypothermia reduces mortality after severe controlled cortical impact in rats. J Cereb Blood Flow Metab 16(2): 253–261
3. Journal of Neurotrauma, Volume 13, No. 11, 1996; 639–734, ISSN: 0897-7151
4. Maas AIR, Dearden M, Teasdale GM *et al* on behalf on the European Brain Injury Consortium (1997) EBIC – Guidelines for management of severe head injury in adults. Acta Neurochir (Wien) 139: 286–294
5. Marion DW, Penrod LE, Kelsey SF *et al* (1997) Treatment of traumatic brain injury with moderate hypothermia. N Engl J Med Feb 20: 336(8): 540–546
6. Quigley MR, Vidovich D, Cantella D *et al* (1997) Defining the limits of survivorship after very severe head injury. J Trauma 42(1): 7–10

Correspondence: Martin Smrčka, Neurosurgical Department, University Hospital Brno, Jihlavská 20, 625 00, Brno, Czech Republic. e-mail: msmrcka@med.muni.cz

Acta Neurochir (2005) [Suppl] 95: 277–280

Contribution of raised ICP and hypotension to CPP reduction in severe brain injury: correlation to outcome

A. Marmarou, A. Saad, G. Aygok, and **M. Rigsbee**

Department of Neurosurgery, Virginia Commonwealth University Medical Center, Richmond, VA, USA

Summary

The aim of this study was to determine to what degree hypotension and ICP contribute to the reduction of cerebral perfusion pressure (CPP), particularly in light of the shift in emphasis to CPP management by the use of pressors. The study population consisted of severely head injured patients extracted retrospectively from the Traumatic Coma Data Bank and compared with 139 patients from the Smith Kline component of the American Brain Injury Consortium database where outcome was available. The percentage time that ICP exceeded 20 mm Hg and CPP less than 60 mm Hg was computed for 5 days post injury. At each hour when CPP was less than 60 mm Hg the contribution of raised ICP and low arterial pressure or both was determined. In the first cohort, hypotension was the predominant factor leading to CPP reduction. With use of the CPP concept of treatment, the major contribution to CPP shifted to ICP and arterial hypotension played less of a role. Overall, CPP management has been associated with improved outcome.

Keywords: Cerebral perfusion pressure; hypotension; head injury.

Introduction

The cerebral perfusion pressure (CPP) is the pressure gradient across the capillary bed and is considered the difference between mean systemic arterial pressure and intracranial pressure [14]. The maintenance of adequate CPP and low ICP remains as a key goal of head injured patient management. In normal brain, cerebral autoregulation fails and cerebral blood flow begins to fall when CPP is less than 50 mm Hg. In injured regions of the brain, when focal tissue pressure may be higher than the mean ICP, a CPP of 50 mm Hg may be too low for optimal perfusion. Therefore, recent guidelines have posed that CPP should be maintained above 60 mm Hg to assure adequate cerebral perfusion, especially in the injured brain [6]. The cerebral perfusion pressure necessary to maintain brain functions varies with the state of metabolism of the brain. In normal brains, CPP may fall as low as 20–30 torr without evidence of neurophysiological dysfunction, although blood flow falls to 50% of normal [5]. With regard to ICP, there is clear evidence that in trauma patients, a sustained ICP of greater than 20 mm Hg is associated with a significantly worse clinical outcome [7, 9, 11]. However, it is not clear to what extent raised ICP and hypotension are each responsible for reduced CPP. The objective of this study was to examine the relative contribution of ICP and hypotension to reductions in CPP during the first 5 days post injury and to determine which factor, raised ICP or hypotension, is most responsible for CPP reduction during the acute course of management. A second objective was to study outcome as assessed by the GOS and determine the relationship of GOS to the specific cause of CPP reduction.

Methods

The population consisted of 387 severely head injured patients (admission GCS of 8 or less with ICP and blood pressure monitoring) extracted retrospectively from the combination of traumatic coma data bank (TCDB) and subsequent patients documented by the MCV Neurocore database. [Total n = 387; males = 296; females 91; age 5 mos – 87 years, mean age 30.1 ± 18.05 years. The mean GCS is 5.4 and the median GCS is 6.0] The percentage time CPP was less than 60 mm Hg was computed for 5 days post injury. For the time that CPP was less than 60 mm Hg, the contribution of raised ICP (>20 mm Hg), low mean arterial pressure (<80 mm Hg) or both was determined based on hourly measures of these parameters recorded by the bedside nurse and was correlated to the outcome.

Results

Of the entire cohort of patients, 30.3% had mass lesions that required surgical removal, and 58.8% had diffuse brain injury. Eighty percent of our patients experienced elevated ICP (>20 mm Hg) at some point during the first five days post trauma. We studied the cerebral perfusion pressure (CPP) of these patients and looked for its effects on their outcome. The percentage time CPP was less than 60 mm Hg was computed for five days post injury. For the time CPP was less than 60 mm Hg, the contribution of raised ICP alone, low mean arterial blood pressure (MAP) alone, or both was determined based on hourly measures of these parameters.

Of the 387 severe head injury patients, 286 (73.9%) experienced CPP reduction in the first five days post injury. Of the patients with CPP reduction, 121 (42.3%) suffered these reductions due to reduced MAP alone, while 34 (11.9%) suffered these reduction due to high ICP alone, and 131 (45.8%) suffered CPP reductions due to both factors together. It is evident from the table that the predominant factor accounting for CPP reduction during the first five days post-admission, in terms of number of patients, is hypotension and not ICP elevation. Concerning the daily incidence of CPP reduction in the first five days post injury, we found that 50.6% of patients experienced CPP reduction in the first day, then this percentage reduced to 37–41% in the next 3 days and slightly increased on the fifth day to 42%.

Contribution of ICP and hypotension to CPP reduction

With regard to the percent responsibility of CPP reduction factors for end hour readings CPP < 60 mm Hg (Fig. 1), MAP reduction alone was the most frequent cause of CPP reduction especially in the first four days (56.5% of reduced CPP readings). Also MAP reduction, either alone or in combination with ICP elevation, was involved in 80% of CPP reduction events especially in the first four days. High ICP alone was the cause of CPP reduction in 20% of instances of CPP reduction. ICP rise either alone or in combination with MAP reduction was the cause of CPP reduction in about 42% of times in the first four days and 60% of the time on the fifth day.

We also studied how the percent contribution of CPP reduction factors to CPP reduction is affected at different levels of CPP. We found that as CPP is reduced, high ICP gradually becomes the major contributor to CPP reduction. In patients who suffered CPP reductions below 40 mm Hg, high ICP was contributing to 80–90% of CPP reduction events. Furthermore,

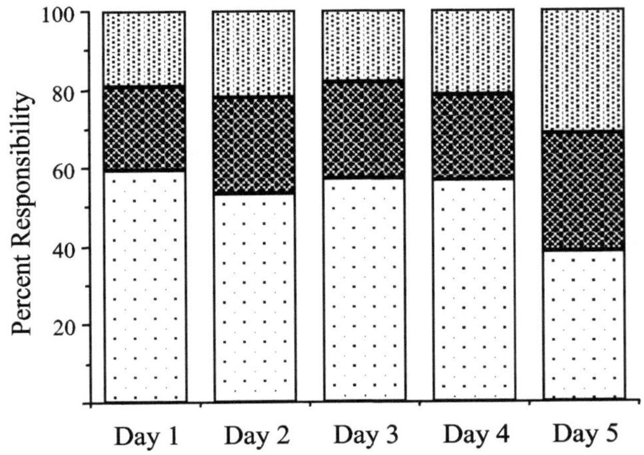

Fig. 1. Percent responsibility of ICP elevation alone, MAP reduction alone, and both factors together for CPP reduction in the first five days in severely head injured patients. It appears that MAP reduction alone is the most common cause of CPP reduction in the first 4 days (around 56.5% of reduced CPP readings). Also MAP reduction, either alone or in combination with ICP rise, was present in about 80% of instances of CPP reduction in the first 4 days and in about 60% in the 5th day (the lower two segments of the columns). While ICP elevation, either alone or in combination with MAP reduction, was present in these instances in about 42% in the first 4 days and 60% in the 5th day (the upper two segments of the columns). ▨ ICP (>20) Alone; ▩ Both; □ MAP (<80) Alone

as CPP decreased below 60 mm Hg, the average value of ICP increased while that of MAP remained nearly unchanged.

The effect of reduced CPP on outcome

On correlating the percent time CPP < 60 mm Hg in the first five days to GOS outcome we observed that survivors suffered significantly shorter periods of CPP reduction (17%) than those who died (45%) {$p < 0.001$}. We concluded that percent time CPP < 60 mm Hg did not correlate with outcome in the form of (Good/Mod/Sev/Veg), but only defined mortality. On the other hand, as CPP decreased below 70 mm Hg, the outcome gradually became worse. We found that outcome in severe head injury patients was more dependent on the lowest level of CPP they suffered than the percentage time below 60 mm Hg. Moreover, in the CPP reduction patients, we found that patients who showed good outcome were those who suffered reductions mainly due to reduced MAP alone. Also we observed that the greater the contribution of high ICP either alone or in concomitant combination with reduced MAP to CPP reduction the worse the outcome.

Table 1. *Grouping patients with CPP < 60 mm Hg according to which factor was responsile for CPP reduction, either ICP > 20 mm Hg alone, MAP < 80 mm Hg alone, or both factors in the first five days and correlating that to prognosis*

	Total	G/M	S/V	Dead	# of end hours/day CPP < 60
ICP > 20 alone	34	9 (26.5%)	12 (35.3%)	13 (38.2%)	3.8 hrs
			73.5%		
MAP < 80 alone	121	48 (39.6%)	59 (48.8%)	14 (11.6%)	4.5 hrs
			60.5%		
Both	131	31 (23.8%)	55 (41.3%)	45 (34.9%)	6.7 hrs
			76.2%		
Total	286	88 (30.8%)	126 (44.1%)	72 (25.1%)	5.8 hrs
			69.2%		

CPP non-reduction group

Total	G/M	S/V	Dead
101	38 (37.6%)	50 (49.5%)	13 (12.9%)
		62.4%	

In Table 1 we show that the outcome was worse when the cause of CPP reduction was a result of both ICP elevation and MAP reduction. Also the table shows that patients who suffered these CPP reductions due to MAP reduction alone, did better than the other two groups. The data also shows that patients in the "Both" group suffered these CPP reduction periods more frequently (6.7 hrs/day) than the other two groups. The overall outcome in patients who did not suffer CPP reductions was better than those who suffered these reductions. Severely head injured patients who did not experience CPP reduction resulted in a 37.6% favorable outcome with only 12.9% mortality. On the other hand, only 30.8% of severely head injured patients who suffered CPP reduction showed favorable outcome, while 69.2% of them showed unfavorable outcome with 25.1% mortality.

The analysis of the relationship of hypotension (%time MAP < 80 mm Hg day one to five) and outcome in patients with no CPP reduction resulted in no clear relation. On the other hand, when we studied this relation for high ICP (%time ICP > 20 mm Hg), we found that the longer the peroids of high ICP the worse the outcome.

The appropriate threshold for CPP

Based on these data we attempted to clarify whether a CPP of 60 mm Hg is an appropriate treatment threshold as posited from the recent change in guidelines. We compared the outcome in patients of CPP threshold above 70 mm Hg to those of CPP threshold of 61–70. There was a generally improved outcome (GOS Good/Mod) with CPP threshold above 70 mm Hg but this did not reach statistical significance. However there was a statistically significant reduction in severely disabled and vegetative patients with CPP > 70 mm Hg.

Discussion

There is agreement among investigators in the field of head trauma that improvement in outcome could be achieved by control of ICP, blood pressure and CPP, but still there is controversy regarding the importance of the role played by each of these factors. Eighty percent of our patients experienced elevated ICP (>20 mm Hg) in the first five days post trauma. Marmarou and Co-workers [8] reported an incidence of 72% high ICP in their severe head injury population in the first 72 hours post trauma. The incidence in our study is markedly higher than the 50% incidence reported by Nordby and Co-workers [11]. We also found that % ICP > 20 mm Hg correlates well with the outcome independent of CPP; the longer the periods of high ICP the worse the outcome. This agrees with what has been reported by many others [7, 9, 11].

Of the 387 severe head injury patients, 286 (73.9%) experienced CPP reduction (<60 mm Hg). O'Sullivan and Co-workers [12] reported an incidence of 62.5% in their small series. In our series, the most common cause for CPP reduction is hypotension (MAP < 80 mm Hg). It was the only cause of this reduction in 56.5% of instances and was present in about 80% of these instances either alone or in combination with elevated ICP. CPP reduction less often was caused by elevated ICP. This is consistent with others who reported that reduction of CPP down to 40 mm Hg was more often caused by decrease in arterial blood pressure than increase in ICP [1]. It is obvious in our study that, as CPP is decreased, high ICP gradually becomes the major contributor to CPP reduction.

We found that percent time CPP < 60 mm Hg does not correlate with outcome in the form of (Good/Mod/Sev/Veg), but only effects mortality. In our study there was a statistically significant difference in %time CPP < 60 mm Hg between survivors and dead, ($p < 0.001$), however we could not find a significant difference between different outcome groups among survivors. This is consistent with Rosner and Rosner [13] who reported that, with CPP management mortality has been reduced to 50% compared to traditional treatment with probably better morbidity. Duthie and colleagues [2] also stated that, episodes of arterial hypotension are associated with increased mortality in head injury patients.

In addition, in patients who suffered CPP reduction, we found that the lower the minimum CPP level they experienced, the worse the outcome. This is consistent with that reported by Mendelow and Co-workers [10]. We also believe that a minimum CPP of 70 mm Hg should be maintained in severely head injured patients despite the fact that there was only a trend toward better outcome. Rosner and Rosner [13] also suggested a minimum CPP of ≥ 70 mm Hg. However, we did not find any correlation between duration of hypotension (%time MAP < 80 mm Hg) and outcome. This is in contradiction to that reported by Grande and Co-workers [3]. Also, Jones and Co-workers [4] reported that the most significant predictor for mortality and outcome in severe head injury patients is the duration of hypotensive insults.

Conclusions

1. ICP elevation was observed in 80% and CPP reduction in 73.9% of head injured patients.
2. Reduced MAP was the main contriibutor (80%) to CPP reduction below 60 mm Hg, however, this reduction was poorly correlated to the outcome.
3. GOS was more clearly related to the severity of CPP reduction than the percent time below 60 mm Hg. This applied for a single event of CPP reduction.
4. As CPP is reduced further, the relative contribution of ICP was greater than that of low MAP.
5. The percent time CPP below 60 mm Hg effected mortality but was similar between Good/Mod and Sev/Veg groups.
6. In patients with no CPP reduction, outcome is correlated to the percent time that ICP is above 20 mm Hg.
7. CPP above 70 mm Hg was slightly better than CPP > 60 mm Hg.

References

1. Czosnyka M, Price DJ, Williamson M (1994) Monitoring of cerebrospinal dynamics using continuous analysis of intracranial pressure and cerebral perfusion pressure in head injury. Acta Neurochir (Wien) 126(2–4): 113–119
2. Duthie SE, Goulin GD, Zornow MH, *et al* (1994) Effects of THAM and sodium bicarbonate on intracranial pressure and mean arterial pressure in an animal model of focal cerebral injury. J Neurosurg Anesthesiol 6(3): 201–208
3. Grande PO, Asgeirsson B, Nordstrom CH (1993) A new potential therapy for treatment of post-tramatic brain edema based on haemodynamic principles for brain volume regulation. In: Nakamura N, Hashimoto T, Yasue M (eds) Recent advances in neurotraumatology. Springer, Berlin
4. Jones PA, Andrews PJ (1994) Measuring the burden of secondary insults in head injured-patients during intensive care. J Neurosurg Anesthesiol 6(1): 4–14
5. Lassen NA (1974) Control of the cerebral circulation in health and disease. Cric Res 34: 746–760
6. Marion DW, Spiegel TP (2000) Changes in the management of severe raumatic braininjury: 1991–1997. CritCare Med 28(1): 16–18
7. Marmarou A, Anderson RL, Ward JD, *et al* (1991) Impact of ICP instability and hypertension onoutcome in patients with severe head trauma. J Neurosurg 75[Suppl]: S59–S66
8. Marmarou A, Anderson RL, Ward JD, *et al* (1991) NINDS traumatic coma data bank: intracranial pressure monitoring methodology. J Neurosurgery 75[Suppl]: S21–S27
9. Marshall LF, Marshall SB, Klauber MR, *et al* (1991) A new classification of head injury based on computerized tomography. J Neurosurg 75[Suppl]: S14–S20
10. Mendelow AD, Allcutt DA, Chabers IR, *et al* (1993) Intracranial and cerebral perfusion pressure monitoring in the head injured patients: which index? In: Nakamura N, Hashimoto T, Yasue M (eds) Recent advances in neurotraumatology. Springer, Berlin
11. Nordby HK, Gunnerod N (1985) Epidural monitoring of the intracranial pressure in severe head injury characterized by nonlocalizing motor responses. Acta Neurochir 74: 21–26
12. O'Sullvian MG, Statham PF, Miller JD, *et al* (1994) Role of intracranial pressure monitoring in severely head-injured patients without signs of intracranial hypertension on initial computerized tomography. J Neurosurg 80: 46–50
13. Rosner MJ, Rosner SD (1993) Cerebral perfusion pressure management in head injury. In: Nakamura N, Hashimoto T, Yasue M (eds) Recent advances in neurotraumatology. Springer, Berlin
14. Schweitzer JS, Bergsneider M, Becker DP (1994) Intracranial pressure monitoring. In: Cottrell JE, Smith DS (eds) Anesthesia and neurosurgery, 3rd ed. Mosby, Chicago, pp 117–135

Correspondence: Anthony Marmarou, Department of Neurosurgery, Virginia Commonwealth University Medical Center, 1001 East Broad Street, Suite 235, Richmond, VA, 23219, USA. e-mail: amarmarou@vcu.edu

Acta Neurochir (2005) [Suppl] 95: 281–285

Mortality from traumatic brain injury

J. Lu[1], **A. Marmarou**[1], **S. Choi**[1], **A. Maas**[2], **G. Murray**[3], and **E. W. Steyerberg**[2]
the Impact and Abic Study Group

[1] Department of Neurosurgery, Virginia Commonwealth University Medical Center, Richmond, VA, USA
[2] Department of Neurosurgery, Erasmus University Medical Center Rotterdam, The Netherlands
[3] Department of Community Health Sciences, University of Edinburgh, Edinburgh, Scotland, UK

Summary

It is the general sense that mortality has been decreasing in recent years compared to earlier studies described by the NIH traumatic coma data bank. We studied mortality during the period of 1984 to 1996 to determine if indeed mortality from severe traumatic brain injury was decreasing and to identify factors which might account for the reduction. The study population (N = 1839) consisted of severely head injured patients extracted retrospectively from the TCDB (635), MCV (382), and 822 patients from clinical trial databases conducted in the United States. Mortality was obtained from each of the databases for the age range form 16 to 65. Penetrating injury and treatment groups in the clinical trial databases were excluded. Mortality in the year 1984 equaled 39% and gradually decreased to a level of 27% in 1996. When adjusting for age, motor score and pupil reaction, the mortality of the period from 1984 to 1987 was significantly higher ($p < 0.05$) than that of the period 1988 to 1996.

During the period 1984 through 1996, mortality from severe brain injury steadily declined. Factors other than age, motor score and pupil reactivity over time are responsible for this reduction. This reduction over time is an important factor for prognostic modeling of TBI.

Keywords: Traumatic brain injury; clinical trial; outcome from traumatic brain injury.

Introduction

Traumatic brain injury is associated with a significant mortality and morbidity despite intensive therapy. In the late 70's, investigators established a standardized protocol for management of severe head injury with emphasis of early diagnosis and evacuation of intracranial mass lesions (Becker, Miller *et al.* 1977). In addition, this protocol included artificial ventilation, control of increased intracranial pressure and aggressive therapy in a specialized intensive care unit. The mortality from traumatic brain injury (TBI) under these conditions averaged 30% with overall poor outcome (S/V/D) of about 40%. It was proposed that vigorous surgical and medical therapy, by preventing secondary insults contributed to the outcome. It was generally assumed that mortality has steadily declined over the past decade although no definitive studies have documented this reduction. The objective of this study was to examine the mortality and to determine if there was indeed reduction of mortality and identify risk factors, which might account for the reduction. The study was based on several TBI database.

Methods

Study population

The data for the study were from Traumatic Brain Injury Data Bank (TCDB), Medical College of Virginia/Virginia Commonwealth University (MCV/VCU), Steling Winthrop PEG-Superoxide Dismutase Clinical Trial (PEGSOD), Ciba-Geigy Selfotel Clinical Trial part of U.S. potion (CIBA) and SmithKline Beecham Bradykinin Antagonist Clinical Trial (SKB). A total of 1839 patients with an age range of 16 to 65 years and GCS 3 to 8 with no penetrating injury were included in this study.

TCDB was founded and supported by National Institute of Neurological Disorders and Stroke (NINDS), a total of 1030 severe head injury patients were collected for the study during 1984 to 1987. A total of 163 patients with penetrating injury were excluded and 76 patients dead on arrival were also excluded. The remaining 635 patients with an age range from 16 to 65 years old were included in this study.

MCV continued the patient survey after the end of the TCDB study. During 1988 to 1992, 482 severe head trauma patients were treated at MCV hospital. A total of 382 patients with the age range of 16–65 were included in this study and 65 patients with penetrating injury were excluded.

Steling Winthrop Inc. conducted two phase III multi-center studies to evaluate the safety and effectiveness of PEG-Superoxide Dismutase on 1562 severe closed head injury patients during 1992 to 1995 (trial 005 with 463 patients, and trial 006 with 1099 patients). A total of 911 patients were assigned to the treatment arms and excluded from this analysis, the remaining 624 patients with a matched age range as compared with other studies were included in this study.

Ciba-Geigy Selfotel trial was a multi-center, randomized, double blinded trial on intubated patients with severe closed traumatic brain injury. The study inclusion criteria were GCS 4-8 or GCS 3 with at least one reactive pupil, and CT classification II to VI. A total 266 patients were enrolled in the U.S. and Israel during 1995 (American Brain Injury Consortium Database). After exclusion of those patients who received the treatment of the study drug, 133 patients with an age range of 16–65 years were included in this study. The remaining data for this trial were not available at the time when this study was conducted.

SKB Bradykinin Antagonist Trial was a multi-center, double blinded, and placebo-controlled study, which was conducted during 1996 in the United States. A total of 139 patients with severe head injury and at least one reactive pupil with GCS 3 were included in the clinical trial. Sixty-nine patients were randomized to the treatment arm, and the remaining 65 placebo patients with age 16–65 were included in this analysis.

Statistical analysis

As stated, the primary variable in this study outcome was the mortality rate of the patient population from 1984 to 1996. In the study, the effect of the following covariables on the mortality were examined: age, gender, race, cause of injury, place of injury, admission hours, neurological evaluation at admission, opening intracranial pressure, blood pressure, and certain characteristics of the admission CT examination.

Two-sided Pearson's Chi-Square tests were performed on the unordered categorical variables and two-sided Wilcoxon Rank Sum tests were used on the ordered categorical variables; and two-sided Student-T tests were applied on the continues variables. Logistic regression model was used for comparing the differences of the mortalities and the adjustment for covariables such as age, motor scores and pupillary responses. The statistical software SAS (version 8.2) was used for the data analysis.

Results

Study population

From the time period of 1984 to 1997, 635 severe head injury patients with the age range of 16 to 65 years old were selected from TCDB; from the period of 1988 to 1996, 1204 patients with severe brain trauma and the same age range were selected from MCV/VCU hospital, the randomized clinical trials of PEG-SOD, CIBA U.S. and SKB during 1988 to 1992, 1992 to 1995, 1995 and 1996, respectively. Those patients who were randomized to the treatment arm from the clinical trials or injured from gun short were excluded from this analysis. The total sample size is 1839.

Demographic characteristics

The demographic background of the study population is show in Table 1. Both study samples from the time period of 1984 to 1987 and 1988 to 1996 have a similar age distribution, and the largest population falls into the young adolescence group with an age range of 20 to 30 years old. Male is seen dominant in both periods, with 79.5 and 77.9 percent, respectively. In comparison with the TCDB data, the data collected from MCV/VCU hospital and other clinical trials showed over 40 percent increase in the Black population and a 15 percent decrease in the Caucasian population, and the Chi-Square Test showed significant difference with p-value less than 0.0001.

The data showed that the cause of head injury has changed significantly over the years, with 35 percent increase in automobile accidents and more than 60 percent increase in assault between the period of 1984 to 1987 and that of 1988 to 1996 (Chi-Square Test, $p < 0.0001$). Within the same time periods, however, the incident rate of fall to the severe head injury was relatively stable. In an earlier time period, TCDB recorded a large proportion of patients injured by motorcycle and work related injuries (table 1, in the category of others), which was not specified in the clinical trails during a later period. Although the information about injury place was only specified from the TCDB and MCV data, the increased incidents of severe head trauma on the street and highway in a later time period seemed consistent with the increase in automobile accidents during the same time period, which is also statistically significant (Chi-Square Test, $p < 0.05$).

The shortened interval from the time of injury to the time of study hospital admission was statistically significant (Student-T Test, $p < 0.0001$), when the TCDB data (a mean of 2.52 hours) was compared with the MCV and other clinical trial data (a mean of 1.82 hours).

Admission status

Table 2 illustrates certain patient clinical conditions upon arrival at the study hospital. Patients from TCDB showed a slightly higher percentage of no motor responses, while patients from the MCV hospital and other clinical trials showed a relative higher percentage of motor responses such as withdrawal or normal flexion, which did not reach the statistical significance (Wilcoxon Rank Sum Test, $p = 0.14$). With regard to the pupillary response, the patients from both periods had a similar rate of bilateral normal pupillary response; the patients from TCDB had a slightly higher rate of bilateral abnormal pupillary response and the patients from the MCV hospital and

Table 1. *Demographic characteristics*

Characteristic	Total Cohort	1984–1987		1988–1996		P-Value
		N	%	N	%	
Mean age (yrs)	1839	30.03 ± 12.16		30.60 ± 12.16		0.34[a]
Age group						
– 16–<20	360	122	19.21	238	19.77	0.289[b]
– 20–<30	691	262	41.26	429	35.63	
– 30–<40	396	124	19.53	272	22.59	
– 40–<50	212	62	9.76	150	12.46	
– 50–<60	120	43	6.77	77	6.40	
– 60–65	60	22	3.46	38	3.16	
Gender						
– Male	1444	505	79.53	939	77.99	0.445[c]
– Female	395	130	20.47	265	22.01	
Race						
– Caucasian	1244	528	84.75	716	71.46	<0.0001[c]
– Black	306	92	14.77	214	21.36	
– Asian/Oriental	54	0	0.00	54	5.39	
– Other	21	3	0.48	18	1.80	
Cause of injury						
– Automobile accident	1078	344	54.60	734	73.69	<0.0001[c]
– Fall	205	87	13.81	118	11.85	
– Assault	113	32	5.08	81	8.13	
– Other	230	167	26.51	63	6.33	
Injury place						
– Street/High way	797	487	77.18	310	84.93	0.028[c]
– Home	58	38	6.02	20	5.48	
– Work/School	51	35	5.55	16	4.38	
– Recreational	29	23	3.65	6	1.64	
– Other	61	48	7.61	13	3.56	
Time from injury to admission (hrs)	1835	2.52 ± 3.20		1.82 ± 1.82		<0.0001[a]

[a] Student-T test, [b] Wilcoxon Rank-Sum Test, [c] Chi-Square Test.

Table 2. *Admission status*

Characteristic	Total Cohort	1984–1987		1988–1996		P-Value
		N	%	N	%	
Motor score						
– None	286	112	18.76	174	15.17	<0.135[b]
– Extensor	287	100	16.75	187	16.30	
– Abnormal flexion	237	86	14.41	151	13.16	
– Withdraw/Normal flexion	446	129	21.61	317	27.64	
– Localize	425	151	25.29	274	23.89	
– Obeys Commands	63	19	3.18	44	3.84	
Papillary response						
– Both side normal	885	244	62.72	641	61.22	0.918[b]
– One side abnormal	142	24	6.17	118	11.27	
– Both side abnormal	409	121	31.11	288	27.51	
Mean systolic blood pressure	1814	140.16 ± 31.44		141.53 ± 29.76		0.270[a]
Mean artery blood pressure	1698	104.94 ± 20.56		102.72 ± 21.68		0.041[a]
Mean opening intracranial pressure	735	19.22 ± 19.01		17.75 ± 14.38		0.205[a]
Time from injury to start ICP monitoring (hrs)	656	6.90 ± 4.38		7.35 ± 4.25		0.190[a]

[a] Student-T test, [b] Wilcoxon Rank-Sum Test, [c] Chi-Square Test.

Table 3. *Radiological examinations*

Characteristic	Total Cohort	1984–1987		1988–1996		P-Value
		N	%	N	%	
Cisterns						
– Absent/Compressed	375	206	35.64	169	33.67	0.497[c]
– Present	705	372	64.36	333	66.33	
Subarachnoid hemorrhage						
– Yes	547	125	21.04	422	37.44	<0.0001[c]
– No	1174	469	78.96	705	62.56	

[a] Student-T test, [b] Wilcoxon Rank-Sum Test, [c] Chi-Square Test.

other clinical trial centers had a small increase in the rate of a unilateral abnormal pupillary response, although the difference is not statistically significant.

There was no significant difference in mean systolic blood pressure between the patients from TBCD and the rest of trial centers. However, the difference in mean artery blood pressure was statistically significant (Student-T Test, $p < 0.05$).

The data from TCDB, MCV hospital and SKB trial included the information on the opening intracranial pressure (ICP). There was no significant difference in mean opening ICP between the two ($P > 0.05$), nor was there difference in the mean interval between the time of injury and the time of starting ICP monitoring ($P > 0.05$).

CT database

Table 3 demonstrates the admission CT diagnoses on cisterns and traumatic subarachnoid hemorrhage. The cisterns on CT images were specified from all data resources except for the PEGSOD trail. There was no significant difference in the presence of cisterns between the patients from TCDB and those from the MCV hospital and other clinical trial centers. However, the significant difference was seen in traumatic subarachnoid hemorrhage (TSAH) between these two general patients populations. The patients from the MCV hospital, SKB, PEGSOD, and CIBA trails had a significant higher rate of TSAH than the TCDB patients (Chi-Square Test, $p < 0.0001$).

Mortality

Of 1839 patients with severe head injury patients who were selected in this study, 526 patients died and the mortality rate was 28.6 percent for the entire cohort. Figure 1 showed the yearly mortality rate from 1984 to 1996 over the study population. From 1984 to 1996, the mortality rate of severe head injury patients was reduced from 39 percent to 27 percent. In particular, a sharp reduction after 1987 is noted in the figure.

Figure 2 showed a significant difference in the head injury mortality rate between the time period of 1984 to 1987 (37%) and 1988 to 1996 (24%) (Chi-Square Test, $p < 0.0001$). After the factors of patient age, admission motor score and pupillary response were adjusted, the difference of the mortalities between these two periods remained significant ($p < 0.05$). In a separated logistic regression model, after controlling covariances of age, gender, race, cause of injury, admission pupillary response and motor score, the significant difference of the mortalities between these two periods persisted ($p < 0.05$).

Discussion

This study shows that mortality from severe brain injury has reduced significantly from 39% in 1984 to 27% in 1996, representing close to a 31% reduction in death. More recent studies reporting mortality are consistent with this level of reduction and this phenomenon occurred in the absence of a neuroprotective pharmacologic agent administered during this period or a dramatic change in surgical management of head injury. What then is the cause of this reduction? It is noted that factors of patient age, admission motor score and pupillary response were adjusted for co-variance and yet the difference of the mortalities between these two periods remained significant ($p < 0.05$). Moreover, it is considered that age, motor score and pupil reactivity are strong predictors of outcome [2–5] and as these factors were not statistically different between the two periods. Further, it is difficult to argue that the reduction was due to devices such as seat belts or air bags. These effectively have re-

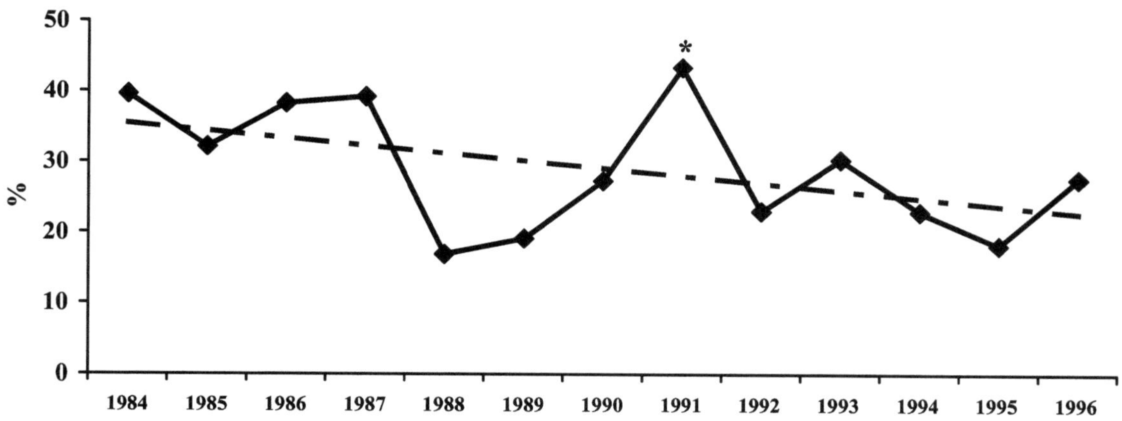

Fig. 1. Mortality from 1984 to 1996 (N = 1839). *MCV hospital only, 46 patients

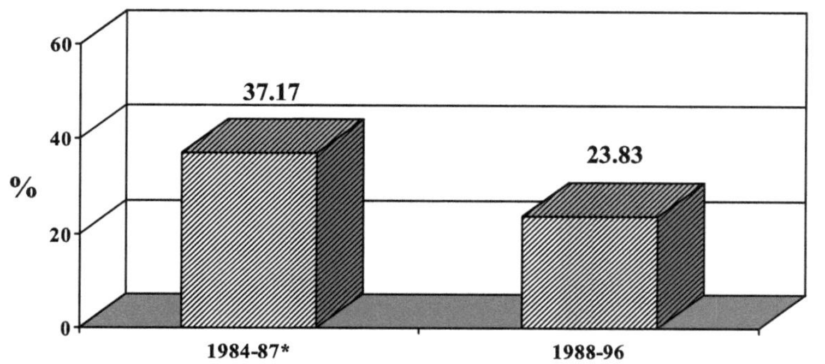

Fig. 2. Severe head injury mortality (N = 1837). *$p < 0.0001$; $p < 0.05$, adjusting for age, motor score, papillary response

duced the incidence of TBI but have not affected the severity of injury of those patients admitted to the hospital.

We hypothesize that there are at least two main factors resulting in a reduction of mortality. The factors include a more rapid and effective emergency medical service resulting in a reduction of secondary insults such as hypoxia and hypotension [6] and a more aggressive treatment of cerebral perfusion pressure [7–9]. The difference in mortality in two periods presented in this paper should be tested in the studies conducted after 1996.

Acknowledgment

This work was supported by the National Institutes of Health Grants NIH 2R01 NS19235, NIH P50NS12587 and NIH R01 NS042691.

References

1. Becker DP, Miller JD, Ward JD, Greenberg RP, Young HF, Sakalas Rl (1977) The outcome from severe head injury with early diagnosis and intensive management. J Neurosurg 47(4): 491–502
2. Choi SC, Muizelaar JP, Barnes TY, Marmarou A, Brooks DM, Young HF (1991) Prediction tree for severely head-injured patients. J Neurosurg 75(2): 251–255
3. Choi SC, Narayan RK, Anderson RL, Ward JD (1988) Enhanced specificity of prognosis in severe head injury. J Neurosurg 69(3): 381–385
4. Choi SC, Ward JD, Becker DP (1983) Chart for outcome prediction in severe head injury. J Neurosurg 59(2): 294–297
5. Hukkelhoven CW, Steyerberg EW, Rampen AJ, Farace E, Habbema JD, Marshall LF, Murray GD, Maas AI (2003) Patient age and outcome following severe traumatic brain injury: an analysis of 5600 patients. J Neurosurg 99(4): 666–673
6. Klauber MR, Marshall LF, Toole BM, Knowlton SL, Bowers SA (1985) Cause of decline in head-injury mortality rate in San Diego County, California. J Neurosurg 62(4): 528–531
7. Rosner MJ (1996) Role of cerebral perfusion pressure in acute brain trauma. Crit Care Med 24(7): 1274, author reply 1275–1276
8. Rosner MJ, Daughton S (1990) Cerebral perfusion pressure management in head injury. J Trauma 30(8): 933–940; discussion 940–941
9. Rosner MJ, Rosner SD, Johnson AH (1995) Cerebral perfusion pressure: management protocol and clinical results. J Neurosurg 83(6): 949–962

Correspondence: Anthony Marmarou, Department of Neurosurgery, Virginia Commonwealth University Medical Center, 1001 East Broad Street, Suite 235, Richmond, VA, 23219, U.S.A. e-mail: amarmarou@vcu.edu

Brain compliance

Acta Neurochir (2005) [Suppl] 95: 289–292

Assessment of different data representations and averaging methods on the Spiegelberg compliance device

Y. H. Yau[1], I. R. Piper[2], C. Contant[3], L. Dunn[2], and I. R. Whittle[1] on behalf of *the Brain IT Group*

[1] Department of Clinical Neurosciences, Western General Hospital, Edinburgh, UK
[2] Institute of Neurological Sciences, Southern General Hospital, Glasgow, UK
[3] Department of Neurosurgery, Baylor College of Medicine, Houston, USA

Summary

The Spiegelberg Compliance Device (Spiegelberg KG, Hamburg, Germany)[1] has been available for the automated measurement and calculation of minute by minute intracranial compliance. Widespread practical use has been somewhat limited by the instability of values; especially at low intracranial pressures.

We looked at two aspects of a methodology in an attempt to increase the value of the Spiegelberg device in the clinical setting. Firstly, we discussed the difference in representing measured values as elastance (dp/dv) instead of compliance (dv/dp); and secondly we proposed the use of an averaging algorithm called the Exponentially Weighted Moving Average (*ewma*), which could be applied as a flexible method to follow trends and rapid changes in the elastance (or compliance).

Clinical data from sixteen patients were gathered and statistical analysis was focused on three particular aspects, the coefficient of variation which indicates the variability of data values, the correlation between the elastance (or compliance) time series and the underlying ICP signal and the percentage of outliers greater than 2.5 standard deviations from the mean. Our results showed that expressing elastance (dp/dv) instead of compliance (dv/dp) yielded fewer outliers and had a better correlation to ICP, and the *ewma* method had a better correlation to ICP than the Spiegelberg method.

Keywords: Elastance; compliance; averaging methods; brain IT; Spiegelberg compliance device.

Introduction

The value of volume-pressure parameters in the context of changing intracranial pressure has been well established [3, 4]. Recent introduction of the Spiegelberg Compliance Device has provided possible benefit of having a continuous measurement of the volume-pressure parameters [6, 10]. Technical aspects of the device have been described previously [1, 7–9, 11] and the lack of uptake of the device by clinicians has been attributed to instability of measurements and therefore indeterminate practical usefulness [2].

We aimed to use early clinical data to assess theoretical concerns over the methodology of the Spiegelberg Compliance Device with a view to improving practicality. The aims of this study were:

1. to compare elastance (dp/dv) with compliance (dv/dp) in optimising signal to noise ratio of data; and
2. to improve the algorithm for averaging raw dp and dv values generated by the Spiegelberg compliance device (*ewma*).

Compliance vs. elastance

The issue of compliance (dv/dp) versus elastance (dp/dv), is best seen by examining graphs of the relationship of each measure to a change in dp. The two graphs depicted in Fig. 1 illustrate the relationship between compliance, elastance and changes in dp.

As can be quickly seen, the compliance rises rapidly as the value of dp drops, but the slope of the elastance curve is constant across all values of dp. One advantage of compliance is that it is very sensitive to changes from high compliance to lower compliance. However, when the ICP is low, the compliance can become extremely unstable. Additionally, when the compliance is low, large changes in dp result in small changes in compliance. From the theoretical point of view, there are clear problems with compliance. Elastance, on the other hand, has a constant slope across the values of dp, so that it is an interval measure. That is, the change in elastance between 0.2 and 0.25 has the same meaning as the change between 1.5 and 1.55. Clearly, this is not the case with the compliance measure. Also, as an

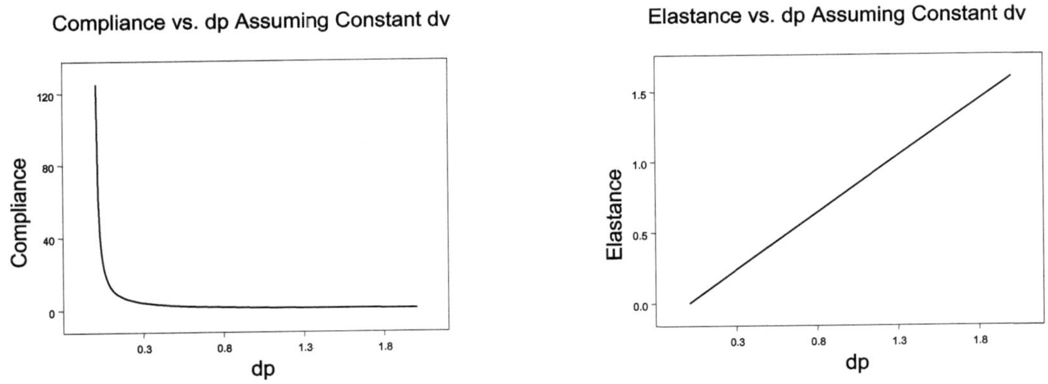

Fig. 1

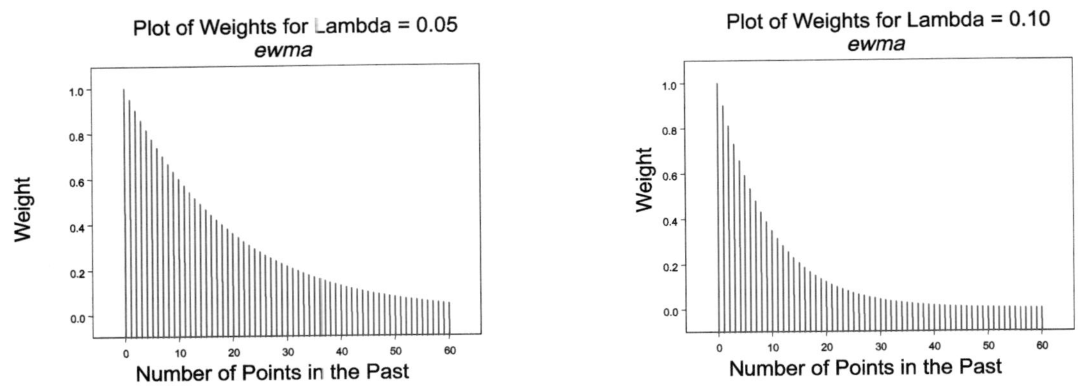

Fig. 2

interval measure, selection of a "cutpoint" becomes more meaningful. Finally, the direction of change in the measured parameter is also important. An increase in elastance indicates a "worse" patient state (like increased ICP), which is an "algorithm" many clinicians are already familiar with. Overall, there appears to be stronger arguments in favour of moving towards a display of "elastance" rather than "compliance".

Exponentially-weighted moving average

The current compliance measure from the Spiegelberg device is calculated by a "rolling" average of up to 200 pulses. The average of *dv* is calculated, and divided by the average of *dp*. Each point in the average is given equal weight, so that the average calculated at time *t* is actually centered at $t - \frac{1}{2}$(*interval length*).

The exponentially-weighted moving average (*ewma*) is calculated as follows:

$$\mu_t = \lambda X_t + (1 - \lambda)\mu_{t-1}$$

Where μ_t is the value of the *ewma* at time *t*. Now, if one expands μ_t by replacing μ_{t-1} with its value $\mu_{t-1} = \lambda X_{t-1} + (1 - \lambda)\mu_{t-2}$, and continuing the process back as many times as desired, the series is found to be equivalent to:

$$\mu_t = \frac{\lambda X_t \left(\sum_{k=2}^{r} (1 - \lambda)^k X_{t-k-1} \right)}{\sum_{k=2}^{r} (1 - \lambda)^k}$$

Therefore, the weight for each value in the past is $(1 - \lambda)^k$. The graphs in Fig. 2 illustrate these weights for a λ of 0.05 and for 0.10.

The larger the value of λ, the more rapidly the weights drop to zero. In one sense, this can be thought of as defining the "memory" of the *ewma*. Where the weights become small quickly, the system has a "short" memory, for slowly decreasing weights, the system has a "long" memory.

One of the advantages of the *ewma* is that the value calculated at time *t* is most heavily weighted by the recent values of the process, even for "long" memory *ewma's*. Smoothing still occurs, but there is more emphasis on the most recent data.

Materials and methods

Sixteen patients underwent continuous intraventricular monitoring using the Spiegelberg compliance device were evaluated. There were 11 males and 5 females (median age 58 years with range of 26 to 77 years): 2 postoperative following tumour excision, 2 after subarachnoid haemorrhage and the remaining 12 with hydrocephalus. A special software version of the compliance monitor captured the raw volume additions *dv* and ICP responses *dp* to yield data sets for analysis. An analysis program was developed to allow comparison of the coefficient of variation (CV), percentage of outliers exceeding 2.5 standard deviations from the mean (%Outliers) and the time-series correlation ($x_n - x_{n-1} \ldots$) with the ICP signal (Corr ICP). Comparison of the standard Spiegelberg averaging method with the *ewma* method, which gives exponentially decreasing weight to the preceding data, was also made.

Results

The findings are summarised in Tables 1 and 2.

Representing changing compliance as elastance (*dp*/*dv*) yields fewer outliers and shows an improved correlation with the underlying ICP than compared with the standard compliance representation (*dv*/*dp*). The *ewma* method for (*dp*/*dv*), although showing higher variability compared to the greater smoothed Spiegelberg method, shows a better correlation to the underlying ICP.

Discussion

The results from this pilot clinical study demonstrate the display of elastance (*dp*/*dv*) as being preferable to that of compliance (*dv*/*dp*) in the monitoring of "intracranial volume decompensation". Use of an "exponentially weighted moving average" (*ewma*) algorithm is feasible and provides a flexible method for following both trends and rapid changes in compliance and elastance. There may be an advantage to changing λ as a function of the ICP. That is, it may be that when the ICP is low, a small value of λ is acceptable as the probable change in elastance or compliance is low, and a very stable value can be used. When the ICP rises, it may be advantageous to increase the value of λ so that the *ewma* for elastance or compliance is more responsive to change. While this modification would somewhat reduce the nice mathematical properties of the *ewma*, the gain by tuning the system to current situation may be overwhelming. The answer to this specific question will have to await further analyses.

Continuing data recruitment is necessary to further define the early results presented above. The concept and infrastructure of the *Brain IT* organisation (www.brainit.org) [5] should provide a basis for quicker collaborative data accrual and validation.

Table 1

Median Values	CV	%Outliers	Corr ICP
dV/dP	182	0.82	−0.009
ewma dV/dP	169	0.83	−0.006
dP/dV	175	0.32	0.087
ewma dP/dV	160	0.27	0.082
Spiegelberg	22	0.35	0.018

Table 2

ANOVA single factor	CV	%Outliers	Corr ICP
ewma dP/dV vs *ewma dV/dP*	$p = 0.65$	$p < 0.001$	$p < 0.001$
ewma dP/dV vs Spiegelberg	$p < 0.01$	$p = 0.20$	$p < 0.05$

Acknowledgments

This work has been supported in part by the Scottish Office Department of Health (Grant K/MRS/50/C2662). Technical collaboration with Spiegelberg company has made data analysis feasible.

References

1. Czosnyka M, Czosnyka Z, Pickard J, *et al* (1997) Laboratory testing of the Spiegelberg brain pressure monitor: A technical report. J Neurol Neurosurg Psychiat 63: 732–735
2. Kiening KL, Schoening WN, Stover JF, Unterberg AW (2003) Continuous monitoring of intracranial compliance after severe head injury: relation to data quality, intracranial pressure and brain tissue PO_2. Br J Neurosurg 17: 340–346
3. Maset AL, Marmarou A, Ward JD, *et al* (1987) Pressure-volume index in head injury. J Neurosurg 67: 832–840
4. Miller JD, Pickard JD (1974) Intracranial volume pressure studies in patients with head injury. Injury 5: 265–268
5. Piper I, Citerio G, Chambers I, Contant C, Enblad P, Fiddes H, Howells T, Kiening K, Nilsson P, Yau YH for the BrainIT Group (2003) The BrainIT Group: Concept and Core Dataset Definition. Acta Neurochir (Wien) 145(8): 615–628
6. Piper I, Dunn L, Contant C, Yau Y, Whittle I, Citerio G, Kiening K, Schoening W, Ng S, Poon W, Enblad P, Nilsson P (2000) Multi-centre assessment of the Spiegelberg compliance monitor: preliminary results. Acta Neurochir [Suppl] 76: 491–494
7. Piper IR, Spiegelberg A, Whittle I, Signorini D, Mascia L (1999) A comparative study of the Spiegelberg compliance device with a manual volume-injection method: a clinical evaluation in patients with hydrocephalus. Br J Neurosurg 13: 581–586
8. Yau YH, Piper IR, Clutton RE, Whittle IR (2000) Experimental

evaluation of the Spiegelberg intracranial pressure and intracranial compliance monitor. Technical note. J Neurosurg 93: 1072–1077
9. Yau YH, Piper IR, Clutton RE, Whittle IR (2002) An experimental evaluation of a new intraparenchymal continuous compliance probe: preliminary studies. Acta Neurochir [Suppl] 81: 181–182
10. Yau Y, Piper I, Contant C, Citerio G, Kiening K, Enblad P, Nilsson P, Ng S, Wasserberg J, Kiefer M, Poon W, Dunn L, Whittle I (2002) Multi-centre assessment of the Spiegelberg compliance monitor: interim results. Acta Neurochir [Suppl] 81: 167–170
11. Yau YH, Piper IR, Contant CF, Dunn LT, Whittle IR (2002) Clinical experience in the use of the Spiegelberg automated compliance device in the assessment of patients with hydrocephalus. Acta Neurochir [Suppl] 81: 171–172

Correspondence: Y. H. Yau, Department of Clinical Neurosciences, Western General Hospital, Edinburgh, EH4 2GA, United Kingdom. e-mail: yhyau@doctors.org.uk

Acta Neurochir (2005) [Suppl] 95: 293–297

Assessment of the relationship between age and continuous intracranial compliance

K. L. Kiening[1], W. Schoening[2], A. W. Unterberg[1], J. F. Stover[3], G. Citerio[4], P. Enblad[5], P. Nilsson[5], and the Brain-IT Group

[1] Department of Neurosurgery, Heidelberg Medical Center, Ruprecht-Karls-University, Heidelberg, Germany
[2] Department of Surgery, Charité, Humboldt-University, Berlin, Germany
[3] Department of Surgery, Division of Surgical Intensive Care Medicine, University Hospital Zuerich, Switzerland
[4] Department of Anesthesiology, Ospedale San Gerardo, Monza, Italy
[5] Department of Neurosurgery, University Hospital, Uppsala, Sweden

Summary

The aim of this open, descriptive and prospective study was to determine if the new monitoring parameter "continuous intracranial compliance (cICC)" decreases with age in patients with traumatic brain injury (TBI).

30 patients with severe and moderate TBI (Glasgow Coma Scale score ≤ 10) contributing to a European multicenter study, organized by the Brain-IT group, underwent computerized monitoring of blood pressure, intracranial pressure (ICP), cerebral perfusion pressure and cICC.

Regression analyses of individual median ICP and median cICC versus patients' age revealed no significant dependency. Median cICC declined significantly with increasing ICP (when median ICP = 10, 20 and 30 mmHg, cICC = 0.64, 0.56 and 0.42 ml/mmHg respectively, $p < 0.05$). These three ICP groups were then subdivided according to age (0–20, 21–40, 41–60 and 61–80 years). Median cICC declined with age in both high ICP groups (median ICP = 20, 30 mmHg). Percentage cICC values below a set pathological threshold of lower than 0.05 ml/mmHg across the four age groups were 28% (0–20 yrs), 59% (21–40 yrs), 60% (41–60 yrs) and 70% (61–80 yrs) respectively.

The observed phenomenon of decreased intracranial volume challenge compensation with advancing age may contribute to the well-known fact of a worse outcome in elderly patients after TBI.

Keywords: Multimodal cerebral monitoring; head injury; intracranial pressure; outcome.

Introduction

The tight functional relationship between increases in intracranial pressure (ICP) and intracranial volume is strongly influenced by the underlying neuropathology and the actual speed of volume expansion within the closed intracranial cavity. Due to the exponential function of the intracranial pressure-volume curve, small increases in intracranial volume can result in massive increases in ICP once compensatory mechanisms have been exhausted as shown experimentally [15] and clinically [10, 11]. The rapid development of hematomas and edema formation in face of a rather late assessment when only measuring ICP predispose to undetected perturbations. Thus, it is thought that continuous assessment of changes in intracranial compliance (ICC) – determined by repeatedly challenging the intracranial compartment with transient injections and withdrawal of a defined volume-reveals ensuing pathological increase in ICP before the ICP is actually increased [9]. This, in turn, could make the ICC technique an early warning system.

The newly developed Aesculap®-Spiegelberg compliance system (ASCS) (Aesculap®, Tuttlingen, Germany) allows for the first time to perform quantitative computerized online measurements of continuous intracranial compliance (cICC), using a fully automated and "closed" external volume load. The ASCS has been shown to be feasible and safe in both experimental [18] as well as clinical settings, in hydrocephalus and TBI research [7, 12, 13, 17, 19].

As observed in a preliminary study in seven patients with severe TBI [6], cICC seemed to be influenced by the age of these patients which may interfere with correct interpretation of the measured cICC data. In the present open, descriptive, and prospective study we focused on the relationship between changes in cICC and age in a multi-center study enrolling thirty TBI patients.

Patients, materials and methods

Demographic characteristics

From 1998 to 2002, a total of thirty patients (3 females/27 males) with closed TBI requiring extended computerized monitoring of ICP and cICC on the intensive care units (ICU) at the Charité, Virchow Medical Center in Berlin, Germany (n = 12), Ospedale San Gerardo in Monza, Italy (n = 9) and the University Hospital in Uppsala, Sweden (n = 9) contributed to a European multi-center study organized by the "Neuro-Intensive Care Monitoring Research Group" ("Brain-IT"; http://www.brainit.org) [12], primarily designed to define the inherent variability of cICC in TBI, subarachnoid hemorrhage, hydrocephalus and brain tumor. Based on the Glasgow coma score (GCS), patients presented with a severe (GCS ≤ 8; n = 26) or moderate (GCS 9–10; n = 4) TBI. All patients with initial moderate TBI deteriorated later, i.e. within 48 hrs post trauma to unconsciousness (GCS ≤ 7) due to progressing contusions, thus making invasive intracranial monitoring indispensable. The underlying intracranial pathology was graded according to the Marshall CT-classification [8] from the "worst" CT scan by the assigned center investigators (Berlin: K. Kiening, Uppsala: P. Enblad, Monza: G. Citerio). Major concomitant injuries were middle facial or limb fractures resulting in a median injury severity score (ISS) of 27 (range: 18–57) [2]. Any space occupying lesions (solid intacerebral contusions/hematoama, epidural/subdural hematomas) with a midline shift more than 5 mm were evacuated immediately (n = 13). Hereby, the skull flap was re-fixed tightly and the dura was sutured "water tight" following subdural procedures to ensure a closed intracranial compartment and thus guarantee valid cICC measurements.

Ethical permission

Permission to enroll patients in the European multi-center study organized by the BrainIT group [12] was granted by the local institutional ethics committees at the Charité (Berlin), Ospedale San Gerardo (Monza) and University Hospital (Uppsala).

General management of patients

General intensive care and specifically the management of pathologically elevated ICP adhered to the "Guidelines for the Management of Severe Traumatic Brain Injury" brought forward by the Brain Trauma Foundation of The American Association of Neurological Surgeons on behalf of the Joint Section on Neurotrauma and Critical Care [4].

Multimodal cerebral monitoring

Invasive mean arterial blood pressure (MAP), ICP, cerebral perfusion pressure (CPP), and cICC were recorded simultaneously using local multimodal computerized cerebral monitoring systems digitizing parameter signals with a frequency of 1/min. Data communication via the Internet between BrainIT data base located in Glasgow, UK and the analyzing center (Berlin) was performed using "WS_ftp Pro-software" (Ipswich, Inc., Lexington, MA). The ASCS catheter was positioned via a conventional frontal burr hole into the frontal horn of lateral ventricles (either left or right). Correct catheter location was assured by CT scan.

Data analyses

After electronical re-transmission of the entire data set of the investigated 30 patients to Berlin, artifacts were removed off-line according to specific remarks documented by nurses and physicians. Thus, a total of 150, 407 minutes artifact-free monitoring time was achieved. From this dataset, individual median ICP and cICC were calculated. 9,133 minutes were taken to determine age-related changes in cICC at distinct ICP levels (ICP = 10, 20, 30 mmHg).

(1) In a first step, regression analyses of individual median ICP and median cICC vs. patient ages were performed followed by **(2)** median cICC determination for each patient at different ICP levels (ICP = 10, 20, 30 mmHg). **(3)** Finally, each ICP group was further subdivided into age blocks (0–20, 21–40, 41–60, 61–80 years) in which median cICC was determined.

Statistics

Results are either given as medians or expressed in box plots showing mean, median, 5th, 25th, 75th, and 95th-percentiles. Sigmaplot 2001 7.0® for Windows (SPSS Science™, Chicago, IL) was used for regression analyses (regression line, 95-confidence interval, Spearman rank correlation coefficient $\{r_s\}$, p value) and the generation of box plots. Differences assessed by ANOVA on ranks (Sigmastat 3.0. Jandel Scientific, Corte Madera, CA) were rated significant at $p < 0.05$.

Results

Median ICP and median cICC did not correlate with age of individual patients (Figs. 1a and b). Median cICC was significantly reduced with increasing ICP in a stepwise fashion (Fig. 2). At ICP 10 mmHg, median cICC was reduced stepwise from 0.64 to 0.42 ml/mmHg at ICP 30 mmHg, which was below the presumed critical cICC of 0.5 ml/mmHg in TBI [7]. In the low ICP group 10 mmHg, the lowest median cICC (0.35 ml/mmHg) was found in the youngest patients (0–20 yrs), which was significantly elevated with increasing age (Fig. 3).

In both high ICP groups at 20 and 30 mmHg, the highest median cICC values were observed in the youngest age group (0–20 years), median $cICC_{ICP=20\,mmHg} = 0.76$ ml/mmHg and median $cICC_{ICP=30\,mmHg} = 0.58$ ml/mmHg, being significantly elevated compared to all other age groups (Fig. 3). By taking the presumed critical cICC threshold of 0.5 ml/mmHg, percentage reduction in this threshold continued through ascending ICP groups of 20, and 30 mmHg with age (0–20 yrs: 28%; 21–40 yrs: 59%; 41–60 yrs: 60%; 61–80 yrs: 70%) (Fig. 1).

Discussion

It is thought that aging is associated with a loss in brain tissue compressibility [16] and/or reduced

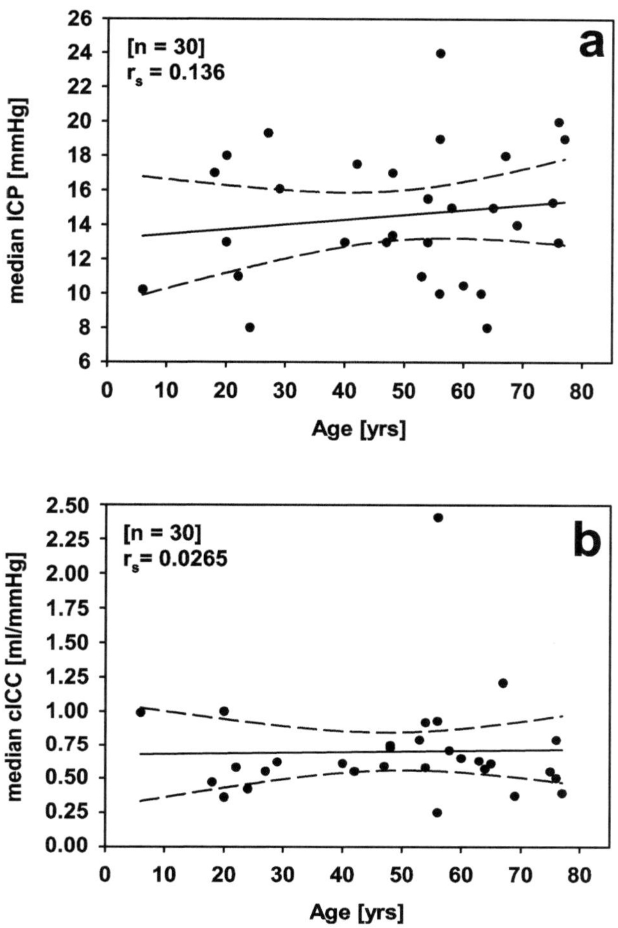

Fig. 1. Linear regression analyses of individual median ICP (a) and median cICC vs. age (b) without significant correlation

cerebro-spinal fluid (CSF) absorption [1, 5], resulting in sustained "stiffness" of the brain and consequently, reduced ICC. Not surprising in line with the hydrocephalus literature [5], ICP does not increase with age in TBI, suggesting that this may be due to an increase pointing to the fact of an increased CSF outflow (R_{csf}) and a simultaneously reduced CSF production (Fig. 1a). Likewise overall cICC does not correlate with age (Fig. 1b). However, by analyzing cICC in more detail (Fig. 3), it became clear that at higher levels of ICP levels 20 and 30 mmHg, the compromised intracranial compliance (Fig. 2), cICC decreases with age, supporting the findings of Czosnyka and colleagues [5] who reported a non-linear increase of elastance coefficient (reciprocal to ICC) with age in hydrocephalic patients. For TBI the observed phenomenon was most pronounced in the 0–20 years age group, where a 28% reduction in cICC below the critical threshold of 0.5 ml/mmHg, which compares favorably to that of the older age group (61–80 years) of 70%. The combination of age and severity of underlying TBI, pushes ICP to higher levels ($\geq$ 20 mmHg), resulting in impaired cICC (Fig. 3). The subsequent increase in resistance of the ageing brain may possibly contribute to a worse outcome in elderly patients [3, 14]. In conclusion ICP monitoring alone may not be sufficient in elderly patients based on the presence cICC threshold data for age.

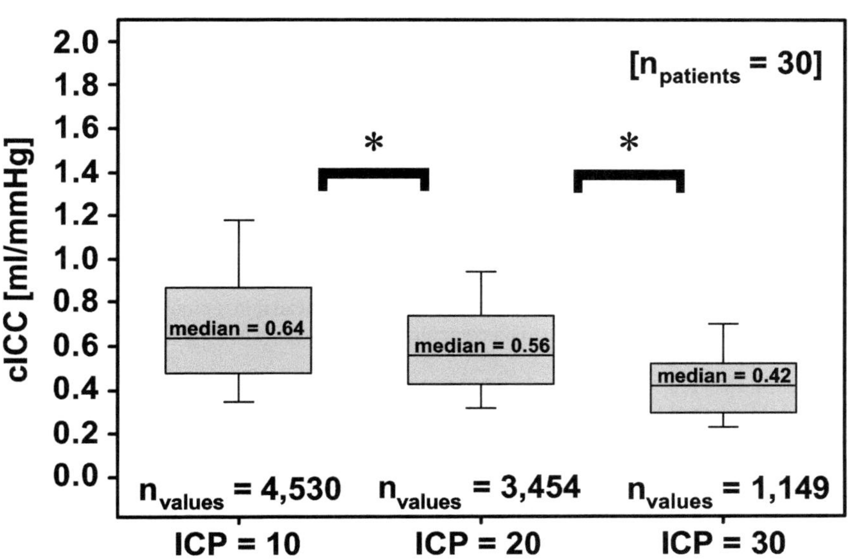

Fig. 2. Changes in continuous intracranial compliance (cICC) determined at three ICP levels (ICP 10, 20, 30 mmHg). With increasing ICP, cICC was significantly decreased (ANOVA on ranks, $p < 0.05$)

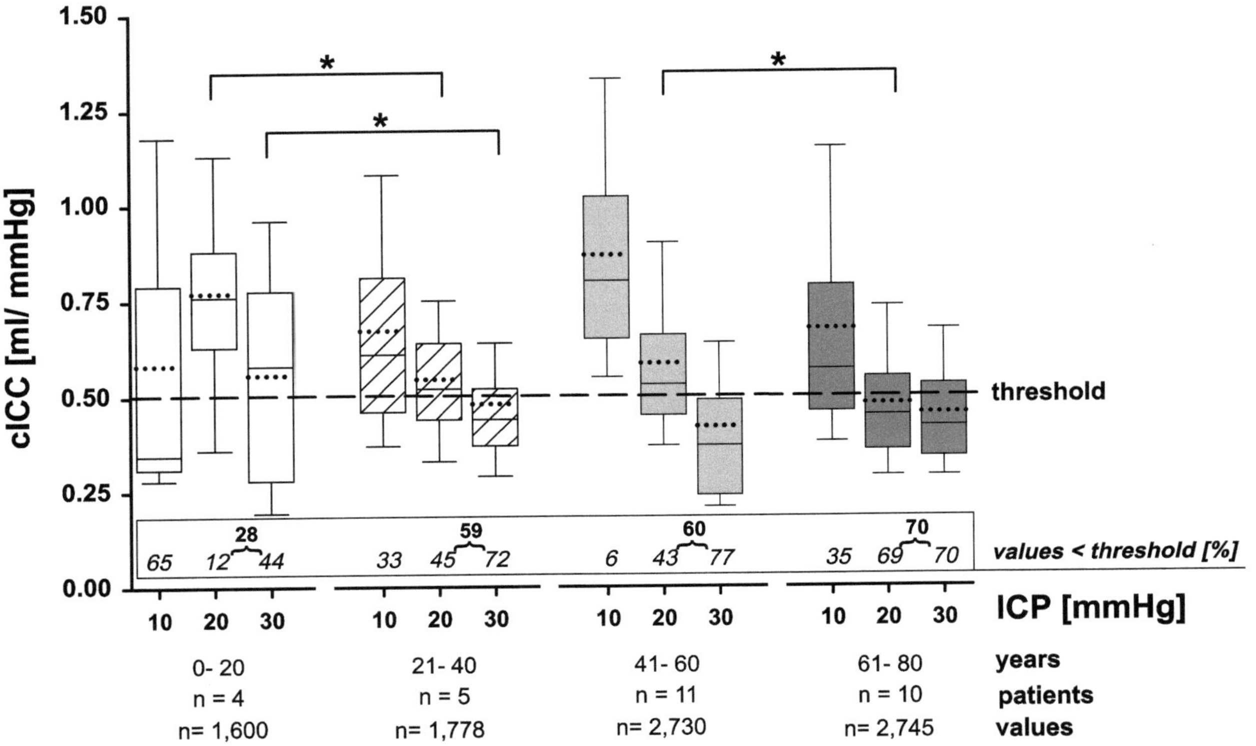

Fig. 3. Distribution of cICC determined at three ICP levels (10, 20, 30 mmHg) for age intervals (0–20, 21–40, 41–60, 61–80 years). At high ICP levels (20 and 30 mmHg), the incidence of pathological cICC levels < 0.5 ml/mmHg expressed in percent to the total number of cICC values determined at each ICP level was significantly increased in patients older than 21 years, being mostly sustained in patients between 61 and 80 years (ANOVA on ranks, $p < 0.05$). The box plots reveal the distribution of the data showing the 5th and 95th percentile (lower and upper error bar), the 25th and 75th percentile (lower and upper rim of the box), the median (solid line), and mean (dotted line). The discongruence of mean and median clearly shows that the data is nonparametric

Acknowledgments

This work was supported by a grant of the "Kuratorium ZNS" (project no.: 99011) and a Charité research grant from the "Humboldt University, Berlin". The Brain-IT group was funded by a grant of the European Union (project number: QLRI 2000-00454).

References

1. Albeck MJ, Skak C, Nielsen PR, Olsen KS, Borgesen SE, Gjerris F (1998) Age dependency of resistance to cerebrospinal fluid outflow. J Neurosurg 89: 275–278
2. Baker SP, O'Neill B, Haddon W Jr, Long WB (1974) The Injury Severity Score: a method for describing patients with multiple injuries and evaluating emergency care. J Trauma 14: 187–196
3. Braakman R, Gelpke GJ, Habbema JD, Maas AI, Minderhoud JM (1980) Systematic selection of prognostic features in patients with severe head injury. Neurosurgery 6: 362–370
4. Brain Trauma Foundation, American Association of Neurological Surgeons, Joint Section on Neurotrauma and Critical Care (2000) Guidelines for the management of severe traumatic brain injury. J Neurotrauma 17: 451–553
5. Czosnyka M, Czosnyka ZH, Whitfield PC, Donovan T, Pickard JD (2001) Age dependence of cerebrospinal pressure-volume compensation in patients with hydrocephalus. J Neurosurg 94: 482–486
6. Kiening KL, Schoening WN, Lanksch WR, Unterberg AW (2002) Intracranial Compliance as bed-side monitoring in severely head-injured patients. Acta Neurochir [Suppl] 81: 177–180
7. Kiening KL, Schoening WN, Stover JF, Unterberg AW (2003) Continuous monitoring of intracranial compliance after severe head injury: relation to data quality, intracranial prerssure and brain tissue PO_2. Br J Neurosurg 17: 340–346
8. Marshall LF, Marshall SB, Klauber MR, Van Berkum Clark M, Eisenberg H, Jane JA, Luerssen TG, Marmarou A, Foulkes MA (1992) The diagnosis of head injury requires a classification based on computed axial tomography. J Neurotrauma 9: S287–292
9. Maset AL, Marmarou A, Ward JD, Choi S, Lutz HA, Brooks D, Moulton RJ, DeSalles A, Muizelaar JP, Turner H, *et al* (1987) Pressure-volume index in head injury. J Neurosurg 67: 832–840
10. Miller JD, Leech P (1975) Effects of mannitol and steroid therapy on intracranial volume-pressure relationships in patients. J Neurosurg 42: 274–281
11. Miller JD, Pickard JD (1974) Intracranial volume pressure studies in patients with head injury. Injury 5: 265–268
12. Piper I, Dunn L, Contant C, Yau Y, Whittle I, Citerio G, Kien-

ing K, Schoening W, Ng S, Poon W, Enblad P, Nilsson P (2000) Multi-centre assessment of the Spiegelberg compliance monitor: preliminary results. Acta Neurochir [Suppl] 76: 491–494
13. Piper IR, Spiegelberg A, Whittle I, Signorini D, Mascia L (1999) A comparative study of the Spiegelberg compliance device with a manual volume-injection method: a clinical evaluation in patients with hydrocephalus. Br J Neurosurg 13: 581–586
14. Signorini DF, Andrews PJ, Jones PA, Wardlaw JM, Miller JD (1999) Predicting survival using simple clinical variables: a case study in traumatic brain injury. J Neurol Neurosurg Psychiat 66: 20–25
15. Sullivan HG, Miller JD, Becker DP, Flora RE, Allen GA (1977) The physiological basis of intracranial pressure change with progressive epidural brain compression. An experimental evaluation in cats. J Neurosurg 47: 532–550
16. Uftring SJ, Chu D, Alperin N, Levin DN (2000) The mechanical state of intracranial tissues in elderly subjects studied by imaging CSF and brain pulsations. Magn Reson Imaging 18: 991–996
17. Yau Y, Piper I, Contant C, Citerio G, Kiening K, Enblad P, Nilsson P, Ng S, Wasserberg J, Kiefer M, Poon W, Dunn L, Whittle I (2002) Multi-centre assessment of the Spiegelberg compliance monitor: interim results. Acta Neurochir [Suppl] 81: 167–170
18. Yau YH, Piper IR, Clutton RE, Whittle IR (2000) Experimental evaluation of the Spiegelberg intracranial pressure and intracranial compliance monitor. Technical note. J Neurosurg 93: 1072–1077
19. Yau YH, Piper IR, Contant CF, Dunn LT, Whittle IR (2002) Clinical experience in the use of the Spiegelberg automated compliance device in the assessment of patients with hydrocephalus. Acta Neurochir [Suppl] 81: 171–172

Correspondence: Karl L. Kiening, Department of Neurosurgery, Heidelberg-Medical Center, Ruprecht-Karls-University, Heidelberg, Im Neuenheimer Feld 400, 69120 Heidelberg, Germany. e-mail: karl.kiening@med.uni-heidelberg.de

Acta Neurochir (2005) [Suppl] 95: 299–301

Cerebral haemodynamic assessment in patients with thalamic haemorrhage: a pilot study with continuous compliance monitoring

S. C. P. Ng[1], **W. S. Poon**[1], and **M. T. V. Chan**[2]

[1] Division of Neurosurgery, Department of Surgery, Prince of Wales Hospital, The Chinese University of Hong Kong, Shatin, Hong Kong, China
[2] Department of Anaethesia and Intensive Care, Prince of Wales Hospital, The Chinese University of Hong Kong, Shatin, Hong Kong, China

Summary

Objective. Thalamic brain haemorrhage is a common disabling and potentially fatal condition. However, management is mainly supportive, very rarely do neurosurgeons resort to evacuation of the haematoma. We hypothesised that cerebral haemodynamic abnormalities in the forms of lost pressure autoregulatory response (PAR) and/or impaired cerebral vasoreactivity (CVR) to carbon dioxide may indicate the haematoma should be aspirated to prevent further brain damage.

Material and methods. Patients with thalamic haemorrhage were selected on clinical ground for intracranial pressure (ICP) monitoring and intensive care management. Spiegelberg double lumen intraventricular balloon catheter was inserted as any other fluid-filled ICP monitoring technique, on the side of the haematoma. Data of ICP, arterial blood pressure (ABP), cerebral perfusion pressure (CPP) and intracranial compliance were collected on a minute basis. Hourly averages were used for analysis. To assess PAR and CVR, blood flow velocity (BFV) in both middle cerebral arteries were measured continuously by transcranial Doppler (TCD) ultrasonography.

Results. Six patients with medium (15–25 ml) to large (>25 ml) haematoma volume were subjected to ultrasoud-guided aspiration. 51 CVR and 53 PAR tests were performed. 80820 sets of data were prospectively collected. Progressive reduction in ICP and sustained improvement in compliance and BFV were observed after aspiration. Impairment in PAR and/or CVR was consistent with clinical deterioration in four patients. Such disturbance was normalised after aspiration. Increments in BFV and improvement in compliance were demonstrated.

Conclusions. Cerebral haemodynamic abnormalities in thalamic haematomas can be demonstrated by the non-invasive TCD ultrasonograpy. These abnormalities can be corrected by aspiration of the haematoma, and hence improve intracranial compliance.

Keywords: Thalamic haemorrhage; intracranial compliance; intracranial pressure monitoring; pressure autoregulatory response; cerebral vasoreactivity; transcranial Doppler ultrasonography.

Introduction

Spontaneous intracerebral haemorrhage is twice more common in Southern Chinese than in Caucasians. Thalamic haemorrhage accounts for 32% of these patients with ICH [2] and carries a 35% management mortality [1].

In general, thalamic haemorrhage is complicated by intraventricular haemorrhage (IVH) with or without hydrocephalus. It causes more extensive and persistent in the metabolic depression and more pronounced cerebral blood flow (CBF) reduction, and therefore a worse outcome [3]. Nevertheless, management is mainly supportive and there is no rigid guideline for surgical intervention in treating these deep-seated haematoma.

In order to improve the outcome, apart from monitoring the intracranial pressure (ICP), we may need more intensive monitoring to help clinical decision making in evacuating or aspirating the haematoma.

Cerebral vasoreactivity (CVR) to carbon dioxide (CO_2) and pressure autoregulatory response (PAR) to cerebral perfusion pressure (CPP) alterations are physiological mechanisms that can have profound influence on CBF. These cerebral haemodynamic assessments in patients with traumatic brain injury, carotid stenosis and hydrocephalus have gained clinical importance. Continuous intracranial compliance monitoring provides information on the volume change resulting from unit change in pressure and helps to identify process which leads to raised ICP.

In this pilot study with intracranial compliance monitoring, we hypothesised that cerebral haemodynamic abnormalities in the forms of lost PAR and/or impaired CVR may indicate the haemoatoma should be aspirated to prevent further brain damage.

Table 1. *Collected data of ICP, ABP, CPP, intracranial compliance and BFV on both pathology and contra-lateral sides are compared: 1 day and 2 days after aspiration versus before aspiration (p1) and the last day in ICU (data are presented as mean ± s.d.)*

	Before	1 day after	2 days after	Last day in ICU
ABP (mmHg)	92.5 ± 7.0	91.1 ± 6.9 (p1 = 0.066, p2 = 0.390)	91.4 ± 3.2 (p1 = 0.773, p2 = 0.768)	94.5 ± 6.9
ICP (mmHg)	16.8 ± 5.6	13.0 ± 4.2 (p1 = 0.126, p2 = 0.060)	12.8 ± 4.3 (p1 = 0.093, p2 = 0.032*)	10.4 ± 3.3
CPP (mmHg)	78.5 ± 6.6	77.2 ± 8.4 (p1 = 0.635, p2 = 0.108)	78.0 ± 5.8 (p1 = 0.939, p2 = 0.239)	84.1 ± 7.8
Compliance (ml/mmHg)	0.39 ± 0.17	0.58 ± 0.26 (p1 < 0.005*, p2 = 0.031*)	0.83 ± 0.27 (p1 < 0.001*, p2 = 0.769)	0.84 ± 0.24
BFV/pathology (cm/s)	43.7 ± 20.8	54.1 ± 19.5 (p1 < 0.001*, p2 = 0.026*)	61.7 ± 28.3 (p1 = 0.006*, p2 = 0.737)	60.5 ± 20.9
BFV/contra-lateral (cm/s)	45.4 ± 14.5	55.3 ± 15.0 (p1 = 0.001*, p2 = 0.176)	59.7 ± 15.1 (p1 < 0.001*, p2 = 0.899)	58.3 ± 13.8

Material and methods

All patients with thalamic haemorrhage aged between 45 to 75 years were included. Patients selected for ICP monitoring and intensive care management were based on clinical ground. They included impaired conscious levels (Glasgow Come Scale score, GCS < 13 points), evidence of IVH, hydrocephalus and haematoma volume was greater than 15 ml.

ICP and intracranial compliance were measured on the side of predominant haemorrhage by Spiegelberg continuous compliance monitor. Data of ICP, compliance, arterial blood pressure (ABP) and cerebral perfusion pressure (CPP) were collected at one-minute intervals. Hourly averages were used for analysis.

During assessment of CVR and PAR, blood flow velocities (BFV) in both middle cerebral arteries were monitored continuously. To evaluate CVR, the initial end-tidal carbon dioxide concentration ($EtCO_2$) was lowered or raised by moderate hyperventilation or hypoventilation for at least 6 mmHg, and the baseline ABP was kept. To evaluate PAR, normoventilaton was kept and induced a 20 to 25% increase or decrease in baseline mean ABP.

Results

Four males and two female aged 53.2 ± 7.3 years (mean ± standard deviation, s.d.) with medium (15–25 ml) to large (>25 ml) haematoma volume were subjected to ultrasould-guided aspiration. Median admission G.C.S. score was 12.5 (range: 3–13) and intensive care unit (ICU) stay was 9.4 ± 2.3 days. 51 CVR and 53 PAR tests were performed. 80820 sets of valid physiological data were prospectively collected.

On admission and during (ICU) stay, BFV on the pathology side was not different from that on the contra-lateral side. Comparing data collected from the day before aspiration, after aspiration and the last day in ICU (Table 1), ICP did not changed on the day after aspiration but was reduced on the last day in ICU (16.8 ± 5.6 *vs* 10.4 ± 3.3 mmHg, p = 0.015), compliance was improving (0.39 ± 0.17 *vs* 0.58 ± 0.26 *vs* 0.84 ± 0.24 ml/mmHg, p < 0.005) and BFV on both sides were increasing (43.7 ± 20.8 *vs* 54.1 ± 19.5 *vs* 60.5 ± 20.9 cm/s, 45.4 ± 14.5 *vs* 55.3 ± 15.0 *vs* 59.3 ± 13.8 cm/s, p < 0.005). No significant changes were found in ABP and CPP during the ICU stay. Further analysis of the data after aspiration revealed although no changes in ICP in a 48-hour interval after aspiration, progressive improvement was observed in compliance (0.58 ± 0.26 *vs* 0.83 ± 0.27 ml/mmHg, p = 0.031) and BFV on the pathology side (54.1 ± 19.5 *vs* 61.7 ± 28.3 cm/s, p = 0.026). These improved compliance and BFV were maintained during the ICU stay whereas ICP took a longer time to achieve significant improvement (12.8 ± 4.3 *vs* 10.4 ± 3.3 mmHg, p = 0.032).

Table 2 shows the analysed results of data collected before and after aspiration from four patients with lost PAR and/or impaired CVR. Such disturbance was consistent with clinical deterioration and able to be normalised after aspiration. No significant changes were found in ICP, ABP and CPP whereas improvement in compliance (0.41 ± 0.15 *vs* 0.80 ± 0.23

Table 2. *Comparison of ICP, ABP, CPP, intracranial compliance and BFV on both pathology and contra-lateral sides collected before and after aspiration in four patients with lost PAR and/or impaired CVR (data are presented as mean ± s.d.)*

	Before	After	p-value
ABP (mmHg)	94.8 ± 10.0	91.1 ± 9.2	0.410
ICP (mmHg)	15.3 ± 2.4	13.3 ± 3.0	0.126
CPP (mmHg)	79.5 ± 9.6	77.8 ± 8.8	0.706
Compliance (ml/mmHg)	0.41 ± 0.15	0.80 ± 0.23	0.018*
BFV/pathology (cm/s)	43.7 ± 22.9	56.1 ± 20.7	0.003*
BFV/contra-lateral (cm/s)	46.6 ± 15.4	58.3 ± 15.9	<0.001*

ml/mmHg, $p = 0.018$) and increments in BFV (43.7 ± 22.9 *vs* 56.1 ± 20.7 cm/s, $p = 0.003$; 46.4 ± 15.4 *vs* 58.3 ± 15.9 cm/s, $p < 0.001$) were demonstrated.

Discussion

Cerebral haemodynamic abnormalities in patients with thalamic haemorrhage can be demonstrated by non-invasive TCD ultrasonography coupled with blood pressure challenge and CO_2 challenge. These abnormalities can be corrected by aspiration of the haematoma, and hence improves intracranial compliance. Whether this can be translated into clinical outcome benefits, such as Glasgow outcome scale scores, the number of days requiring CSF drainage and the need for CSF diversion requires further study in a larger number of patients.

Acknowledgment

This study was supported by a grant earmarked by Direct Grant for research, the Chinese University of Hong Kong.

References

1. Hsiang JNK, Zhu XL, Wong LKS, Kay R, Poon WS (1998) Putaminal and thalamic haemorrhage in ethnic Chinese living in Hong Kong. Surg Neruology 50: 526–532
2. Kreel L, Kay R, Woo J, *et al* (1991) The radiological (CT) and clinical sequelae of primary intracerebral haemorrhage. Brit J Radiol 64: 1096–1100
3. Tanaka A, Yoshinaga S, Nakayama Y, Kimura M, Tomonaga M (1996) Cerebral blood flow and clinical outcome in patient with thalamic haemorrhages: a comparison with putaminal haemorrhages. J Neruo Sci 144: 191–197

Correspondence: Wai S. Poon, Division of Neurosurgery, Department of Surgery, The Chinese University of Hong Kong, Shatin, Hong Kong, China. e-mail: wpoon@surgery.cuhk.edu.hk

Acta Neurochir (2005) [Suppl] 95: 303–306

Non-invasive measurement of intracranial compliance using cine MRI in normal pressure hydrocephalus

M. Mase[1], **T. Miyati**[2], **K. Yamada**[1], **H. Kasai**[3], **M. Hara**[3], and **Y. Shibamoto**[3]

[1] Department of Neurosurgery and Restorative Neuroscience, Nagoya City University Graduate School of Medical Sciences, Nagoya, Japan
[2] Department of Radiological Technology, School of Heath Sciences, Faculty of Medicine, Kanazawa University, Kanazawa, Japan
[3] Department of Central Radiology, Nagoya City University Hospital, Nagoya, Japan

Summary

The aim of this study is to clarify biophysics of normal pressure hydrocephalus (NPH) based on non-invasive intracranial compliance measurement using magnetic resonance imaging (MRI). Patients with NPH after subarachnoid hemorrhage (NPH group, n = 5), brain atrophy or asymptomatic ventricular dilation (VD group, n = 5), and healthy volunteers (control group, n = 12) were included in this study. Net blood flow (bilateral internal carotid and vertebral arteries, and jugular veins) and cerebrospinal fluid (CSF) flow in subarachnoid space at the C2 level of cervical vertebra were measured using phase-contrast cine MRI. CSF pressure gradient and intracranial volume changes during a cardiac cycle were calculated based on Alperin's method. Compliance index (Ci = delta V/delta P) was obtained from the maximum pressure gradient and volume changes. Pressure volume response (PVR) was measured in the NPH group during a shunt operation. Ci in the NPH group was the lowest among the three studies groups. No difference was found between the control and VD groups. There was a linear correlation between Ci and PVR. In conclusion, intracranial compliance can be determined by cine MRI non-invasively. It is well known that NPH has relatively low intracranial compliance, this non-invasive method can be used for the diagnosis of NPH.

Keywords: Intracranial compliance; elastance; MRI; normal pressure hydrocephalus; cerebrospinal fluid flow dynamics.

Introduction

The diagnosis of normal pressure hydrocephalus (NPH) continues to be difficult in some cases and many diagnostic examinations have been attempted. However, in electing NPH patients for shunt surgery, there remains no unambiguously useful and non-invasive examinations which are easily performed or widely accepted in medical centers.

Intracranial compliance is one of the most important parameters to estimate the compensatory capacity of the cranial cavity in patients with various intracrainal pathologies. It has been reported that the intracranial compliance was relatively low in NPH [4, 8]. We have already shown that cerebrospinal fluid (CSF) flow dynamics on magnetic resonance imaging (MRI) was affected by changes of intracranial compliance [5–7, 9].

In the present study, we directly measured intracranial compliance as compliance index (Ci) non-invasively using cine MRI based on Alperin's method [1], and studies the biophysis by analyzing the changes of Ci in these patients.

Materials and methods

Compliance index

The compliance index (Ci) was calculated from the ratio of intracranial volume change to craniospinal CSF pressure gradient change during a cardiac cycle with retrospective cardiac gated phase-contrast (PC) cine MRI [1]. We set the vertical slice plane at the dense (C2) level against the cerebrospinal axis (Fig. 1a and b), and obtained velocity-mapped phase images (32 cardiac phases) with different velocity encoding gradients for CSF flow, cord motion and transcranial blood flow respectively (Fig. 1c and d). Measurements were done by using a gradient echo pulse sequence (T1-FFE) with shortest echo time, 20 degrees flip angle, 3 mm slice thickness, 2 signals averaged, 150×120 mm^2 rectangular field of view, and 256×256 or 256×128 acquisition matrix on a 1.5T MR system (Gyroscan ACS II; Philips). The velocity-encoding value for the CSF flow measurement was ± 15 cm/sec in patients suspected of having NPH, and ± 10 cm/sec in the others. The velocity-encoding value for measuring blood flow was ± 90 cm/sec.

To obtain the net transcranial flow, we measured the flow (or displacement) during a cardiac cycle, CSF flow [$V_c(t)$], displacement of cord [$V_s(t)$], arterial inflow (the sum of both internal carotid arteries and vertebral arteries) [$V_a(t)$] and venous outflow (the sum of both

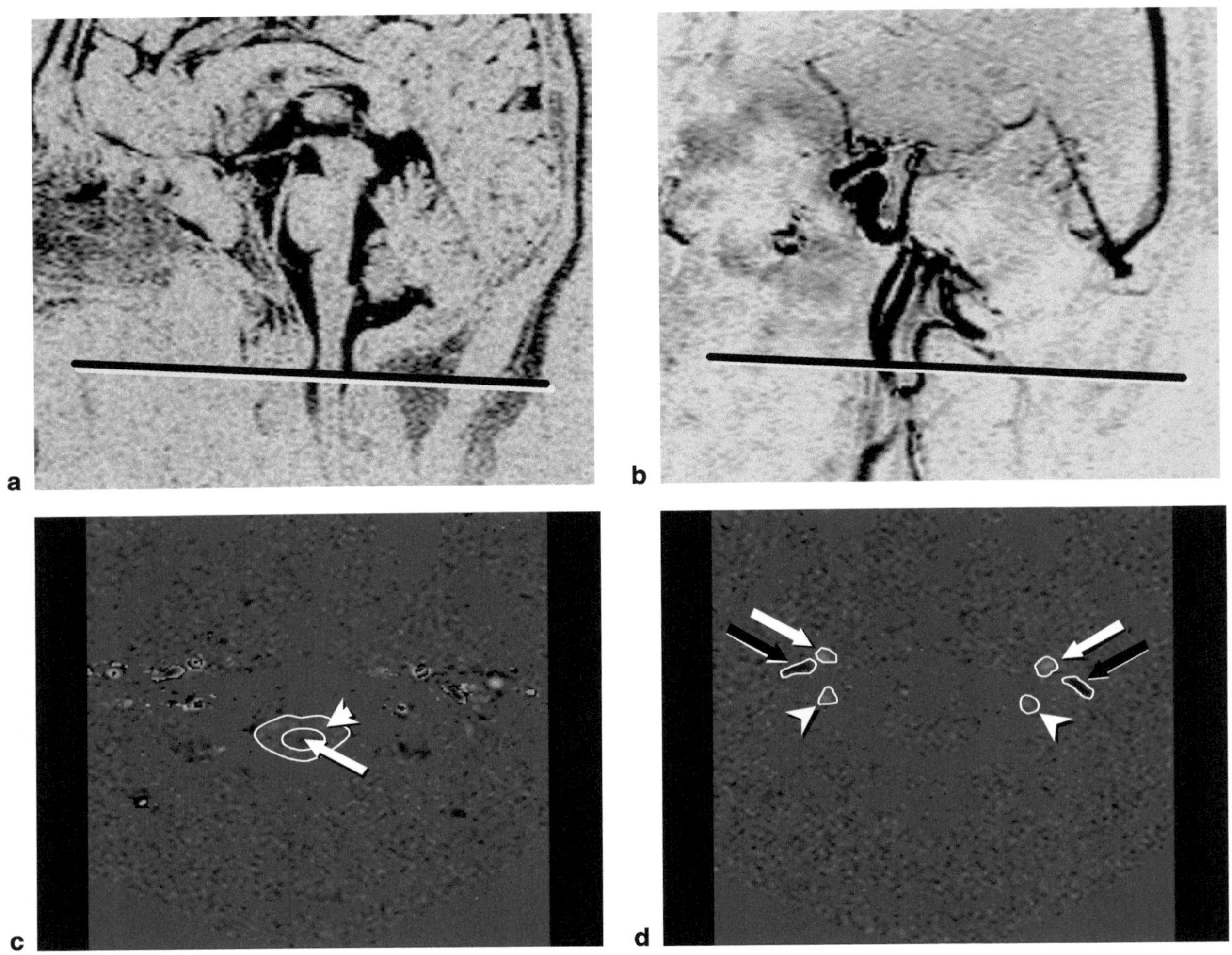

Fig. 1. Slice planes of MRI for CSF flow (a) and vascular flow (b) were set at the C2 level of cervical vertebra. Region of interest (*ROI*) on velocity-mapped phase images with different velocity encoding gradients for CSF flow (white arrow head) and cord motion (white arrow) (c) as well as for transcranial blood flow (white arrows: internal carotid artery, black arrows: jugular veins, white arrow heads: vertebral arteries) (d) were shown

internal jugular veins) $[V_v(t)]$. Firstly, we corrected the baseline offset due to eddy currents by a subtraction process. Secondly, to match the difference in inflow and outflow capacity to the cranium in a cardiac cycle, the flow wave of the venous outflow was scaled down and the net transcranial flow (intracranial volume change) $[V_n(t)]$ in each cardiac phase was calculated using the equation:

$$V_n(t) = [V_a(t) - V_v(t) - V_c(t) - V_s(t)]\Delta t$$

And then, we determined the craniospinal CSF pressure gradient change $[\square P(t)]$ during the cardiac cycle, which was calculated from the above measured CSF flow velocity using a simplified Navier-Stokes equation [11]:

$$\nabla P = \frac{\partial P}{\partial z} = -\rho\frac{\partial w}{\partial t} + \mu\left(\frac{\partial^2 w}{\partial x^2} + \frac{\partial^2 w}{\partial y^2}\right)$$

where x, y, and z are coordinates, and z-direction is a canal axis. P and w are the pressure and the axial velocity, and ρ and μ are the fluid density (CSF: 1.0007 g/cm^3) and viscosity (CSF: 1.1 cP), respectively.

Finally, Ci was obtained from dividing the peak-to-peak intracranial volume change by the peak-to-peak CSF pressure gradient change.

Clinical subjects

Ci analysis was performed in patients with NPH after subarachnoid hemorrhage (n = 5), asymptomatic ventricular dilation or brain atrophy (VD, n = 5), and healthy volunteers (control group, n = 12).

During the shunt operations, intracranial pressure (ICP) and pulse pressure of the ICP pulse wave in the lateral ventricle were recorded (n = 4). Pressure-volume response (PVR) was determined as the ICP changes after a bolus injection of saline (1–5 ml) into the lateral ventricle, as an invasive measure of compliance [5]. $PaCO_2$ was kept within a normal range during the measurements (38.2 ± 3.6 mmHg). Relationship between the PVR and Ci was investigated.

The purpose and procedures of all our investigations were fully explained to all patients, and the studies were performed only after obtaining informed consent from each patient.

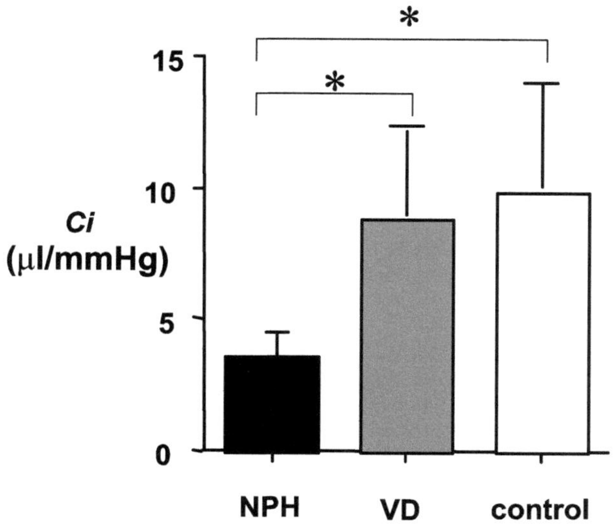

Fig. 2. Compliance index (*Ci*) of each group. Ci of the NPH group is significantly smaller than those of VD and control groups

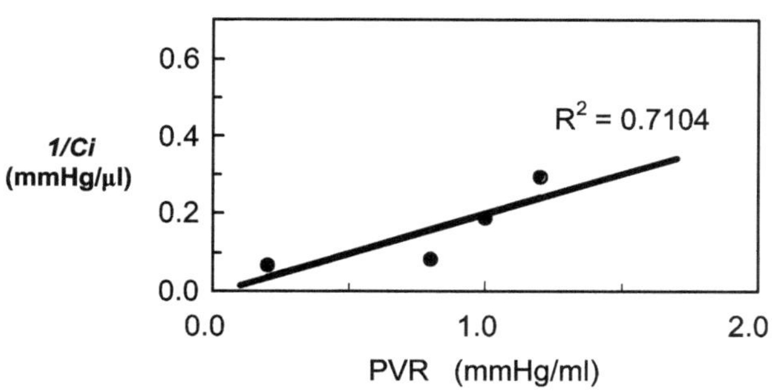

Fig. 3. Relationship between 1/Ci and PVR. There was a linear correlation between 1/Ci and PVR

Results

The Ci obtained with MRI was significantly smaller in the NPH group (mean ± SD: 3.52 ± 0.78 μl/mmHg) than those in the control (9.80 ± 4.04 μl/mmHg) and VD (8.77 ± 3.38 μl/mmHg) groups. There was no significant difference in the Ci between the VD and control groups (Fig. 2). There was a positive linear correlation between the reciprocal of Ci and the PVR (Fig. 3).

Discussion

Intracranial compliance is one of the most important parameters to understand the intracranial condition of patients with various intracrainal pathologies. Especially in cases with no abnormal findings using conventional imaging techniques (CT or MRI), such as diffuse brain swelling or slit ventricle syndrome. Therefore, intracranial compliance can be very useful for determining therapeutic strategies and evaluating the effects of treatments. In genereal, the intracranial compliance is determined by measurement the change in intracranial pressure (ICP) after intrathecal or intraventricular infusion of saline (pressure volume response, PVR). However, this procedure is invasive and has some risks of infection or other complications, and therefore it is not widely accepted. In the present study, we could directly measure intracranial compliance non-invasively (Ci) using cine MRI based on Alperin's method [1]. Significant corrections between Ci and PVR suggests Ci is accurate and reliable. And Ci may become a very important clinical index for estimating intracranial condition non-invasively.

In many trials CSF flow dynamics were measured by using MRI to evaluate the changes in intracranial conditions in normal-pressure hydrocephalus (NPH) [2, 4–7, 9, 10]. However, the diagnosis of NPH based on the CSF flow study using MRI has not been established yet. The most reliable test could the infusion test for evaluating intracranial compliance and out flow resistance of craniospinal cavity, or the tap-test (analyze symptoms after withdrawal of CSF) [3], nevertheless both are invasive. In the present study, we showed that patients with NPH had significantly lower Ci, which meant that compensatory capacity decreased. This corresponds with the other reports that NPH has low compliance [4–6, 8]. Analysis of Ci may be useful for more accurate diagnosis of NPH. Although further investigation is needed, CSF flow study including Ci using MRI promises to become an essential component of preoperative examination.

Conclusion

Ci (compliance index) analysis measured by phase contrast cine MRI enables the determination of intracranial state and dynamics precisely and non-invasively. Ci of patients with NPH was significantly lower than that of healthy controls or patients with asymptomatic brain atrophy. Ci may become one of useful indices for the diagnosis of NPH.

References

1. Alperin NJ, Lee SH, Loth F *et al* (2000) MR-Intracranial pressure (ICP): a method to measure intracranial elastance and non-invasively by means of MR imaging: baboon and human study. Radiology 217: 877–885
2. Bradley WG, Scalzo D, Queralt J *et al* (1996) Normal pressure

hydrocephalus; evaluation with cerebrospinal fluid flow measurements at MR imaging. Radiology 198: 523–529
3. Damasceno BP, Carelli EF, Honorato DC (1997) The predictive value of cerebrospinal fluid tap-test in normal pressure hydrocephalus. Arch Neuropsychiatr 55: 179–185
4. Kim DS, Choi JU, Huh R *et al* (1999) Quantitative assessment of cerebrospinal fluid hydrodynamics using a phase-contrast cine MR image in hydrocephalus. Child's Nerv Syst 15: 461–467
5. Mase M, Yamada K, Banno T *et al* (1998) Quantitative analysis of CSF flow dynamics using MRI in normal pressure hydrocephalus. Acta Neurochir (Wien) [Suppl] 71: 350–353
6. Mase M, Yamada K, Banno T *et al* (1988) Differential diagnosis of normal pressure hydrocephalus based on CSF flow dynamics study using MRI. Current Tr Hyd (Tokyo) 8: 13–18
7. Mase M, Banno T, Yamada K *et al* (1998) Quantitative measurement of CSF flow velocity in the aqueduct using MRI – experimental simulation and normal value. Current Tr Hyd (Tokyo) 8: 6–12
8. Matsumoto T, Nagai H, Kasuga Y *et al* (1986) Changes in intracranial pressure (ICP) pulse wave following hydrocephalus. Acta Neurochir (Wien) 82: 50–56
9. Miyati T, Mase M, Banno T (2003) Frequency analysis of CSF flow on cine MRI in normal pressure hydrocephalus. Eur Radiol 13: 1019–1024
10. Ohara S, Nagai H, Matsumoto T (1988) MR imaging of CSF pulsatory flow and its relation to intracranial pressure. J Neurosurg 69: 675–682.
11. Urchuk SN, Plewes DB (1995) MR measurement of time-dependent blood pressure variations. J Magn Reson Imaging 5: 621–627

Correspondence: M. Mase, Department of Neurosurgery and Restorative Neuroscience, Nagoya City University Graduate School of Medical Sciences, 1-Kawasumi, Mizuho-ku, Mizuho-cho, Nagoya, 467-8602, Japan. e-mail: mitmase@med.nagoya-cu.ac.jp

Acta Neurochir (2005) [Suppl] 95: 307–310

Pulse and mean intracranial pressure analysis in pediatric traumatic brain injury*

M. Aboy[1], **J. McNames**[1], **W. Wakeland**[2], and **B. Goldstein**[3]

[1] Biomedical Signal Processing Laboratory, Department of Electrical and Computer Engineering at Portland State University, Portland, OR, USA
[2] Systems Science Ph.D. Program, Portland State University, Portland, OR, USA
[3] Oregon Health and Science University, Portland, OR, USA

Summary

Objective. We investigated the relationship between the intracranial pulse pressure (ICP_{PP}) and the mean intracranial pressure (ICP_M) in pediatric patients with traumatic brain injury (TBI).

Methods. We screened ICP records of 42 patients admitted to the Pediatric Intensive Care Unit at Doernbecher Children's Hospital (OHSU) for segments in which the ICP_M varied at least 5 mmHg. We found 54 ICP segments in 9 pediatric TBI patients (ages 0.2–17.8 years, mean = 9.9). ICP was continuously monitored ($f_s = 125$ Hz). We used an automatic algorithm to detect ICP beat components. We then calculated the ICP_{PP} and ICP_M for each beat and created density plots of ICP_{PP} vs. ICP_M.

Results. The coefficient of linear correlation was $r > 0.70$ in 43/54 segments ($p < 0.01$). We found that an underlying linear relationship exits between ICP_{PP} and ICP_M in most 1-hour records of pediatric patients with TBI. This finding is consistent with the data in adult studies, suggesting that children with TBI demonstrate similar changes in brain compliance. However, density plots revealed that there are also nonlinear ICP_{PP}-ICP_M patterns present that are not captured by linear metrics.

Conclusion. Although there is an underlying linear relationship between ICP_{PP} and ICP_M, nonlinear patterns are also present. Further research is required to determine if specific nonlinear ICP_{PP}-ICP_M patterns correlate with clinically significant information.

Keywords: Intracranial hypertension; intracranial pressure; waveform analysis; pulse pressure.

Introduction

Traumatic brain Injury (TBI) is the leading cause of death and disability in children in the United States [1]. Elevated intracranial pressure (ICP) following TBI may result in secondary injury due to decreased cerebral perfusion pressure (CPP) and cerebral ischemia. ICP monitoring and therapeutic interventions to control elevated ICP (>20 mmHg) have resulted in improved outcomes [9].

Several investigators and research groups have studied the relationship between the ICP pulse pressure (ICP_{PP}) and the mean ICP (ICP_M) in adult patients and dogs. Castel and Cohadon studied the ICP waveform in three groups of neurosurgical patients and noted that a linear relationship exists between ICP_{PP} and ICP_M [4]. Avezatt and Van Eijndoven examined this relationship in dogs and described a linear relationship between ICP_{PP} and ICP_M below a breakpoint where the slope changes, which they attributed to vasoparesis and failure of autoregulation [2]. They proposed to use the ratio ICP_{PP}-ICP_M as an index of brain compliance. The rationale behind this definition is that a change in this relationship during patient monitoring indicates a change either in the volume-pressure relationship or in the net volume change per cardiac cycle. Price defined the *pulse wave index* as the ratio of ICP_{PP} and ICP_M [10]. More recently, the correlation coefficient between ICP_{PP} and ICP_M has been termed the *pressure-volume compensatory reserve index* (RAP). This index measures the degree of correlation between ICP_{PP} and ICP_M over short periods of time, and indicates the relationship between ICP and changes in volume of the intracerebral space. In general, increased ICP_{PP} has been associated with decreasing intracranial compliance [3, 5–7].

Due to the unavailability of automatic ICP component detection algorithms, none of these previous studies calculated ICP_{PP} and ICP_M on a beat-by-beat

* This research was supported in part by the Thrasher Research Grant. This work was presented at the International ICP 2004 Conference and at the Annual International IEEE Engineering in Medicine and Biology Conference 2003.

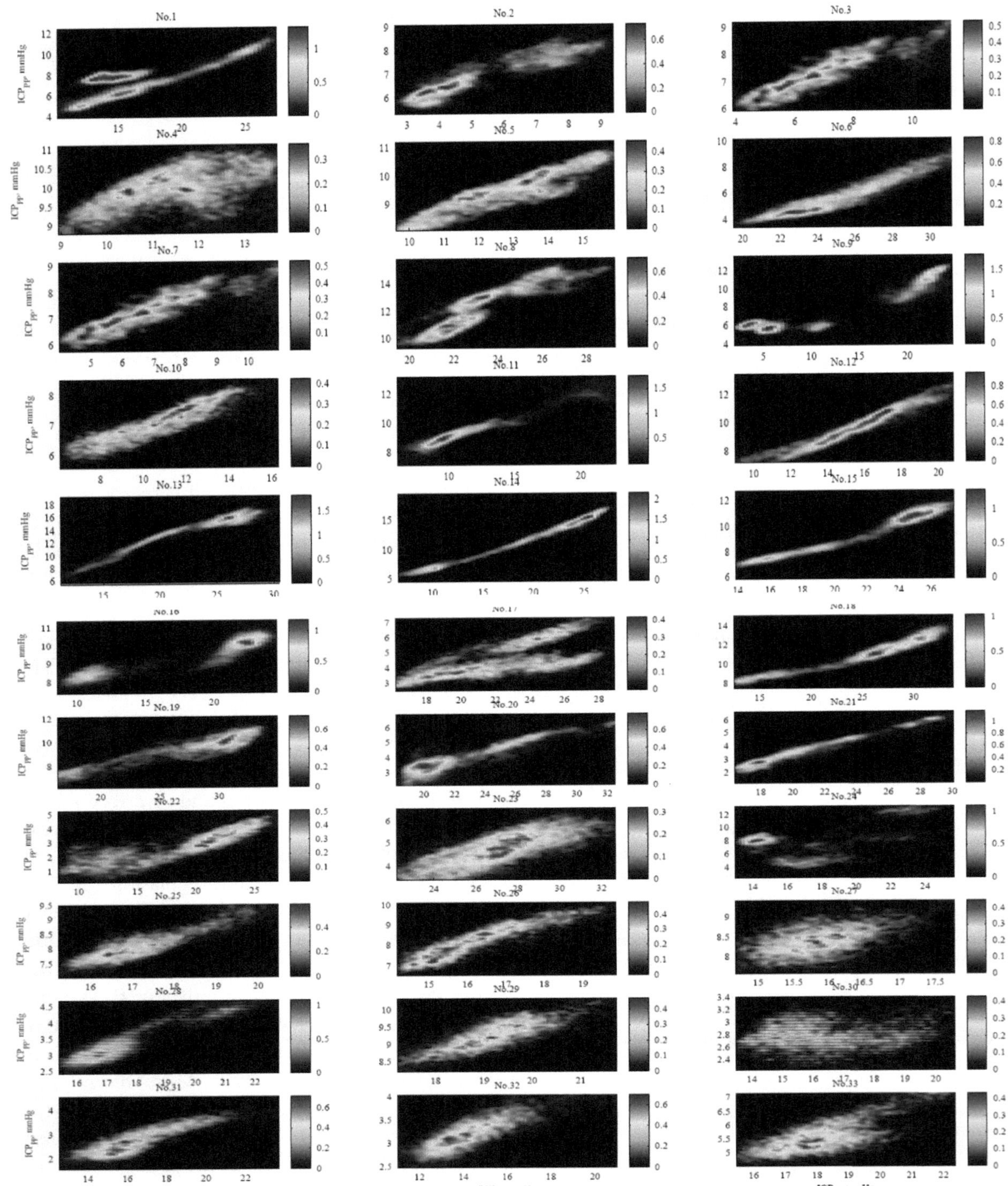

Fig. 1. Density plots showing the relationship between ICP_{PP} and ICP_M for 33 episodes

basis. Instead, ICP_{PP} and ICP_M are often estimated by indirect measures based on moving averages or frequency domain techniques. The most common methodology used to calculate the RAP index, for instance, consists of estimating the ICP_{PP} as the squared root of the power of the fundamental harmonic component, and the ICP_M as the moving average mean ICP over specific time window length [6].

We investigated the relationship between ICP_{PP} and ICP_M in pediatric subjects with TBI. This study had three objectives: 1) to investigate the ICP relationship on a previously unstudied patient population, 2) to determine whether there is further information that can be derived from the relationship between ICP_{PP}-ICP_M that is not captured by linear metrics such as the coefficient of linear correlation, and 3) to describe a new methodology to calculate ICP_{PP} and ICP_M and visualize their relationship.

Materials and methods

Patient population and data acquisition

We examined ICP records of 42 patients admitted to the Pediatric Intensive Care Unit at Doernbecher Children's Hospital (Oregon Health and Science University, OHSU) for segments in which the ICP_M varied at least 5 mmHg over a 1-hour period. The study was reviewed and approved by the Institutional Review Board at OHSU, and the requirement of informed consent was waived. We found 54 1-hour ICP segments from 9 different subjects that met this condition (age range of 0.2 to 17.8, mean age 9.9 years).

ICP was monitored continuously using a ventricular catheter or parenchymal fiber-optic pressure transducer (Integra Neurocare, Integra LifeSciences, Plainsboro, NJ). The ICP monitor was connected to a Philips Merlin patient monitor (Philips, Best, Netherlands) that sampled the ICP signal at 125 Hz. An HPUX workstation automatically acquired these signals through a serial data network and stored them on CD-ROM [8].

Detection algorithm and ICP_{PP} /ICP_M definitions

An automatic beat detection algorithm was used to detect each ICP beat. The algorithm performs minima detection to identify the time location corresponding to the start of each beat a_k,

$$\boldsymbol{a} = (a_1\ a_2 \ldots a_{k-1}\ a_k\ a_{k+1} \ldots)^T. \tag{1}$$

Details of the beat detection process follow. The pressure signal is preprocessed by three bandpass elliptic filters with different cutoff frequencies. The output of the first bandpass filter is used to estimate the heart rate based on the estimated power spectral density (PSD). The estimated heart rate is then used to calculate the cutoff frequencies of the other two filters. Minima detection and the associated decision logic are performed based on rank-order (percentile-based) nonlinear filters, that incorporate relative amplitude and slope information to coarsely estimate the minima preceding each beat (a_k). A nearest neighbor algorithm is used to combine the information extracted from the relative amplitude and slope. Finally, an interbeat-interval stage uses this classification together with the estimated heart rate to make the final classification and detection of signal components. Since detection is made on the filtered signal, a second nearest neighbor algorithm is used to find the minima in the raw signal that are closest to the detected components.

The maximum in each ICP beat b_k was determined using the $\boldsymbol{a}$ components identified by the beat detection algorithm in the previous step as follows,

$$b_k \triangleq \arg \max_{a_k \leq n \leq a_{k+1}} x(n)$$
$$\boldsymbol{b} = (b_1\ b_2 \ldots b_{k-1}\ b_k\ b_{k+1} \ldots)^T \tag{2}$$

where $x(n)$ denotes the ICP signal. Based on these definitions, the ICP_M was calculated as the ICP beat mean $\bar{x}_k$, and ICP_{PP} was calculated as the beat pulse pressure p_k,

$$\bar{x}_k \triangleq \frac{1}{a_{k+1} - a_k + 1} \sum_{k=a_k}^{a_{k+1}} x(k)$$
$$\bar{\boldsymbol{x}} = (\bar{x}_1\ \bar{x}_2 \ldots \bar{x}_{k-1}\ \bar{x}_k\ \bar{x}_{k+1} \ldots)^T. \tag{3}$$

Analysis

We determined the ICP_M and ICP_{PP} for each beat in the dataset, created density plots ICP_{PP} vs. ICP_M using a Gaussian Kernel with width of 0.05 mmHg, performed least-squares linear regression, and calculated the coefficient of linear correlation. Statistical significance was assessed using the nonparametric binary sign test.

Results and discussion

We found that in 43 out of 54 segments ($p < 0.001$) the coefficients of correlation were greater than 0.70. This result indicates that in pediatric patients with TBI there is an underlying relationship with a significant linear component between ICP_{PP} and ICP_M. This finding is consistent with the results previously reported on adult subject population and dogs.

Visual inspection of the density plots reveals that even though there is an underlying linear component between ICP_{PP} and ICP_M, there are also other nonlinear phenomena present (Fig. 1). In some episodes we observed underlying linear relationships with nonlinear clusters (e.g., 1, 9, 24). Others exhibit two parallel close-to-linear relationships (e.g., 8, 17) and slope changes (e.g., 13, 15, 18).

These nonlinearities have not been previously described. This may be due to the fact the ICP_{PP} vs ICP_M relationship has been visualized primarily using simple scatter plots. Density plots are more informative since they have an extra dimension that provides information about the density of points. In fact, most of the episodes presented on Fig. 1 appear to be linear if visualized with scatter plots.

The detection of these nonlinearities may also be due to the methodological differences in the calculation of ICP_{PP} and ICP_M. In this study the ICP_{PP} and ICP_M were calculated exactly for each beat. This eliminates the correlation introduced by moving average or frequency domain techniques.

Further research on prospective clinically annotated data is required to determine if these nonlinear patterns contain clinically significant information.

References

1. Adelson P, Kochanek P (1998) Head injury in children. Child Neurol 13: 2–15
2. Avezaat C, Van Eijndhoven, Wyper D (1979) Cerebrospinal fluid pulse pressure and intracranial volume-pressure relationships. J Neurol, Neurosurg, Psychiat 42: 687–700
3. Cardoso ER, Rowan JO, Galbraith S (1983) Analysis of the cerebrospinal fluid pulse wave in intracranial pressure. J Neurosurg 59: 817–821
4. Castel J, Cohadon F (1976) The pattern of cerebral pulse: automatic analysis. In: Intracranial pressure III. Springer, Berlin, pp 303–307
5. Chopp M, Portnoy HD (1980) Systems analysis of intracranial pressure. J Neurosurg 53: 516–527
6. Czosnyka M, Guazzo E, Whitehouse M, Smielewski P, Czos-

nyka Z, *et al* (1996) Significance of intracranial pressure waveform analysis after head injury. Acta Neurochirur (Wien) 138: 531–541
7. Doyle JD, Mark PWS (1992) Analysis of intracranial pressure. J Clin Monitoring 8(1): 81–90
8. Goldstein B, McNames J, McDonald B, Ellenby M, Lai S, Sun Z, Sclabassi R (2003) Physiologic data acquisition system and database for the study of disease dynamics in the intensive care unit. Crit Care Med 31(2): 433–441
9. Luerssen TG (1997) Intracranial pressure: Current status in monitoring and management. Semi Pediatr Neurol 4(3): 146–155
10. Price D, Dugdale R, Mason J (1980) The control of ICP using three asynchronous closed loops. In: Shulman, Marmarou K, Miller JA, Becker D, Hochwald G, Brock M (eds), Intracranial pressure IV. Springer, Berlin, pp 395–399

Correspondence: Wayne Wakeland, Systems Science Ph.D. Program, PO Box 751, Portland, OR 97208, USA. e-mail: wakeland@pdx.edu

Acta Neurochir (2005) [Suppl] 95: 311–314

Preliminary report on spiegelberg pre and post-operative monitoring of severe head-injured patients who received decompressive craniectomy

J. Abdullah[1], **I. Zamzuri**[1], **S. Awang**[1], **S. Sayuthi**[1], **A. Ghani**[1], **A. Tahir**[1], and **N. N. Naing**[2]

[1] Department of Neurosciences, School of Medical Sciences, Universiti Sains Malaysia, Kelantan, Malaysia
[2] Biostatistic and Research Methodology Unit, School of Medical Sciences, Universiti Sains Malaysia, Kelantan, Malaysia

Summary

The monitoring of craniospinal compliance is uncommonly used clinically despite it's value. The Spiegelberg compliance monitor calculates intracranial compliance ($C = \Delta V/\Delta P$) from a moving average of small ICP perturbations (ΔP) resulting from a sequence of up to 200 pulses of added volume ($\Delta V = 0.1$ ml, total $V = 0.2$ ml) made into a double lumen intraventricular balloon catheter. The objective of this study was thus to determine the effectiveness of the decompressive craniectomy done on the worst brain site with regard to compliance (CI), pressure volume index (PVI), jugular oximetry ($SjVo_2$), autoregulation abnormalties, brain tissue oxygen (TiO_2) and cerebral blood flow (CBF). This is a prospective cohort study of 17 patients who were enrolled after consent and approval of the ethics committee between the beginning of the year 2001 and end of the year 2002. For pre and post assessment on compliance and PVI, all 12 patients who survived were reported to become normal after decompressive craniectomy. There is no significant association between pre and post craniectomy assessment in jugular oxymetry ($p > 0.05$), autoregulation ($p > 0.05$), intracranial brain oxymetry ($p = 0.125$) and cerebral blood flow ($p = 0.375$). Compliance and PVI improved dramatically in all alive patients who received decompressive craniectomy. Compliance and PVI monitoring may be crucial in improving the outcome of severe head injured patients after decompressive craniectomy.

Keywords: Craniospinal; compliance; pressure volume index; head injury.

Introduction

The monitoring of craniospinal compliance is uncommonly used clinically despite it's value. Intracranial hypertension (ICH) is common in patients with head injury and leads to further secondary brain damage. Craniospinal compliance offers detection and prediction of raised intracranial pressure (ICP) [1, 2]. The Spiegelberg compliance monitor calculates intracranial compliance ($C = \Delta V/\Delta P$) from a moving average of small ICP perturbations (ΔP) resulting from a sequence of up to 200 pulses of added volume ($\Delta V = 0.1$ ml, total $V = 0.2$ ml) made into a double lumen intraventricular balloon catheter. Once a stable average has developed, the device produces a minute by minute measure of intracranial compliance [3]. The scientific value of the Spiegelberg monitoring device in head injured patients after decompressive craniectomy is not known. The objective of this study was thus to determine the effectiveness of the decompressive craniectomy done on the worst brain site with regard to compliance (CI), pressure volume index (PVI), jugular oximetry ($SjVo_2$), autoregulation abnormalties, brain tissue oxygen (TiO_2) and cerebral blood flow (CBF).

Patients and methods

This is a prospective cohort study of 17 patients who were enrolled after consent and approval of the ethics committee between the beginning of the year 2001 and end of the year 2002. Intracranial pressure was monitored until an acceptable upper limit of 35 mm Hg in adults by using a compliance monitor (Spiegelberg GmbH & Co., Hamburg, Germany) ($CI > 1$ and $PVI > 0.5$ was taken as normal) [4]. Compliance monitoring was placed in the non-dominant frontal horn. All measurements were made up to 4 days of trauma. All patients received decompressive craniectomy, when there was persistent ICH for more than 30 minutes not responding to medical therapy. Craniectomy was done on the worst brain site seen on computed tomography (CT) scan of the brain. A cerebral perfusion pressure (CPP) of more than 70 mm Hg was maintained using one or more inotropic supports with a central venous pressure between 5 and 10 mg. The Saber 2100 CBF sensor (Flowtronics, Phoenix, AZ) measured regional cortical blood flow (ml/100 g/min) in the cortex where the area concerned is approximately 1 cm deep in the cerebral hemisphere and 3 cm by implication. The sensor measures an area approximately 8 mm hemispheric due to the size of the gyrus. Its sensor plates are 24-k gold, whereas its wires are solid silver and Teflon-coated. Regional cortical blood flow was maintained by expanding the plasma volume, by using inotropic support, or implementing other medical methods to maintain a value higher than 50 ml/100 g/min at all times [5]. Brain tissue PO_2 (Licox, Kiel-Mielkendorf,

Germany) was kept at a minimal level of 15 mm Hg, with a fractional inspired O_2 level of 1 for a duration of 24 hours [6]. The sensor was placed in the worse brain seen on CT Scan of the brain. Jugular venous saturation was maintained from a baseline of 65 to 75% by using an optical catheter in the dominant jugular bulb (Oximetrix3 system; Abbott Laboratories, Chicago, IL) [7]. All patients with impaired cerebral vasomotor reactivity test during which the flow velocity in the middle cerebral artery was monitored intermittently by using a 2-MHz pulsed TCD ultrasonography probe (model Multi-Dop 2; DWL Electronishe System GmbH, Sipplingen, Germany) inserted at depths of between 4 and 6 cm [8]. All patients were positioned with their heads elevated at 30°. All received a decompressive craniectomy defined as a large 12 cm diameter cranial bone involving the frontal, parietal, temporal and part of the occipital bone with duraplasty [9].

For statistical analysis, mean with standard deviation was applied for numerical variables and frequency and percentage were used for categorical variables for summarization of results. Mc Nemar's test was applied to determine the association between pre and postoperative assessments after decompressive craniotomy for severe head injury. The level of significance was set at 0.05. SPSS version 11.0 was used in data analysis. Disability Rating Scale (DRS) was noted in these patients but were not measured as an outcome value due to the small number of observations in each group in this study.

Results

Seventeen head injured patients with a GCS of less than 8 were managed in our neurosciences Intensive Care Unit (ICU). All of them were males. Mean age was 21.35 ± 2.26 years. The range was 18–27 years. GCS was reported as 5.35 ± 1.73 score. 8 patients had a GCS of between 6–8 (47.1%) and 9 had a GCS of between 3–5 (52.9%). CT scan revealed at least one contusion. A total of 4 patients (23.5%) were found to have one contusion only whereas 8 (47.1%) and 5 (29.4%) were found to have 2 and more than two contusion respectively. All patients had a Borovich oedema scale of 4.

Precraniectomy compliance was normal in only 1 (5.9%) patients whereas the rest 16 (94.1%) were abnormal. Contrast to this postcraniectomy compliance was found normal in 12 patients (70.6%) whereas only 5 (29.4%) were found abnormal. All patients were having abnormal PVI before the operation, however after the decompressive craniectomy only 12 (70.6%) were found normal. Precraniectomy $SjVo^2$ was found to be normal in 6 patients (35.3%) whereas the rest 11 (64.7%) were abnormal. $SjVo^2$ after craniectomy was found to be normal 8 (47.1%) whereas 9 (52.9%) were abnormal. Autoregulation precraniectomy was abnormal in 10 (58.8%) of patients whereas the rest 7 (41.2%) were found to be abnormal. For autoregulation after craniectomy, 9 (52.9%) were reported normal and the rest 8 (47.1%) were reported abnormal. TiO_2 was reported normal in 2 patients (11.8%) whereas the rest 15 (88.2%) were reported abnormal. After craniectomy TiO_2 was found normal in 7 patients (41.2%) whereas the rest 10 (58.8%) were abnormal. CBF precraniectomy was normal in 2 patients (11.8%) whereas the rest 15 (88.2%) were reported abnormal. After craniectomy 6 patients (35.3%) were reported normal whereas the rest 11 (64.7%) were found normal. DRS was reported as 6 (35.3%) as good, 6 (35.3%) as bad and 5 (29.4%) as death.

Among all those who died ($n = 5$), two (40%) were found as having less than or equal to 2 contusions whereas three (50%) as having more than 2 contusions. Compliance, PVI, autoregulation, TiO_2 were found to be abnormal among those who died in both pre and postoperative measurements. $SjVo_2$ and CBF were reported as all being abnormal in preoperative assessment but improved to be normal in 4 (80%) of those who died in postoperative assessment. Those who died were all in the GCS of 3 to 5. 8 patients who survived (66.7%) had a GCS of between 6 to 8 and 4 patients (33.3%) were found to have a GCS of 3 to 5.

For pre and post assessment on compliance and PVI, all 12 patients who survived were reported to become normal after decompressive craniectomy. There is no significant association between pre and post craniectomy assessment in jugular oxymetry ($p > 0.05$), autoregulation ($p > 0.05$), intracranial brain oxymetry ($p = 0.125$) and cerebral blood flow ($p = 0.375$) (Table 1). Compliance and PVI improved dramatically in all alive patients who received decompressive craniectomy. Compliance and PVI monitoring may be crucial in improving the outcome of severe head injured patients after decompressive craniectomy.

Discussion

It is well known that the relationship between ICP and intracranial volume is often described by an exponential function which is relatively flat at low ICPs, that rises as pressure increases above 20 mmHg. At the low ICP end of this curve, compliance is high. When the patients volume-pressure status changes so as to move up the curve, compliance will fall rapidly. Measurement of craniospinal compliance in brain injured patients may, offer the potential for early detection of raised intracranial pressure (ICP) before it rises to levels damaging to brain parenchyma. The most commonly used methods of measuring craniospinal compliance were developed by Miller [2] and Mar-

Table 1. *Comparison of pre-craniectomy and post-craniectomy assessments on jugular oxymetry, autoregulation, brain tissue oxymetry and cerebral blood flow amongst alive patients with severe head injury*

Pre-craniectomy[a] assessment n (%)	Post-craniectomy[a] assessment n (%)		p-value[b]
	Normal	Abnormal	
Jugular oximetry			
– Normal	4 (66.7)	2 (33.3)	>0.05
– Abnormal	3 (50.0)	3 (50.0)	
Autoregulation			
– Normal	9 (90.0)	1 (10.0)	>0.05
– Abnormal	0 (0.0)	2 (100.0)	
Brain tissue oxymetry			
– Normal	1 (50.0)	1 (50.0)	0.125
– Abnormal	6 (60.0)	4 (40.0)	
Cerebral blood flow			
– Normal	1 (50.0)	1 (50.0)	0.375
– Abnormal	4 (40.0)	6 (60.0)	

[a] Assessment expressed in frequency and percentage.
[b] p-value obtained by using Mc Nemar's test.

marou [10], and depend upon the rapid injection of known volumes of fluid into the CSF space with immediate measurement of the resultant increase in CSF pressure. Pressure volume index (PVI); is the volume that when added to the CSF space would produce a 10-fold rise in ICP. Miller and co-workers found a measure of the craniospinal volume pressure relationship, the volume pressure response (VPR) which was calculated from the intracranial pressure response resulting from a rapid bolus injection of saline into the CSF space [1, 2]. The PVI, which assumes a monoexponential pressure versus volume relationship, is a single index that characterizes the patients' entire volume-pressure relationship and is, numerically, a measure of the slope of the logarithm of the ICP versus intracranial volume relationship and defines the whole volume pressure status of the patient with a single index which will be effected by a large decompressive craniectomy with duraplasty [1, 2]. In severe head injured patients, a continuous measure of absolute compliance (or its inverse: elastance is measured by the VPR) provides, more useful information that the VPR correlated better with the degree of brain midline shift, as imaged on CT, than it did to absolute ICP alone. VPR could serve as an indicator for decompressive craniectomy with critical levels being between 3–5 mmHg/ml [2, 11]. In our group of patients $SjVo_2$, TiO_2, CBF and autoregulation test did not improve significantly compared to other studies. Compliance and PVI improved after decompressive craniotomy in all alive patients and thus may be crucial in improving the outcome of severe head injured patients [12–16]. An animal study has shown that intracranial compensatory volume could still be impaired despite having normal ICP and more so when patient has intracranial hypertension [17]. We conclude the a larger group of patients needed to be studied before a direct relationship between PVI, CI and VPR can be done with prediction values followed by mathematical modeling of the size of craniectomy needed to decrease intracranial pressure and thus improve PVI, CI and VPR.

References

1. Miller JD, Garidi J (1972) Intracranial volume/pressure relationships during continuous monitoring of ventricular fluid pressure. In: Brock M, Dietz H (eds) Intracranial pressure. Springer, Berlin, pp 270–274
2. Miller JD, Garibi J, Pickard JD (1973) Induced changes of cerebrospinal fluid volume: effects during continuous monitoring of ventricular fluid pressure. Arch Neurol 28: 265–269
3. Piper I, Spiegelberg A, Whittle I, Signorini D, Mascia L (1999) A comparative study of the Spielberg compliance device with a manual volume-injection method: a clinical evaluation in patients with hydrocephalus. J Neurosurg 13(6): 581–586
4. Chambers IR, Siddique MS, Banister K *et al* (2001) Clinical comparison of the Spiegelberg parenchymal transducer and ventricular fluid pressure. J Neurol Neurosurg Psychiatry 71: 383–385
5. Dickman CA, Carter LP, Baldwin HZ *et al* (1991) Continous regional cerebral blood flow monitoring in acute craniocerebral trauma. Neurosurgery 28: 467–472
6. Hoffman WE, Charbel FT, Edelman G (1996) Brain tissue oxygen, carbon dioxide, and pH in neurosurgical patients at risk for ischaemia. Anesth Analg 82: 582–586
7. Chan KH, Miller JD, Dearden NM *et al* (1992) The effect of changes in cerebral perfusion upon middle cerebral artery blood flow velocity and jugular bulb venous oxygen saturation after severe brain injury. J Neurosurg 77: 55–61
8. Czosnyka M, Smielewski P, Kirkpatrick P *et al* (1991) Continuous assessment of the cerebral vasomotor reactivity in head injury. Neurosurg 41: 11–19
9. Albanese J, Leome M, Alliez JR, Kaya JM, Antonini F, Alliez B, Martin C (2003) Decompressive craniectomy for severe traumatic brain injury: Evaluation of the effects at one year. Crit Care Med 31(10): 2535–2582
10. Marmarou A, Shulman K, LaMorgese J (1975) Compartmental analysis of compliance and outflow resistance of the cerebrospinal fluid system. J Neurosurg 43: 523–534
11. Hase U, Reulen HJ, Meinig G, Schurmann K (1978) The influence of the decompressive operation on the intracranial pressure and the pressure volume relation in patients with severe head injuries. Acta Neurochir (Wien) 45: 1–13
12. Al-Rawi PG, Hutchinson PJA, Gupta AK *et al* (2000) Multiparameter brain tissue monitoring-correlation between parameters and identification of CPP thresholds. Zentralbl Neurochir 61: 74–79

13. Cruz J (1998) The first decade of continuous monitoring of jugular bulb oxyhemoglobin saturation: management strategies and clinical outcome. Crit Care Med 26: 344–351
14. Martin NA, Patwardhan RV, Alexander MJ *et al* (1997) Characterization of cerebral hemodynamic phases following severe head trauma: hypoperfusion, hyperemia, and vasospasm. J Neurosurg 87: 9–19
15. Newell DW, Aaslid R, Lam A *et al* (1994) Comparison of flow and velocity during dynamic autoregulation testing in humans. Stroke 25: 793–797
16. Steiner LA, Czosnyka M, Piechnik SK *et al* (2002) Continuous monitoring of cerebrovascular pressure reactivity allows determination of optimal cerebral perfusion pressure in patients with traumatic brain injury. Crit Care Med 16: 429–439
17. Dalci K, Enblad P, Piper I, Contant C, Nilsson P (2004) A model for studies of intracranial volume pressure dynamics in traumatic brain injury. J Neurotrauma 21(3): 317–327

Correspondence: Jafri Malin Abdullah, Department of Neurosciences, School of Medical Sciences, Universiti Sains Malaysia, Health Campus, 16150 Kubang Kerian, Kelantan, Malaysia. e-mail: deptneurosciencesppspusm@yahoo.com

Acta Neurochir (2005) [Suppl] 95: 315–318

Age-dependence of cerebrospinal parameters

K. König, H. E. Heissler, M. Zumkeller, and **E. Rickels**

Medizinische Hochschule Hannover, Neurochirurgische Klinik, Hannover, Germany

Summary

Objective. The records of 159 patients were reviewed who routinely underwent a classical volume-pressure testing procedure due to suspected cerebrospinal fluid circulatory disorders.

Methods. Regression analyses were carried out to explain the cerebrospinal parameters' relationships to age. Least median squares (LMS) algorithms achieved robust estimation of simple linear model parameters. Also, method of weighted regression was used because of unequal variances in the observations.

Results. Cerebrospinal elastance and resistance to outflow revealed significant LMS regressions: $y = 0.171x + 7.460$ ($n = 159$, $p < 0.0001$) and $y = 0.096x + 1.871$ ($n = 97$, $p < 0.0001$), respectively. Similarly, weighting data resulted in models such as $y = 0.151x + 6.830$ ($p < 0.0001$) for elastance and $y = 0.087x + 1.730$ ($p < 0.0001$) for resistance estimates. The intracranial pressure at rest showed no age-related dependency. Both clear clinical and morphological signs were found in 20% of patients.

Conclusion. Expecting no time-variant properties we surprisingly found a significant linear relationship in cerebrospinal parameters and age. Thus, parameter magnitudes are not likely dominated by a pathological process only but also determined by temporal system alterations.

Keywords: Cerebrospinal system; aging; elastance; resistance; regression analysis.

Introduction

Recent publications on age-related changes in cerebrospinal parameters as measured by volume-pressure tests give cause to revise relevance and clinical significance of cerebrospinal parameters.

Materials and methods

As a routine diagnostic procedure in patients with suspected cerebrospinal fluid (CSF) circulatory disorders, commercial epidural pressure transducers were implanted for long-term ICP measurement. To complete the impression of individual pressure dynamics, each patient underwent a diagnostic test for the assessment of individual volume-pressure relationship. Patients were examined in the sequence of their appearance in our clinic.

Examination setup

In all patients a lumbar volume-pressure test was performed. This test examines the individual CSF pressure responses to transient (bolus) and continuous infusion of intrathecal volume loading [5, 9]. For testing transient behaviour, a series of 1, 2, 4 and 8 ml isotonic saline was injected. Bolus test volume of 8 ml was administered in random order. For the reasons of standardization, the 8 ml bolus should have been infused within five to seven seconds. Eventually a constant infusion of 90 ml/h of isotonic saline was used to perform continuous volume loading. Throughout the examination waveforms of epidural intracranial pressure and electrocardiogram were registered by a commercial patient data monitoring system (HP 78354A, Philips Medizin Systeme, Böblingen, Germany) and routed to a personal computer for analysis.

Parameter extraction

For elastance estimation, only the pressure response to 8 ml was considered. To minimize the influence of endogenous volumetric processes during injection that potentially could have corrupted the corresponding pressure response, special attention was paid to intracranial pressures readings both directly before and after volume loading. Only these readings were used for volume-pressure response calculations. A constant rate infusion test was performed for resistance to outflow estimation. During infusion onset of a pressure plateau was assessed on-line by computation of the running mean pressure. After stabilization of the mean pressure readings within confidence limits the pressure level was defined the plateau pressure and its difference to baseline levels used for resistance calculation.

Regression analysis

Data subsets were used for regression analyses to estimate the formal relationship between variables. As computations are sensitive to outliers a so-called robust estimation algorithm was chosen. Dallal [3] introduced a regression algorithm in minimizing model error using least median deviations instead of least squares. In all regression analyses age was the explanatory variable. Data was eligible for analysis if the patient completed volume-pressure testing procedure and had almost constant heart rate over the entire examination. Linear regression was carried out on 97 resistance estimates. In 159 patients measures of cerebrospinal elastance and resting pressure were assigned for analysis.

Statistical computations were performed using JMP Statistical Software (SAS Institute, Cary, NC, USA) and LMSMVE-software package [3]. For dispersion estimates the standard error of the mean

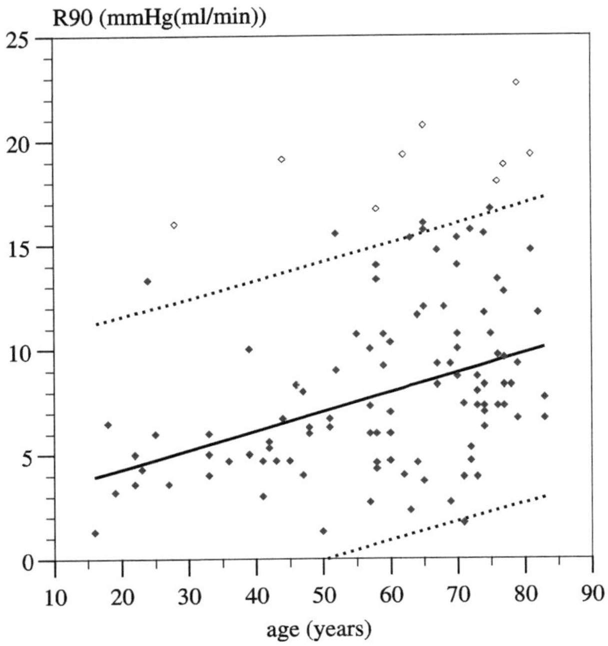

Fig. 1. Regression between resistance to CSF outflow (R90) versus age in patients with suspected CSF circulatory disorders. Regression line (solid); 95% prediction interval (dashed); outliers (open diamonds)

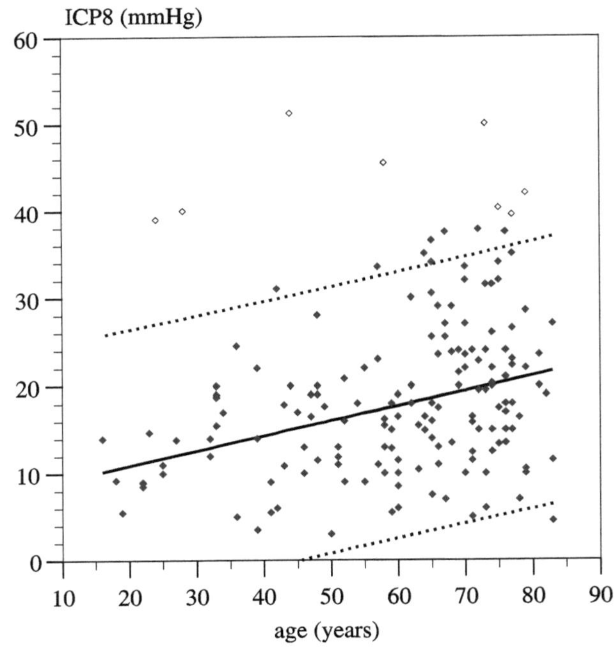

Fig. 2. Regression between cerebrospinal elastance as measured by the difference of intracranial pressure before and at the end of a rapid intrathecal infusion of 8 ml saline (ΔICP8) versus age in patients with suspected CSF circulatory disorders. Regression line (solid); 95% prediction interval (dashed); outliers (open diamonds)

was used. Significance was indicated by a two-tailed p-value of less than 0.05.

Results

Age

Patient age was distributed between 16 and 83 years of age (median 65 years). The densest 50% of observations were found between 63 and 78 years.

Resistance to CSF outflow

Regression analysis resulted in a significant positive slope (0.096 ± 0.018 mmHg/(ml/min) per year, $n = 97$, $p < 0.0001$) in linear model approach. Regression function was given by $y = 0.096x + 1.871$. Correction for heteroscedasticity yielded the model $y = 0.087x + 1.730$ ($p < 0.0001$). For this parameter median age was 64 year and 50% data density was highest between 57 and 73 years.

Cerebrospinal elastance

Analysis of elastance parameter and age revealed a significant positive slope (0.171 ± 0.037 mmHg/year, $n = 159$, $p < 0.0001$. The regression equation was $y = 0.171x + 7.460$. Recalculation for inhomogeneous variances resulted in $y = 0.151x + 6.830$ ($p < 0.0001$). Median age was 65 years and densest 50% of observations were ranging from 62 to 77 years of age.

Resting pressure

Resting pressure and age were not related as calculated by regression analysis. The average resting pressure was 15.4 ± 0.8 ($n = 159$).

Discussion

There is little knowledge published on age-related changes in cerebrospinal parameters. König [7] found a highly significant relationship between cerebrospinal elastance, resistance and age in hydrocephalic patients. Also, Albeck [1] reported from a significant regression coefficient when applying a linear model to resistance to CSF outflow and age data. Unfortunately although the slopes resembled well the patients have been completely different: Albeck's data [1] described patients with no known CSF disorders. Another remarkable feature is the overall resistance level which is more than twice as high in young patients (about 20 years) and about 1.6-times in the elderly (about 80 years) when compared to the present study. The marked difference in intercept could potentially be traced back on the different methods for the resistance calculations used. When, however, normalizing the present study's resistance data to an infusion rate of 1 ml/min the effective pressure differences used for resistance calculation were found to be in the same range. Lundar [8] determined ventricular outflow resistance by infusions rates of 1.5 ml/min and 5 ml/min. Carrying out regression analyses from his data tables published regression coefficients were 0.092 ± 0.035 mmHg/(ml/min) per year ($n = 26$, $p = 0.014$) and 0.049 ± 0.021 mmHg/(ml/min) per year ($n = 26$, $p = 0.026$), respectively. There were differences in the steepness of regression lines that assumed resistance values to be over-estimated when using an infusion rate of 1.5 ml/min. Resistance values were ranging from 4.4 to 20.7 mmHg/(ml/min) and age from 1 to 74 years. The assumption of a linear relationship between resistance and age must be rejected for a patient sample adopted from Kosteljanetz [6]. The patients in his population differ in aetiology or concurrent disease. Furthermore Kosteljanetz [6] and co-workers used Marmarou's [9] method for resistance calculation. Resistance ranged from 1 to 34.9 mmHg/(ml/min) and age from 40 to 80 years. The lacking relationship is possibly due to the missing data in patients under 40 years. Czosnyka [2] found for both elastance and resistance estimates best-fit-models such as $y = 1/(a - bx)$ in 46 hydrocephalic patients with heterogeneous aetiology. This working group found also proportionality in parameters and age. In the present study a linear relationship was found for cerebrospinal elastance. Calculation of the volume-pressure response (VPR) instead of widely used pressure-volume index (PVI) was preferred because the VPR had an almost constant error over the complete range of response magnitude whereas the PVI error was varying [4, 9–11].

Linear regression of resting pressure and age did not reach statistical significance [2, 7]. As it was the pressure after lumbar puncture and before volume loading it cannot really be considered a representative pressure reading but an artefact [1].

The data presented suffered from heterogeneity, i.e. variability in the elderly is pronounced in comparison to the young patient group. Furthermore, age is not distributed normally and thus violated the application of classical regression analysis. Albeck [1], Czosnyka [2], and König [7] pursued a comparable mathematical approach and there were hints that their datasets were featured by the same inconsistency as mentioned. To overcome these limitations in this study computations were also carried out as weighted linear regression resulting in a flattening and downward shift of graphs without loss in significance. Furthermore, bootstrap analyses were performed to verify the results of regression analyses.

In the individual case estimating resistance as a function of age is governed by wide confidence intervals that propose a simple dichotomized scale suggesting a critical and non-critical range. The terminus normal is intentionally avoided. Patients whose outflow resistance exceeds the upper confidence limit are most likely characterised by severe CSF circulatory impairment. In patients within limits the influence of age should be considered, e.g. a resistance value of more than 12 mmHg/(ml/min) becomes unlikely in patients younger than 55 years showing a CSF outflow perturbation. At the same resistance level, however, both aging and a specific pathophysiology are likely in patients older than 55 years. In similar fashion the positive slope in the elastance versus age diagram reflects the commonly accepted structural changes of the cerebrospinal system, i.e. a loss in brain tissue and a complementary increase in the CSF compartment resulting in a hardening of the system.

The results of regression analyses suggested that aging of biomechanical constituents was mirrored in the magnitude of cerebrospinal system parameters and it seemed reasonable to correct parameters for age, i.e. despite the uncertainties in reasoning about the correct model approach, the models introduced were better estimates of system parameters than the corresponding mean values or ranges.

References

1. Albeck MJ, Skak C, Nielsen PR, Olsen KS, Børgesen SE, Gjerris F (1998) Age dependency of resistance to cerebrospinal fluid outflow. J Neurosurg 89: 275–278
2. Czosnyka M, Czosnyka Z, Whitfield PC, Donovan T, Pickard JD (2001) Age dependence of cerebrospinal pressure-volume compensation in patients with hydrocephalus 94: 482–486
3. Dallal GE, Rousseeuw PJ (1992) LMSMVE A program for least median of squares regression and robust distances. Comput Biol Med 25: 384–391
4. Heissler HE, Holl K, Gaab MR, Dietz H (1993) Interfering factors in cerebrospinal parameter estimation. In: Avezaat CJJ, van Eijndhoven JHM, Maas AIR, Tans JTJ (eds) Intracranial pressure VIII. Springer, Berlin

5. Katzman R, Hussey F (1970) A simple constant-infusion manometric test for measurement of CSF absorption. I. Rationale and method. Neurology 20: 534–544
6. Kosteljanetz M, Nehen AM, Kaalund J (1990) Cerebrospinal fluid outflow resistance measurements in the selection of patients for shunt surgery in the normal pressure hydrocephalus syndrome. A controlled trial. Acta Neurochir (Wien) 104: 48–53
7. König K, Heissler HE, Zumkeller M, Fischer S (1996) Age-related Alterations in Cerebrospinal Parameters. Zentralbl Neurochir [Suppl] 74
8. Lundar T, Nornes H (1990) Determination of ventricular fluid outflow resistance in patients with ventriculomegaly. J Neurol Neurosurg Psychiatry 53: 896–898
9. Marmarou A, Shulman K, Rosende RM (1978) A nonlinear analysis of the cerebrospinal fluid system and intracranial pressure dynamics. J Neurosurg 48: 332–344
10. Miller JD, Garibi J, Pickard JD (1973) Induced changes of cerebrospinal fluid volume. Effects during continuous monitoring of ventricular fluid pressure. Arch Neurol 28: 265–269
11. Schettini A, Walsh EK (1991) Contribution of brain distortion and displacement to CSF dynamics in experimental brain compression. Am J Physiol 260: R172–178

Correspondence: K. König, Medizinische Hochschule Hannover, Neurochirurgische Klinik 7249, 30623 Hannover, Germany. e-mail: koenig.kathrin@mh-hannover.de

Biophysics

Acta Neurochir (2005) [Suppl] 95: 321–326

A computer model of intracranial pressure dynamics during traumatic brain injury that explicitly models fluid flows and volumes*

W. Wakeland[1,2] and **B. Goldstein**[3]

[1] Systems Science Ph.D. Program, Portland State University, Portland, OR, USA
[2] Biomedical Signal Processing Laboratory, Portland State University, Portland, OR, USA
[3] Complex Systems Laboratory, Oregon Health and Science University, Portland, OR, USA

Summary

A model of intracranial pressure (ICP) dynamics that uses fluid volumes as primary state variables is presented, along with clinical data for two subjects with elevated ICP. The data includes annotations to indicate the precise timing of clinical changes in cerebral spinal fluid drainage, head of bed elevation, and minute ventilation. The response to changes in the clinical parameters was used to calibrate the model to correspond to specific subjects by estimating values for key characteristics such as hematoma volume and CSF uptake resistance. The error in mean ICP predicted by the model was less than 2 mmHg when cerebral spinal fluid is drained and the head of bed elevation was increased. The error in mean ICP predicted by the model exceeded 5 mmHg during an episode when the head of bed was decreased and also during a reduction in minute ventilation. The estimated values for hematoma volume and other subject characteristics were plausible but could not be verified empirically.

Keywords: Intracranial pressure; mathematical model; simulation; clinical data.

Introduction

There are several clinical treatment options for elevated ICP following severe traumatic brain injury (TBI). These include drainage of cerebral spinal fluid (CSF) via a ventriculostomy catheter, raising the head-of-bed (HOB) elevation to 30° to promote jugular venous drainage, and mild hyperventilation [1]. However, it is currently impossible to accurately and quantitatively predict how a subject will respond to a particular combination of therapeutic interventions. In addition, under certain conditions treatments may exhibit side effects that can worsen the subject's overall condition. Tools are needed that help clinicians contend with these complex interactions. Unfortunately, the complexity of most of the simulation models reported in the literature limits their clinical usefulness.

Materials

Data was acquired from two subjects. Subject 1 was a 3-year-old female who was in a motor vehicle accident. She suffered a severe closed head injury and was found to be pneic when paramedics arrived at the scene. Her GCS at the scene was 4. She was transferred to a local community hospital where CT scan revealed a large right-sided epidural hematoma. In the emergency room she was noted to have a right pupil that was fixed, dilated and unresponsive to light, and her GCS was 3. She was tracheally intubated and taken to the operating room where the epidural hematoma was evacuated. She was then transported by helicopter to the pediatric intensive care unit (PICU) where a ventriculostomy catheter was placed. In the PICU she had equal, round, and mid-position pupils that were reactive to light. She had no spontaneous movements, no cough or gag reflex, and exhibited extensor posturing with painful stimulus. She required intravenous infusions of isotonic solutions, dopamine, dobutamine and epinephrine to maintain her blood pressure. She showed signs of severe anoxic/ischemic brain and multi-organ injury. On hospital day 2, her right pupil again became fixed, dilated and unresponsive to light. Repeat CT scan showed severe cerebral edema and cerebellar infarct. Due to the irreversible nature of her brain injury, supportive therapy was withdrawn by the family.

Subject 2 was an 8-year-old female who was a restrained passenger in a motor vehicle accident. On arrival to a local emergency hospital, she was moaning, and had a Glasgow Coma Scale (GCS) score of 8. She was tracheally intubated and transferred to the PICU. A cranial CT scan demonstrated diffuse axonal injury with no intra- or extra-cranial hemorrhage. The GCS in the PICU was 6. Pupils were 2 mm and reactive bilaterally. An intraventricular catheter was placed for ICP monitoring. Initial ICP ranged between 15–17 mmHg. The patient also had a left mandibular fracture and facial avulsion. The patient remained on a ventilator in the PICU and her ICP rose to a maximum of 38 mmHg by day 3. She was treated with mild hyperventilation ($PaCO_2$ 32–35 mmHg) and intravenous mannitol

* This research was supported in part by the Thrasher Research Fund.

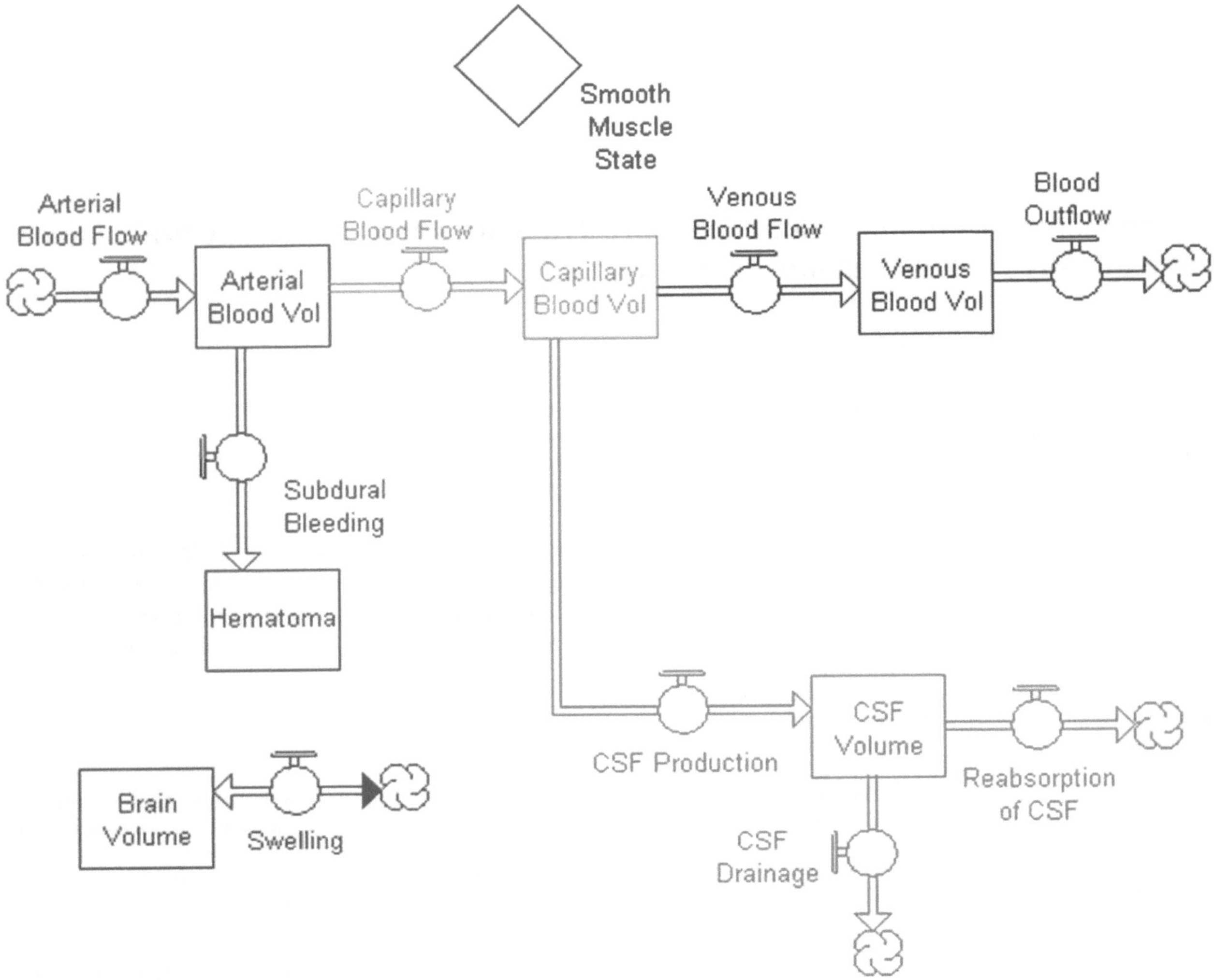

Fig. 1. Model state variables and flows. Fluid volumes are represented by rectangles, and fluid flows are represented by double arrows with a circular "valve-like" symbol superimposed on them to indicate that flows are variable. Brain swelling, the rate of change of brain volume, is also represented as a flow. The state of auto regulation is shown as a diamond, representing a submodel in this simulation software

(0.25 gram/kg IV every 3 hours). On day 5 she followed commands and was purposeful in all movements and was successfully extubated. She was transferred to the general pediatric ward on day 7, and then to a rehabilitation center on hospital day 9.

Methods

We developed a mathematical model that treats fluid volumes as the primary state variables. This is different from ICP dynamic models in the literature that treat pressures as the state variables [2, 5, 7]. To calibrate the model, we estimated a number of model parameters, including the volume and rate of change of an intracranial hematoma, CSF absorption resistance, and autoregulatory characteristics. Intra- and extracranial flows and pressures were calculated from the values of the state variables and the parameters, both fixed and estimated. Figure 1 shows the model state variables and flows, and how they are related.

Blood pressures are computed from the blood volumes and their respective compliances as shown in Equations 1–3.

$$P_{a\ ic} = P_{ic} + V_a/C_a \quad (1)$$

$$P_{c\ ic} = P_{ic} + V_c/C_c \quad (2)$$

$$P_{v\ ic} = P_{ic} + V_v/C_v \quad (3)$$

where:
$P_{a\ ic}$ = pressure in the intracranial arteries
P_{ic} = intracranial pressure (ICP)
V_a = arterial blood volume
C_a = arterial complicance
$P_{c\ ic}$ = pressure in the intracranial capillaries
V_c = capillary blood volume
C_c = capillary compliance
$P_{v\ ic}$ = pressure in the intracranial veins
V_v = venous volume
C_v = venous compliance

P_{ic} is computed as shown in Equation 4, based on the total intracranial volume and the pressure volume index (PVI) as defined by Marmarou [3]: the additional volume needed to cause a 10-fold increase in P_{ic}.

$$P_{ic} = P_{icb} * 10^{(V_{tc}-V_{bc})/PVI} \tag{4}$$

where: P_{icb} = baseline ICP for the subject
V_{tc} = total cranial volume
V_{bc} = base cranial volume

To model cerebrovascular autoregulation (CAR), the resistance at the arterioles changes rapidly, but within physiological limits, in order to adjust the cerebral blood flow to match metabolic needs, as indicated in Equations 5–7.

$$S_{sm} = \int^{t} \left[(F_{cap} - f_i) + \frac{1}{S_{sm}^3} + \frac{1}{(1 - S_{sm})^3} \right] * g \tag{5}$$

$$R_{cap} = \gamma * S_{sm} \tag{6}$$

$$F_{cap} = \frac{(P_{aic} - P_{cic})}{R_{cap}} \tag{7}$$

where: S_{sm} = smooth muscle state (~ resistance, mmHg/ml/min; 0 = fully dilated, 1 = max. constriction)
F_{cap} = capillary blood flow
f_i = indicated blood flow
g = cerebrovascular autoregulation gain (dimensionless)
R_{cap} = capillary resistance
γ = conversion factor (dimensionless)

The cubic terms in Equation 5 force the smooth muscle state to remain in the range [0, 1]. The CAR logic responds to changes in indicated blood flow needs due to diurnal variation, and changes in ICP, respiratory rate, arterial blood pressure, and HOB elevation.

The model also incorporates logic to reflect intracranial pathology such as intracranial hemorrhage and cerebral edema, and to reflect therapeutic interventions such as CSF drainage, elevation of the HOB, and changes in $PaCO_2$ using a proxy variable of minute ventilation (defined as the respiratory frequency × tidal volume). For this preliminary report, we will only present data on changes in respiratory frequency when tidal volume was held constant. The flow rate for CSF drainage, F_{CSF} is modeled as shown in Equation 8.

$$F_{csf} = \frac{(P_{ic} - P_t)}{r_c} \tag{8}$$

where: F_{csf} = rate of cerebrospinal fluid drainage
P_t = pressure at the terminus of the drainage system (a function of its elevation)
r_c = resistance of the catheter

The pressure differential, ΔP, due to changing HOB elevation is calculated as shown in Equation 9.

$$\Delta P = d * \sin(\theta) * \frac{\text{mmHg}}{\text{mmH}_2\text{O}} \tag{9}$$

where: ΔP = pressure difference
d = distance between the heart and the head
θ = head of bed angle

ΔP is subtracted from both the systemic arterial pressure and the systemic venous pressure to determine the external arterial and venous pressure at the cranial vault.

The impact of changing the $PaCO_2$ using a proxy variable of respiratory frequency (i.e., the ventilator rate) is much more complex (cf. [6]). We use the logic shown in Equations 10 and 11 to calculate the indicated blood flow that serves as input to the CAR logic.

$$f_i = k_1 + \alpha * (PaCO_2 - S) \tag{10}$$

$$PaCO_2 = MOV[(k_2 - \beta * VR), t_c] \tag{11}$$

where: k_1 = baseline blood flow
α = flow multiplier
$PaCO_2$ = actual CO_2 pressure
S = setpoint for CO_2 pressure
$MOV(x, t)$ = moving average of variable x with averaging period t
k_2 = offset for CO_2 pressure
β = conversion factor from respiration rate to CO_2 pressure
VR = respiration rate in breaths per minute
t_c = time constant

The model was calibrated to specific subjects based on recorded clinical data, including ICP, arterial blood pressure, and central venous pressure. The data used to calibrate the model was clinically annotated to indicate the exact timing of events such as changes in ventilator rate, CSF drainage and amounts, and changes in HOB elevation from 0° to 40° in 10° increments. Details of the data acquisition system were reported previously by Goldstein *et al.* [3]. Data for this report included both retrospective, non-annotated data as well as prospectively collected physiologic signals following an experimental protocol that included random changes to the height of the CSF drainage system, HOB elevation, and minute ventilation.

Results

Figure 2 shows the clinical data from the two subjects that was used the calibrate the model. Figure 2a shows the ICP signals recorded during three episodes of CSF drainage for Subject 1. These signals have been lowpass filtered to remove the pulsatile component, and synchronized in the time domain so that the point of drainage occurs at minute 5. Figure 2b shows the ICP signal recorded for Subject 2 when the HOB was changed from 30° to 0° and then back to 30°. Figure 2c shows the ICP signal recorded for Subject 2 when the ventilation rate was changed from 15 breaths per minute (bpm) to 12 bpm and then back to 15 bpm.

The model was calibrated to fit Subject 1 using the clinical data shown in Figure 2a. The estimated parameters were: *Initial intracranial hematoma volume* (24 mL), *hematoma volume increase rate* (0 mL/min.), *CSF drainage volume* (6.5 mL), and *CSF absorption resistance* (160 mmHg/mL/min). Figure 3a shows the ICP response to treatment in the model compared with the actual data. The error is less than 2 mmHg throughout the episode. On the other hand, the estimated *CSF absorption resistance* is much higher than typically reported and would normally be consistent with a higher steady state ICP value. In this case however, periodic CSF drainage therapy was employed to maintain ICP below its steady state value.

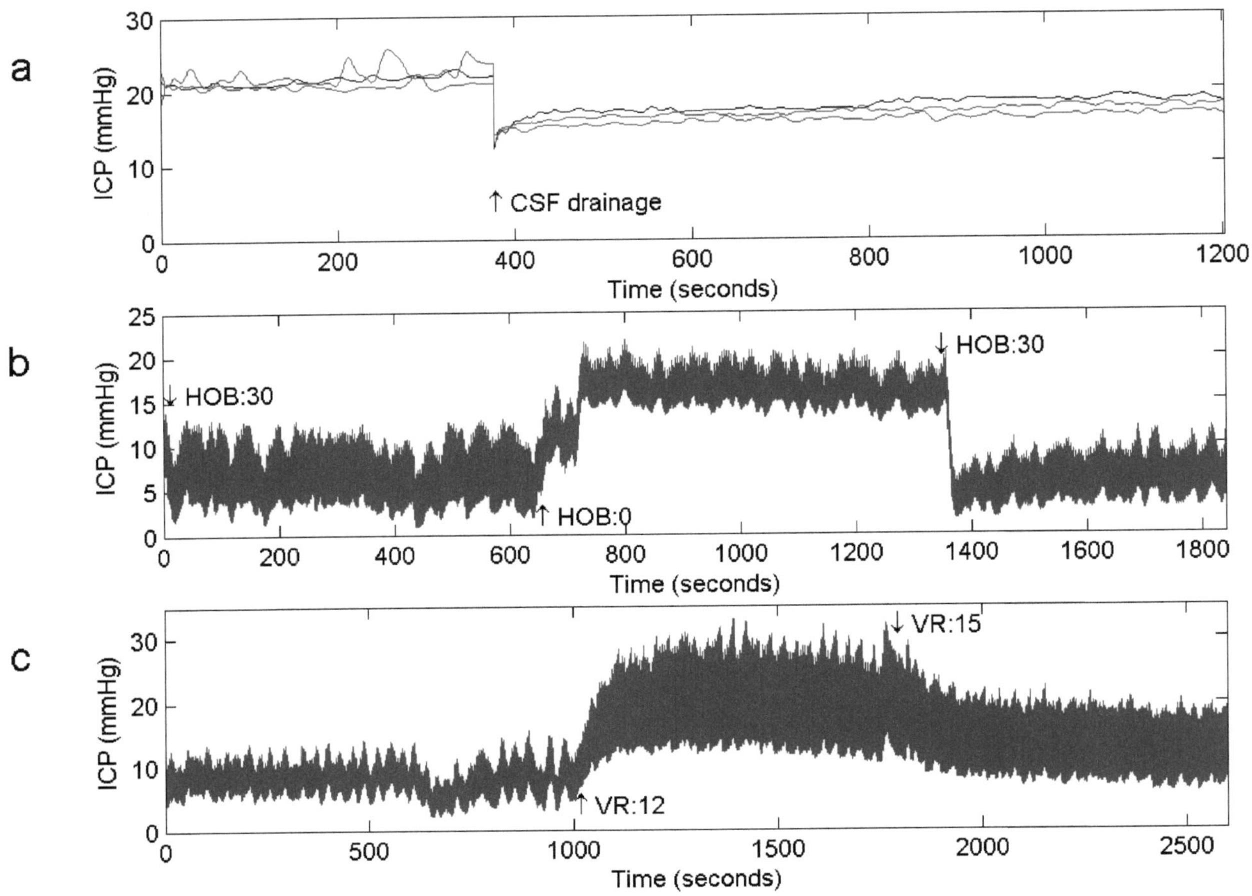

Fig. 2. Clinically recorded ICP signals. (a) ICP before and after CSF drainage for subject 1. (b) ICP during head of bed change for subject 2, from 30° to 0° and then from 0° to 30°. (c) ICP during respiration change for subject 2, from 15 breaths per minute (bpm) to 12 bpm and then from 12 bpm to 15 bpm

Next, the model was calibrated to fit Subject 2, using the data shown in Figure 2b. The estimated parameters included: *Initial intracranial hematoma volume* (6 mL), *hematoma increase rate* (.5 mL/min.), *CSF absorption resistance* (24 mmHg/mL/min), and estimated *distance from heart to brain* (45 cm). Figure 3b shows the ICP calculated by the model for this episode, compared to clinical data. During the first stage, the error is greater than 5 mmHg at some time points. During the second stage, the error is less than 2 mmHg.

The model was then further calibrated to Subject 2 based on the data collected during changes in respiratory frequency (Figure 2c). The previously-estimated parameters were retained, and the new data was used to estimate parameters associated with the CAR logic, including α (75 ml/mmHg), S (34 mmHg), k_2 (64 mmHg), β (2 mmHg-breaths/min.) and t_c (2.5 minutes). Figure 3c shows the ICP calculated by the model for this episode, compared to the clinical data. The error is greater than 5 mmHg.

Discussion

Our main finding was that this first generation ICP dynamic model was able to accurately replicate the subject's response to therapeutic intervention using CSF drainage and HOB elevation, but not following changes in respiratory frequency. The mean ICP calculated by the model for Subject 1 is very similar to the actual mean ICP recorded during their CSF drainage episodes. The large value for the estimated CSF absorption resistance implies a significant impediment to flow and/or absorption. This may be due to the initial injury, subsequent cerebral edema, or a combination. The model was also able to replicate Subject 2's overall

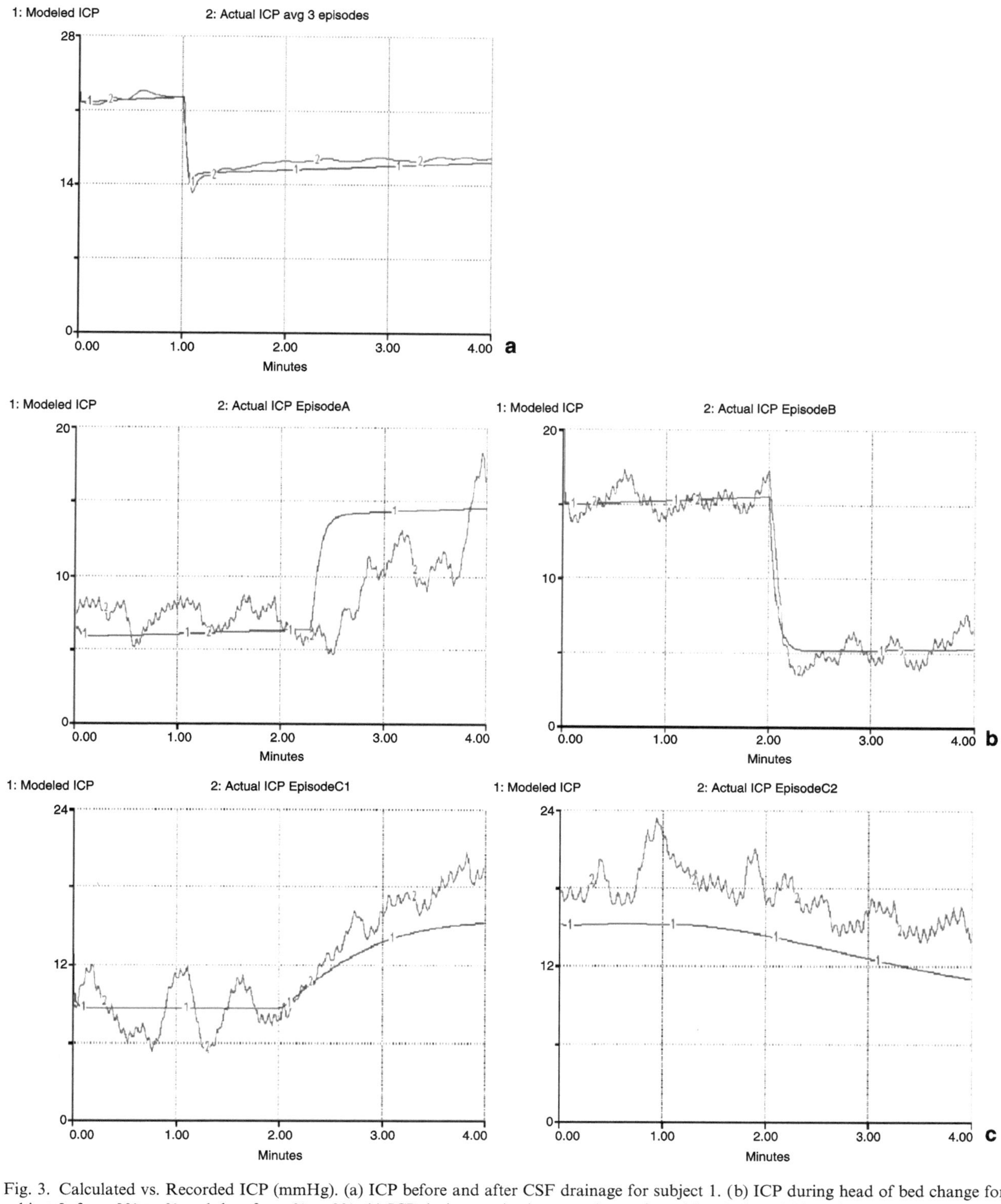

Fig. 3. Calculated vs. Recorded ICP (mmHg). (a) ICP before and after CSF drainage for subject 1. (b) ICP during head of bed change for subject 2, from 30° to 0° and then from 0° to 30°. (c) ICP during respiration rate change for subject 2, from 15 breaths per minute (bpm) to 12 bpm and then from 12 bpm to 15 bpm

ICP response to changes in HOB elevation, especially the response to raising the HOB. The modeled ICP response when HOB was lowered was less accurate. This is most likely due to the multiple autoregulatory mechanisms with differential responses triggered when the HOB is lowered and ICP begins to increase. These mechanisms currently are not incorporated in the model. The model was also not able to fully replicate the ICP signal for Subject 2 during changes in respiratory frequency. This may be due to the simplified CAR logic – indicating the need to enhance this portion of the model, or that respiratory frequency is a poor proxy for $PaCO_2$.

Other researchers have also recently focused their attention on multiple autoregulatory mechanisms. For example, Ursino and Magosso [8] demonstrate that three distinct CAR mechanisms are "... necessary to provide good reproduction of autoregulation to cerebral perfusion pressure changes." These mechanisms include reactions to CO_2 changes, the reaction to tissue hypoxia, and a mechanism that responds to pressure changes in the large pial arteries and to flow rate in the small pial arteries. This latter mechanism might correspond to myogenic or neurogenic regulation, endothelium dependent factors, or other processes. Simulation results are validated against data reported in the literature. The reaction to tissue hypoxia was noted as being particularly complex, as we found in the present study. Another example is a recently-published mathematical model of ICP dynamics by Lakin, *et al.* [4] that incorporates, in addition to CAR mechanisms, the regulation of systemic vascular pressures by the sympathetic nervous system, and the regulation of CSF production in the choroid plexus. Simulation results are inspected for reasonableness but not compared to clinical data. According to the authors, the additional autoregulatory mechanisms they describe are needed to accurately represent normal physiology as well as pathological conditions. The present study supports their assertion.

Acknowledgments

The authors greatly appreciate the support and assistance received from Mateo Aboy, Louis Macovsky, James McNames, and the Thrasher Research Foundation.

References

1. Adelson P *et al* (2003) Guidelines for the acute medical management of severe traumatic brain injury in infants, children and adolescents. Ped Crit Care Med 4(3)
2. Czosnyka M, Piechnik S, Richards H, Kirkpatrick P, Smielewski P, Pickard J (1997) Contribution of mathematical modeling to the interpretation of bedside tests of cerebrovascular autoregulation. J Neurosurg 63: 721–731
3. Goldstein B *et al* (2003) A physiologic data acquisition system and database for the study of disease dynamics in the intensive care unit. Crit Care Med 31: 433–441
4. Lakin W, Stevens S, Tranmer B, Penar P (2003) A whole-body mathematical model for intracranial pressure dynamics. Math Biology 46: 347–383
5. Marmarou A, Shulman K, Rosende R (1978) A nonlinear analysis of compliance and outflow resistance of the cerebrospinal fluid system. J Neurosurg 48: 332–344
6. Phillis (ed) (1993) The regulation of cerebral blood flow. CRC Press, Boca Raton
7. Ursino M, Lodi CA (1997) A Simple mathematical model of the interaction between intracranial pressure and cerebral hemodynamics. J App Physiol 82(4): 1256–1259
8. Ursino M, Magosso M (2001) Role of tissue hypoxia in cerebrovascular regulation: a mathematical modeling study. Ann of Biomedical Engr 29: 563–574

Correspondence: Wayne Wakeland, Systems Science Ph.D. Program, PO Box 751, Portland, OR 97208, USA. e-mail: wakeland@pdx.edu

Acta Neurochir (2005) [Suppl] 95: 327–332

Plateau waves: changes of cerebrovascular pressure transmission

M. L. Daley[1], **C. W. Leffler**[2], **M. Czosnyka**[3], and **J. D. Pickard**[3]

[1] Department of Electrical and Computer Engineering, The University of Memphis, Memphis, TN, USA
[2] Department of Physiology, The University of Tennessee Health Science Center, Memphis, TN, USA
[3] Academic Neurosurgical Unit, Addenbrooke's Hospital, Cambridge, UK

Summary

Objective. To test the validity of the hypothesis that active vasodilatation and vasoconstriction underlie the occurrence of intracranial pressure (ICP) plateau waves by evaluating corresponding changes of cerebrovascular pressure transmission of arterial blood pressure (ABP) to ICP.

Methods. Digitized recordings of ICP and ABP sampled at 30 Hz were obtained from nine patients with traumatic brain injury. For each 16.5 s recording interval mean values of ICP, ABP, cerebral perfusion pressure (CPP), and the corresponding highest modal frequency (HMF) of cerebrovascular pressure transmission were calculated.

Results. Mean ICP and HMF significantly increased ($P < 0.003$) and mean CPP decreased significantly ($P < 0.00036$) at onset of the wave. Conversely at termination, mean ICP and HMF significantly decreased ($P < 0.026$) and mean CPP significantly increased ($P < 0.028$). In addition, the strong negative correlations between mean ICP and mean CPP ($r = -0.87$) and mean HMF and CPP ($r = -0.87$) were demonstrated.

Conclusion. The findings that HMF increased at onset and decreased at the termination of plateau wave support the validity of the vasodilatatory/constriction cascade model that postulates active vasodilation at the onset and active vasoconstriction of the cerebrovascular bed at the termination of a plateau wave.

Keywords: Plateau waves; vasodilation; vasoconstriction; cerebrovascular pressure transmission; highest modal frequency.

Introduction

Plateau waves consist of a sudden rapid elevation of intracranial pressure (ICP) to 50–100 mmHg for 5 to 20 min [2, 5, 8]. After a sustained period of elevation, termination of the wave is characterized by a rapid decrease of ICP. Changes of cerebral blood volume are thought to produce corresponding changes of ICP [6]. The onset of the wave has been described by the vasodilatory cascade model, a closed loop unstable process linking active vasodilation, increased cerebral blood volume (CBV), elevated ICP and decreased cerebral perfusion pressure (CPP) [7, 8]. The relatively rapid termination of the wave has been described by a vasoconstriction cascade model in which active vasoconstrictive events lead to decreases in CBV and ICP and an increase of CPP [7, 8].

Since the vasodilatory/constriction cascade model postulates intact pressure regulation prior to onset of the plateau wave [7, 8], we surmised that during the period before the wave the critical vibration frequencies of cerebrovascular pressure transmission should be dampened. For this state of cerebrovascular tone the highest modal frequency (HMF), the highest vibration mode by which energy is transferred from ABP to ICP, will be low. However, at the onset of the wave, active vasodilation will reduce vascular tone of the cerebrovascular bed resulting in an increase of both the ability of the cerebrovascular bed to vibrate and of the HMF. For termination of the wave, the vasodilatory/constriction cascade model describes a reduction in CBV due to active vasoconstriction as the basis for the reduction of ICP. Thus, tone of the cerebrovascular bed will increase resulting in a reduction of the HMF. The purpose of this study was to test the validity of the postulated descriptive changes of the cerebrovascular bed of the vasodilatory/vasoconstriction cascade model by the evaluation of changes of the HMF of cerebrovascular pressure transmission at the onset and termination of a plateau wave.

Materials and methods

Patients

Measurement of ICP and ABP is a standard monitoring routine in the management of severely head injured patients in Addenbrooke's

Hospital, Cambridge, United Kingdom. Computer recording of the data did not influence clinical decision; the data were processed anonymously. Local Ethical Committee was informed but individual permissions were not required at the time of recording. The Institutional Review Board at The University of Memphis approved the data analysis protocol.

Pressure recording analysis

Monitoring included invasive ABP from the radial or dorsalis pedis artery. ICP was monitored using an intraparenchymal probe (Camino ICP transducer or Codman ICP MicroSensors). ICP and ABP recordings sampled at 30 Hz and CPP was calculated as the difference between mean ABP and mean ICP. The length of the pressure recordings obtained from each patient varied from 16.2 min to 117 min. To reconstruct the sampled recordings to a more accurate description of the actual pressure recordings, a moving average filter was applied to every 3 sample values of the sampled recordings.

Numerical methods

Because the pressure recordings are analyzed over a brief period of 16.5 s during relatively steady-state conditions, our method of analysis assumes that cerebrovascular pressure transmission can be described by a linear time invariant process. Selection of the appropriate model equation is essential for constructing a valid identification model [4]. We chose a cerebrovascular pressure transmission equation based on a mathematical model that has been successfully used to interpret bedside tests of cerebrovascular autoreguation [1] and is of the form:

$$\begin{aligned} &Y((n+3)T) + a_2Y((n+2)T) + a_1Y((n+1)T) + a_0Y(nT) \\ &\quad = b_1U(nT) + b_2U((n+1)T) + b_3U((n+2)T) \end{aligned} \tag{1}$$

Specifically, in the above equation, Y(nT) and U(nT) represent ICP and ABP, respectively, and T represents the sampling epoch of 33 ms. For each 500 paired samples of pressure values representing 16.5 s segments, the autoregressive moving average (ARMAX) numerical technique was applied using MATLAB System Identification Toolbox software (Mathworks, Inc., Natick, MA) to obtain the minimum least square error set of constants a_0, a_1, a_2 and b_1, b_2, b_3 for equation #1 above. These constants are used to derive the equivalent continuous-time differential equation description of cerebrovascular pressure transmission of the form:

$$\begin{aligned} &d^3Y(t)/dt^3 + aa_2d^2Y(t)/dt^2 + aa_1dY(t)/dt + aa_0Y(t) \\ &\quad = bb_1U(t) + bb_2dU(t)/dt + bb_3d^2U(t)/dt^2 \end{aligned} \tag{2}$$

The eigenvalues of this differential equation are the modal radian frequencies of the cerebrovascular pressure transmission and are the roots of the polynomial equation:

$$\lambda^3 + aa_2\lambda^2 + aa_1\lambda^1 + aa_0 = 0 \tag{3}$$

The highest modal radian frequency is defined as the eigenvalue with the greatest absolute value and is converted to modal frequency by division by 2*pi.

The ABP recording obtained from either the radial artery or dorsalis pedis artery does not have the same time relationship to the ICP recording as does an ABP recording obtained from one of the major arteries entering the intracranial subarachnoid space. To account for the time difference between recordings, the computation is done on 10 paired recordings. These paired recordings are obtained by shifting the start of the 16.5 s ICP recording in steps of 33 ms over a range of ±165 ms relative to the same 16.5 s ABP recording. As a result, 10 sets of constants each derived by the ARMAX technique were determined and the best fit set which produced minimum least square error was selected to describe the cerebrovascular pressure transmission between ABP and ICP and the corresponding value of HMF.

Results

The initial numerical step in implementation of an identification technique is to remove the mean value from each data set [8]. As a result, computation of HMF is based solely on the variation of ICP with respect to ABP. Examples of the ICP and ABP recordings with the mean removed and the corresponding simulated ICP waveform computed by the autoregressive moving average technique described above are given in Fig. 1. An example of ICP and ABP recordings containing a plateau wave along with serial mean values of ICP, ABP, and CPP with each corresponding value of HMF is given in Fig. 2. Serial values of HMF demonstrate a sudden significant increase ($P < 0.001$, $n = 20$) from a mean baseline value of 2.42 (±0.12) Hz before the plateau wave to a mean value of 3.97 (±0.50) Hz during the plateau wave. After more than 5 min with CPP below 50 mmHg, termination of the wave occurs.

Mean values of ABP, ICP, and CPP were computed from the 500 paired pressure values representing 16.5 s of paired ABP and ICP recordings. Three recordings contained a complete wave with periods corresponding to before, during, and after the wave. Five recordings contained the onset of a wave, and one recording contained a termination of a wave. For each phase of each set of pressure recording, at least 10 serial mean values representing 16.5 s of paired ABP and ICP recordings were used to compute the grand mean (±standard deviation) of ICP, ABP, CPP, and HMF. The grand mean value (±standard deviation) for the three periods are shown in Table 1.

Individual grand mean values of HMF, ICP, and CPP for each period in each recording were plotted (see Fig. 3). Grand mean ICP was found to inversely correlate with grand mean CPP with $r = -0.85$ with regression line parameters of slope and intercept of −0.78 and 75.5 mmHg, respectively (see Fig. 3a). Also, grand mean HMF inversely correlated with grand mean CPP with $r = -0.87$ with regression line parameters of slope and intercept of −0.083 Hz/mmHg and 7.68 Hz, respectively (see Fig. 3b).

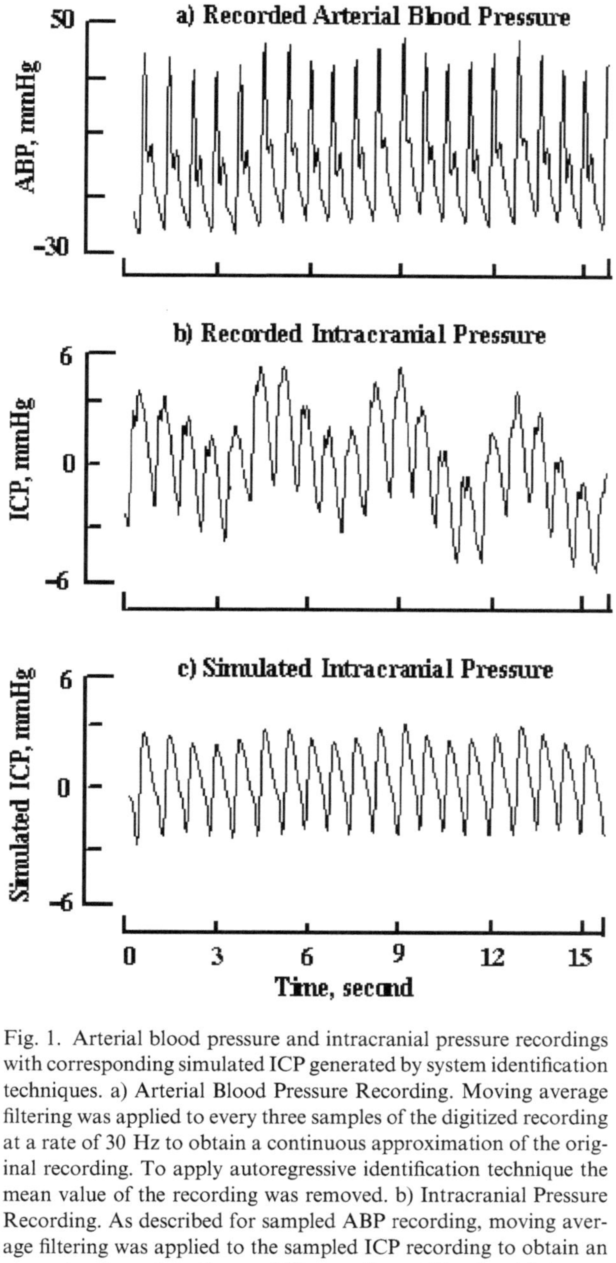

Fig. 1. Arterial blood pressure and intracranial pressure recordings with corresponding simulated ICP generated by system identification techniques. a) Arterial Blood Pressure Recording. Moving average filtering was applied to every three samples of the digitized recording at a rate of 30 Hz to obtain a continuous approximation of the original recording. To apply autoregressive identification technique the mean value of the recording was removed. b) Intracranial Pressure Recording. As described for sampled ABP recording, moving average filtering was applied to the sampled ICP recording to obtain an approximation of continuous ICP recording. c) Simulated Intracranial Pressure. The Autoregressive Moving Average (ARMAX) identification technique was applied to recordings shown in a) and b) above. The ARMAX identification technique determines by least square criteria the set of best fit constants for equation #1 given in Materials and Methods section. By using the ABP recording shown above as the input function to eqn #1, the simulated ICP recording is produced. Of note is that the identification technique is based on variation of pressure and is not dependent on mean values

Discussion

The vasodilatory/constrictive cascade model postulates the occurrence of dynamic structural and functional changes of the cerebrovascular bed associated with the onset and termination of a plateau wave. To evaluate the postulated descriptive changes of the cascade model, a novel measure, the HMF of cerebrovascular pressure transmission, was derived from paired ABP and ICP recordings obtained before, during, and after a plateau wave. Mean HMF significantly increased at onset and significantly decreased at termination of plateau wave (see Table 1). Such a finding provides additional support for the active vasodilatory and constrictive changes of the cerebrovascular bed at onset and termination of a plateau wave described by the cascade model.

To some, our findings may seem obvious, intuitive, and established knowledge ever since ICP and CPP monitoring were initiated. What is original in this study is use of the methodology which employs modal frequency analysis on clinical pressure recordings. Recently, demonstrated in a laboratory study that during active cerebrovascular pressure regulation the relationship between HMF and CPP is an inverse one in uninjured preparations [3]. In this study the relationship between HMF and CPP was also found to be inverse (see Fig. 3b) which is consistent with our laboratory finding that pressure regulation is intact when the HMF is inversely related to CPP [3].

Use of HMF analysis enables both a structural and functional understanding of changes in cerebrovascular pressure transmission which may occur during a plateau wave. In particular, the rapid increase of HMF at the onset of the wave (see Table 1 and Fig. 2f) is consistent with the structural and functional changes of the cerebrovascular bed described by the vasodilatory cascade model [7, 8]. Structurally, dilation results in an increase of the compliance of the arterial-arteriolar bed and a decrease of intracranial compliance primarily due to decreased compliance of the cerebrovascular venous bed. Functionally, vasodilation results in a significant reduction of arterial-arteriolar resistance and increased intravascular pressure within the cerebrovascular bed. At termination of the wave, the structural and functional changes of vasoconstriction are the reverse of those that occur during vasodilation. Compliance of the arterial-arteriolar bed is reduced because of increased vascular tone, intracranial compliance increases due to increased compliance of the cerebrovascular venous bed. Functionally, vasoconstriction results in an increase of arterial-arteriolar resistance and a decrease of intravascular pressure with the cerebrovascular bed. These changes are consistent with the decrease of

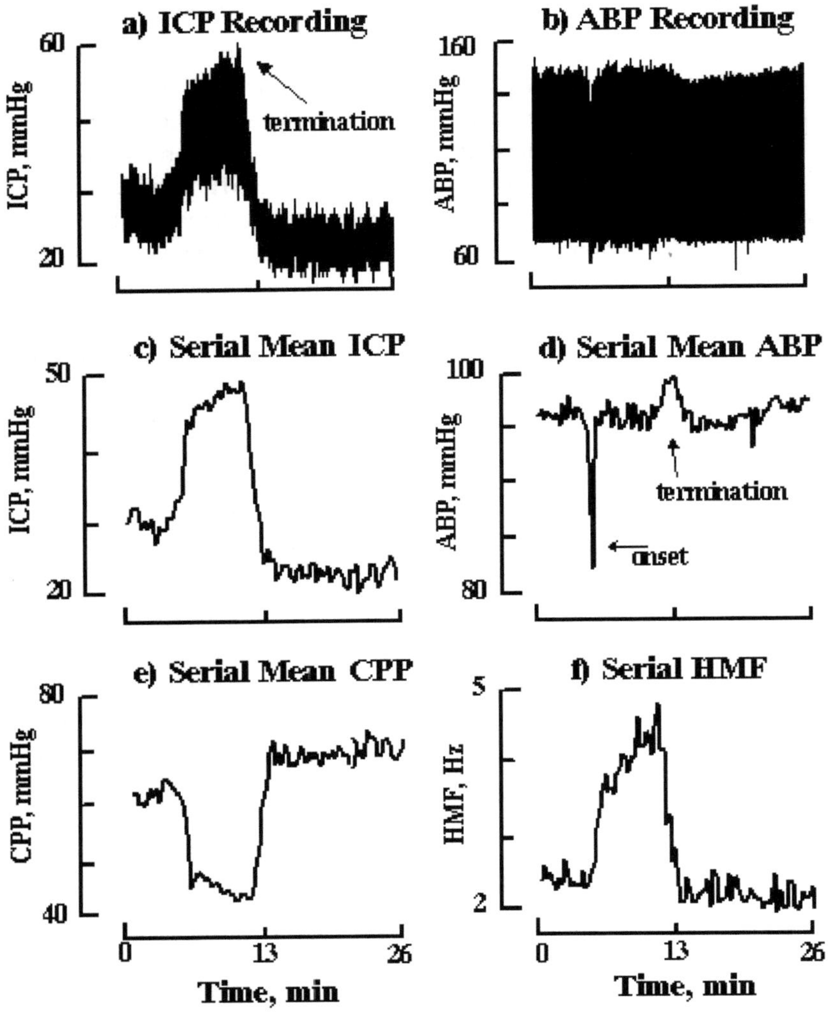

Fig. 2. Plateau wave: ICP and ABP recording, serial mean ICP, ABP, and CPP values with corresponding highest modal frequency derived from serial segments of 500 samples of ABP and ICP recordings. (a) ICP Recording. ICP recording illustrates complete plateau wave over a 26 min period. (b) ABP Recording. Corresponding ABP recording over same period as ICP recording in a). (c) Serial Mean ICP. For each 500 serial samples of ICP, mean ICP was computed and plotted at 16.5 s intervals. Standard error within symbol for each plotted point. (d) Serial Mean ABP. Mean ABP computed and plotted as described in c) above. Standard error within symbol for each plotted point. (e) Serial Mean CPP. For each 500 samples of difference between ABP and ICP, mean CPP was computed and plotted. Standard error within symbol for each plotted point. (f) Serial HMF. For each serial 500 samples of paired ABP and ICP recordings, an autoregressive moving average technique was used to determine the best set of constants for cerebrovascular pressure transmission described by equation #1. Given these constants, the highest modal frequency of cerebrovascular pressure transmission for each series of 500 consecutive samples representing 16.5 s was determined and plotted. Onset of wave associated with brief salient decrease in ABP. Termination of plateau period associated with brief increase in ICP and ABP

the HMF observed at the termination of the wave (see Table 1 and Fig. 2f).

Our reported strong correlative linear relationships between mean ICP, mean CPP, and mean HMF all support the contention that all of the patients had some degree of pressure regulation. Rosner has previously reported that ICP correlated indirectly with CPP for a group of patients who demonstrated plateau waves [7]. Here we report a similar finding for our patient group with the correlation between mean ICP and mean CPP at $r = -0.85$ (see Fig. 3a). With cerebral arterial pressure regulation, an increase in CPP would cause active vasoconstriction resulting in a reduction in CBV and ICP; whereas, a decrease in CPP would cause active vasodilation resulting in an increase of CBV and ICP [8]. The strong negative correlation ($r = -0.87$) between mean HMF and mean CPP further supports the contention that pressure regula-

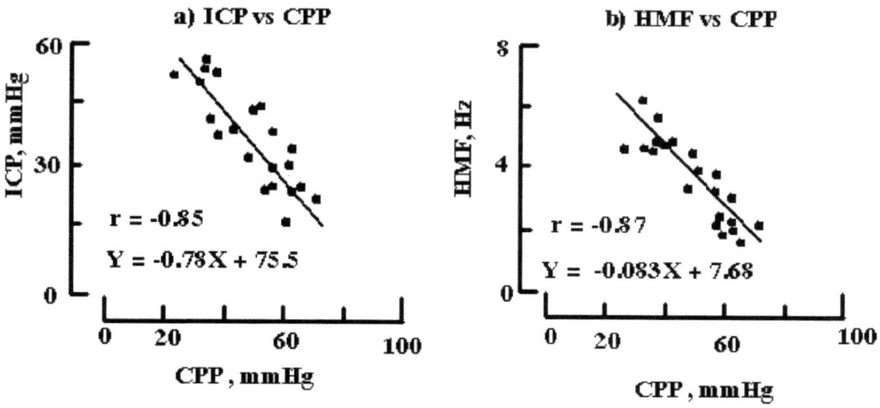

Fig. 3. Inverse relationship between intracranial pressure, highest modal frequency, and cerebral perfusion pressure. (a) ICP vs. CPP. Mean paired values of ICP and CPP for all recordings segments (N = 21) are plotted. Standard error bar is within each symbol. Linear regression line parameters of slope and intercept are −0.78 and 75.5 mmHg, respectively, with correlation value of −0.85 at $P < 0.001$. (b) Highest Modal Frequency vs. CPP. As described in a) above mean values of HMF and CPP were determined and plotted. Linear regression parameters of slope and intercept are −0.083 Hz/mmHg and 7.68 Hz with correlation value of −0.87 at $p < 0.001$

Table 1. *Grand mean (±S.D) of ICP, HMF, CPP, & ABP before, during, and after plateau wave*

Phase of wave and No. recordings	ICP (mmHg)	HMF (Hz)	CPP (mmHg)	ABP (mmHg)
Before	28.3 (±5.5)	2.69 (±1.01)	58.2 (±7.6)	86.5 (±5.2)
(n = 8)	P < 0.0003[a]	P < 0.00003	P < 0.00036	
During	48.5 (±6.6)	4.71 (±0.70)	37.2 (±8.41)	86.4 (±7.6)
(n = 9)	P < 0.017	P < 0.026	P < 0.028	
After	21.9 (±8.9)	2.32 (±2.32)	67.1 (±7.84)	89.3 (±10.1)
(n = 4)				

[a] Level of significance between grand means was determined using Student's t-test

tion was at least partially intact in all patients (see Fig. 3b). With pressure regulation, increases of CPP cause active vasoconstriction which produces structural and functional changes of the cerebrovascular bed that are consistent with the reduction of the HMF. The converse also occurs with active vasodilation associated with reduced CPP. In contrast, with loss of pressure regulation, an increase of CPP would cause passive vasodilation (distension) and an increase of the HMF, while a decrease of CPP would cause passive vasoconstriction (compression) and a decrease of the HMF. The HMF would be directly related to CPP.

In summary, predicted changes in the cerebrovascular bed during a plateau wave described by the vasodilatory/constriction cascade model were evaluated by examining changes in the HMF of cerebrovascular pressure transmission of ABP to ICP. The predicted structural and functional changes of the cerebrovascular bed associated with the cascade model were validated by the finding that mean HMF values increased significantly at the onset and decreased significantly at termination of the wave. In addition, the strong inverse linear correlation between HMF and CPP supports the contention that all patients who demonstrate plateau waves have some degree of cerebrovascular pressure regulation at onset and termination of the plateau wave.

Acknowledgments

This project was partially (MC and JDP) supported by the UK Government Technology Foresight Initiative, and the Medical Research Council (Grant No G9439390 ID 65883). This research was also supported in part by the NHLB National Institutes of Health.

Dr. M. Czosnyka is on leave from Warsaw University of Technology, Poland.

References

1. Czosnyka M, Piechnik P, Richards HK, Kirkpatrick P, Smielewski P, Pickard JD (1997) Contribution of mathematical modeling to the interpretation of bedside tests of cerebrovascular autoregulation. J Neurol Neurosurg Psychiat 63: 721–731
2. Czosnyka M, Smielewski P, Piechnik S, Schmidt E, Al-Rawi PG, Kirkpatrick PJ, Pickard JD (1999) Hemodynamic characterization of intracranial pressure plateau waves in head-injured patients. J Neurosurg 92: 11–19
3. Daley ML, Pourcyrous M, Timmons SD, Leffler CW (2004) Assessment of cerebrovascular autoregulation: changes of highest modal frequency of cerebrovascular pressure transmission with cerebral perfusion pressure. Stroke 35: 1952–1956

4. Ljung L (1987) Chapter 1: Introduction. In: System identification: theory for the user. Prentice Hall, Upper Saddle River, NJ, pp 1–12
5. Lundberg N (1960) Continuous recording and control of ventricular fluid pressure in neurological practice. Acta Psychiat Neurol Scand (Copenhagen) (thesis)
6. Risberg J, Lundberg N, Ingvar DH (1969) Regional cerebral blood volume during acute transient rises of the intracranial pressure (plateau waves). J Neurosurg 31(3): 303–310
7. Rosner MJ (1986) The vasodilatory cascade and intracranial pressure. In: Miller JD, Teasdale GM, Rowan JO (eds) Intracranial pressure VI. Springer, Berlin, pp 137–141
8. Rosner MJ, Becker DP (1984) Origin and evolution of plateau waves. Experimental observations and a theoretical model. J Neurosurg 60(2): 312–324

Correspondence: Michael L. Daley, Department of Electrical and Computer Engineering, The University of Memphis, Engineering Science Building, Rm. 208B, Memphis, TN 38152-3180, USA. e-mail: mdaley@memphis.edu

Acta Neurochir (2005) [Suppl] 95: 333–336

Brain tissue biomechanics in cortical contusion injury: a finite element analysis

A. Peña[1,2], **J. D. Pickard**[1,2,3], **D. Stiller**[4], **N. G. Harris**[1,2,3], and **M. U. Schuhmann**[5]

[1] Academic Neurosurgery Unit, University of Cambridge, Cambridge, UK
[2] Wolfson Brain Imaging Centre, University of Cambridge, Cambridge, UK
[3] Cambridge Centre for Brain Repair, University of Cambridge, Cambridge, UK
[4] In-vivo MRI Lab, Boehringer Ingelheim KG, Biberach, Germany
[5] Department of Neurosurgery, University of Leipzig, Leipzig, Germany

Summary

The controlled cortical impact model has been used extensively to study focal traumatic brain injury. Although the impact variables can be well defined, little is known about the biomechanical trauma as delivered to different brain regions. This knowledge however could be valuable for interpretation of experiment (immunohistochemistry etc.), especially regarding the comparison of the regional biomechanical severity level to the regional magnitude of the trauma sequel under investigation. We used finite element (FE) analysis, based on high resolution T2-weighted MRI images of rat brain, to simulate displacement, mean stress, and shear stress of brain during impact. Young's Modulus E, to describe tissue elasticity, was assigned to each FE in three scenarios: in a constant fashion ($E = 50$ kPa), or according to the MRI intensity in a linear ($E = [10, 100]$ kPa) and inverse-linear fashion ($E = [100, 10]$ kPa). Simulated tissue displacement did not vary between the 3 scenarios, however mean stress and shear stress were largely different. The linear scenario showed the most likely distribution of stresses. In summary, FE analysis seems to be a suitable tool for biomechanical simulation, however, to be closest to reality tissue elasticity needs to be determined with a more specific approach, e.g. by means of MRI elastography.

Keywords: Experimental traumatic brain injury; controlled cortical impact; magnetic resonance imaging; finite element analysis; brain tissue biomechanics; Young's modulus E; tissue elasticity.

Introduction

Since its original description in the late eighties [5, 6] the controlled cortical impact injury model (CCI) has been extensively used in different forms to investigate focal traumatic brain injury. The rationale behind these studies is that this type of injury might partially replicate human brain contusions. A current Medline search with the keywords "CCI" and "controlled cortical impact" will yield hundreds of publications. One virtue of the model is, that it uses a known impact interface, a measurable impact velocity and a predefined depth and duration of cortical compression and thus is regarded as biomechanically well-defined. It has been assumed that the quantifiable single mechanical input would permit to conduct a standardized statistical analysis (e.g. correlation) between the amount of deformation and the resultant pathology and functional changes [6]. However, although the "external" biomechanical parameters can be well defined and described, the "internal" biomechanics of the brain during impact according to the regionally different tissue elasticity, the amount of compression/pressure and the regional distribution of tissue pressures and strains is unknown and has not been systematically investigated at large. Looking at the distribution of e.g. post-traumatic changes in CBF maps, MRI images or (immuno-) histopathological stains, it remains often unclear to which extent the observed findings are directly related to the local severity of the primary biomechanical trauma or rather to the many secondary and tertiary reactions which widely unfold immediately following the biomechanical action. A deeper insight into this complex regional interaction between a predominant "biomechanical" trauma and a predominant secondary/tertiary "neurochemical and molecular" trauma would be of significant value. For example, such a distinction could help identifying those areas around contusions, which do not receive a lethal biomechanical impact and therefore might be the target for salvage strategies summarized under the term "neuroprotection". Furthermore, it has become evident that "CCI" differs between labs, and it is unknown how the different variables (brain size, impact location, speed, depth, duration) actually influence comparability of results between labs. Therefore a biomechanical description of the applied trauma seems of high interest to better interpret own results and to create a basis for a biomechanical standardization of the brain trauma. Our study is a first attempt to achieve those goals by using a finite element analysis of the CCI, based on high resolution T2-weighted MRI images of rat brain, for a biomechanical simulation of the resulting trauma inside the brain tissue.

Materials and methods

Several high resolution T2 weighted image series were obtained from a 260 g male Sprague Dawley rat in a Bruker Biospec 47/20 scanner at 4.7 T running Paravision® software (Bruker GmbH, Ettlingen, Germany). We used a CPMG sequence with a TR 5000 ms, and different TE of 15, 30, 45, 60, 75, 90, 105, 120 ms, 8 echoes, FOV = 2.56 × 2.56 cm, 2 averages, 1 mm slices. We obtained at TE 45 ms the best contrast and delineation between white and grey matter and between hippocampus, basal ganglia and ventricles and chose a slice corresponding to −3.6 mm to Bregma. (see Fig. 1a) as basis for the finite element analysis. Maps of stress (force per unit area) were calculated following the indentation. The Navier-Cauchy governing equations:

$$\mu\nabla^2 u_i + (\lambda + \mu)\nabla\nabla u_i + \rho b_i = 0 \qquad (1)$$

where ∇ is the Laplacian operator, λ and μ are Lame's elasticity constants, u_i is the displacement vector, ρ is the density and b_i is the body force vector, were solved using the FE method. Stress was quantified in terms of the magnitude of the mean stress (p) and the von Mises shear stress (q). The equations (1) were solved using the FE software FEMLAB 2.0 with the Structural Mechanics Module (COMSOL Ltd, John Eccles House, The Oxford Science Park, Oxford OX4 4GP, UK). Second order triangular elements were used.

In order to ensure the convergence of the numerical solution, the analysis was repeated at two levels of mesh refinement. Delaunay tringulation was used to generate the meshes. Analyses were done using an Intel Pentium 4 computer with 512MB RAM and on HP-UX Unix workstations. Because of the uncertainty regarding the tissue elasticity, we considered three scenarios relating the magnitude of Young's modulus (E) and the MRI intensity (I). The first is the *linear*, in which E is directly proportional to I. The second is the *inverse linear*, in which E is inversely proportional to I. Finally, the constant scenario in which E is constant and thus independent of I.

For simulating the impact, we used standard previously-defined parameters: impactor diameter 5 mm, impact speed 4 m/s, impact depth 2.5 mm [8, 9].

Results

Maps were calculated with *deformed* coordinates – corresponding to the time of impact, when the brain is indented – and *un-deformed* coordinates – corresponding to the time thereafter, when the brain has resumed its previous shape. At each scenario maps were computed for displacement, mean stress (corresponding to compression or pressure) and shear stress (corresponding to tissue distortion). Figure 2 shows the resulting maps with un-deformed coordinates.

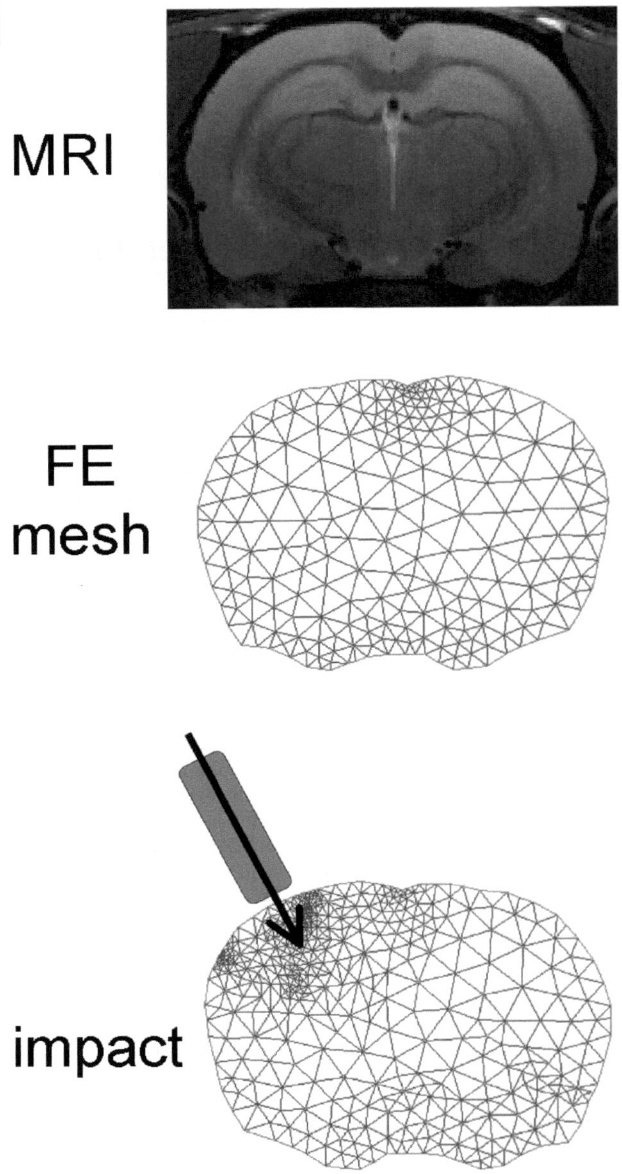

Fig. 1. Shows the high resolution T2 weighted MR image, which formed the basis for construction of the finite element mesh and the assignment of tissue properties (elasticity) to the single triangular shaped finite elements. On the left side of the brain, an impact with a round 5 mm indentor travelling at 4 m/s to an impact depth of 2.5 mm was simulated

Tissue displacement (Fig. 2a) was similar in all three scenarios of tissue elasticity with only little variation.

Mean stress distribution (Fig. 2b), however, varied considerably. With constant elasticity (all brain regions were equally hard or soft), there was a rather constant mean stress distribution inside the tissue with highest compression in the "contusion core" at and

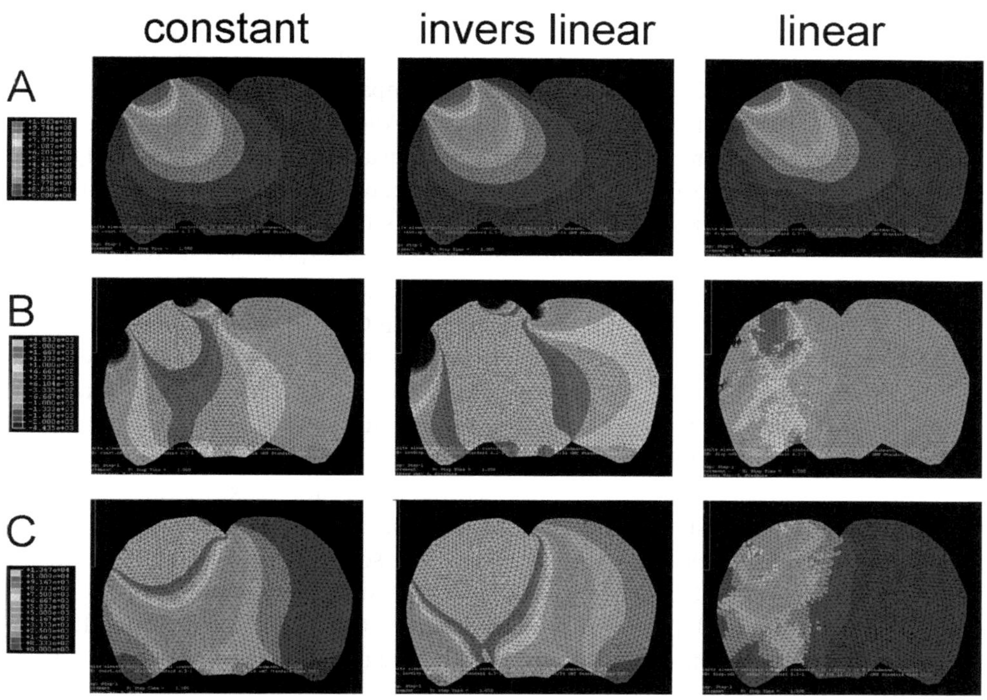

Fig. 2. Shows the resulting computations in three rows: (A) tissue displacement, (B) mean stress (positive values correspond to tissue compression, negative values correspond to tissue stretch) and (C) shear stress (tissue distortion). All of these based on the undeformed coordinates. In the first column are results for constant brain elasticity (E = 50 kPa), in the second for an inverse-linear relationship between MRI intensity and elasticity (the brighter the pixels the softer the tissue) and in the third for a linear relationship between MRI intensity and elasticity (the brighter the pixels the harder the tissue)

immediately below the impact site and higher compression along the axis of impact direction. In the cortex laterally to the impact site the map showed negative compression (tissue stretch – negative pressures). In the inverse-linear scenario (the brighter pixels in the MRI, the softer the tissue, e.g. the corpus callosum was harder than cortex and hippocampus) the biomechanical impact appeared stronger, with a larger zone of highest mean stress – compression – down to brain above the skull base and also a pronounced tissue compression on the contralateral side. In the linear scenario (the brighter pixels in the MRI, the harder the tissue, e.g. the corpus callosum was softer than cortex and hippocampus) the biomechanical burden appeared lower and more scattered than in the constant scenario, with highest levels only at the borders of the impact zone, smaller areas of negative compression – stretch – laterally and some areas of stress discontinuity below the corpus callosum. At the ipsilateral base of the brain there was an increase of compression like a côntre-coup effect. No changes in mean stress appeared on the contralateral side.

Regarding shear stress (Fig. 2c) there were similar findings in all three scenarios of tissue elasticity. With constant elasticity (all brain regions were equally hard or soft) there appeared a broad area of maximum shear stress – tissue distortion, which was larger than the volume of maximum mean stress. Centrifugally there was a rather constant decrease of shear stress distribution inside the tissue with higher distortion again along the axis of impact direction. In the inverse-linear scenario (the brighter pixels in the MRI, the softer the tissue, e.g. the corpus callosum was harder than cortex and hippocampus) the biomechanical distortion appeared again much stronger, with a very large zone of highest shear stress almost down to the skull base and also shear stress on the contralateral side. In the linear scenario (the brighter pixels in the MRI, the harder the tissue, e.g. the corpus callosum was softer than cortex and hippocampus) the biomechanical impact again appeared lighter, with scattered areas of shear stress discontinuity. Higher distortion was seen mainly at the borders of the impact zone, and below the corpus callosum. No effects were seen on the contralateral side.

Discussion

We used a two-dimensional approach in the CCI model of focal brain contusion to explore the ability of finite element modelling for simulating primary biomechanical trauma – and how it relates to "mechanical" variables such as displacement, mean stress and shear stress. High resolution MRI appears to be a adequate to be used to create the underlying finite element mesh.

Obviously the knowledge of brain elasticity is crucial. In our constant model we used an E = 50 kPa for the whole brain according to the literature [4]. This is clearly unrealistic. Regional tissue properties can be expected to influence regional elasticity. For example, the corpus callosum with a dense axonal and oligodendrocytic structure should be different in elasticity than the cortex. Areas of packed neuronal density like basal ganglia should be different from the hippocampus with its distinct neuronal layers. The next obvious step is to model brain tissue based on a functional relationship between MR intensity and Young's modulus. Our analysis was based on high resolution T2 weighted MR imaging. It is unknown if such relationship exists, however, for modelling purposes we assumed a linear (the brighter the pixel the harder the tissue) and inverse-linear (the brighter the pixel the softer the tissue) relationship between T2 signal intensity and tissue elasticity, in order to capture the influence of regional variations of tissue elasticity on the distribution of the internal stresses following impact.

As expected, our three scenarios had very small influence on the tissue displacement, after the indentation of the impactor. In contrast, we could observe dramatic differences for mean and shear stress, in other words, between the amount of tissue compression or stretch and tissue distortion between both models. Most literature describes a predominantely unilateral reaction of glial, microglial, and neuronal tissue following trauma [1–3, 7]. Therefore it can be assumed, that the simulation with the linear relationship between tissue intensity in T2 images and brain elasticity, which showed a unilaterally confined compression and tissue distortion as well as influences of regional tissue elasticity on stress patterns, is closer to reality than the inverse-linear relationship.

Our three different models with constant and varying elasticity demonstrate that the key feature of a modelling approach is a precise knowledge of regional tissue elasticity. Only then a simulation will be reliable enough to use it for standardisation of the CCI trauma and to investigate the effects of varying trauma severity with regard to impact depth and impact velocity on a theoretical basis. And – most importantly – only then this method will gain a potential usefulness with regard to the differentiation between a contribution of the primary trauma or of secondary neurochemical effects if it comes to the interpretation of e.g. regionally inhomogeneous patterns of neuronal, glial, microglial reactions or changes in CBF or glucose consumption.

MR elastography, which is a novel technique to obtain maps of tissue elasticity in vivo, might be a possible way to assign tissue properties on a pixel-by-pixel basis, such that individual finite elements can reflect the local elastic properties making the simulation much more accurate.

In summary, we successfully used high resolution MRI as a basis for finite element modelling of the biomechanical primary trauma in the CCI model. We demonstrated that regional tissue elasticity has a large influence on how the forces created by the initial tissue displacement are transferred into the surrounding tissues, and these in turn affect the regional distribution patterns of mean stress and shear stress following CCI. The linear scenario showed the distribution of stresses that most resembles experimental data.

References

1. Baldwin SA, Scheff SW (1996) Intermediate filament change in astrocytes following mild cortical contusion. Glia 16(3): 266–275
2. Chen S, Pickard JD *et al* (2003) Time-course of cellular pathology after controlled cortical impact injury. Exp Neurol 182: 87–102
3. Dunn-Meynell AA, Levin BE (1997) Histological markers of neuronal, axonal and astrocytic changes after lateral rigid impact traumatic brain injury. Brain Res 761(1): 25–41
4. Fung YC (1994) Biomechanics: mechanical properties of living materials. Prentice-Hall, Englewood Cliffs, NJ
5. Lighthall JW (1988) Controlled cortical impact: a new experimental brain injury model. J Neurotrauma 5(1): 1–15
6. Lighthall JW, Dixon CE *et al* (1989) Experimental models of brain injury. J Neurotrauma 6(2): 83–97
7. Newcomb JK, Zhao X *et al* (1999) Temporal profile of apoptotic-like changes in neurons and astrocytes following controlled cortical impact injury in the rat. Exp Neurol 158(1): 76–88
8. Schuhmann MU, Mokhtarzadeh M *et al* (2003) Temporal profiles of cerebrospinal fluid leukotrienes, brain edema and inflammatory response following experimental brain injury. Neurol Res 25(5): 481–491
9. Schuhmann MU, Stiller D *et al* (2003) Metabolic changes in the vicinity of brain contusions: a proton magnetic resonance spectroscopy and histology study. J Neurotrauma 20(8): 725–743

Correspondence: Martin U. Schuhmann, Klinik und Poliklinik fuer Neurochirurgie, Universitaetsklinikum Leipzig, Liebigstrasse 20, 04103 Leipzig, Germany. e-mail: mus@uniklinik-leipzig.de

Acta Neurochir (2005) [Suppl] 95: 337–339

Modifications of spontaneous cerebral blood flow oscillations during cardiopulmonary bypass

J. Nicolet[2], **T. Gillard**[2,3], **G. Gindre**[2,3], **F. Cervenansky**[3], **C. Duale**[2], **J. E. Bazin**[2], **C. De Riberolles**[1], **P. Schoeffler**[2], and **J. J. Lemaire**[3]

[1] Cardiovascular Surgery Department, University Hospital and Auvergne University, Clermont-Ferrand, France
[2] Anaesthesiology Department, University Hospital and Auvergne University, Clermont-Ferrand, France
[3] ERIM-EA3295, University Hospital and Auvergne University, Clermont-Ferrand, France

Summary

Spontaneous slow waves are present in the systemic circulation including the intracranial compartment. They are supposed to reflect the cerebral autoregulation. We hypothesised that in the absence of cardio respiratory variability, during cardiopulmonary bypass (CPB), we should reveal extreme physiologic controls.

Material/methods. Ten patients were included. Arterial blood pressure (ABP, radial invasive), extracorporeal circuitry pressure and cerebral blood flow velocity (CBFV, middle cerebral artery) were recorded. We analysed the slow waves in the B (8 to 50) and the UB (>50 to 200) bands (in milli-Hz). The analysis, before and during CPB, was performed in the tine domain (correlation coefficient, entropy, mean quantity of mutual information, relative entropy) and in the frequency domain (spectrogram, frequency spectrum, coherence).

Results. CPB dramatically changed monitored signals decreasing their entropy and revealing a dominant CBFV 70 mHz-frequency and a dominant ABP 9 mHz-frequency. There was no association between the signals ($p < 0.05$). Before CPB we found complex patterns where B and UB waves were present.

Conclusion. We hypothesised that CPB provoked a highly protective mechanism, reducing the fluctuations of CBF, by a deactivation of B waves, revealing monotonous UB waves.

Keywords: Slow waves; vasomotor control; cardiovascular bypass; cerebral autoregulation.

Introduction

Spontaneous slow waves are physiologically present in the systemic circulation including the intracranial compartment [3]. They are influenced by pathological conditions, mainly brain stem ischemia and intracranial hypertension and by artificial modifications of the autonomic nervous system (ANS), mainly sympathetic blockade. The slow-wave physiology is still debated on both control system and regulation. The control system is based on the ANS and the cardiovascular pulsatility, on the vessel wall movements (active vasomotion, passive mechanical properties) and on the responses to metabolic substrates. The regulation is based on high pass filtering, linear and non linear links, in particular supposed to reflect, through frequency dependence, the cerebral autoregulation [2, 7, 10, 11]. We hypothesised that in the absence of cardio respiratory variability, during cardiopulmonary bypass (CPB), we should reveal extreme physiologic controls, allowing a better understanding of this complex domain.

Material

Ten consecutive patients, mean age 62.9 (±17) years, scheduled in our institution for elective cardiac surgery (8 for valve replacement, 5 for coronary artery bypass graft), were included after ethical approval, between April and July 2002. The mean EuroScore [6] was 5. According to clinical and ultrasonic examination patients who showed carotid or cerebrovascular disease were excluded. Six patients had a history of hypertension pharmacologically controlled (ß adrenergic blockade, 3 cases; diuretic, 4 cases; angiotensin convertase inhibitor, 6 cases; 4 patients had more than one drug). Arterial blood pressure (ABP, radial invasive), extracorporeal circuitry pressure (ECCP) and cerebral blood flow velocity (CBFV, middle cerebral artery) were recorded (Fig. 1) at a rate of 100 samples/sec (ASC, Soluscience, France). Anaesthesia was realised with sufentanil (0.5 μg/kg/h) and continuous concentration-targeted intravenous infusion of propofol (2–5 μg/ml). After heparinisation, CPB was carried out through an ascending aortic cannula using a peristaltic pump (RP 06, Rhône-Poulenc, France) with a flow of 2.6 $l/min/m^2$. ABP variations were treated according to routine management protocols (sufentanil, midazolam, propofol, urapidil, phenylephrin or noradrenalin) with the aim of a mean ABP superior or equal to 55–60 mmHg (56.3 ± 6.5 mmHg and 55 ± 9.6 mmHg respectively before and during CPB). End tidal-CO_2, pH and base deficit were adjusted to 40 mmHg, 7.4 and below 4 mEq/l (37 °C), respectively. No deep hypothermia was induced (minimal temperature during CPB was 32 °C).

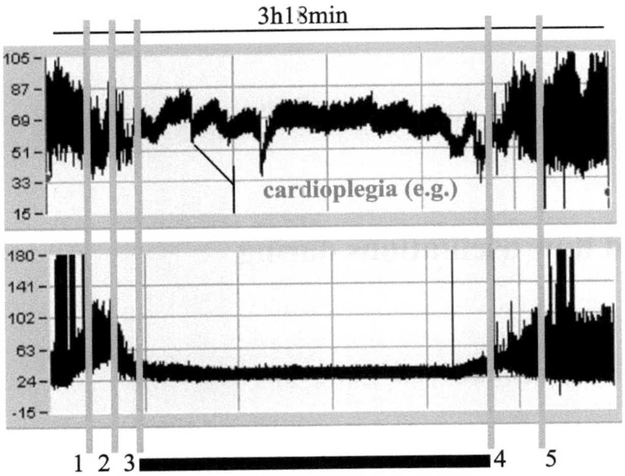

Fig. 1. ABP (top) and CBFV (bottom) recordings (X-axis: time, outset = 10AM21/Y-axis: signal mean value) during cardiac surgery (aortic valve replacement and coronary bypass graft): 1-aortic cannulation, 2-cardiopulmonary bypass (CPB) outset, 3-aortic cross clamp, 4-removal of aortic clamp, 5-end of CPB

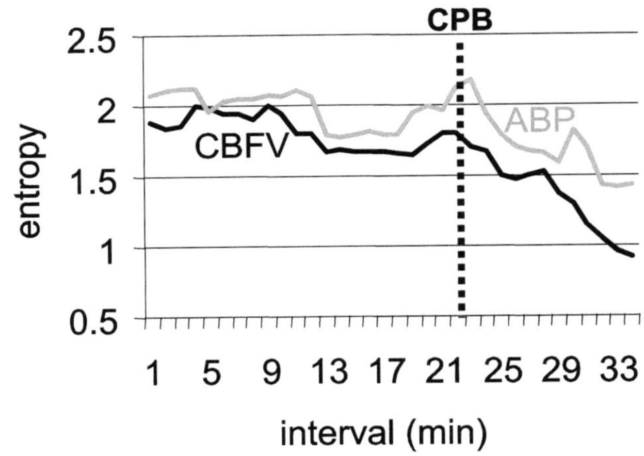

Fig. 2. Example of the evolution of the entropy (Shannon) on CBFV (black line) and ABP (grey line) signals (X-axis: time expressed in intervals/Y-axis: mean value of the entropy): before (left) and after (right) cardiopulmonary bypass (CPB)

Method

We analysed, before and during CPB, the signals and the slow waves included in the B (8 mHz to 50 mHz) and the UB (above 50 mHz up to 200 mHz) bands. The periods during which artefacts were noted (on the CBFV signal), due to the electric scalpel, were removed. In the temporal domain we calculated: the correlation coefficient (Pearson) between two signals; the entropy (Shannon, histogram method) which can be simplified as the complexity of the signal (the higher entropy, the irregular and non stationary signal); the mean quantity of mutual information (MQMI) which represents the mutual entropy between two signals (the lower value, the lower link); the relative entropy (wavelet analysis) or time-entropy representation. In the frequency domain we calculated: the spectrogram (time-frequency representation); the frequency spectrum, allowing the distribution of frequencies function of the power (close to the amplitude), and the partial spectrum which represents the participation of a signal in the constitution of another one; the coherence function between two signals (from 0, no link, to 1, direct link) and this for the waves (B and UB bands) with the maximal amplitude if applicable (presence of the waves on both signals).

Results

There was no link between the signals except between ABP and ECCP signals (B band) and between ABP and CBFV signals (UB band) respectively during and before CPB. The results are summarised Table 1. Before CPB we found complex patterns where B and UB waves (high amplitude) were present. During CPB, the analysis of partial spectra (CBFV vs. ABP and CBFV vs. ECCP) did not reveal influences of ABP or ECCP signals on the spectral distribution of CBFV. CPB dramatically changed CBFV and ABP signals, decreasing the entropy (Fig. 2), in particular on CBFV (2.01 ± 0.32 and 1.3 ± 0.18, respectively be-

Table 1. *Estimation of the links between CBFV, ABP and ECCP signals, before and during CPB (cardiopulmonary bypass)*

–		ABP versus CBFV		ABP versus ECCP	CBFV versus ECCP
		during CPB	before CPB	during CPB	before CPB
Correlation coefficient		0.09 ± 0.24*	0.24 ± 0.15*	0.58 ± 0.1*	0.06 ± 0.01*
MQMI		0.08 ± 0.08	0.2 ± 0.06	0.17 ± 0.2	0.08 ± 0.1
Coherence function	B band	n.a.	n.a.	0.93 ± 0.2	n.a.
	UB band		0.57 ± 0.2	n.a.	

MQMI Mean quantity of mutual information.
ECCP Extracorporeal circuitry pressure; *ABP* arterial blood pressure; *CBFV* cerebral blood flow velocity.
Frequency bands: B, from 8 mHz to 50 mHz; UB above 50 mHz up to 200 mHz.
* Difference, $p < 0.05$; n.a., not applicable. Data are expressed as mean ± SD.

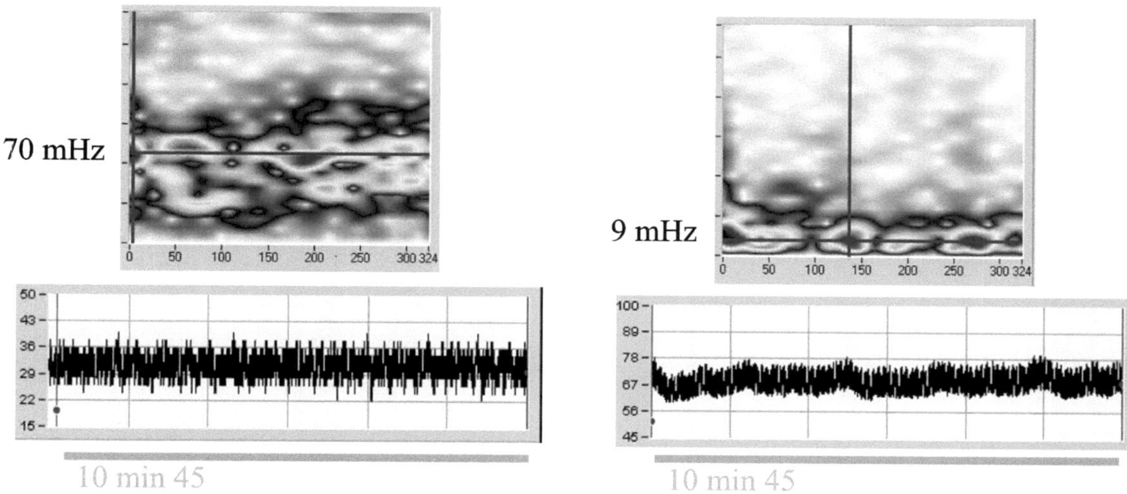

Fig. 3. Example of CBFV (left, bottom) and ABP (right, bottom) signals (Y-axis: mean value of the signal) and their respective spectrograms (left and right, top; Y-axis, frequency), during cardiopulmonary bypass: the duration of the analysed period (X-axis, time, for both signals and spectrograms) was 10 min 45 sec

fore and after CPB). CPB revealed, during about 50% of the time, a dominant 70 mHz – frequency (B wave) on CBFV signal and a dominant 9 mHz – frequency (UB wave) on ABP (Fig. 3).

Discussion

We hypothesised that CPB provoked a highly protective mechanism, reducing the fluctuations of CBF, by a deactivation of B waves, revealing monotonous UB waves. The 50%-duration of the activity could be due to a progressive control tuning. The interpretation of the influence of anaesthetic and opiate drugs is complex, as these drugs can interfere with ANS controls [4, 9]. Nevertheless the results could argue in a frequency dependence of autoregulation [2, 7, 10, 11] and vasomotor sensors linked to the intracranial enclosure [1, 8]. Due to the moderate hypothermia we may emphasise that the temperature did not have a major influence on the cerebral autoregulation [5], conversely to the mean ABP which was at the lower limit of the autoregulation. UB and B bands could be linked to two different sectors, B activity supported by the main vessels and UB activity by the small vessels closed to the brain (and others organs).

References

1. Agassadian K, Fazan VPS, Adanina V, Talman WT (2002) Direct projections from the cardiovascular nucleus tractus solitarii to pontine preganglionic parasympathetic neurons: a Link to cerebrovascular regulation. J Comp Neurol 452: 242–254
2. Giller CA, Roseland A, Lam M (1993) Periodic variations in transcranial Doppler mean velocities. J Neuroimaging 3: 160–162
3. Lemaire JJ, Khalil T, Cervenansky F, Gindre G, Boire JY, Bazin JE, Irthum B, Chazal J (2002) Slow pressure waves in the cranial enclosure. Acta Neurochir (Wien) 144: 243–254
4. Monitto CL, Kurth CD (1993) The effect of fentanyl, sufentanil, and alfentanil on cerebral arterioles in piglets. Anesth Analg 76: 985–989
5. Neri E, Sassi C, Barabesi L, Massetti M, Pula G, Buklas D, Tassi R, Giomarelli P (2004) Cerebral autoregulation after hypothermic circulatory arrest in operations on the aortic arch. Ann Thorac Surg 77: 72–80
6. Nashef SA, Roques F, Michel P, Gauducheau E, Lemeshow S, Salamon R (1999) The EuroSCORE study group: European system for cardiac operative risk evaluation (EuroScore). Eur J Cardiothorac Surg 16: 9–13
7. Panerai RB, Rennie JM, Kelsall AW, Evans DH (1998) Frequency-domain analysis of cerebral autoregulation from spontaneous fluctuations in arterial blood pressure. Med Biol Eng Comput 36: 315–322
8. Sandor P (1999) Nervous control of the cerebrovascular system: doubts and facts. Neurochemistry Int 35: 237–259
9. Wang X, Huang ZG, Gold A, Bouairi E, Cory Evans C, Andresen MC, Mendelowitz D (2004) Propofol modulates γ-aminobutyric acid–mediated inhibitory neurotransmission to cardiac vagal neurons in the nucleus ambiguus. Anaesthesiology 100: 1198–1205
10. Zhang R, Zuckerman JH, Levine BD (2000) Spontaneous fluctuations in cerebral blood flow: insights from extended-duration recordings in humans. Am J Physiol Heart Circ Physiol 278: H1848–H1855
11. Zhang R, Zuckerman JH, Iwasaki K, Wilson TE, Crandal G, Levine B (2002) Autonomic neural control of dynamic cerebral autoregulation in humans. Circulation 106: 1814–1820

Correspondence: Jean-Jacques Lemaire, Service de Neurochirurgie A, Hôpital Gabriel Montpied, BP 69, Clermont – Ferrand Cedex 1, France. e-mail: jjlemaire@chu-clermontferrand.fr

Acta Neurochir (2005) [Suppl] 95: 341–344

Concept of "true ICP" in monitoring and prognostication in head trauma

M. Czosnyka[1], L. Steiner[1,2], M. Balestreri[1,3], E. Schmidt[1,4], P. Smielewski[1], P. J. Hutchinson[1], and **J. D. Pickard[1]**

[1] Academic Neurosurgical Unit, Addenbrooke's Hospital, Cambridge, UK
[2] Department of Anaesthesia, University Hospital Basel, Basel, Switzerland
[3] Department of Anaesthesia and Intensive Care (IInd) Policlinico San Matteo, Pavia, Italy
[4] Department of Neurosurgery, Hopital Purpan, Toulouse, France

Summary

Objective. To propose a new coefficient, which contains information about both the absolute ICP and the position of the 'working point' on the pressure-volume curve.

Method. ICP was monitored continuously in 187 sedated and ventilated patients. The RAP coefficient was calculated as the running (3 minutes) correlation coefficient between slow changes in pulse amplitude and mean ICP. RAP has value 0 on the flat part of the Pressure-Volume Curve and +1 on the ascending exponential part. Then RAP decreases to zero or becomes negative when ICP increases further and affects cerebrovascular pressure-reactivity (which flattens the pressure-volume curve). Variable $tICP = ICP * (1 - RAP)$ has been called 'trueICP'. It magnifies the critical values of ICP when cerebrovascular reactivity is exhausted and dampens those states where absolute ICP is elevated but vascular reactivity is not affected.

Results. Both Mean ICP and RAP were independently correlated with outcome (ANOVA:ICP-GOS: $F = 22$; $p < 0.00001$, RAP-GOS: $F = 9$; $p < 0.001$). 'TrueICP' had stronger association with outcome: $F = 28$; $p < 0.000001$. Mortality in those patients having 'trueICP' above the threshold of 19 mm Hg was above 80%, while the mortality in those having cICP below 19 mm Hg was only 20% ($F = 80$; $p < 10^{-8}$). 'TrueICP' was also suitable for continuous monitoring: sustained rise in tICP above 19 mm Hg was strongly associated with fatal complications.

Conclusion. The proposed variable is a powerful predictor of fatal outcome following head injury. It is sensitive to both the rising absolute ICP and the critical loss of cerebrovascular regulation.

Keywords: Intracranial pressure; head injury; monitoring; outcome.

Introduction

Mean intracranial pressure is an important prognostic variable in severe head injury [7]. We hypothesised that absolute ICP would be even more predictive if one could take into account where on the pressure-volume curve [4] the 'working' pressure point is positioned.

A pressure-volume curve according to Lofgren *et al.* [5] shows three distinctive zones. In normal conditions, without elevation of ICP and with good compensatory reserve the 'working point' lies in the first flat zone of the pressure-volume curve; Fig. 1. Patients after TBI 'work' usually in the second zone ('poor compensatory reserve'). In this zone, any change in intracerebral volume cases deep, exponential pressure response. Real trouble starts when ICP rises as high that they enter the third zone, where cerebral perfusion pressure (CPP) is too low to keep normal vascular regulatory mechanisms (autoregulation and regulation of cerebral blood volume) working. This conceptual threshold of 'critical' ICP is patient- and time-dependent.

Moving correlation coefficient between changes in mean ICP and pulse amplitude of ICP (called RAP) can be used to indicate passage between three zones of the pressure-volume curve [2].

A simple model explains (Fig. 1) the pulse amplitude of ICP resulting from pulsatile changes in cerebral blood volume transformed by the pressure-volume curve [1]. The pulse amplitude of ICP is low and does not depend on mean ICP in the first zone, making the value of RAP close to 0. Pulse amplitude increases linearly with mean ICP in the zone of poor compensatory reserve, resulting in RAP close to +1. In the third zone, pulse amplitude starts to decrease with rising ICP resulting (theoretically) in low or even negative values for RAP.

We proposed a new coefficient, which contains information on both the absolute ICP and the pressure-volume compensatory reserve. The coefficient tICP

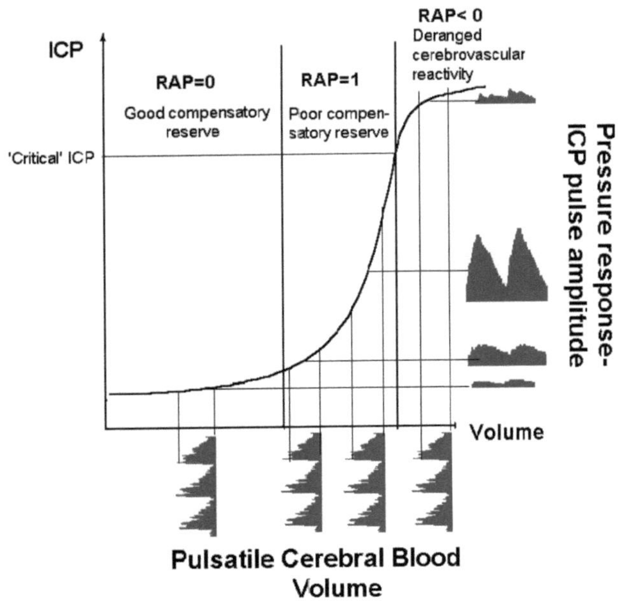

Fig. 1. In a simple model, pulse amplitude of ICP (expressed along the y-axis on the right side of the panel) results from pulsatile changes in cerebral blood volume (expressed along the x-axis) transformed by the pressure-volume curve. This curve has three zones: a flat zone, expressing good compensatory reserve, an exponential zone, depicting poor compensatory reserve and a flat zone again, seen at very high ICP (above the 'critical' ICP) depicting derangement of normal cerebrovascular responses. The pulse amplitude of ICP is low and does not depend on mean ICP in the first zone, resulting in values of RAP close to 0. The pulse amplitude increases linearly with mean ICP in the zone of poor compensatory reserve, resulting in RAP close to +1. In the third zone, the pulse amplitude starts to decrease with rising ICP making RAP theoretically negative. Adopted from [1, 5]

has been called 'true ICP' and is mathematically expressed as:

$$tICP = ICP * (1 - RAP)$$

It magnifies the critical values of ICP when cerebrovascular pressure reactivity is exhausted (i.e. when RAP is close to 0 or negative at high levels of ICP). Value of 'trueICP' is attenuated in those states where absolute ICP is elevated but the vascular reactivity reserve is not affected (i.e. when RAP is close to +1).

Material and methods

One hundred and eighty-seven patients admitted after head injury to Addenbrooke's Hospital (1992–98) with a median Glasgow Coma Score (GCS) of 6 (range 3 to 13, 10% of patients with initial GCS > 9) were studied. There were 44 women and 143 men, their ages ranging from 6 to 75 years (mean age 36 years; only 7 patients were younger than 15). 31% had subdural hematomas on their initial CT of which 64% were evacuated surgically. Intracerebral hematomas were found in 28% (50% were removed surgically) and extradural hematomas in 10% of these patients. 11% had diffuse brain injury, 57% brain swelling and 38% presented with a midline shift. Subarachnoid blood was found in 25% of patients, with only 12 demonstrating mean flow velocity above 120 cm/s. The patients were paralyzed, sedated and ventilated to achieve mild hypocapnia. Falls in ABP, which reduced CPP below 60 mmHg, were managed with alternating colloid/normal saline infusions and supplementary inotropic agents if necessary (dopamine 2–15 μg/kg per minute). If ICP rose to above 20 mm Hg, boluses of mannitol (200 ml of 20%, over a period of 20 min or longer) were administered.

Intracranial pressure was monitored continuously using microtransducers (Camino Direct Pressure Monitor, Camino Laboratories, San Diego, CA or Codman MicroSensor, Johnson&Johnson Professional, Rynham, MA), inserted intraparenchymally into the frontal region. Arterial pressure was monitored directly from the radial or dorsalis pedis artery (System 8000, S&W Vickers Ltd, Sidcup, UK or Solar 6000 System, Marquette, USA). All data were digitized, sampled (30 Hz) and saved on bedside computer (time period from 30 min to 3 hours daily).

The RAP index (correlation coefficient [R] between AMP amplitude [A] and mean pressure [P]) was derived by linear correlation between 40 consecutive, time-averaged data points of AMP and ICP acquired over 6.4 s. This index indicates the degree of correlation between AMP and mean ICP over short periods of time (~ 4 minutes). Its clinical significance has been discussed before [2]. Following head injury and subsequent brain swelling RAP is usually close to +1. With further increase in ICP, AMP decreases and RAP values fall below zero. This occurs when the cerebral autoregulatory capacity is exhausted and the pressure-volume curve bends to the right as the capacity of cerebral arterioles to dilate in response to a CPP decrement is exhausted, and they tend to passively collapse. This indicates terminal cerebrovascular derangement with a decrease in pulse pressure transmission from the arterial bed to the intracranial compartment.

Statistical analysis

For statistical evaluation averaged values from artifact-free whole monitoring periods of RAP, ICP and 'trueICP' were taken. Analysis of variance was used to investigate statistical difference between patients who survived and died. F-Snodecker coefficient was used to examine separation between these two groups. Ch-square statistics was used to examine different mortality rates. Mean values ± 95% confidence limits for mean values were given.

Results

Relationship with outcome

Mortality rate was 23% in the studied group of patient. Both mean ICP and RAP were independently associated with outcome (ICP: 17 ± 3 mmHg for survivors versus 25 ± 4.2 mm Hg for those who died $F = 22$; $p < 0.00001$. RAP: 0.48 ± 0.03 for survivors and 0.33 ± 0.08 for those who died; $F = 9$; $p < 0.001$). 'Corrected ICP' had stronger association with outcome: $F = 28$; $p < 0.000001$. It was 7.5 ± 1.5 mmHg in survivors and 17.5 ± 2 mmHg in those who died.

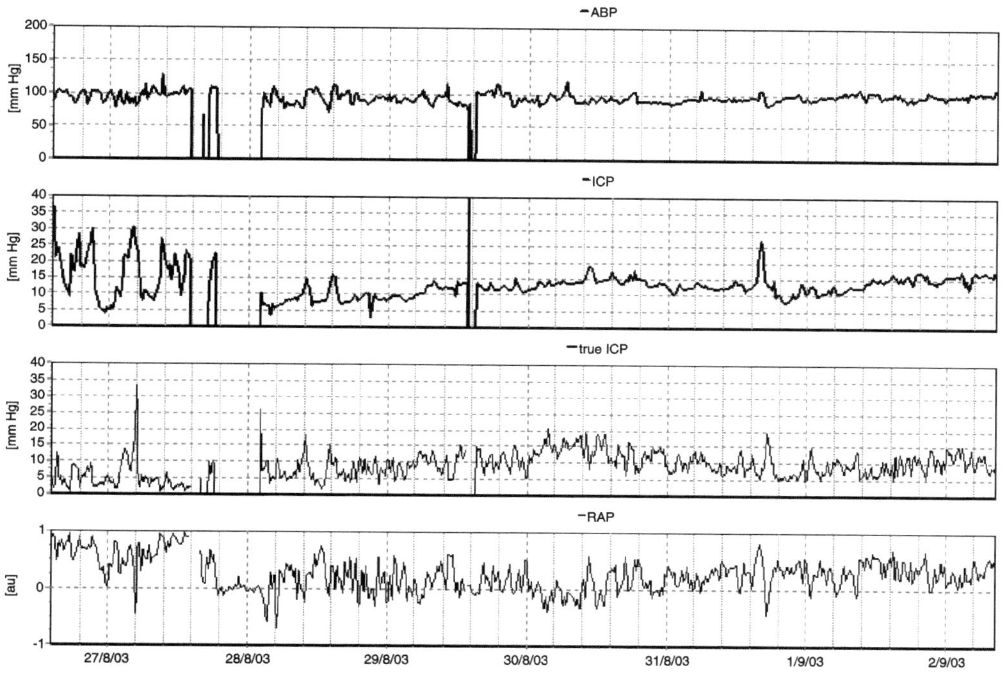

Fig. 2. Time-trend of mean arterial pressure, ICP, 'True ICP' and RAP coefficient recorded in patient after severe head injury with good outcome. X-axis: day of monitoring

Mortality in those patients having tICP above the threshold of 18 mm Hg was above 80%, while the mortality in those having tICP below 18 mm Hg was only 20% ($p < 10^{-5}$).

The critical threshold maximizing separation (F coefficient) between patients who died and survived is 31 mmHg for ICP and 18 mmHg for 'trueICP'.

Use of 'trueICP' in continuous monitoring

Use of 'trueICP' for continuous monitoring can best illustrated using two examples chosen from our own material.

First patient achieved good outcome, in spite of severity of initial injury (Admission GCS was 5); see Fig. 2. After first few hours left temporal haematoma was removed. Before that event, ICP was unstable, fluctuating up to 30 mmHg. Compensatory reserve was poor, making RAP close to +1. TrueICP remained (in spite of one, relatively short elevation) low. After removal of haematoma ICP settled, RAP was low (good compensatory reserve and 'trueICP' stayed low.

In contrast, the second case, Fig. 3, contains data from the patient who died following injury. ICP was moderately elevated over all period but after first day vascular response deteriorated (RAP decreased), causing increase in 'trueICP' well above critical threshold. Patient died.

Discussion

Continuous analysis of intracranial pressure adds information to simple recording of mean trend values. 'TrueICP' shows a potential to warn against these elevations of ICP, which may lead to deterioration of cerebrovascular control mechanisms. Undoubtedly, such an index is nothing more than mathematical derivative of two other variables, already used in brain monitoring. But it brings a new quality to already reach family of 'brain pressures' (we know ICP, CPP but also critical closing pressure, cerebral venous pressure, non-invasive ICP and CPP, intrahemispherical pressure gradients, etc.). First of all it is scaled in millimetres of mercury. And secondly-sustained rise in 'trueICP' above 15–18 mmHg almost always indicates life-threatening situation after head injury. The name 'trueICP' is probably not ideal, as it may suggest that traditional ICP may be 'untrue'. But, on the other hand, it shows when the rise in ICP is becoming dangerous. We know that 'critical' thresholds for both

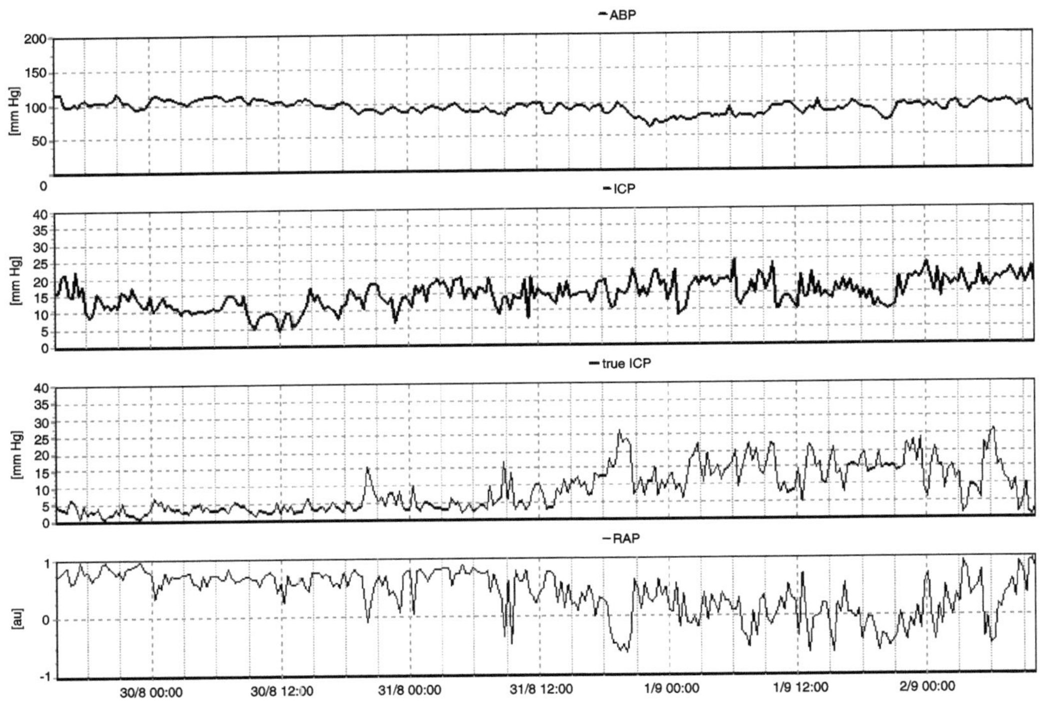

Fig. 3. Example of time-trend of mean arterial pressure, ICP, 'True ICP' and RAP coefficient recorded in patient after severe head injury who died third day after injury. X-axis: date and time

ICP and CPP are subject- and time-dependent. Facilitating information about ICP with compensatory reserve derived from the pulse amplitude of ICP is in fact nothing knew. The lesson from the concepts given by 'old masters' [1, 3–9] has been learnt and thanks to new computer technology [10] is becoming available at the bedside.

References

1. Avezaat CJJ, van Eijndhoven JHM (1984) Cerebrospinal fluid pulse pressure and craniospinal dynamics. Thesis. A Jongbloed and Zoon Publishers, The Hague
2. Balestreri M, Czosnyka M, Steiner LA, Schmidt E, Smielewski P, Matta B, Pickard JD (2004) Intracranial hypertension: what additional information can be derived from ICP waveform after head injury? Acta Neurochir (Wien) 146(2): 131–141
3. Eijndhoven JHM van, Avezaat CJJ (1986) Cerebrospinal fluid pulse pressure and the pulsatile variation in cerebral blood volume: an experimental study in dogs. Neurosurgery 19: 507–522
4. Langfitt TW, Weinstein JD, Kassell NF (1965) Cerebral vasomotor paralysis produced by intracranial pressure. Neurology (Minneap) 15: 622–641
5. Lofgren J, von Essen C, Zwetnow NN (1973) The pressure-volume curve of the cerebrospinal space in dogs. Acta Neurol Scand 49: 557–574
6. Marmarou A, Maset AL, Ward JD, Choi S, Brooks D, Lutz HA, Moulton RJ, Muizelaar JP, DeSalles A, Young HF (1987) Contribution of CSF and vascular factors to elevation of ICP in severely head injured patients. J Neurosurg 66: 883–890
7. Miller JD, Becker DP, Ward JD, Sullivan HG, Adams WE, Rosner MJ (1977) Significance of intracranial hypertension in severe head injury. J Neurosurg 47(4): 503–516
8. Nornes H, Aaslid R, Lindegaard KF (1977) Intracranial pulse pressure dynamics in patients with intracranial hypertension. Acta Neurochirur (Wien) 38: 177–186
9. Portnoy HD, Chopp M, Branch C, Shannon M (1982) Cerebrospinal fluid pulse waveform as an indicator of cerebral autoregulation. J Neurosurg 56: 666–678
10. Smielewski P, Czosnyka M, Zabolotny W, Kirkpatrick P, Richards HK, Pickard JD (1997) A computing system for the clinical and experimental investigation of cerebrovascular reactivity. Int J Clin Monitor Comput 14: 185–198

Correspondence: Marek Czosnyka, Academic Neurosurgery, Box 167, Addenbrooke's Hospital, Cambridge CB2 2QQ, UK. e-mail: MC141@medschl.cam.ac.uk

Acta Neurochir (2005) [Suppl] 95: 345–349

Fuzzy pattern classification of hemodynamic data can be used to determine noninvasive intracranial pressure*

B. Schmidt[1], **S. F. Bocklisch**[2], **M. Päßler**[2], **M. Czosnyka**[3], **J. J. Schwarze**[1], and **J. Klingelhöfer**[1]

[1] Department of Neurology, Medical Centre Chemnitz, Chemnitz, Germany
[2] Department of Systems Theory, Technical University Chemnitz, Chemnitz, Germany
[3] Academic Neurosurgical Unit, Addenbrooke's Hospital, Cambridge, UK

Summary

Objective. The authors previously introduced a method in which intracranial pressure (ICP) was estimated using parameters (TCD characteristics) derived from cerebral blood flow velocity (FV) and arterial blood pressure (ABP). Some results suggested that this model might be influenced by the patient's state of cerebral autoregulation and other clinical parameters. Hence, it was the aim of the present study to improve the method by modifying the previously used global procedure in certain subgroups of patients.

Methods. In 103 traumatic brain injured patients (3–76 years, mean: 31 ± 16 years) signal data of FV, ABP and ICP were used to generate samples of TCD characteristics together with time corresponding ICP. Fuzzy Pattern Classification was used to identify cluster subsets (classes) of the sample space. On each class a local estimator of ICP was defined. This approach provides a non-invasive assessment of ICP (nICP) as follows: Using FV and ABP the TCD characteristics were computed and related to the matching classes. nICP was calculated as a weighted sum of local ICP estimations.

Results. ICP A and B waves and long-term trends could be visibly assessed. The median absolute difference between ICP and nICP was 5.7 mmHg.

Conclusions. The class structure of the model facilitates nICP assessment in heterogeneous patient groups and supports a stepwise extension of the target patient group without affecting the former validity.

Keywords: Intracranial pressure; cerebral autoregulation; cerebral blood flow; transcranial Doppler ultrasonography; fuzzy pattern classification.

Introduction

In patients with acute cerebral diseases elevated intracranial pressure (ICP) constitutes a frequent and serious complication. Until now reliable monitoring of ICP has only been provided by invasive methods such as for instance the implantation of intraventricular ICP probes. On the other hand, various approaches have been made in the past to analyse the relationship between ICP and parameters derived from cerebral blood flow velocity (FV) and arterial blood pressure (ABP) [1, 4, 6–8]. The authors previously introduced a data based mathematical model in which a linear relationship between selected hemodynamic parameters (TCD characteristics) and the quotient ICP/ABP was assumed [11, 12]. Since the introduction of the so-called linear method of nICP assessment various clinical studies have been performed in order to assess the accuracy of this method and its potential benefit for clinical application [13–15]. The results of these studies showed that this procedure was generally able to visibly assess ICP pulse waves, B waves, plateau waves [9, 10] and ICP trends. Furthermore, the mean absolute difference between non-invasively and invasively assessed ICP was between 4 and 8 mm Hg depending on the underlying population. However, there is a general problem with this method: By definition the linear nICP procedure depends on the relationship to the underlying patients reference group which was used for the model construction. Homogeneity of this group with respect to defined clinical parameters increases the accuracy of the model. For the same reason clinical parameters of the patients reference and target groups should be similar. Unfortunately, there are a lot of varying parameters which might influence the relationship between TCD characteristics and the ABP-ICP relationship such as the underlying disease and its severity, state of cerebral autoregulation (CA), arterial CO_2 pressure, cerebrovascular state, age and others.

* This study was supported by the Deutsche Forschungsgemeinschaft (KL960/1-2).

These problems can be eliminated by the introduction of fuzzy pattern method which could be used to cover heterogeneous patient data by homogeneous subgroups and to smoothly fit together locally defined nICP estimators. Hence, the main objective of the current study was to clarify the general feasibility of assessing nICP by using a fuzzy pattern classification.

Materials and methods

Patient population

The investigations were performed on data of 103 traumatic brain injured (TBI) patients (3–76 years, mean: 31 ± 16 years) treated in Addenbrooke's Hospital, Cambridge, UK. At the time of data recording all the patients were sedated, paralyzed and mechanically ventilated. No patient showed vasospasm (exclusion criteria was MCA FV > 90 cm/s) nor evidence of stenosis of the intracranial or extracranial arteries.

Monitoring

TCD measurements were taken by 2 MHz pulsed Doppler devices (PCDop842, Scimed, Bristol, UK or Neuroguard, Medasonics, CA). Flow patterns of the MCA were continuously recorded in the hemisphere ipsilateral to the ICP measurement. Blood pressure was measured with a standard manometer line inserted into the radial or femoral artery. ICP was measured using implanted intraparenchymal microsensors (Camino Laboratories, San Diego, CA, USA).

Such a way of monitoring was a routine clinical practice and did not require separate Local Ethical Committee approvals at the time of recording.

Computer – assisted recording

PC was used for recording and analyzing FV, ABP and ICP signals. The computer for signal recording was fitted with data acquisition systems (DTA2814, Data Translation, Marlboro, CA, USA). The sampling frequencies used ranged from 25 Hz to 50 Hz. The signals were recorded daily for the duration of 20–120 min. Home-written software [16] was used for data recording.

Assessment of state of cerebral autoregulation

Ten-second time averaged values of FV, ICP, ABP, and CPP (CPP = ABP – ICP) were continuously calculated. The autoregulation index Mx [5] was computed as the correlation coefficient of 36 consecutive samples of CPP and FV mean values. Positive association between CPP and FV (Mx > 0) indicates passive dependence of cerebral blood flow (CBF) on CPP. Zero or negative value of Mx indicates CBF being independent on CPP, i.e. intact (or even over-regulating) CA. The patients were stratified into two different groups, one with unequivocally preserved CA (Mx < –0.2, N = 50), and the other with clearly impaired CA (Mx > 0.2, N = 53). Patients with equivocal results (–0.2 < Mx < 0.2) were excluded from the study.

Non-invasive ICP assessment – linear procedure

The procedure is controlled by parameters called TCD characteristics, which are derived from FV and ABP signals. They essentially consist of coefficients of an ABP → FV impulse response function

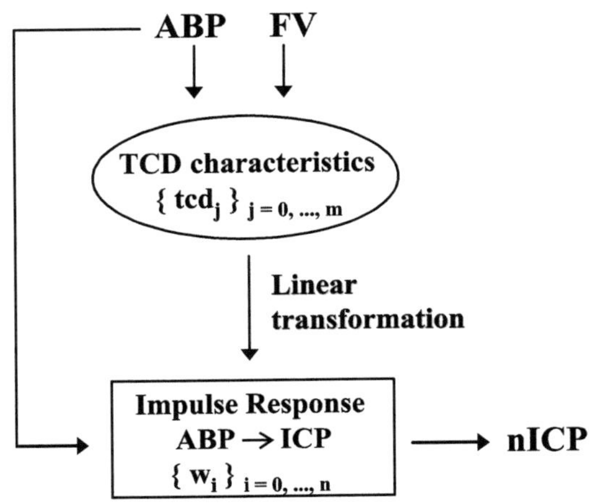

Fig. 1. nICP assessment procedure. From FV and ABP signals, TCD characteristics consisting of an ABP → FV impulse response function, its coefficients divided by ABP, and PI and PL, are computed. A linear transformation on the TCD characteristics gives an estimation of the ABP → ICP impulse response function, which transfers ABP into the nICP signal. The rules of linear transformation have been formerly derived by a multiple regression analysis on patients reference data

transformation and additional ICP related parameters such as pulsatility index. The TCD characteristics are used to calculate the ABP → ICP impulse response function. The model assumes linear relationship between TCD characteristics and the coefficients of ABP → ICP impulse response. This linear function constitutes the essential part of the nICP assessment procedure and is constructed using data recordings of FV, ABP and invasively assessed ICP from a well-defined group of reference patients. In Fig. 1 a flow chart of the nICP assessment is presented.

Non-invasive ICP assessment using fuzzy pattern classification

Fuzzy Pattern Classification method allows to analyze the inner structure of a parameter space by means of identification of cluster subsets [2, 3]. To each of these subsets a smooth membership function is assigned which is defined on the whole parameter space and which expresses the affiliations of the space points to this cluster set in terms of fuzzy logic values, i.e. any value between 0 (no affiliation) and 1 (full affiliation). A fuzzy pattern class consists of a cluster subset and its associated membership function (Fig. 2a). By definition classes are not sharply bounded and may overlap with other classes. Applied to nICP assessment this method was used to classify the sample space of TCD characteristics. On each of the classes a local estimator of nICP ($nICP_i$) was defined. The final nICP estimation was achieved by assembling the local estimators (Fig. 2b). House written software of Technical University (Fuzzy Toolbox; GWT, Dresden, Germany) and Medical Centre Chemnitz [11] was used for model construction.

The nICP was quantitatively compared to invasively measured ICP using mean absolute errors (ΔICP) which were defined as absolute values of ten-second averaged differences ICP–nICP. Fuzzy versions of sensitivity and specificity of distinction between increased ICP (>20 mm Hg) and normal ICP (<15 mm Hg) were defined as fuzzy sensitivity: ratio #(cases ICP > 20 mm Hg *and* nICP > 18 mm Hg)/#(cases ICP > 20 mm Hg), and fuzzy specificity: ratio #(cases ICP < 15 mm Hg *and* nICP < 17 mm Hg)/ #(cases ICP < 15 mm Hg).

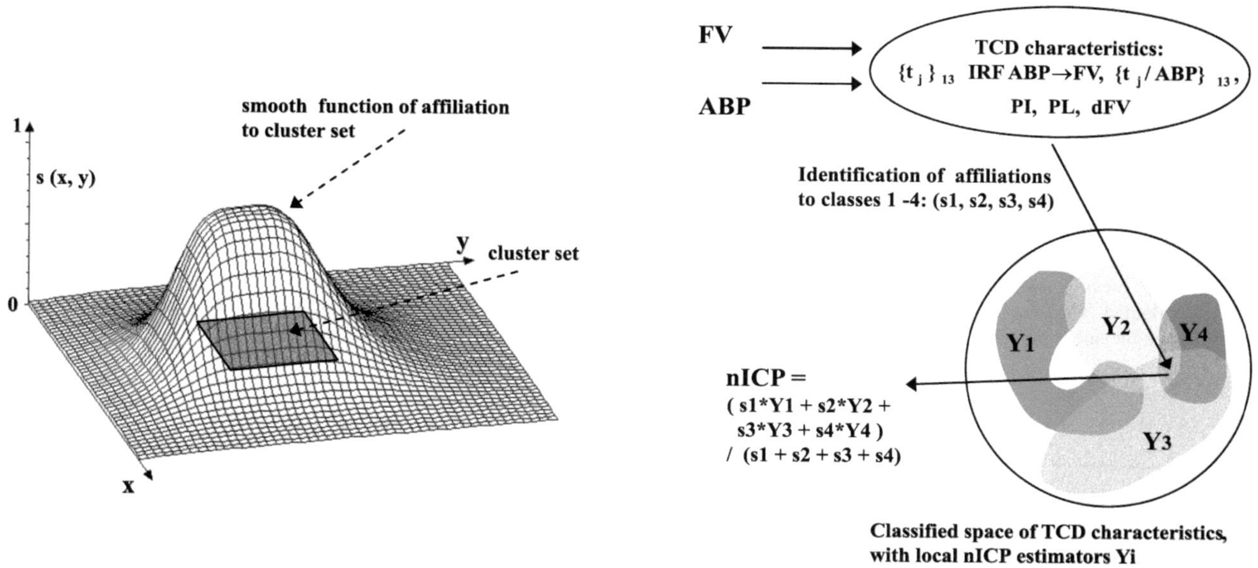

Fig. 2. (a) Fuzzy pattern class in a two-dimensional parameter space. A fuzzy pattern class consists of the underlying cluster subset together with a related smooth function of membership to the cluster set. The membership function may take on all values between 0 and 1. The class is fuzzy in the sense that even values outside of the cluster set have non-zero grades of membership to it. (b) Assessment of nICP in a four-class fuzzy pattern model. From measured FV and ABP samples of TCD characteristics are calculated and their grades of membership to the classes are expressed in terms of s1–s4. These are used to calculate nICP as the weighted sum of class-local estimators Y1–Y4. (*PI* pulsatility index; *PL* pulse length, *dFV* diastolic FV)

Results

The nICP was calculated from data collected in 145 different recordings of 103 patients. Invasively assessed reference ICP exceeded 20 mm Hg in 60 patients and 30 mm Hg in 25 patients with a mean ($\pm$SD) of 22 ± 13 mm Hg.

Three different fuzzy pattern classification models were constructed and their results compared to results of the linear nICP procedure. Twenty-nine TCD characteristics (13 coefficients ABP $\rightarrow$ FV IRF $\{t_j\}$, 13 ratios t_j/ABP, pulsatility index, pulse length, diastolic FV) were used for nICP assessment by the linear procedure as well as for data classification in the fuzzy pattern models.

Model 1: Twenty-seven classes were generated, 14 of which were associated with intact CA and 13 with disturbed CA data. On each class a constant nICP estimator was defined as the mean ICP of all samples of the underlying cluster set.

Model 2: Seventeen classes were generated. CA information was not evaluated. On each class a constant nICP estimator was defined as the mean ICP of all samples of the underlying cluster set.

Model 3: Three classes were generated. CA information was not evaluated. On each class a linear nICP estimator was defined, calculated by a multiple regression process on the class data.

Mean nICP could be assessed by all models, while nICP pulse waves could only be calculated by the linear procedure and by model 3. Accuracy was assessed in terms of ΔICP using invasively measured ICP as reference. Compared to the linear model the three fuzzy pattern models showed slight but not statistically significant improvements of accuracy. In model 1 median of ΔICP was 5.68 mm Hg compared to 5.93 mm Hg in the linear model. Restricted to patients with ICP below 30 mm Hg the mean ΔICP ($\pm$SD) was 5.97 ± 4.68 mm Hg. The sensitivity and specificity were 0.79 and 0.69 in model 1. In Table 1 an overview of the results in the different models is provided. All models had in common that trends of ICP changes usually were repeated by the non-invasive assessment. Examples of nICP assessments during ICP plateau and B waves [9, 10] are shown in Fig. 3.

Discussion

In the studied fuzzy pattern classification models the TCD characteristics were used as classifying parameters in order to allow a direct comparison to the linear

Table 1. *Accuracy of nICP assessment in the linear model and in fuzzy pattern models*

nICP model	Median ΔICP [mm Hg]	Mean ΔICP for ICP < 30 [mm Hg]	Mean ΔICP [mm Hg]	Fuzzy sensitivity	Fuzzy specificity	Standard deviation of ICP-nICP
Linear model	5.93	6.05 ± 4.13	7.98 ± 7.59	0.66	0.71	10.38
Fuzzy pattern model 1	5.68	5.97 ± 4.68	7.50 ± 6.44	0.79	0.69	9.67
Fuzzy pattern model 2	5.87	5.68 ± 4.00	7.86 ± 7.06	0.75	0.59	10.08
Fuzzy pattern model 3	6.09	6.48 ± 4.39	7.67 ± 6.31	0.73	0.67	9.71

Median ΔICP, mean ± standard deviation of ΔICP of all patients and of those with ICP < 30 mm Hg, as well as fuzzy sensitivity and specificity were evaluated. The fuzzy pattern models showed similar accuracy of nICP assessment as the linear model. Results of model 1 were slightly better than in the linear model. Mean values of ΔICP did not differ significantly in the different models.

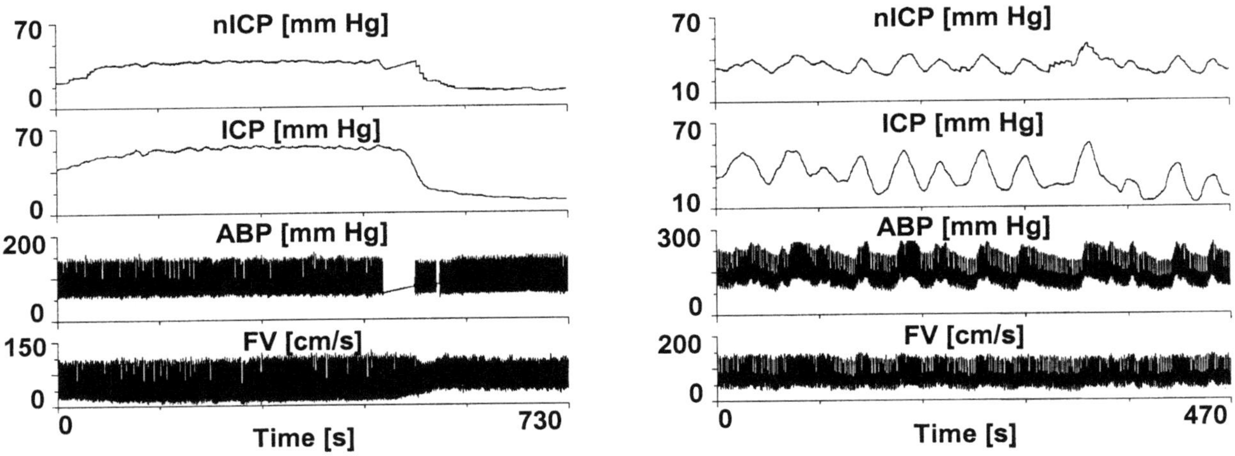

Fig. 3. Curves of nICP, ICP, ABP, and FV assessed by fuzzy pattern model 1. In a) nICP was assessed during generation of an ICP plateau wave. The effect of ABP signal artefact during disconnected arterial line to nICP can be observed at the end of nICP plateau; b) shows corresponding B waves of ICP and nICP with slightly reduced nICP dynamics

nICP assessment method. The results showed similar accuracy of the linear and the fuzzy pattern models. The highest accuracy of nICP assessment was achieved in model 1 by using additional information about the state of CA, although even the improvement on the linear model was not statistically significant. Obviously, the classification of TCD characteristics did not provide a clear potential of accuracy improvement. However, the results may be interpreted as a prove of the general feasibility of nICP assessment by using fuzzy pattern models. Nevertheless, the question arises whether improvements could be achieved by using other types of classifying parameters. Fuzzy pattern classification applied to nICP assessment provides a great variety of different models. In the construction of the current model we have not yet integrated formerly-mentioned clinical parameters such as patient's primary disease, arterial CO2 pressure, vascular state and age. The reason was because the studied population was rather homogeneous with regard to these parameters, age being the only exception. Included patients suffered from TBI and were mildly hyperventilated, patients with cerebral vasospasm and stenosis had been excluded. It may be assumed that in a target group with varying diseases and cerebrovascular irregularities the fuzzy pattern approach would be clearly preferable. On the other hand the general question arises for the potential of improvement of this data driven model. Median ΔICP of almost 6 mm Hg might appear to be much. However, it should be taken into consideration that intraventricular or intraparenchymal ICP assessment which is used as reference method is erroneous as well. Moreover, implanted ICP probes assess a local pressure value while Doppler ultrasonographic assessed FV is influenced by the pressure distribution in the whole MCA flow area. This constitutes a systematic difference between the invasive and the non-invasive method. As a third point non-optimal insonation angles during Doppler measurements may cause serious ICP estimation errors. Due to these effects it seems unlikely that a median ΔICP of less than 5 mm Hg could ever be achieved. However, considering the procedure's property of properly reflecting the invasively measured ICP trends, the current accuracy appears sufficient for clinical usage of the nICP assessment method.

An important point of fuzzy pattern classification is that it allows the coverage of heterogeneous patient data by homogeneous data subsets and locally defined nICP estimators. Besides a possible accuracy improvement this modular structure facilitates continuous addition of new patient data without affecting validity of the complete model. The modular structure, therefore, supports the ability of the model to learn.

Conclusion and outlook

Fuzzy pattern classification provides a promising nICP assessment method for the application in heterogeneous patient groups. In currently ongoing investigations models are constructed which include patients with vasospasm.

References

1. Aaslid R, Lundar T, Lindegaard KF, Nornes H (1993) Estimation of cerebral perfusion pressure from arterial blood pressure and transcranial Doppler recordings. In: Miller JD, Teasdale GM, Rowan JO, Galbraith SL, Mendelow AD (eds) Intracranial pressure VI. Springer, Berlin, pp 226–229
2. Baur M, Bocklisch SF (2001) Similarity based local model approach for nonlinear modelling. In: Proc 6th European Control Conference. ECC01, Porto, Portugal, pp 3905–3910
3. Bocklisch SF, Päßler M (2000) Fuzzy time series analysis. In: Hampel *et al* (eds) Fuzzy control – theory and practice. Physica-Verlag, Heidelberg
4. Chan KH, Miller JD, Dearden NM, Andrews PJ, Midgley S (1992) The effect of changes in cerebral perfusion pressure upon middle cerebral artery blood flow velocity and jugular bulb venous oxygen saturation after severe brain injury. J Neurosurg 77: 117–130
5. Czosnyka M, Smielewski P, Kirkpatrick P, Menon DK, Pickard JD (1996) Monitoring of cerebral autoregulation in head-injured patients. Stroke 27: 1829–1834
6. Czosnyka M, Matta BF, Smielewski P, Kirkpatrick P, Pickard JD (1998) Cerebral perfusion pressure in head-injured patients: a noninvasive assessment using transcranial Doppler ultrasonography. J Neurosurg 88(5): 802–808
7. Klingelhöfer J, Conrad B, Benecke R, Sander D (1987) Relationship between intracranial pressure and intracranial flow patterns in patients suffering from cerebral diseases. J Cardiovasc Ultrasonogr 6: 249–254
8. Klingelhöfer J, Conrad B, Benecke R, Sander D, Markakis E (1988) Evaluation of intracranial pressure from transcranial Doppler studies in cerebral disease. J Neurol 235: 159–162
9. Lundberg N (1960) Continuous recording and control of ventricular fluid pressure in neurosurgical practice. Acta Psych Neurol Scand [Suppl] 149: 1–193
10. Newell DW, Aaslid R, Stooss R, Reulen HJ (1992) The relationship of blood flow velocity fluctuations to intracranial pressure B waves. J Neurosurg 76: 415–421
11. Schmidt B, Klingelhöfer J, Schwarze JJ, Sander D, Wittich I (1997) Noninvasive prediction of intracranial pressure curves using transcranial Doppler ultrasonography and blood pressure curves. Stroke 28: 2465–2472
12. Schmidt B, Schwarze JJ, Czosnyka M, Sander D, Wittich I, Klingelhöfer J (1998) A method for a simulation of continuous intracranial pressure curves. Comp Biomed Res 31(4): 231–243
13. Schmidt B, Czosnyka M, Schwarze JJ, Sander D, Gerstner W, Lumenta CB, Pickard JD, Klingelhöfer J (1999) Cerebral vasodilatation causing acute intracranial hypertension – a method for non-invasive assessment. J Cereb Blood Flow Metab 19: 990–996
14. Schmidt B, Czosnyka M, Schwarze JJ, Sander D, Gerstner W, Lumenta CB, Klingelhöfer J (2000) Evaluation of a method for noninvasive intracranial pressure assessment during infusion studies in patients with hydrocephalus. J Neurosurg 92: 793–800
15. Schmidt B, Czosnyka M, Raabe A, Yahya H, Schwarze JJ, Sackerer D, Sander D, Klingelhöfer J (2003) Adaptive noninvasive assessment of cerebral autoregulation and ICP. Stroke 34: 84–89
16. Zabolotny W, Czosnyka M, Smielewski P (1994) Portable software for intracranial pressure recording and waveform analysis. In: Nagai H, Kamiya K, Ishii S (eds) Intracranial Pressure IX. Springer, Berlin, pp 439–440

Correspondence: Bernhard Schmidt, Pembaurstrone. 14a, 81243 München, Germany. e-mail: BSchmidt@LRZ.TUM.DE

Acta Neurochir (2005) [Suppl] 95: 351–355

Clinical experience with the noninvasive ICP monitoring system

Y. L. Zhao[1], **J. Y. Zhou**[2], and **G. H. Zhu**[3]

[1] Department of Neurosurgery, Beijing Tiantan Hospital, Beijing, China
[2] Department of Neurology, Beijing Xuanwu Hospital, Beijing, China
[3] Chongqing Welcom Co. Ltd., Beijing, China

Summary

The Noninvasive ICP (Intracranial Pressure) Monitoring System NIP-200/210 has been used in several hospitals with more than 2000 patients since March 2002. It is based on the N2 wave response to flash visual evoked potentials (FVEP). According to our data, the mean latency period for the FVEP-induced N2 wave in healthy controls was 126.61 $\pm$ 14.64 ms, in which that of females was shorter than that of males (123.95 $\pm$ 10.345 ms vs. 130.75 $\pm$ 14.632 ms; $p < 0.05$). There was no significant difference between the left or right side response (126.71 $\pm$ 14.91 ms vs. 124.468 $\pm$ 15.043 ms, $p > 0.05$). No significant difference in latency was found across age groups in our patient pool. In general, the N2 wave was stable and easily identified in most of the patients or healthy controls. When the data obtained with the NIP-200/210 Noninvasive ICP Monitoring System was compared with that from invasive techniques, the results were quite consistent (correlation index 0.651–0.97, standard error 8–15%). From our clinical trial results, we conclude that the latency periods for the FVEP-induced N2 wave reflected ICP values. However this technique is not suitable in patients with bifrontal hematoma, retinal concussion, or contusion of the optical nerve, because an FVEP value cannot be measured accurately in these cases. In our clinical trials, we used the FVEP technique to determine the effectiveness of mannitol in decreasing the ICP. The data revealed that ICP values decreased significantly within 20 minutes after a mannitol injection, and reached a minimum level at 40 minutes. For a single bolus of mannitol, the duration of the ICP decrease ranged from 30–210 minutes. Elevated ICP is one of the most important clinical issues in neurosurgery and neurology. The present noninvasive technique is safe and easy to perform, with a minimal risk of complications.

Keywords: Intracranial pressure; monitoring; visual evoked potential.

Introduction

Intracranial hypertension is an acute, severe symptom in nervous diseases and is one of the most common direct causes for a patient's death. Thus, a key step for rescuing such patients is to monitor their intracranial pressure promptly and accurately, and to deliver effective treatment.

Currently, there are various ICP monitoring methods used in clinical practice, but most of them are invasive. Although the invasive methods can measure ICP with relative accuracy, these methods require sophisticated technology and complicated procedures, and often induce additional complications such as intracranial infection, leakage of cerebrospinal fluid and intracranial hemorrhage. Their application is thus limited, and currently in most of the hospitals in China, doctors assess ICP levels through by clinical observation, which can result in the inappropriate usage of dehydrating agents. It is imperative to find a noninvasive technique for monitoring ICP in order to incorporate this factor into the clinical diagnosis.

Flash visual evoked potential (FVEP) can accurately reflect injuries of the visual pathway. Because the latency period for FVEP is prolonged in parallel with increases in ICP, a study was designed to apply both FVEP and invasive methods to the monitoring of ICP in patients, and to analyze the correlation and consistency between these methods. The feasibility of FVEP as a noninvasive means for monitoring ICP is discussed.

Clinical data and methods

Clinical data

Subjects: 152 participants in this study (89 males and 63 females) were enrolled from patients in the Department of Neurosurgery between March and April 2002 who had signed informed consent forms. Blood pressure, respiration rate, heart rate and body temperature were recorded for each patient, and tests of consciousness, pupil size and response to light were carried out before determining blood pressure. For patients under treatment with diuretics or anesthetics, the dosage and the schedule (day, hour) of these medications were recorded.

The following criteria were used to exclude patients from this study: (1) bilateral visual pathways under pressure from a hypophyseal tumor; (2) hypoxia (O2 saturation less than 95%); (3) obvious liver dysfunction (abdominal dropsy, severe hypoproteinemia, jaundice); (4) uremia; (5) severe acidosis; (6) diseases obviously affecting the visual acuity, such as severe cataracts, glaucoma or optic atrophy.

Methods

Instruments: the NIP-200 noninvasive ICP monitoring system (produced by Chongqing HaiWeiKang Medical Instrument Co., Ltd) was used to induce the flash visual stimulation.

Method: According to the operating instructions for the monitor, the sunflower-shaped galactic disk electrodes (8 mm in diameter) were placed 3 cm above the occipital tuberosity separately on the left and right sides, the reference electrode was located on the midline of forehead hairline, and the ground electrode was located on the glabella. The impedance between the electrodes was lower than 50 kΩ. The FVEP stimulation was produced by a light emitting diode array that was arranged in a pair of LED light glasses; dispersed yellow light was emitted in a pulsed-wave mode that was triggered by computer. The brightness of light glasses was 20000 cd/m^2.

Invasive examination: either the lumbar puncture method or the cerebral epidural manometric method (LCY-3.10 intelligent intracranial pressure monitor) were carried out to measure the pressure, depending on the diagnosis and condition of the individual patients. All invasive examinations were performed after carrying out the FVEP.

Statistical analysis

All data were expressed as means ± standard deviation, and were analyzed with a paired t-test and linear correlation analysis using SAS 6.12 software.

Results

A total of 152 patients were examined, including 89 males and 63 females, with an average age of 45 years (13–82 years). The indications included: 32 cases with cerebral hemorrhage, 27 cases with subarachnoid hemorrhage, 24 cases with meningitis, 16 cases with brain trauma (subdural hematoma, contusion and laceration of brain, etc.), 9 cases with brain tumor, 8 cases with hydrocephalus, and 36 cases with various other indications (cerebral infarction, benign intracranial hypertension, encephalitis, headache, etc.).

Recognition of the FVEP waveform

Figure 1a shows the FVEP waveform of a normal person, where the waves that appear below the baseline are negative waves (Negative, N). N126 is the large peak that occurs early on: it is stable and easy to distinguish, so the latency period of the N126 wave in NIP-200 noninvasive ICP monitoring system was adopted as a baseline reflecting the indices of ICP changes. Figures 1b–d show the FVEP waveforms for intracranial hypertension patients.

Comparisons between the results from FVEP and those of the invasive ICP methods

(1) Figure 2 displays the results for a correlation analysis of the ICP values showing a linear correlation between FVEP and the invasive methods, with a correlation coefficient (r) of 0.97. This indicates that the ICP values from the FVEP examination are remarkably well correlated with those from the invasive methods.
(2) A paired t-test gives a t-value of 0.37, which demonstrates there is no significant difference between the results from noninvasive and invasive ICP examinations ($p > 0.05$).
(3) The average relative error (δ) for the FVEP ICP values from the non-invasive ICP technique is 13.2%. 95% confidence limit for prediction is around 8 mm Hg.

There were no patterns in the alterations of the N126 wave amplitude, and there are no correlations of the amplitude with the ICP.

Discussion

The FVEP reflects the integrity of the visual pathway from the retina to the occipital lobe cortex [1, 2]. When the ICP increases, ischemia and anoxia are induced in the neurons and nerve fibers and the level of lactic acid increases, which results in a decrease in cerebrospinal fluid pH and elicits a nerve conduction blockade. The conduction velocity of the electrical signal decreases so as to prolong the latency period of FVEP peak, which is positively correlated with the ICP level. Based on this mechanism, it should be valid to employ FVEP as an indicator of changes in the ICP. Furthermore, FVEP is less affected by visual acuity and can easily be performed for patient monitoring without their active cooperation, for example, including patients with severe diseases and especially in comatose patients.

Changes in the FVEP waveform accompanying ICP increases

We have adopted the general international conventions for naming the waveform, and the portion above

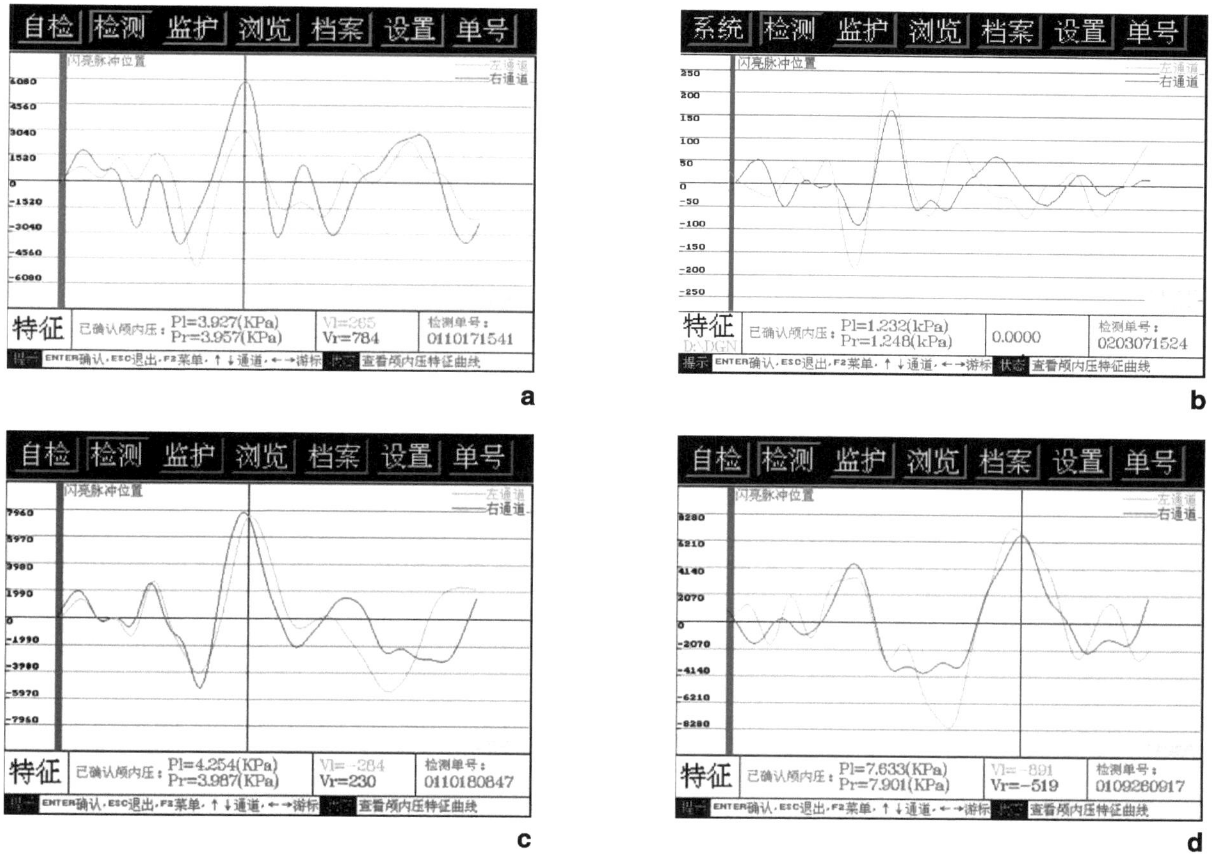

Fig. 1. (a) FVEP of a normal person. The marker is the N126 wave, showing that the average ICP is 126.5 mmH_2O. (b) FVEP of a patient with cerebral hemorrhage. The latency period of N126 is prolonged (147.25 ms). The FVEP indicates an ICP of 363 mmH_2O, while the cerebral epidural pressure is 380 mmH_2O. (c) FVEP of a patient with a subarachnoid hemorrhage. The latency period of the N126 peak is 155.91 ms. The average ICP from FVEP is 400 mmH_2O, while the pressure from lumbar puncture is 375 mmH_2O. (d) FVEP of a patient with a cerebral hemorrhage (hemorrhage volume is about 80 mL, patient is deceased). The FVEP displays double peaks; the marker indicates the N126 wave, and the average ICP of noninvasive methods is 741.5 mmH_2O

the baseline is referred to as a negative wave N. A review of the literature (2–9) shows that various researchers use different names to refer to the waveforms that reflect ICP changes, such as N2, N3, P1, P2, P100 wave, etc. The reasons for these differences may be related to the instruments, the use of different conventions for positive vs. negative waves, delays in the instrument's response time, different times to commence sampling, etc. For example, the N2 and N3 waves could refer to the same signal on different FVEP instruments. In order to facilitate comparisons of these findings, we propose that a typical graphic output for the instrument used should be included in publications in this field, and also the adoption of a standard peak to establish a benchmark for latency times, such as the N125 peak. In our preliminary experiments (manuscript in preparation), we have adopted the N125 wave latency period for the NIP-200 noninvasive ICP monitoring system as an index for changes in the ICP.

As seen from Figures 1b–d, the latency period for the N126 peak is prolonged with increases the ICP, and there is a linear correlation with these ICP values, but there were no patterns in the wave amplitude changes. The results of this study also showed that when the ICP exceeded 900 mmH_2O, the FVEP displayed 2 large waves (double peaks), as shown in Figure 1d, of which the N126 should be the latter peak. A dying patient's FVEP wave shows a similar change.

The correlation analysis of FVEP and ICP

Correlations between the FVEP and ICP have already been reported by researchers from different parts of the world. Donald *et al.* [3] used FVEP and

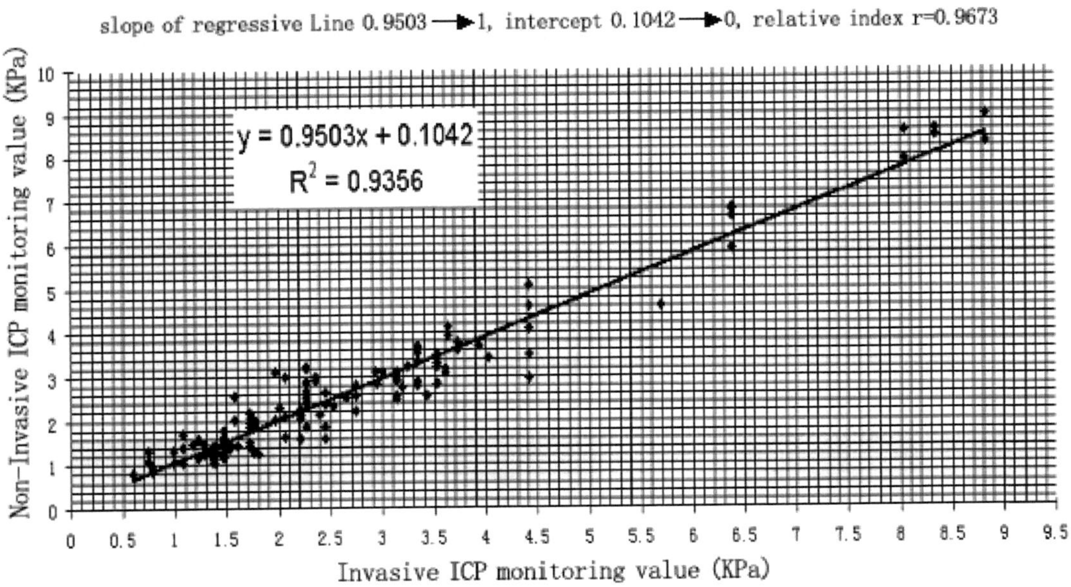

Fig. 2. Correlation analysis of ICP values from FVEP and from invasive examinations

the subdural ICP monitoring technique to conduct a comparative study of the patients with hydrocephalus and brain trauma. They discovered that the N2 latency period increased continuously in tandem with ICP increases, and the N2 latency period showed a positive correlation with ICP and a negative correlation with cerebral perfusion pressure. York *et al.* [4] obtained the same results, and suggested that FVEP could be used to examine the ICP in patients with hydrocephalus and brain trauma. They proposed that the prolongation of the latency period was related to a drop in the blood flow and oxygen pressure in the brain.

Burrows *et al.* [5] also confirmed this relationship in extracorporeal circulatory support in infants. Liu Jianjun *et al.* [6] studied the correlation between FVEP and brainstem auditory evoked potentials (BAEP) with ICP using the acute ICP rabbit model, and discovered that the latency periods for the N2 and P1 waves in the FVEP were positively correlated with increases in ICP, but there was no obvious correlation between BAEP and ICP. Our findings also indicated that the ICP values obtained from FVEP were remarkably well correlated with those from invasive methods, with a correlation coefficient (r) of 0.9. The paired t-test across both methods proved that the difference has no statistical significance ($p > 0.05$), and the average relative error is only 13.22%.

At present, there are many studies of noninvasive ICP monitoring methods, including the relatively mature methods such as FVEP, TCD, etc., and anterior fontanelle can be used in infants [1, 2, 10]. Compared with other methods, FVEP itself is an effective means for monitoring brain function and doing follow-up with seriously injured patients.

Both our findings and other literature reports indicate a good linear correlation between the prolongation of the FVEP peak with changes in ICP, and FVEP has the advantages of being simple, rapid, and convenient for bedside use. Thus, it is strongly recommended for clinical practice as a noninvasive ICP monitoring method to inform the design of treatment strategies. We have used the NIP-200 noninvasive ICP system to monitor dynamically a number of patients with intracranial hypertension, and subsequently obtained significant results (manuscript in preparation). However, there remain several important issues to be resolved before FVEP-based ICP noninvasive monitoring is ready for wider application, including enhancements in the accuracy, analysis of potential variations in the FVEP wave with different clinical indications, and recognizing and controlling any other factors that may influence the FVEP-based ICP system as a noninvasive monitoring method.

References

1. Martinez-Manas RM, Santamarta D, Campos JMD, *et al* (2000) Camino intracranial pressure monitor: prospective study

of accuracy and complications. J Neurol Neurosurg Psychiatry 69: 82
2. Zhang Dan, Peng Guoguang (2000) The advancements of noninvasive ICP monitoring technology. Foreign Med Sci (Cerebrovasc Dis) 8: 234
3. Donald H, Pullam YMW, Rosenfeld JG (1981) Relationship between visual evoked potentials and intracranial pressure. J Neurosurg 55: 909
4. York D, Legan M, Benner S, *et al* (1984) Further studies with a noninvasive method of intracranial pressure estimation. Neurosurgrey 14: 456
5. Burrows FA, Hillier SC, Mcleod ME, *et al* (1990) Anterior fontanel pressure and visual evoked potentials in neonates and infants undergoing profound hypothermic circulatory arrest. Anesthsiology 73: 632
6. Liu Jianjun, Yue Yun, Cai Wei, *et al* (1998) The feasibility of noninvasively monitoring ICP through visual evoked potential and brain stem auditory evoked potential. Chin J Anesthesiol 18: 173
7. Davenport A, Bramley PN (1993) Cerebral function analyzing monitoring and visual evoked potientials as a noninvasive method of detecting dysfunction in patients with acute hepatic and renal failure treated with intermittent machine hemofiltration. Renal Failure 15: 515
8. Gumerlock MK, York D, Durkis D (1994) Visual evoked responeses as a monitor of intracranial pressure during hyperosmolar blood-brain barrier disruption. Acta Neurochir (Wien) [Suppl] 60: 132
9. Jordan KG (1993) Continuous EEG and evoked potential monitoring in the neuroscience intensive care unit. J Clin Neurophysiol 10: 445
10. Desch LW (2001) Longitudinal stability of visual evoked potentials in children and adolescents with hydrocephalus. Dev Med Child Neurol 43: 113–117
11. Zhong J, Dujovny M, Park HK, *et al* (2003) Advances in ICP monitoring techniques. Neurol Res 25: 339–350

Correspondence: Yuanli Zhao, Department of Neurosurgery, Beijing Tiantan Hospital, Beijing, 100050, China. e-mail: zhaolang88@hotmail.com

Acta Neurochir (2005) [Suppl] 95: 357–361

Innovative non-invasive method for absolute intracranial pressure measurement without calibration

A. Ragauskas[1], **G. Daubaris**[1], **A. Dziugys**[1], **V. Azelis**[2], and **V. Gedrimas**[2]

[1] Kaunas University of Technology, Kaunas, Lithuania
[2] Kaunas University of Medicine, Kaunas, Lithuania

Summary

A new absolute ICP (aICP) measurement method was designed which does not need calibration. In this study we compared a new method with invasive aICP method in ICU on the patients with closed severe traumatic brain injury.

A new method is based on two-depth TCD technique for aICP and external absolute pressure aPe comparison using the eye artery (EA) as natural "balance". The intracranial segment of EA is compressed by aICP and the extracranial segment is compressed by aPe applied to the tissues surrounding the eye. The blood flow parameters in both EA segments are approximately the same when aPe = aICP. Two-depth TCD device is used as an indicator of balance aPe = aICP when the pulsatility index of blood flow velocity waveform in intracranial and extracranial segments are the same.

Fifty seven simultaneous invasive and non-invasive aICP measurements were performed in aICP range from 3.0 to 37.0 mmHg. Bland Altman plot of the differences between simultaneous invasive and non-invasive aICP measurements shows the negligible mean difference (mean = 0.94 mmHg) with a standard deviation of 6.18 mmHg.

This validation study shows that it is possible to measure aICP non-invasively without calibration of the system with 95% confidence interval of 12 mmHg.

Keywords: Non-invasive absolute intracranial pressure; calibration; transcranial Doppler; traumatic brain injury; eye artery.

Introduction

Most of existing absolute intracranial pressure (ICP) measurement is usually invasive. It would be very helpfull to measure absolute ICP without implantation of invasive transducers into the human brain [1]. Non-invasive absolute ICP measurement is extremelly important for the early diagnosing of a brain injury in emergency cases, for military and aerospace medicine and for a lot of other applications outside intensive care unit (ICU). It is expected that the non-invasive absolute ICP measurement technology could provide a solution to some of the problems associated with the current invasive ICP monitoring. The problems include the increased risk of infection, limitations on the duration of invasive ICP monitoring, delays in obtaining ICP diagnosis in emergency due to the need of implanting the sensors.

Non-invasive techniques for the measurement of ICP were reported as early as 1966 in newborns and infants. Because these techniques rely on an open fontanelle, they are not applicable in older children and adults. Other methods of non-invasive ICP measurement attempt to find the biophysical objects or physiological characteristics that would be related to the ICP and that could be measured or monitored non-invasively. Some of the proposed non-invasive technologies are based on dielectric properties [9] of cranium or intracraniospinal structures. Most of proposed approaches are based on ultrasonic measurements of ICP through the associated physiological parameters such as skull diameter, blood flow parameters in intracranial or intraocular vessels, pulsations of cerebral ventricles, acoustic properties of the cranium, skull bones, transintracranial parenchymal acoustic path, brain tissue resonance parameters, etc. [2, 3, 11, 12, 14, 15]. Very promissing for absolute ICP measurement is dynamic magnetic resonance imaging (MRI) technique [7]. Other approaches include tympanic membrane displacement [8], ophthalmodynamometry [4], quantitative pupillometry [13], othoacoustic emission [15], pulsatility of the ocular circulation [5], prediction of ICP based on transcranial Doppler (TCD) and arterial blood pressure (ABP) simultaneous measurements [10].

However, there are a few common problems limiting the introduction of the proposed methods into clinical application:

1. Biophysical object: which biophysical parameter of what object inside the human cerebrospinal system could be a stable and a repeatable function f of absolute ICP, i.e., $f(aICP)$?

2. Functional dependence $f(aICP)$: is that function linear and independent of such influential factors as ABP, cerebrovascular autoregulation impairment, anatomical and physiological individuality of the patient, etc.?

3. General individual calibration problem: how to calibrate the system "individual patient – non-invasive ICP meter" non-invasively?

Current clinical studies failed to solve these problems.

Calibration is a metrological term. By definition a calibration is determination of the correct value of measuring instrument by comparison with a "golden standard" instrument which is much more accurate. Non-invasive absolute ICP "golden standard" does not exist therefore the non-invasive calibration of the system "individual patient – non-invasive ICP meter" is impossible. This is the main reason why it is impossible to apply the existing approaches to non-invasive ICP measurement for creation of reliable non-invasive absolute ICP meter. The only solution of non-invasive absolute ICP meter's calibration problem is to apply for aICP measurement the direct comparison of aICP and absolute extracranial pressure aPe [6] by the natural physiological "scales", preliminary balanced by human anatomy (Fig. 1 a).

The objectives of this study were to show that the eye artery (EA) can be applied for non-invasive absolute ICP measurement as a pressure transducer and natural "scales", to show that special two-depth TCD technology is suitable for the pressure balance indication of natural "scales" and also to show that proposed pressure balance method (Fig. 1 a, b) does not need calibration of the system "individual patient – non-invasive absolute ICP meter".

Material and methods

In order to test our hypothesis about possibility to use the natural "scales" based on two EA segments for aICP measurement without calibration an anatomical study, numerical simulation and clinical validation study have been performed.

Anatomical study

In order to determine the technical requirements for specially designed compact battery powered two-depth TCD device (HemoPilot R2) for aICP measurement application we performed an anatomical study of the eye arteries. Study population was 34 cadavers, 67 eyes, 25 male and 19 female, mean age 23 years (8 months ... 60 years). We determined the following parameters:

- diameters of EA: right 1.6 ± 0.5 mm, left 1.5 ± 0.5 mm,
- lengths of the optic nerve canal: right 7.2 ± 2.5 mm, left 6.7 ± 2.4 mm,
- distances between internal carotid artery and dura mater crossing point by EA inside optic nerve canal: right 6.2 ± 2.2 mm, left 5.5 ± 2.3 mm.

The anatomical study shows that intracranial segment of EA affected by aICP is long enough for TCD measurements with standard TCD volume 3.0 mm. Anatomically identified distance between two TCD depths is from 4.5 to 5.0 mm. The typical depth of EA extracranial segment is from 36 mm to 40 mm and the typical depth of EA intracranial segment is from 46 mm to 53 mm. These anatomical findings were used for a numerical simulation of the pressure balance method and for the validation study in ICU.

Numerical simulation of absolute ICP value measurement method

The purpose of numerical simulation was to analyze the possible systematic error of the proposed non-invasive aICP measurement method and to show that this method does not need the individual calibration of the system "individual patient – non-invasive ICP meter".

The mathematical model of blood flow in the EA is based on the equations resulting from the Navier-Stokes theory (pulsatile incompressible flow in a flexible tube).

The mathematical model of pulsating blood flow through the EA is based on the assumptions that the behaviour of EA and blood flow is axis-symmetric and that ABP is constant across the artery. Therefore, blood flow can be modelled by one-dimensional equations resulting from the integration of three-dimensional non-compressible Navier-Stokes equations over the cross section of EA. The artery is considered as an axis-symmetric elastic tube. As a result, time evolution of blood volume yield in the artery is described by a system of continuity and momentum equations as follows:

$$\frac{\partial A}{\partial t} + \frac{\partial Q}{\partial z} = 0 \tag{1}$$

$$\frac{\partial Q}{\partial t} = -\frac{A}{\rho}\frac{\partial p}{\partial z} - \frac{8\pi\mu}{\rho A}Q - \frac{\partial}{\partial z}\left(\frac{Q^2}{A}\right) \tag{2}$$

where t is time, z is the axis along the EA, $A(z,t)$ is the inner cross-section area of the EA, $Q(z,t)$ is the blood volume yield, $p(z,t)$ is an ABP, μ and ρ are the blood viscosity and density, respectively.

The system is completed by the equilibrium equation of forces acting on the arterial wall:

$$p - p_{outer} = M\frac{\partial^2 r}{\partial t^2} + \gamma\frac{\partial r}{\partial t} + p_{elastic} + T\frac{\partial^2 r}{\partial z^2} \tag{3}$$

where $r(z,t)$ is the inner radius of the artery, p_{outer} – pressure acting on the outer wall of the artery and is a function of z, M is related to the density of the the wall, $T(r,z)$ – longitudinal tensile force, γ is the damping coefficient for wall motion, $p_{elastic}$ is pressure created by the elastic force of the arterial wall and is defined as follows:

$$p_{elastic} = p_0 \exp\left(k\frac{A - A_0}{A_0}\right) \tag{4}$$

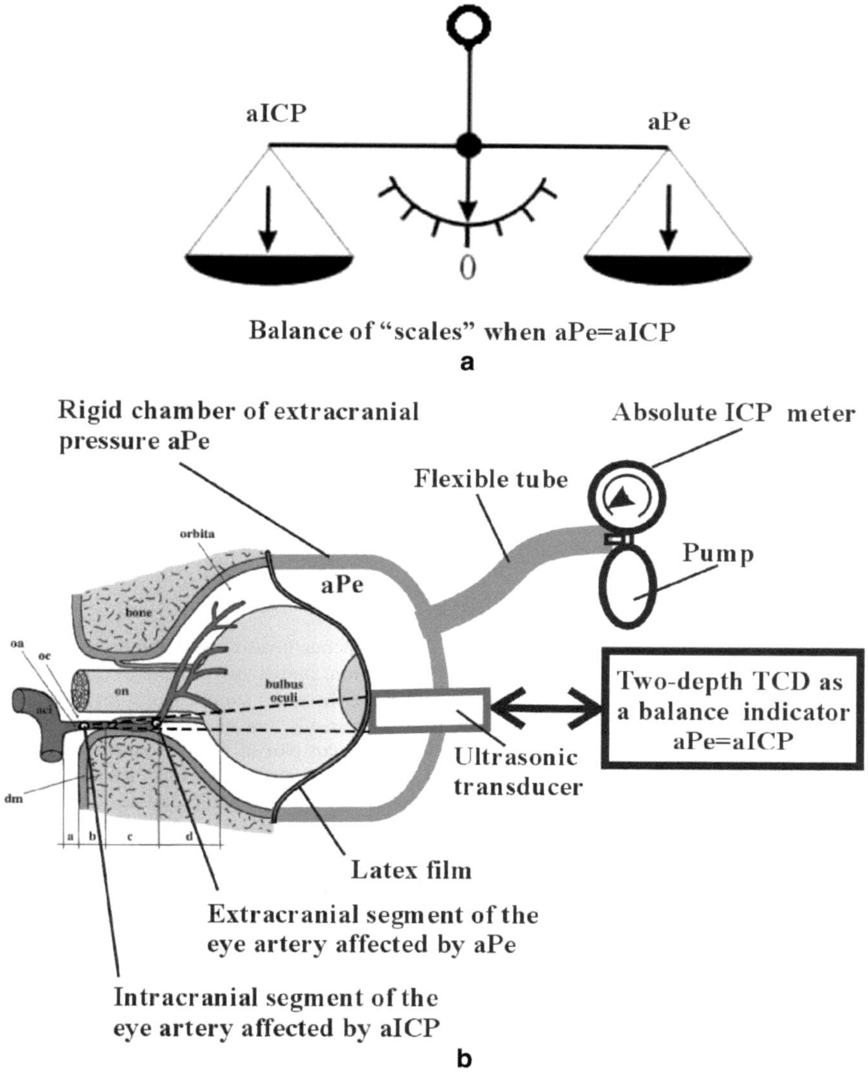

Fig. 1. Innovative non-invasive method for absolute ICP (*aICP*) measurement: (a) preliminary balanced "scales" as an aICP measurement instrument without calibration problem, (b) structure of proposed non-invasive aICP meter

where r_0 is the artery inner radius, while arterial pressure is p_0 and $k(z)$ is arterial stiffness.

The simulation area is divided into three segments: intracranial, inside the optic nerve canal and extracranial. That defines the different outer pressures: $p_{outer} =$ aICP inside the brain, $p_{outer} =$ aPe. Under assumption that the EA does not pulsate inside the optic nerve canal, the inner radius of the EA inside that canal is constant. As a result, the equilibrium equation (3) is not solved inside the optic nerve canal. All segments are coupled during calculations.

Boundary conditions are defined by inlet pressure p_{in}(t) or blood yield Q_{in}(t). Results of numerical simulation are presented in Fig. 2. It is shown on Fig. 2 a and b, that the ICP dependent systematic error of the proposed non-invasive method is within limits $[-3.0 \ldots +1.0]$ mmHg when the aICP values are within clinically important limits $[10 \ldots 40]$ mmHg. The calculated absolute systematic error is small enough and it is not important in clinical practice. This result confirms that the proposed method of two EA segments pressure balancing could be used for aICP measurement without a calibration problem. It is important to notice that the pressure balance aICP = aPe is not affected by changing EA blood flow parameters, by changing vascular resistances of the EA or the eye, by absolute values of ICP or Pe and by individual anatomy of patient.

Clinical study

In order to validate the proposed aICP measurement method the clinical study of simultaneous invasive and non-invasive aICP measurements has been performed under neurosurgical ICU conditions. The main goal of the study was to determine the systematic error of non-invasive method compared with the invasive method. In order to minimize the systematic errors of invasive ICP monitor (Codman ICP Express) the simultaneous invasive and non-invasive aICP measurements were performed as soon as possible after implantation of invasive ICP transducers. Study population was: pathology –

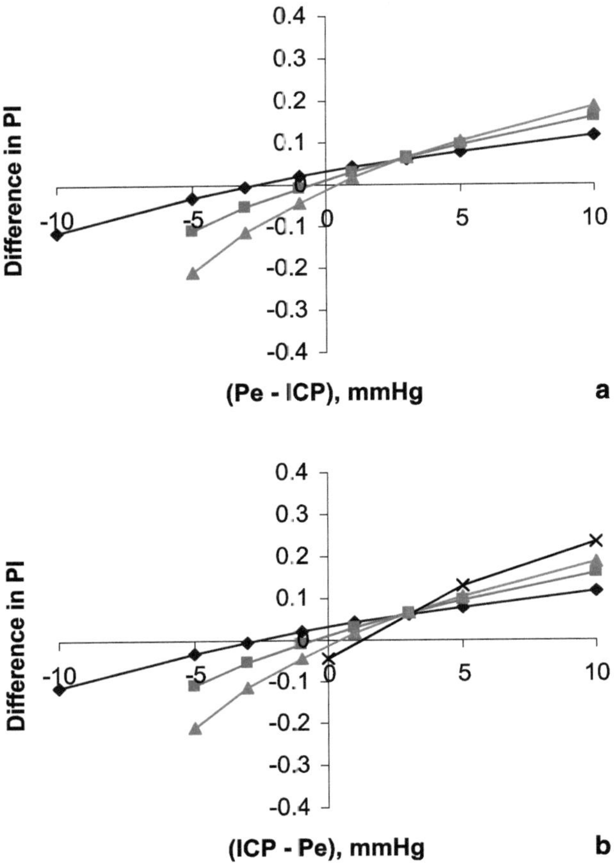

Fig. 2. Relationships between aPe, aICP and the differences of Goesling pulsatility indexes (*PI*) in intracranial and extracranial segments of the eye artery (numerical simulation): (a) model with constant mean blood flow in the eye artery. —◆— ICP = 10 mmHg; —■— ICP = 20 mmHg; —▲— ICP = 30 mmHg, (b) model with Pe dependent mean blood flow in the eye artery, —◆— ICP = 10 mmHg; —■— ICP = 20 mmHg; —▲— ICP = 30 mmHg; —×— ICP = 40 mmHg

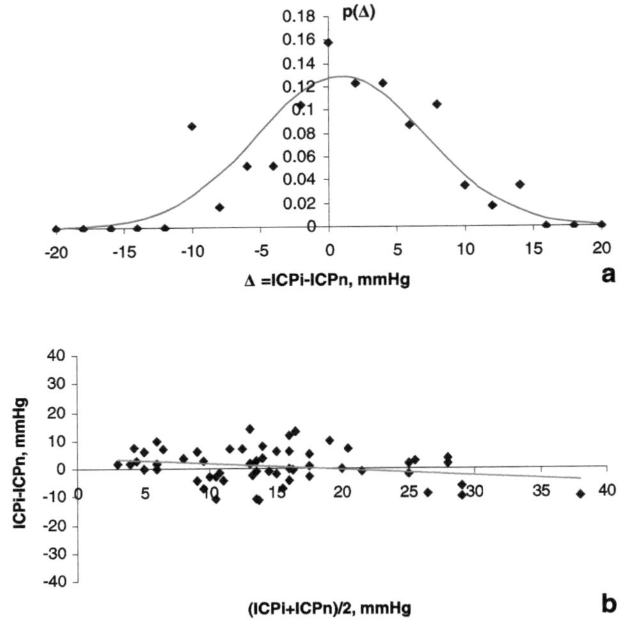

Fig. 3. Statistics of simultaneous invasive and non-invasive aICP measurement data (validation study): (a) probability mass function p(Δ) of the difference between simultaneous invasive and non-invasive aICP measurements (mean = 0.939 mmHg, SD = 6.18 mmHg), (b) Bland Altman plot of 57 simultaneous invasive and non-invasive aICP measurements in ICU: solid line is a negligible systematic error which is the evidence that proposed non-invasive method does not need calibration

closed severe traumatic brain injury 100%, 57 male/female 37/20, age mean 27.6 (18–70 years), implanted invasive ICP transducers.

The results of clinical study are summarized in Fig. 3. Probability mass function p(Δ) (Fig. 3 a) of the difference between simultaneous invasive and non-invasive aICP measurements shows negligible systematic error mean = 0.939 mmHg of the difference Δ = ICPi – ICPn between invasively (ICPi) and non-invasively (ICPn) measured absolute ICP values. The hypothesis on zero value of differences between invasive and non-invasive measurements can be proved by *t*-test statistics $|T| = 1.1471 < 2.0$. Bland Altman plot (Fig. 3 b) of simultaneous invasive and non-invasive aICP measurements shows that the negligible systematic error (solid line on Fig. 3 b) practically does not depend on absolute ICP in the tested aICP range. That is the evidence that proposed non-invasive absolute ICP measurement method does not need the correction of negligible systematic error, i.e., does not need calibration. That was predicted by mathematical modeling and numerical simulation.

The random error (Fig. 3) of clinical data (SD = 6.18 mmHg) are caused by both invasive and non-invasive meters. The random error of non-invasive aICP is mainly because of limited resolution and accuracy of two-depth TCD device. That is a new challenge for two-depth TCD technology development. Twice value of the standard deviation gives the predicted 95% confidence limit for prediction of invasive ICP around 12 mmHg.

The results of the clinical study support our hypotheses that the balance between aICP and aPe applied to the eye can be achieved and identified applying two-depth TCD technology with clinically non-significant systematic errors of non-invasive aICP measurement and with relatively small 95% confidence interval.

Acknowledgment

The research was supported by the US Dept. of Defence (Agreement DAMD17-00-2-0065).

References

1. Czosnyka M, Pickard JD (2004) Monitoring and interpretation of intracranial pressure. J Neurol Neurosurg Psychiatry 75: 813–821
2. Hassler W, Steinmetz H, Gawlowski J (1988) Transcranial Doppler ultrasonography in raised intracranial pressure and in intracranial circulatory arrest. J Neurosurg 68: 745–751
3. Kushiwaki H *et al* (1997) Human dural thickness measured by ultrasonographic method: reflection of intracranial pressure. J Ultrasound Med 16:11: 725–730

4. Motschmann M, Muller C, Kuchenbecker J *et al* (2001) Ophthalmodynamometry: a reliable method for measuring intracranial pressure. Strabismus 9: 13–16
5. Querfurth HW, Lagreze WD, Hedges TR, Heggerick PA (2002) Flow velocity and pulsatility of the ocular circulation in chronic intracraqnial hypertension. Acta Neurol Scand 105(6): 431–440
6. Ragauskas A, Daubaris G, Dziugys A (1999) Method and apparatus for determining the pressure inside the brain. US patent 5,951,477
7. Raksin PB, Alperin N, Sivaramakrishnan A *et al* (2003) Non-invasive intracranial compliance and pressure based on dynamic magnetic resonance imaging of blood flow and cerebrospinal fluid flow: review of principles, implementation, and other non-invasive approaches. Neurosurg Focus 14(4): article 4
8. Reid A, Marchbanks RJ, Bateman DE *et al* (1989) Mean intracranial pressure monitoring by a non-invasive audiological technique: a pilot study. J Neurol Neurosurg Psychiatry 52: 610–612
9. Russegger L, Ennemoser O (1990) Atraumatic measurement of intracranial pressure. Wien Klin Wochenschr 102: 543–547 (in German)
10. Schmidt B, Klingelhofer J, Schwarce JJ *et al* (1997) Non-invasive prediction of intracranial pressure curves using transcranial Doppler ultrasonography and blood pressure curves. Stroke 28: 2465–2472
11. Schoser BG, Riemenschneider N, Hansen HC (1999) The impact of raised intracranial pressure on cerebral venous hemodynamics: a prospective venous transcranial Doppler ultrasonography. J Neurosurg 91: 744–749
12. Shakhnovich AR, Shakhnovich VA, Galushina AA (1999) Non-invasive assessment of the elastance and reserve capacity of the craniovertebral contents via flow velocity measurements in the straights sinus by TCD during body tilting test. J Neuroimaging 9: 141–149
13. Taylor WR, Chen JW, Meltzer H *et al* (2003) Quantitative pupillometry, a new technology: normative data and preliminary observations in patients with acute head injury. Technical Note. J Neurosurg 98: 205–213
14. Ueno T, Shuer LM, Yost WT *et al* (1998) Development of a non-invasive technique for the measurement of intracranial pressure. Biol Sci Space 12: 270–271
15. US Patents (1980–2004) 4204547, 4984567, 5117835, 5388583, 5617873, 5919144, 6086533, 6117089, 6129682, 6231509, 6390989, 6413227, 6475147, 6589189, 6702743, 6740048, 6746410, 6773407

Correspondence: Arminas Ragauskas, Head of Telematics Scientific Laboratory, Kaunas University of Technology, Lithuania, Studentu 50-451a, 3031 Kaunas, Lithuania. e-mail: telematics@tef.ktu.lt

Acta Neurochir (2005) [Suppl] 95: 363–365

The correlation of midline shifts of human brain with large brain haematoma using a finite element approach

A. Y. S. Cheng[1], **M. C. Y. Pau**[1], **W. S. Poon**[2], and **G. K. C. Wong**[2]

[1] Department of Physics and Materials Science, City University of Hong Kong, Hong Kong, China
[2] Division of Neurosurgery, Prince of Wales Hospital, Chinese University of Hong Kong, Hong Kong, China

Summary

Objectives. We report on using a computational (finite element) model to simulate a human skull-brain structure to quantify the distortion of brain.

Methods. We simulated various effects of brain haematoma causing the distortion of brain. Midline shifts of the human brain in relation to size and location of haematoma were compared with the theoretical prediction.

Results. Prediction of midline shifts in lobar space-occupying lesions was more accurate that in deep-seated ones (such as thalamic lesions).

Discussion. More accurate boundary conditions of space-occupying lesions and better knowledge of physical materials properties of brain tissues can improve predictions of brain deformation using mathematical models.

Keywords: Cerebral hemorrhage; simulation; finite element method; brain shift; space occupying brain lesion.

Introduction

Frameless neuronavigation has proven its usefulness in neurosurgical operation for accurate spatial delineation. However, it suffers from the inaccuracy caused by the intraoperative brain shift. Mathematical understanding and correction of brain shift would be therefore useful.

Neuro-surgical CT and MRI scan images of patients can now be imported into our contour digitization program. This program has been modified to take up the DICOM 3 format, which is a common format for medical image transfer.

The finite element model software from Unigraphics is the same as the 'Nastran' FEA software, except that it has better user interface. The modeling of a human skull using this kind of method for a specific patient has not been reported in the literature so far.

Method

After image extraction, the skull, the left and right hemispheres and the haematoma are all separately extracted as independent objects. The boundary condition for the FEA model presumes the inner boundary of the skull being a rigid support structure. The nodes on this area are all constrained in the displacement and rotation in all 6 degrees of freedom. When the haematoma is located inside one side of the brain, the model is only built for that side. The other side is extracted and processed as a reference for comparison.

The co-ordinates of outline of the extracted haematoma are used to compute the location of its mass centroid. A circular seed is used to model the growth of the haematoma. A loading in a constant and uniform form is applied as the initial loading conditions for the analysis.

The physical material properties of a human brain can be found readily in the literature [1, 2] and they were summarized in the Table 1 below. The Young modulus taken for the calculation is probably appropriate for dead brain, without volume of circulating blood which can be easily displaced outside craniospinal system. This volume makes the compliance of 'living' brain much lower than 'dead' incompressible brain. Therefore the value of the Young modulus intuitively should be lower for the living brain. Unfortunately, there is no data concerning such a modification in the literature (editorial comment).

The FEA model generated with 3D mesh using the contour outlined previously. For the simplicity in the calculations, the whole model was treated as a 5 mm thick solid in our investigation.

After validating the method described above, the variations of the

Table 1. *Physical parameters used in the Finite Element Analysis (FEA) model*

Physical Parameters of Brain	
Young's Modulus (N/m^2)	2.52357×10^5
Poisson ratio ν	0.48
Mass density (kg/m^3)	1.007×10^3
FEA Model Characteristics	
Element Type	3 mm Tetra
Diameter of tumour seed (mm)	10
Initial Loading on seed (mN)	1×10^6

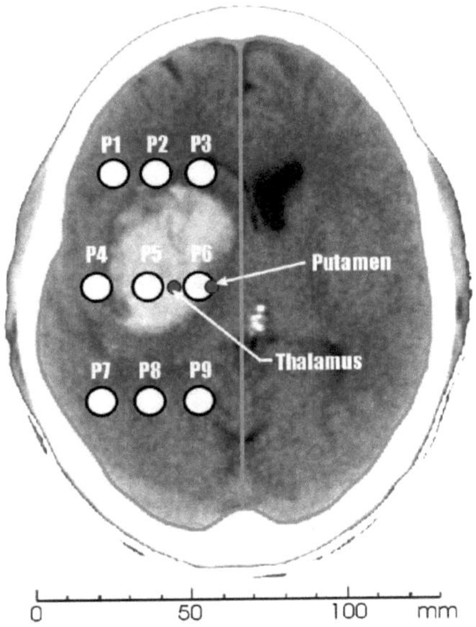

Fig. 1. The positions of the seed used in this study

Table 2. *The maximum shifts along the midline line for the 11 cases*

Calculated Cases	Max Displacement (mm)	Strain Energy (%)
Haematoma	16.82	1.331
P1	16.66	0.592
P2	16.90	0.578
P3	22.11	0.770
P4	16.23	0.624
P5	16.95	0.672
P6	20.32	0.670
P7	16.87	0.679
P8	16.28	0.651
P9	21.37	0.855
Thalamus	19.27	0.906
Putamen	28.25	0.872

shape of the mid-line shifts as a function of haematoma sites were studied. This was achieved by repositioning the seed to various positions. To cover most of specific cases found clinically, we decided to pick 9 representative positions for our calculations. They are depicted as positions P1 to P9 as shown in Fig. 1. Two clinically useful positions were also studied. They are haematoma in thalamus and putamen.

Result

The maximum displacements in the model and the corresponding maximum percentage strain energies for all the 11 different cases were summarized in Table 2.

We also studied the effect of changing the size of the haematoma on the maximum mid-line shifts. The goodness of fit of the relationship was shown graphically in Fig. 2. The variations of the maximum mid-line shifts followed a linear correlation with a linear proportional constant of 0.89 mm increase in mid-line shifts for every 100 mm^2 increase in size of the space-occupying lesion. The space-occupying lesion sizes were measured by graphical method from the presented output of the FEA software. The accuracy was about ± 0.1 mm since the used a pixel scale was 0.208 mm.

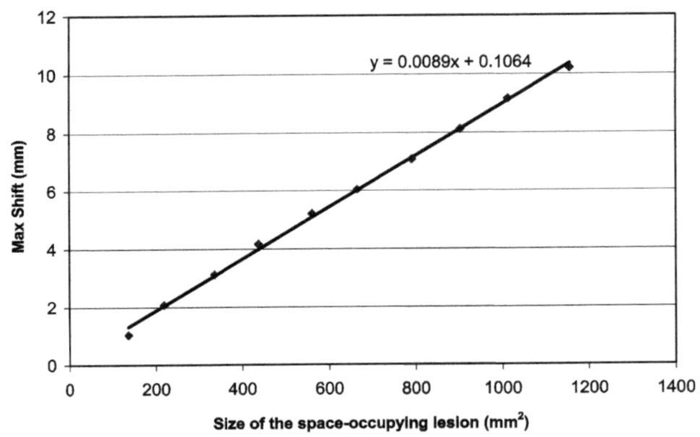

Fig. 2. The linear correlation between the sizes of the space-occupying lesion and the maximum mid-line shifts of the human brain was shown in this plot

Conclusion

Using Nastran, a linear structural finite element analysis software, computed model can be validated by comparing simulations and clinical results. The boundary conditions and the method of building the brain model were both adequate for the FEA computation.

For large size of space-occupying lesions within 40 mm diameter, the human brain basically behaves as a linear elastic solid model. The mid-line shifts can be predicted to sub-mm accuracy in our method without having to go to elaborate 3D model of the whole brain.

From the results, we can conclude that the locations and sizes of the space-occupying lesions were the major factors that dominated the magnitude and shape of brain mid-line shifts. The results also suggested that the idea of utilizing a finite element approach to interpret the midline shifts in the presence of large space-occupying lesion have considerable merit. It will lead to the development of quantitative models that could predict brain shifts during operations, which is a very important tool for 3D navigated brain surgery in the 21st century.

References

1. Hickling R, Wenner ML (1973) Mathematical Model of a head subjected to an axisymmetric impact. J Biomechanics 6, 115–132
2. Charalambopoulos A, Dassios G, Fotiadis DI, Massalas CV (1998) Mathl Comput Modelling 27(2): 81–101

Correspondence: W. S. Poon, Department of Surgery, Prince of Wales Hospital, Shatin, New Territories, Hong Kong SAR, China. e-mail: wpoon@surgery.cuhk.edu.hk

Acta Neurochir (2005) [Suppl] 95: 367–370

Clinical study of continuous non-invasive cerebrovascular autoregulation monitoring in neurosurgical ICU

A. Ragauskas[1], **G. Daubaris**[1], **V. Petkus**[1], **V. Ragaisis**[2], and **M. Ursino**[3]

[1] Kaunas University of Technology, Kaunas, Lithuania
[2] Kaunas Medical University Neurosurgical Clinic, Kaunas, Lithuania
[3] University of Bologna, Bologna, Italy

Summary

Ultrasonic "time-of-flight" monitor (Vittamed) was used for continuous monitoring of intracranial blood volume (IBV) pulse, respiratory, slow waves and cerebrovascular autoregulation (CA). The objectives are to compare of invasively and non-invasively monitored slow intracranial waves and CA of ICU patients and to evaluate the phase shift between ABP and IBV respiratory waves as a possible estimator of CA.

CA monitoring has been performed in 13 patients with severe TBI (age mean/range 30.5/(18–64)). Data were collected from 87 one-hour sessions of simultaneous invasive and non-invasive wave monitoring and from 53 one-hour sessions of invasive and non-invasive CA monitoring.

High correlation ($R > 0.9$) has been obtained between invasively and non-invasively recorded intracranial slow waves. Bland Altman difference between invasively and non-invasively recorded intracranial slow waves is clinically not significant (mean $= -0.07$, $SD = 0.089$, $\alpha = 0.05$). Agreement has been confirmed between invasive and non-invasive CA monitoring data in a wide range of $R = [-0.85; +0.96]$.

Hypothesis of the coincidence of invasive and non-invasive CA assessment is accepted ($p < 0.05$).

Phase shift monitoring of permanent respiratory ABP waves and IBV waves permit continuous non-invasive CA estimation without unnatural physical or pharmacological stimulations of CA system.

Keywords: Traumatic brain injury; ICP; ABP; slow waves; cerebrovascular autoregulation; non-invasive monitoring.

Introduction

It is assumed that a negative correlation or the phase shift close to 180° in the time courses of 'slow waves' (0.03 Hz to 0.008 Hz) of CBF and CPP reflects an active change in cerebrovascular resistance and therefore an intact CA system [2, 3, 5]. A positive correlation or the phase shift close to zero of CBF and CPP shows a non-reacting cerebral vessels and impaired CA system [2, 3, 5].

Control of CA in order to prevent secondary brain insults and to guide ICP/CA targeted therapy requires continuous uninterrupted and real-time CA monitoring. Transcranial Doppler (TCD) technology together with non-invasive ABP wave monitoring has been proposed for non-invasive CA assessment [5].

A convenient way to assess CA continuously under ICU conditions is to monitor the moving Pearson's correlation coefficient between invasively measured ABP and ICP waves [2, 8] at B wave frequencies [4]. It is necessary to note that the moving Pearson's correlation coefficient in the case of CA monitoring reflects the cosine of phase shift between the waves under comparison (+1 is equivalent to 0 degrees phase shift, −1 is equivalent to 180 degrees phase shift). The limitation of slow B wave moving correlation method are the intermittent nature of B waves. Moving averaging of monitoring data in order to reduce the uncertainty of CA estimation is the cause of 3 min to 10 min instrumental delay between actual CA changes and the reflection of such changes in the CA monitoring [2]. Such a data delay time can be difficult in the acute patients.

ABP and ICP respiratory waves are permanent and up to 10 times more frequent comparing with slow B waves. The main advantage of natural or ventilatory supported respiratory wave application for CA assessment is the possibility of continuous uninterrupted CA monitoring with up to 10 times shorter monitoring data delay comparing with B wave method.

Slow, respiratory and pulse ICP waves are the consequences of the variations of intracranial blood volume (IBV). All IBV waves can be monitored con-

Table 1. *Study population parameters*

Total	13 ICU patients
Gender: male/female	10(76.9%)/3(23.1%)
Age: Mean/Range	30.5/(18–64) years
Pathology: Closed severe traumatic brain injury	100%
Glasgow coma scale before sedation	3–7
Mean ABP range, mmHg	35–140
Mean ICP range, mmHg	3–80
Mean CPP range, mmHg	54–142
Cerebrovascular autoregulation simultaneous invasive and non-invasive monitoring	53 one hour sessions
Slow intracranial B wave, respiratory wave and pulse wave simultaneous invasive and non-invasive monitoring	87 one hour sessions

tinuously, non-invasively and in real-time applying ultrasonic "time-of-flight" technique [6, 7].

The aim of this study was to verify clinically continuous CA monitoring on neurosurgical ICU patients with severe TBI. Simultaneous continuous invasive and non-invasive monitoring of CA were compared Additionally, the phase shift between ABP and IBV respiratory waves as an estimator of CA state was preliminary evaluated.

Materials and methods

Clinical data were collected from neurosurgical ICU of Kaunas Medical University, Lithuania. ABP, ICP and IBV monitoring data have been obtained following Clinical Research Protocol No. 99124006, AIBS No. 990135, HSRRB log No. A-9676. Invasive ICP monitors Camino V420 and Codman ICP Express have been used simultaneously with the non-invasive IBV monitor (Vittamed) in the ICU on TBI coma patients. Datex CardioCup II ABP monitor was used for invasive ABP wave monitoring. Study population parameters are shown in Table 1.

ICP, ABP and IBV data were simultaneously and collected (50 Hz sampling frequency) at the bedside by a personal computer running Vittamed software. In order to select ICP, ABP and IBV slow B waves the digital real-time bandpass filtering (bandwidth 0.03 Hz to 0.008 Hz) procedure has been performed. The digital real-time bandpass filtering (bandwidth 0.1 Hz to 0.45 Hz) procedure has been used for respiratory wave selection. Moving Pearson's correlation coefficient method with three minute averaging window has been applied simultaneously for ABP/ICP and for ABP/IBV slow B wave.

The phase shifts between ABP/IBV waves in the case of intact CA were estimated applying monitoring data within subinterval of $R = [-1.0; -0.4]$ and in the case of impaired CA – within subinterval of $R = [+1.0; +0.4]$. A mathematical model of CA [21] has been used for simulation of the ABP/IBV wave phase shift frequency dependences in the cases of intact CA and impaired CA.

Results and discussion

Comparison of invasive and non-invasive slow B wave monitoring clinical data

The typical clinical results of simultaneous invasive and non-invasive monitoring of CA are illustrated in Fig. 1. It shows a good agreement between invasive and non-invasive moving correlation coefficient R individual variability during one hour invasive and non-invasive CA monitoring sessions.

It was determined that the coefficient R(ICP; IBV) exceeded 0.9 when the amplitude of ICP B waves was above 3 mmHg. Correlation between invasive and non-invasive CA monitoring data also exceeded 0.9 when the amplitudes of ICP B wave was above 3 mmHg and the amplitude of ABP slow wave was above 5 mmHg.

In order to assess the similarity between the invasive ICP slow waves and non-invasive IBV slow waves data two types of data sets were evaluated for each monitoring session. The first type data sets that contain all measurement points were checked by calculating the standard deviation and mean of differences between the normalized invasive and non-invasive data and also by applying the correlation analysis. The second type data sets that contain randomly selected data points were evaluated by performing *t*-test of the paired data. The values of *t*-criterion corresponding to each data set were calculated using the normalized data. Normalization was performed dividing the original measurement data by standard deviation. The calculated values of *t*-criterion were checked in terms of inequality $|t| < T_{crit}(\alpha = 0.05, n = 50 \ldots 70)$, where, t is the calculated value of *t*-criterion of paired samples, T_{crit} is a critical value of *t*-statistics, α is a significance level of the test, n is a number of randomly selected points. $N = 83$ samples of paired data were found which satisfied condition of accepted hypothesis on the equality of the mean values of measurements under comparison. The investigated number of samples was $N = 87$. The obtained proportion of samples with positive decision was higher than the proportion of interest ($\pi = 0.95$) and that confirms the hypothesis on the coincidence of invasively and non-invasively measured slow waves $\pi = N_+/N = 0.954 > 0.95$.

The Bland Altman evaluation of invasively and non-invasively measured slow B waves and the distribution of the differences between these waves shows the good agreement between invasive and non-

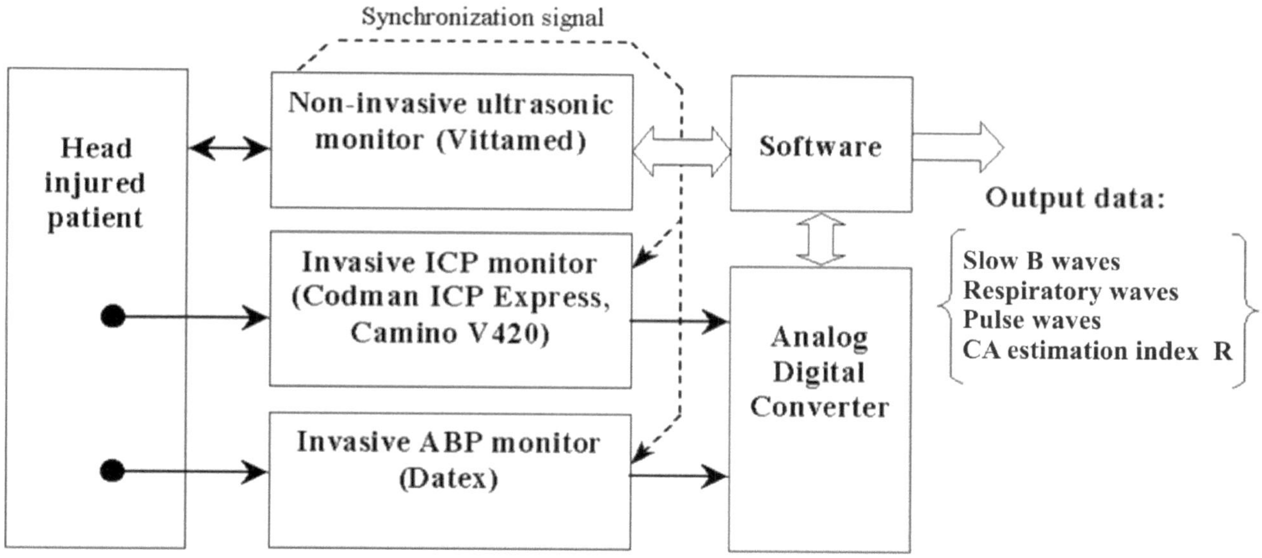

Fig. 1. Phase shift between ABP and IBV waves in the cases of intact and impaired CA comparing with intact CA/impaired CA phase shift data of mathematical simulation: *Tpa* time constant of pial arteries, *Tar* time constant of arterioles. Slow B wave data are presented in frequency band 0.01 Hz...0.04 Hz, respiratory wave data – in frequency band 0.1 Hz...0.45 Hz and pulse wave data – in frequency band 0.65 Hz...3.0 Hz

invasive data (SD = 0.089, $p < 0.0001$). The Bland Altman evaluation of invasive and non-invasive CA monitoring data also shows good agreement (SD = 0.05, $p < 0.0001$). Both invasive and non-invasive methods of CA monitoring seem to give the same diagnostic information about CA state with the uncertainty of experimental data which is not important for clinical practice.

Relationship between ABP/IBV respiratory wave phase shift and CA

Our data suggest that the phase shift not only between slow intermittent B waves of ABP and IBV, but also the phase shift between permanent respiratory waves is able to differentiate between intact and impaired CA (Fig. 2). In order to support the clinical results, the phase shift between ABP and ICP waves has been simulated on a computer using the mathematical model described in [9], with different values of autoregulation gains and time constants. The simulations were performed at the frequencies 0.01 Hz, 0.1 Hz and 1.0 Hz (Fig. 2). The different time constants of pial artery reactivity (Tpa) and reactivity of arterioles (Tar) has been used in the case of intact CA. The impairement of CA was simulated by reducing all gains in CA model up to zero. Mathematical simulation shows that the phase shift between respiratory waves under intact CA and impaired CA conditions is from 26° to 73° in the wide range of Tpa = [1.5 s; 10 s] and Tar = [15 s; 2 s]. That is in agreement with our empirical data (Fig. 2). It is evident (Fig. 2) that the phase shift between ABP and IBV can be used as an estimator of CA impairement under ICU conditions when the frequences of natural or ventilatory supported patient respiration can change in wide frequency band from 0.1 Hz to 0.45 Hz. This is an attractive alternative to slow B wave application for CA monitoring because natural or ventilatory supported respiration is a permanent physiological process which can be monitored continuously, uninterruptedly and with the instrumental delay time of monitoring data up to 10 times less comparing with B wave method (Ragauskas A., Daubaris G. Method and apparatus for non-invasive continuous monitoring of cerebrovascular autoregulation state. US patent pending, 2004).

ABP and IBV pulse wave phase shift clinical data shows no difference in the cases of intact and impaired CA (Fig. 2). That is in agreement with simulation data.

In conclusion, our study shows that the good correlation between invasive and non-invasive CA assessment using intermittent B wave. Phase shift changes of permanent respiratory ABP waves and IBV waves

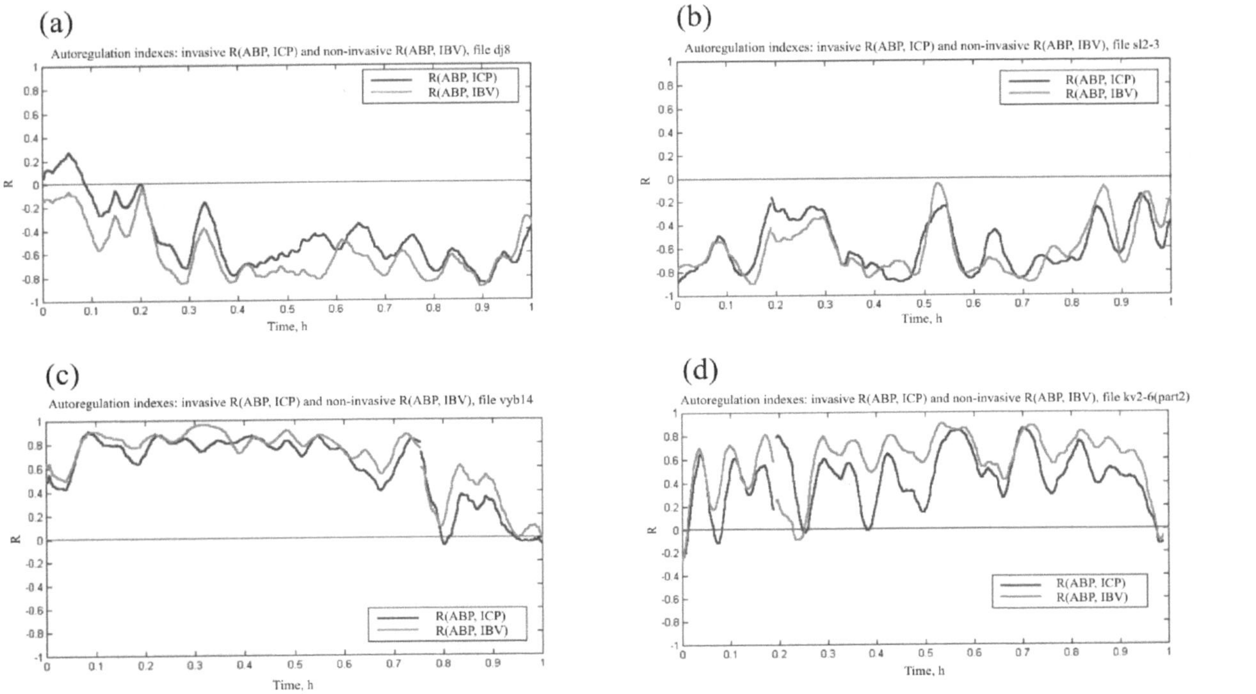

Fig. 2. Typical simultaneous invasive and non-invasive CA monitoring data (1 hour monitoring sessions): a), b) variability of intact CA, c), d) variability of impaired CA

also permit continuous non-invasive CA monitoring under ICU conditions without physical or pharmacological stimulations of CA system.

Acknowledgment

The research was supported by Lithuanian State Science and Studies Foundation and partially supported by the US Dept. of Defence (Agreement DAMD17-00-2-0065).

References

1. Czosnyka M, Smielewski P, Kirkpatrick P *et al* (1997) Continuous assessment of the cerebral vasomotor reactivity in head injury. Neurosurgery 41: 11–17
2. Czosnyka M, Smielewski P, Piechnik S, Pickard JD (2002) Clinical significance of cerebral autoregulation. Acta Neurochir [Suppl] 81: 117–119
3. Czosnyka M, Pickard JD (2004) Monitoring and interpretation of intracranial pressure. J Neurol Neurosurg Psychiatry 75: 813–821
4. Lemaire JJ, Khalil T, Cervenansky F *et al* (2002) Slow pressure waves in the cranial enclosure. Acta Neurochir (Wien) 144: 243–254
5. Panerai RB (1998) Assessment of cerebral pressure autoregulation in humans – a review of measuring methods. Physiol Meas 19: 305–338
6. Petkus V, Ragauskas A, Jurkonis R (2002) Investigation of intracranial media ultrasonic monitoring model. Ultrasonics 40: 829–833
7. Ragauskas A, Daubaris G, Ragaisis V, Petkus V (2003) Implementation of non-invasive brain physiological monitoring concepts. Med Eng Phys 25: 667–678
8. Reinhard M, Roth M, *et al* (2003) Cerebral autoregulation in carotid artery occlusive disease assessed from spontaneous blood pressure fluctuations by the correlation coefficient index. Stroke 32: 2138–2144
9. Ursino M, Magosso E (2001) Role of tissue hypoxia in cerebrovascular regulation: a mathematical modeling study. Ann Biomed Eng 29: 563–574

Correspondence: Arminas Ragauskas, Head of Telematics Sc. Lab., Kaunas University of Technology, Lithuania Studentu 50-451a, LT-3031, Kaunas, Lithuania. e-mail: telematics@ktu.lt

Experimental aspects

Acta Neurochir (2005) [Suppl] 95: 373–376

Increased intracranial pressure induces a rapid systemic interleukin-10 release through activation of the sympathetic nervous system

C. Woiciechowsky[1] and **H.-D. Volk**[2]

[1] Department of Neurosurgery, Charité Campus Virchow-Klinkum, Berlin, Germany
[2] Department of Medical Immunology, Charité Campus Mitte, University Medicine of Berlin, Berlin, Germany

Summary

There is a bi-directional communication between the immune and central nervous system. In this context, it is known that patients with traumatic brain injury suffered from systemic immunodepression and an increased risk to develop infectious complications. We investigated the role of an increased intracranial pressure (ICP) and sympathetic activation on systemic immune changes. A sustained increase in ICP was achieved by inflation of a subdural balloon. At different time points, plasma levels of the anti-inflammatory cytokine, interleukin (IL)-10, were measured. Furthermore, the effect of a sympathetic blockade by co-administration of the β_2-adreoreceptor antagonist, propranolol, was evaluated. Finally, we examined the impact of epinephrine infusion on blood IL-10 levels. We showed that an increase in ICP with activation of the sympathetic nervous system was able to induce systemic release of IL-10. This effect was blocked by administration of the β_2-adreoreceptor antagonist. Furthermore, epinephrine infusion directly induced systemic release of IL-10. Our data suggested that sympathetic activation with release of epinephrine may induce systemic immunodepression with risk of infectious complications in brain-injured patients.

Keywords: Intracranial pressure; interleukin-10; immunodepression; catecholamines.

Introduction

Brain injury is an independent risk factor for infectious complications in patients with polytrauma [18]. It is reported that early pneumonia occurs in 40% of patients with closed head injury [10, 15]. Moreover, brain injury is associated with the appearance of different cytokines (e.g. IL-1, IL-6, IL-8, IL-10) in the cerebrospinal fluid and plasma [3, 6, 12, 13, 25]. Interestingly, high levels of pro-inflammatory cytokines in the brain and an elevated intracranial pressure (ICP) can stimulate neuroimmune pathways like the hypothalamic pituitary adrenal (HPA) axis and the sympathetic nervous system (SNS). This correlates with the severity of injury and outcome [1, 2, 23]. Since monocytes and macrophages are the main targets for the immunomodulatory action of glucocorticoids and catecholamines. Therefore, alterations of these immunologically important cells should reflect the cerebral impact on the post-injury immunodepression [11, 20, 21]. An important mediator of monocytic deactivation is interleukin (IL)-10, an anti-inflammatory cytokine that inhibits the production of pro-inflammatory cytokines (for example tumor necrosis factor, TNF) and is a major depressor of specific cellular immunity through its reduction of monocytic MHC class II expression and IL-12 production [7, 14].

With this background, we investigated the importance of epinephrine infusion and increased ICP with sympathetic activation for the systemic IL-10 release in a rat model.

Our results provide strong evidence that catecholamine-mediated systemic release of IL-10 may be an important neuroimmunological mechanism contributing to immunodepression after injury and stress.

Material and methods

All experimental protocols were approved by the Animal Protection Board of the Senate of Berlin. Thirteen male Sprague-Dawley rats (body weight 350–400 g; Harlan Winkelmann, Borchen, Germany) were anaesthetized using an isoflurane (1.5–2.5%/v/v), N_2O/O_2 (0.5 l/min, 0.25 l/min) gas-mixture. A Forgarty-catheter (Baxter Healthcare, Deerfield, Illinois, USA) was placed subdurally through a bore hole. ICP was elevated to 60 mm Hg by inflation of the balloon of this catheter. Six of these animals also received an intravenous infusion of propranolol (Sigma-Aldrich, Seelze, Germany) at a dosage of 7.5 mg/h during the entire observation period. ICP and mean arterial pressure (MAP) were monitored continuously using an intraparenchymal ICP probe (Codman/Johnson & Johnson, Raynham, Massachusetts, USA) and a catheter in the femoral artery connected to a Transpac transducer (Abbott, Ludwigshafen, Ger-

Table 1. *An increase in ICP followed by an increase in MAP (as indication of sympathetic activation) induces a systemic release of IL-10*

Time (min)	ICP (mmHg)	MAP (mmHg)	IL-10 (pg/ml)
0	13.6 ± 1.1	96.6 ± 4.5	<40,0
10	42.5 ± 6.7	102.1 ± 5.9	58.4 ± 18.4
30	64.8 ± 3.6	131.8 ± 8.2	757.6 ± 237.8*
50	65.3 ± 7.4	152.2 ± 11.7	1072.6 ± 333.5*

* $p < 0.05$ versus 0 h value, Mann-Whitney U test.

Table 2. *The application of the β_2 adrenoreceptor antagonist propranolol blocks the ICP triggered systemic release of IL-10*

Time (min)	ICP (mmHg)	MAP (mmHg)	IL-10 (pg/ml)
0	9.0 ± 0.5	83.2 ± 3.3	<40,0
10	38.7 ± 7.9	95.8 ± 1.5	<40,0
30	60.2 ± 0.9	109.6 ± 2.9	<40,0
50	59.8 ± 0.8	115.0 ± 4.8	64.4 + 24.4

many), respectively. Arterial blood samples were collected before the increase in ICP and 10, 30 and 50 minutes thereafter. IL-10 levels were analyzed using a commercial rat IL-10 ELISA kit (Biosource, Laboserv, Giesen, Germany). The detection limit was 40 pg/ml.

In another 10 anaesthetized animals, the experiments were repeated. Intravenous epinephrine (suprarenin®, Aventis, Bad Soden, Germany) was also infused through a femoral catheter (1 mg/ml; 25 µl/min infusion rate). MAP was monitored continuously, and arterial blood samples were collected before increase of ICP and thereafter every 30 minutes up to 2 hours.

Results

In our model, mean (± standard deviation) ICP was increased up to 65.8 ± 7.4 mm Hg after the subdural balloon was inflated for 50 min (Table 1). This leads to a clear increase of the MAP up to 152 ± 12 mm Hg. Correspondingly, there was a significant increase in IL-10 plasma levels of 757.6 ± 237.8 pg/ml and 1072.6 ± 333.5 pg/ml, 30 min and 50 min after elevation of ICP ($P < 0.05$ versus baseline, Mann-Whitney U test), respectively. This systemic release of IL-10, however, was prevented by simultaneous infusion of propranolol (Table 2). As expected, propranolol administration also significantly attenuated the rise in the MAP. This was considered as an effective blockade of the sympathetic activation.

Furthermore, direct infusion of epinephrine at the dosage of 25 µg/min also leads to an elevation of the MAP. This was associated with significant increase in plasma IL-10 levels with a peak at 60 minutes (Fig. 1).

To study the source of IL-10, we analyzed IL-10 mRNA expression by real-time reverse transcription-polymerase chain reaction (RT-PCR) in different tissues including peripheral blood leukocytes. A significant increase of IL-10 mRNA levels were observed in liver (>10 fold) and spleen (2–4 fold) but not in lung, peripheral blood and brain (data not shown).

Discussion

Our findings demonstrate that increased ICP leads to a systemic release of IL-10. It is mediated by sympathetic activation and direct action of catecholamines on immune cells. Therefore, direct epinephrine infusion produces the same effect, proven the importance

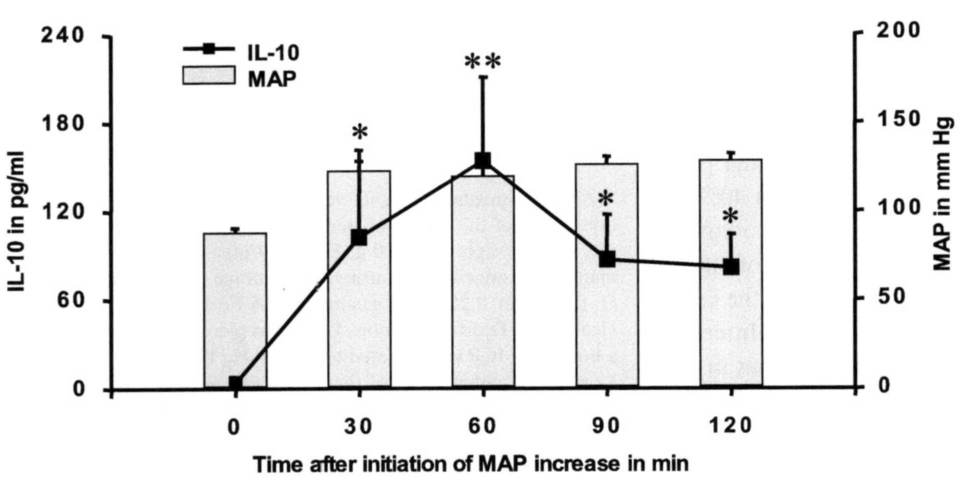

Fig. 1. Intravenous epinephrine infusion at the dosage of 25 µg/min leads to an increase of mean arterial pressure *(MAP)* and an increase of blood IL-10 levels. *$P < 0.05$, **$P < 0.05$ versus baseline, Mann-Whitney U test

of neurotransmitters for the modulation of the systemic immune capacity.

In previous studies, we have demonstrated that neurosurgical procedures and traumatic brain injury are associated with immunodepression which leads to monocytic deactivation [1, 2, 23, 24]. This immunodepression was linked to brain cytokine induced stimulation of the HPA axis [1].

However, our studies revealed that stimulation of the HPA axis was not the unique mechanism involved in immunodepression following brain injury. Thus, patients with infratentorial tumors with brainstem compression and patients with severe brain injury showed a marked systemic release of the anti-inflammatory cytokine, IL-10, early after the acute event or operation. Furthermore, these patients showed signs of sympathetic activation like increases in systolic blood pressure. Therefore, we assumed that sympathetic activation may be responsible for some of the reported effects [23].

Under these circumstances the rapid catecholamine mediated systemic release of the immunoinhibitory cytokine IL-10 was assumed to play a key role in brain mediated immunodepression [4, 8, 9, 17, 19]. To confirm this clinical hypothesis, different animal studies have been performed. In our model of raised ICP and epinephrine infusion, we confirmed for the first time that ICP induced "sympathetic storm" leads to a systemic release of the immunoinhibitory cytokine IL-10, particularly in the liver. This may be in part responsible for the observed systemic immunodepression after severe brain injury. These results correspond to data from the literature, where it is known that IL-10 acts an immunosuppressive. Futheremore, there is evidence that catecholamines released non-synaptically from the sympathetic axon terminals are able to inhibit production of proinflammatory (TNF-α, interferon-γ, IL-12, IL-1) and increase antiinflammatory cytokines (IL-10). These molecules are enhanced in concentration in the close proximity of immune cells, indicating a fine-tuning control of the production of cytokines by sympathetic innervation under stressful conditions [5, 22]. Furthermore, it is described that these effects are mediated via β_2-adrenoceptors expressed on immune cells and coupled to cAMP levels [16].

References

1. Asadullah K, Woiciechowsky C, Döcke WD, Egerer K, Kox W, Vogel S, Sterry W, Volk HD (1996) Very low monocytic HLA-DR expression indicates high risk of infection – immunomonitoring for patients after neurosurgery and patients during high dose steroid therapy. Eur J Emerg Med 2: 184–190
2. Asadullah K, Woiciechowsky C, Döcke WD, Liebenthal C, Wauer H, Volk HD, Kox W, Vogel S, von Baehr R (1995) Immunodepression following neurosurgical procedures. Crit Care Med 23: 1976–1983
3. Beamer NB, Coull BM, Clark WM, Hazel JS, Silberger JR (1995) Interleukin-6 and interleukin-1 receptor antagonist in acute stroke. Ann Neurol 37: 800–805
4. Bogdan C, Paik J, Vodovotz Y, Nathan C (1992) Contrasting mechanisms for suppression of macrophage cytokine release by transforming growth factor-beta and interleukin-10. J Biol Chem 267: 23301–23308
5. Elenkov IJ, Wilder RL, Chrousos GP, Vizi ES (2000) The sympathetic nerve – an integrative interface between two supersystems: the brain and the immune system. Pharmacol Rev 52: 595–638
6. Fassbender K, Rossol S, Kammer T, Daffertshofer M, Wirth S, Dollman M, Hennerici M (1994) Proinflammatory cytokines in serum of patients with acute cerebral ischemia: kinetics of secretion and relation to the extent of brain damage and outcome of disease. J Neurol Sci 122: 135–139
7. Fiorentino DF, Zlotnik A, Mosmann TR, Howard M, O'Garra A (1991) IL-10 inhibits cytokine production by activated macrophages. J Immunol 147: 3815–3822
8. Fiorentino DF, Zlotnik A, Vieira P, Mosmann TR, Howard M, Moore KW, O'Garra A (1991) IL-10 acts on the antigen-presenting cell to inhibit cytokine production by Th1 cells. J Immunol 146: 3444–3451
9. Howard M, O'Garra A (1992) Biological properties of interleukin 10. Immunol Today 13: 198–200
10. Hsieh AH, Bishop MJ, Kubilis PS, Newell DW, Pierson DJ (1992) Pneumonia following closed head injury. Am Rev Respir Dis 146: 290–294
11. Kapcala LP, Chautard T, Eskay RL (1995) The protective role of the hypothalamic-pituitary-adrenal axis against lethality produced by immune, infectious, and inflammatory stress. Ann NY Acad Sci 771: 419–437
12. Kossmann T, Hans VH, Imhof HG, Stocker R, Grob P, Trentz O, Morganti Kossmann C (1995) Intrathecal and serum interleukin-6 and the acute-phase response in patients with severe traumatic brain injuries. Shock 4: 311–317
13. Kushi H, Saito T, Makino K, Hayashi N (2003) IL-8 is a key mediator of neuroinflammation in severe traumatic brain injuries. Acta Neurochir (Wien) [Suppl] 86: 347–350
14. Mosmann TR (1994) Interleukin-10. In: Thomson A (ed) The cytokine handbook. London, Academic Press, pp 223–237
15. Piek J, Chesnut RM, Marshall LF, van Berkum Clark M, Klauber MR, Blunt BA, Eisenberg HM, Jane JA, Marmarou A, Foulkes MA (1992) Extracranial complications of severe head injury. J Neurosurg 77: 901–907
16. Platzer C, Docke W, Volk H, Prosch S (2000) Catecholamines trigger IL-10 release in acute systemic stress reaction by direct stimulation of its promoter/enhancer activity in monocytic cells. J Neuroimmunol 105: 31–38
17. Pretolani M, Goldman M (1997) IL-10: a potential therapy for allergic inflammation? Immunol Today 18: 277–280
18. Rodriguez JL, Gibbons KJ, Bitzer LG, Dechert RE, Steinberg SM, Flint LM (1991) Pneumonia: incidence, risk factors, and outcome in injured patients. J Trauma 31: 907–912
19. Sherry RM, Cue JI, Goddard JK, Parramore JB, DiPiro JT (1996) Interleukin-10 is associated with the development of sepsis in trauma patients. J Trauma 40: 613–616
20. van der Poll T, Coyle SM, Barbosa K, Braxton CC, Lowry SF

(1996) Epinephrine inhibits tumor necrosis factor-alpha and potentiates interleukin 10 production during human endotoxemia. J Clin Invest 97: 713–719
21. van-der Poll T, Jansen J, Endert E, Sauerwein HP, van DS (1994) Noradrenaline inhibits lipopolysaccharide-induced tumor necrosis factor and interleukin 6 production in human whole blood. Infect Immun 62: 2046–2050
22. Vizi ES, Elenkov IJ (2002) Nonsynaptic noradrenaline release in neuro-immune responses. Acta Biol Hung 53: 229–244
23. Woiciechowsky C, Asadullah K, Nestler D, Eberhardt B, Platzer C, Schöning B, Glöckner F, Lanksch WR, Volk HD, Döcke WD (1998) Sympathetic activation triggers systemic interleukin-10 release in immunodepression induced by brain injury. Nat Med 4: 808–813
24. Woiciechowsky C, Asadullah K, Nestler D, Schöning B, Glöckner F, Robinson PN, Volk HD, Vogel S, Lanksch WR (1997) Different release of cytokines into the cerebrospinal fluid following surgery for intra- and extra-axial brain tumours. Acta Neurochir (Wien) 139: 619–624
25. Woiciechowsky C, Schoning B, Cobanov J, Lanksch WR, Volk HD, Docke WD (2002) Early IL-6 plasma concentrations correlate with severity of brain injury and pneumonia in brain-injured patients. J Trauma 52: 339–345

Correspondence: PD Dr. med. Christian Woiciechowsky, Klinik für Neurochirurgie, Charité, Campus Virchow-Klinikum, Universitätsmedizin Berlin, Augustenburger Platz 1, D-13353 Berlin, Germany. e-mail: christian.woiciechowsky@charite.de

Acta Neurochir (2005) [Suppl] 95: 377–380

Experimental model for investigating hyponatremia after subarachnoid hemorrhage in rats

T. Mori[1], **Y. Katayama**[1], **J. Kojima**[1,2], **N. Moro**[1], **H. Kawai**[1], **M. Yoneko**[2], and **T. Kawamata**[1]

[1] Department of Neurological Surgery, Nihon University School of Medicine, Tokyo, Japan
[2] Omiya Research Laboratory, Nikken Chemicals Co. Ltd., Saitama, Japan

Summary

Hyponatremia is a common complication in patients with aneurysmal subarachnoid hemorrhage (SAH). Such patient demonstrates excessive natriuresis and an increased risk of symptomatic cerebral vasospasm. However, the precise mechanisms underlying SAH induced hyponatremia remain unclear. In the present study, in order to establish an experimental model of hyponatremia following SAH, we induced SAH in rats, and evaluated the serum sodium (Na) levels, Na excretion and physiological parameters. Twenty-four male Wistar rats were used. SAH was induced by an endovascular puncture method. The mean arterial pressure (MAP), intracranial pressure (ICP), and cerebral blood flow (CBF) were monitored continuously. The urine was collected cumulatively for 12 hours after SAH, and the urine Na concentration was determined with a spectrophotometer. The serum Na levels were measured at 12 hrs, 2 and 4 days following the SAH induction. The mean ($\pm$ standard deviation) baseline ICP was 3.5 ± 2.6 mmHg, and increased to 67.4 ± 17.6 mmHg immediately following induction of SAH. CBF decreased rapidly, and then gradually recovered to 70–80% of baseline. The urine volume and total Na excretion were significantly increased in comparison to those of the sham ($P < 0.05$). The serum Na level was significantly decreased at 4 days following SAH ($P < 0.05$). The present results demonstrated for the first time that rats with SAH exhibited excessive natriuresis. The endovascular puncture model is suitable for investigating hyponatremia that occurs concomitantly with natriuresis and diuresis after SAH.

Keywords: Animal model; subarachnoid hemorrhage; hyponatremia; natriuresis.

Introduction

Subarachnoid hemorrhage (SAH) is a common life-threatening neurological disease [2] and the overall outlook for patients with SAH still remains poor [2]. Electrolyte abnormalities, especially hyponatremia, are frequently observed during the acute period of SAH [4, 11]. Concurrent hyponatremia in SAH may increase the severity of neurological symptoms because of an increased risk of cerebral ischemia. It has been reported that a relationship exists between hyponatremia and the occurrence of cerebral vasospasm [3, 9, 13]. The cause of the hyponatremia after SAH is now believed to involve cerebral salt wasting (CSW) [6]. From the acute phase of SAH, patients suffer from excessive natriuresis, concomitantly developing osmotic diuresis. The induced hyponatremia and decrease total blood volume are indicative of dehydration, and dehydration is reported to increase the risk of cerebral vasospasm. We focused our attention on these events after SAH and indicated the efficacy of inhibition of natriuresis [7, 8]. In the present study, we attempted to establish an experimental model of hyponatremia following SAH.

Materials and methods

SAH model

All experimental procedures were approved by the Committee on Animal Research at Nihon University School of Medicine, Tokyo, Japan. Male Wistar rats (weight 250–300 g) were used for the experiments ($n = 24$). The animals were anesthetized with 2% halothane in a mixture of 70% N_2O and 30% O_2 under spontaneous respiration. During the surgical procedure, anesthesia was maintained with 1% halothane. We followed the time course of MAP, ICP, CPP, and CBF for 90 min after induction of injury. Each rat was placed in a prone position on a heated pad and its body temperature was maintained at between 37 and 38 C. A laser Doppler probe for assessing the CBF was positioned within the temporal skull vault in the area of the middle cerebral artery. For ICP monitoring, a tip of 1.5 mm in diameter was inserted into the brain parenchyma, with the tip situated at a depth of 1 mm from the brain surface. The catheter was connected to a transducer for continuous monitoring of the ICP. The rat was rotated to the supine position and the right femoral artery was cannulated for continuous blood pressure monitoring. The induction of SAH was performed according to the methods described by Veelken *et al.* [12]. Briefly, the common carotid artery including its bifurcation was exposed and dissected. A nylon monofilament (0.235 mm in diameter) was introduced into the right internal carotid artery through the external carotid artery to perforate the internal bifurcation. Sham control animals were subjected to similar operations to expose the carotid arteries without perforation ($n = 6$). The urine was collected cumulatively for 12 hours post SAH, and the urinary specific gravity was measured. The urine sodium (Na) concentration was determined with a spectrophotometer. The serum Na levels were measured at 12 hours, 2 and 4 days following the SAH induction.

Results

Mortality and blood distribution

Three of the 18 rats died within 12 hours after SAH. These animals were confirmed to have diffuse SAH especially in the basal cistern macroscopically. Four animals were observed to have subdural hemorrhage without SAH. The endovascular puncture model occasionally produced subdural hemorrhage which was dependent on the site of puncture.

Physiological data

The mean $\pm$ standard deviation (SD) values of the physiological variables before SAH induction were 104.3 $\pm$ 11.4 mmHg for the MAP and 3.5 $\pm$ 2.6 mmHg for the ICP. After SAH, the MAP increased immediately to 122 $\pm$ 11 mmHg and then returned to baseline values within 5 min. The ICP rose rapidly to 67 $\pm$ 18 mmHg after SAH and remained at around 40 mmHg up to 90 min. The CBF decreased suddenly to reach 40% of baseline, and then gradually recovered to normal values at 90 min after SAH.

Urine volume and sodium levels

At 12 hours following SAH, the urine volume was significantly increased in comparison to that of the sham ($P < 0.05$). The sodium excretion following SAH was also increased. These data indicated that the urine volume increased concomitantly with the sodium excretion (Fig. 1). There were no significant differences in urine specific gravity and urinary Na concentration between the groups (Table 1). In the SAH group, the serum Na level decreased gradually and showed a significant difference on day 4 ($P < 0.05$; Fig. 2).

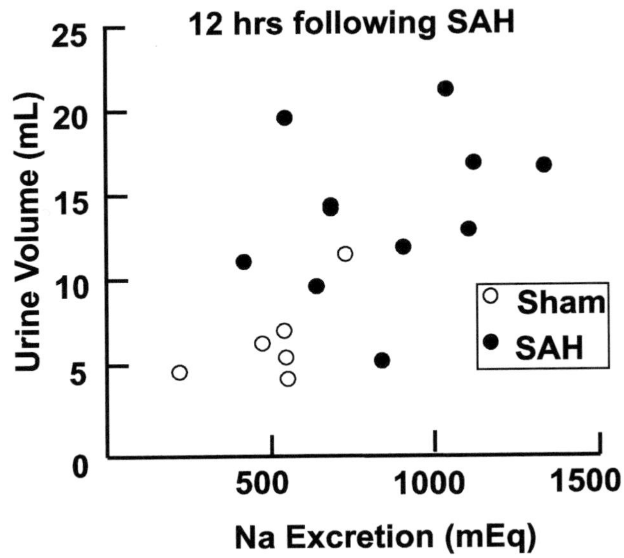

Fig. 1. Scatter plot showing the overall relationship between urinary sodium (Na) excretion and urine volume after SAH: The strong correlation ($p < 0.05$, $r = 0.726$) indicates that the urine volume increased concomitantly with the sodium excretion

Table 1. *Urine specific gravity (U-S.G.) and urinary sodium concentration (U-Na conc) 12 hrs following SAH*

	Sham	SAH
U-S.G.	1.015 $\pm$ 0.02	1.022 $\pm$ 0.04
U-Na conc.	82.3 $\pm$ 12.2	68.1 $\pm$ 10.1

(mEq/mL).

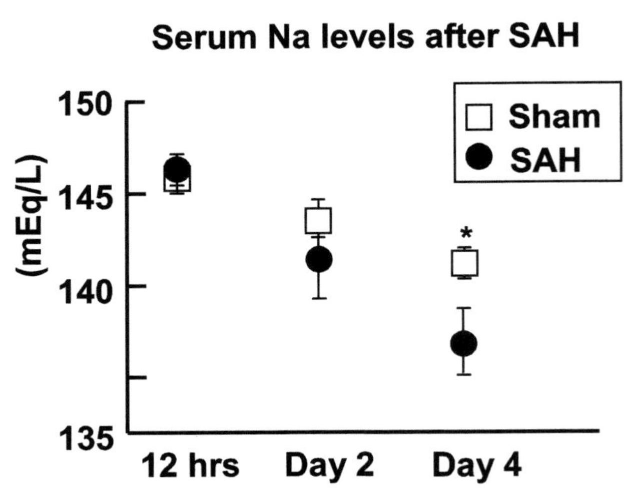

Fig. 2. Bar graphs demonstrating serum sodium levels at 12 hrs, 2 and 4 days after SAH. The serum sodium level was significantly decreased at 4 days following SAH. Asterisk indicates a statistically significant difference between two groups ($p < 0.05$)

Discussion

Hyponatremia after SAH

Hyponatremia is a common finding in acute brain disease. Especially in patients with SAH, it is observed in about 30% of cases. Such patients tend to suffer from excessive natriuresis and diuresis which decrease the total blood volume, and they eventually become exposed to the risk of cerebral vasospasm. It has been reported that a relationship exists between hyponatremia and the occurrence of cerebral vasospasm [3, 9, 13]. Evidence has been accumulated to indicate that the most common cause of hyponatremia in acute brain disease is CSW [6]. Among the diagnostic criteria of CSW, excessive natriuresis and total blood volume depletion are crucially important. We reported previously that maintenance of an adequate intravascular volume is vital for the management of SAH in order to minimize cerebral ischemia due to cerebral vasospasm [7, 8]. It is important to clarify the pathology of CSW following SAH. In the present study, our aim was thus to establish an experimental model of hyponatremia after SAH.

SAH model in rats

A variety of animal models have been established and various species have been employed in the study of SAH, including dogs, cats, monkeys, and more recently rats [1]. Recent animal studies of SAH have focused on the investigation of delayed ischemia. To the best of our knowledge, there are no published papers giving detailed descriptions of hyponatremia following SAH in rats. We chose the endovascular puncture model to investigate hyponatremia after SAH because this model does not require craniotomy to produce and bleeding induced by arterial injury. It is important that the mechanism of hemorrhage closely simulates aneurysmal rupture. SAH in rats demonstrates an extensive distribution of blood throughout the subarachnoid space, reduces the cortical blood flow, and produces a consistent and significant elevation of ICP. However, one serious problem for SAH research is the inability to control the severity of hemorrhage, giving rise to a high mortality [10]. Veelken *et al.* attempted to control the SAH severity by occluding the ipsilateral common carotid artery [12].

Cause of CSW

Natriuretic peptides have been investigated in hyponatremia during acute brain disease and they represent suitable candidates for the purposed natriuretic factors in CSW [6]. We have measured the levels of atrial natriuretic peptide (ANP), brain natriuretic peptide (BNP) and antidiuretic hormone (ADH), but there were no significant differences between the control group and SAH group (data not shown). Further studies are needed to clarify the detailed mechanism by which natriuresis occurs following SAH.

Conclusion

The present results demonstrated for the first time that rats with SAH exhibited excessive natriuresis. The endovascular puncture model is considered suitable for investigating hyponatremia concomitant with natriuresis and diuresis after SAH.

References

1. Alkan T, Korfali E, Kahveci N (2002) Experimental Subarachnoid haemorrhage models in rats. Acta Neurochir (Wien) [Suppl] 83: 61–69
2. Dorsch NWC (2002) Therapeutic approaches to vasospasm in subarachnoid hemorrhage. Curr Opin Crit Care 8: 128–133
3. Hasan D, Wijdicks EF, Vermeulen M (1990) Hyponatremia is associated with cerebral ischemia in patients with aneurysmal subarachnoid hemorrhage. Ann Neurol 27: 106–108
4. Kuroiwa Y, Uede T, Ishiguro M *et al* (1996) Pathogenesis of hyponatremia following subarachnoid hemorrhage due to ruptures cerebral aneurysm. Surg Neurol 46: 500–507
5. Ljunggren B, Säveland H, Brandt L, Uski T (1984) Aneurysmal subarachnoid hemorrhage: Total annual outcome in a 1.46 million population. Surg Neurol 22: 435–438
6. McGirt MJ, Blessing R, Nimjee SM *et al* (2004) Correlation of serum brain natriuretic peptide with hyponatremia and delayed ischemic neurological deficits after subarachnoid hemorrhage. Neurosurgery 54: 1369–1373
7. Mori T, Katayama Y, Kawamata T *et al* (1999) Improved efficiency of hypervolemic therapy with inhibition of natriuresis by fludrocortisone in patients with aneurysmal subarachnoid hemorrhage. J Neurosurg 91: 947–952
8. Moro N, Katayama Y, Kojima J *et al* (2003) Prophylactic management of excessive natriuresis with hydrocortisone for efficient hypervolemic therapy after subarachnoid hemorrhage. Stroke 34: 2807–2811
9. Okuchi K, Fujioka M, Fujikawa A, Nishimura A *et al* (1996) Rapid natriuresis and preventive hypervolaemia for symptomatic vasospasm after subarachnoid haemorrhage. Acta Neurochir (Wien) 138: 951–956
10. Prunell GF, Mathiesen T, Diemer NH (2003) Experimental subarachnoid hemorrhage: subarachnoid blood volume, mortality rate, neuronal death, cerebral blood flow, and perfusion pressure in three different rat models. Neurosurgery 52: 165–167

11. Segatore M (1993) Hyponatremia after aneurysmal subarachnoid hemorrhage. J Neurosci Nurs 25: 92–99
12. Veelken JA, Laing RJ, Jakubowski J (1995) The Sheffield model of subarachnoid hemorrhage in rats. Stroke 26: 1279–1283
13. Wijdicks EF, Vermeulen M, Hijdra A *et al* (1985) Hyponatremia and cerebral infarction in patients with ruptured intracranial aneurysms: is fluid restriction harmful? Ann Neurol 17: 137–140

Correspondence: Tatsuro Mori, Department of Neurological Surgery, Nihon University School of Medicine, 30-1 Oyaguchi-kamimachi, Itabashi-ku, Tokyo 173-8610, Japan. e-mail: forestat@med.nihon-u.ac.jp

Acta Neurochir (2005) [Suppl] 95: 381–387

Effects of venom defibrase on brain edema after intracerebral hemorrhage in rats

G. Wu and **F. P. Huang**

Department of Neurosurgery, Huashan Hospital, Fudan University, Shanghai, China

Summary

We evaluated the effects of defibrase DF-521 batroxobin on reducing brain edema formation and the expression of ICAM-1, complement C3d and C9 in the perihematomal area after intracerebral hemorrhage (ICH) in rats. A rat ICH model, involving infusion of autologous blood into the right basal ganglia, were used in this study. The animals were sacrificed at 24 and 72 hours after ICH to determine the water content of the brain tissue with wet/dry weight measurement. While the expression of ICAM-1 and complement C3d was detected using immuno-histochemistry, and C9 was detected semi-quantitatively with Western blot analysis in the perihematomal area. Perihematomal brain edema was reduced after intraperitoneally injection of DF-521 batroxobin 24 and 72 hours after intracerebral hemorrhage. Immunohistochemistry showed that there were less ICAM-1 positive cells were found around the hematoma after intraperitoneally injection of DF-521 batroxobin 24 and 72 hours after ICH. Immuno-histochemistry also showed that C3d deposition reduced significantly, and the Western blot analysis also showed the content of C9 protein declined around the hematoma in DF-521 batroxobin treatment group at 72 hours after ICH. Defibrase DF-521 batroxobin down-regulate ICAM-1 and complement C3d and C9 expression in the perihematomal area, and attenuate brain edema formation in ICH rats.

Keywords: Intracerebral hemorrhage (*ICH*); brain edema; intercellular adhesion; molecule-1 (*ICAM-1*); defibrase; complement.

Introduction

Spontaneous intracerebral hemorrhage (ICH) represents one of the most common and lethal type of stroke. The incidence of ICH is continuously increasing in China [7]. Brain edema plays an important role in secondary brain injury after ICH [3], there has been no effective treatment for ICH. The current therapeutic approaches for ICH include hematoma evacuation and injection of fibrinolytic substance to lyse the hematoma. However, under many circumstances fibrinolytic substance may worsen thrombin-induced brain injury [16]. Clinically, there has been no effective treatment that lyses the clot and attenuates the brain edema.

Venom defibrase is an enzyme purified from Viperidae, which can inhibit thrombosis and effectively lyses thrombus. This enzyme is widely used in ischemic stroke treatment [5]. Experimental research has shown that the mechanisms underlying the beneficial effects of venom Defibrase in attenuating ischemic brain injury include clearance of free radical [4], down-regulation of c-fos gene expression [17], and reduction of the complement infiltration [13]. However, the effect of venom defibrase on ICH-induced brain edema has not been investigated.

Df-521 batroxobin (Tobishi Biomedical Technology) is a mono-component venom defibrase agent. We evaluate the effect of Df-521 batroxobin on brain edema formation and expression of intracellular adhesion molecule-1 (ICAM-1) and complement C3d and C9 following ICH in a rat model.

Materials and methods

Animal preparation and intracerebral infusion

Male Sprague-Dawley rats (Shanghai Experimental Animal Center), weighing 300 to 420 g each, were used in this study. The animals were anesthetized with 400 mg/kg of intraperitoneal chloral hydrate injection. The right femoral artery of the rats was catheterized for blood pressure monitoring, blood sampling and used as the source for intracerebral blood infusion. Blood pH, arterial oxygen tension (PaO_2), and glucose levels were monitored during infusion. Rectal temperature was maintained at $37 \pm 0.5\,°C$. The rats were then placed in a stereotactic head frame (Kopf Instrument, Tujunga, CA), and a 1-mm cranial burr hole was drilled on the right coronal suture 4 mm lateral to the midline. One hundred microliters of autologous blood withdrawn from the right femoral artery or saline was infused into the right caudate nucleus by a micro infusion pump (Harvard Apparatus Inc) at a rate of 13 μl/min through a needle (coordinates: 0.2 mm anterior, 5.5 mm ventral, and 4 mm lateral to the bregma). The needle was removed 10 min later and the skin incision was closed.

Experimental groups

There were three experimental groups. In the first group ($n = 5$), rats received intracerebral injections of 100 µl autologous blood and intraperitoneal injections of 1 ml saline, intracerebral injections of 100 µl autologous blood and intraperitoneal injections Df-521 batroxobin (8 IU/kg), or intracerebral injections of 100 µl saline (as naive control). After 24 and 72 hours, the rats were killed for brain water and ion contents. In the second group ($n = 4$ rats per group), rats received intracerebral injections of 100 µl autologous blood and intraperitoneal injections 1 ml saline or Df-521 batroxobin (8 IU/kg). After 24 and 72 hours, the rats were killed for ICAM-1 and complement C3d immunohistochemical studies. In the third group ($n = 3$ rats per group), rats received intracerebral injections of 100 µl autologous blood and intraperitoneal injections 1 ml saline or Df-521 batroxobin (8 IU/kg). After 72 hours, the rats were killed for complement C9 Western blot analysis.

Measurement of brain water and ion contents

Animals were re-anesthetized 24 or 72 hours later with 400 mg/kg of intraperitoneally injected chloral hydrate and killed by decapitation. Their brains were removed and a 3-mm thick coronal brain slice was cut with a blade approximately 4 mm from the frontal pole. The brain slice was separated into ipsilateral and contralateral cortex, and ipsilateral and contralateral basal ganglia. The cerebellum served as a control specimen. Brain samples were immediately weighed on an electronic analytical balance for wet weight, and were dried at 95 °C for 24 hours to obtain the dry weight. Water content was determined as (wet weight – dry weight)/wet weight. The dehydrated samples were then digested in 1 ml of 1 M nitric acid for more than 7 days. Sodium and potassium contents were measured using atomic absorption spectrophotometer (Hitachi Z-8000); ion contents were expressed in µg per kg of dehydrated brain tissue.

Immunohistochemical studies

The rats were re-anesthetized 24 or 72 hours later with 600 mg/kg of intraperitoneally injected chloral hydrate and underwent transcardiac perfusion with 4% paraformaldehyde in 0.1 M pH 7.4 phosphate-buffered saline (PBS). The brains were removed and kept in 4% paraformaldehyde for 6 hours, and immersed in 25% sucrose for 3 to 4 days at 4 °C. Brains were then embedded in OCT compound and sectioned in 6-µm slices on a cryostat. Sections were incubated using the avidin-biotin complex technique. Primary antibodies were rabbit anti-human C3d Pab (DAKO), and mouse-anti-rat ICAM-1 Pab (DAKO). Normal rabbit or mouse IgG was used as a control.

Western blot analysis

Rats were reanesthetized with 600 mg/kg of intraperitoneally injected chloral hydrate at 72 hours after intracerebral infusion and underwent transcardiac perfusion with saline. A coronal brain slice was cut as described for water content measurements and were divided into ipsilateral and contralateral part. The brain tissues were immersed in 1 ml of 0.05 mol/L Tris/HCl (pH 7.4; 0.01% PMSF) and Homogenized in ice. The samples then were centrifuged at 1000× G for 20 minutes 2 hours later at 4 °C. 15 µg protein of ipsilateral brain tissue was run on 7.5% polyacrylamide gels with a 5% stacking gel after boiling for 5 minutes at 95 °C. The protein was transferred to pure nitrocellulose membrane, and membranes were probed with a 1:400 dilution of the primary antibody (rabbit anti-rat C9 PAB, Gift from Dr. P. Morgan, University of Wales) and a 1:250 dilution of the secondary antibody (HRP-Goat anti-rabbit IgG). Finally, the antigen-antibody complexes were visualized with diaminobenzidine. The relative densities of C9 protein bands were analyzed using the NIH image software.

Statistical analysis

All data in this study are presented as the mean ± standard deviation (SD). Data were analyzed using Student's t-test. A P value less than 0.05 was considered statistically significant.

Results

All physiological variables including mean arterial pressure (MAP), body temperature, PaO_2, and blood glucose were within normal ranges.

As shown in Fig. 1, Df-521 batroxobin treatment significantly reduced brain edema in the ipsilateral basal ganglia 24 hours (79.5 ± 0.7% versus 81.8 ± 1.1% in ICH group; $P < 0.01$; (Fig. 1a) and 72 hours (81.1 ± 0.3% versus 83.1 ± 0.4% in ICH group; $P < 0.05$; (Fig. 1b) after ICH. In contrast, the cerebellar water content was unaffected. The reduction of water content was associated with reduction of sodium ion accumulation (7.5 ± 1.2 versus 10.7 ± 1.4 µg/kg dry wt in ICH group; $P < 0.01$) and reduction of potassium ion loss (24.0 ± 2.2 versus 19.6 ± 2.3 µg/kg dry wt in ICH group; $P < 0.05$) (Fig. 1c and d).

ICAM-1 was detected in the ipsilateral brain tissue, at both 24 and 72 hours after ICH in non-Df-521 batroxobin treated group, and there were more ICAM-1 immunoreactive particles at 72 hours compared to 24 hours. However, compared to ICH group, there were smaller number of ICAM-1immunoreactive particles in the ipsilateral brain tissue in Df-521 batroxobin treated animals (Fig. 2A–D).

Three days after ICH, there was greater C3d immunoreactivity around the hematoma in ICH group compared to the Df-521 batroxobin-treated group (Fig. 2E and F). Western blot analysis showed that complement C9 content in the Df-521 batroxobin treated group was significantly decreased in the ipsilateral brain tissue (gray value 185.5 ± 14.9 versus 225.8 ± 3.2 in ICH group, $P = 0.01$) at 72 hours after intracerebral infusion of 100 µl of autologous blood (Fig. 3).

Discussion

The major novel finding of this study is that Venom defibrase can also decrease ICH-induced brain damage. We found that Df-521 batroxobin treatment re-

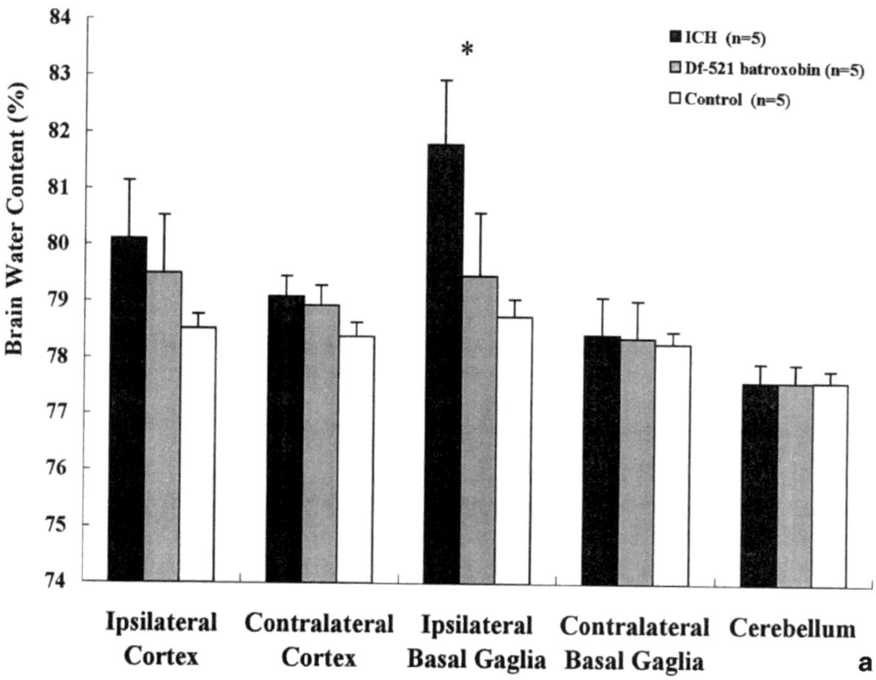

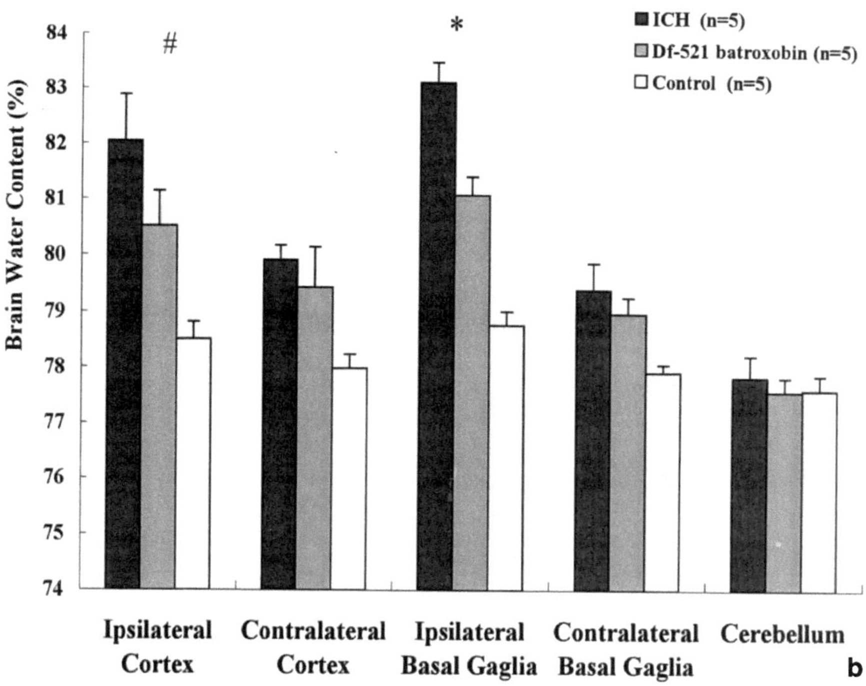

Fig. 1. Bar graph of brain water contents at 24 hours (a), 72 hours (b), brain sodium (c) and potassium (d) contents at 24 hours after intracerebral infusion of 100 μL autologous blood. ICH = intracerebral hematoma group. Values are mean ± SD; $n = 5$. *$P < 0.01$, #$P < 0.05$ *vs* ICH group

duced the water content in the ipsilateral brain tissues, which coincided with reductions of sodium accumulation and potassium loss. Df-521 batroxobin also decreased ICAM-1 expression at both 24 and 72 hours after ICH, and reduce the expression of complement C3d and C9 at 72 hours after ICH. We suggest that Df-521 batroxobin decreases ICH-induced edema formation at least partially by its effects on the expression

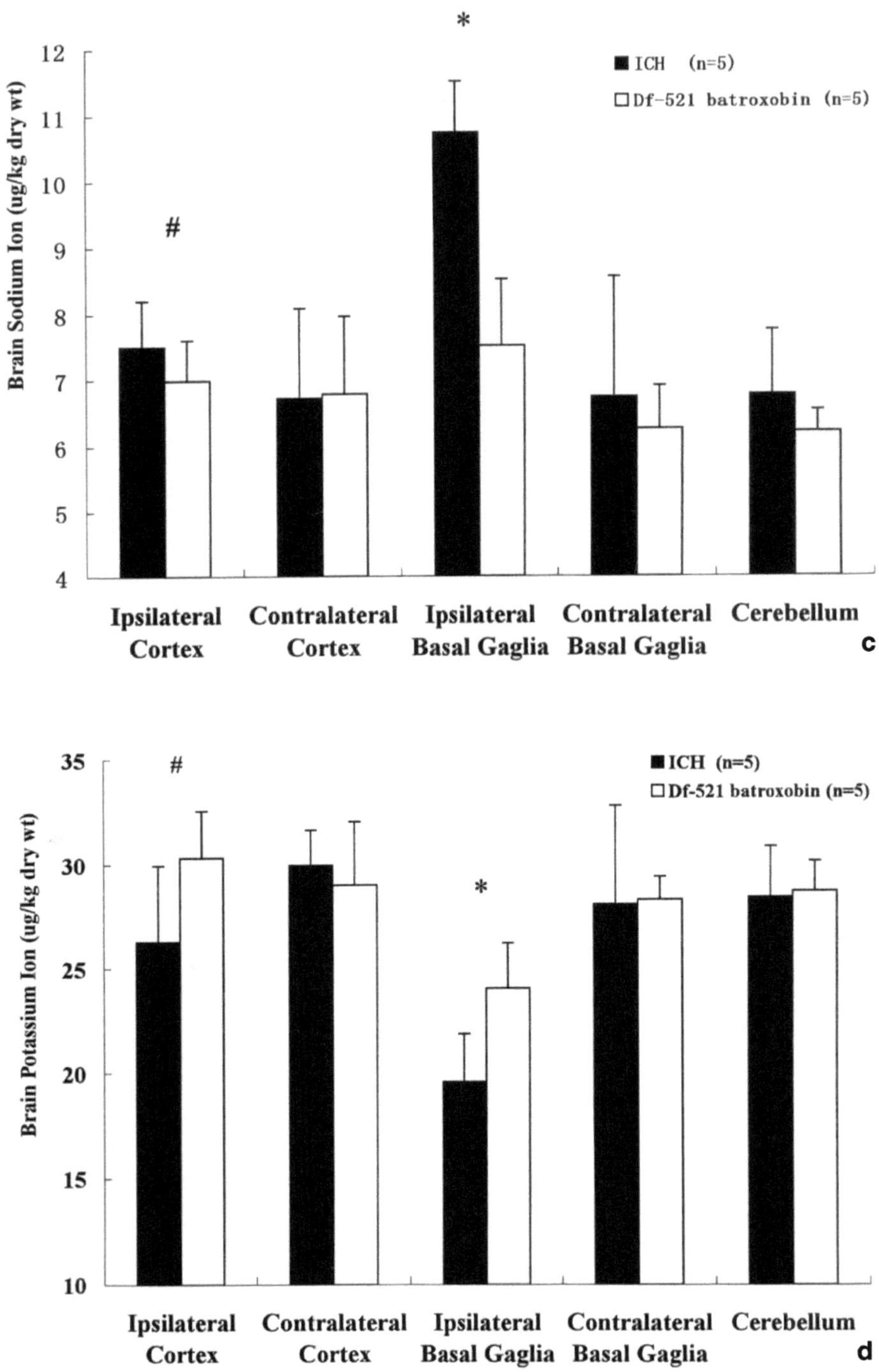

Fig. 1 (*continued*)

of ICAM-1 and complements – two components in inflammatory responses.

Many studies have indicated that inflammatory reactions play important roles in many neurological diseases such as cerebral ischemia [2, 15]. Cerebral ischemia causes infiltration of leukocyte, activation of microglia and upregulation of ICAM-1, which contributes ischemic brain damage. Although there is no typical brain ischemia in ICH models, inflammatory reactions do occur around the hematoma [9]. These

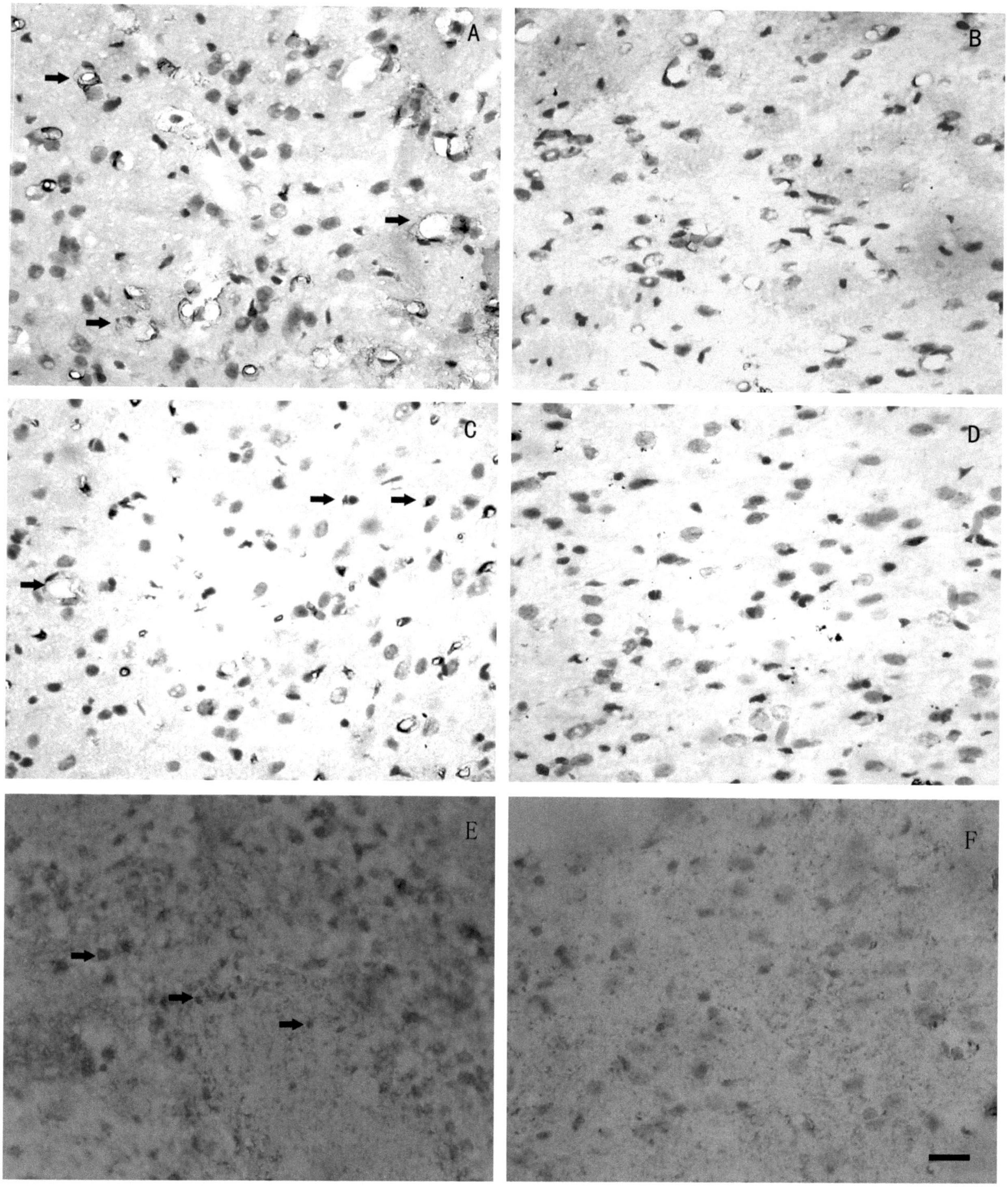

Fig. 2. ICAM-1 immunoreactivity 24 (A and B), 72 (C and D) hours and Complement C3d immunoreactivity 72 hours (E and F) in ipsilateral brain tissue after intracerebral infusion of 100 μL blood. Fewer ICAM-1 immunoreactive particles were found in the ipsilateral brain tissue in Df-521 batroxobin treated animals (A and B) than in ICH group (C and D). Also the complement C3d-immunoreactive minute particles are less in ipsilateral brain tissue of Df-521 batroxobin treated (B) rats than the ICH ones. Examples of positive cells are indicated by arrows. Bar = 40 mm

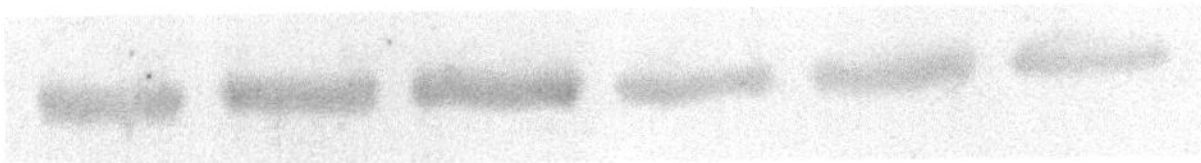

Fig. 3. Western blot analysis of complement C9 protein levels in ipsilateral brain tissue 72 hours after intracerebral infusion of 100 μl of autologous blood in ICH group (lanes 1–3) and Df-521 batroxobin treated group (lanes 4–6). Equal amounts of protein (15 μg) were used in Western blot analysis

hematoma release inflammatory mediators that may contribute to edema formation. ICH also induces inflammation by producing up-regulation of adhesion molecules in local brain tissues, which induces leucocytes infiltration [6].

ICAM-1 is widely expressed in various types of cells, including activated T cells, B cells, dendricytes and microglia. Tissue injury and inflammatory cytokines, such as IL-1 and TNF, also up-regulates ICAM-1 expression in endothelial cells and some hematopoietic cells [19]. The coagulation reaction in hematoma after ICH releases large number of thrombin, which induces the expression of cytokine such as IL-1 and TNF [18]. This hematoma induced cytokine production may account for the increased ICAM-1 expression found after ICH [9]. The increased ICAM-1 on the endothelial cells can promote infiltration of leukocytes and interactions between microglia and neurons, which may contribute to the brain edema formation and neuronal death due to the release of cytotoxic factors from leukocytes and microglia and direct cell-cell interactions. The infiltrating leukocytes can alter cerebral vasoreactivity by releasing vasoactive mediators, such as superoxidase, endothelin and prostagladins. In addition, some leukocytes can release proteolytic enzymes such as elastase. These factors produced by infiltrating leukocytes might damage endothelial cell membranes and the basal lamina, thus altering BBB and leading to formation of brain edema [8].

In our study, Df-521 batroxobin treatment decreased ICAM-1 expression at both 24 and 72 hours after ICH. Because ICAM-1 may play a significant role in brain edema formation and neuronal injury, we propose that the beneficial effects of Df-521 batroxobin may be, at least partially, due to its inhibition of ICAM-1 expression. We also found that there were much more ICAM-1 positive microvessels and cells at 72 hours after ICH compared with 24 hours after ICH, which may be determined by the different levels of hematoma induced cytokine production at different time points.

Complement activation has been shown in many diseases including cerebral ischemia [1] and subarachnoid hemorrhage [14]. Complements mediated brain injury may be caused by the formation of membrane attack complex (MAC), their cytotoxicity [18] and complement mediated inflammatory response [19]. Complement, that is excluded from the brain under normal conditions, enters the brain soon after ICH due to disruption of blood-brain barrier. After ICH, C3d, a fragment of C3 was detected in and around the hematoma [1]. Deposition of C9 was also shown on the cell membrane of perihematoma [10]. These observations suggest that the complement system is activated and MAC is formed after ICH. In ICH, MAC leads to red cells lysis. Subsequently hemoglobin, iron and other productions such as bilirubin are released, which causes delayed brain edema after ICH [11]. In addition, the activated complement system stimulates microgia and macrophage to release TNF-α and IL-1, which induces ICAM-1 expression, leukocytes infiltration, and edema formation. Our study showed that Df-521 batroxobin decreased the expression of complement C3d and C9. Based on the information indicating important roles of complement, the Df-521 batroxobin decreases brain edema formation and may be due to, at least partially, its inhibitory effects on the expression of the complements.

However, our study does not exclude other mechanisms by which venom defibrase decreases edema formation. For example, many studies have shown that venom defibrase can reduce c-fos expression in brain tissues [17]. Since marked expression of c-fos may correlate with T cell activation, venom defibrase might also attenuate edema formation by affecting c-fos expression.

Conclusion

Our study shows that venom defibrase Df-521 batroxobin attenuates brain edema formation induced by ICH. This beneficial effect of the enzyme may result from its anti-inflammatory capacity via decreasing expression of ICAM-1 and preventing complement activation and deposition. Thus, venom defibrase, a drug widely used in treatment of ischemic diseases, may also be a promising drug for treating ICH.

References

1. Bellander BM, von Holst H, Fredman P *et al* (1996) Activation of the complement cascade and increase of clusterin in the brain

following a cortical contusion in the adult rat. J Neurosurg 85: 468–475
2. Berti R, Williams AJ, Moffett JR *et al* (2002) Quantitative real-time RT-PCR analysis of inflammatory gene expression associated with ischemia-reperfusion brain injury. J Cereb Blood Flow Metab 22: 1068–1079
3. Broderick J, Brott T, Tomsick T *et al* (1994) Management of intracerebral hemorrhage in a large metropolitan population. Neurosurgery 34: 882–887
4. Chen Q, Zeng YM, Wang SL *et al* (1999) The effect of batroxobin on brain ischemia-reperfusion in gerbils. J Chin Pharmacol Toxicol 13: 277–280 (Chinese)
5. Chen QT, Zhao YB, Zheng YW *et al* (1996) Clinical study of cerebral infarction with Batroxobin (DF-521). J Stroke Nervous Syst 4: 213–218 (Chinese)
6. Del Bigio MR, Yan HJ, Buist R *et al* (1996) Experimental intracerebral hemorrhage in rats. Magnetic resonance imaging and histopathological correlates. Stroke 27: 2312–2319
7. Ding L, Peng QL, Gao QX *et al* (1995) Epidemiologic research of stroke in 100 thousand people in Changsha city. J Hunan Med Uni 20: 219–222 (Chinese)
8. Feuersteun GZ, Liu T, Barone FC (1994) Cytokines, inflammation, and brain injury: role of tumor necrosis factor-alpha. Cerebrovasc Brain Metab 6: 341–360 (Review)
9. Gong C, Hoff JT, Keep RF (2000) Acute inflammatory reaction following experimental intracerebral hemorrhage in rat. Brain Res 871: 57–65
10. Hua Y, Xi G, Keep RF *et al* (2000) Complement activation in the brain after experimental intracerebral hemorrhage. J Neurosurg 92: 1016–1022
11. Huang FP, Xi G, Keep RF *et al* (2002) Brain edema after experimental intracerebral hemorrhage: role of hemoglobin degradation products. J Neurosurg 96: 287–293
12. Lucchesi BR (1993) Complement activation, neutrophils, and oxygen radicals in reperfusion injury. Stroke 24[Suppl]: 141–147
13. Luo XG, LI SY, QU HC *et al* (1998) Study of complement deposition and the effect of defibrase on brain ischemia-reperfusion in rats. J Chin Neuroimunol Neurol 5: 203–206 (Chinese)
14. Østergaard JR, Kristensen BØ, Svehag SE *et al* (1987) Immune complexes and complement activation following rupture of intracranial saccular aneurysms. J Neurosurg 66: 891–897
15. Pang L, Ye W, Che XM *et al* (2001) Reduction of inflammatory response in the mouse brain with adenoviral-mediated transforming growth factor-ss1 expression. Stroke 32: 544–552
16. Rohde V, Rohde I, Thiex R *et al* (2002) Fibrinolysis therapy achieved with tissue plasminogen activator and aspiration of the liquefied clot after experimental intracerebral hemorrhage: rapid reduction in hematoma volume but intensification of delayed edema formation. J Neurosurg 97: 954–962
17. Wu WP, Kuang PG, LI ZZ *et al* (1995) Effect of batroxobin on c-fos in brain tissue after brain ischemia-reperfusion. J Stroke Nervous Syst 12: 73–75 (Chinese)
18. Xi G, Wagner KR, Keep RF *et al* (1998) Role of blood clot formation on early edema development after experimental intracerebral hemorrhage. Stroke 29: 2580–2586
19. Zhang RL, Chopp M, Jiang N *et al* (1995) Anti–intercellular adhesion molecule–1 antibody reduces ischemic cell damage after transient but not permanent middle cerebral artery occlusion in the Wistar rat. Stroke 26: 1438–1443

Correspondence: Fengping Huang, Department of Neurosurgery, Huashan Hospital, No. 12 Middle Wulumuqi Road, Shanghai, China, 200040. e-mail: huangfengping@hotmail.com

Acta Neurochir (2005) [Suppl] 95: 389–392

Complement inhibition attenuates brain edema and neurological deficits induced by thrombin*

Y. Gong[1], G. H. Xi[1], R. F. Keep[1,2], J. T. Hoff[1], and Y. Hua[1]

[1] Department of Neurosurgery, University of Michigan, Ann Arbor, Michigan, USA
[2] Department of Physiology, University of Michigan, Ann Arbor, Michigan, USA

Summary

The present study examined whether thrombin activates the complement cascade in the brain and whether N-acetylheparin, an inhibitor of complement activation, attenuates brain injury induced by thrombin. There were three sets of studies. In the first set, rats had an intracerebral infusion of either five-unit thrombin or a needle insertion. Brains were sampled at 24 hours for Western blot analysis and immuno-histochemistry. In the second set, rats received either five-unit thrombin+saline, five-unit thrombin+25 μg N-acetylheparin or five-unit thrombin+100 μg N-acetylheparin infusion. Brains were sampled 24 hours later for water content measurement. In the third set, rats received either five-unit thrombin+saline or five-unit thrombin+100 μg N-acetylheparin. Behavioral tests sensitive to unilateral striatal damage were carried out for two weeks. Western blotting demonstrated that complement C9 and clusterin levels increase 24 hours after thrombin infusion ($P < 0.01$). Both C9 and clusterin positive cells were found around the injection site. High-dose (100-μg) but not low-dose (25-μg) N-acetylheparin attenuated thrombin-induced brain edema ($81.5 \pm 0.4\%$ vs. $83.7 \pm 0.3\%$ in the vehicle, $P < 0.05$). Behavior was also significantly improved by N-acetylheparin ($P < 0.05$). In conclusion, thrombin-induced edema formation and neurological deficits were both reduced by N-acetylheparin. This suggests that inhibition may be a novel treatment for the thrombin-induced brain injury that occurs in intracerebral hemorrhage.

Keywords: Cerebral hemorrhage; thrombin; complement; N-acetylheparin; brain edema.

Introduction

Mechanisms of edema formation after intracerebral hemorrhage (ICH) have been identified during the past decade [14]. We now know that several processes are responsible for edema formation around the clot. These include hydrostatic pressure during the clot formation, clot retraction, coagulation cascade activation with thrombin production, erythrocyte lysis with hemoglobin induced toxicity, and complement cascade activation in the brain parenchyma [14].

Thrombin, in particular, has been shown to play a major role in early edema formation after ICH. Indeed, a thrombin inhibitor (argatroban) reduced ICH-induced edema formation when given after six hours in a rat ICH model [9] and there is some evidence to support human efficacy [11].

The present study, therefore, examined whether thrombin can activate the complement cascade and whether complement inhibition can reduce thrombin-induced brain injury.

Materials and methods

Animal preparation

The protocols for these animal studies were approved by the University of Michigan Committee on the Use and Care of Animals. Adult male Sprague-Dawley rats (275–325 g, Charles River Laboratories, Portage, Michigan) were anesthetized with pentobarbital (40 mg/kg, i.p.). Aseptic precautions were utilized in all procedures. A polyethylene catheter (PE-50) was then inserted into the right femoral artery in order to monitor arterial blood pressure and to obtain blood samples for analysis of blood gases, blood pH, hematocrit, blood glucose concentration. Body temperature was maintained at 37.5 °C by using a feedback-controlled heating pad.

Intracerebral infusion

Before intracerebral infusion, the rats were positioned in a stereotactic frame (Kopf Instrument, Tujunga, CA), the scalp was incised along the sagittal midline using a sterile technique. A cranial burr hole (1 mm) was drilled near the right coronal suture 4.0 mm lateral to the midline and a 26-gauge needle was inserted stereotaxically into the right basal ganglia (coordinates: 0.2 mm anterior, 5.5 mm ventral, and 4.0 mm lateral to the bregma). Thrombin, saline or throm-

* This study was supported by grants NS-17760 (J.T.H), NS-39866 and NS-47245 (G.X) from the National Institutes of Health.

bin plus N-acetylheparin were infused at 5 μl/min into the right basal ganglia using a microinfusion pump. After infusion, the needle was removed and the skin incisions were closed with sutures. Animals were allowed to recover.

Experiment groups

There were three sets of studies. In the first set, rats had intracerebral infusion of either five-unit thrombin or a needle insertion. Brains were sampled at 24 hours for Western blot analysis and immunohistochemistry. In the second set, rats received either five-unit thrombin+saline, five-unit thrombin+25 μg N-acetylheparin or five-unit thrombin+100 μg N-acetylheparin infusion. Brains were sampled 24 hours later for water content measurement. In the third set, rats received either five-unit thrombin+saline or five-unit thrombin+100 μg N-acetylheparin. Behavioral tests sensitive to unilateral striatal damage were carried out for two weeks.

Brain water content measurement

The rats were sacrificed by decapitation under deep pentobarbital anesthesia (60 mg/kg i.p.). The brains were removed immediately and a 3 mm thick coronal brain slice 4 mm from the frontal pole was cut. That slice was divided into four samples, ipsilateral and contralateral basal ganglia, and ipsilateral and contralateral cortex. Cerebellum was obtained as a control. Tissue samples were weighed on an electronic analytical balance to obtain the wet weight (WW). The tissue was then dried in a gravity oven at 100 °C for more than 24 hours to determine the dry weight (DW). Tissue water contents (%) were calculated as $[(WW - DW)/WW] \times 100$.

Western blot analysis

The rat brains were perfused with saline, and a coronal brain slice was cut as described for brain water content measurements. The brain tissue was immersed in 0.5 ml of Western blot sample buffer and then sonicated for Western blot analysis [15]. Primary antibodies were rabbit anti-complement C9 polyclonal antibody (gift from Dr. P. Morgan, University of Wales) and rabbit anti-clusterin polyclonal antibody (gift from Dr. M. Griswold, Washington State University). The relative densities of complement C9 and clusterin protein bands were analyzed using the NIH image software.

Immuno-histochemistry

Immuno-histochemical studies were performed according to our previous report [15]. Brain sections were incubated according to the avidin-biotin complex technique. Primary antibodies were rabbit anti-complement C9 polyclonal antibody (gift from Dr. P. Morgan, University of Wales) and rabbit anti-clusterin polyclonal antibody (gift from Dr. M. Griswold, Washington State University). Normal rabbit IgG was used as negative control.

Behavioral tests

Animals were placed in a cylindrical enclosure to record preferential use of the non-impaired forelimb for weight shifting movements during spontaneous vertical exploration. The percentage independent use of the non-impaired forelimb (ipsilateral to the injection side), that for the contralateral forelimb, or for both forelimbs together in rapid succession for stepping movements along the walls of the cylinder were calculated. A single score was then used to reflect forelimb use asymmetry: percentage ipsilateral limb use minus percentage contralateral limb use (low score = better function). In addition, a vibrissae-stimulated forelimb placing test (10 trials per side for each rat) was used to examine sensorimotor/proprioceptive capacity (high score = better function) [6]. All behavior was scored by experimenters who were blind to both neurological and treatment conditions. These tests are highly correlated with extent of striatal injury without being influenced by repeated testing.

Statistical analysis

All data in this study are presented as mean ± standard deviation. Data were analyzed with ANOVA using the Scheffe F test or Mann-Whitney U rank test. Significance levels were measured at $P < 0.05$.

Results

Mean arterial pressure, blood pH, arterial oxygen and carbon dioxide tensions, hematocrit, and blood glucose were controlled within normal ranges.

Western blot analysis demonstrated that C9 had a seven-fold increase 24 hours after intracerebral infusion of five-unit thrombin (3329 ± 433 vs 458 ± 395 pixels in saline control, $P < 0.01$). Immuno-histochemistry showed that C9 was deposited on neuronal membranes, indicating activation of the complement cascade and the formation of membrane attack complex. Clusterin, an inhibitor of membrane attack complex formation, also increased after thrombin infusion (4925 ± 686 vs 2453 ± 264 pixels in saline control, $P < 0.01$) and was expressed in neurons.

High-dose (100 μg) but not low-dose (25 μg) N-acetylheparin attenuated thrombin-induced brain edema (81.5 ± 0.4% vs 83.7 ± 0.3% in the vehicle, $P < 0.05$; Fig. 1). Forelimb use asymmetry score and forelimb placing score were also significantly improved by high-dose N-acetylheparin ($P < 0.05$; Fig. 2).

Discussion

In the present study, we demonstrated that thrombin can increase complement C9 and clusterin levels in the brain. Complement inhibition with N-acetylheparin, a heparin congener without anticoagulant properties, reduced thrombin-induced brain edema and neurological deficits.

The complement system is involved in various immune reactions, including cell lysis and the inflammatory response [12]. Complement is normally excluded from the brain parenchyma by the blood-brain barrier (BBB), but entry can occur after ICH as a part of the extravasated blood and later as a result of BBB disruption. There is evidence that the complement cascade is

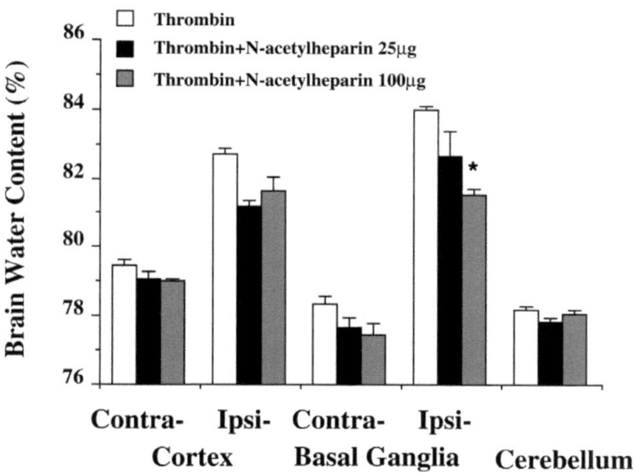

Fig. 1. Brain water content 24 hours after intracerebral infusion of 5 units thrombin, thrombin 5 units + 25 μg N-acetyheparin, and thrombin 5 units + 100 μg N-acetyheparin. Values are mean ± SD, $n = 5$, $^{*}P < 0.05$ vs. thrombin

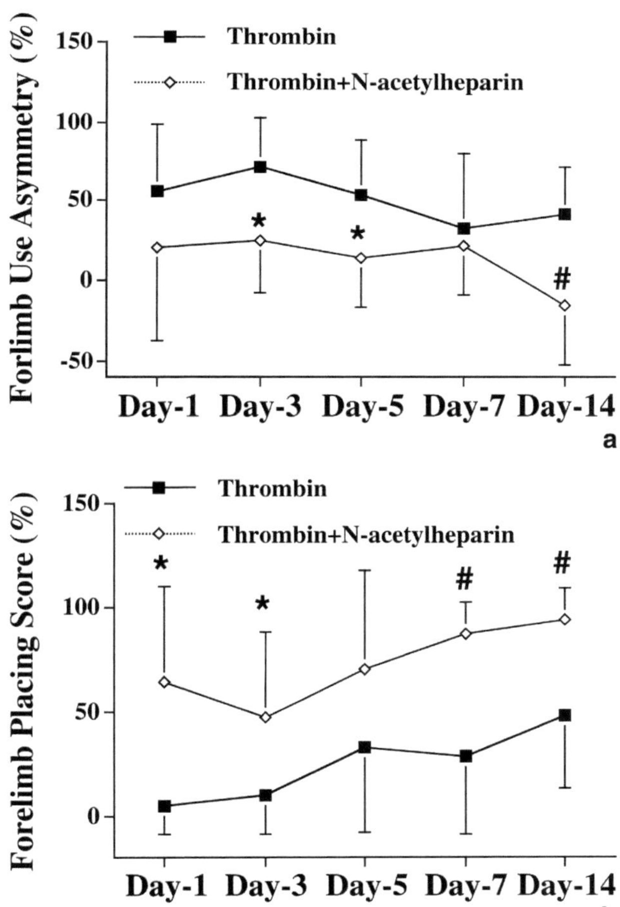

Fig. 2. Forelimb use asymmetry (low score = better function; a) and forelimb placing score (high score = better function; b) at days 1, 3, 5, 7 and 14 after incerebral infusion of 5 units thrombin or 5 units thrombin + 100 μg N-acetyheparin. Values are mean ± SD, $n = 8$, #P < 0.01 and $^{*}P < 0.05$, Mann-Whitney U rank test

activated in brain parenchyma after ICH. Our previous studies have found that inhibition and depletion of the complement cascade attenuate perihematomal brain edema [7, 13].

The current study indicates that thrombin can activate the complement cascade in the brain. Intracerebral infusion of thrombin resulted in a seven-fold increase in complement C9 and a deposition of complement C9 on neuronal membranes. Complement-related brain injury may be due to membrane attack complex (MAC) formation and the classic inflammatory response. MAC consists of C5b-9 complement forms which are assembled following complement activation [3]. The presence of C9 on cell membranes is an indicator of MAC formation. Its formation causes the production of a pore in the cell membrane that leads to cell lysis. Thus, MAC formation may be involved in the lysis of erythrocytes within the clot after ICH. However, MAC insertion may also occur in neurons, glia and endothelial cells, causing neuronal death and BBB leakage. Our studies have also shown that MAC is assembled after ICH [7]. Recent studies have demonstrated that MAC not only causes cell lysis, but also modulates cellular functions such as the release of cytokines, oxygen radicals, and matrix proteins [5].

Clusterin, an inhibitor of MAC formation, was also up-regulated and found in neurons after intracerebral thrombin infusion. We have also found that it is up-regulated in the brain parenchyma after ICH [7]. The balance between complement activation and clusterin up-regulation may help to determine the extent of brain injury.

The effects of coagulation cascade on complement activation are not well studied. However, several studies suggest that there is a very close relationship between thrombin and complement. About 50% of C3 is cleaved during clot formation [4]. Thrombin can cleave and activate C3 [1]. Thrombin cleaved C3a like fragments are chemotatic for leukocytes and to induce enzyme release from neutrophils [8]. Thrombin can also cleave C5 to produce C5a like fragments which are leukotactic [8].

Although the primary role of thrombin in hemostasis is through cleaving fibrinogen to fibrin and inducing platelet aggregation, other important cellular activities of thrombin may be related to thrombin receptor activation [16]. Three protease-activated receptors (PARs), PAR-1, PAR-3 and PAR-4, have been identified as thrombin receptors [2]. Little is known about the interaction between the activation of PARs and the complement system. Lidington *et al.* [10] found that thrombin stimulates decay-accelerating factor (DAF) production through PAR-1. However, whether the effects of thrombin on the complement system found in this study are PAR mediated is still uncertain.

In conclusion, thrombin-induced edema formation and neurological deficits were both reduced by N-acetylheparin suggesting that complement inhibition may be a novel treatment for ICH.

References

1. Bokisch VA, Muller-Eberhard HJ, Cochrane CG (1969) Isolation of a fragment (C3a) of the third component of human complement containing anaphylatoxin and chemotactic activity and description of an anaphylatoxin inactivator of human serum. J Exp Med 129: 1109–1130
2. Coughlin SR (1999) How the protease thrombin talks to cells. Proc Nat Acad Sci USA 96: 11023–11027
3. Esser AF (1991) Big MAC attack: complement proteins cause leaky patches. Immunol Today 12: 316–318; discussion 321
4. Fareed J, Messmore HL, Fenton JW, Brinkhous KM (1981) Perspectively in hemostasis. New York, Pergamon Press
5. Hansch GM, Shin Ml (1998) Complement attack phase. In: Rother K, Till GO, Hansch GM (eds) The complement system. Berlin: Springer, pp 115–145
6. Hua Y, Schallert T, Keep RF, Wu J, Hoff JT, Xi G (2002) Behavioral tests after intracerebral hemorrhage in the rat. Stroke 33: 2478–2484
7. Hua Y, Xi G, Keep RF, Hoff JT (2000) Complement activation in the brain after experimental intracerebral hemorrhage. J Neurosurg 92: 1016–1022
8. Hugli TE (1977) Complement factors and inflammation: Effects of thrombin on components of C3 and C5. In: Lundblad RL, Fenton JW, Mann KG (eds) Chemistry and biology of thrombin. Ann Arbor, Ann Arbor Science Publishers, pp 345–360
9. Kitaoka T, Hua Y, Xi G, Hoff JT, Keep RF (2002) Delayed argatroban treatment reduces edema in a rat model of intracerebral hemorrhage. Stroke 33: 3012–3018
10. Lidington EA, Haskard DO, Mason JC (2000) Induction of decay-accelerating factor by thrombin through a protease-activated receptor 1 and protein kinase C-dependent pathway protects vascular endothelial cells from complement-mediated injury. Blood 96: 2784–2792
11. Matsuoka H, Hamada R (2002) Role of thrombin in CNS damage associated with intracerebral haemorrhage. CNS Drugs 16: 509–516
12. Morgan BP, Gasque P, Singhrao S, Piddlesden SJ (1997) The role of complement in disorders of the nervous system. Immunopharmacology 38: 43–50
13. Xi G, Hua Y, Keep RF, Younger JG, Hoff JT (2001) Systemic complement depletion diminishes perihematomal brain edema. Stroke 32: 162–167
14. Xi G, Keep RF, Hoff JT (2002) Pathophysiology of brain edema formation. Neurosurg Clin North America 13: 371–383
15. Xi G, Keep RF, Hua Y, Xiang JM, Hoff JT (1999) Attenuation of thrombin-induced brain edema by cerebral thrombin preconditioning. Stroke 30: 1247–1255
16. Xi G, Reiser G, Keep RF (2003) The role of thrombin and thrombin receptors in ischemic, hemorrhagic and traumatic brain injury: deleterious or protective? J Neurochem 84: 3–9

Correspondence: Ya Hua, R5550 Kresge I, University of Michigan, Ann Arbor, Michigan 48109-0532, USA. e-mail: yahua@umich.edu

Acta Neurochir (2005) [Suppl] 95: 393–397

Reduction of neural and vascular damage by transplantation of VEGF-secreting neural stem cells after cerebral ischemia

W. Zhu, Y. Mao, and **L. F. Zhou**

Department of Neurosurgery, Shanghai Neurosurgical Center, Huashan Hospital, Fudan University, Shanghai, China

Summary

We determined the role of VEGF-transfected neural stem cells (NSCs) transplantation in rat brain subjected to ischemia.

Fetal NSCs were cultured from E14 days SD rats and transfected with VEGF121 gene by using lipofectamine technique. Temporary middle cerebral artery occlusion (tMCAO) models were established and randomly divided into 1: control group, 2: PBS transplantation group, 3: NSCs transplantation group and 4: VEGF-secreting NSCs transplantation group. Grafts were transplanted into the penumbra zones 3 days after tMCAO model established. Neurological Severity Score (NSS) was checked in all groups 2–12 weeks after transplantation. By using immunofluorescent staining, VEGF expression of transplanted cells, differentiation and migration of transplanted NSCs after transplantation were detected.

VEGF gene-transfected neural stem cells expressed gene products during the first 2 weeks. NSS in this group was significantly lower compared with that in other 3 groups 12 weeks after transplantation. VEGF gene-transfected NSCs migrated and expressed VEGF in hosts' brains, some of them differentiated to neurons 12 weeks after transplantation.

VEGF-transfected NSCs expressed gene products during the early time after transplantation, which reduce brain injury through protecting the vascular system against ischemic attack.

Keywords: Neural stem cells; Vascular endothelial growth factor; transplantation; cerebral ischemia; gene transfer.

Introduction

Vascular endothelial growth factor (VEGF) belongs to neurotrophic factor family, which has a broad range of effects on neural and endothelial cells, including neuroprotection. The expression of VEGF and its receptors are regulated by a variety of acute brain insults, including cerebral ischemia. The increased synthesis of VEGF triggered by cerebral ischemia should be an intrinsic neuroprotective response of the damaged brain. VEGF regulates neovascularization in ischemic brain, decreases infarct volume and brain edema after temporary cerebral ischemia [1, 2]. This raised the possibility that exogenously supplying VEGF might be an useful strategy to up-regulate neovascularization, ameliorate brain edema in ischemic region and protect injured neurons from dying after acute ischemic attack.

Biological delivery of VEGF is an effective approach for administering it locally into the mammalian brain. However, the poor diffusion properties, short half-life in the brain interstitium and the blood brain barrier have previously require intraventricular or intracerebral injection [1]. Such procedures often induce additional trauma, and the duration of action is limited. *Ex vivo* gene transfer of neurotrophic factors to different types of carrier cells might be a more useful and less invasive approach.

Neural stem cells (NSCs), is capabile to develop into integral cyto-architectural components of many region throughout the host brain as neurons, astrocytes and oligodendrocytes. They may also be used as cellular vectors for the stable in vivo expression of foreign therapeutic genes. Since they display significant migratory capacity as well as ability to integrate widely throughout the brain when implanted into germinal zones, NSCs may help reconstitute enzyme and cellular deficiencies in a global fashion. Furthermore, NSCs appear to possess a tropism for degenerating CNS region. The intrinsic properties of NSCs may overcome some of the limitations of standard viral and cellular vectors, and provides more promising strategies against cerebral ischemia.

The main objective of the present study was to explore the possibility that VEGF, secreted by grafts of *ex vivo* transfected NSCs, could ameliorate neural and vascular damage induced by transient focal ischemia.

Materials and methods

The procedures for the use of laboratory animals were approved by the institutional animal care and use committee in the Shanghai Medical School, Fudan University, Shanghai, China. Embryonic NSCs were isolated from the hippocampus of E14 fetal Sprague-Dawley rats. NSCs were cultured according to Vescovi's methods [3]. Briefly, diluted VEGF121 plasmid (2 μg pcDNA3-VEGF121 plasmid in 375 μl NSCs culture medium) was mixed with lipofectamine (12 μl lipofactamine in 375 μl NSCs culture medium) gently. The mixture was incubated at room temperature for 45 min. Antibiotics (0.7 ml) was then added to the free medium complex. The neurosphere of NSCs was digested. Forty thousand cells were centrifugated and resuspended in the DNA/lipofectamine medium complex. They were incubated for 6 h at 37 °C in a carbon dioxide incubator. The cells were then centrifugalized from the transfection mixture. A further 6 ml culture medium with antibiotics was added and gently mixed.

From 24 h to 14 days after transfection, VEGF121 expression of NSCs was evaluated by reverse transcription-polymerase chain reaction (RT-PCR). VEGF expression of transfected NSCs was also assessed by immunoflurescent staining 48 h after transfection.

The middle cerebral artery was temporarily occluded (tMCAO) according to Hayashi's method [2]. After surgery, anesthesia was continued and reperfusion was performed by extracting the surgical thread 120 min later. During the surgery, arterial pressure and body temperature were monitored. Ventilation was adjusted according to arterial oxygen and carbon dioxide tensions and pH values.

Experimental groups were designed as follows:

Group 1: rats were subjected tMCAO alone without donor cell administration ($n = 10$);

Group 2: rats received 10 μl of PBS injected into right striatum 3 days after tMCAO ($n = 10$);

Group 3: rats were treated with 10 μl of NSCs (2×10^5 cells) 3 days after tMCAO ($n = 10$);

Group 4: rats were treated with 10 μl of VEGF-secreting NSCs (2×10^5 cells) 3 days after tMCAO ($n = 10$).

Another 10 rats were given 10 μl of VEGF-secreting NSCs (2×10^5 cells) 3 days after tMCAO, and fluorescent study was performed to detect VEGF expression of donor cells 7 days after the transplantation. Transplantation of donors into the striatum was carried out according to the method described by Fukunaga [4]. Briefly, 10 μl PBS or semisuspended BrdU labeled cells were injected over a 10-minute period into right striatum (posterior = 0.5 mm, right to midline = 3.5 mm, vertical to dura = 5.0 mm from the bregma). This position approximates the ischemic penumbra zone in the striatum.

In all animals, behavior tests were performed just before donor transplantation and at 2, 4, 6, 8, 10, and 12 weeks after transplantation by an investigator who was blinded to the experimental groups. Neurological Severity Scores (NSS) [5] was used to evaluate neurological function of each animal. Before tMCAO surgery, all animals were trained to be familiar with the testing environment. Once the rats scored 0 and passed all the tests, they were subjected to tMCAO.

Rats' brains were fixed by transcardial perfusion with saline, followed by perfusion and immersion in 4% paraformaldehyde for 6 hours, then transferred to 20% sucrose solution (in 0.1 M PBS). Sucrose-impregnated brain slabs were covered with O.C.T. compound embedding medium and frozen on dry ice. 10 μm thick cryostat section were cut and mounted on Poly-L-lysine-coated microscope slides. To detect VEGF expression of grafted cells in vivo, double immunofluorescent staining of anti-BrdU and anti-VEGF antibodies were used 1 week after transplantation. To visualize the cellular colocalization of BrdU-specific and neuron-specific markers in the same transplanted cells, double immunofluorescent staining of anti-BrdU and anti-neurofilament antibodies were used 12 weeks after transplantation. Sections in 3 and 4 groups were also immunohistochemical study for anti-Factor VIII antibody to investigate the capillary vessel structure at transplanted area.

For statistical procedures, SPSS 11.5 software was used and data in different groups were analyzed by a one-way analysis of variance. All data are presented as mean ± standard deviations and $P < 0.05$ was considered statistically significant.

Results

Neurosphere of NSCs were isolated, cultured and were transferred and adhered to poly-L-ornithine-coated glass coverslips. NSCs showed nestin positive with immunofluorescent staining. These neurosheres also showed neurofilament-positive, GFAP-positive or GalC-positive, respectively, when they were induced to differentiate.

From 24 hours after VEGF gene transfection, NSCs continued to express VEGF mRNA until about 2 weeks, and reached its peak at 48–72 hours after transfection. Immunofluorescent study for VEGF showed that both undifferentiated and differentiated NSCs were stained positively.

on the 7th day after grafting, VEGF-transfected NSCs migrated out from the injection site. The central core of heavily labeled cells expressed VEGF, showed double labeling of BrdU and VEGF (Fig. 1). These findings indicated that the *ex vivo* VEGF-transfected NSCs also expressed gene products in vivo, as they migrated with the common NSCs after being transplanted into the penumbra zone of ischemia.

At 12th week after grafting, BrdU positive cells were found scattered in an area of the striatum surrounding the injection site, and only very few labeled cells remained at the injection site. Some of the immigrated cells also showed neurofilament-positive (Fig. 2), indicating that they had differentiated to neurons.

The ischemic lesion in rats of VEGF gene-transfected NSCs group appeared to be less severe compared with that in rats of NSCs grafted group, with more Factor-VIII positive cells and less destroyed vessel structure at 12th week after transplantation. (Fig. 3)

There was no significant differences in NSS among groups before transplantation. At the 8th week after transplantation, NSS in group 3 and 4 was significantly lower than that in group 1 and 2. At the 12th week, NSS in group 4 was significantly lower than that in other 3 groups, NSS in group 3 was also significantly lower than that in group 1 and 2. There was no

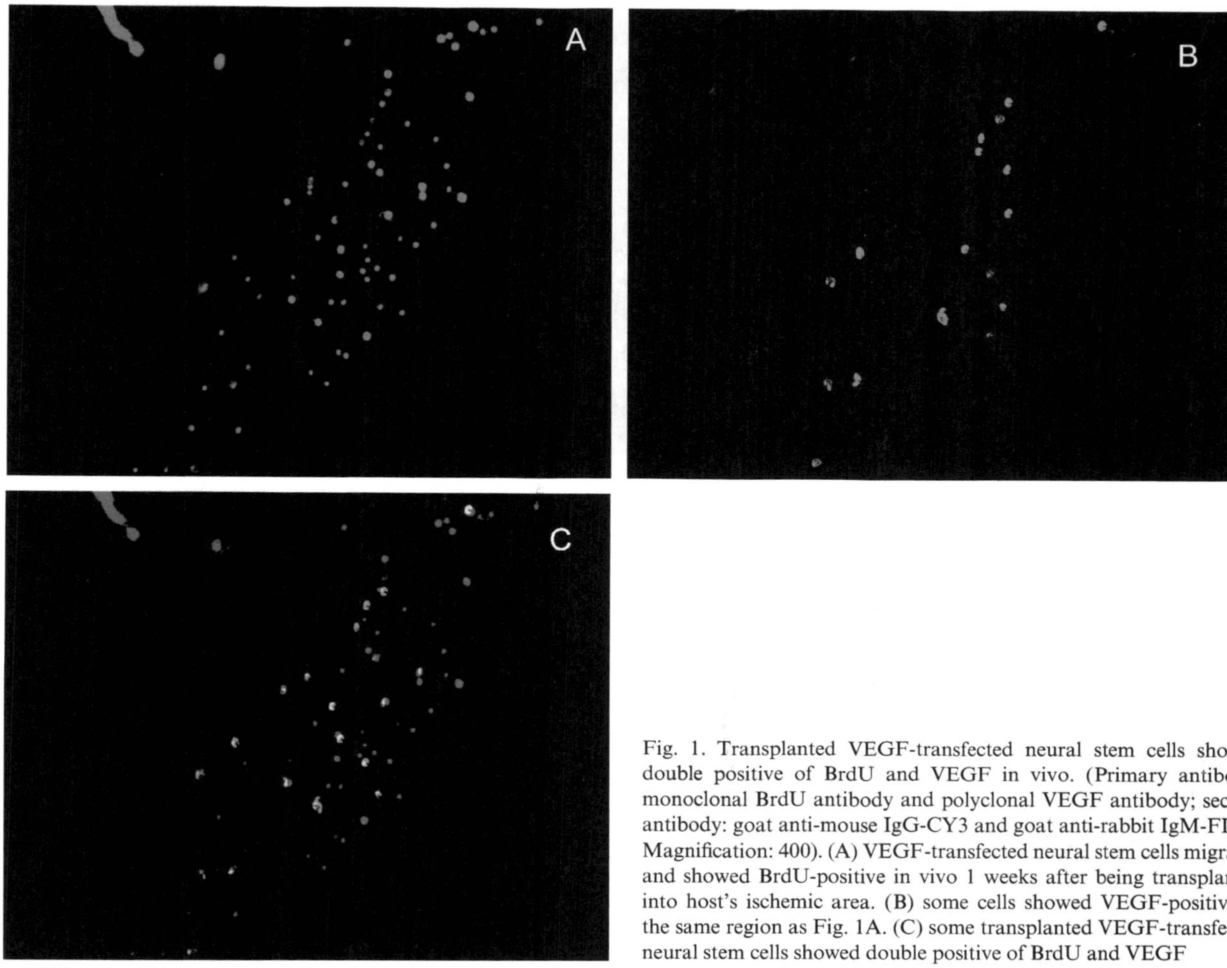

Fig. 1. Transplanted VEGF-transfected neural stem cells showed double positive of BrdU and VEGF in vivo. (Primary antibody: monoclonal BrdU antibody and polyclonal VEGF antibody; second antibody: goat anti-mouse IgG-CY3 and goat anti-rabbit IgM-FITC. Magnification: 400). (A) VEGF-transfected neural stem cells migrated and showed BrdU-positive in vivo 1 weeks after being transplanted into host's ischemic area. (B) some cells showed VEGF-positive at the same region as Fig. 1A. (C) some transplanted VEGF-transfected neural stem cells showed double positive of BrdU and VEGF

difference between 1 and 2 groups during the 12 weeks after transplantation. (Table 1)

Discussion

Because of their inherent biologic properties, NSCs represent valuable therapeutic tools against a variety of neuropathologies, including cerebral ischemic disease. Their ability to develop into integral cytoarchitectural components of ischemic region in the host's brain, makes them capable of replacing the missing or dysfunctional neural cells.

Veizovic *et al.* implanted NSCs into the striatum and cortex of MCAO rats, stable and long-lasting resolution of sensory dysfunction was observed 8 weeks after transplantation [6]. However, Andsberg *et al.* [7] did not find any significant protection during the early period after implantation of NSCs into the penumbra area of tMCAO rats.

Because NSCs require time to migrate and differentiate into neural cells and integrate with the host cytoarchitecture, limited restoration of neurological function is achieved during the early period after transient ischemia. However, NSCs secret cytokine which can ameliorate neural death [3]. Therefore, transfecting NSCs with foreign therapeutic gene in vitro, transplanting them to the penumbra zone, delivering these gene products to the ischemic area of the host's brain might be an effective way to reduce neurological damage in the early period.

Several lines of evidences supported that the reduction of neural damage in the tMCAO rats was due to the transplantation of VEGF-secreting NSCs. Firstly, there was no systematic difference between the experimental groups in the severity of ischemic insult before donor transplantation. Secondly, NSS in VEGF-secreting NSCs group was significantly lower than those in other 3 groups 12 weeks postgrafting. Thirdly,

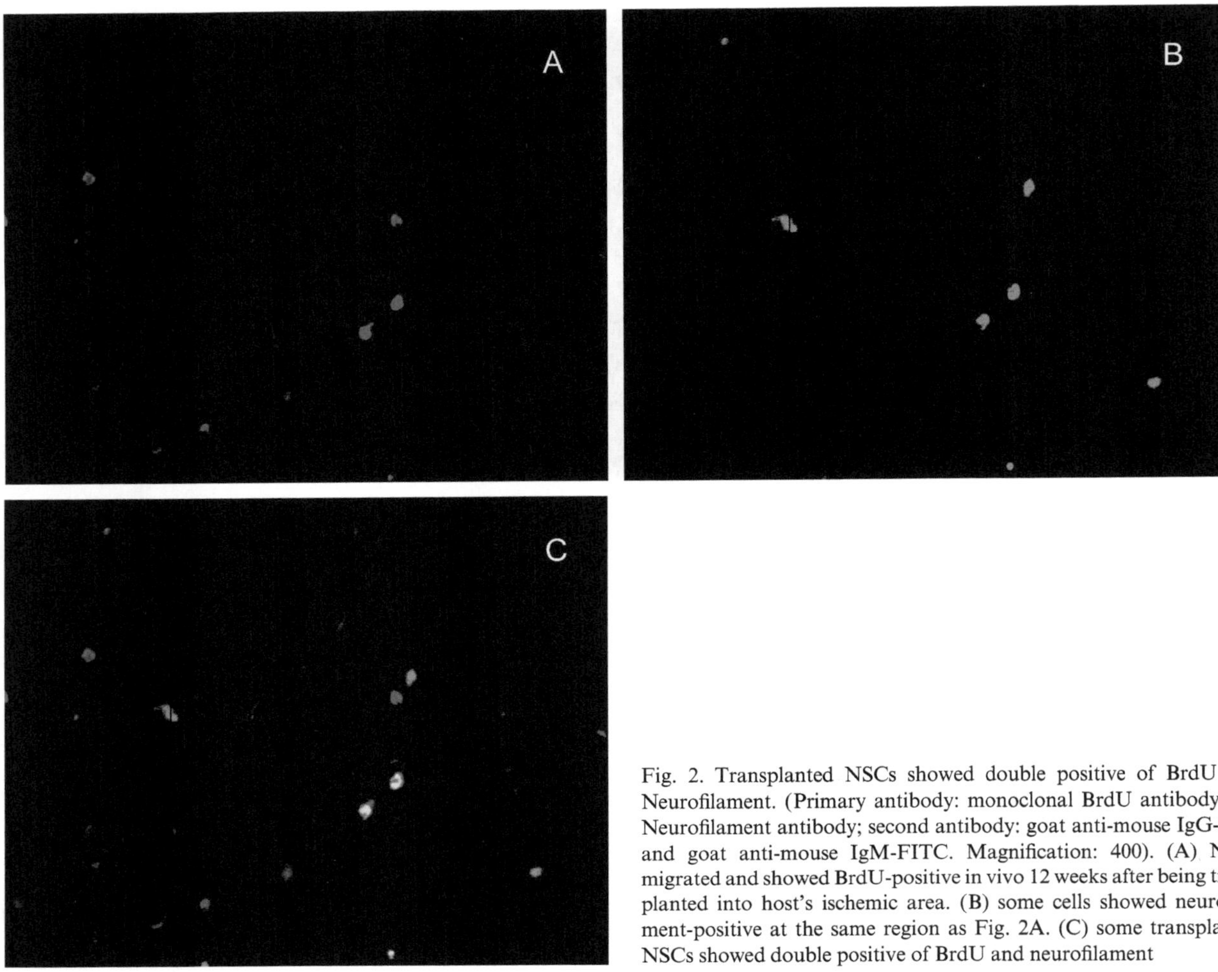

Fig. 2. Transplanted NSCs showed double positive of BrdU and Neurofilament. (Primary antibody: monoclonal BrdU antibody and Neurofilament antibody; second antibody: goat anti-mouse IgG-CY3 and goat anti-mouse IgM-FITC. Magnification: 400). (A) NSCs migrated and showed BrdU-positive in vivo 12 weeks after being transplanted into host's ischemic area. (B) some cells showed neurofilament-positive at the same region as Fig. 2A. (C) some transplanted NSCs showed double positive of BrdU and neurofilament

in the VEGF-secreting NSCs group, the rats which exhibited very poor grafts survival also showed no reduction of ischemic damage. (Data not shown)

It is unclear as to the mechanism or factors that promote damaged neurological function with VEGF-secreting NSCs transplantation after stroke. One possibility is that the NSCs integrated into the tissue, replaced the damaged neural cells, and reconstructed neural circuity. Although we have no direct evidence that NSCs developed direct contacts with other neurons, NSS in NSCs group declined more quickly compared with control and PBS graft groups at the 8th week after transplantation. A more important reason is that VEGF121 secreted by transplanted NSCs contributed to recovery of function lost as a result of lesions [2]. VEGF secreted by host's inherent cells declined greatly 3 days after tMCAO [8]. Exogenously administrated VEGF which is delivered by transplanted NSCs contributed to inhibit activation of the apoptosis signal pathway of endothelial cells. In addition, with the migration of NSCs, VEGF made its effect on larger ischemic area. However, in present study, the NSS in group 4 was significantly lower at the 12th week after transplantation, indicating that exogenously administrated VEGF might be an initiating factor of vascular and neurological protection.

Hayashi [8] investigated VEGF expression in ischemic rats after tMCAO, and showed that VEGF expression was increased at 1 hour after tMCAO and reached its peak at 3 hour, then decreased to the baseline level within 1 day. On the contrary, flk-1 expression, one of the main VEGF receptor expressed by endothelial cells, reached its peak 1–3 days after tMCAO and remained at high level until 3 weeks after tMCAO. In our present study, gene-transfected NSCs expressed VEGF121 in about 2 weeks. When being transplanted into the hosts' brain 3 days after tMCAO, these cells were most effective because VEGF receptor of endo-

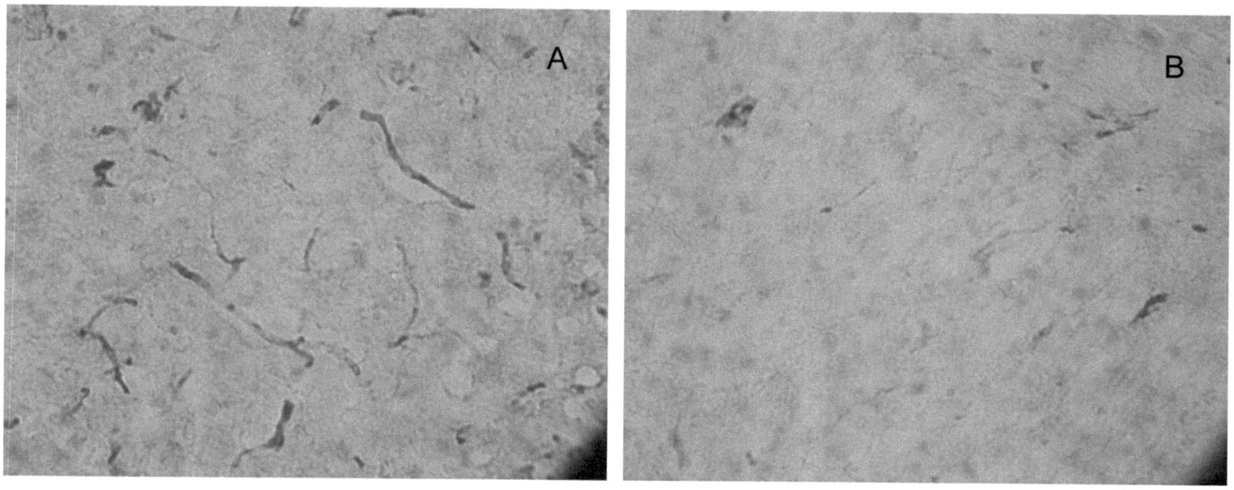

Fig. 3. Immunohistochemistry staining showed larger amount of capillary vessels in transplanted regions of VEGF-secreting NSCs transplantation group. (Primary antibody was monoclonal Factor-VIII antibody; ABC kit. Magnification: 400). (A) VEGF-secreting NSCs transplantation group. (B) NSCs transplantation group

Table 1. *Neurological severity scores in 4 groups from 0–12 weeks after transplantation*

Group	Case	0 week	2 weeks	4 weeks	6 weeks	8 weeks	10 weeks	12 weeks
1	10	6.8 ± 1.3	6.2 ± 1.2	5.5 ± 1.5	5.7 ± 2.0	5.2 ± 1.2	5.4 ± 1.4	5.4 ± 0.7
2	10	6.8 ± 1.7	6.2 ± 1.3	6.0 ± 1.9	6.3 ± 1.9	5.4 ± 0.9	5.1 ± 1.7	5.1 ± 0.7
3	10	6.7 ± 1.5	6.4 ± 1.7	5.7 ± 1.4	5.5 ± 1.5	*4.5 ± 0.9*	4.5 ± 1.6	*4.4 ± 0.7*
4	10	7.0 ± 1.7	5.8 ± 1.5	5.0 ± 1.1	4.6 ± 0.9	*4.0 ± 0.7*	4.0 ± 1.1	***3.8 ± 0.4***
P values		0.98	0.82	0.51	0.16	0.01*	0.15	<0.001*

thelial cells is still being expressed at that time. If gene-transfected NSCs, which have the ability of permanently secreting VEGF, were transplanted, the effect of VEGF would be relatively poor in the later time. Additionally, permanently secreting of VEGF could lead to more severe brain edema.

References

1. Harrigan MR, Ennis SR, Sullivan SE *et al* (2003) Effects of intraventricular infusion of vascular endothelial growth factor on cerebral blood flow, edema, and infarct volume. Acta Neurochir (Wien) 145(1): 49–53
2. Hayashi T, Abe K, Itoyama Y (1998) Reduction of ischemic damage by application of vascular endothelial growth factor in rat brain after transient ischemia. J Cereb Blood Flow Metab 18: 887–895
3. Vescovi AL, Snyder EY (1999) Establishment and properties of neural stem cell clones: plasticity in vitro and in vivo. Brain Pathol 9: 569–598
4. Fukunaga A, Uchida K, Hara K *et al* (1999) Differentiation and angiogenesis of central nervous system stem cells implanted with mesenchyme into ischemic rat brain. Cell Transplant 8: 435–441
5. Chen J, Li Y, Wang L *et al* (2001) Therapeutic benefit of intravenous administration of bone marrow stromal cells after cerebral ischemia in rats. Stroke 32: 1005–1011
6. Veizovic T, Stroemer P, Beech J *et al* (1999) Stem Cell grafts resolve sensory dysfunction following MCAO in the rats. J Cere Blood Flow Metab 19: 616–621
7. Andsberg G, Kokaia Z, Bjorklund A *et al* (1998) Amelioration of ischaemia-induced neuronal death in the rat striatum by NGF-secreting neural stem cells. Eur J Neurosci 10: 2026–2036
8. Hayashi T, Noshita N, Sugawara T *et al* (2003) Temporal profile of angiogenesis and expression of related genes in the brain after ischemia. J Cereb Blood Flow Metab 23: 166–180

Correspondence: Ying Mao, Department of Neurosurgery, Huashan Hospital, Fudan University, 12th Wulumuqi Zhong Rd (200040), Shanghai, China. e-mail: weszhu398@yahoo.com

Acta Neurochir (2005) [Suppl] 95: 399–402

Ischemia-induced endothelial cell dysfunction

R. F. Keep[1,2], **A. V. Andjelkovic**[1,3], **S. M. Stamatovic**[1], **P. Shakui**[1], and **S. R. Ennis**[1]

[1] Department of Neurosurgery, University of Michigan, Ann Arbor, Michigan, USA
[2] Department of Molecular and Integrative Physiology, University of Michigan, Ann Arbor, Michigan, USA
[3] Department of Pathology, University of Michigan, Ann Arbor, Michigan, USA

Summary

Hemorrhagic transformation upon reperfusion therapy has focused attention on ischemia-induced endothelial dysfunction. This study examined whether hyperglycemia may induce hemorrhagic transformation by enhancing endothelial mitochondrial damage during ischemia and whether preconditioning (PC) stimuli may limit ischemia-induced endothelial damage. In vivo, rats received 2.8 M D-glucose or arabinose (1 ml/100 g; i.p.) prior to undergoing two hours of middle cerebral artery occlusion and transcardiac fixation for electron microscopy. In vitro, brain endothelial cells were exposed to a PC impulse (short-term oxygen glucose deprivation; OGD) prior to an injurious event (5 hours OGD). Endothelial injury was assessed by measuring lactate dehydrogenase release. Hyperglycemia during cerebral ischemia resulted in marked changes in endothelial morphology and mitochondrial swelling. Thus, in the ischemic hemisphere, there was no evidence of endothelial mitochondrial swelling in normoglycemic rats (mean profile width 0.22 ± 0.04 vs. 0.17 ± 0.01 μm in contralateral hemisphere) but there was marked swelling in hyperglycemic rats (0.44 ± 0.02 μm). In vitro, cells preconditioned with one hour of OGD one day prior to 5 hours of OGD, showed reduced lactate dehydrogenase release ($p < 0.05$). In conclusion, hyperglycemia may have specific adverse effects on endothelial cell mitochondria during ischemia. Preventing those effects may help to ameliorate blood-brain barrier disruption on reperfusion. Insights into how to prevent endothelial injury may come from determining the mechanisms involved in endothelial preconditioning.

Keywords: Cerebral ischemia; hyperglycemia; cerebral endothelial cells; blood-brain barrier; mitochondria.

Introduction

The blood-brain barrier (BBB) is formed by the cerebral endothelial cells and their linking tight junctions. BBB disruption occurs during cerebral ischemia with and without reperfusion [1]. The increased barrier permeability may contribute to increased intracranial pressure by causing vasogenic edema. In cases of marked disruption, however, it can also lead to hemorrhagic transformation with potentially devastating effects. Indeed, such transformation is a major factor limiting the use of tissue plasminogen activator (tPA) as a stroke therapy.

One of the factors influencing the occurrence of hemorrhagic transformation after a stroke is hyperglycemia [2, 5]. Animal ischemia reperfusion models have also shown that hyperglycemia markedly increases the occurrence of hemorrhagic transformation [4, 7]. The mechanisms involved in this effect of hyperglycemia are still uncertain, but we have found that hyperglycemia affects endothelial function during ischemia [7, 8]. This suggests that hyperglycemia-enhanced endothelial damage during ischemia may lead to hemorrhagic transformation if the brain is reperfused. To examine this hypothesis further, the current study examined the effects of hyperglycemia on endothelial morphology during ischemia.

We have recently shown that ischemic preconditioning (PC) can limit BBB disruption during a subsequent ischemic event [10]. Thus, there are endogenous mechanisms that can limit BBB injury. To help identify those mechanisms, we have examined whether the in vivo PC phenomenon can be mimicked in vitro using cultured cerebral endothelial cells. An in vitro system allows examination of the effects of PC on endothelial cells in the absence of other cell types.

Materials and methods

In vivo experiments

The animal protocols were approved by the University of Michigan Committee on the Use and Care of Animals. Adult male Sprague-Dawley rats (Charles River Laboratories) weighing 275–350 g were used for all experiments. Animals were anesthetized with pentobarbital (65 mg/kg, i.p.) and body temperature was main-

Table 1. *Endothelial mitochondrial dimensions after normo- and hyperglycemic MCA occlusion*

	Normoglycemic contralateral	Normoglycemic ipsilateral	Hyperglycemic contralateral	Hyperglycemic ipsilateral
Length (μm)	0.36 ± 0.03	0.36 ± 0.04	0.31 ± 0.01	0.55 ± 0.03*†
Width (μm)	0.18 ± 0.02	0.22 ± 0.04	0.16 ± 0.01	0.42 ± 0.02**††

Effect of hyperglycemia on the average mitochondrial dimensions in the cerebral endothelium ipsi- and contralateral to an MCA occlusion. Values are means ± S.E., n = 3 animals per group. * and ** indicate a significant difference from contralateral at the $P < 0.01$ and $p < 0.001$ levels. † and †† indicate a significant difference between the normo- and hyperglycemic animals at the $P < 0.01$ and $p < 0.001$ levels. Comparisons by ANOVA with a Newman-Keuls multiple comparisons test.

tained at 37 °C. The rats then received 2.8 M D-glucose (1 ml/100 g; i.p.), to induce hyperglycemia, or arabinose, an osmotic control, 20 minutes before undergoing two hours of permanent middle cerebral artery (MCA) occlusion using the suture method [9]. The rats were then transcardially perfused with phosphate-buffered fixative, pH 7.4 containing 2% paraformaldehyde and 2.5% glutaraldehyde. Brains were removed and the cerebral cortex ipsi- and contralateral to the MCA occlusion was sampled using a punch scheme [7]. Tissue was sampled from the core of the ipsilateral and contralateral MCA territory and processed for electron microscopy. Ultrathin sections were taken and stained for examination with a Philips CM100 transmission electron microscope equipped with a Kodak 1.6 Megaplus high-resolution digital camera. For each sample 10–12 randomly selected capillary profiles were selected and the measurements made of mitochondrial lengths and widths (longest dimension perpendicular to the length). An average of 52 ± 3 profiles were measured per sample.

In vitro experiments

These experiments were performed on an immortalized mouse brain endothelial cell line (bEND.3; American Type Culture Collection VA). Cells were plated and grown in the recommended cell media (DMEM, 10%FBS, 1xAA, 2 mM glutamine) in the presence of 10% CO_2 to confluence. They were then exposed to one hour of oxygen glucose deprivation (OGD) as a PC stimulus or no PC. The OGD was induced by transferal into a temperature-controlled (37 ± 1 °C) anaerobic chamber (Coy Laboratory, MI) and replacement of the media by deoxygenated glucose-free DMEM solution. Then, at different time intervals (4 to 72 hours), the cells were exposed to 5 hours of OGD (to induce injury) and reoxygenated for 24 hours. During the injury-inducing OGD and subsequent reoxygenation, lactate dehydrogenase (LDH) release into the media was monitored to assess cell damage using a commercially available kit (CytoTox 96; Promega, WI).

Results

Two hours of permanent MCA occlusion in normoglycemic rats had little effect on the structure of the cerebral endothelium. For example, there was no significant swelling of the mitochondria in the ipsilateral (ischemic) hemisphere compared to contralateral (Table 1). By contrast, in hyperglycemic animals, the cerebral endothelium was grossly abnormal in the ischemic hemisphere. Particularly noticeable were the presence of trapped erythrocytes within the lumen of the blood vessels (despite perfusion fixation) and grossly swollen mitochondria. Thus, the average width of a mitochondrial profile was 175% greater in the ipsilateral hemisphere compared to contralateral and the average profile length also increased 77% (Table 1). This enhanced endothelial damage during ischemia may contribute to the greater occurrence of reperfusion-induced hemorrhagic transformation in hyperglycemic animals (75% of rats compared to 9% of normoglycemic animals; [7]).

In vitro, bEND.3 cells were preconditioned with one hour of OGD and then exposed to 5 hours of OGD 4, 24 or 72 hours later. Those cultures where the interval was 24 or 72 hours, showed markedly reduced LDH release at the end of the 5 hours of OGD and at the end of 24 hours of reoxygenation (Table 2).

Discussion

Currently, a very small percentage of patients obtain tPA therapy for stroke (~ 2%). An expansion of the use of tPA-induced reperfusion therapy could occur if the number of patients reaching hospitals within the current therapeutic time window (three hours) increases, if methods are found to prolong that therapeutic time window or if the occurrence of hemorrhagic transformation with t-PA could be reduced. In terms of the latter, hyperglycemia is a risk factor for hemorrhagic transformation [2, 5] and the adverse effects of hyperglycemia in humans can be mimicked in animals [4, 7]. The mechanisms by which hyperglycemia enhances hemorrhagic transformation are, however, still uncertain. The results of this study suggest that adverse effects of hyperglycemia occur during ischemia (rather than being solely a reperfusion event). Evidence was found for enhanced endothelial injury. In particular, the endothelial mitochondria of the ischemic hemisphere in hyperglycemic animals became grossly swollen and there was evidence that some blood vessels may cease to be perfused (as evinced by trapped erythrocytes). This suggests a possible scenario whereby

Table 2. *Effects of preconditioning on OGD-induced LDH release from bEND.3 cells*

	Media LDH (U/ml)			
	No PC	PC (4 h)	PC (24 h)	PC (72 h)
5 hour OGD	0.86 ± 0.09	0.73 ± 0.09	0.23 ± 0.03**	0.22 ± 0.05**
24 hours reoxygenation	0.65 ± 0.07	0.54 ± 0.06	0.18 ± 0.04**	0.18 ± 0.03**

bEND.3 cells were preconditioned (*PC*) with one hour OGD. Four, 24 or 72 hours later (*PC* (4 h), (24 h) and (72 h), respectively), the cells were exposed to 5 hours of OGD with or without 24 hours reoxygenation. Media LDH was measured at the end of the 5 hours OGD or at the end of the reoxygenation (fresh media at the start of reoxygenation). LDH release was compared to non-preconditioned cells (*No PC*) by ANOVA with a Dunnett multiple comparison test. Values are means ± S.E., n = 3 separate cultures. ** indicates a significant difference from the No PC group at the $p < 0.01$ level.

hyperglycemia during ischemia causes extensive endothelial damage, perhaps by affecting mitochondrial function, which causes activation of the coagulation cascade. Normally, activation of the coagulation cascade would prevent perfusion of the damaged vessel. However, reperfusion of the damaged vessel (by removal of the suture before the clot can withstand the restoration of pressure or by tPA treatment) may lead to hemorrhagic transformation.

The mechanism underlying hyperglycemia-enhancement of endothelial damage during ischemia remains to be elucidated. The mitochondrial swelling that occurs suggests that those organelles may be specifically involved in the injury, possibly through activation of the mitochondrial permeability transition pore. This pore is permeable to large molecular weight particles (such as cytochrome c) and its opening can result in mitochondrial swelling [6].

An understanding of the mechanisms involved in ischemia-induced endothelial injury is necessary for the rational design of therapeutic agents to limit that injury. Insights into those mechanisms can potentially be derived by examining the endogenous protective mechanisms, such as those upregulated by PC to limit ischemic injury. We have previously shown that ischemic PC limits ischemia-induced BBB disruption in vivo [10] but, in such studies, it is always problematic to determine whether PC affects the endothelium directly or has indirect effects via parenchymal cell protection. The current study, therefore, examined whether cerebral endothelial cells could be protected in vitro in the absence of other cell types. As shown previously in neurons [3], the current results demonstrate that cerebral endothelial cells are robustly protected by PC. Whether there are differences in the cellular mechanisms underlying protection in these two different cell types is currently undergoing investigation. The fact that there is a delay in the onset of protection with PC suggests that the phenomenon requires protein synthesis.

In conclusion, endothelial injury *during* ischemia is an understudied phenomenon. Such endothelial injury is markedly exacerbated by hyperglycemia and may lead to hemorrhagic transformation upon reperfusion. Endothelial injury in vivo [10] and in vitro (this study) can be limited by preconditioning. Understanding the mechanisms involved may help in the design of therapeutic agents to limit endothelial injury.

Acknowledgments

This work was supported by grant NS 37409 from the National Institutes of Health.

References

1. Betz AL, Dietrich WD (1998) Blood-brain barrier dysfunction in cerbral ischemia. In: Ginsburg MD, Bogousslavsky J (eds) Cerebrovascular disease: pathophysiology, diagnosis, and management. Blackwell, Malden, pp 358–370
2. Bruno A, Levine SR, Frankel MR *et al* (2002) Admission glucose and clinical outcomes in the NINDS rt-PA Stroke Trial. Neurology 59: 669–674
3. Dawson VL, Dawson TM (2000) Neuronal ischaemic preconditioning. TiPS 21: 423–424
4. de Courten-Myers G, Kleinholz M, Holm P *et al* (1992) Hemorrhagic infarct conversion in experimental stroke. Ann Emerg Med 21: 120–126
5. Demchuk AM, Morgenster LB, Krieger DW *et al* (1999) Serum glucose level and diabetes predict tissue plasminogen activator-related intracerebral hemorrhage in acute ischemic stroke. Stroke 30: 34–39
6. Hansson MJ, Persson T, Friber H *et al* (2003) Powerful cyclosporin inhibition of calcium-induced permeability transition in brain mitochondria. Brain Res 960: 99–111
7. Kawai N, Keep RF, Betz AL (1997) Hyperglycemia and the vascular effects of cerebral ischemia. Stroke 28: 149–154
8. Kawai N, Stummer W, Ennis SR, Betz AL, Keep RF (1999)

Blood-brain barrier glutamine transport during normo- and hyperglycemic focal ischemia J Cereb Blood Flow Metab 19: 79–86
9. Longa EZ, Weinstein PR, Carlson S, Cummins R (1990) Reversible middle cerebral artery occlusion without craniectomy in rats. Stroke 20: 84–91
10. Masada T, Hua Y, Xi G, Ennis SR, Keep RF (2001) Attenuation of ischemic brain edema and cerebrovascular injury following ischemic preconditioning in the rat. J Cereb Blood Flow Metab 21: 22–33

Correspondence: Richard F. Keep, Department of Neurosurgery, University of Michigan, R5550 Kresge I, Ann Arbor, MI 48109-0532, USA. e-mail: rkeep@umich.edu

Acta Neurochir (2005) [Suppl] 95: 403–406

Systemic use of argatroban reduces tumor mass, attenuates neurological deficits and prolongs survival time in rat glioma models*

Y. Hua[1], L. L. Tang[1], M. E. Fewel[1], R. F. Keep[1,2], T. Schallert[1,3], K. M. Muraszko[1], J. T. Hoff[1], and G. H. Xi[1]

[1] Department of Neurosurgery, University of Michigan, Ann Arbor, Michigan, USA
[2] Department of Physiology, University of Michigan, Ann Arbor, Michigan, USA
[3] Department of Psychology and Institute for Neuroscience, University of Texas at Austin, Austin, Texas, USA

Summary

Our previous studies showed that intracerebral infusion of argatroban, a specific thrombin inhibitor, reduces brain edema and neurological deficits in a C6 glioma model. The present study investigated whether systemic argatroban administration can reduce glioma mass and neurological deficits and extend survival time in C6 and F98 gliomas. Rat C6 or F98 glioma cells were infused into the right caudate of adult male Fischer 344 rats. Osmotic minipump loaded with argatroban (0.3 mg/hour) or vehicle was implanted into abdomen immediately after glioma implantation. Tumor mass was determined at day 9. Over the period of the experiment, the animals underwent behavioral testing (forelimb placing and forelimb use asymmetry). In addition, survival time was tested in the F98 glioma model. In C6 glioma, argatroban reduced glioma mass ($p < 0.05$) and neurological deficits ($p < 0.05$) at day 9. In F98 glioma, agratroban prolonged the survival time ($p < 0.05$) and reduced the body weight loss (84 ± 15 gram vs. 99 ± 2 gram in the vehicle group, $P < 0.05$). In conclusion, systemic use of argatroban reduced tumor mass and neurological deficits, and prolonged survival time. These results suggest that thrombin plays a key role in glioma growth and thrombin inhibition with argatroban may be a novel treatment for gliomas.

Keywords: Glioma; brain edema; thrombin; argatroban; neurological deficits; rats.

Introduction

It has long been known that thrombin has a pivotal role in the coagulation cascade. However, the discovery and cloning of a series of thrombin receptors [2], has suggested that thrombin has a wide range of potential actions. Several pieces of evidence led us to examine whether thrombin might have a major role in malignant gliomas. (A) In contrast to normal brain, the vasculature of these tumors is often highly permeable with the potential for prothrombin entry from blood into tumor. (B) Brain edema in and around gliomas contributes to the high mortality (within months) of patients with malignant gliomas by causing herniation-related death. Our previous studies indicate that thrombin plays an important role in edema formation after intracerebral hemorrhage (ICH) [4, 6]. (C) Thrombin is a potent mitogen, which enhances the synthesis and secretion of nerve growth factor in glial cells and stimulates astrocyte and tumor cell proliferation [1]. (D) Angiogenesis is essential for rapid glioma growth because of the need for oxygen and metabolites. Although many factors regulate angiogenesis, thrombin may play an important role [7].

Our previous study found that intracerebral infusion of argatroban reduces brain edema and improves function outcome in a C6 glioma model [3]. The present study was to determine whether systemic use of argatroban can reduce tumor mass and tumor-induced behavioral deficits and prolong survival time.

* This study was supported by Mitsubishi Pharma Corporation.

Materials and methods

Animal preparation and C6, F98 glioma models

The protocols in this study were approved by the University of Michigan Committee on the Use and Care of Animals. Adult male Fischer 344 rats (200–225 g, Charles River Laboratories, Portage, Michigan) were anesthetized with pentobarbital (50 mg/kg, i.p.). Aseptic precautions were utilized in all surgical procedures. After anesthesia was achieved, a polyethylene catheter (PE-50) was inserted into the right femoral artery in order to monitor arterial blood pressure and to obtain blood for analysis of blood gases, blood pH, hematocrit, and blood glucose concentration. They then were placed in a Kopf stereotaxic frame. Body temperature was maintained at 37 ± 1 °C using a rectal temperature probe and a feedback regulator.

A small burr hole was drilled in the skull overlying the right caudate nucleus. The tip of a 26-gauge stainless steel cannula was then lowered into the right caudate (0.2 mm anterior, 5.5 mm ventral, 3.5 mm lateral to the bregma) and 1×10^6 C6 or F98 glioma cells were infused. Behavioral deficits in these animals were examined at 3, 6, 9 and 12 days. Animals were sacrificed at 3 and 12 days for measurement of tumor mass and edema formation, while other animals were sacrificed at 12 days for immunohistochemistry and measurement of thrombin activity.

Alzet osmotic mini-pump implantation and argatroban treatment

To examine whether thrombin contributed to brain edema formation, glioma growth, glioma-related neurological deficits and survival time, animals were treated with argatroban, a thrombin inhibitor. At the time of C6 or F98 glioma cell implantation, rats were implanted intraperitoneally with an osmotic minipump (Alzet, Cupertino, California). The pump was preloaded with argatroban (60 mg/ml and delivers 5 μl/h) or vehicle.

Brain water content

The rats were sacrificed by decapitation under deep pentobarbital anesthesia. The brains were removed immediately and cut into three parts: front part (frontal pole to 3 mm back), the middle part (3–7 mm frontal pole) and the rear part (7 mm to occipital pole). The brain samples were then divided in to cortex or basal ganglia (ipsilateral or contralateral). The brain water content was measured by wet/dry method.

Tumor mass

We used the weight difference between the ipsilateral (tumor side) and contralateral hemisphere to estimate the tumor mass.

Behavioral tests

Animals were placed in a cylindrical enclosure to record preferential use of the non-impaired forelimb for weight shifting movements during spontaneous vertical exploration. The % independent use of the non-impaired forelimb (ipsilateral to the tumor), the % independent use of the contralateral forelimb, and the % use of both forelimbs together in rapid succession for stepping movements along the walls of the cylinder were assessed. A single score was then used to reflect forelimb use asymmetry: % ipsilateral limb use minus % contralateral limb use (low score = better function). In addition, a vibrissae-stimulated forelimb placing test (10 trials per side for each rat) was used to examine sensorimotor/proprioceptive capacity (high score = better function) [4]. All behavior was scored by experimenters who were blind to both neurological and treatment conditions. These tests are highly correlated with extent of striatal injury without being influenced by repeated testing.

Cell culture

C6 (passage number 36 to 42) and F98 (passage number 9–11) glioma cell lines were obtained from American Type Culture Collection (Manassas, Virginia). C6 glioma cells were grown at 37 °C in air with 5% CO_2 in Ham's F-10 medium with 2.5% fetal bovine serum and 15% horse serum. F98 glioma cell were obtained from American Type Culture Collection (ATCC, Manassas VA) and grown in DMEM (ATCC) medium with 10% fetal bovine serum at 37 °C in air with 5% CO_2. Cells were maintained in a monolayer culture.

Statistics

Data were analyzed by Student t test, Mann-Whitney U rank test or Trend Peto-Peto-Wilcoxon test. Differences were considered significant at the $p < 0.05$ level.

Results

Brain water content was measured at day 9 after infusion of C6 glioma cells with and without argatroban treatment. Argatroban did not reduce brain edema in the glioma significantly (80.7 ± 1.5% vs. 81.2 ± 0.7% in the vehicle, $p > 0.05$). Minor hemorrhages were found in all gliomas with argatroban treatment that may prevent argatroban causing the significant reduction in edema we found with intracerebral treatment [3].

We used the weight difference between the ipsilateral (tumor side) and the contralateral hemisphere to estimate the tumor mass. Argatroban (0.3 mg/h/rat) reduced C6 glioma mass at day 9 (43.3 ± 31 mg vs. 80.1 ± 35.2 mg in the vehicle group, n = 8, $p < 0.05$, Fig. 1).

Argatroban treatment also improved F98 glioma-related neurological deficits including forelimb use asymmetry and forelimb placing (Fig. 2). For forelimb use asymmetry test, we only could test the glioma rats to day 9 because the rats could not perform at day 12. Early argatroban systemic treatment reduced rat body weight loss (84 ± 15 gram vs. 99 ± 2 gram in the vehicle group, $p < 0.05$). Argatroban prolonged survival

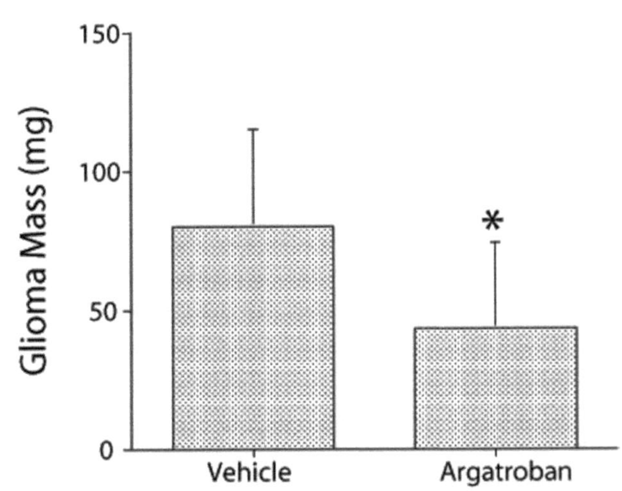

Fig. 1. Bar graph showing glioma mass at day 9 after C6 glioma cell implantation in the argatroban-treated group (0.3 mg/h/rat) and vehicle-treated group. Values are mean ± SD, n = 8, *$p < 0.05$ vs. vehicle

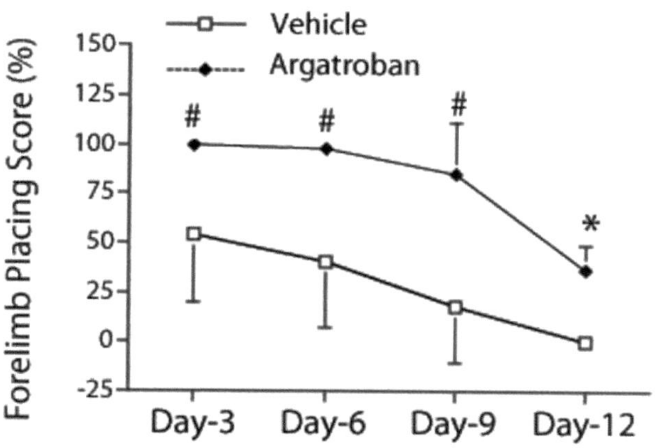

Fig. 2. Forelimb placing score at days 3, 6, 9 and 12 after intracerebral infusion of F98 glioma cells with systemic treatment of argatroban or vehicle. The lower the score the greater the neurological deficits. Values are mean $\pm$ SD, n = 15, *$p < 0.05$, #$p < 0.01$ vs. vehicle by Mann-Whitney U rank test

time in F98 glioma model (18 $\pm$ 3 days vs. 14 $\pm$ 4 days in the vehicle group, $p < 0.05$).

Discussion

The current study demonstrates that systemic use of argatroban (0.3 mg/h/rat) reduced C6 glioma mass. It also improved F98 glioma-induced behavioral deficits and prolonged survival time in that glioma model. These results suggested that thrombin plays an important role in glioma growth and argatroban may be a useful treatment for glioma.

Our recent studies have found that intracerebral infusion of argatroban reduced tumor mass, edema formation and neurological deficits in a C6 glioma model [3]. The effect of thrombin on tumor mass may result from a direct effect on tumor cell proliferation. Thrombin enhances the synthesis and secretion of nerve growth factor in glial cells [10] and stimulates astrocyte proliferation [1]. In addition, thrombin may act as a growth factor for tumor cells and induces proliferative response in T-47D mammary tumor cells [9]. In T98G and TM-1 human glioma cells, thrombin also induced proliferation. This mitogenic effect was abolished by hirudin, a thrombin inhibitor [12]. Another possibility is thrombin may increase tumor mass by promoting angiogenesis [18].

Other experimental data suggest a role for coagulation and fibrinolysis systems in tumor development, progression and metastasis [11, 13]. Clinical data also suggest that targeting the coagulation system might influence the course of malignant disease. The anticoagulant drug warfarin prolongs the length of survival of patients with small cell lung cancer. Heparin has a benefit in patients with several types of cancer [19]. Our results suggest that drugs directly targeting thrombin may have similar actions and that warfarin and heparin may be having their effects by reducing thrombin production or activity.

Argatroban was chosen as the thrombin inhibitor in this study. It is a small molecule (molecule weight: 508.7) and a direct thrombin inhibitor binding to the catalytic site of thrombin molecule. Argatroban is an effective inhibitor of thrombin both bound to fibrin and thrombin free in solution [5, 8]. We have found that systemically delivered argatroban is effective in reducing intracerebral hemorrhage-induced edema formation [6].

In the first part of this study C6 glioma cells were injected into the caudate. This takes advantage of the fact that a number of behavioral tests have been devised to assess caudate damage induced by ischemia, hemorrhage or neurotoxins [4, 15, 16]. This model is stable and reproducible for edema, tumor mass and neurological deficit measurements. There is a very serious limitation for the survival studies, however, since C6 glioma can be immunogenic in allogeneic hosts [17]. For this reason, we developed the F98 glioma model in the rat for survival time studies.

Argatroban reduced neurological deficits in the F98 glioma model. Traditional preclinical investigations of brain tumor therapies have focused primarily on inhibiting tumor cell proliferation or survival with little regard to behavioral assessment. However, in clinical trials brain function has been an essential index of treatment benefit. A model that includes sensitive functional outcome measures would appear to represent a significant advantage in brain tumor research.

The current study provides evidence that thrombin could be a therapeutic target for malignant gliomas. It does not address the question of whether thrombin plays a role in other forms of brain tumor, i.e. is it specific for gliomas or does thrombin stimulate tumor growth in, for example, low grade astrocytomas? But, in relation to this point, there is evidence that the effect of thrombin is not confined to gliomas from examination of tumors outside the brain [14]. In particular, the effects of thrombin on angiogenesis suggest that it may affect growth in a variety of solid tumors which depend on the formation of new blood vessels [7, 11, 14]. In summary, we found that early systemic use of argatroban reduced glioma mass, improved glioma-related neurological deficits and prolonged survival time in two rat glioma models. Clarification of the effects of argatroban on glioma proliferation, angiogenesis and edema formation should help to develop new therapeutic strategies for glioma treatment.

References

1. Cavanaugh KP, Gurwitz D, Cunningham DD, Bradshaw RA (1990) Reciprocal modulation of astrocyte stellation by thrombin and protease nexin-1. J Neurochem 54: 1735–1743
2. Coughlin SR (2000) Thrombin signalling and protease-activated receptors. Nature 407: 258–264
3. Hua Y, Keep RF, Schallert T, Hoff JT, Xi G (2003) A thrombin inhibitor reduces brain edema, glioma mass and neurological deficits in a rat glioma model. Acta Neurochir (Wien) [Suppl] 86: 503–506
4. Hua Y, Schallert T, Keep RF, Wu J, Hoff JT, Xi G (2002) Behavioral tests after intracerebral hemorrhage in the rat. Stroke 33: 2478–2484
5. Hursting MJ, Alford KL, Becker JC, Brooks RL, Joffrion JL, Knappenberger GD, Kogan PW, Kogan TP, McKinney AA, Schwarz RP, Jr (1997) Novastan (brand of argatroban): a small-molecule, direct thrombin inhibitor. Sem Thrombos Hemostas 23: 503–516
6. Kitaoka T, Hua Y, Xi G, Hoff JT, Keep RF (2002) Delayed argatroban treatment reduces edema in a rat model of intracerebral hemorrhage. Stroke 33: 3012–3018
7. Maragoudakis ME, Tsopanoglou NE (2000) On the mechanism(s) of thrombin induced angiogenesis. Adv Exp Med Biol 476: 47–55
8. McKeage K, Plosker GL (2001) Argatroban. Drugs 61: 515–522
9. Medrano EE, Cafferata EG, Larcher F (1987) Role of thrombin in the proliferative response of T-47D mammary tumor cells. Mitogenic action and pleiotropic modifications induced together with epidermal growth factor and insulin. Exp Cell Res 172: 354–364
10. Neveu I, Jehan F, Jandrot-Perrus M, Wion D, Brachet P (1993) Enhancement of the synthesis and secretion of nerve growth factor in primary cultures of glial cells by proteases: a possible involvement of thrombin. J Neurochem 60: 858–867
11. O'Reilly MS, Pirie-Shepherd S, Lane WS, Folkman J (1999) Antiangiogenic activity of the cleaved conformation of the serpin antithrombin. Science 285: 1926–1928
12. Ogiichi T, Hirashima Y, Nakamura S, Endo S, Kurimoto M, Takaku A (2000) Tissue factor and cancer procoagulant expressed by glioma cells participate in their thrombin-mediated proliferation. J Neuro Oncol 46: 1–9
13. Ornstein DL, Meehan KR, Zacharski LR (2002) The coagulation system as a target for the treatment of human gliomas. Sem Thromb Hemostas 28: 19–28
14. Rickles FR, Patierno S, Fernandez PM (2003) Tissue factor, thrombin, and cancer. Chest 124: 58S–68S
15. Schallert T, Leasure JL, Kolb B (2000a) Experience-associated structural events, subependymal cellular proliferative activity, and functional recovery after injury to the central nervous system. J Cereb Blood Flow Metab 20: 1513–1528
16. Schallert T, Fleming SM, Leasure JL, Tillerson JL, Bland ST (2000b) CNS plasticity and assessment of forelimb sensorimotor outcome in unilateral rat models of stroke, cortical ablation, parkinsonism and spinal cord injury. Neuropharmacology 39: 777–787
17. Trojan J, Johnson TR, Rudin SD, Ilan J, Tykocinski ML (1993) Treatment and prevention of rat glioblastoma by immunogenic C6 cells expressing antisense insulin-like growth factor I RNA. Science 259: 94–97
18. Yamahata H, Takeshima H, Kuratsu J, Sarker KP, Tanioka K, Wakimaru N, Nakata M, Kitajima I, Maruyama I (2002) The role of thrombin in the neo-vascularization of malignant gliomas: an intrinsic modulator for the up-regulation of vascular endothelial growth factor. Int J Oncol 20: 921–928
19. Zacharski LR, Ornstein DL (1998) Heparin and cancer. Thromb Haemostas 80: 10–23

Correspondence: Guohua Xi, M.D., R5550 Kresge I, University of Michigan, Ann Arbor, Michigan 48109-0532, USA. e-mail: guohuaxi@umich.edu

Acta Neurochir (2005) [Suppl] 95: 407–410

Spinal contribution to CSF pressure lowering effect of mannitol in cats

M. Klarica[1], **R. Varda**[1], **M. Vukić**[2], **D. Orešković**[3], **M. Radoš**[1], and **M. Bulat**[1]

[1] Department of Pharmacology and Croatian Institute for Brain Research, School of Medicine, Zagreb, Croatia
[2] Department of Neurosurgery, School of Medicine, Zagreb, Croatia
[3] Rudjer Bošković Institute, Zagreb, Croatia

Summary

Objectives. After application of hyperosmolar mannitol the cerebrospinal (CSF) pressure is usually lowered within 30 min but this effect cannot be explained either by changes in intracranial blood volume and flow or by changes in brain volume. We assume that this effect of mannitol my be consequence of CSF volume decrease primarily in the spinal CSF due to high compliance of the spinal dura.

Methods. To explore such a possibility we planned to separate spinal and cerebral CSF. In chloralose anaesthetized cats dorsal laminectomy of C_2 vertebrae was performed and a plastic semi ring was positioned extradurally separating cranial and spinal CSF. CSF pressures were recorded via cannulas positioned in lateral ventricle and lumbar subarachnoid space at L_3 vertebrae, respectively.

Results. After intravenous bolus of 20% mannitol (0.5 or 1.0 g/kg/3 min) in control animals without cervical stenosis, the fall of both ventricular and lumbar CSF pressures was equal over time. At 15 min after mannitol application in cats with cervical stenosis an slight increase of ventricular and a fall of lumbar CSF pressures were observed, while at 30 min a gradient of these pressures of 5.5 and 7 cm H_2O at lower and higher dose of mannitol, respectively, were registered. However, after removal of cervical stenosis these gradients disappeared.

Conclusion. The observed changes of CSF pressures in spinal and intracranial space indicate that spinal subarachnoid space contributes a great deal to overall fall of CSF pressure and volume in the early period after mannitol application probably due to high compliance of the spinal dura.

Keywords: Mannitol; intracranial pressure; cervical stenosis; spinal dura compliance.

Introduction

Mannitol has been widely used in clinical practice for reduction of intracranial pressure (ICP) but the exact mechanism of its action still remains controversial. It is generally accepted that decrease of the volume of intracranial contents (brain, blood, cerebrospinal fluid (CSF)) may change ICP. It is observed that the ICP usually is lowest at 30 min after i.v. bolus of hyperosmolar mannitol but this effect does not correlate with changes of percentage of brain tissue water [3, 9] or cerebral blood volume [7, 8]. We assume that this effect of mannitol may be consequence of CSF volume reduction primarily in the spinal CSF due to high compliance of the spinal dura. To test this hypothesis we have developed a new model of cervical stenosis in cats separating the cranial and spinal CSF, and in this model ventricular and lumbar CSF pressures were measured after application of mannitol.

Material and methods

Adult cats weighing 2.5–5.0 kg were anaesthetised with chloralose (100 mg/kg, i.p.) and positioned in sphinx position in stereotaxic frame (D. Kopf, USA) with their heads elevated 12 cm above the stereotaxic table. Mean arterial blood pressure and arterial blood gases controlled during experiment. The CSF pressures were recorded via cannulas positioned in left lateral ventricle and lumbar subarachnoid space at L_3 vertebrae, respectively (Fig. 1). Interaural line was taken as zero reference point for pressures measurements and the pressures were registered via transducers (Statham P 23 ID, Gold Electronics, USA) and polygraph (Model R511A, Beckman, USA) (Fig. 1).

Since effect of mannitol on normal CSF pressure is small (reduction about 18%) [1] to enhance this pressure we infused mock CSF at a rate of 52 µl/min (Harvard M-975 infusion pump) over 10 min via a cannula positioned in right lateral ventricle. Such an infusion increases CSF pressure to 22–25 cm H_2O. During the last 3 min of intraventricular infusion 20% solution of mannitol (0.5 g/kg) was intravenously injected and CSF pressures recorded over 60 min (Fig. 2). The animals so treated served as control.

In the other group of experiments, the laminectomy of cervical C_2 vertebrae and exposition of dura was performed. Immediately after mock CSF infusion and mannitol application (see above), a plastic semi ring, covering dorsal and lateral parts of dura and gently pressing on the cord, was positioned and CSF pressures recorded over 60 min (Fig. 2). It was shown in preliminary experiments that after infusion (7 µl/min) of 2% phenolsulfophthalein (Fluka, Switzerland) in saline over 20 min and after waiting of 60 min, tissue above plastic

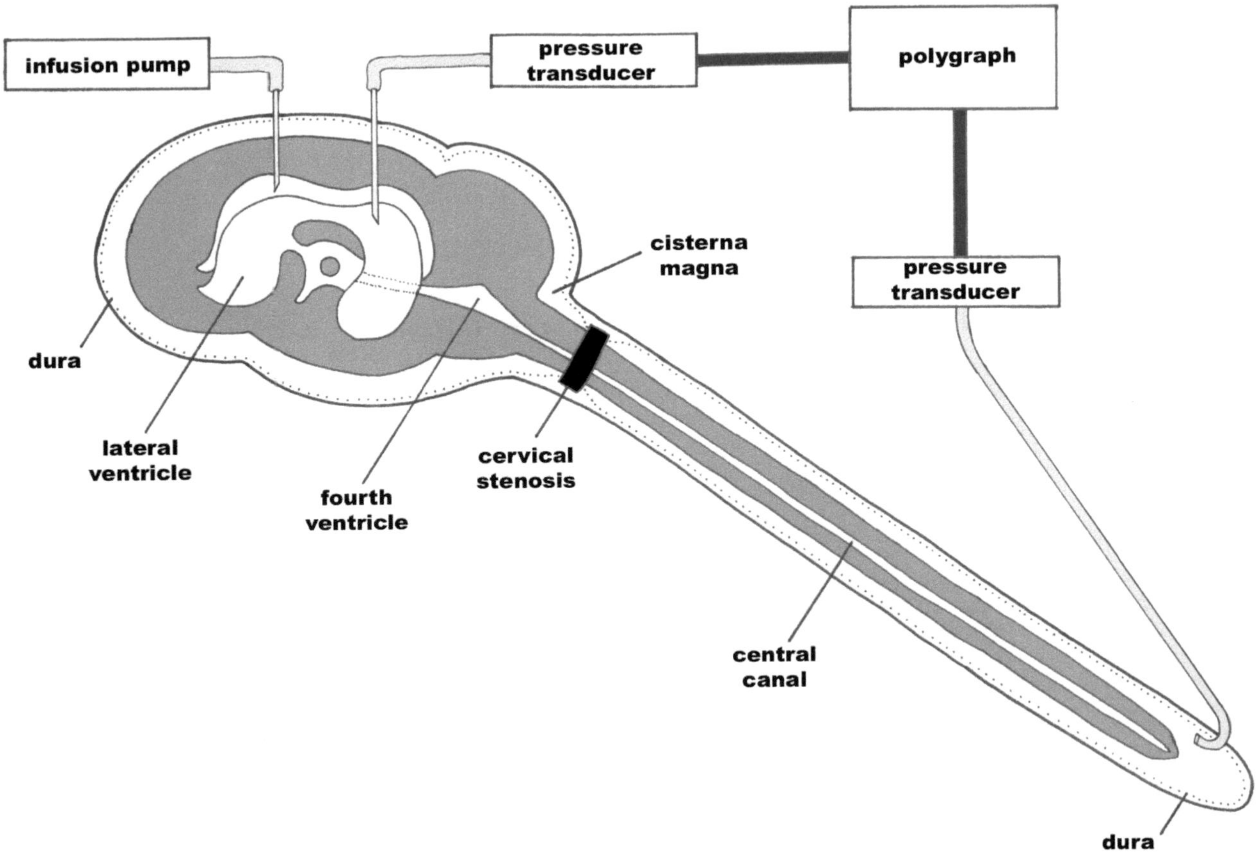

Fig. 1. Scheme of experimental model showing position of cannulas in lumbar subarachnoid space and left lateral ventricle for CSF pressure recording, cannula in right lateral ventricle for infusion of mock CSF and the site of cervical stenosis

semi ring adjacent to CSF showed intensive pink colour while below plastic semi ring in the spinal region no pink colour was detected indicating a block of communication of CSF above and below plastic semi ring.

Results are shown as mean ± S.E.M. Significance between data was estimated by the Student's t-test.

Results

After i.v. mannitol (0.5 g/kg) application change of blood pressure from control 91 ± 7 mmHg to 103 ± 30 at 30 min and 95 ± 17 mmHg at 60 min was observed. Control plasma osmolarity was 307 ± 1 milliosmoles (mosm) per litre, immediately after mannitol application (0.5 g/kg) osmolarity was 317 ± 4 mosm/L, while at 30 and 60 min after mannitol osmolarities were 324 ± 4 and 313 ± 2 mosm/L, respectively.

Figure 2 shows that in control animals without stenosis (n = 5) CSF pressures in lateral ventricle and lumbar subarachnoid space fall to the same degree and after 20 min they reached a relatively constant value of about 8 cm H_2O. Similar fall in CSF pressures is observed in animals with cervical stenosis (n = 4) during first 15 min. However, after that time the ventricular CSF pressure reaches value about 11.5 cm H_2O while lumbar CSF pressure falls to about 6 cm H_2O, so that significant gradient of pressure between ventricular and lumbar CSF of 5 to 7 cm H_2O ($p < 0.05$) is maintained until the end of experiment. When the plastic semi ring was removed at the end of experiment this gradient of CSF pressure disappeared and ventricular and lumbar CSF pressure reached a value as in control animals. In three animals with cervical stenosis mannitol in dose of 1.0 g/kg was applied and gradient of CSF pressure was somewhat higher than in previous case reaching a value of 9 cm H_2O (mean value at 30 min was 7 cm H_2O).

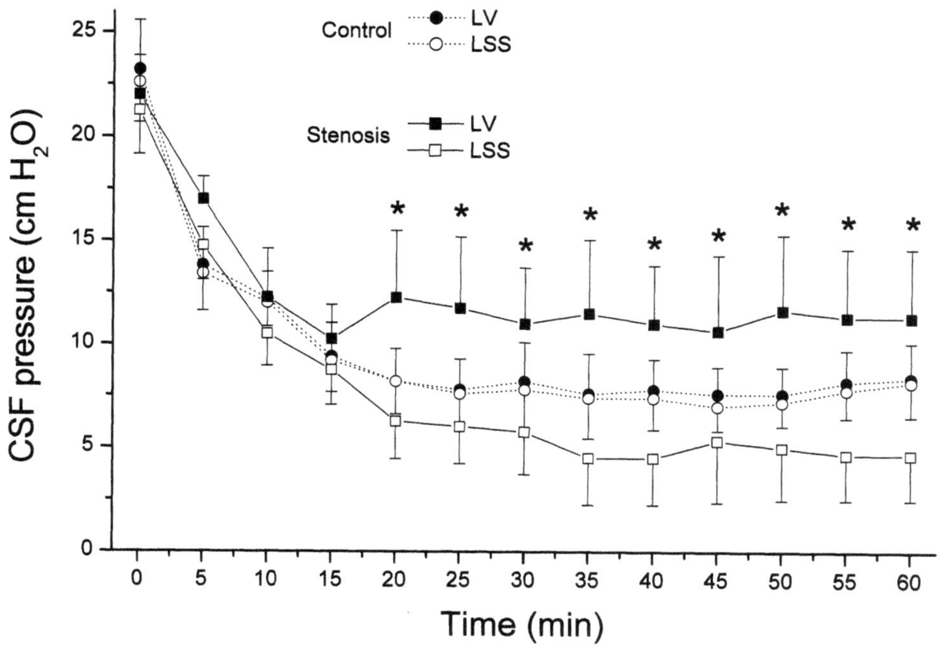

Fig. 2. The CSF pressures (cm H_2O) in lateral ventricle (*LV*; ●) and lumbar subarachnoid space (*LSS*; ○) in control cats without cervical stenosis (Control; n = 5), and CSF pressures in LV (■) and LSS (□) in animals with cervical stenosis (Stenosis; n = 5). Immediately before these pressure recordings the mock CSF was infused (52 µl/min) into right lateral ventricle, mannitol (20%; 0.5 g/kg) injected over last 3 min of intraventricular infusion, and thereafter cervical stenosis executed (Stenosis). Significant statistical difference between LV and LSS CSF pressures in animals with cervical stenosis is indicated (*$p < 0.05$)

Discussion

The large fall of CSF pressures in both groups of animals during first 15 min after intraventricular infusion of mock CSF and mannitol application (Fig. 2) can be ascribed to increased absorption of CSF under high CSF pressure and osmotic withdrawal of water into blood capillaries due to increased plasma osmolarity of about 10 mosm/L (see Results). After 15 min in animals with cervical stenosis a separation of ventricular and lumbar CSF pressure is manifested until the end of experiment, the ventricular CSF pressure being significantly higher than lumbar CSF pressure ($p < 0.05$). At the same time in control animals without cervical stenosis the ventricular and lumbar CSF pressures were equal what should be expected due to free communication of these two compartments.

Exchange of water and dissolved substances between the CSF and interstitial fluid is free process [12]. After intravenous application of hyperosmolar mannitol the hydrostatic pressure in interstitial fluid decreases what is followed by decrease of CSF pressure [11] indicating the absorption of water from interstitium (and CSF) into blood capillaries. We assume that replacement of water in interstitium by water coming from CSF is reason why a decrease of percentage of interstitial water was not observed in early period after mannitol application [3, 9]. During mannitol action the osmolarity of CSF is significantly increased [2] primarily due to retention of electrolytes (Na, Cl and K). Since permeability of cerebral capillaries to electrolytes is very poor, the increase of their concentration and osmotic pressure in CSF and interstitium should limit the intravascular osmotic effect of mannitol on interstitial and CSF water volume.

In our model of cervical stenosis the transmission of CSF pressure and volume is blocked between cranial and spinal CSF so that the effect of mannitol on these two CSF compartments can be studied separately. Our observation in animals with cervical stenosis that lumbar CSF is significantly lower than ventricular CSF pressure (Fig. 2) requires explanation. It was shown that mannitol increases cerebral blood volume and flow [8] what should oppose to intraventricular CSF pressure decrease. Furthermore, intracranial dura with adjacent arachnoidea is tightly apposed to cranial bones preventing the collapse of subarachnoid space and opposing to absorption of intracranial CSF water volume during osmotic action of mannitol, what should not contribute to fall of ventricular pressure.

On the other hand, the spinal dura with adjacent arachnoidea is not fixed to vertebrae so that subarachnoid space filled with CSF can easily change volume under different conditions [6, 10]. So we assume that during mannitol osmotic action water from interstitium of spinal cord and spinal CSF is predominantly absorbed in spinal capillaries due to high compliance of spinal dura. The importance of compliance of spinal dura in regulation of CSF pressure was previously stressed [4, 5].

The observed changes of CSF pressures in spinal and intracranial space indicate that spinal subarachnoid space contributes a great deal to overall fall of CSF pressure and volume in the early period after mannitol application probably due to high compliance of the spinal dura. We suppose that our observation is missing point which enables interpretation of main effects of mannitol on ICP, cerebral blood volume and flow, brain water and CSF.

Acknowledgments

This work has been supported by the Ministry of Science and Technology, Croatia.

References

1. Aurer LM, Haselberger K (1987) Effect of intravenous mannitol on cat pial arteries and veins during normal and elevated intracranial pressure. Neurosurgery 21: 142–146
2. Donato T, Shapira Y, Artru A, Powers K (1994) Effect of mannitol on cerebrospinal fluid dynamics and brain tissue edema. Anesth Analg 78: 58–66
3. Hartwell RC, Sutton LN (1993) Mannitol, intracranial pressure, and vasogenic edema. Neurosurgery 32: 444–450
4. Loeffgren J, Zwetnow NN (1973) Cranial and spinal components of the cerebrospinal fluid pressure-volume curve. Acta Neurol Scand 49: 575–585
5. Marmarou A, Shulman K, LaMorgese J (1975) Compartmental analysis of compliance and outflow resistance of the cerebrospinal fluid system. J Neurosurg 43: 523–534
6. Martins AN, Wiley JK, Myers P (1972) Dynamics of the cerebrospinal fluid and the spinal dura mater. J Neurol Neurosurg Psychiatr 35: 468–473
7. Muizelaar JP, Wei EP, Knotos HA, Becker DP (1983) Mannitol causes compensatory cerebral vasoconstriction and vasodilatation in response to blood viscosity changes. J Neurosurg 59: 822–828
8. Ravussin P, Archer DP, Tyler JL, Meyer E, Abou-Madi M, Diksic M, Yamamoto L, Trop D (1986) Effects of rapid mannitol infusion on cerebral blood volume. A positron emission tomographic study in dogs and man. J Neurosurg 64: 104–113
9. Takagi H, Saitoh T, Kitahara T, Morii S, Ohwada T, Yada K (1983) The mechanism of ICP reducing effect of mannitol. In: Ishii S, Nagai H, Brock M (eds) Intracranial pressure V. Berlin, Springer, pp 729–733
10. Tunturi AR (1977) Elasticity of the spinal cord dura in the dog. J Neurosurg 47: 391–396
11. Wiig H, Reed RK (1983) Rat brain interstitial fluid pressure measured with micropipettes. Am J Physiol 244: H239–H246
12. Zmajević M, Klarica M, Varda R, Kudelić N, Bulat M (2002) Elimination of phenolsulfophthalein from the cerebrospinal fluid via capillaries in central nervous system in cats by active transport. Neurosci Lett 321: 123–125

Correspondence: Marijan Klarica, Department of Pharmacology, School of Medicine, University of Zagreb, POB 916, Salata 11, HR-10 000, Zagreb, Croatia. e-mail: mklarica@hiim.hr

Acta Neurochir (2005) [Suppl] 95: 411–414

Cell column chromatography: a new research tool to quantify cerebral cell volume changes following chemically-induced anoxia/re-oxygenation

F. Xiao[1], S. Pardue[1], T. Nash[1], T. C. Arnold[1], J. S. Alexander[2], D. L. Carden[1], R. Turnage[3], A. Jawahar[4], and **S. A. Conrad[1]**

[1] Department of Emergency Medicine, Louisiana State University Health Science Center, Shreveport, Louisiana, USA
[2] Department of Physiology, Louisiana State University Health Science Center, Shreveport, Louisiana, USA
[3] Department of Surgery, Louisiana State University Health Science Center, Shreveport, Louisiana, USA
[4] Department of Neurosurgery, Louisiana State University Health Science Center, Shreveport, Louisiana, USA

Summary

Objective. The roles of individual types of cerebral cells in contributing to brain edema are undefined. The objective of this study was to determine the role of cerebral cell-column chromatography in quantifying cell volumes of individual cerebral cell lines, under chemically-induced anoxia/re-oxygenation (A/R).

Methods. Cerebral endothelial cells (4 experiments) or type II astrocytes (4 experiments) were cultured to confluence on microcarrier beads. A chromatographic cell-column of 1.5 cm height was filled with non-treated cell-covered beads. The column was perfused at 1 ml/min with a balanced perfusate for one hour (Baseline). The perfusate was then switched to that containing 5 mM thioglycolic acid for one hour (Anoxia). Then the column was perfused with the normal perfusate for another two hours (Re-oxygenation). The total free space in the column, reversely reflecting cell volumes, was determined by averaged transit time (TTa) of a non-permeable flow tracer blue dextran. Decreased TTa means that cells swell, and vice versa.

Results. TTa in endothelial cell columns increased with a peak at 60 minutes of re-oxygenation. TTa in astrocyte columns decreased with a nadir at 30 minutes of re-oxygenation.

Conclusion. Cell column chromatography can be used to determine the cerebral cell volume changes following chemically-induced anoxia/re-oxygenation.

Keywords: Brain edema; cell column chromatography; cell volume; anoxia; re-oxygenation; cytotoxic edema; vasogenic edema.

Introduction

The pathophysiological mechanisms of brain damage caused by cardiac arrest (CA) and resuscitation are complicated and multifactorial [20], including the primary insult of CA (complete temporary *global* brain ischemia) and secondary inflammatory responses during and after resuscitation, also known as the post-resuscitation syndrome. A characteristic feature of this syndrome is brain edema [24]. Brain edema formation correlates with morphological neuronal injury, including endothelial pinocytosis and opening of the endothelial cell-to-cell tight junctions [6, 7]. Diffuse brain edema formation has been documented in CA patients with CT and MRI scanning and has been shown to predict a poor neurological outcome [8, 16]. Increased ICP has also been reported in CA patients [14, 16]. We have recently reported that experimental CA of eight-minute asphyxiation results in brain edema formation following one hour of restoration of spontaneous circulation (ROSC) [25–27]. The mechanisms underlying CA-elicited brain edema are not clear, but increasing evidence indicates an association of brain edema formation with N-methyl-D-aspartate (NMDA) receptor activation and aquaporin 4 (AQP4), a brain water-selective transport channel [24, 26, 28]. Furthermore, the contributions of each type of cerebral cells to CA-elicited brain edema are undefined, partly due to the complexity of in vivo experiments and the paucity of in vitro systems that accurately measure cerebral cell water content and volume. Cell column chromatography is an in vitro technique that has been successfully used to determine alterations in endothelial cell permeability [9, 10, 18]. The objective of this study was to determine if a cell column chromatography technique, modified from that in our previous studies, can be used to quantify cerebral cell volumes during chemically-induced anoxia/re-oxygenation (A/R) [10, 18].

Methods and materials

Microcarrier bead cultures

Cerebral endothelial cells (4 experiments) or type II astrocytes (4 experiments) were seeded on Cytodex-3 microcarrier beads (Pharmacia, Uppsala, Sweden) at a density of 2×10^4 cells per cm^2. Microcarrier cultures were stirred at 60 rpm and 50% of the medium exchanged 3 times a week, as previously described [10, 18]. Cultures were used for these assays after confluence occurred.

Cell-column methods

Chromatographic cell-columns were made from water-jacketed glass columns (0.65 cm diameter; Rainin, Emeryville, CA). Non-treated cell-covered beads were poured into a cell column to a height of approximately 1.5 cm (approximately 1.5×10^8 cells). The column was equilibrated and perfused with HBSS + 15 mM HEPES (pH 7.4, Sigma). Perfusion through the column was maintained by a peristaltic pump (Minipuls II; Gilson, Middleton, WI) at 1 ml/min (chosen to approximate the gravity flow rate). After equilibrating columns in this buffer for one hour as baseline, the column perfusate was then switched to HBSS + HEPES containing 5 mM thioglycolic acid for one hour of anoxia. After one hour of thioglycollate-induced anoxia, the perfusate was switched back to normal perfusate and the column was perfused for another two hours of re-oxygenation. TTa was determined at baseline, 30 and 60 minutes of anoxia, and 30, 60, 90, and 120 minutes of re-oxygenation.

A/R injury

Perfusing the cell column with thioglycollate of 5 mM for 60 minutes was used to create anoxic injury since we observe in our previous studies that, at this concentration, after 30 minute treatment there was only 7% of oxygen tension remained in the media, and after 60 minute treatment there were no changes in viability of cells, determined by the trypan blue exclusion test [18]. Re-oxygenation state was established by washing out thioglycollate-containing perfusate with thioglycollate-free perfusate.

Cell volume determination

Changes in cell volume were indirectly inferred from the average transit time (TTa) of blue dextran through the cell columns, at baseline, 30 and 60 minutes of anoxia, 30, 60, 90, and 120 minutes of re-oxygenation. Blue dextran with a molecular weight of 2×10^6 cannot penetrate the monolayer through the cell membrane or cell-cell junctions and follows the fluid phase, i.e., it is a flow tracer. To determine TTa at each time point, a bolus of non-permeable flow tracer blue dextran at a concentration of 10 mg/ml was applied by a rotary injection valve (Rainin) using a 50 µl loop. 100 µl of outflow fluid were collected every second after dye injection until the total injected dye was recovered and the optical absorbance (OD) of each sample at 620 nm was determined using a 96-well plate reader (Titertek Multiskan MCC/340, Vienna, VA). According to the formula, transit time (TT) = cross sectional perfusion area (free space)/flow rate, since flow rate was kept constant in these experiments, any increase in cell volume reflected by decreased cell column free space, was observed as an increase in fluid velocity through cell columns, resulting in more rapid transit times, i.e., decreased TT. Average TT (TT_a) of the complete recovery of the impermeable tracer blue dextran in each experiment was determined using the formula, $TT_a = \Sigma T_{(n)} \times OD_{(n)}/\Sigma OD_{(n)}$, where $T_{(n)}$ = the time in seconds at each n time point after dye injection; and $OD_{(n)}$ = OD of the recovered blue dextran at each n time point after dye injection. This technique permits extremely rapid and sensitive analysis of cell volume changes.

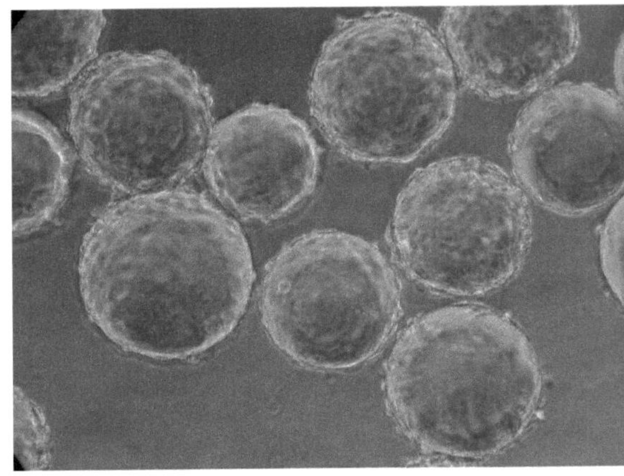

Fig. 1. A representative light microscopic picture of type II astrocyte-covering microcarrier beads

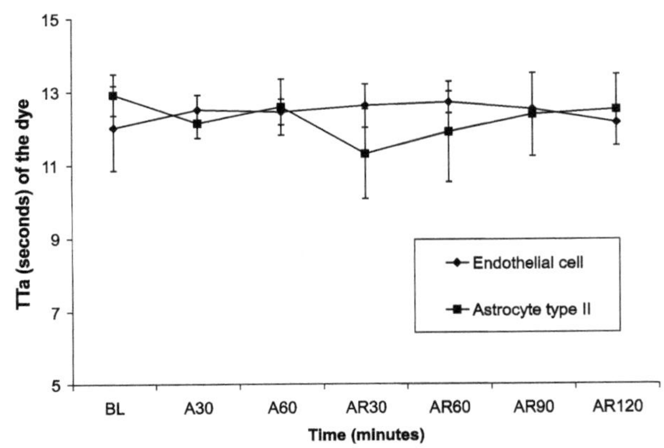

Fig. 2. Brain AQP4 expression following CA treated with ifenprodil. TTa increased in columns containing cerebral endothelial cells with a peak at 60 minutes of re-oxygenation. TTa decreased in columns containing Type II astrocytes with a nadir at 30 minutes of re-oxygenation. *BL* Baseline, *A* anoxia, *R* re-oxygenation

Results

Figure 1 showed a representative light microscopic picture of type II astrocytes-covered microcarrier beads. Although there were no statistical differences in TTa at any time point compared to baseline measurements, the following trends were observed (Fig. 2). TTa increased after anoxia in columns containing cerebral endothelial cells with a peak at 60 minutes of re-oxygenation. TTa decreased after anoxia in columns containing Type II astrocytes with a nadir at 30 minutes of re-oxygenation.

Discussion

Brain edema, occurs in CA patients, detected with CT or MRI scanning, and correlates to brain histology and predicts a poor neurological outcome [4, 8, 16]. If patients survive the time of maximal brain edema, chances for recovery improve, with or without neurological deficits [19]. We recently documented in several studies that experimental CA results in brain edema as determined by brain wet-to-dry weight ratio following one hour of ROSC [25–28]. The reduction in brain edema by the neuroprotective agents correlates with the reduction in infarct volume following focal cerebral ischemia [22]. The mechanisms underlying brain edema elicited by CA are unclear. By defining the mechanisms underlying CA-elicited brain edema, effective new therapeutic strategies to preserve cellular, tissue and neurological function following CA may become feasible.

Brain edema can be localized either intracellular (cytotoxic edema, cerebral cell swelling) or interstitially (vasogenic edema) [12]. Because of the dominant effect of hypoxia on cells, cerebral ischemia results primarily in cytotoxic edema. After reperfusion, vasogenic edema plays a critical role in brain edema formation. Vasogenic brain edema is a consequence of the damage to the blood-brain barrier (BBB). BBB shields brain parenchyma from plasma constituents, including ions and large molecules. Cerebral capillaries are the major sites of the BBB [3]. The capillary complex is composed of the lining endothelial cells, the surrounding pericytes, the basal lamina, and the astrocytic end-feet near the vessels. Tight junctions formed between the endothelial cells are major sites for restriction of molecular trafficking for molecules that are hydrophobic, large, or charged. Vasogenic edema is important because of the increased risk for hemorrhage from the damaged vessel and the excess fluid [19]. There has been increasing evidence that a considerable overlap exists between these two types of brain edema, especially in ischemic brain edema [11, 13]. At the present time, it is difficult to differentiate cytotoxic from vasogenic brain edema in any clinical setting. Nevertheless, brain edema, probably with both cytotoxic and vasogenic components in many if not all clinical scenarios [11, 13], has been documented to be well related to severity of brain injury and to be a major predictor of neurological outcome in patients of CA [8, 16], stroke [5, 23], traumatic brain injury [1, 15, 17], acute liver failure [2], and diabetic ketoacidosis [21]. Furthermore, the contributions of each type of cerebral cells, including endothelial cells, astrocytes, and neurons, to brain edema are not known. This line of the studies is very limited in vivo due to a lack of a technique that can reproducibly measure cerebral cell volume.

In this study, we have described a new in vitro approach to measure cerebral cell volume. The method was modified from previously reported studies in that cell column chromatography was used to determine endothelial cell monolayer permeability [9, 10, 18]. The method combines aspects of cell culture on microcarrier beads, column chromatography, and flow indicator optical analysis. The monolayer formed in this culture system appear to be confluent and provide stable and reproducible characteristics of cell volume changes for at least four hours of column perfusion, allowing serial measurements on the same population of cells subjected to several treatments. The use of flow tracer blue dextran, which is non-permeable to monolayer cells covering microcarrier beads due to its large molecular weight of 2×10^6, allows us to determine the intra-column free space, reflecting total cell volume.

We found in this study that type II astrocytes and endothelial cells behave differently under a condition of chemically-induced anoxia/re-oxygenation. Astrocytes swell following anoxia with a peak of swelling at 30 minutes of re-oxygenation, and endothelial cells shrink following anoxia with a peak of shrinking at 60 minutes of re-oxygenation. Due to facts that the astrocyte is one dominant type of glial cells in brain and play roles in providing structural support and modulating the micro-environment for neurons, regulating neuronal growth factors and neurogenesis, and maintaining BBB, studies on astrocyte swelling will shine a light on the mechanisms responsible for both cytotoxic and vasogenic brain edema. The finding from this study that cerebral endothelial cells shrink following chemically-induced anoxia/re-oxygenation suggests that inter-endothelial tight junction might be physically opened, resulting in vasogenic brain edema. While BBB is formed by tight junctions between endothelial cells, the disruption of relationship of end-feet to the endothelial cells leads to disruption of BBB and subsequent leakage.

In in vitro studies of hypoxia, a hypoxia chamber with an anoxic gas mixture has been widely used. However, this system needs a long incubation time in the medium to produce an anoxic condition. In addition, it is very difficult to handle the cells in anoxic chambers. The current and previous studies suggest that thioglycollate-induced anoxia/re-oxygenation is a useful model for investigating the effect of ischemia/reperfusion injury on cerebral cell volume. Because thioglycollate produces a rapid, nontoxic, and controllable hypoxic environment, this in vitro model of hypoxia/anoxia may be used to study the effects of different levels of hypoxia on specific cells [18].

In summary, this study suggests that cell column chromatography may be a useful technique to determine cerebral parenchymal cell (type II astrocytes) and vascular cell (endothelial) volume changes following chemically-induced anoxia/re-oxygenation. Cerebral type II astrocyte swelling and vascular cell contraction are evident using cell column chromatography following chemical-induced anoxia/re-oxygenation.

References

1. Aldrich EF, Eisenberg HM, Saydjari C, Luerssen TG, Foulkes MA, Jane JA, Marshall LF *et al* (1992) Diffuse brain swelling in severely head-injured children. A report from the NIH Traumatic Coma Data Bank. J Neurosurg 76: 450–454
2. Alper G, Jarjour IT, Reyes JD, Towbin RB, Hirsch WL, Bergman I (1998) Outcome of children with cerebral edema caused by fulminant hepatic failure. Pediatr Neurol 18: 299–304
3. Brightman MW, Reese TS (1969) Junctions between intimately apposed cell membranes in the vertebrate brain. J Cell Biol 40: 648–677
4. Clasen RA, Huckman MS, Von Roenn KA, Pandolfi S, Laing I, Lobick JJ (1981) A correlative study of computed tomography and histology in human and experimental vasogenic cerebral edema. J Comput Assist Tomogr 5: 313–327
5. Davalos A, Toni D, Iweins F, Lesaffre E, Bastianello S, Castillo J (1999) Neurological deterioration in acute ischemic stroke: potential predictors and associated factors in the European cooperative acute stroke study (ECASS) I. Stroke 30: 2631–2636
6. Dietrich WD, Busto R, Halley M, Valdes I (1990) The importance of brain temperature in alterations of the blood barrier following cerebral ischemia. J Neuropathol Exp Neurol 49: 486–497
7. Dietrich WD, Halley M, Valdes I, Busto R (1991) Interrelationships between increased vascular permeability and acute neuronal damage following temperature-controlled brain ischemia in rats. Acta Neuropathol 81: 615–625
8. Fujioka M, Okuchi K, Sakaki T, Hiramatsu K, Miyamoto S, Iwasaki S (1994) Specific changes in human brain following reperfusion after cardiac arrest. Stroke 25: 2091–2095
9. Haselton FR, Mueller SN, Howell RE, Levine EM, Fishman AP (1989) Chromatographic demonstration of reversible changes in endothelial permeability. J Appl Physiol 67: 2032–2048
10. Haselton FR, Woodall JH, Alexander JS (1996) Neutrophil-endothelial interactions in a cell-column model of the microvasculature: effects of fMLP. Microcirculation 3: 329–342
11. Joo F (1987) A unifying concept on the pathogenesis of brain oedemas. Neuropathol Appl Neurobiol 13: 161–176
12. Klatzo I (1967) Neuropathological aspects of brain edema. J Neuropathol Exp Neurol 26: 1–14
13. Klatzo I (1987) Pathophysiological aspects of brain edema. Acta Neuropathol (Berl) 72: 236–239
14. Laake JH, Haug FM, Wieloch T, Ottersen OP (1999) A simple in vitro model of ischemia based on hippocampal slice cultures and propidium iodide fluorescence. Brain Res Brain Res Protoc 4: 173–184
15. McCarron MO, Hoffmann KL, DeLong DM, Gray L, Saunders AM, Alberts MJ (1999) Intracerebral hemorrhage outcome: apolipoprotein E genotype, hematoma, and edema volumes. Neurology 53: 2176–2179
16. Morimoto Y, Kemmotsu O, Kitami K, Matsubara I, Tedo I (1993) Acute brain swelling after out-of-hospital cardiac arrest: pathogenesis and outcome. Crit Care Med 21: 104–110
17. Muttaqin Z, Uozumi T, Kuwabara S, Arita K, Kurisu K, Ohba S, Kohno H *et al* (1993) Hyperaemia prior to acute cerebral swelling in severe head injuries: the role of transcranial Doppler monitoring. Acta Neurochir (Wien) 123: 76–81
18. Park JH, Okayama N, Gute D, Krsmanovic A, Battarbee H, Alexander JS (1999) Hypoxia/aglycemia increases endothelial permeability: role of second messengers and cytoskeleton. Am J Physiol 277: C1066–C1074
19. Rosenberg GA (1999) Ischemic brain edema. Prog Cardiovasc Dis 42: 209–216
20. Safar P (2000) On the future of reanimatology. Acad Emerg Med 7: 75–89
21. Scibilia J, Finegold D, Dorman J, Becker D, Drash A (1986) Why do children with diabetes die? Acta Endocrinol (Copenh) [Suppl] 279: 326–333
22. Takano K, Tatlisumak T, Formato JE, Carano RA, Bergmann AG, Pullan LM, Bare TM, Sotak CH, Fisher M (1997) Glycine site antagonist attenuates infarct size in experimental focal ischemia. Postmortem and diffusion mapping studies. Stroke 28: 1255–1262
23. Wijdicks EF, Diringer MN (1998) Middle cerebral artery territory infarction and early brain swelling: progression and effect of age on outcome. Mayo Clin Proc 73: 829–836
24. Xiao F (2002) Bench to bedside: brain edema and cerebral resuscitation: the present and future. Acad Emerg Med 9: 933–946
25. Xiao F, Rodrignez J, Arnold TC, Zhang S. Ferrara D, Ewing J, Alexander JS, Carden DL, Conrad SA (2004) Near-infrared spectroscopy: a tool to monitor cerebral hemodynamic and metabolic changes after cardiac arrest in rats. Resuscitation 63: 213–220
26. Xiao F, Pardue S, Arnold T, Carden D, Alexander JS, Monroe J, Sharp CD, Turnage R, Conrad S (2004) Effect of ifenprodil, a polyamine site NMDA receptor antagonist, on brain edema formation following asphyxial cardiac arrest in rats. Resuscitation 61: 209–219
27. Xiao F, Zhang Z, Arnold TC, Alexanderl JS, Huang J, Carden DL, Conrad SA (2002) Mild hypothermia induced before cardiac arrest reduces brain edema formation in rats. Academic Emerg Med 9: 105–114
28. Xiao F, Arnold TC, Zhang S, Brown C, Alexander JS, Carden DL, Conrad SA (2004) Cerebral cortical aquaporin-4 expression in brain edema following cardiac arrests in rats. Academic Emerg Med 11: 1001–1007

Correspondence: Feng Xiao, MD, Department of Emergency Medicine, Louisiana State University Health Sciences Center in Shreveport, 1501 Kings Highway, Shreveport, LA 71130, USA. e-mail: fxiao@lsuhsc.edu

Acta Neurochir (2005) [Suppl] 95: 415–419

Ifenprodil treatment is associated with a down-regulation of brain aquaporin 4 following cardiac arrest in rats

F. Xiao[1], **S. Pardue**[1], **T. C. Arnold**[1], **J. Monroe**[1], **J. S. Alexander**[2], **D. L. Carden**[1], **R. Turnage**[3], and **S. A. Conrad**[1]

[1] Department of Emergency Medicine, Louisiana State University Health Science Center, Shreveport, Louisiana, USA
[2] Department of Physiology, Louisiana State University Health Science Center, Shreveport, Louisiana, USA
[3] Department of Surgery, Louisiana State University Health Science Center, Shreveport, Louisiana, USA

Summary

Objectives. Ifenprodil, a NMDA receptor polyamine site antagonist, reduces experimental cardiac arrest (CA)-elicited brain edema, which is associated with an up-regulation of aquaporin 4 (AQP4), a brain water-selective channel. However, the interacting roles of NMDA receptors and AQP4 in CA-elicited brain edema are unknown. The objective of this study was to test our hypothesis that ifenprodil treatment is associated with a down-regulation of brain AQP4.

Methods. Twenty-five rats were assigned to normal controls (group 1, $n = 6$) or subjected to eight minutes of asphyxial CA treated with placebo (group 2, $n = 9$) or ifenprodil (group 3, $n = 10$). Ifenprodil at 10 mg/kg or normal saline of equal volume was given intraperitoneally, 5 minutes before CA. The density of AQP4 protein and actin bands was scanned and expressed as the ratios of the optical density of AQP4 relative to that of actin. The ANOVA analysis was used to compare the group differences.

Results. The ratios of the optical density of AQP4 to that of actin were 0.88 ± 0.06 in group 1, 1.11 ± 0.08 in group 2 ($p < 0.05$ vs. group 1), and 0.78 ± 0.04 in group 3 ($p < 0.01$ vs. group 2; NS vs. group 1).

Conclusion. Ifenprodil given before CA is associated with a down-regulation of brain AQP4 in rats.

Keywords: Cardiac arrest; brain edema; ifenprodil; aquaporin 4; NMDA receptor; AQP4 phosphorylation.

Introduction

Sudden cardiac arrest (CA) claims approximately 1,200 lives daily in the United States [9, 23, 39, 40]. CPR attempts have achieved suboptimal results. Even when the restoration of spontaneous circulation (ROSC) is achieved by current treatment, about 40% of those who arrive at the hospital are never awakened, and about 30% of those who are awakened suffer permanent brain damage, including cognitive and motor deficits [18]. Recent data demonstrate that although the implementation of the survival of chain increases overall survival rate of out-of-hospital CA victims from 5–10% to 15–40%, 44% of the survivors suffer from some extents of neurological, psychosocial, or cognitive deficits [3, 4, 30]. Furthermore, survival in patients following in-hospital CA has not improved in last 40 years [2]. More interestingly, the outcomes after out-of-hospital CA in pediatric population have not changed in last 30 years [37, 38]. Poor neurological outcome after CA illustrates improved understanding of pathophysiology of *global* cerebral ischemia/reperfusion (I/R) elicited by CA.

Brain edema occurs in experimental CA and CA patients and predicts a poor neurological outcome [8, 14, 21, 33–36]. It has been documented that ifenprodil (α-4-hydroxyphenyl-β-methyl-4-benzyl-1-piperidineethanol), a polyamine modulatory site blocker of N-methyl-D-aspartate (NMDA), reduces brain edema following focal cerebral I/R and traumatic brain injury (TBI) [1, 6, 10]. We recently documented that ifenprodil reduces experimental CA-elicited brain edema and CA-elicited brain edema is associated with an up-regulation of AQP4 in brain [34, 36]. Aquaporin 4 (AQP4), a brain water-selective transport channel, contributes to brain edema formation elicited by focal cerebral I/R [20, 29]. However, the interacting roles of NMDA receptors and AQP4 and their sequence in the signaling pathway underlying CA-elicited brain edema remain unclear [32]. Therefore, the objective of this study was to test our hypothesis that ifenprodil treatment is associated with a down-regulation of brain AQP4. The role of ifenprodil on phosphorylation of AQP4 was also investigated.

Methods and materials

1. Study design

This study was approved by the Institutional Animal Care and Use Committee. Twenty-five Sprague-Dawley rats weighing 350–450 grams were assigned to normal controls (group 1, $n = 6$) or subjected to CA treated with placebo (group 2, $n = 9$) or ifenprodil (group 3, $n = 10$). Ifenprodil at 10 mg/kg or normal saline of equal volume was given intraperitoneally, 5 minutes before CA.

2. Animal protocol

Halothane was used for anesthesia induction (3%) and maintained as low as possible throughout the surgical procedures. The rat was instrumented with a tracheostomy for mechanical ventilation and femoral arterial/venous cannulations for blood pressure monitoring or medicine administration. Vecuronium bromide at 2 mg/kg prn was used to facilitate ventilation. After baseline measurements of blood pressure and rectal temperature, the rat was ventilated with intermittent positive pressure ventilation (IPPV) of 100% O_2 for one minute followed by room air for four minutes. Asphyxia was then induced by stopping IPPV and clamping the ventilator tubing for eight minutes. The animal was resuscitated by resuming ventilation, CPR, and epinephrine (0.01 mg/kg) and NaCO3 (2 mEq/kg) administered IV push without interrupting CPR. Resuscitation was continued until ROSC which was defined as a spontaneous systolic BP ≥ 60 mmHg in placebo-treated animals (group 2) and ≥ 40 mmHg in ifenprodil-treated animals (group 3). The discrepancy in defining ROSC between these two groups is from our observation that ifenprodil-treated animals maintained systolic BP < 60 mmHg for about two minutes after resuscitation. We also noticed that all ifenprodil-treated animals survived the experiment if they had a systolic BP ≥ 40 mmHg during resuscitation [34]. Rectal temperature was controlled at $37.5 \pm 0.5\,°C$ in all animals throughout the experiments. At one hour of ROSC, the experiment was terminated with 3% halothane and the brain was perfused transcardially with 100 ml of normal saline under a fixed infusion rate, until clear fluid drainage occured [34, 35]. The whole brain was removed and a part of the brain was frozen for brain AQP4 expression and phosphorylation determination.

3. AQP4 protein expression in brain using western blotting

Brain samples were homogenized for 30 s in 10 mM Tris (pH 7.4) and 2 mM EDTA, containing protease inhibitors (10 uM pepstatin, 10 uM leupeptin, and 0.1% phenylmethyl-sulfonyl fluoride) (Sigma). Whole cell homogenates (40 µg) were loaded on a 12% polyacrylamide gel. After electrophoresis, proteins were electrophoretically transferred for 2 h to PDVF membranes. The membranes were blocked with 20% milk in PBS (overnight) and incubated with primary antibodies (rabbit anti-rat AQP4, 1:1000, Chemicon, Temecula, CA) overnight. Membranes were rinsed in PBS with 2% milk, and horse-radish peroxidase conjugated secondary antibodies (anti-rabbit Ig, Vector Lab, CA) were bound for 1 h. Membranes were rinsed and visualization was preformed using a chemiluminesence kit (Pierce Biochemicals, Rockford, IL). Films were scanned and densotometric measurement was preformed using Imagej software (NIH). The optical density of AQP4 band was normalized to that of actin, which was determined using Ponseau S staining in the same gels. The level of AQP4 protein was expressed as the ratios of the optical density of AQP4 relative to that of actin.

4. Total phosphorylation of AQP4 using immunoprecipitation

To determine the effect of ifenprodil on AQP4 phosphorylation, immunoprecipitation was preliminarily performed to measure the levels of total phosphorylated AQP4. Brain samples challenged with ifenprodil were homogenized in 150 mM NaCl, 1% NP-40 and 50 mM Tris (pH 8.0) containing both protease inhibitors (as used in western blot) and a phosphatase inhibitor cocktail (Sigma). Samples were precleared for 1 h on ice using normal rabbit serum and precipitated with washed staph aureus (Sigma). AQP4 antibody (4 µg) was added for 1 hour (Chemicon, Temecula, CA). Protein A conjugated agarose (Sigma) was added and pelleted. The pellets were washed in the same buffer and re-suspended with Lammli sample buffer and loaded on a 12% agarose gel. After electrophoresis, the gels were fixed, washed, and stained for 2 h with Pro-Q diamond phosphoprotein stain for total phosphorylated AQP4 (Molecular Probes, Eugene, OR). To remove non-specific bindings of the stain, the gels were destained 3 h with 3 changes of destain solution (15% 1,2-propanediol, 50 mM sodium acetate, pH 4.0. Sigma). Gels were then scanned and densotometric measurement was performed using Imagej software (NIH) to determine the optical density of total phosphorylated AQP4.

5. Statistical analysis

All data obtained from this proposal were entered into SPSS 11.0 for Windows. One-way ANOVA with a post-hoc analysis of Scheffe's procedure was used for group comparisons. The results were presented as mean $\pm$ STD and significant differences were defined as $p < 0.05$.

Results

There were no statistical differences in animal body weight, surgical preparation time, and baseline measurements of blood pressure, temperature, ABG, K+, Na+, and Ca+. There were no statistical differences between groups in physiological variables at any time point of the experiments, including blood pressure, temperature, ABG, K+, Na+, and Ca+, except that mean blood pressure at 30 minutes of ROSC were 55 ± 5 mmHg in group 2 and 68 ± 17 mmHg in group 3 ($p < 0.05$).

Figure 1 showed a representative illustration of brain AQP4 protein expression determined by western blotting. Brain AQP4, expressed as the ratio of the optical density of AQP4 to actin, were 0.88 ± 0.06 in group 1, 1.11 ± 0.08 in group 2 ($p < 0.05$ vs. group 1); and 0.78 ± 0.04 in group 3 ($p < 0.01$ vs. group 2) (Fig. 2). Total phosphorylated AQP4, expressed as absolute staining intensity, were 1004 ± 149 in group 1 ($n = 3$), 3303 ± 794 in group 3 ($n = 3$); and 2201 ± 769 in group 3 ($n = 3$) (Fig. 3). However, there

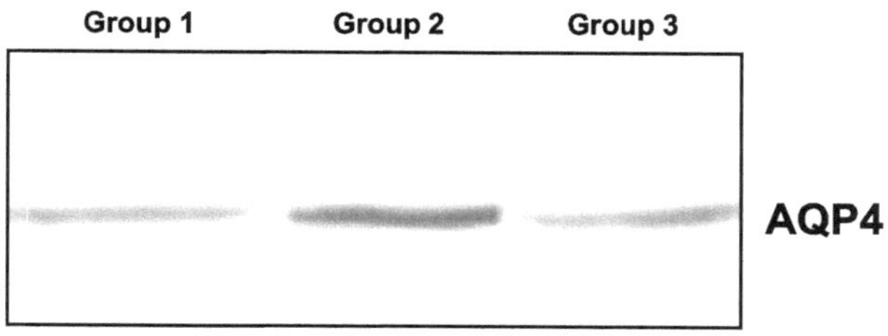

Fig. 1. A representative illustration of brain AQP4 expression determined by western blot. Group 1 – normal control animals; group 2 – cardiac arrest treated with placebo; and group 3 – cardiac arrest treated with ifenprodil

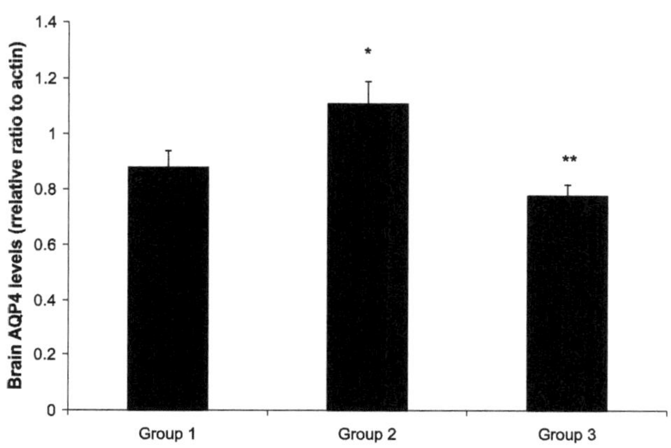

Fig. 2. Brain AQP4 expression following CA treated with ifenprodil. Brain AQP4, expressed as the ratio of the optical density of AQP4 to actin, were 0.88 ± 0.06 in control animals (group 1), 1.11 ± 0.08 in CA animals treated with placebo (group 2); and 0.78 ± 0.04 in CA animals treated with ifenprodil (group 3). * – $p < 0.05$ vs. group 1; ** – $p < 0.01$ vs. group 2

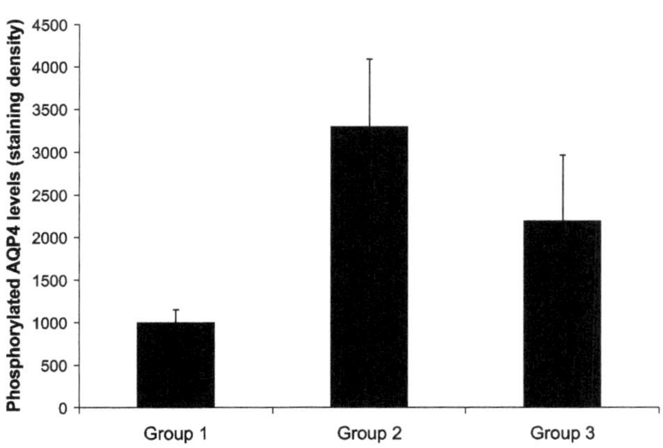

Fig. 3. Phosphorylated AQP4 following CA treated with ifenprodil. Total phosphorylated AQP4, expressed as absolute staining intensity, were 1004 ± 149 in control animals (group 1), 3303 ± 794 in CA animals treated with placebo (group 2); and 2201 ± 769 in CA animals treated with ifenprodil (group 3)

were no statistical differences between groups in total phosphorylated AQP4 due to their small sample sizes.

Discussion

Brain edema occurs in CA patients, detected with CT or MRI scanning, and correlates to brain histology and predicts a poor neurological outcome [5, 8, 14, 21]. Increased ICP has also been reported in CA patients [7, 21]. If patients survive the time of maximal brain edema, chances for recovery improve, with or without neurological deficits [26]. We recently documented in several studies that experimental CA results in brain edema as determined by brain wet-to-dry weight ratio following one hour of ROSC [33–36]. The reduction in brain edema by the neuroprotective agents correlates with the reduction in infarct volume following focal cerebral ischemia [28]. The mechanisms underlying brain edema elicited by CA are unclear. By defining the mechanisms underlying CA-elicited brain edema, effective new therapeutic strategies to preserve cellular, tissue and neurological function following CA may become feasible.

The fact that glutamate receptor antagonists are neuroprotective following *focal* cerebral I/R suggests that glutamatergic transmission is enhanced during I/R injury [24]. NMDA receptors are the most important and the most studied glutamate receptors. NMDA receptors include NMDA receptor-1 (NMDAR1) and NMDA receptor-2A-C (NMDAR2A-C). NMDAR1 is ligand-operated Ca^{2+} channels, whereas, NMDAR2 does not constitute a functional Ca^{2+} permeable channel. Although NMDA receptors are preponderant on neurons and glia, they are present on cerebral endothelial cells as well [15, 16]. Ifenprodil inhibiting NMDA receptor modulatory sites has been shown to attenuate brain edema and injury size in rats after focal cerebral ischemia [1, 6]. We recently reported that ifenprodil pretreatment reduces brain edema following CA in rats [34].

AQPs, a family of the membrane water-channel proteins, regulate transcellular water permeability. Ten AQPs have thus far been identified, each with a distinct distribution in the kidneys, lung, eye and brain. AQP4 is the most abundant AQP in CNS including astrocytes, ependymal cells, endothelial cells as well as neurons, suggesting that AQP4 plays an important role in the regulation of brain water exchange in the CNS [25, 31]. Manley *et al.* reported that AQP4-

deficient mice have an improved neurological outcome and decreased infarct size following *focal* cerebral ischemia [20]. Taniguchi *et al.* reported an up-regulation of AQP4 mRNA in the peri-infarcted cortex seven days after focal cerebral ischemia that correlated with the generation and resolution of brain edema monitored by MRI [29]. We recently found that CA-elicited brain edema is associated with an up-regulation of AQP4 protein expression in rat brain [36]. However, the role of AQP4 in CA-elicited brain edema remains unclear.

NMDA receptor mediates protein phosphorylation in studies using the hippocampal slice cultures, cultured neurons, and in vivo *focal* cerebral I/R [12, 13, 17, 19, 27]. The literatures support that AQP4 is regulated through a mechanism involving protein phosphorylation [11, 22] and AQP4 contains both tyrosine and serine residues (I59283, Swiss Protein Data Base). Due to the fact that AQP4 is widely distributed in brain and responsible for transcellular water transport [25, 31], we hypothesized that NMDA receptor causes CA-elicited brain edema through activating AQP4. In this study, we documented that ifenprodil pre-treatment inhibits asphyxial CA-induced over-expression of brain AQP4. Our study also suggests that ifenprodil pre-treatment inhibits CA-elicited AQP4 phosphorylation.

We conclude that ifenprodil pretreatment inhibits brain AQP4 over-expression elicited by asphyxial CA. This study also suggests that ifenprodil may play a role in regulating AQP4 phosphorylation. The results derived from this study provide indirect evidence of an interaction between the NMDA receptor and AQP4 following asphyxial CA. Further studies to define the mechanism by which NMDA receptor-mediated AQP4 expression modulates CA-induced brain edema are indicated.

References

1. Baskaya MK, Rao AM, Donaldson D, Prasad MR, Dempsey RJ (1997) Protective effects of ifenprodil on ischemic injury size, blood-brain barrier breakdown, and edema formation in focal cerebral ischemia. Neurosurgery 40: 364–370
2. Brindley PG, Markland DM, Mayers I, Kutsogiannis DJ (2002) Predictors of survival following in-hospital adult cardiopulmonary resuscitation. CMAJ 167: 343–348
3. Bunch TJ, White RD, Gersh BJ, Meverden RA, Hodge DO, Ballman KV, Hammill SC, Shen WK, Packer DL (2003) Long-term outcomes of out-of-hospital cardiac arrest after successful early defibrillation. N Engl J Med 348: 2626–2633
4. Bunch TJ, White RD, Khan AH, Packer DL (2004) Impact of age on long-term survival and quality of life following out-of-hospital cardiac arrest. Crit Care Med 32: 963–967
5. Clasen RA, Huckman MS, Von Roenn KA, Pandolfi S, Laing I, Lobick JJ (1981) A correlative study of computed tomography and histology in human and experimental vasogenic cerebral edema. J Comput Assist Tomogr 5: 313–327
6. Dogan A, Rao AM, Baskaya MK, Rao VL, Rastl J, Donaldson D, Dempsey RJ (1997) Effects of ifenprodil, a polyamine site NMDA receptor antagonist, on reperfusion injury after transient focal cerebral ischemia. J Neurosurg 87: 921–926
7. Dolinak D, Smith C, Graham DI (2000) Global hypoxia per se is an unusual cause of axonal injury. Acta Neuropathol (Berl) 100: 553–560
8. Fujioka M, Okuchi K, Sakaki T, Hiramatsu K, Miyamoto S, Iwasaki S (1994) Specific changes in human brain following reperfusion after cardiac arrest. Stroke 25: 2091–2095
9. Gillum RF (1989) Sudden coronary death in the United States: 1980–1985. Circulation 79: 756–765
10. Gotti B, Duverger D, Bertin J, Carter C, Dupont R, Frost J, *et al* (1988) Ifenprodil and SL 82.0715 as cerebral anti-ischemic agents. I. Evidence for efficacy in models of focal cerebral ischemia. J Pharmacol Exp Ther 247: 1211–1221
11. Han Z, Wax MB, Patil RV (1998) Regulation of aquaporin-4 water channels by phorbol ester-dependent protein phosphorylation. J Biol Chem 273: 6001–6004
12. Hardingham GE, Chawla S, Cruzalegui FH, Bading H (1999) Control of recruitment and transcription-activating function of CBP determines gene regulation by NMDA receptors and L-type calcium channels. Neuron 22: 789–798
13. Hu BR, Fux CM, Martone ME, Zivin JA, Ellisman MH (1999) Persistent phosphorylation of cyclic AMP responsive element-binding protein and activating transcription factor-2 transcription factors following transient cerebral ischemia in rat brain. Neuroscience 89: 437–452
14. Kelsen J, Obel A (2003) Fatal cerebral hypoxemia after cardiac arrest. NEJM 348: 817
15. Koenig H, Trout JJ, Goldstone AD, Lu CY (1992) Capillary NMDA receptors regulate blood-brain barrier function and breakdown. Brain Res 588: 297–303
16. Krizbai IA, Deli MA, Pestenacz A, Siklos L, Szabo CA, Andras I, Joo F (1998) Expression of glutamate receptors on cultured cerebral endothelial cells. J Neurosci Res 54: 814–819
17. Laake JH, Haug FM, Wieloch T, Ottersen OP (1999) A simple in vitro model of ischemia based on hippocampal slice cultures and propidium iodide fluorescence. Brain Res Brain Res Protoc 4: 173–184
18. Longstreth WT Jr, Inui TS, Cobb LA, Copass MK (1983) Neurologic recovery after out-of-hospital cardiac arrest. Ann Intern Med 985: 588–592
19. Mabuchi T, Kitagawa K, Kuwabara K, Takasawa K, Ohtsuki T, Xia Z, Storm D, Yanagihara T, Hori M, Matsumoto M (2001) Phosphorylation of cAMP response element-binding protein in hippocampal neurons as a protective response after exposure to glutamate in vitro and ischemia in vivo. J Neurosci 21: 9204–9213
20. Manley GT, Fujimura M, Ma T, Noshita N, Filiz F, Bollen AW, Chan P, Verkman AS (2000) Aquaporin-4 deletion in mice reduces brain edema after acute water intoxication and ischemic stroke. Nat Med 6: 159–163
21. Morimoto Y, Kemmotsu O, Kitami K, Matsubara I, Tedo I (1993) Acute brain swelling after out-of-hospital cardiac arrest: pathogenesis and outcome. Crit Care Med 21: 104–110
22. Nakahama K, Nagano M, Fujioka A, Shinoda K, Sasaki H (1999) Effect of TPA on aquaporin 4 mRNA expression in cultured rat astrocytes. Glia 25: 240–246

23. National Center for Health Statistics (1995) Advance report of final mortality statistics, 1995. Monthly Vital Statistics Report. 45: 82
24. Nellgard B, Wieloch T (1992) Postischemic blockade of AMPA but not NMDA receptors mitigates neuronal damage in the rat brain following transient severe cerebral ischemia. J Cereb Blood Flow Metab 12: 2–11
25. Nielsen S, Nagelhus EA, Amiry-Moghaddam M, Bourque C, Agre P, Ottersen OP (1997) Specialized membrane domains for water transport in glial cells: high-resolution immunogold cytochemistry of aquaporin-4 in rat brain. J Neurosci 17: 171–180
26. Rosenberg GA (1999) Ischemic brain edema. Prog Cardiovasc Dis 42: 209–216
27. Sala C, Rudolph-Correia S, Sheng M (2000) Developmentally regulated NMDA receptor-dependent dephosphorylation of cAMP response element-binding protein (CREB) in hippocampal neurons. J Neurosci 20: 3529–3536
28. Takano K, Tatlisumak T, Formato JE, Carano RA, Bergmann AG, Pullan LM, Bare TM, Sotak CH, Fisher M (1997) Glycine site antagonist attenuates infarct size in experimental focal ischemia. Postmortem and diffusion mapping studies. Stroke 28: 1255–1262
29. Taniguchi M, Yamashita T, Kumura E, Tamatani M, Kobayashi A, Yokawa T, Maruno M, Kato A, Ohnishi T, Tohyama M, Yoshimine T (2000) Induction of aquaporin-4 water channel mRNA after focal cerebral ischemia in rat. Brain Res Mol Brain Res 78: 131–137
30. van Alem AP, Waalewijn RA, Koster RW, de Vos R (2004) Assessment of quality of life and cognitive function after out-of-hospital cardiac arrest with successful resuscitation. Am J Cardiol 93: 131–135
31. Venero JL, Vizuete ML, Machado A, Cano J (2001) Aquaporins in the central nervous system. Prog Neurobiol 63: 321–336
32. Xiao F (2002) Bench to bedside: brain edema and cerebral resuscitation: the present and future. Acad Emerg Med 9: 933–946
33. Xiao F, Rodrignez J, Arnold TC, Zhang S. Ferrara D, Ewing J, Alexander JS, Carden DL, Conrad SA (2004) Near-infrared spectroscopy: a tool to monitor cerebral hemodynamic and metabolic changes after cardiac arrest in rats. Resuscitation 63: 213–220
34. Xiao F, Pardue S, Arnold T, Carden D, Alexander JS, Monroe J, Sharp CD, Turnage R, Conrad S (2004) Effect of ifenprodil, a polyamine site NMDA receptor antagonist, on brain edema formation following asphyxial cardiac arrest in rats. Resuscitation 61: 209–219
35. Xiao F, Zhang Z, Arnold TC, Alexander1 JS, Huang J, Carden DL, Conrad SA (2002) Mild hypothermia induced before cardiac arrest reduces brain edema formation in rats. Academic Emerg Med 9: 105–114
36. Xiao F, Arnold TC, Zhang S, Brown C, Alexander JS, Carden DL, Conrad SA (2004) Cerebral cortical aquaporin-4 expression in brain edema following cardiac arrests in rats. Academic Emerg Med 11: 1001–1007
37. Young KD, Gausche-Hill M, McClung CD, Lewis RJ (2004) A prospective, population-based study of the epidemiology and outcome of out-of-hospital pediatric cardiopulmonary arrest. Pediatrics 114: 157–164
38. Young KD, Keidel JS (1999) Pediatric cardiopulmonary resuscitation: a collective review. Ann Emerg Med 33: 195–205
39. Zheng ZJ, Croft JB, Giles WH, Mensah GA (2001) Sudden cardiac death in the United States, 1989 to 1998. Circulation 104: 2158–2163
40. Zheng ZJ, Croft JB, Giles WH (2002) MMWR 51(06): 123–126

Correspondence: Feng Xiao, Department of Emergency Medicine, Louisiana State University Health Sciences Center in Shreveport, 1501 Kings Highway, Shreveport, LA 71130, USA. e-mail: fxiao@lsuhsc.edu

Acta Neurochir (2005) [Suppl] 95: 421–424

Intracerebral hemorrhage induces edema and oxidative stress and alters N-methyl-D-aspartate receptor subunits expression

T. Nakamura[1,2], **R. F. Keep**[1], **Y. Hua**[1], **J. W. Park**[1], **T. Itano**[3], **S. Nagao**[2], **J. T. Hoff**[1], and **G. H. Xi**[1]

[1] Department of Neurosurgery, University of Michigan, Ann Arbor, MI, USA
[2] Department of Neurological Surgery, Kagawa University Faculty of Medicine, Kagawa, Japan
[3] Department of Neurobiology, Kagawa University Faculty of Medicine, Kagawa, Japan

Summary

Intracerebral hemorrhage (ICH) induces brain edema formation via a variety of mechanisms including toxicity due to thrombin and erythrocyte lysis. However, the roles of oxidative damage and excitotoxicity have not been fully elucidated and they are examined in this rat ICH study. Adult male Sprague-Dawley rats received an intracaudate injection of 100 μl autologous whole blood and 5 U of thrombin. Rats were sacrificed at 1 hour, 1 and 3 days, and then the brains processed using Western blotting to quantify N-methyl-D-aspartate receptor (NR) subunit expression. At 3 days, animals were also sacrificed for assessment of protein oxidation using Western blot analysis for dinitrophenyl (DNP) and brain water content. Compared to the contralateral side, ipsilateral basal ganglia NR1 and NR2A subunit expression transiently increased at 1 hour after ICH and thrombin injection. From 24 hours there was a marked down-regulation. At 3 days, marked edema and DNP up-regulation were observed in ICH and thrombin injection groups. The present NR expression up-regulation at 1 hour may reflect the acute cell response after ICH. The down-regulation of NR subunits and up-regulation of DNP may be associated with cell damage, towards which thrombin may contribute.

Keywords: Cerebral hemorrhage; N-methyl-D-aspartate receptor; thrombin; oxidative stress.

Introduction

N-methyl-D-aspartate receptor (NR) subunits play an important role in the mechanism of excitotoxic injury after ischemic and traumatic brain injury [7, 10]. However, the influence of hemorrhagic stroke on NRs is unclear. Intracerebral hemorrhage (ICH) induces perihematomal brain edema and causes neurological deficits. The mechanisms of brain injury after ICH are not fully umderstood but are different from ischemic stroke. One of several effects of thrombin, via activation of protease-activated receptor-1 (PAR-1), is potentiation of NR [3]. However, roles of excitotoxicity and oxidative damage have not been fully elucidated and they are examined in this rat ICH study. The current study examines whether ICH induces NR alteration in the rats. This study also examines whether thrombin may alter NR expression. Protein oxidation was examined using an anti-dinitrophenyl (DNP) antibody following ICH and thrombin injection models.

Materials and methods

Animal protocols were approved by the University of Michigan Committee on the Use and Care of Animals. Male Sprague-Dawley rats (Charles River Laboratories, Portage, MI) were used for all experiments. The animals were anesthetized with pentobarbital and the right femoral artery was catheterized to sample blood for intracerebral infusion. The rats were positioned in a stereotaxic frame and a 26-gauge needle was inserted stereotaxically into the right basal ganglia (0.2 mm anterior, 5.5 mm ventral, and 3.5 mm lateral to the bregma). Autologous whole blood (100 μl), rat thrombin (5 U/50 μl saline) or saline (100 μl) was infused at a rate of 10 μl/min with the use of a microinfusion pump.

Animals were anesthetized before undergoing intracardiac perfusion with saline. The brains were then removed and a 3 mm thick coronal brain slice was cut approximately 4 mm from the frontal pole. The slice was separated into ipsilateral and contralateral basal ganglia. Briefly, 50 μg proteins for each were separated by sodium dodecyl sulfate polyacrymide gel electrophoresis and transferred to a Hybond-C pure nitrocellulose membrane (Amersham, Piscataway, NJ). The membranes were blocked in nonfat milk. Membranes were probed using the primary antibody (mouse anti-NR1 antibody, rabbit anti-NR2A antibody, mouse anti-β actin antibody or mouse anti-DNP antibody) and the second antibody (BIO-RAD, Hercules, CA). The antigen-antibody complexes were visualized with a chemiluminescence system and exposed to film. The relative densities of bands were analyzed with the NIH Image.

Reverse transcription reaction was performed at 42 °C for 30 min and terminated at 99 °C after 5 min. Polymerase chain reaction was performed with reverse transcriptase reaction mixture. Oligonucleotide primer for rat NR1 was synthesized by Sigma Genosis (Woodlands, TX). The primer sequences for NR1 were 5′-

TCAGCGACGACCACGAGGGAG-3′ (sense primer) and 5′-TTGTAGATGCCCACTTGCACCA-3′ (antisense primer), generating a 603 bp PCR product. Rat GAPDH primers were used to amplify GAPDH mRNA, a housekeeping gene used as a control. Samples were subjected to 31 cycles (94 °C for 30 sec; 60 °C for 30 sec; and 72 °C for 1 min). PCR was analyzed by electrophoresis on 1% agarose gel. Gels were observed with ethidium bromide staining and ultraviolet transillumination. Photographs were taken with black and white film (Polaroid, Cambridge, MA). The relative densities of the bands were analyzed with the NIH Image.

Animals were reanesthetized and decapitated 3 days after ICH for brain water content measurements. A coronal brain slice (approximately 3 mm thick) 4 mm from the frontal pole was cut with a blade. The brain slice was dissected into the basal ganglia. The cerebellum also served as a control. A total of three samples from each brain were obtained, i.e. the bilateral basal ganglia and the cerebellum. Brain samples were immediately weighed on an electric analytical balance to obtain the wet weight. Brain samples were then dried at 100 °C for 24 hours to obtain the dry weight. The formula for calculation was as follows: (Wet Weight – Dry Weight)/Wet Weight.

All data in this study is presented as mean ± SD, and analyzed using Student's t-test or analysis of variance, followed by Scheffe's post hoc test for multiple comparisons. Significance levels are measured at $p < 0.05$.

Results

NR1 protein levels in the ipsilateral basal ganglia significantly increased at 1 hour after ICH ($p < 0.01$). At 1 and 3 days after ICH, however, the NR1 protein levels in the ipsilateral basal ganlgaia were significantly decreased compared to the contralateral basal ganglia ($p < 0.05$). Similarly, NR2 protein levels significantly increased at 1 hour after ICH ($p < 0.05$) and decreased at 1 and 3 days after ICH ($p < 0.05$, data not shown). NR1 levels at 1 hour after thrombin-infusion were significantly increased in the ipsilateral basal ganglia ($p < 0.05$). At 1 and 3 days after infusion of thrombin, there were significant decreases in NR1 protein levels in the ipsilateral basal ganglia ($p < 0.05$, Fig. 1B).

NR1 mRNA levels were significantly increased in the ipsilateral basal ganglia at 1 hour after ICH ($p < 0.05$). At 1 day and 3 days after ICH, however, the significant differences of the NR1 mRNA levels were not found in the ipsilateral basal ganglia compared to the contralateral basal ganglia (Fig. 1C).

The brain water content was significantly increased in the ipsilateral basal ganglia in the ICH group ($p < 0.01$) and thrombin injection group at 3 days after injection, compared with the saline-injected control (Fig. 2A). DNP upregulation was observed in the ipsilateral basal ganglia at 3 days after ICH ($p < 0.01$) and thrombin injection ($p < 0.01$) groups compared to the saline injection group (Fig. 2B).

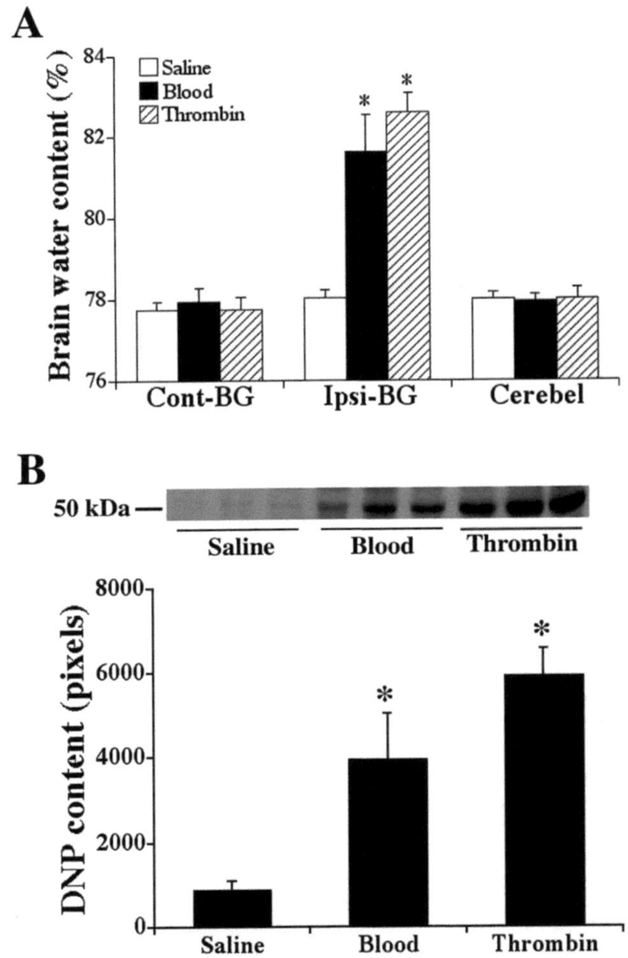

Fig. 1. (A) Bar graphs showing the brain water content 3 days after saline, blood or thrombin injection ($n = 5$ in each group). Cont-BG: contralateral basal ganglia; Ipsi-BG: ipsilateral basal ganglia; Cerebell: cerebellum. (B) Western blot analysis of DNP protein levels in the basal ganglia at 3 days after injection. Equal amounts of protein (50 µg) were used. Values are means ± SD. * indicates $p < 0.01$ compared with that saline injection group ($n = 3$ in each group)

Discussion

In the present study, we showed that protein levels of the NR1 and NR2A subunits were transiently increased at 1 hour and there was a marked down-regulation from 1 day after ICH. The mRNA level of NR1 was also increased at 1 hour after ICH, but there was no down-regulation at 1 and 3 days after ICH. In addition, intracerebral injection of thrombin resulted in similar results as ICH. DNP up-regulation and marked edema were observed at 3 days after ICH and thrombin injection.

NR expression up-regulation may reflect the acute cell response after ICH, which is, in part, activated by

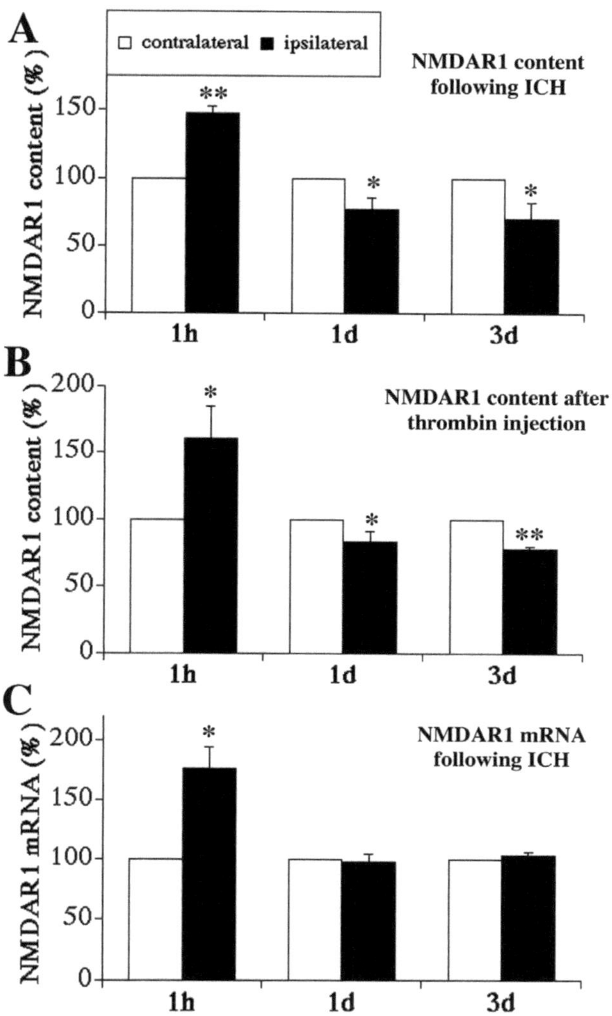

Fig. 2. (A) The graph showing the results of Western blot analysis of the time course of NMDAR1 (NR1) expression in contralateral and ipsilateral basal ganglia following ICH. (B) Western blot analysis of the time course of NR1 expression in contralateral and ipsilateral basal ganglia after thrombin injection. (C) Time course of NR1 gene expression in the contralateral and ipsilateral basal ganglia after ICH. Values are means ± SD. *: $p < 0.05$ and **: $p < 0.01$ compared with the contralateral group ($n = 3$ in each group)

glutamate release. Several studies have shown increase in NR mRNA expression following ischemic brain injury [4, 9]. The present study suggests that increases in NR subunit levels in acute phase were due to up-regulation of gene expression, like previous studies in ischemic model. Interestingly, NR began to decrease from 24 hours after ICH. Our results suggest that decreases in NR subunit levels were not due to down-regulation of gene expression. Kumar *et al.* also showed decreases in NR subunits were not induced by down-regulation of gene expression [6]. One possibility for decreased NR is enzyme degeneration. Transient activation of calpain-like protease has been shown in traumatic brain injury [5] and seizure models [1]. The role of proteases, including thrombin, and their receptors may be important in ICH. Evidence indicates that brain injury following ICH is, at least in part, due to the action of thrombin formed during clot formation. The possibility that thrombin may enter brain tissue after ICH raises the potential for activation of protease receptors with detrimental consequences. Gingrich *et al.* demonstrated the potentiation of NR function by serine protease thrombin [3]. From this evidence, thrombin-induced brain injury may be associated in part with NR. The alterations of NR subunit expressions following ICH may be associated with cell damage, to which thrombin may contribute.

Some reports have shown the relation of excitotoxicity and oxidative stress in other models [7, 10]. The interactions of reactive oxgen species (ROS) and glutamate receptors have also been shown [2]. These reports suggest glutamate receptor activation as a source of ROS and the role of ROS in accentuating excitotoxicity. Oxidative glutamate toxicity toward neurons lacking functional NRs could be a component of the excitotoxicity-initiated cell death pathway [8]. In the present study, the down-regulation of NR subunits and up-regulation of DNP have been observed. These regulations may be associated with edema and cell damage, to which thrombin may contribute.

Acknowledgments

This study was supported by grants NS-17760 (J.T.H.), NS-039866 and NS-047245 (G.X.) from the National Institutes of Health.

References

1. Bi X, Bi R, Baudry M (2000) Calpain-mediated truncation of glutamate ionotropic receptors. Methods for studying the effects of calpain activation in brain tissue. Methods Mol Biol 144: 203–217
2. Facchinetti F, Dawson VL, Dawson TM (1998) Free radicals as mediators of neuronal injury. Cell Mol Neurobiol 18: 667–682
3. Gingrich MB, Junge CE, Lyuboslavsky P, Traynelis SF (2000) Potentiation of NMDA receptor function by the serine protease thrombin. J Neurosci 20: 4582–4595
4. Heurteaux C, Lauritzen I, Widmann C, Lazdunski M (1994) Glutamate-induced over expression of NMDA receptor messenger RNAs and protein triggered by activation of AMPA/kainate receptors in rat hippocampus following forebrain ischemia. Brain Res 659: 67–74
5. Kampfl A, Posmantur RM, Zhao X, Schmutzhard E, Clifton

GL, Hayes RL (1997) Mechanisms of calpain proteolysis following traumatic brain injury: implications for pathology and therapy: implications for pathology and therapy: a review and update. J Neurotrauma 14: 121–134
6. Kumar A, Zou L, Yuan X, Long Y, Yang K (2002) N-Methyl-D-aspartate receptors: transient loss of NR1/NR2A/NR2B subunits after traumatic brain injury in a rodent model. J Neurosci Res 67: 781–786
7. Said SI, Pakbaz H, Berisha HI, Raza S (2000) NMDA receptor activation: critical role in oxidant tissue injury. Free Radic Biol Med 28: 1300–1302
8. Schubert D, Piasecki D (2001) Oxidative glutamate toxicity can be a component of the exicitotoxicity cascade. J Neurosci 21: 7455–7462
9. Small DL, Poulter MO, Buchan AM, Morley P (1997) Alteration in NMDA receptor subunit mRNA expression in vulnerable and resistant regions of in vitro ischemic rat hippocampal slices. Neurosci Lett 232: 87–90
10. Won MH, Kang T, Park S, Jeon G, Kim Y, Seo JH, Choi E, Chung M, Cho SS (2001) The alterations of N-Methyl-D-aspartate receptor expressions and oxidative DNA damage in the CA1 area at the early time after ischemia-reperfusion insult. Neurosci Lett 301: 139–142

Correspondence: Takehiro Nakamura, Department of Neurological Surgery, Kagawa University Faculty of Medicine, 1750-1 Ikenobe, Miki-cho, Kita-gun, Kagawa 761-0793, Japan. e-mail: tn_mius@yahoo.co.jp

Acta Neurochir (2005) [Suppl] 95: 425–427

Aging enhances intracerebral hemorrhage-induced brain injury in rats*

Y. Gong[1], G. H. Xi[1], R. F. Keep[1,2], J. T. Hoff[1], and Y. Hua[1]

[1] Department of Neurosurgery, University of Michigan, Ann Arbor, Michigan, USA
[2] Department of Physiology, University of Michigan, Ann Arbor, Michigan, USA

Summary

Age is an important factor affecting oxidative stress and plasticity after brain injury. The present study investigated the effects of aging on brain injury after intracerebral hemorrhage (ICH). Aging (18-month) and young (3-month) male Sprague-Dawley rats received an intracerebral infusion of 100-μl autologous blood. Age-related changes in brain edema and neurological deficits were examined and heat shock protein 27 (HSP27) and heat shock protein 32 (HSP32) levels were determined by Western blotting. Perihematomal brain swelling was more severe in aged rats compared to young rats at three days after ICH ($P < 0.05$). The behavioral tests used were forelimb placing test and forelimb use asymmetry test. There were more severe neurological deficits and a slower recovery in aged rats compared to those in young rats after ICH ($P < 0.05$). In addition, perihematomal HSP27 and HSP32 protein levels were higher ($p < 0.05$) in aged rats. In conclusion, ICH causes more severe brain swelling and neurological deficits in aged rats. Clarification of the mechanisms of brain injury after ICH in the aging brain should help develop new therapeutic strategies for hemorrhagic brain injury.

Keywords: Cerebral hemorrhage; aging; brain edema; neurological deficits.

Introduction

Intracerebral hemorrhage (ICH) is a common and often fatal stroke subtype and results in severe neurological deficits in survivors. Brain edema is an important component of brain injury after ICH, causing both acute herniation-related deaths and long-term neurological deficits [6]. ICH is primarily a disease of the elderly, but most current experimental ICH models have used young animals. Age is an important factor affecting brain injury after ischemic stroke in animals and humans [1], which may also affect brain injury after ICH. Recently, we have developed behavioral tests for rodent models of ICH [2], allowing assessment of the impact of age on ICH-induced brain damage. In addition, heat shock proteins are brain injury markers, some of which are induced after ICH [3]. In the present study, we examined the effects of aging on neurological deficits, brain edema formation and the levels of heat shock proteins after ICH.

Materials and methods

Animal preparation and intracerebral infusion

Animal use protocols were approved by the University of Michigan Committee on the Use and Care of Animals. Thirty-four young (3-month old) and thirty-four aged (18-month old) male Sprague-Dawley rats (Charles River Laboratories) were used in this study. The rats were anesthetized with pentobarbital (45 mg/kg i.p.). The right femoral artery was catheterized for continuous blood pressure monitoring and blood sampling. Blood was obtained from the catheter for analysis of blood pH, PaO_2, $PaCO_2$, hematocrit and blood glucose. Core temperature was maintained at 37 °C with use of a feedback-controlled heating pad. All rats received an injection of 100-μl autologous whole blood into the right caudate nucleus at a rate of 10-μL/min through a 26-gauge needle (coordinates: 0.2 mm anterior, 5.5 mm ventral, and 3.5 mm lateral to bregma) with the use of a microinfusion pump.

Experimental groups

These experiments were divided into three parts. All rats had an intracerebral infusion of 100-μL autologous whole blood. In the first part, 4 groups of aged and young rats ($n = 7$) were killed 1 or 3 days after blood injection for brain water content measurements. In the second part, aged and young rats ($n = 8$) underwent behavioral testing the day before ICH and at days 1, 3, 5, 7, 14, and 28 after ICH. In the third part, brains were sampled 1, 3, 7 and 28 days after ICH for heat shock protein analysis using Western blotting ($n = 3$ each time point).

* This study was supported by grants NS-17760 (J.T.H), NS-39866 and NS-47245 (G.X) from the National Institutes of Health and by grant 0435354Z from the American Heart Association (Y.H).

Brain swelling

Brain water content was determined by wet/dry weight method [6]. Brain swelling after ICH is usually examined by measuring % water content ((water content × 100)/wet weight). Contralateral brain and cerebellar water content, however, was found to be less in aged rats than in younger rats. Instead, therefore, we calculated tissue swelling. For each sample the ratio of wet weight to dry weight was determined and the % swelling was calculated as the % change in this ratio between the ipsilateral and contralateral side.

Behavioral tests

Two types of test [2] were employed: (A) Forelimb Placing Test. Forelimb placing was scored using a vibrissae-elicited forelimb-placing test. (B) Forelimb Limb-Use Asymmetry Test.

Western blot analysis

Western blot analysis was performed at day 1, 3, 7, 28 after ICH using as previously described methods [8]. In brief, protein concentration was determined by Bio-Rad protein assay kit. Samples (50 µg protein) were run on a polyacrylamide gel and then transferred to pure nitrocellulose membrane (Amersham). For HSP27 and HSP32 measurements, membranes were probed with 1:2500 dilution of rabbit anti-rat HSP27 or anti-rat HSP32 (StressGen), followed by a 1:2500 dilution of the secondary antibody (peroxidase-conjugated goat anti-rabbit antibody, Bio-Rad). The antigen–antibody complexes were visualized with a chemiluminescence system (Amersham) and exposed to photosensitive film. The relative densities were analyzed using the NIH image software (NIH Image Version 1.61).

Statistical analysis

Mann-Whitney U tests and Students t test were used. Values are mean ± SD. Statistical significance was set at $P < 0.05$.

Results

Physiological parameters in all animal groups were recorded during intracerebral infusions. There was a significant difference between young and aged rats in body weight. However, the other physiological variables, including mean arterial blood pressure, blood pH, blood gases, hematocrit, and blood glucose, were not different between the groups.

Forelimb placing and forelimb use asymmetry were used to assess neurological deficits. Rats were tested from day 1 to day 28 after ICH. There were marked neurological deficits by day 1 after ICH but there was progressive recovery of function over four weeks. The initial (day 1) behavioral deficit, as assessed by the forelimb placing score (Fig. 1) and the forelimb use asymmetry scores, was greater in aged compared to young rats ($P < 0.05$). This difference with age was maintained for the 4 weeks after ICH (Fig. 1). ICH has no effect on the ipsilateral (non-impaired) forelimb placing. The enhanced neurological deficits in aged rats were associated with enhanced brain swelling.

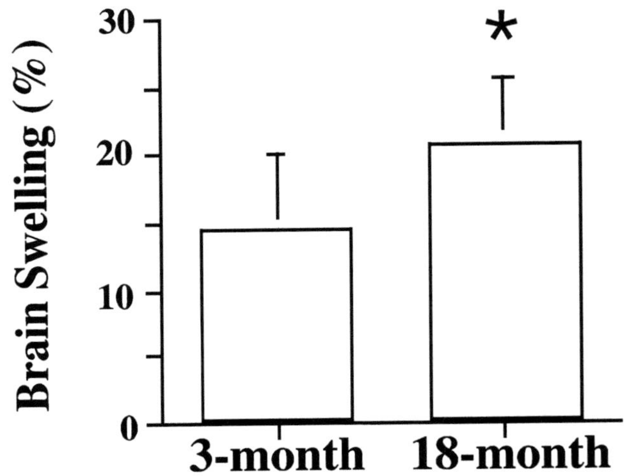

Fig. 1. Forelimb placing test in young and aged rats after ICH. Values are mean ± SD, n = 8. *$p < 0.05$, #$p < 0.01$ vs. 3-month old rats

Perihematomal swelling was more severe in the aged rats compared to young rats at three days after ICH (Fig. 2).

Further evidence for enhanced injury in aged animals came from analysis of HSP27 and 32, two brain injury markers. After ICH, young rats had increased HSP27 and HSP32 levels by day 1. These levels peaked at days 3–7 and there was a decrease at day 28, particularly for HSP32. The time course for HSP32 in aged rats was similar, but the peak induction at days 3 and 7 was greater than in young rats ($p < 0.05$). For HSP27 not only were the levels greater at days 3 and 7 in aged rats ($p < 0.05$), but the upregulation was prolonged with there still being a marked upregulation at day 28.

Discussion

In the present study, we found that intracerebral hemorrhage caused greater neurological deficits, more severe brain swelling and a greater induction of heat shock proteins in aged rats compared to young rats. These results suggest that age is a significant factor in determining brain injury after ICH.

Accumulating evidence indicates that multiple mechanisms are involved in brain edema after ICH [7]. These include coagulation cascade activation with thrombin production, red blood cell lysis with hemoglobin-induced toxicity and complement cascade activation in the brain parenchyma. The age dependent injury may reflect differences in one or more of these mechanisms.

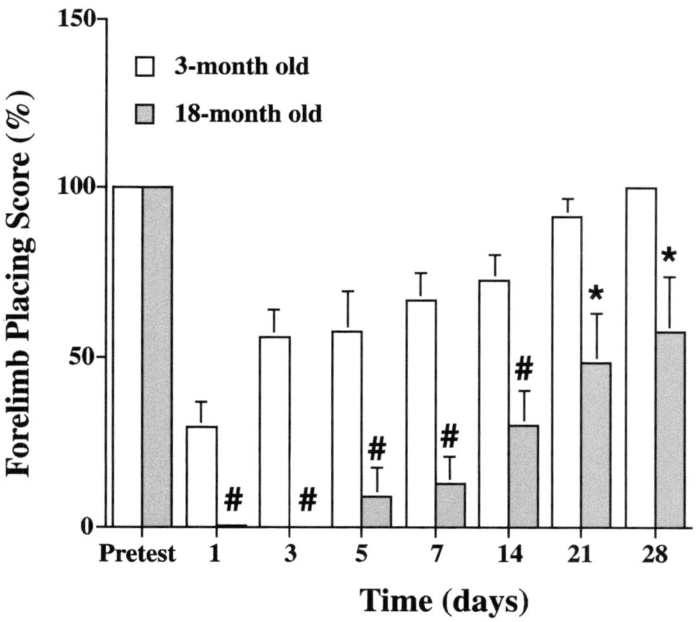

Fig. 2. Brain swelling in the ipsilateral basal ganglia at day 3 in the young and aged rats after ICH. Values are mean ± SD, n = 7. *p < 0.05 vs. 3-month old rats

Heat shock proteins are very sensitive markers for brain injury [4]. HSP27 and HSP32 were chosen as two brain injury markers for the current study. The upregulation of both perihematomal HSP27 and HSP32 levels was found to be higher in the aged rats. It should also be noted that HSP32 (also called hemeoxygenase 1; HO-1) is a key enzyme for heme degradation. By increasing the release of iron from hemoglobin, HSP32 may participate in brain injury. Indeed, hemeoxygenase inhibitors reduce brain edema after ICH [5]. Greater upregulation of HSP32 in aged rats may not only be a response to greater injury but it may participate in that injury.

Neurological deficits following ICH result from the balance of injury and repair. Thus, the greater neurological deficits in aged rats could reflect alterations in either the initial injury or the ability to recover functions. Figure 1 and our previous study have shown that there is time-dependent recovery after ICH may result from plasticity [2]. Our present behavioral data showed that the temporal profiles of recovery in aged and young rats are identical. This result suggests that differences in acute injury, rather than less plasticity, results in the greater brain swelling and neurological deficits in aged rats.

References

1. Davis M, Mendelow AD, Perry RH, Chambers IR, James OF (1995) Experimental stroke and neuroprotection in the aging rat brain. Stroke 26: 1072–1078
2. Hua Y, Schallert T, Keep RF, Wu J, Hoff JT, Xi G (2002) Behavioral tests after intracerebral hemorrhage in the rat. Stroke 33: 2478–2484
3. Matz PG, Weinstein PR, Sharp FR (1997) Heme oxygenase-1 and heat shock protein 70 induction in glia and neurons throughout rat brain after experimental intracerebral hemorrhage. Neurosurgery 40: 152–160
4. Sharp FR (1995) Stress proteins are sensitive indicators of injury in the brain produced by ischemia and toxins. J Toxicol Sci 20: 450–453
5. Wagner KR, Hua Y, de Courten-Myers GM, Broderick JP, Nishimura RN, Lu SY, Dwyer BE (2000) Tin-mesoporphyrin, a potent heme oxygenase inhibitor, for treatment of intracerebral hemorrhage: in vivo and in vitro studies. Cell Mol Biol 46: 597–608
6. Xi G, Keep RF, Hoff JT (1998) Erythrocytes and delayed brain edema formation following intracerebral hemorrhage in rats. J Neurosurg 89: 991–996
7. Xi G, Keep RF, Hoff JT (2002) Pathophysiology of brain edema formation. Neurosurg Clin N Am 13: 371–383
8. Xi G, Keep RF, Hua Y, Xiang JM, Hoff JT (1999) Attenuation of thrombin-induced brain edema by cerebral thrombin preconditioning. Stroke 30: 1247–1255

Correspondence: Ya Hua, R5550 Kresge I, University of Michigan, Ann Arbor, Michigan 48109-0532, USA. e-mail: yahua@umich.edu

Acta Neurochir (2005) [Suppl] 95: 429–431

The antioxidant effect of N-acethylcysteine on experimental contusion in rats

U.-W. Thomale[1], **M. Griebenow**[1], **S.-N. Kroppenstedt**[1], **A. W. Unterberg**[2], and **J. F. Stover**[3]

[1] Department of Neurosurgery, Charité, Virchow Medical Center, University of Berlin, Berlin, Germany
[2] Department of Neurosurgery, Ruprecht-Karls University, Heidelberg, Germany
[3] Department Surgery, Division of Surgical Intensive Care Medicine, University Hospital, Zurich, Switzerland

Summary

N-acethylcysteine (NAC) is known to have direct and indirect antioxidant abilities. We investigated the potential protective effect of NAC on ICP, brain edema and contusion volume after Controlled Cortical Impact (CCI) injury. A moderate CCI injury was induced on the left hemisphere in 48 Sprague Dawley rats. The animals were treated with intraperitoneal injection of NAC (163 mg/kg/KG) or physiological saline. Measurements of intracranial pressure (ICP) were performed and brains were removed at 24 hours. Gravimetric analysis of post-traumatic edema and morphometric measurements (TTC staining) of contusion volume were carried out in 24 animals, respectively. ICP measurements increased significantly over time with no significant differences between both groups. The relative difference in water content in NAC treated animals ($1.45 \pm 0.1\%$) did not differ significantly versus placebo ($1.47 \pm 0.2\%$). The contusion volume was diminished by 19% in the NAC group (53.52 ± 5.3 mm^3) versus placebo (66.28 ± 4.7 mm^3) without showing statistical significance. The antioxidant properties of NAC did not affect intracranial pressure or posttraumatic brain edema formation, although the moderate reduction of contusion volume might reveal beneficial effects on focal contusion.

Keywords: N-acetylcysteine; traumatic brain injury; brain edema; controlled cortical impact injury; free radicals.

Introduction

The significant role of free radicals on secondary posttraumatic brain injury has been stressed by different authors [5, 8, 10]. In particular, the generation of oxygen free radicals such as hydroxyl or superoxide radicals, has been described to occur following CNS damage. Free radicals may cause secondary brain damage due to peroxidation of membrane lipids resulting in breakdown of the blood-brain-barrier (BBB), cell membranes or denaturation of proteins and nucleic acids. N-acetylcysteine (NAC) is a precursor molecule of glutathione, which is one of the antioxidant molecules acting as a physiological defense mechanism against free radical mediated cell damage. NAC, a well-known compound from clinical routine, may eliminate free radical species and other oxidants by direct scavenging abilities [1] as well as by promoting glutathione synthesis [3]. Following focal cerebral ischemia, NAC has previously been shown to be neuroprotective for neuronal survival, neurological outcome score and infarct volume [4, 6, 12]. Following Controlled Cortical Impact Injury, a restoration of GSH levels and mitochondrial dysfunction following NAC treatment has been reported [17]. In this study, we investigate the effect of NAC on ICP, brain edema formation and contusion volume following Controlled Cortical Impact injury in rats.

Material and methods

Approval by the local animal care and use committee of the Berlin state department was obtained. 48 male Sprague Dawley rats (300–350 g), anesthetized with isoflurane (1.9 vol% in N_2O and O_2; mixture of 2:1), were traumatized over the intact dura of the left hemisphere using the Controlled Cortical Impact (CCI) injury model (bolt diameter: 5 mm; velocity: 7 m/sec; penetration depth: 1 mm; contact time: 300 msec). Rectal temperature was kept at 37 °C. Intracranial pressure (ICP) was monitored by positioning a microsensor (Codman, Johnson and Johnson) in the right frontal hemisphere. Rats received either n-acetylcysteine (NAC, Hexal AG, Holzkirchen, Germany; 163 mg/kg body weight i.p.) or physiological saline solution immediately, 2 and 4 hours after CCI. ICP measurements were performed before and after trauma, following the first drug administration and before brain removal at 24 hours following CCI.

To assess edema formation, brains [n = 24] were dissected in midline and weighed before and after drying at 100 °C for 24 hours to quantify wet weight (WW) and dry weight (DW). Water content and delta water content were calculated (water $\text{content}_{L/R}$ [%] $= [(WW_{L/R} - DW_{L/R})/WW_{L/R}] * 100$); (delta water content [%] = water content_L [%] − water content_R [%]). To determine contusion volume brains [n = 24] were cut in 1.3 mm slices. The brain

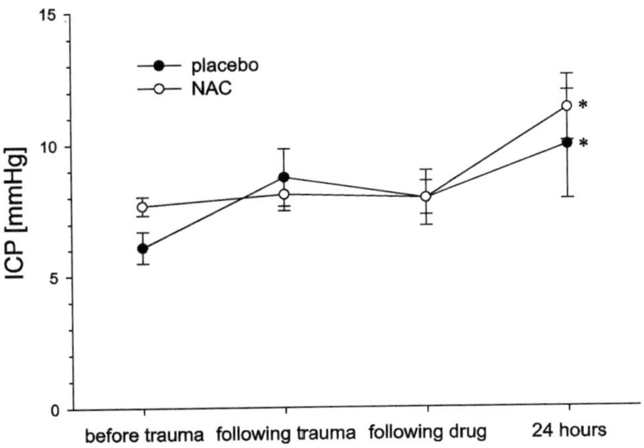

Fig. 1. ICP measurements showed increased values over time. At 24 hours pressure values were significantly increased versus baseline in both groups. No statistical differences could be determined between the groups

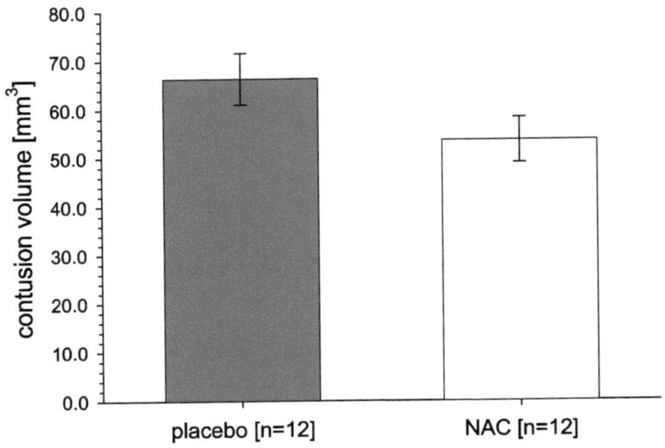

Fig. 2. Contusion volume was reduced by 19% in NAC treated animals in comparison to the placebo group without reaching the level of statistical significance ($p = 0.09$)

sections were incubated in 2% triphenyltetrazolium-chloride (TTC) solution at 37 °C for 30 minutes and were recorded photographically. The contusion area was measured morphometrically and the volume was calculated. All data is expressed as means ± SEM, with differences rated significant at $p < 0.05$ (Sigma Stat® 3.0; Jandel Scientific, Germany).

Results

Arterial blood gases remained within physiological limits at all time points. MABP and CPP were in normal range without statistical differences between the groups. *ICP* significantly increased in both groups over time. Differences between the groups did not reach statistical significance (Fig. 1). The *water content* at 24 hours after CCI in the traumatized hemisphere was significantly increased in all animals compared to non-traumatized hemispheres ($p < 0.01$). In the NAC group, delta water content (1.45 ± 0.1%) was similar to the placebo group (1.47 ± 0.2%). *Contusion volume* in NAC treated animals (53.5 ± 4.7 mm^3) was reduced by 19% compared to placebo treated animals (66.3 ± 5.3 mm^3; $p = 0.09$; Fig. 2).

Discussion

The present study could not show significant differences in elevated posttraumatic intracranial pressure or brain edema formation following NAC administration versus placebo animals. However, a tendency to reduction in contusion volume in NAC treated animals was observed.

Free radical formation is described as a factor responsible for secondary traumatic brain injury. Antioxidants and free radical scavengers have been tested as neuroprotective compounds following traumatic brain injury in experimental and clinical settings [9, 11, 13, 14]. For n-acetylcysteine (NAC), antioxidant properties are described due to its sulphydryl group, which regenerates the radical scavenger glutathione. Furthermore, direct reaction with hydroxyl radicals are described [1, 3]. NAC has been tested as neuroprotective drug following ischemic brain damage and reperfusion injury [4, 6, 12]. Despite following a similar protocol of drug administration, our study did not show any effect of NAC treatment on post-traumatic edema nor on intracranial pressure. Brain edema formation following CCI has been described as cytotoxic and vasogenic origin [15, 16]. Vasogenic edema is characterized by a biphasic opening of the blood brain barrier at 6 and 24 hours following injury [2], while vasogenic edema after CCI correlates with the release of hydroxyl radicals [13]. We hypothesize, that the lack of edema reduction by NAC may be explained by the small period of vasogenic edema formation within the treatment window and is less represented compared to models of ischemia and reperfusion.

The beneficial effect of NAC treatment has been observed on infarct volume, hippocampal neuronal cell loss and neurological outcome in various models of ischemic brain damage [3, 6, 12]. Interestingly the efficacy of NAC inversely correlates in a model testing different ischemia intensities [7]. After CCI, mitochondrial dysfunction is reportedly restored after NAC administration [17]. Our data has not shown statistical significance in the reduction of absolute contusion volume. Nevertheless, we quantified a 19% decrease in contusion volume 24 hours after trauma. A similar mechanism to ischemic injury may be responsible for the increase in neuronal survival after NAC administration. Direct and indirect free radical scavenging activity of NAC and its metabolite glutathione as well as indirect anti-inflammatory effects may be responsible for protective properties in brain injury [6, 12].

Conclusion: The antioxidant abilities of NAC result in promising effects following experimental cerebral ischemia but are only restricted in the treatment of focal traumatic brain lesions in rats. The tendency for reduction in contusion volume warrants further investigations on neuroprotection at a cellular level.

References

1. Aruoma OI, Halliwell B, Hoey BM, Butler J (1989) The antioxidant action of N-acetylcysteine: its reaction with hydrogen peroxide, hydroxyl radical, superoxide, and hypochlorous acid. Free Radic Biol Med 6: 593–597
2. Baskaya MK, Rao AM, Dogan A, Donaldson D, Dempsey RJ (1997) The biphasic opening of the blood-brain barrier in the cortex and hippocampus after traumatic brain injury in rats. Neurosci Lett 226: 33–36
3. Corcoran GB, Wong BK (1986) Role of glutathione in prevention of acetaminophen-induced hepatotoxicity by N-acetyl-L-cysteine in vivo: studies with N-acetyl-D-cysteine in mice. J Pharmacol Exp Ther 238: 54–61
4. Cuzzocrea S, Mazzon E, Costantino G, *et al* (2000) Beneficial effects of N-acetylcysteine on ischaemic brain injury. Br J Pharmacol 130: 1219–1226
5. Hall ED, Andrus PK, Yonkers PA (1993) Brain hydroxyl radical generation in acute experimental head injury. J Neurochem 60: 588–594
6. Khan M, Sekhon B, Jatana M, *et al* (2004) Administration of N-acetylcysteine after focal cerebral ischemia protects brain and reduces inflammation in a rat model of experimental stroke. J Neurosci Res 76: 519–527
7. Knuckey NW, Palm D, Primiano M, Epstein MH, Johanson CE (1995) N-acetylcysteine enhances hippocampal neuronal survival after transient forebrain ischemia in rats. Stroke 26: 305–310; discussion 311
8. Kontos HA, Wei EP (1986) Superoxide production in experimental brain injury. J Neurosurg 64: 803–807
9. Marshall LF, Maas AI, Marshall SB, *et al* (1998) A multicenter trial on the efficacy of using tirilazad mesylate in cases of head injury. J Neurosurg 89: 519–525
10. Povlishock JT, Kontos HA (1992) The role of oxygen radicals in the pathobiology of traumatic brain injury. Hum Cell 5: 345–353
11. Sarrafzadeh AS, Thomale UW, Kroppenstedt SN, Unterberg AW (2000) Neuroprotective effect of melatonin on cortical impact injury in the rat. Acta Neurochir (Wien) 142: 1293–1299
12. Sekhon B, Sekhon C, Khan M, Patel SJ, Singh I, Singh AK (2003) N-Acetyl cysteine protects against injury in a rat model of focal cerebral ischemia. Brain Res 971: 1–8
13. Smith SL, Andrus PK, Zhang JR, Hall ED (1994) Direct measurement of hydroxyl radicals, lipid peroxidation, and blood-brain barrier disruption following unilateral cortical impact head injury in the rat. J Neurotrauma 11: 393–404
14. Stoffel M, Berger S, Staub F, Eriskat J, Jacob K, Baethmann A (1997) The effect of dietary alpha-tocopherol on the experimental vasogenic brain edema. J Neurotrauma 14: 339–348
15. Stroop R, Thomale UW, Pauser S, *et al* (1998) Magnetic resonance imaging studies with cluster algorithm for characterization of brain edema after controlled cortical impact injury (CCII) Acta Neurochir (Wien) [Suppl] 71: 303–305
16. Unterberg AW, Stroop R, Thomale UW, *et al* (1997) Characterisation of brain edema following "controlled cortical impact injury" in rats. Acta Neurochir (Wien) [Suppl] 70: 106–108
17. Xiong Y, Peterson PL, Lee CP (1999) Effect of N-acetylcysteine on mitochondrial function following traumatic brain injury in rats. J Neurotrauma 16: 1067–1082

Correspondence: Ulrich-Wilhelm Thomale, Department of Neurosurgery, Charité, Virchow Medical Center, Humboldt University, Augustenburger Platz 1, D-13353 Berlin, Germany. e-mail: uthomale@charite.de

Acta Neurochir (2005) [Suppl] 95: 433–436

The investigation of cerebrospinal fluid formation by ventriculo-aqueductal perfusion method in cats

D. Orešković[1], **M. Vukić**[2], **M. Klarica**[3], and **M. Bulat**[3]

[1] Department of Molecular Biology, Rudjer Bošković Institute, Zagreb, Croatia
[2] Department of Neurosurgery, School of Medicine, Zagreb, Croatia
[3] Department of Pharmacology and Croatian Institute for Brain Research, School of Medicine, Zagreb, Croatia

Summary

Objectives. The perfusion of cerebrospinal fluid (CSF) spaces by artificial CSF (aCSF) containing an indicator, is an indirect method used to calculate CSF formation. To evaluate this method, we have developed a ventriculo-aqueductal perfusion method, which enables a direct measurement of CSF formation in the ventricles.

Methods. In chloralose anaesthetized cats, the aqueduct of Sylvius was cannulated so that the outflow end of the plastic cannula was positioned extracranially. Both lateral ventricles were also cannulated, with one cannula for infusion of aCSF containing blue dextrane and the other for measurement of CSF pressure.

Results. During ventriculo-aqueductal perfusion (direct method) under physiological CSF pressure, the outflow rate from aqueductal cannula did not differ significantly from the inflow rate, i.e. no CSF formation was observed. When the indirect method based on dilution of blue dextran in the outflowing perfusate was used, the formation of approximately 5 μl/min of CSF was obtained.

Conclusion. Results of the direct method indicate that net CSF formation inside brain ventricles does not exist. The opposite results obtained by the indirect method questions this method as a reliable study of CSF formation.

Keywords: Cerebrospinal fluid (CSF); ventriculo-aqueductal perfusion; CSF formation; CSF pressure.

Introduction

The CSF formation rate (Vf) in animals has been extensively studied by the ventriculo-cisternal perfusion technique, which is still regarded as the most precise method [1, 4–6]. This method and the equation for the calculation of the CSF formation rate have been established by Heisey *et al.* [5], who assumed that the dilution of the indicator substance is a consequence of newly formed CSF, i.e. that a higher CSF formation rate will result in a higher degree of dilution of the indicator substance in outflow perfusate. Therefore, this method is indirect and any uncertainty in the interpretation of the degree of dilution of the indicator substance in the perfusate caused by some other factors will result in questionable and often contradictory conclusions regarding CSF formation rates [3, 7–9, 12].

For this reason we have developed a direct ventriculo-aqueductal perfusion method for measurement Vf at the physiological CSF pressure assuming that the inflow rate should be smaller than the outflow rate if the net CSF formation in ventricles exists [11]. The purpose of this study was to compare results obtained by these two methods. Namely, if both methods are reliable and precise, no differences should exist between calculated Vf (indirect method) and measured Vf (direct method).

Materials and methods

Adult cats of both sexes weighing 2.0–3.8 kg were anaesthetized with chloralose (100 mg/kg, i.p.) and positioned in sphinx position in stereotaxic frame (D. Kopff, USA) with their heads elevated 12 cm above stereotaxic table. The aqueduct of Sylvius was approached via transcerebellar route and cannulated so that outflow end of the plastic cannula was positioned extracranially, and thereafter the skull was hermetically closed [11]. Both lateral ventricles were also cannulated, with one cannula used for 26.06 μl/min infusion (by Palmer pump) of aCSF containing blue dextrane, (m.w. 2 millions; 1 mg/ml) and the other for measurement of CSF pressure (Fig. 1). From the external end of the aqueductal cannula, 30-min samples of the perfusate were collected in a plastic tube and the optical density of the dextran blue in perfusate was measured using a spectrophotometer (Perkin-Elmer 55B) at wavelength of 635 nm. All experiments were performed at physiological CSF pressure (6.5–8.0 cm H_2O), where the external auditory meatus was taken as the zero pressure level.

The CSF formation rate (Vf) was calculated by the indirect

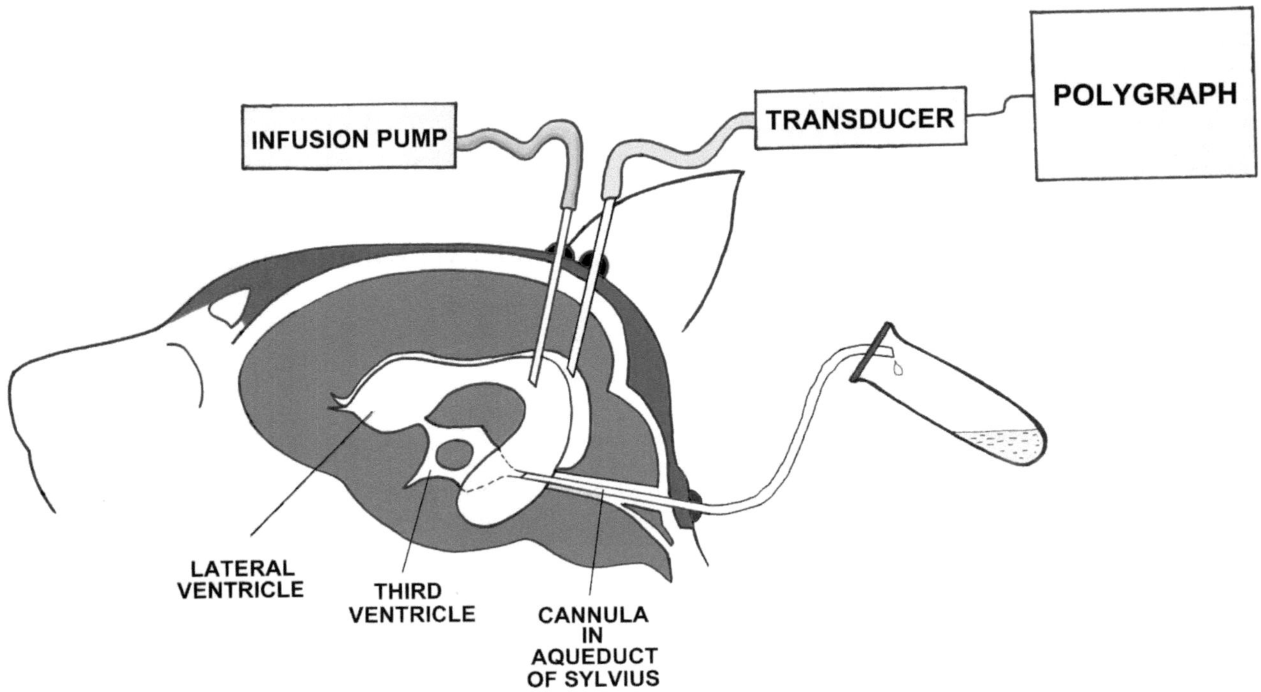

Fig. 1. Scheme of experimental model showing position of cannulas in left lateral ventricle for infusion of aCSF, in right lateral ventricle for CSF pressure recording, and in aqueduct of Sylvius for collection of outflow perfusate

method using the equation derived by Heisey *et al.* [4]: $Vf = Vi(Ci - Co)/Co$, where Vi is the inflow perfusate rate, Ci is the concentration of the blue dextran in the inflow perfusate and Co is the concentration of the blue dextran in the outflow perfusate.

The Vf was obtained by the direct method as the difference between measured outflow and inflow volumes of perfusate at physiological CSF pressure. The volume of samples was determined by weighing (Mettler, Toledo, Switzerland). The Vf was achieved by both methods for each cat under the same experimental conditions. Results are shown as mean values ± S.E.M ($n = 5$) and differences between data estimated by the Student's t-test.

Results

The results of the outflow concentration of blue dextran ($Co/Ci \times 100$) and calculated CSF formation rate (Vf) in five cats at different time intervals (30 min) at physiological CSF pressure are shown in Fig. 2. The outflow concentration of blue dextran was stable during the examined period of 120 min and the calculated rate of Vf was about 5 μl/min.

Simultaneously, the outflow and inflow rates of perfusate measured by direct ventriculo-aqueductal perfusion method in the same cats were 27.06 ± 2.66 μl/min and 26.06 ± 0.22 μl/min, respectively, and these did not differ significantly ($p > 0.05$). This suggests that by the direct method, CSF formation in ventricles is not observed.

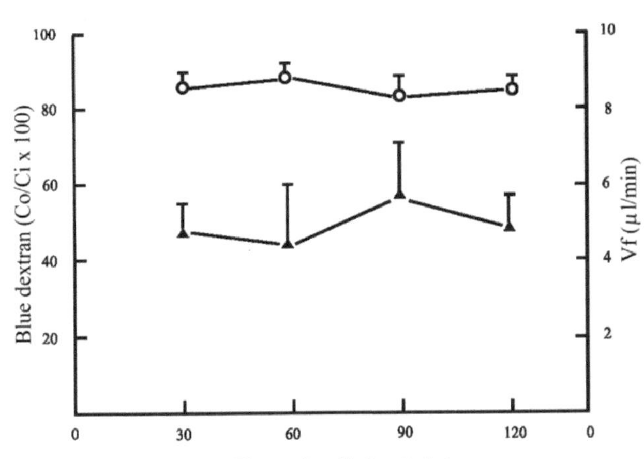

Fig. 2. Outflow concentration of blue dextran in the percentage of inflow concentration ($Co/Ci \times 100$; open circles) and the rate of calculated CSF formation (Vf; closed triangles) during ventriculo-aqueductal perfusion (26.06 μl/min over 120 min) with aCSF containing blue dextran at physiological cerebrospinal fluid pressure in five cats. The vertical lines show the standard error of mean value

Discussion

It is a generally accepted hypothesis that CSF is produced mainly by the secretory activity of the choroid plexuses within the brain ventricles. At physiological CSF pressure, CSF supposedly circulates slowly from the ventricles towards the cortical subarachnoid space, to be passively absorbed into the venous sinuses across the arachnoid villi. If CSF is mainly formed inside the ventricles and absorbed from the subarachnoid space, it should circulate at a physiological CSF pressure, through the aqueduct of Sylvius or, as in our cat model [11], through the plastic cannula positioned in the aqueduct (Fig. 1). Therefore, if Vf is measured directly during ventriculo-aqueductal perfusion with aCSF, outflow rate of perfusate is increased in comparison to inflow rate. Furthermore, if during such perfusion, the aCSF contains an indicator substance as dextran blue, the dilution of dextran blue in the outflow perfusate, in comparison to inflow perfusate, can be used for calculation of Vf according to the equation developed by Heisey *et al.* (see Materials and Methods).

Our approach gives the opportunity of measuring the CSF formation rate (Vf) on the same experimental animal in two ways, i.e. directly, by measuring outflow volume of perfusate, and indirectly, by calculations based on the dilution of indicator substance in outflow samples.

Our study has shown that by using the direct method, no significant differences are observed between outflow and inflow rate during two hours of ventriculo-aqueductal perfusion (26.06 μl/min) at physiological CSF pressure. This means that the same volume of perfusate as has been infused by the pump in lateral ventricle, has been collected at the end of the aqueductal cannula, i.e. nett CSF formation inside the brain ventricles is not observed. This data largely corresponds to our previous observations when perfusion has been performed using a rate of 12.96 μl/min [11], or when CSF is freely collected from the aqueductal cannula over one [11] or two hours [10]. Furthermore, these results are in line with a new hypothesis [2, 11] which suggests that CSF is not formed at one site (brain ventricles) and absorbed at the other one (villi arachnoidales). According to this hypothesis the osmotic and hydrostatic forces in plasma operating across the capillary walls are regulators of cerebral interstitial fluid and CSF volume.

However, when the dilution of blue dextran in the outflow perfusate is used to calculate CSF formation rate according to Heisey *et al.* [5], a nett CSF formation of nearly 5 μl/min is obtained (Fig. 2). This discrepancy between results obtained by two mentioned methods again questions the possibility of using the dilution of indicator substance as a parameter for calculation of Vf. The variable mixing of CSF across the whole CSF system and the indicator in perfusate, the different spread of perfusate through the CSF system and possible different absorption of the indicator or water from the perfusate into surrounding tissue, could lead to contradictory experimental results obtained by that method [3, 7–9, 12]. For previously mentioned reasons, our results indicate that the perfusion method with dilution of indicator in the perfusate is not reliable for calculation of the rate of the CSF formation and that data attained by this method should be reevaluated.

Acknowledgments

We thank Mrs Katarina Karlo for her skilled technical assistance. This work has been supported by the Ministry of Science and Technology, Republic of Croatia.

References

1. Artru AA (1988) Dose-related changes in the rate of cerebrospinal fluid formation and resistance of reapsorption of cerebrospinal fluid following administration of thiopental, midazolam and etomidate in dogs. Anesthesiology 69: 541–546
2. Bulat M (1993) Dynamics and statics of the cerebrospinal fluid: the classical and a new hypothesis. In: Avezaat CJJ, Ejindhoven JHM, Maas AIR, Tans JTJ (eds) Intracranial pressure VIII. Berlin, Springer, pp 726–730
3. Curran RE, Mosher MB, Owens ES, Fenstermacher JD (1970) Cerebrospinal fluid production rates determined by simultaneous albumin and inulin perfusion. Exp Neurol 29: 546–553
4. Harnish PP, Samuel K (1988) Reduced cerebrospinal fluid production in the rat and rabbit by diatrizoate ventriculo-cisternal perfusion. Invest Radiol 23: 534–536
5. Heisey SR, Held D, Pappenheimer JR (1962) Bulk flow and diffusion of the cerebrospinal fluid system of the goat. Am J Physiol 203: 775–781
6. Lorenzo AV, Page LK, Waters V (1970) Relationship between cerebrospinal fluid formation, absorption and pressure in human hydrocephalus. Brain 93: 679–692
7. Martins AN, Newby N, Doyle TF (1977) Source of error in measuring cerebrospinal fluid formation by ventriculo-cisternal perfusion. J Neurol Neurosurg 40: 645–650
8. Orešković D, Bulat M (1993) Hydrostatic force in regulation of CSF volume. In: Avezaat CJJ, Ejindhoven JHM, Maas AIR, Tans JTJ (eds) Intracranial pressure VIII. Berlin, Springer, pp 731–734
9. Orešković D, Klarica M, Lupret V, Vukić M (2000) The character of the cerebrospinal fluid production. Neurosci Res Commun 26: 69–76
10. Orešković D, Klarica M, Vukić M (2001) Does the secretion

and circulation of the cerebrospinal fluid really exist? Med Hypotheses 55: 622–624
11. Orešković D, Klarica M, Vukić M (2002) The formation and circulation of cerebrospinal fluid inside the cat brain ventricles: a fact or an illusion? Neurosci Lett 327: 103–106
12. Orešković D, Klarica M, Vukić M, Maraković J (2003) Evaluation of ventriculo-cisternal perfusion method to study cerebrospinal fluid formation. Croat Med J 44: 161–164

Correspondence: Darko Orešković, Department of Molecular Biology, Rudjer Bošković Institute, Bijenička c. 54, HR-10 000 Zagreb, Croatia. e-mail: doresk@irb.hr

Satellite symposium on neurochemical monitoring

Acta Neurochir (2005) [Suppl] 95: 439

Introduction for the neurochemical satellite symposium

P. J. Hutchinson[1] and **W. S. Poon**[2]

[1] Academic Neurosurgery Unit, University of Cambridge, Cambridge, UK
[2] Division of Neurosurgery, The Chinese University of Hong Kong, Hong Kong, China

Of the advances in monitoring the injured brain, techniques which enable derangements in neurochemistry to be characterised are evolving at a rapid rate. Methods are now available to interrogate brain chemistry are several levels including nucleic acids, amino acids, peptides, proteins and fundamental substrates and products of metabolism such as glucose, lactate and pyruvate. These methods can be divided into multi-modality techniques such as microdialysis and brain tissue oxygen sensors which involve invasive monitoring, and imaging techniques such as positron emission tomography and magnetic resonance spectroscopy. Both approaches have advantages and disadvantages. Multi-modality monitoring techniques enable continuous monitoring of the brain over prolonged periods of time but in focal areas of the brain. Imaging, on the other hand, is able is provide a global map of the whole brain but only as a snapshot for the duration of the scan. Combining monitoring and imaging techniques is therefore an attractive proposal.

This neurochemical satellite symposium to the ICP2004 meeting in Hong Kong addressed the current state of the art of monitoring and imaging brain chemistry, covering the current role of microdialysis in brain injury (methodology and pathophysiology) including the biochemical signature of the Lund protocol, brain tissue oxygenation, near infrared spectroscopy and imaging. In addition, new and exciting techniques – circulating nucleic acid and proteomic technology were reviewed. The theme of this symposium was to describe the contribution that each of these techniques has made and has the future potential to make in terms of increasing our understanding of the pathophysiology of brain injury. In addition, each technique was appraised in terms of its ability to translate from purely a research technique to one that can be used as a clinical monitor for the management of patients on an intention to treat analysis.

Acta Neurochir (2005) [Suppl] 95: 441–445

Microdialysis in traumatic brain injury – methodology and pathophysiology

P. J. Hutchinson

Academic Department of Neurosurgery and Wolfson Brain Imaging Centre, University of Cambridge, Addenbrooke's Hospital, UK

Summary

The application of clinical microdialysis to monitor changes in cerebral extracellular chemistry is now well established in several neurosurgical units worldwide. In neuro-intensive care the technique has been predominantly applied to patients with traumatic brain injury and subarachnoid haemorrhage. There is no doubt that microdialysis has increased and continues to increase our understanding of the pathophysiology of these conditions. Current studies are addressing the potential role of microdialysis as a clinical monitoring technique assisting in the management of patients on an intention to treat basis. This involves establishing the relationship between microdialysis and outcome, and the effect of therapeutic manoeuvres on the chemistry. This manuscript describes the place of microdialysis in traumatic brain injury in terms of the fundamental principles, methodology, pathophysiology and clinical application.

Keywords: Traumatic brain injury; head injury; microdialysis; cerebral metabolism; hypoxia; ischemia.

Introduction

Following trauma to the head, one of the fundamental processes that results in progression of injury is cerebral ischemia. Both the primary insult (ictus) and secondary insults such as hypoxia, hypotension and seizures contribute to this process. Ischemia results in reduced oxygen delivery, energy failure, and cerebral edema. One of the goals of traumatic brain injury management is to detect and treat these secondary insults in order to reduce injury progression and improve outcome.

In neuro-intensive care, measurement of intracranial pressure (ICP) is the cornerstone of multimodality monitoring and is recommended by US and European guidelines [5, 16]. Monitoring of cerebral blood flow (CBF; by extrapolation of flow velocity from trans-cranial Doppler or by functional imaging) and brain oxygenation (by jugular venous oxygen catheters and brain tissue oxygen sensors) is also widely practised both for research and clinical purposes [15]. Measurement of the effect of fundamental derangements of pathophysiology such as increased ICP, reduced CBF and reduced cerebral oxygenation in terms of the impact on cerebral metabolism is now feasible with the application of cerebral microdialysis. This has enabled monitoring of brain injury in terms of changes in the extracellular concentration of fundamental substrates and metabolites.

The microdialysis technique was first described by Ungerstedt and Pycock in 1974 [29], applied to the human brain initially to monitor neurotransmitters in Parkinson's disease [20] and subsequently used to monitor patients with traumatic brain injury (TBI) and subarachnoid haemorrhage in neuro-intensive care.

Principle of microdialysis

The principle of cerebral microdialysis is the perfusion of a fine catheter lined with semi-permeable renal dialysis membrane [28], introduced into the cerebral cortex either directly or via a cranial access device. The catheter comprises an inlet tube, a shaft, a membrane and an outlet tube. The inlet tube is perfused with a physiological solution such as normal saline or Ringer's solution at very low flow rates (typically 0.1–2.0 μl/min i.e. approximately 0.15–3.0 ml/day) using a precision pump. The solution passes down the concentric shaft to the membrane where molecules diffuse from the extracellular space into the perfusion fluid. The solution then passes up the shaft via the outlet tube into collecting vials. Standard membranes have a 20 kDa cut-off and are applied to measure the concen-

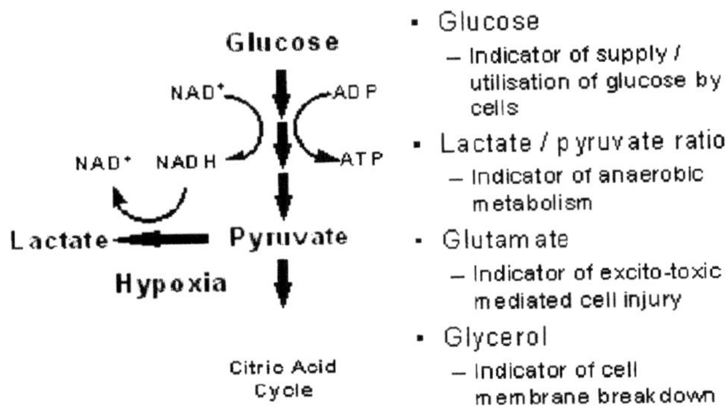

Fig. 1. Fundamental substrates and metabolites monitored by microdialysis. Glucose is metabolised to pyruvate (glycolysis) which enters the citric acid cycle. However, under conditions of hypoxia, pyruvate is preferentially metabolised to lactate. The lactate/pyruvate ratio is therefore a marker of anaerobic metabolism. Glutamate and glycerol are also markers of evolving injury

tration of glucose as an indicator of glucose supply and uptake, the lactate/pyruvate ratio as an indicator of anaerobic metabolism, glutamate as a marker of excitotoxicity and the release of glycerol from damaged cell membranes (Fig. 1). Molecular cut-off membranes of 100 kDa can also be applied which enable the detection of larger proteins. The collecting vials are changed at set intervals of 10–60 minutes and analysed for glucose, pyruvate, lactate and glutamate at the bedside using automated enzyme analysers and then stored for delayed off-line analysis using, for example HPLC, to detect other amino acids.

Methodology of microdialysis

There are a number of important considerations to be taken into account in terms of the clinical application of microdialysis to the human brain. These include the concepts of recovery, the choice of perfusion fluid, and critical appraisal of the technique.

Recovery

The microdialysis catheter lies within the extracellular space mimicking a blood vessel allowing communication with the extracellular fluid along a logarithmic concentration gradient towards and away from the catheter. It is important to recognise, therefore that the level of substance detected in the dialysis fluid does not necessarily equate to the true extracellular concentration. The term "recovery" is applied to the proportion of substance in the extracellular fluid that is detected in the dialysate. "Relative recovery" depends on the length of the dialysis membrane, the rate of flow of the perfusion fluid, the speed of diffusion of the substance and the properties of the membrane. Relative recovery is defined as concentration of a particular substance in the fluid as it leaves the dialysis membrane expressed as a percentage of the concentration in the extracellular fluid surrounding the membrane. It approaches 100% when the flow rate approaches zero. Absolute recovery is the total amount of substance recovered during a defined time period, usually the sampling period. It approaches a maximum value at higher flow rates because the concentration gradients between the environment of perfusate/dialysate are then maximal. In practice, long membranes (10–30 mm) and slow flow rates (0.3–2 μl/min) are used to increase recovery rates. Longer 30 mm membranes enable significantly higher recovery rates than shorter 10 mm membranes but are more difficult to implant and may monitor heterogeneous volumes of brain. Slow flow rates while increasing recovery, reduce the volume of dialysate available for analysis in a given unit time.

The concept of recovery can be applied to determine the true extracellular concentration of a particular substance by varying the flow rate, while measuring the changes in concentration of substance coming out of the catheter, and extrapolating to zero flow. Using this technique the relative recovery for the 10 mm 20 kDa cut-off CMA70 catheter has been shown to be approximately 70% at 0.3 μl/min and 30% at 1.0 μl/min for glucose, pyruvate, lactate and glutamate [12].

Perfusion fluid

Ideally, the composition of the perfusion fluid should be as close to the normal physiological values of the extracellular fluid. Early clinical microdialysis studies utilised normal saline. However, concerns raised from animal studies that depletion of calcium will impair neurotransmitter release have lead to introduction of other solutions such as CMA perfusion fluid. Despite these concerns the absence of calcium from the perfusate does not appear to affect the recovery of glucose, lactate, pyruvate and glutamate [12]. For the larger cut-off molecular membranes it is recommended that dextran is added to the perfusion fluid to maintain dialysate volume.

Critical appraisal

There are a number of points which require critical evaluation in terms of the application of clinical cerebral microdialysis.

(i) Microdialysis-induced tissue trauma. Animal studies have raised concerns regarding tissue trauma induced by catheter insertion. There is evidence of uncoupling of regional cerebral blood flow from local brain glucose metabolism for the first 24 hours after probe insertion which is [2, 4] probably a consequence of relative size of the microdialysis catheter. Other animal studies have shown reticular fibre deposition and gliosis after several days which is likely to be a consequence of lack of sterility of the catheter [3]. These factors must be considered in the design of studies and interpretation of the results. However, the recovery rates of substances stabilise after 30 minutes, histological studies show only occasional microhaemorrhages [3] and the blood brain barrier remains intact [18]. Post-mortem studies in sheep and human brains have revealed minimal or no disturbance to the cerebral parenchyma as a result of microdialysis catheter implantation [13, 31].

(ii) The effect of vasogenic oedema on recovery also needs to be considered. The expansion of the extracellular space in vasogenic oedema will potentially lead to a greater volume of distribution of substances in this environment, resulting in a change in substance concentration. The expansion of the extracellular space in vasogenic oedema will potentially lead to a lower tortuosity factor (i.e. ease of passage of molecules through the extracellular fluid) and hence a higher diffusion coefficient than in normal brain.

(iii) The origin of neurotransmitters needs to be established. It is important to determine whether neurotransmitters detected in the dialysate reflect true synaptic release or non-specific overflow from synaptic and non-synaptic sources [27]. Currently the smallest catheters available are about 200 μm external diameter. The synaptic cleft is approximately 0.02 μm across. Consequently, it is not possible to monitor neurotransmitter release in this region. However, evidence suggests that many systems function via overflow mechanisms as well as classic vesicular release from the presynaptic membrane, diffusion across the synaptic cleft and binding to post-synaptic membrane receptors.

Pathophysiology of traumatic brain injury

There is now increasing experience of microdialysis in patients with severe traumatic brain injury. This has enabled biochemical signatures of adverse events to be defined, characterised by a reduction in cerebral glucose, increase in lactate, increase in lactate/pyruvate ratio, increase in glutamate and increase in glycerol (Fig. 2).

One of the initial studies performed in Sweden dem-

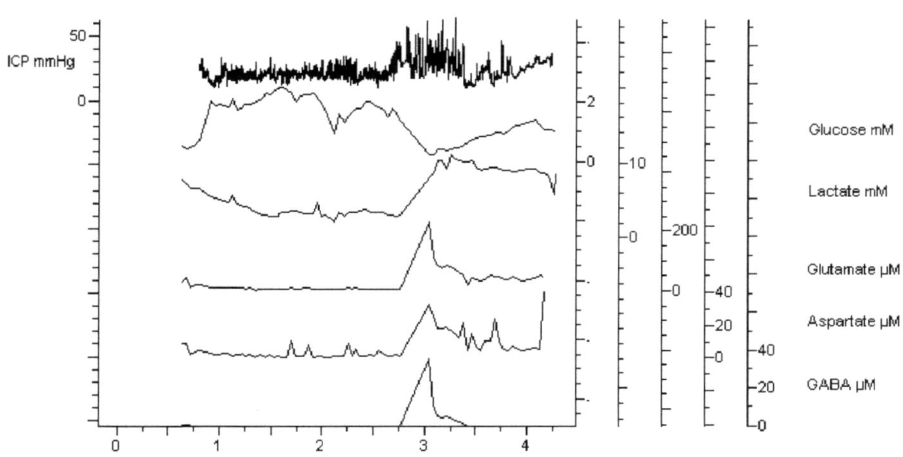

Fig. 2. Characteristic metabolic signature of physiological insult in a TBI patient showing reduction in glucose, increase in lactate and amino acid release in association with intracranial hypertension

onstrated the feasibility of the technique and identified the lactate/pyruvate ratio as a marker of energy disturbance within the brain [22]. Subsequent studies have examined the impact of adverse events including ischemia and hypoxia [8, 9, 23] with a brain tissue oxygen threshold of approximately 10 mmHg resulting in significant derangements in microdialysis parameters. Cross-validation with Positron Emission Tomography (PET) has shown a good correlation between oxygen metabolism and microdialysis markers of the brain redox state (PET oxygen extraction fraction versus microdialysis lactate/pyruvate ratio) [11]. Seizures have been shown to be associated with increased extracellular glutamate [30]. Metabolic derangements in relation to focal injury have been addressed in patients with acute subdural haematoma and contusions [6, 10, 24].

Other studies have sought to determine the impact of therapeutic interventions on cerebral chemistry. These interventions include hyperventilation, barbiturate coma, hyperoxia and manipulation of cerebral perfusion pressure (CPP). In terms of hyperventilation, even brief periods have been shown to cause an increase in the lactate/pyruvate ratio and release of glutamate in penumbral areas [19]. Barbiturate coma has been shown to associated with a reduction in extracellular glutamate and lactate [7]. There is conflicting data on the effect of hyperoxia on cerebral metabolism with one study demonstrating improvement in biochemical markers [26] and another showing slight reduction in lactate but with no change in the redox status [17]. The effect of manipulation of CPP is also unclear with augmentation from 70 mmHg to 90 mmHg increasing the level of brain tissue oxygen and reducing oxygen extraction fraction but not producing any significant changes in microdialysis variables [14]. Reduction in CPP from 73 mmHg to 62 mmHg, however, appeared to be associated with normalisation of cerebral metabolism [25]. What is clear, however, is that considerable biochemical heterogeneity exists between patients [21].

Clinical application

In terms of the continuing clinical evaluation of microdialysis in TBI, there are two fundamental requirements. Firstly, that the changes in the biochemistry relate to both tissue outcome (detected by late imaging) and patient outcome. Secondly, that therapeutic manoeuvres can be applied to "manipulate" the chemistry in a favourable direction. In order to address the application of microdialysis to patients with TBI on an intention to treat basis, a consensus statement has been produced to assist with the implementation of the technique [1]. This statement addresses: (a) catheter placement recommending that for diffuse injury one catheter may be placed in the right frontal region and for focal lesions catheter placement in the pericontusional tissue (Fig. 3) with option for a second catheter in "normal" tissue (b) insertion artefact with the first hour of monitoring regarded as unreliable (c) the lactate/pyruvate ratio as a sensitive marker of the brain redox state and secondary ischaemic injury (with glucose, glycerol and glutamate as additional markers of evolving ischemia) and (d) in terms of clinical use in traumatic brain injury, microdialysis in association with other brain monitoring techniques may assist in delivery of targeted therapy for prevention of secondary ischaemic injury.

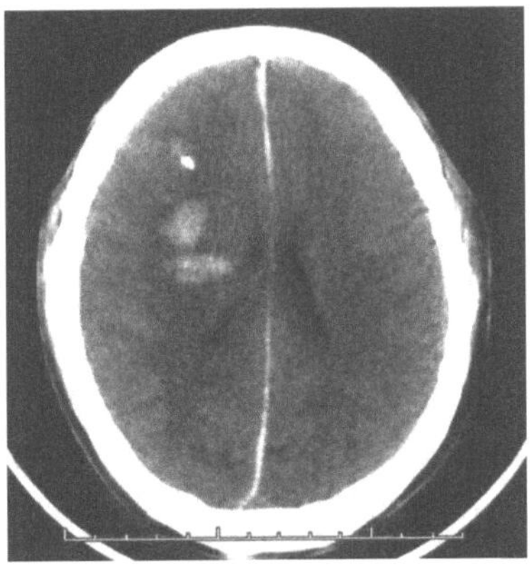

Fig. 3. CT scan showing peri-contusional placement of microdialysis catheter, visible due to gold-tip

Conclusion

There is increasing experience of the application of microdialysis to patients with TBI. There is no doubt that this technique has increased our understanding of the pathophysiology of this condition. In terms of the clinical application, microdialysis is now employed routinely in several centres. Current research is being directed to determine the role of microdialysis as a monitor of secondary injury on an individual intention to treat basis.

Acknowledgments

P. J. Hutchinson is supported by an Academy of Medical Sciences/Health Foundation Senior Surgical Scientist Fellowship.

References

1. Bellander BMCE, Enblad P, Hutchinson P, Nordstrom CH, Robertson C, Sahuquillo J, Smith M, Stocchetti N, Ungerstedt U, Unterberg A, Olsen NV (2004) Consensus meeting on microdialysis in neurointensive care. Int Care Med (in press)
2. Benveniste H (1989) Brain microdialysis. J Neurochem 52: 1667–1679
3. Benveniste H, Diemer NH (1988) Early postischaemic 45Ca accumulation in rat dentate hilus. J Cereb Blood Flow Metab 8: 713–719
4. Benveniste H, Drejer J, Schousboe A, Diemer NH (1987) Regional cerebral glucose phosphorylation and blood flow after insertion of a microdialysis fibre through the dorsal hippocampus in the rat. J Neurochem 49: 729–734
5. Brain Trauma Foundation (2003) American Association of Neurological Surgeons. Guidelines for the management of severe head injury. J Neurotrauma 13: 641–734
6. Bullock R, Zauner A, Woodward JJ, Myseros JS, Choi SC, Ward JD, Marmarou A, Young HF (1998) Factors affecting excitatory amino acid release following severe human head injury. J Neurosurg 89: 507–518
7. Goodman JC, Valadka AB, Gopinath SP, Cormio M, Robertson CS (1996) Lactate and excitatory amino acids measured by microdialysis are decreased by pentobarbital coma in head-injured patients. J Neurotrauma 13: 549–556
8. Goodman JC, Valadka AB, Gopinath SP, Uzura M, Robertson CS (1999) Extracellular lactate and glucose alterations in the brain after head injury measured by microdialysis. Crit Care Med 27: 1965–1973
9. Hlatky R, Valadka AB, Goodman JC, Contant CF, Robertson CS (2004) Patterns of energy substrates during ischaemia measured in the brain by microdialysis. J Neurotrauma 21: 894–906
10. Hlatky R, Valadka AB, Goodman JC, Robertson CS (2004) Evolution of brain tissue injury after evacuation of acute traumatic subdural haematomas. Neurosurgery 55: 1318–1323
11. Hutchinson PJ, Gupta AK, Fryer TF, Al-Rawi PG, Chatfield DA, Coles JP, O'Connell MT, Kett-White R, Minhas PS, Aigbirhio F, Clark JC, Kirkpatrick PJ, Menon DK, Pickard JD (2002) Correlation between cerebral blood flow, substrate delivery, and metabolism in head injury: a combined microdialysis and triple oxygen positron emission tomography study. J Cereb Blood Flow Metab 22: 735–745
12. Hutchinson PJ, O'Connell MT, Al-Rawi PG, Maskell LB, Kett-White CR, Gupta AK, Richards HK, Hutchinson DB, Kirkpatrick PJ, Pickard JD (2000) Clinical cerebral microdialysis – a methodological study. J Neurosurg 93: 37–43
13. Hutchinson PJA, O'Connell MT, Al-Rawi PG, Maskell LB, Gupta AK, Hutchinson DBA, Pickard JD, Kirkpatrick PJ, Wharton SB (1999) Neuropathological findings after microdialysis catheter implantation (letter). NeuroReport 10(3): i
14. Johnston AJ, Steiner LA, Coles JP, Chatfield DA, Fryer TD, Smielewski P, Hutchinson PJ, O'Connell MT, Al-Rawi PG, Aigbirhio FI, Clark JC, Pickard JD, Gupta AK, Menon DK (2005) Effect of cerebral perfusion pressure augmentation on regional oxygenation and metabolism after head injury. Crit Care Med 33: 189–195
15. Kett-White R, Hutchinson PJ, Czosnyka M, Boniface S, Pickard JD, Kirkpatrick PJ (2002) Multi-Modal Monitoring of Acute Brain Injury. Adv Tech Stand Neurosurg 27: 87–134.
16. Maas AI, Dearden M, Teasdale GM, Braakman R, Cohadon F, Iannotti F, Karimi A, Lapierre F, Murray G, Ohman J, Persson L, Servadei F, Stocchetti N, Unterberg A (1997) EBIC-guidelines for management of severe head injury in adults. European Brain Injury Consortium. Acta Neurochir Wien 139: 286–294
17. Magnoni S, Ghisoni L, Locatelli M, Caimi M, Colombo A, Valerian V, Stocchetti N (2003) Lack of improvement in cerebral metabolism after hyperoxia in severe head injury: a microdialysis study. J Neurosurg 98: 952–958
18. Major O, Shdanova T, Duffek L, Nagy Z (1990) Continuous monitoring of blood-brain barrier opening to 51Cr-EDTA by microdialysis following probe injury. Acta Neurochir [Suppl] 51: 46–48
19. Marion DW, Puccio A, Wisniewski SR, Kochanek P, Dixon CE, Bullian L, Carlier P (2002) Effect of hyperventilation on extracellular concentrations of glutamate, lactate, pyruvate, and local cerebral blood flow in patients with traumatic brain injury. Crit Care Med 30: 2619–2625
20. Meyerson BA, Linderoth B, Karlsson H, Ungerstedt U (1990) Microdialysis in the human brain: extracellular measurements in the thalamus of parkinsonian patients. Life Sci 46: 301–308
21. Nelson DW, Bellander BM, Maccallum RM, Axelsson J, Alm M, Wallin M, Weitzberg E, Rudehill A (2004) Cerebral microdialysis of patients with severe traumatic brain injury exhibits highly individualistic patterns as visualised by cluster analysis with self-organising maps. Crit Care Med 2428–2436
22. Persson L, Hillered L (1992) Chemical monitoring of neurosurgical intensive care patients using intracerebral microdialysis. J Neurosurg 76: 72–80
23. Sarrafzadeh AS, Kiening KL, Callsen TA, Unterberg AW (2003) Metabolic changes during impending and manifest cerebral hypoxia in traumatic brain injury. Br J Neurosurg 17: 340–346
24. Stahl N, Schalen W, Ungerstedt U, Nordstrom CH (2001) Bedside biochemical monitoring of the penumbra zone surrounding an evacuated acute subdural haematoma. Acta Neurol Scand 108: 211–215
25. Stahl N, Ungerstedt U, Nordstrom CH (2001) Brain energy metabolism during controlled reduction of cerebral perfusion pressure in severe head injuries. Intensive Care Med 27: 1215–1223
26. Tolias CM, Reinert M, Seiler R, Gilman C, Scharf A, Bullock MR (2004) Normobaric hyperoxia-induced improvement in cerebral metabolism and reduction in intracranial pressure in patients with severe head injury: a prospective historical cohort-matched study. J Neurosurg 435–444
27. Ungerstedt U (1984) Measurement of neurotransmitter release by intracranial dialysis. In: Marsden CA (ed) Measurement of Neurotransmitter Release in vivo. John Wiley & Sons Ltd, Chichester, p 81–105
28. Ungerstedt U (1991) Microdialysis-principles and applications for studies in animals and man. J Intern Med 230: 365–373
29. Ungerstedt U, Pycock C (1974) Functional correlates of dopamine neurotransmission. Bull Schweiz Akad Med Wiss 30: 44–55
30. Vespa P, Prins M, Ronne-Engstrom E, Caron M, Shalmon E, Hovda DA, Martin NA, Becker DP (1998) Increase in extracellular glutamate caused by reduced cerebral perfusion pressure and seizures after traumatic brain injury: a microdialysis study. J Neurosurg 89: 971–982
31. Whittle IR, Glasby M, Lammie A, Ball H, Ungerstedt U (1998) Neuropathological findings after intracerebral implantation of microdialysis catheters. NeuroReport 9: 2821–2825

Correspondence: P. J. Hutchinson, Academic Department of Neurosurgery, University of Cambridge, Box 167, Addenbrooke's Hospital, Cambridge, UK, CB2 2QQ. e-mail: pjah2@cam.ac.uk

Acta Neurochir (2005) [Suppl] 95: 447–451

Brain tissue oxygenation monitoring in acute brain injury

I. Ng[1], **K. K. Lee**[2], and **J. Wong**[1]

[1] Acute Brain Injury Research Laboratory, Department of Neurosurgery, National Neuroscience Institute, TTSH Campus, Singapore
[2] Acute Brain Injury Research Laboratory, Department of Nursing Administration, National Neuroscience Institute, TTSH Campus, Singapore

Summary

Cerebral ischemia is one of the most important causes of secondary insults following acute brain injury. While intracranial pressure monitoring in the intensive care unit constitutes the cornerstone of neurocritical care monitoring, it does not reflect the state of oxygenation of the injured brain. The holy grail of neuromonitoring is a modality that would reflect accurately real time the status of oxygenation in the tissue of interest, is robust, artefact free and that which provides information that can be used for therapeutic interventions and to improve outcome. Such a device could conceivably be used to augment the sensitivity of current multi-modality monitoring systems in the neurocritical management of brain injured patients. This article examines the availability of data in the literature to support clinical use of local tissue oxygen probes in intensive care.

Keywords: Brain tissue oxygen; acute brain injury; clinical application; prognostic value; outcome.

Introduction

Cerebral hypoxia, as a result of injury, is recognised as one of the most important causes of secondary insults in patients with acute brain injury. While intracranial pressure monitoring and cerebral perfusion maintenance form the cornerstone of neuromonitoring in the neurocritical care setting, they do not reflect the state of oxygenation in the internal milieu of the brain. As such, the holy grail of neuromonitoring has been the reliable identification of inadequate cerebral oxygenation to prevent secondary brain injury and unfavourable outcome in patients with acute cerebral lesions. In the past two decades, techniques of real time monitoring cerebral oxygenation via implanted brain tissue sensors has allowed the real-time tracking of tissue oxygenation in the region of interest. While large amounts of investigative work on local brain tissue oxygenation (PtiO2) have been published in the literature, the role of monitoring local tissue oxygenation in the clinical setting has been undefined. In this article, we examine the current evidence available to make the case for using this technology routinely in the neurocritical care management of acute brain injury.

Brain tissue oxygenation (PtiO2) monitoring

The measurement of PtiO2 can be performed with two commercially available disposable and sterile probes: the polarographic Clarke-type probe (Licox, Integra Neuroscience, Plainsboro, NJ) which measures PtiO2 and brain temperature; the Neurotrend (Diametrics Medical, St Paul, Mn) which is a multiparameter sensor that measures brain pH, carbon dioxide, oxygen and temperature based on a colorimetric method using optical fluorescence. Both systems may be introduced through a dedicated two or three-way bolt that allows the introduction of catheters for monitoring of intracranial pressure, PtiO2, or brain temperature, or for microdialysis. Alternatively, the probes may be tunnelled and positioned locally in the penumbra of a lesion after intracranial surgery. The differences in the two commercially available systems are summarised in Table 1.

For any real time monitoring device to be applied to neurocritical care, it must fulfil the following criteria:

- Catheter insertion must be safe;
- Data should be accurate, real-time and artifact free;
- Data can be used to predict prognosis;
- Detect adverse events reliably;
- Data can be used to fine tune treatment;
- Impact on outcome.

Table 1. *Comparison of Licox and Neurotrend system*

Comparison of Licox and Neurotrend system		
	Licox	Neurotrend
Parameters measured	PtiO2, brain temperature	PtiO2, PtiCo2, pH, brain temperature
Method	Polaragraphic	Colorimetric
Bolt	3-way bolt * A new 2-way bolt is currently available	2-way bolt
Ischemic Threshold	10 mm Hg	19 mm Hg
Calibration	Smart card	Gas

Accuracy and safety of tissue oxygen probes

Various studies [4, 26, 31–32] have demonstrated good accuracy of the probes in in-vitro and in-vivo conditions with local tissue oxygen probes. The data quality is excellent with few artifacts and the catheters demonstrated very little drift, although the Neurotrend system tends to over-estimate PtiO2 at very low levels.

In a review of nine studies conducted between 1996 and 2000 on the safety parameters in 250 patients, Littlejohns *et al.* [16] found few complications with the insertion of local oxygen probes. There were 2 small hematomas, and no infections reported. Nevertheless, the determination of bleeding caused by the probe may be difficult as suggested by Dings *et al.* [5] as three catheters (Intracranial Pressure, temperature, and oxygen probes) are inserted through a single bolt. Zauner *et al.* [32] and Van den Brink *et al.* [28] demonstrated additionally with microscopic studies that local damage with the use of the brain tissue oxygen sensors is minimal. Our findings are similar to previous studies and suggest that brain tissue catheter probe insertion should not carry any risk higher than the standard ICP insertion (Table 2). Nevertheless, the inadvertent placement of the catheter near or within areas of hemorrhage may lead to the sensor displaying artifactually low readings of oxygen tension. It is therefore recommended that the position of the catheter needs verification with CT scanning [27]. Increasing the fraction of inspired oxygen (FiO2) momentarily to verify that the sensor is functioning can test the response of the catheter properly before use [10].

Several studies have made a comparison between PtiO2 measurement and SjvO2 monitoring [8, 15, 29]. Kiening *et al.* [15] compared the use of PtiO2 and jugular venous oximetry (SjvO2) and found that the PtiO2

Table 2. *Our experiences in the use of Licox & Neurotrend system from Jan 2001 to Nov 2004*

Our Experiences in the use of Licox & Neurotrend System from Jan 2001 to Nov 2004		
	Licox (From 2002 Jan)	Neurotrend (From 2001 Jan)
Number of patients monitored	42	35
Total catheters used	49	50
No. of patient with artifact free data collected	37	14
Technical complications:		
– Catheter/bolt breakage	0	10
– Catheter/sensor malfunction	1	4
Adverse events		
• Infection	0	0
• Hematomas	0	2

monitoring technique seemed to be superior as it provided 95% artifact-free "good-quality" data as compared to 43% in the SjvO2. It could also provide twice the duration of monitoring as SjvO2. In contrast to the frequent re-calibration required by jugular bulb oximetry, monitoring PtiO2 only requires one calibration prior to catheter insertion without subsequent re-calibration. Additionally, monitoring PtiO2 provides nearly zero artefacts, which would suggest PtiO2 monitoring might be more suitable for continuous and routine use.

Technical complications, including catheter or bolt displacements/breakage and defective catheters are 13.6% [4]. The rate of displacement was noted to be higher with the surface sensors than with the inserted probes [18]. Our complication rates for insertion of the catheters are shown in Table 2.

Ischemic threshold and its prognostic value

PtiO2 levels in uninjured brain tissue range from 20–35 mm Hg in animal experiments and severely brain injured patients without intracranial hypertension or reduced CPP [17]. However, the depth and duration of tissue hypoxia were related to outcome and proved to be an independent predictor of unfavorable outcome and death. The likelihood of death increased with increasing duration of monitoring where PtiO2 is less than 15 mm Hg or any occurrence of a PtiO2 less than 6 [26]; Similar results have also been reported by other authors [1, 29]. Of great clinical significance was that early tissue hypoxia was observed in 50% of patients and persisted despite aggressive treatment for intracranial pressure and cerebral perfusion pressure [27]. Low PtiO2 is reported in >50% of patients in the first 24 hours after injury, and most critical in the first 8 to 12 hours with subsequent recovery, consistent with the results of cerebral blood flow studies reported by Bouma *et al.* [2].

Clinical studies performed using the Licox system suggested that the critical threshold for ischemia was between 5 and 10 mm Hg [4, 8, 15, 26, 29]. The critical threshold appears to be higher for Neurotrend catheters, 19–23 mm Hg [6–7, 32]. While the differences between the ischemic thresholds are not clear, it is likely to be due to differences in technology and/or the fact that Neurotrend catheters tend to over-estimate PtiO2 under ischemic situations [26].

Reliable detection of adverse events

Validation of PtiO2 with PET has been performed. While there was no correlation between PtiO2 and regional end-capillary venous oxygen tension, there was a significant association between change of PtiO2 and change in end-capillary oxygen tension; PtiO2 was also noted to be consistently lower than PvO2, suggesting that it reflects tissue oxygenation downstream of capillaries, i.e. it reflects the demand-supply relationship within the extracellular space of interest and is closely correlated to cerebral blood flow [9]. Comparison of PtiO2 and jugular venous oximetry showed that when inserted in the penumbra of a mass lesion, PtiO2 reflected the oxygenation in the tissue of interest and did not correlate with SjO2, which reflected a more global state of oxygenation. However, if the PtiO2 probe was inserted in a relatively "normal area" of brain, then PtiO2 was closely correlated with SjvO2. Certainly, when PtiO2 is intended to reflect global oxygenation states, the catheter should be inserted in non-lesioned brain tissue [23].

Data can be used to fine tune treatment

Conventional approaches to the management of severe brain injury patients have concentrated on a reduction in ICP and maintenance of adequate CPP to prevent secondary ischaemic insults. Nevertheless, conventional monitoring will miss 40% of adverse events in the neurocritical care unit [21]. In a prospective interventional study of 353 patients with severe TBI, Cruz [3] found that outcome at 6 months after injury was significantly better in the 178 patients who had monitoring and management of cerebral extraction of oxygen along with CPP as compared with 175 patients undergoing monitoring and management of CPP alone. However, monitoring of SjvO2 is prone to many technical problems and vigilance is required to obtain reliable data [15]. Despite frequent recalibration and special attention, problems such as poor light intensity, head movements and spontaneous waves contribute to the issues of poor data quality [25].

Hyperventilation showed detrimental effects on cerebral oxygenation, although ICP and CPP values were normalized. Schneider *et al.* [24] found that PtiO2 decreased to as low as 10 mm Hg during hyperventilation even though it may effectively decrease ICP and increased CPP. In Van den Brink's [28] study of 23 patients, 17 of them had PtiO2 decrease during the first 24 hours after injury due to decreases in PaCO2 associated with hyperventilation. Imberti *et al.* [12] found that PtiO2 detected with greater frequency cerebral ischemia brought about by hyperventilation. The above findings would suggest that optimized hyperventilation may be more effectively monitored with PtiO2 monitoring. However, SjvO2 reflects the global state of oxygenation while PtiO2 would reflect more on local states, as such, both two systems should be used together to provide complimentary information, as neither alone identifies all episodes of ischemia [8, 11–12, 15].

The use of inotropic support to augment CPP has also been investigated. Kiening *et al.* [14] discovered that when CPP < 60 mm Hg, PtiO2 decreased. However, when CPP was >60 mm Hg, the effect on PtiO2 was minimal. Johnston *et al.* [13] found that while there were no significant differences between norepinephrine and dopamine on cerebral oxygenation or metabolism either at baseline or following a CPP intervention, however, the response to a CPP intervention with dopamine seemed to be more variable than the response achieved with norepinephrine. As such, norepineprhine may be a more suitable inotrope in terms of perfusion and oxygenation augmentation.

Several studies have demonstrated an increasing brain tissue oxygen in response to increasing arterial oxygen tensions [29–30]. Additionally, Menzel *et al.* [20] and Reinert *et al.* [21] further showed that increasing FiO2 significantly increases PtiO2 with a concurrent decrease in brain microdialysate lactate suggesting that aerobic metabolism is improved. While the authors did not examine the outcome of these patients, it is conceivable that the administration of 100% oxygen to patients with acute brain injury can indeed improve cerebral oxygen delivery and metabolism during the critical early period.

Improvement of outcome

More recently, PtiO2 guided treatment to supplement ICP/CPP therapy has been suggested by several authors. Meixensberger *et al.* [18] studied 93 patients with severe traumatic brain injury and found that cerebral hypoxic events can be reduced significantly by increasing CPP. However, no statistical differences can be observed in outcome although there was a trend towards lower mortality in the group with PtiO2 guided therapy.

Despite the preceding discussion on PtiO2 and supportive evidence for its prognostic value and its potential use in the intensive care to fine tune therapy, the presence of a consistently high PtiO2 does not by itself guarantee a good outcome. Since PtiO2 reflects the oxygen-demand relationship only, it is conceivable that irreversible damage may have occurred at the time of injury and that the possibility of impaired cerebral oxygen utilization despite adequate oxygen delivery may exist. The use of a prospective randomized trial to evaluate the usefulness of PtiO2 monitoring is logical but we must be mindful of the fact that we are evaluating a monitoring modality and not a therapeutic modality. Optimal algorithms for oxygen targeted therapy utilizing the information afforded by the use of local tissue oxygen probes is still lacking and would need to be studied first before we embark on a multi-centre trial to assess the overall efficacy of targeted therapy in improving outcome. As such, a reasonable strategy may be to ascertain if PtiO2 monitoring can lead to an increased detection of reversible secondary deficits as compared to conventional modalities. Current cohort studies have provided compelling evidence for use of PtiO2 in intensive care but given the known limitations of monitoring, it may be prudent to combine PtiO2 and SjO2 as part of multi-modality neuromonitoring to detect cerebral hypoxia.

Conclusion

With a mounting body of experimental and clinical data, the routine clinical use of PtiO2 may be justified by the fact that it is safe, accurate, robust and has an excellent prognostic value. Additionally, the information obtained may be used to fine-tune treatment. However, evidence for efficacy in improvement in outcome is lacking. Nevertheless, we are evaluating a monitoring modality and its efficacy can only be reliably assessed in concert with targeted therapy. As such, optimal algorithms for oxygen targeted therapy will have to be elucidated before a multi-centre trial to test its efficacy can be undertaken.

Support

These studies were supported by the National Medical Research Council NRN01/002 awarded to Dr Ivan Ng.

References

1. Bardt T, Unterberg A, Haertl R, Kiening K, Schneider G, Lanksch W (1998) Monitoring of brain tissue PO2 in traumatic brain injury: effect of cerebral hypoxia on outcome. Acta Neurochir [Suppl] 71: 153–156
2. Bouma GJ, Muizelaar JP, Stringer WA *et al* (1992) Ultra-early evaluation in regional cerebral blood flow in severely head injured patients using xenon-enhanced computerized tomography. J Neurosurg 77: 360–368
3. Cruz J (1998) The first decade of continuous monitoring of jugular bulb oxyhemoglobin saturation: Management strategies and clinical outcome. Crit Care Med 26: 344–351
4. Dings J, Meixensberger J, Jager A, Roosen K (1998) Clinical experiences with 118 brain tissue oxygen partial pressure catheter probes. Neurosurgery 43: 1082–1095
5. Dings J, Meixensberger J, Jager A, Roosen K (1997) Brain tissue-PO2 monitoring; catheter stability and complications. Neurol Res 19: 241–245
6. Doppenberg EMR, Zauner A, Watson JC, Bullock R (1998) Determination of the ischemic threshold for brain oxygen threshold. Acta Neurochir [Suppl] 71: 166–169
7. Doppenberg EMR, Watson JC, Broaddus WC, Holloway KL, Young HF, Bullock R (1997) Intra-operative monitoring of substrate delivery during aneurysm and hematoma surgery: initial experience in 16 patients. J Neurosurg 87: 809–816
8. Gopinath SP, Valadka AB, Uzura M *et al* (1999) Comparison of jugular venous oxygen saturation and brain tissue PO2 as monitors of cerebral ischemia after head injury. Crit Care Med 27: 2337–2345
9. Gupta AK, Hutchinson PJ, Fryer T, Al-Rawi PG, Parry DA, Minhas PS, Kett-White R, Kirkpatrick PJ, Mathews JC, Downey S, Aigbirhio F, Clark J, Pickard JD, Menon DK (2002) Measurement of brain tissue oxygenation performed using positron emission tomography scanning to validate a novel monitoring method. J Neurosurg 96: 263–268
10. Haitsma IK, Maas AIR (2002) Advanced monitoring in the intensive care unit: brain tissue oxygen tension. Curr Opin Crit Care 8: 115–120
11. Hilton G (2000) Cerebral oxygenation in the traumatically brain-injured patient: Are ICP and CPP enough? Journal of Neuroscience Nursing 32: 278–282

12. Imberti R, Bellinzona G, Langer M (2002) Cerebral tissue PO2 and SjvO2 changes during moderate hyperventilation in patients with severe traumatic brain injury. J Neurosurg 96: 97–102
13. Johnston AJ, Steiner LA, Chatfield DA, Coles JP, Hutchinson PJ, Al-Rawi PG, Menon DK, Gupta AK (2004) Effect of cerebral perfusion pressure augmentation with dopamine and norepinephrine on global and focal brain oxygenation after traumatic brain injury. Intensive Care Medicine 30: 791–797
14. Kiening K, Schneider G, Bardt T *et al* (1997) Bifrontal measurements of brain tissue-pO2 in comatose patients. Acta Neurochir Res 19: 233–240
15. Kiening KL, Unterberg AW, Bardt TF, Schneider G, Lanksch WR (1996) Monitoring of cerebral oxygenation in patients with severe head injuries: brain tissue PO2 versus jugular vein oxygen saturation. J Neurosurg 85: 751–757
16. Littlejohns LR, Bader MK, March K (2003) Brain tissue oxygen monitoring in severe brain injury, I: Research and usefulness in critical care. Critical Care Nurse 23: 17–27
17. Maas A, Fleckenstein W, de Jong D, van Santbrink H (1993) Monitoring cerebral oxygenation: experimental studies and preliminary clinical results of continuous monitoring of cerebrospinal fluid and brain tissue oxygen tension. Acta Neurochir [Suppl] 59: 50–57
18. Meixensberger J, Jaeger M, Väth A, Dings J, Kunze E, Roosen K (2003) Brain tissue oxygen guided treatment supplementing ICP/CPP therapy after traumatic brain injury. J Neurol Neurosurg Psychiatry 74: 760–764
19. Meixensberger J, Dings J, Kuhnigk H, Roosen K (1993) Studies of tissue pO2 in normal and pathological human brain cortex. Acta Neurochir [Suppl] 59: 58–63
20. Menzel M, Doppenberg EMR, Zauner A, Soukup J, Reinert MM, Bullock R (1999) Increased inspired oxygen concentration as a factor in improved brain tissue oxygenation and tissue lactate levels after severe human head injury. J Neurosurg 91: 1–10
21. Reinert M, Barth A, Rothen HU, Schaller B, Takala J, Suiler RW (2003) Effects of cerebral perfusion pressure and increased fraction of inspired oxygen on brain tissue oxygen, lactate and glucose in patients with severe head injury. Acta Neurochirurgica 145: 341–350
22. Robertson C (1993) Desaturation episodes after severe head injury: influence on outcome. Acta Neurochir [Suppl] 59: 98–101
23. Sarrafzadeh AS, Kiening KL, Bardt TF *et al* (1998) Cerebral oxygenation in contusioned vs nonlesioned brain tissue: monitoring of PtiO2 with Licox and Paratrend. Acta Neurochir [Suppl] 71: 186–189
24. Schneider GH, Sarrafzadeh AS, Kiening KL, Bardt TF, Unterberg AW, Lanksch WR (1998) Influence of hyperventilation on brain tissue – PO2, PCO2, and pH in patients with intracranial hypertension. Acta Neurochir (Wien) 71: 62–65
25. Sheinberg M, Kanter MJ, Robertson CS *et al* (1992) Continuous monitoring of jugular venous oxygen saturation in head-injured patients. J Neurosurg 76: 212–217
26. Valadka AB, Shankar P, Gopinath SP, Charles F, Uzura M, Robertson CS (1998) Relationship of brain tissue PO2 to outcome after severe head injury. Crit Care Med 26: 1576–1581
27. Van den Brink WA, van Santbrink H, Steyerberg EW, Avezaat CJJ, Suazo JAC, Hogesteeger C, Jansen WJ, Kloos LMH, Vermeulen J, Maas AIR (2000) Brain oxygen tension in severe head injury. Neurosurgery 46: 868–876
28. Van den Brink WA, Haitsma IK, Avezaat CJJ, Houtsmuller AB, Kros JM, Maas AIR (1998) Brain parenchymal/pO2 catheter interface: a histopathologic study in the rat. J Neurotrauma 15: 813–824
29. Van Santbrink H, Mass AI, Avezaat CJ (1996) Continuous monitoring of partial pressure of brain tissue oxygen in patients with severe head injury. Neurosurgery 38: 21–31
30. Zauner A, Clausen T, Alves OL, Rice A, Levasseur J, Young HF, Bullock R (2002) Cerebral metabolism after fluid-percussion injury and hypoxia in a feline model. J Neurosurg 97: 643–649
31. Zauner A, Doppenberg EMR, Woodward JJ *et al* (1997) Continuous monitoring of substrate delivery and clearance; initial experience in 24 patients with severe acute brain injuries. Neurosurgery 41: 1082–1093
32. Zauner A, Bullock R, Di X, Young HF (1995) Brain oxygen, CO2, pH, and temperature monitoring: evaluation in the feline brain. Neurosurgery 37: 1168–1176

Correspondence: Ivan Ng, Department of Neurosurgery, National Neuroscience Institute, 11 Jalan Tan Tock Seng Singapore 308433. e-mail: ivan_ng@nni.com.sg

Acta Neurochir (2005) [Suppl] 95: 453–457

Near infrared spectroscopy in brain injury: today's perspective

P. G. Al-Rawi

Academic Neurosurgery Unit, Addenbrooke's Hospital, Cambridge, UK

Summary

The technique of near infrared spectroscopy (NIRS) is based on the principle of light attenuation by the chromophores oxy-haemoglobin (HbO_2), deoxyhaemoglobin (Hb) and cytochrome oxidase. Changes in the detected light levels can therefore represent changes in concentrations of these chromophores.

Clinical use of NIRS in the brain has been well established in neonates where transillumination is possible. While it has become a useful research tool for monitoring the adult brain, clinical application has been hampered by the fact that it must be applied in reflectance mode. This has resulted in a number of concerns, most significantly the issue of signal contamination by the extracranial tissue layers. Algorithms have been applied to try to overcome this problem, and techniques such as time resolved, phase resolved and spatially resolved spectroscopy have been developed.

There has been renewed interest in NIRS as an easy to use, non-invasive technique for measuring tissue oxygenation in the adult brain. Recent technical advances have led to the development of compact, portable instruments that detect changes in optical attenuation of several wavelengths of light.

Near infrared spectroscopy is an evolving technology that holds significant potential for technical advancement. In particular, NIRS shows future promise as a clinical tool for bedside cerebral blood flow measurements and as a cerebral imaging modality for mapping structure and function.

Keywords: Near infrared spectroscopy; cerebral oxygenation; brain injury; cerebral oximetry.

Introduction

The use of in-vivo tissue near infrared spectroscopy in humans was first described more than 25 years ago by FF Jöbsis [28]. The technique is based on the concept that light of wavelengths 680–1000 nm is able to penetrate human tissue and is absorbed by the chromophores oxy-haemoglobin (HbO_2), deoxyhaemoglobin (Hb) and cytochrome oxidase. Changes in the detected light levels can therefore represent changes in concentrations of these chromophores. The non-invasive nature of the technique led to its first clinical application for monitoring the cerebral oxygenation status of premature infants [7]. Since then it has become an established research tool with numerous applications [3, 8, 15, 17, 36, 42, 53, 56–58].

While its clinical use for monitoring the brain has been well established in neonates, where transillumination is possible due to the thin skull and small dimensions, clinical application of NIRS for monitoring the adult brain has been hampered by the fact that it must be applied in reflectance mode [62]. This has resulted in concerns about quantification, the volume and type of tissue being illuminated, and most significantly the issue of signal contamination by the extracranial tissue layers [19, 20, 22, 30, 35, 38, 49]. A number of algorithms have been applied to try to overcome these issues, and techniques such as time resolved, phase resolved and spatially resolved spectroscopy have been developed [1, 11, 41, 44]. Initially, NIRS was not quantified and only provided a trend of increased or decreased oxygenation.

Over the last 2–3 years, there has been renewed interest in NIRS as an easy to use, non-invasive technique for measuring tissue oxygenation in the adult brain. Recent technical advances have led to the development of instruments that detect changes in optical attenuation of several wavelengths of light, with the potential to derive a tissue oxygen saturation from spontaneous Hb and HbO_2 signal changes [1, 16, 23].

Principles

The theory behind NIRS has been described in detail previously [9]. The basic principle is that near infrared light (delivered via optodes placed on the skin) penetrates the scalp and brain tissue and is subjected to scatter and absorption within the various tissues. While some of these media can be considered to have a fixed concentration (e.g. bone, melanin, bilirubin, lipids, water), and hence fixed scattering and absorption, the concentration of Hb and HbO_2 may vary with time or with oxygenation status. Cytochrome oxidase, the terminal enzyme of the respiratory chain, also contributes to the near infrared spectrum of cortical tissue [59].

Assuming constant scattering properties and applying knowledge of the absorption spectra for oxy- and deoxyhaemoglobin, measurements of light attenuation are converted into concentrations of these chromophores using a modified Beer-Lambert law, which includes a differential pathlength factor to account for scattering within the tissue, and can be defined as:

$$\Delta A = L \cdot \Delta \mu a$$

where,

A = light attenuation
L = differential pathlength, and
μa = absorption scattering coefficient

This concept works very well in the infant head where the skull is thin enough to allow transillumination of light from one side to the other. In the adult, the relative thickness of scalp, skull and brain prevents transmission spectroscopy and NIRS must be used in reflectance mode, with the optodes situated on the same side of the head. This results in the introduction of unknown, nonlinear variables for light absorption and scattering coefficients [45]. Various algorithms have been applied to attempt quantification, but claims as to their accuracy remain controversial [33, 49].

Instrumentation

NIRS instrumentation has continued to evolve. Early machines were large and cumbersome, showed considerable drift and were prone to both movement and light artefact. They were really only useful as trend monitors. Evaluation of the technology culminated in the mid-nineties with numerous publications assessing several devices in different clinical scenarios. Current commercial machines detect changes in optical attenuation of a number of wavelengths of light, and are compact and portable, enabling non-invasive measurements of cerebral oxygenation at the bedside [1, 12, 40, 61]. However, it should be recognised that values obtained from different instruments are likely to give different cerebral tissue oxygen saturations because the data is being derived from different volumes of tissue [11]. Of the current techniques on offer, spatially resolved spectroscopy appears to be the most promising.

Spatially resolved spectroscopy

Spatially resolved spectroscopy has been described in previous publications [1, 39]. Spatial resolution relies upon the measurement of the attenuation gradient as a function of source-detector separation. Using a modified diffusion equation, a product of the absorption and scattering coefficients is calculated [52]. Treating the tissue as homogeneous, the scattering coefficient can be assumed to be a constant (k) in the near infrared wavelength. In order to increase the accuracy of the calculation however, a wavelength dependency for the scattering coefficient is derived in the form of:

$$k(1 - h\lambda)$$

where, λ = wavelength and h is the normalised slope of the scattering coefficient along λ [39]. From here, the relative absorption coefficients and thus the relative concentrations of HbO_2 and Hb can be obtained.

Clinical application

Early experience with NIRS has shown that in patients with head injury, changes in oxy-haemoglobin correlate well with changes in jugular bulb oximetry (SjO_2), transcranial Doppler (TCD) and laser Doppler [32]. Additionally, NIRS can potentially provide a more sensitive indicator of desaturation events than SjO_2 [32]. However, studies have demonstrated that despite using large optode distances there was still a major contribution from the extracranial tissues to the NIRS signal [19]. In a study using the NIRO 500 (Hamamatsu Photonics, KK), it was demonstrated that once the extracranial component was subtracted, a threshold for severe critical ischaemia could be defined in patients undergoing carotid endarterectomy [31].

More recently, it has been shown that spatially resolved spectroscopy (as used in the NIRO 300, Hamamatsu Photonics, KK) reflects cerebral oxygenation in the adult brain to a high degree of sensitivity and specificity, and a threshold for cerebral ischaemia has been defined [1, 2].

Using spatially resolved reflectance spectroscopy, the NIRO 300 (Hamamatsu Photonics KK), gives simultaneous measurement of Tissue Oxygen Index (TOI) and haemoglobin concentration changes [1,

52]. TOI is the ratio of oxygenated to total tissue haemoglobin and can be expressed as:

$$\frac{HbO_2}{HbO_2 + Hb} \times 100$$

However, the accuracy of the calculated TOI as a quantitative measure of brain tissue saturation still needs to be assessed in vivo and compared with known cerebral tissue saturation and haemoglobin concentrations. Normal ranges of TOI need to be defined and further validation studies to investigate NIRO 300 reliability in different clinical scenarios are required. In particular, the local changes that frequently occur in the extracranial tissues of head injured patients, e.g. the development of oedema, will cause increased extracerebral attenuation of the near infrared light and hence lead to potential inaccuracies.

Cerebral blood volume and cerebral blood flow

Since NIRS can measure Hb and HbO_2 it can also measure total haemoglobin (THb). This makes it possible to estimate cerebral blood volume (CBV). Assuming the whole blood Hb concentration & the large to small cerebral vessel haematocrit ratio remains constant for the duration of the measurement, changes in THb may be assumed to imply a change in CBV:

$$\Delta\text{CBV} = \Delta[\text{THb}] \times (0.89/\text{Hb})$$

Calculation of absolute cerebral blood volume (CBV) and cerebral blood flow (CBF) are based on the generalised application of Fick's Principle, with an assumption made that there is no venous outflow of the chromophore during the time interval of the measurement and using HbO_2 and Hb as endogenous intravascular tracer molecules [13, 14]. Quantified CBV and CBF can therefore be obtained by, for example, using graded arterial hypoxaemia, thus varying the HbO_2 [13, 50, 60]. Since measurements of quantification are made over a longer period of time and are based on changes from a stable baseline, signal averaging and rapid response of the NIRS equipment is not essential.

This technique has been further modified by using indocyanine green (ICG) as the intravascular tracer [6, 27, 46, 47, 51, 54]. Indocyanine green is a strong near-infrared absorber. Bolus injection of ICG dye is given via a central vein and the arterial concentration measured either by repeated blood sampling or with a peripheral dye densitometer. The rate of arrival of ICG in the brain is detected by NIRS. Using a modified Fick principle as previously described, the amount of ICG dye delivered to the brain can be calculated from the area under the curve of the arterial ICG concentration change. CBF, CBV and mean transit time can then be calculated [21, 29]. Published results to date suggest that CBF is underestimated by this method. Further validation of this technique as a bedside measurement of CBF in adults is needed.

More recently the technique of tissue dye densitometry, based on tissue blood measurements of ICG rather than arterial blood, has been employed in order to estimate the absolute cerebral blood volume and total circulating blood volume in neonates, following a single injection of ICG [37].

Near infrared optical imaging

The first imaging spectrometer capable of providing a two-dimensional image of the neonatal brain was described by Van Houten *et al.* [55]. Since then, continuous real-time optical imaging of the brain has been shown to be both feasible and achievable [24, 43]. The major requirement for bedside optical imaging is an NIRS device that can accurately measure photon pathlength from multiple detectors and process this data into tomographic images. This has necessitated development of highly sophisticated image reconstruction algorithms. The detailed principles of these techniques are beyond the scope of this article, and can be found elsewhere [4, 24].

With the focus primarily on the pursuit of three-dimensional optical tomography, instruments are being developed and evaluated for use in the clinical environment [25]. There are clear clinical benefits of a device that can image both oxygenation and function non-invasively at the bedside, and several groups are working towards development of a functional or biochemical optical image of the brain [5, 10, 18, 25, 26, 34, 48].

Conclusion

In conclusion, to date NIRS is still not used routinely and remains primarily a research tool. However NIRS is an evolving technology that holds significant potential for technical advancement. Improvements in the methodology, quantitative accuracy and regional specificity will increase its clinical applicability. In particular, NIRS shows future promise as a low cost non-invasive clinical tool for bedside cerebral blood flow measurements and as a cerebral imaging modality for mapping structure and function. Within the next few years it is likely that more multichannel imaging near infrared spectrometers will be developed enabling detailed spatial information on cerebral oxygenation and perfusion to be obtained at the bedside. However, considerable research efforts and future clinical validation studies are still required.

References

1. Al-Rawi PG, Smielewski P, Kirkpatrick PJ (2001) Evaluation of a near infrared spectrometer (NIRO 300) for the detection of intracranial oxygenation changes in the adult head. Stroke 32: 2492–2500
2. Al-Rawi PG, Smielewski P, Kirkpatrick PJ (2002) Threshold for cerebral ischaemia defined by Tissue Oxygen Index (TOI) using near infrared spectroscopy
   Ref Type: Conference Proceeding
3. Aldrich CJ, D'Antona D, Wyatt JS, Spencer JAD, Peebles DM, Reynolds EOR (1994) Fetal cerebral oxygenation measured by near infrared spectroscopy shortly before birth and acid-base status at birth. Obstetrics Gynaecol 84: 861–866
4. Arridge SR, Schweiger M (1997) Image reconstruction in optical tomography. Phil Trans R Soc Lond B 352: 717–726
5. Benaron DA, Hintz SR, Villringer A, Boas D, Kleinschmidt A, Frahm J, Hirth C, Obrig H, van Houten JC, Kermit EL, Cheong WF, Stevenson DK (2000) Noninvasive functional imaging of human brain using light. J Cereb Blood Flow Metab 20: 469–477
6. Boushel R, Langberg H, Olesen J, Nowak M, Simonsen L, Bülow J, Kjær M (2000) Regional blood flow during exercise in humans measured by near infrared spectroscopy and indocyanine green. J Appl Physiol 89: 1868–1878
7. Brazy JE, Lewis DV, Mitnik MH, Jöbsis FF (1985) Non invasive monitoring of cerebral oxygenation in preterm infants: preliminary observations. Paediatrics 75: 217–225
8. Brown R, Wright G, Royston DA (1993) A comparison of two systems for assessing cerebral venous oxyhaemoglobin saturation during cardiopulmonary bypass in humans. Anaesthesia 48: 697–700
   Ref Type: Journal (Full)
9. Cope M, Delpy DT (1988) System for long-term measurement of cerebral blood and tissue oxygenation on newborn infants by near infra-red transillumination. Med Biol Eng Comp 26: 289–294
10. Danen RM, Wang Y, Li XD, Thayer WS, Yodh AG (1985) Regional imager for low resolution functional imaging of the brain with diffusing near-infrared light. Photochemistry and Photobiology 67: 33–40
11. Delpy DT, Cope M (1997) Quantification in tissue near-infrared spectroscopy. Phil Trans R Soc Lond B 352: 649–659
12. Dullenkopf A, Frey B, Baenziger O, Gerber A, Weiss M (2003) Measurement of cerebral oxygenation state in anaesthetized children using the INVOS 5100 cerebral oximeter. Paediatric Anaesthesia 13: 384–391
13. Edwards AD, Wyatt JS, Richardson CE, Delpy DT, Cope M, Reynolds EOR (1988) Cotside measurements of cerebral blood flow in ill newborn infants by near infrared spectroscopy. Lancet ii, 770–771
   Ref Type: Journal (Full)
14. Elwell CE, Cope M, Edwards AD, Wyatt JS, Delpy DT, Reynolds EOR (1994) Quantification of adult cerebral haemodynamics by near-infrared spectroscopy. App Physiol 77: 2753–2760
15. Elwell CE, Matcher SJ, Tyszczuk L, Meek JH, Delpy DT (1997) Measurement of cerebral venous saturation in adults using near infrared spectroscopy. Adv Exp Med Biol 411: 453–460
16. Fantini S, Franceschini MA, Maier JS, Walker SA, Barbieri B, Gratton E (1995) Frequency-domain multichannel optical detector for non-invasive tissue spectroscopy and oximetry. Opt Eng 34: 32–42
17. Ferrari M, De Marchis C, Giannini I (1986) Cerebral blood volume and haemoglobin oxygen saturation monitoring in the neonatal brain by near infrared spectroscopy. Adv Exp Med Biol 200: 203–211
18. Franceschini MA, Toronov V, Filiaci ME, Gratton E, Fantini S (2000) On line optical imaging of the human brain with 160 ms temporal resolution. Optics Express 6: 49–57
19. Germon TJ, Evans DH, Barnett N, Wall P, Manara AR, Nelson RJ (1999) Cerebral near infrared spectroscopy: emitter-detector separation must be increased. Br J Anaesth 82: 831–837
20. Germon TJ, Young AER, Manara AR, *et al* (1995) Extracerebral absorption of near infrared light influences the detection of increased cerebral oxygen monitored by near infrared spectroscopy. J Neurol Neurosurg Psychiatry 58: 477–479
21. Gora F, Shinde S, Elwell CE, Goldstone JC, Cope M, Delpy DT, Smith M (2002) Noninvasive measurement of cerebral blood flow in adults using near-infrared spectroscopy and indocyanine green: a pilot study. J Neurosurg Anaesthesiol 14: 218–222
22. Harris DN, Cowans FM, Wertheim DA (1994) Near infrared spectroscopy in the temporal region: strong influence of external carotid artery. Adv Exp Med Biol 345: 825–828
23. Hazeki O, Tamura M (1988) Quantitative analysis of haemoglobin oxygenation state of brain *in situ* by near-infrared spectrophotometry. J App Physiol 64: 796–802
24. Hebden JC (2003) Advances in optical imaging of the newborn infant brain. Psychophysiology 40: 501–510
25. Hebden JC, Gibson A, Austin T, Yusof RMd, Everdell N, Delpy DT, Arridge SR, Meek JH, Wyatt JS (2004) Imaging changes in blood volume and oxygenation in the newborn infant brain using three-dimensional optical tomography. Phys Med Biol 49: 1117–1130
26. Hintz SR, Cheong WF, van Houten JP, Stevenson DK, Benaron DA (1999) Bedside imaging of intracranial hemorrhage in the neonate using light: comparison with ultrasound, computed tomography, and magnetic resonance imaging. Pediat Res 45: 54–59
27. Hopton P, Walsh TS, Lee A (1999) Measurement of cerebral blood volume using near-infrared spectroscopy and indocyanine green elimination. J Appl Physiol 87: 1981–1987
28. Jöbsis FF (1977) Non-invasive infrared monitoring of cerebral and myocardial oxygen sufficiency and circulatory parameters. Science 198: 1264–1267
29. Keller E, Nadler A, Alkhadi H, Kollias S, Yonekawa Y (2002) Noninvasive measurement of regional cerebral blood flow and regional cerebral blood volume by near infrared spectroscopy and indocyanine green dye dilution. Unknown
30. Kirkpatrick PJ, Smielewski P, Al-Rawi P, Czosnyka M (1998) Resolving extra- and intracranial signal changes during adult near infrared spectroscopy. Neurol Res 20: S19–S22
31. Kirkpatrick PJ, Lam JMK, Al-Rawi PG, Smielewski P, Czosnyka M (1998) Defining thresholds for critical ischaemia by using near-infrared spectroscopy in the adult brain. J Neurosurg 89: 389–394

32. Kirkpatrick PJ, Smielewski P, Czosnyka M, Menon DK, Pickard JD (1995) Near infrared spectroscopy use in patients with head injury. J Neurosurg 83: 963–970
33. Komiyama T, Quaresima V, Shigematsu H, Ferrari M (2001) Comparison of two spatially resolved near-infrared photometers in the detection of tissue oxygen saturation: poor reliability at very low oxygen saturation. Clin Sci 101: 715–718
34. Kusaka T, Kawada K, Okubo K, Nagano K, Namba M, Okada H, Imai T, Isobe K, Itoh S (2004) Noninvasive optical imaging in the visual cortex in young infants. Human Brain Mapping 22: 122–132
35. Kytta J, Ohman J, Tanskanen P, Randell T (1999) Extracranial contribution to cerebral oximetry in brain dead patients: a report of six cases. J Neurosug Anaesthesiol 11: 252–254
36. Lam JMK, Kirkpatrick PJ, Al-Rawi P, Smielewski P, Pickard JD (1996) Internal and external carotid contribution to near infrared spectroscopy (NIRS) during carotid endarterectomy (CE). Journal of Neurology, Neurosurgery and Psychiatry 61(5): 553
Ref Type: Abstract
37. Leung TS, Aladangady N, Elwell CE, Delpy DT, Costeloe K (2004) A new method for the measurement of cerebral blood volume and total circulating blood volume using near infrared spatially resolved spectroscopy and indocyanine green: application and validation in neonates. Pediatr Res 55: 134–141
38. Litscher G, Schwarz G (1997) Transcranial cerebral oximetry – is it clinically useless at this moment to interpret absolute values obtained by the INVOS 3100 cerebral oximeter? Biomed Tech (Berl) 42: 74–77
39. Matcher SJ, Kirkpatrick PJ, Nahid K, Cope M, Delpy DT (1993) Absolute quantification methods in tissue near infrared spectroscopy. Proc SPIE 2389: 486–495
40. McKeating EG, Monjardino JR, Signorini DF, Souter MJ, Andrews PJD (1997) A comparison of the INVOS 3100 and the Critikon 2020 near-infrared spectrophotometer as monitors of cerebral oxygenation. Anaesthesia 52: 136–140
Ref Type: Journal (Full)
41. Miwa M, Ueda Y, Chance B (1995) Development of time resolved spectroscopy system for quantitative non-invasive tissue measurement. Proc SPIE 2389: 142–149
42. Nollert G, Jonas RA, Reichart B (2000) Optimising cerebral oxygenation during cardiac surgery: a review of experimental and clinical investigations with near infrared spectrophotometry. Thorac Cardiovasc Surg 48: 247–253
43. Obrig H, Villringer A (2003) Beyond the visible; Imaging the human brain with light. J Cereb Blood Flow Metab 23: 1–18
44. Oda M, Yamashita Y, Nishimura G, Tamura M (1996) A simple and novel algorithm for time resolved multiwavelength oximetry. Physics Med Biol 40: 2093–2108
45. Okada E, Firbank M, Schweiger M, *et al* (1997) Theoretical and experimental investigation of near-infrared light propagation in a model of the adult head. Appl Opt 36: 21–31
46. Patel J, Marks K, Roberts I, Azzopardi D, Edwards AD (1998) Measurement of cerebral blood flow in newborn infants using near infrared spectroscopy with indocyanine green. Pediatr Res 43: 34–39
47. Roberts I, Fallon P, Kirkham FJ, Lloyd-Thomas A, Cooper C, Maynard R, Elliot M, Edwards AD (1993) Estimation of cerebral blood flow with near infrared spectroscopy and indocyanine green. Lancet 342: 1425
48. Schroeter ML, Bucheler MM, Muller K, Uludag K, Obrig H, Lohmann G, Tittgemeyer M, Villringer A, Yves von Cramon D (2004) Towards a standard analysis for functional near-infrared imaging. NeuroImage 21: 283–290
49. Schwarz G, Litscher G, Kleinert R, Jobstmann R (1996) Cerebral oximetry in dead subjects. J Neurosurg Anesthesiol 8(3): 189–193
Ref Type: Journal (Full)
50. Skov L, Pryds O, Griesen G (1991) Estimating cerebral blood flow in newborn infants: comparison of near infrared spectroscopy and ^{133}Xe clearence. Pediatr Res 30: 570–573
51. Springett R, Sakata Y, Delpy DT (2001) Precise measurement of cerebral blood flow in newborn piglets from the bolus passage of indocyanine green. Phys Med Biol 46: 2209–2225
52. Suzuki S, Takasaki S, Ozaki T, Kobayashi Y (1999) A tissue oxygenation monitor using NIR spatially resolved spectroscopy. Proc SPIE 3597: 582–592
53. Tamura M, Hoshi Y, Okada F (1997) Localised near-infrared spectroscopy and functional optical imaging of brain activity. Phil Trans R Soc Lond B 352: 737–742
54. Terborg C, Bramer S, Harscher S, Simon M, Witte OW (2004) Bedside assessment of cerebral perfusion reductions in patients with acute ischaemic stroke by near-infrared spectroscopy and indocyanine green. J Neurol Neurosurg Psychiatry 75: 38–42
55. van Houten JP, Benaron DA, Spilman S, *et al* (1996) Imaging brain injury using time-resolved near infrared light scanning. Pediat Res 39: 470–476
56. Vernieri F, Tibuzzi F, Pasqualetti P, Rosato N, Passarelli F, Rossini PM, Silvestrini M (2004) Transcranial Doppler and near-infrared spectroscopy can evaluate the haemodynamic effect of carotid artery occlusion. Stroke 35: 64–70
57. Villringer A, Planck J, Stodieck S, Boetzel K, Schleinkofer L, Dirnagl U (1994) Non invasive assessment of cerebral haemodynamics and tissue oxygenation during activation of brain function in human adults using near infrared spectroscopy. Adv Exp Med Biol 345: 559–565
58. Watanabe E, Nagahori Y, Mayanagi Y (2002) Focus diagnosis of epilepsy using near-infrared spectroscopy. Epilepsia 43: 50–55
59. Wray S, Cope M, Delpy DT, Wyatt JS, Reynolds EOR (1988) Characterisation of the near infrared absorption spectra of cytochrome aa3 and haemoglobin for the non-invasive monitoring of cerebral oxygenation. Biochim Biophys Acta 933: 184–192
60. Wyatt JS, Cope M, Delpy DT, Richardson CE, Edwards AD, Wray S, Reynolds EOR (1990) Quantification of cerebral blood volume in newborn infants by near infrared spectroscopy. J Appl Physiol 68: 1086–1091
61. Yoshitani K, Kawaguchi M, Tatsumi K, Kitaguchi K, Furuya H (2002) A comparison of the INVOS 4100 and the NIRO 300 near-infrared spectrometers. Anesth Analg 94: 586–590
62. Young AER, Germon TJ, Barnett NJ, Manara AR, Nelson RJ (2000) Behaviour of near-infrared light in the adult human head: implications for clinical near-infrared spectroscopy. Br J Anaesth 84: 38–42

Correspondence: P. G. Al-Rawi, Academic Neurosurgery Unit, Box 167, Level 4, A-Block, Addenbrooke's Hospital, Hills Road, Cambridge CB2 2QQ, UK. e-mail: pga20@medschl.cam.ac.uk

Acta Neurochir (2005) [Suppl] 95: 459–464

Imaging of cerebral blood flow and metabolism in brain injury in the ICU

J. D. Pickard[1,2], **P. J. Hutchinson**[1,2], **J. P. Coles**[1,3], **L. A. Steiner**[1,2], **A. J. Johnston**[1,3], **T. D. Fryer**[1], **M. R. Coleman**[1,2], **P. Smielewski**[1,2], **D. A. Chatfield**[1,3], **F. Aigbirhio**[1], **G. B. Williams**[1], **K. Rice**[4], **J. C. Clark**[1], **C. H. Salmond**[1,4], **B. J. Sahakian**[4], **P. G. Bradley**[1,3], **T. A. Carpenter**[1], **R. Salvador**[1], **A. Pena**[1,2], **J. H. Gillard**[1,5], **A. S. Cunningham**[1,3], **S. Piechnik**[1,2], **M. Czosnyka**[2], and **D. K. Menon**[1,3]

[1] Wolfson Brain Imaging Centre, University of Cambridge, Addenbrookes Hospital, Cambridge, UK
[2] Academic Neurosurgical Unit, University of Cambridge, Addenbrookes Hospital, Cambridge,UK
[3] Academic Anaesthesia Unit, University of Cambridge, Addenbrookes Hospital, Cambridge, UK
[4] MRC Biostatistics Unit, University of Cambridge, Addenbrookes Hospital, Cambridge, UK
[5] Academic Radiology Unit, University of Cambridge, Addenbrookes Hospital, Cambridge, UK

Summary

The heterogeneity of the initial insult and subsequent pathophysiology has made both the study of human head injury and design of randomised controlled trials exceptionally difficult. The combination of multimodality bedside monitoring and functional brain imaging positron emission tomography (PET) and magnetic resonance (MR), incorporated within a Neurosciences Critical Care Unit, provides the resource required to study critically ill patients after brain injury from initial ictus through recovery from coma and rehabilitation to final outcome. Methods to define cerebral ischemia in the context of altered cerebral oxidative metabolism have been developed, traditional therapies for intracranial hypertension re-evaluated and bedside monitors cross-validated. New modelling and analytical approaches have been developed.

Keywords: Head injury; positron emission tomography; magnetic resonance; intensive care; cerebral blood flow; cerebral ischemia; multimodality bedside monitoring; cognitive outcome.

Introduction

Heterogeneity is the key feature of brain injury that has made randomised controlled trials so difficult [29, 30]. The initial insult varies widely as do the immediate sequelae, both intracranial and extracranial. Service frameworks have been created in many countries using evidence based guidelines, so far as the evidence exists [2, 31, 36]. During the preadmission phase, intermittent assessments of Glasgow Coma Scale, pulse oximetry and non-invasive blood pressure give an incomplete picture but biological markers that integrate the extent of preadmission brain ischemia, hypoxia and hypotension are in their infancy. There is now increasing evidence for the value of admission to an Intensive Care Unit in general and Neurosciences Critical Care Unit [33] in particular but it is not clear what the vital components of such care are.

Monitoring and timely intervention are key to the prevention of secondary insults but the former has many limitations in terms of temporal and spatial resolution. Monitors of global brain function (e.g. intracranial pressure (ICP), jugular venous oxygen saturation) cannot detect regional abnormalities whereas local monitors (e.g., microdialysis, brain tissue oxygen (PtO2) electrode) can only sample a small region of the brain and are highly dependent on accurate placement. However, such monitors give continuous information and hence both invaluable temporal resolution and prognostic information [11, 12, 20, 46]. Imaging gives excellent spatial resolution but only in snapshots aided by the regional responses to physiological challenges such as changes in cerebral perfusion pressure and arterial blood gases. The study of acute middle cerebral artery stroke with PET has revealed that only about one-third of patients are suitable for studies of thrombolysis: the other patients have already recanalised or the brain affected is irreversibly damaged [29]. The design of large scale Phase III randomised controlled trials (RCT's) needs to be informed by studies using surrogate markers that provide early indications of efficacy and side-effects in homogeneous subgroups of patients, pharmacokinetics and dynamics and creation of cocktails.

Wolfson Brain Imaging Centre (WBIC)

The WBIC was created some ten years ago and incorporates both PET and high field (3T) MR within the envelope of the 21-bedded neurosciences critical care unit. Full cyclotron and radiochemistry facilities were built close by [52]. This structure allows for the safe functional brain imaging of critically ill patients from initial ictus through recovery from coma and rehabilitation to final outcome. Full multimodality monitoring based around the ICM+ software [52] is available both in the NCCU and in the WBIC.

Definition and extent of cerebral ischemia following head injury

Cerebral oxidative metabolism varies after head injury as the result, for example, of depressed conscious level, sedative drugs, fits and hypermetabolism associated with excitotoxicity. Hence, cerebral ischemia cannot be defined simply by local cerebral blood flow values alone derived from thresholds for neuronal survival based upon studies of experimental and clinical stroke. Furthermore, the fate of injured brain depends upon the state of coupling between local cerebral blood flow (CBF) and local glucose metabolism [4].

There has been considerable controversy over the interpretation of PET studies after head injury [13] but, using triple 15O-PET to measure oxygen extraction ratio voxel-by-voxel, Coles *et al.* [6–10] have estimated the volume of ischaemic brain (IBV) within 24 hours of head injury and found it to be significantly higher than the artefactual 'ischaemic' volume created in controls by the partial volume effect (67 ± 69 *vs* 2 ± 3 ml; $P < 0.01$). Furthermore, they found that relative ischemia and hyperaemia coexisted in some patients implying mismatch between perfusion and oxygen use. Increase in IBV correlated with a poor Glasgow Outcome Score 6 months after injury. However, it is not possible to use any single early physiological threshold – CBF, oxygen extraction fraction (OEF) or cerebral metabolic rate of oxygen consumption (CMRO2) – to predict the outcome for a specific area around a contusion.

The same group [26] compared cerebral tissue PtO2 and cerebral venous PvO2 from images of PET-derived OEF. Tissue regions with hypoxic levels of PtO2 (<10 mmHg) had similar levels of PvO2 compared with non-hypoxic areas and hence displayed larger PvO2-PtO2 gradients. Despite similar CBF reductions with hyperventilation, hypoxic regions achieved significantly smaller OEF increases compared with normoxic regions. It has been known for some years that pericontusional tissue shows varying degrees of endothelial swelling, microvascular collapse and perivascular oedema. Hence, increased diffusion barriers may reduce cellular oxygen delivery following head injury and attenuate the ability of the brain to increase oxygen extraction in response to hypoperfusion. Global or regional OEF will underestimate tissue hypoxia due to such mechanisms.

MR is now being used to define the outcome of tissue threatened by periods of ischemia and to determine whether post-traumatic brain swelling reflects vasogenic or cytotoxic oedema. Larger studies are required given the heterogeneity of the patient cohorts [1, 23, 24].

Validation of bedside monitors: jugular venous oxygen saturation (JvO2), transcranial Doppler (TCD) and intracerebral microdialysis

It is unlikely in the foreseeable future that MR or PET scanners will be installed in the majority of intensive care units which will continue to rely on multimodality bedside monitoring. It is essential therefore to define the limitations of such monitoring by comparison with gold standard measurements provided by PET.

Jugular venous oximetry is widely used taking SjO2 values above 50% as the threshold for defining cerebral ischemia [2]. However, in Coles' study, such a JvO2 value was only achieved at an IBV of 170 ± 63 ml which equates to $13 \pm 5\%$ of the brain [7, 8]. Hence, jugular bulb oximetry fails to detect regional ischemia that is clinically important and potentially related to outcome. A threshold JvO2 of at least 60% is more appropriate after head injury.

TCD is commonly used to assess the degree of cerebral vasospasm after subarachnoid haemorrhage based on comparison with angiography. However, the relationship of TCD velocity to the development of delayed cerebral ischemia and outcome is notoriously capricious. Minhas *et al.* [27] have demonstrated that TCD middle cerebral artery velocity alone, even when compared with the internal carotid artery velocity (Lindegaard ratio), does not necessarily indicate whether ischemia is present on PET scanning. A markedly heterogeneous pattern of CBF distribution was observed, with hyperaemia, normal CBF values and reduced flow being observed among patients with delayed neurological deficits. The use of TCD to assess autoregulation using manipulations such as the transient hyperaemic response test is more predictive of development of delayed ischaemic neuronal deficit (DIND) and outcome [21].

Finally, the comparison of PET derived estimates of ischemia have been used to validate the findings from intracerebral microdialysis. Significant ischemia is required before metabolite changes occur and there is a correlation between lactate/pyruvate ratio and OEF [18]. Work is ongoing in a number of centres dissecting out the significance of 'hot-spots' as defined by 18F-fluorodeoxyglucose PET [17, 50].

Efficacy of traditional treatments

RCT's of established treatments such as hyperventilation and induced hypertension are difficult to mount in an ICU setting, not least because of the difficulty of dissecting out the effect of one component of a package of care. PET and MR scanning provide an opportunity to use a surrogate end-point to define whether brain tissue viability is being put at risk by such therapies, at least in the short term.

Although hyperventilation has long been used for the control of ICP after head injury, concern is emerging about the risk of ischemia that it may precipitate in the first 24 hours [13]. Coles *et al.* [6] used PET to demonstrate that hyperventilation increases the volume of severely hypoperfused brain tissue within the injured brain, despite improvements in cerebral perfusion pressure and intracranial pressure. Hyperventilation-induced increases in hypoperfused brain volume were apparently non-linear with a threshold value of between 34 and 38 mmHg. A complimentary study of the relative time course of change in ICP and middle cerebral artery TCD velocity indicated that head-injured patients may adapt differently to hyperventilation than healthy volunteers. Mean flow velocity continued to decrease in head injured patients despite ICP returning to baseline. Hence, potentially harmful reductions in CBF may persist beyond the duration of useful ICP reduction [48]. Finally, early studies with diffusion-weighted imaging (DWI) MR indicate that hyperventilation has a variable effect on apparent diffusion coefficient (ADC) values around contusions that is suggestive of precipitating ischemia in some patients [1].

Induced hypertension has been used to reduce ICP in head injured patients who are still autoregulating and increase CBF in those not autoregulating. However, aggressive maintenance of cerebral perfusion pressure may worsen outcome due to the extracranial complications of the therapy as recognised by the recent change in the Brain Trauma Foundation guidelines [2]. Furthermore, it is unclear whether it is the ischaemic areas that selectively benefit from the rise in CPP, rather than the normal appearing regions. PtO2 may certainly rise in structurally normal brain tissue as well as in focal lesions. An ROI based analysis of static autoregulation using PET showed that CBF remains in the ischaemic range around contusions following a rise in CPP and that the greatest increase was in the more normal areas [47]. A voxel-based analysis revealed that IBV was reduced by induced hypertension but that the effect was small and clinically insignificant in the majority of patients [9]. However, patients with the largest baseline IBV showed substantial and clinically significant reductions. The problem is how to identify this minority of patients and thereby avoid the complications of displaying induced hypertension to the whole group. One complication is that the effects depend on the inotrope used: noradrenaline is more consistent than dopamine [19, 49].

Late structural and functional sequelae of head injury

Diffuse head injury, particularly when accompanied by prolonged intracranial hypertension, may be followed by diffuse loss of brain substance as confirmed by both VBM and magnetic resonance spectroscopic measurements of N-acetyl aspartate (NAA) [14–16, 44, 51]. Perhaps surprisingly, it has proven difficult to correlate late cognitive sequelae with late structural imaging after brain injury. Recent developments in voxel based morphometry (VBM), statistical techniques and more system specific neuropsychological paradigms have shown that it now possible to demonstrate some correspondence between anatomy and cognitive function. For example, Salmond *et al.* [39, 40] have shown that head-injured survivors may display deficits in sustained attention, paired associate learning and reaction time but comparative preservation of spatial working memory. VBM revealed reduced grey matter density in the basal forebrain, hippocampal formation and regions of the neocortex. These cognitive and structural changes are consistent with cholinergic dysfunction for which there is also neuropathological and experimental evidence.

It is not clear from structural imaging why patients enter the vegetative state. The neuropathology of VS indicates thalamocortical damage of widely differing degrees and distribution depending on the primary insult. Functional brain imaging (PET and fMR) has revealed that the coupling between neuronal electrical activity and regional glucose metabolism is preserved even in the minimally conscious, but absent in the vegetative state [5]. Furthermore, although it is possible to demonstrate islands of preserved cortical function to various degrees of stimulus complexity in the vegetative patient using PET or functional MR (fMR), there is a lack of connectivity between areas of the cortex unlike that seen in minimally conscious states (MCS) [22, 25, 32]. It remains to be shown in longitudinal studies whether changes in functional imaging may be predictive of emergence from coma and the vegetative state. One problem is that functional imaging under these circumstances is prone to artefact and is not sufficiently robust as yet for routine clinical use.

Modelling

In order to make sense of the large volumes of data generated by imaging studies, it has been necessary for a number of groups to develop new models and ways of analysis to cope with the issues surrounding heterogeneity and connectivity (e.g. 3, 28, 37, 38, 41–43, 45). For example, ROI-based analysis may not always be appropriate for diffuse processes, hence the concept of voxel-based ischaemic brain volume. Spatial heterogeneity has required transforming traditional global lumped models so that asymmetry of responses can be understood. Finite element modelling has proven to be a powerful tool and has created new and testable hypotheses to explain ventricular dilatation in normal pressure hydrocephalus and diaschisis [34, 35].

Conclusions

Although advanced imaging techniques such as PET and MR are not new, patients with acute brain injury have been denied access to them because of various logistical problems which have now been resolved. New insights have been achieved into longstanding questions about the extent and mechanisms of cerebral ischemia, the efficacy and risks of traditional treatments, cross-validation of conventional multimodality bedside monitoring techniques and the pathophysiology of recovery from coma and cognitive outcome.

Acknowledgments

Wolfson Foundation, Medical Research Council, Smiths Charity, Technology Foresight Challenge.

References

1. Bradley PG, Harding SG, Pena A *et al* (2003) Hyperventilation induced changes in diffusion weighted imaging in early severe head injury. J CBF Metab [Suppl] Brain 03
2. Brain Trauma Foundation (2003) The American Association of Neurological Surgeons. The Joint Section on Neurotrauma and Critical Crae. Guidelines for cerebral perfusion pressure. http://www2.braintrauma.org
3. Bullmore E, Fadili J, Breakspear M, Salvador R, Suckling J, Brammer M (2003) Wavelets and statistical analysis of functional magnetic resonance images of the human brain. Stat Meth Med Res 12: 375–399. Review
4. Chen SF, Richards HK, Smielewski P, Johnstrom P, Salvador R, Pickard JD, Harris NG (2004) Relationship between flow-metabolism uncoupling and evolving axonal injury after experimental traumatic brain injury. J Cereb Blood Flow Metab 24: 1025–1036
5. Coleman MR, Menon DK, Fryer TD, Pickard JD (2005) Neurometabolic coupling in the vegetative and minimally conscious states: preliminary findings. J Neurol Neurosurg Psychiatry 76: 432–434
6. Coles JP, Minhas PS, Fryer TD, Smielewski P, Aigbirihio F, Donovan T, Downey SP, Williams G, Chatfield D, Matthews JC, Gupta AK, Carpenter TA, Clark JC, Pickard JD, Menon DK (2002) Effect of hyperventilation on cerebral blood flow in traumatic head injury: clinical relevance and monitoring correlates. Crit Care Med 30: 1950–1959
7. Coles JP, Fryer TD, Smielewski P, Rice K, Clark JC, Pickard JD, Menon DK (2004) Defining ischemic burden after traumatic brain injury using 15O PET imaging of cerebral physiology. J Cereb Blood Flow Metab 24: 191–201
8. Coles JP, Fryer TD, Smielewski P, Chatfield DA, Steiner LA, Johnston AJ, Downey SP, Williams GB, Aigbirhio F, Hutchinson PJ, Rice K, Carpenter TA, Clark JC, Pickard JD, Menon DK (2004) Incidence and mechanisms of cerebral ischemia in early clinical head injury. J Cereb Blood Flow Metab 24: 202–211
9. Coles JP, Steiner LA, Johnston AJ, Fryer TD, Coleman MR, Smieleweski P, Chatfield DA, Aigbirhio F, Williams GB, Boniface S, Rice K, Clark JC, Pickard JD, Menon DK (2004) Does induced hypertension reduce cerebral ischemia within the traumatized human brain? Brain 127: 2479–2490
10. Cunningham AS, Salvador R, Coles JP, Chatfield DA, Bradley PG, Johnston AJ, Steiner LA, Fryer TD, Aigbirhio FI, Smielewski P, Williams GB, Carpenter TA, Gillard JH, Pickard JD, Menon DK (2005) Physiological thresholds for irreversible tissue damage in contusional regionsfollowing traumatic brain injury. Brain. In Press
11. Czosnyka M, Balestreri M, Steiner L, Smielewski P, Hutchinson PJ, Matta B, Pickard JD (2005) Age, intracranial pressure, autoregulation, and outcome after brain trauma. J Neurosurg 102: 450–454
12. Czosnyka M, Pickard JD (2004) Monitoring and interpretation of intracranial pressure. J Neurol Neurosurg Psychiatry 75: 813–821
13. Diringer M (2002) Hyperventilation in head injury: what have we learned in 43 years? Crit Care Med 30: 2142–2143

14. Garnett MR, Corkill RG, Blamire AM, Rajagopalan B, Manners DN, Young JD, Styles P, Cadoux-Hudson TA (2001) Altered cellular metabolism following traumatic brain injury: a magnetic resonance spectroscopy study. J Neurotrauma 18: 231–240
15. Garnett MR, Blamire AM, Corkill RG, Rajagopalan B, Young JD, Cadoux-Hudson TA, Styles P (2001) Abnormal cerebral blood volume in regions of contused and normal appearing brain following traumatic brain injury using perfusion magnetic resonance imaging. J Neurotrauma 18: 585–593
16. Goetz P, Blamire A, Rajagopalan B, Cadoux-Hudson T, Young D, Styles P (2004) Increase in apparent diffusion coefficient in normal appearing white matter following human traumatic brain injury correlates with injury severity. J Neurotrauma 21: 645–654
17. Hattori N, Huang SC, Wu HM, Liao W, Glenn TC, Vespa PM, Phelps ME, Hovda DA, Bergsneider M (2004) Acute changes in regional cerebral (18)F-FDG kinetics in patients with traumatic brain injury. J Nucl Med 45: 775–783
18. Hutchinson PJ, Gupta AK, Fryer TF, Al-Rawi PG, Chatfield DA, Coles JP, O'Connell MT, Kett-White R, Minhas PS, Aigbirhio FI, Clark JC, Kirkpatrick PJ, Menon DK, Pickard JD (2002) Correlation between cerebral blood flow, substrate delivery, and metabolism in head injury: a combined microdialysis and triple oxygen positron emission tomography study. J Cereb Blood Flow Metab 22: 735–745
19. Johnston AJ, Steiner LA, Coles JP, Chatfield DA, Fryer TD, Smielewski P, Hutchinson PJ, O'Connell MT, Al-Rawi PG, Aigbirihio FI, Clark JC, Pickard JD, Gupta AK, Menon DK (2005) Effect of cerebral perfusion pressure augmentation on regional oxygenation and metabolism after head injury. Crit Care Med 33: 189–195
20. Kett-White R, Hutchinsom PJ, Boniface S, Pickard JD, Kirkpatrick P (2002) Multimodality monitoring of acute brain injury. Adv Tech Stand Neurosurg 27: 87–134
21. Lam JM, Smielewski P, Czosnyka M, Pickard JD, Kirkpatrick PJ (2000) Predicting delayed ischemic deficits after aneurysmal subarachnoid hemorrhage using a transient hyperemic response test of cerebral autoregulation. Neurosurgery 47: 819–825
22. Laureys S, Owen AM, Schiff ND (2004) Brain function in coma, vegetative state, and related disorders. Lancet Neurol 3(9): 537–546. Review
23. Maeda T, Katayama Y, Kawamata T, Koyama S, Sasaki J (2003) Ultra-early study of edema formation in cerebral contusion using diffusion MRI and ADC mapping. Acta Neurochir [Suppl] 86: 329–331
24. Marmarou A (2003) Pathophysiology of traumatic brain edema: current concepts. Acta Neurochir [Suppl] 86: 7–10. Review
25. Menon DK, Owen AM, Williams EJ, Minhas PS, Allen CM, Boniface SJ, Pickard JD (1998) Cortical processing in persistent vegetative state. Wolfson Brain Imaging Centre Team. Lancet 352: 200
26. Menon DK, Coles JP, Gupta AK, Fryer TD, Smielewski P, Chatfield DA, Aigbirhio F, Skepper JN, Minhas PS, Hutchinson PJ, Carpenter TA, Clark JC, Pickard JD (2004) Diffusion limited oxygen delivery following head injury. Crit Care Med 32: 1384–1390
27. Minhas PS, Menon DK, Smielewski P, Czosnyka M, Kirkpatrick PJ, Clark JC, Pickard JD (2003) Positron emission tomographic cerebral perfusion disturbances and transcranial Doppler findings among patients with neurological deterioration after subarachnoid hemorrhage. Neurosurgery 52: 1017–1022
28. Momjian S, Owler BK, Czosnyka Z, Czosnyka M, Pena A, Pickard JD (2004) Pattern of white matter regional cerebral blood flow and autoregulation in normal pressure hydrocephalus. Brain 127: 965–972
29. Medical Research Council (1998) Field Review. Neuroprotection in acute brain injury after trauma and stroke – from preclinical research to clinical trials
30. Narayan RK, Michel ME, Ansell B, Baethmann A, Biegon A, Bracken MB, Bullock MR, Choi SC, Clifton GL, Contant CF, Coplin WM, Dietrich WD, Ghajar J, Grady SM, Grossman RG, Hall ED, Heetderks W, Hovda DA, Jallo J, Katz RL, Knoller N, Kochanek PM, Maas AI, Majde J, Marion DW, Marmarou A, Marshall LF, McIntosh TK, Miller E, Mohberg N, Muizelaar JP, Pitts LH, Quinn P, Riesenfeld G, Robertson CS, Strauss KI, Teasdale G, Temkin N, Tuma R, Wade C, Walker MD, Weinrich M, Whyte J, Wilberger J, Young AB, Yurkewicz L (2002) Clinical trials in head injury. J Neurotrauma 19: 503–557. Review
31. NICE guidelines: Head injury – Triage, assessment, investigation and early management of head injury in infants, children and adults. www.nice.org.uk
32. Owen AM, Menon DK, Johnsrude IS, Bor D, Scott SK, Manly T, Williams EJ, Mummery C, Pickard JD (2002) Detecting residual cognitive function in persistent vegetative state. Neurocase 8: 394–403
33. Patel HC, Menon DK, Tebbs S, Hawker R, Hutchinson PJ, Kirkpatrick PJ (2002) Specialist neurocritical care and outcome from head injury. Intensive Care Med 28(5): 547–553
34. Pena A, Owler BK, Fryer TD, Minhas P, Czosnyka M, Crawford PJ, Pickard JD (2002) A case study of hemispatial neglect using finite element analysis and positron emission tomography. J Neuroimaging 12: 360–367
35. Pena A, Harris NG, Bolton MD, Czosnyka M, Pickard JD (2002) Communicating hydrocephalus: the biomechanics of progressive ventricular enlargement revisited. Acta Neurochir [Suppl] 81: 59–63
36. Pickard JD, Seeley HM, Kirker S, Maimaris C, McGlashan K, Roels E, Greenwood R, Steward C, Hutchinson PJ, Carroll G (2004) Mapping rehabilitation resources for head injury. J R Soc Med 97: 384–389
37. Piechnik SK, Czosnyka M, Richards HK, Whitfield PC, Pickard JD (2001) Cerebral venous blood outflow: a theoretical model based on laboratory simulation. Neurosurgery 49: 1214–1222
38. Piechnik SK, Czosnyka M, Harris NG, Minhas PS, Pickard JD (2001) A model of the cerebral and cerebrospinal fluid circulations to examine asymmetry in cerebrovascular reactivity. J Cereb Blood Flow Metab 21: 182–192
39. Salmond CH, Chatfield DA, Menon DK, Pickard JD, Sahakian BJ (2005) Cognitive sequelae of head injury: involvement of basal forebrain and associated structures. Brain 128: 189–200
40. Salmond CH, Menon DK, Chatfield DA, Pickard JD, Sahakian BJ (2005) Deficits in decision-making in head injury survivors. J Neurotrauma 22: 613–622
41. Salvador R, Pena A, Menon DK, Carpenter TA, Pickard JD, Bullmore ET (2005) Formal characterization and extension of the linearized diffusion tensor model. Hum Brain Mapp 24: 144–155
42. Salvador R, Suckling J, Coleman MR, Pickard JD, Menon D, Bullmore E (2005) Neurophysiological Architecture of Functional Magnetic Resonance Images of Human Brain. Cereb Cortex. In Press
43. Schmidt EA, Czosnyka M, Steiner LA, Balestreri M, Smielewski P, Piechnik SK, Matta BF, Pickard JD (2003) Asymmetry of pressure autoregulation after traumatic brain injury. J Neurosurg 99: 991–998
44. Signoretti S, Marmarou A, Tavazzi B, Lazzarino G, Beaumont A, Vagnozzi R (2001) N-Acetylaspartate reduction as a measure

of injury severity and mitochondrial dysfunction following diffuse traumatic brain injury. J Neurotrauma 18: 977–991

45. Smielewski P, Coles JP, Fryer TD, Minhas PS, Menon DK, Pickard JD (2002) Integrated image analysis solutions for PET datasets in damaged brain. J Clin Monit Comput 17: 427–440
46. Steiner LA, Czosnyka M, Piechnik SK, Smielewski P, Chatfield D, Menon DK, Pickard JD (2002) Continuous monitoring of cerebrovascular pressure reactivity allows determination of optimal cerebral perfusion pressure in patients with traumatic brain injury. Crit Care Med 30: 733–738
47. Steiner LA, Coles JP, Johnston AJ, Czosnyka M, Fryer TD, Smielewski P, Chatfield DA, Salvador R, Aigbirhio FI, Clark JC, Menon DK, Pickard JD (2003) Responses of posttraumatic pericontusional cerebral blood flow and blood volume to an increase in cerebral perfusion pressure. J Cereb Blood Flow Metab 23: 1371–1377
48. Steiner LA, Balestreri M, Johnston AJ, Czosnyka M, Coles JP, Chatfield DA, Smielewski P, Pickard JD, Menon DK (2004) Sustained moderate reductions in arterial CO2 after brain trauma time-course of cerebral blood flow velocity and intracranial pressure. Intensive Care Med 30: 2180–2187
49. Steiner LA, Johnston AJ, Czosnyka M, Chatfield DA, Salvador R, Coles JP, Gupta AK, Pickard JD, Menon DK (2004) Direct comparison of cerebrovascular effects of norepinephrine and dopamine in head-injured patients. Crit Care Med 32: 1049–1054
50. Vespa P, Bergsneider M, Hattori N, Wu HM, Huang SC, Martin NA, Glenn TC, McArthur DL, Hovda DA (2005) Metabolic crisis without brain ischemia is common after traumatic brain injury: a combined microdialysis and positron emission tomography study. J Cereb Blood Flow Metab 25: 763–774
51. Wu HM, Huang SC, Hattori N, Glenn TC, Vespa PM, Hovda DA, Bergsneider M (2004) Subcortical white matter metabolic changes remote from focal hemorrhagic lesions suggest diffuse injury after human traumatic brain injury. Neurosurgery 55: 1306–1315
52. www.wbic.cam.ac.uk; www.neurosurg.cam.ac.uk/icmplus

Correspondence: J. D. Pickard, Academic Neurosurgery Unit, University of Cambridge, Addenbrookes Hospital, Box 167, Cambridge CB22QQ, UK. e-mail: idpsecretary@medschl.can.ac.uk

Acta Neurochir (2005) [Suppl] 95: 465–470

Brain injury and proteomics/peptidomics: is it relevant? An overview

M. U. Schuhmann[1], **G. Heine**[2], **M. Skardelly**[1], **M. Jaeger**[1], and **H. Selle**[2]

[1] Department of Neurosurgery, University of Leipzig, Leipzig, Germany
[2] BioVisioN AG, Hannover, Germany

Summary

Proteomics and peptidomics® are different and supplemental to genomics, since – in contrast to the basically constant genome – the proteome and peptidome are dynamic, constantly changing, and complex networks. Proteomics is traditionally linked to 2D-gel electrophoresis techniques. Concerning peptidomics, three different approaches are currently available, all using mass spectrometry as a key element. The use of proteomics or peptidomics in traumatic brain injury (TBI) research is demanding. From the technical point of view there are high-level requirements concerning the preanalytical phase, specific machinery, sophisticated software and skilled manpower/intellectual input. There are currently no bedside techniques and most methods are suitable for experimental TBI research in specialized laboratories. In screening experiments of CSF following controlled cortical impact in rats we identified several peptides, which, although previously known, were so far not reported in the TBI context or in CSF. Peptidomics and proteomics, as highly complex screening technologies, thus seem to carry a large potential to lead TBI science. Newly "discovered" peptide targets have to be validated with different methodology to establish a real diagnostic or therapeutic value.

Keywords: Biomarkers; cerebrospinal fluid; experimental; peptides; peptidomics; proteomics; screening techniques; traumatic brain injury.

Introduction

Proteomics, defined as the large-scale study of proteins, particularly their occurrence and structure but also their function, can be viewed as "just the next step after genomics" or functional genomics. However, it is much more than genomics. While the genome is a rather constant entity, the proteome is a complex and dynamic network that permanently changes. A mutation qualitatively changes the genotype and thus may determine a predisposition for a disease, whereas proteins define the phenotype and may reflect or even cause a pathogenic process. So proteomics is suited for (I) defining the transition from health to disease, (II) describing the current disease status and (III) its progression with time. The concentration of each single protein is influenced by synthesis, modification and proteolytic processing, as well as by degradation and fluid balance, depicting a wide variety of metabolic and catabolic processes. One organism has radically different protein expression and processing in different parts of its body, at different stages of its life cycle and in response to external factors like a pathologic challenge. Evolution, using the same proteins in multiple pathways and functions has reduced complexity in living systems: The Human Genome Project found that a genome of less than 30.000 genes codes for more than 300.000 proteins.

The large group of peptides, small protein molecules with molecular masses below 15 kDa, further increases the complexity in this field. Peptides are not only mere protein degradation products, but also specific breakdown products from larger proteins, generated by posttranslational processing to become biological active substances. Some protein precursors carry several distinct biological activities that are silent until activated by different proteolytic processes. Other peptides are directly generated as active compounds to act as signaling molecules. Thus, the processing of proteins, changing with compartment, during aging and in disease states, is an important principle that allows metabolic variation in living systems. It has become clear that the hard work is still to come before we reasonably understand the interaction between the genome and a specific proteome/peptidome. Advancing the terms proteomics and its pendant for peptides, peptidomics [16], may thus be defined as "the qualitative and quantitative comparison of proteomes/peptidomes under different conditions to unravel the mechanisms of biological processes".

With regard to the application of proteomics or peptidomics in traumatic brain injury (TBI) research, the question arises which compartment should be addressed first. Blood plasma contains too large amounts of proteins and peptides of extracerebral origin to be an ideal target. Brain tissue will be available in the experimental set-up, but in humans only as an exception. Therefore, CSF has moved into our focus, since it is available in experimental models of TBI as well as from humans. As an ultrafiltrate of plasma from the choroid plexus, it is an indicator of the condition of the blood-CSF barrier. Moreover, the choroid plexus actively synthesizes and secrets proteins and peptides into CSF. The brain extracellular fluid – generated from the large surface of the blood brain barrier and enriched by the extracellular components of the metabolism of all brain cells – drains as "brain lymph" into the CSF spaces. Thus, CSF reflects both blood brain barrier function and brain cell metabolism. Finally, the barriers keep proteins and peptides of the blood plasma outside the CSF or control their passage by active mechanisms [1, 2, 7, 9, 12, 14].

Neurotrauma research has a wide range of different and difficult tasks. The additional application of proteomics and peptidomics can be envisioned to be helpful for the development of effective multi-target neuroprotective strategies, for deeper insights into restoration and regeneration, or for the development of neurochemical monitoring beyond the classical microdialysis parameters. As screening techniques, proteomics/peptidomics are used for the identification of biological markers, that e.g. could indicate significant events or conditions – like the extent of damage to neurons or glial cells – for a diagnostic or prognostic statement. Biological markers could furthermore lead to pathophysiological mechanisms and thus a deeper understanding of TBI, or could indicate a TBI related key event, that might be a target for a therapeutic intervention.

The limitations for basic research in human TBI are related to the complexity and heterogeneity of its clinical presentation. The multitude of influencing variables like distribution and severity of biomechanical trauma, pre-resuscitation secondary injury, pre-existing and confounding co-morbidity, secondary ischemic events during the ICU stay, and side effects of treatment, make a systematic assessment of the multitude of biochemical cascades by means of proteomics and peptidomics in humans an undesirable starting point. Therefore, experimental models with much better controlled boundary conditions and trauma definition are more suitable to explore the possibilities and the power of proteomics and peptidomics.

Available methods

Proteomics

The comprehensive analysis of proteins is based on the two-dimensional (2-D) gel electrophoresis technique developed in 1975 by O'Farrell [11] and Klose [8]. Proteins from complex biological samples are separated in an acrylamide gel according to the protein's isoelectric point (first dimension) and to the molecular mass (second dimension). Several thousand different compounds can be depicted. The automation of the experimental procedure as well as resolution and sensitivity of the technique make it a powerful tool, although the reproducibility is impaired by a variety of experimental parameters like sample preparation, the composition and size of gels, and staining conditions. Sophisticated methods and software tools allow the comparison of patterns from different gels. Spots of interest are then identified by excision, tryptic digestion and analysis by mass spectrometry. The acquired mass spectra are matched with database entries in order to identify the parent protein molecule initially detected in the 2D gel.

A systematic analysis of the proteome of human CSF has already been undertaken, e.g. by the combination of 2D gels and mass-spectrometric identification of proteins [17]. The authors identified 480 spots from gels by mass spectrometry that corresponded to a far lower number of precursors. The use of proteomics has been sparse in TBI research so far. To our knowledge, there is no report about a systematic study with proteomics in patients following TBI. In experimental research one paper was recently published using large format 2D Gel electrophoresis in focal injury in immature rats [6]. The authors used conventional and functional proteomics to characterize protein changes in the dorsal hippocampus 24 h after trauma. Of 1500 protein spots in silver stains, 50 could be identified and six were found to be statistically significantly altered.

Peptidomics

Native peptides are not addressed by the proteomics approach since 2D gels show an insufficient retention of compounds with masses below 10 kDa and sensitivity for dyeing decreases with the molecular mass. Therefore specific techniques had to be developed. Three currently available methods use mass spectrometry as the key technology to screen and analyze samples for their peptide content.

A) The *Ciphergen SELDI Chip®* System (SELDI = Surface Enhanced Laser Desorption Ionisation) addresses proteins as well as peptides. It selects a certain subset of peptides/proteins out of a sample according to the surface affinity characteristics of the employed mass-spectrometric chip. They can be varied by changing e.g. characteristics of ion-exchange, hydrophobicity or antigen/antibody binding. Unbound contents of the sample are washed out and mass spectrometric detection of the bound peptides is performed and data converted into a sample specific signal pattern. Statistical analysis identifies marker candidates by indicating differences between different sample sets.

B) The *Bruker Daltonics®* System uses a similar approach of pre-selection and is likewise designed for high throughput analysis. The specific capture of a group of peptides is accomplished according to the surface affinity characteristics of magnetic beads instead of a chip. Via magnetic force the beads are separated from the rest of the sample and after elution and target preparation a mass spectrometric analysis of the peptides is performed. The final step again consists of statistical work-up, and pattern display. Peptide identification is currently not possible.

C) *BioVisioN's Peptidomics® Technologies* (see Fig. 1). Peptides are extracted from biological samples and separated in an automated process by liquid chromatography into 96 individual fractions. Each fraction is analysed by mass spectrometry generating 96 mass spectra from one sample. These mass spectra are combined for each individual sample to one multidimensional diagram (peptide display) depicting molecular mass, chromatographic fraction and mass-spectrometric signal intensity. For visualisation purposes, the intensity of each mass-spectrometric signal is translated into colour intensity. Each peptide's position in such a display is unambiguously characterised by its molecular mass and chromatographic behaviour, i.e. retention time. A peptide display of CSF consists of more than 6,000 different signals, representing about 2,000 different peptide molecules with a molecular mass below 15 kDa. The individual peptide displays of each group of samples/patients are averaged to a joint peptide mass map that represents the peptidome of each group. These combined datasets are used for the Differential Peptide Display® (DPD) analysis: Comparison of the different peptidomes reveals differences occurring in conjunction with the specific conditions used for definition of the groups of samples/patients. Appropriate software is used for identification of signals correlating with clinical parameters. Furthermore, networks are calculated representing groups of signals correlating with each other, allowing the identification of interactions or interdependencies. Peptides of interest are then identified by tandem mass spectrometry (MS/MS) and, if necessary, by Edman degradation [13, 16].

The Ciphergen and Bruker systems are designed for high throughput use (hundreds of samples per day), not only in research but also in clinical routine. The screening results are substantially influenced by the peptide pre-selection due to surface characteristics of the employed chip or bead determining the analyzed protein/peptide spectrum. In combination with antibody binding this allows also for a target specific capture in addition to a screening approach. The larger the amount of peptides analysed per mass spectrometry cycle, the lower the resolution of the resulting mass spectrum will be. Therefore high abundance proteins/peptides are dominating the obtained patterns and low abundance ones can disappear in the "noise". The techniques cannot discriminate between isobaric but different peptides, which happen to have the same molecular weight and aim rather at pattern recognition than at identification of the peptides by sequence analysis. The latter is important, even if the techniques are used in clinical research. Furthermore, the reproducibility of those screening results between different labs seems to be limited, which casts light on the great influence of the way the pre-analytic sample processing is done [3]. BioVisioN's Peptidomics® technologies allow for a more comprehensive analysis and avoid pre-selection effects to a much greater extend. Due to the separation in 96 fractions it has a much higher resolution and discriminates isobaric peptides. After the detection of significant disease-specific signals and differences, the method is linked to an identification process for the desired marker candidates. This detailed and precise screening method requires more time and specific equipment and staff. Therefore it is confined to a

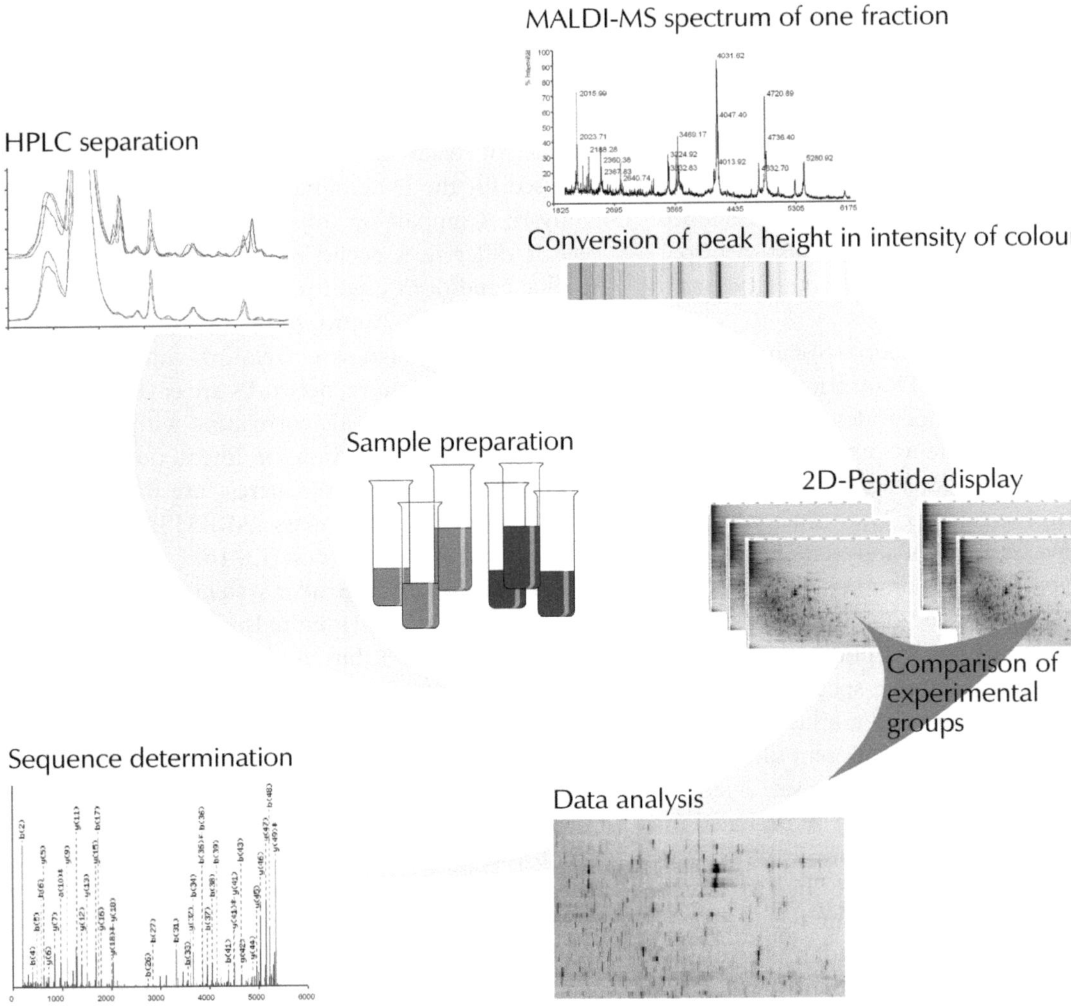

Fig. 1. Workflow of Peptidomics® from sample to peptide identification. For details see text

highly specialised lab and is essentially performed by one company.

Concerning CSF, the Ciphergen SELDI ProteinChip® has meanwhile been used in humans and rats for the analysis of proteins [4, 10]. The BioVisioN Peptidomics® technologies have been adapted to human and rat CSF as well. In a first attempt to systematically screen all CSF peptides in relation to a clinical condition, we analysed CSF from two patients suffering from a primary CNS lymphoma in comparison to CSF from three subjects without a CNS disease undergoing routine myelography. More than 6,000 signals were detected with a mass accuracy better than 500 ppm. Several differences in peptide composition were found [5]. In a second study we used the peptide screening technology for TBI research. In terms of our current knowledge of the literature, this is the first time that TBI pathophysiology has been addressed at the peptide level [15]. Using the CCI model of focal traumatic brain injury in the rat, we screened for consistent, statistically significant post-traumatic changes in CSF over a period of 1 h to 7 d. Out of more than 3000 peptide signals in CSF we identified 8 distinct peptides fulfilling these criteria. Neither these peptides nor their protein precursors have been described in the literature specifically in the context of TBI or as appearing in CSF post trauma (manuscript in preparation).

Discussion

Proteomics and peptidomics techniques used as a screening approach provide the opportunity to see "much at a glance". The described approaches focus not on single proteins or peptides but screen either selected sample portions (SELDI Chip® System and Daltonics®) or the majority of proteins or peptides (classical 2D gel electrophoresis and Peptidmonics® technologies, respectively). Therefore, those techniques open in experimental TBI a new window with a different angle of view, since all other currently available methods (with the exception of genomics) address specific pre-selected targets. Our own work shows, that peptidomics is able to detect differences in CSF that correlate with the disease status and offers therefore the possibility to find new biomarkers or even peptides with therapeutic value. This opens new lines of thought, since none of the peptides we found as candidates in CSF after traumatic brain injury have been described in this specific context before. To evaluate detected peptide patterns we regard it as mandatory to identify the amino acid sequence of the candidate peptides. This allows the development of immuno- or mass spectrometry-based assays that extend the screening approach towards a specific targeted investigation to elucidate the biological significance of marker changes. For experimental TBI research therefore the detailed approach using BioVisioN's Peptidomics® technologies seem most suitable, since it covers a large range of peptide signals and possible marker candidates. In the future, analysis of the peptide composition of brain tissue itself might be possible, as soon as a reliable extractions of peptides from the tissue is established. The Ciphergen SELDI Chip®, Bruker Daltonics® or a comparable high throughput approach seems appropriate for any situation where the heterogeneity of the patient population affords the analysis of large numbers of samples, like in human TBI. However, this will be at the cost of resolution and important changes in peptide composition may be missed.

The new technologies provide a wide-angle view on the highly complex molecular mechanisms following TBI, expressed as changes in protein and peptide composition of the sample. This could help to overcome the past series of bench-to-bedside failures of mono-drug approaches to a multi-layer disease. It could further help to identify groups of targets instead of one. It enables us to follow a multitude of signals and their related or unrelated changes over time. In the case of therapy, it might help to observe all the effects of the therapeutic approach, the expected but also the surprising ones. Such screening in traumatic brain injury and control of therapeutic attempts might help to develop multi-drug-cocktails which influence several pathways in an integrated fashion at the same time. Thus we conclude, that proteomics and peptidomics are relevant tools in brain injury research.

Acknowledgment

The authors wish to thank a large team that has contributed to the adaptation of Peptidomics to CSF and Thomas Brinker, M.D. Ph.D. for his support of the experimental trauma work.

References

1. Banks WA, Kastin AJ (1996) Passage of peptides across the blood-brain barrier: pathophysiological perspectives. Life Sci 59: 1923–1943
2. Chodobski A, Szmydynger Chodobska J (2001) Choroid plexus: target for polypeptides and site of their synthesis. Microscopy Research and Technique 52: 65–82
3. Diamandis EP (2004) Analysis of serum proteomic patterns for early cancer diagnosis: drawing attention to potential problems. J Natl Cancer Inst 96: 353–356
4. Gineste C, Ho L, Pompl P, Bianchi M, Pasinetti GM (2003) High-throughput proteomics and protein biomarker discovery in an experimental model of inflammatory hyperalgesia: effects of nimesulide. Drugs 63 [Suppl] 1: 23–29
5. Heine G, Zucht HD, Schuhmann MU, Burger K, Jurgens M, Zumkeller M, Schneekloth CG, Hampel H, Schulz-Knappe P, Selle H (2002) High-resolution peptide mapping of cerebrospinal fluid: a novel concept for diagnosis and research in central nervous system diseases. J Chromatogr B Analyt Technol Biomed Life Sci 782: 353–361
6. Jenkins LW, Peters GW, Dixon CE, Zhang X, Clark RSB, Skinner JC, Marion DW, Adelson PD, Kochanek PM (2002) Conventional and functional proteomics using large format two-dimensional gel electrophoresis 24 hours after controlled cortical impact in postnatal day 17 rats. J Neurotrauma 19: 715–740
7. Kastin AJ, Pan W, Maness LM, Banks WA (1999) Peptides crossing the blood-brain barrier: some unusual observations. Brain Res 848: 96–100
8. Klose J (1975) Protein mapping by combined isoelectric focusing and electrophoresis of mouse tissues. A novel approach to testing for induced point mutations in mammals. Humangenetik 26: 231–243
9. Maness LM, Kastin AJ, Banks WA (1998) Relative contributions of a CVO and the microvascular bed to delivery of blood-borne IL-1alpha to the brain. Am J Physiol 275: E207–212
10. Mannes AJ, Martin BM, Yang HY, Keller JM, Lewin S, Gaiser RR, Iadarola MJ (2003) Cystatin C as a cerebrospinal fluid biomarker for pain in humans. Pain 102: 251–256
11. O'Farrell PH (1975) High resolution two-dimensional electrophoresis of proteins. J Biol Chem 250
12. Reiber H (2003) Proteins in cerebrospinal fluid and blood: barriers, CSF flow rate and source-related dynamics. Restor Neurol Neurosci 21: 79–96
13. Schrader M, Schulz-Knappe P (2001) Peptidomics technologies for human body fluids. Trends Biotechnol 19: S55–60
14. Schreiber G, Aldred AR (1991) Origin and funcion of proteins in the cerebrospinal fluid. In: Felgenhauer K, Holzgrefe M, Prange HW (eds) CNS barriers and modern CSF diagnostics. VCH, Weinheim, pp 229–246

15. Schuhmann MU, Heine G, Skardelly M, Tammen H, Appel A, Brinker T (2002) Differential Peptide display and analysis in CSF and plasma following traumatic brain injury. J Neurotrauma 19: 1305
16. Schulz-Knappe P, Zucht HD, Heine G, Jurgens M, Hess R, Schrader M (2001) Peptidomics: the comprehensive analysis of peptides in complex biological mixtures. Comb Chem High Throughput Screen 4: 207–217
17. Sickmann A, Dormeyer W, Wortelkamp S, Woitalla D, Kuhn W, Meyer HE (2002) Towards a high resolution separation of human cerebrospinal fluid. J Chromatogr B 770: 167–196

Correspondence: Martin U. Schuhmann, Klinik und Poliklinik fuer Neurochirurgie, Universitaet Leipzig, Liebigstr. 20, 04103 Leipzig, Germany. e-mail: mus@uniklinik-leipzig.de

Acta Neurochir (2005) [Suppl] 95: 471–474

Circulating nucleic acid analysis: diagnostic applications for acute pathologies

R. W. K. Chiu[1], **T. H. Rainer**[2], and **Y. M. D. Lo**[1]

[1] Department of Chemical Pathology, The Chinese University of Hong Kong, Hong Kong, China
[2] Department of Accident and Emergency Medicine Academic Unit, The Chinese University of Hong Kong, Hong Kong, China

Summary

Much research interest has been shown in recent years for the development of molecular diagnostic strategies based on the analysis of DNA/RNA molecules that are present in the plasma/serum of human subjects. Reported applications include the diagnosis, prognostication or monitoring of malignancies and pregnancy-associated complications. While researchers have speculated that cell death is a potential mechanism that leads to the release of DNA/RNA into the circulation, studies have demonstrated that indeed increased amounts of plasma DNA and RNA could be detected in patients sustaining acute traumatic injuries. The degree of plasma DNA elevation correlated with the severity of injury. Similarly, plasma DNA concentrations have been shown to correlate with indices of prognostic significance in patients with acute stroke. It is expected that new diagnostic markers based on plasma RNA detection could be developed for the evaluation of acute pathologies.

Keywords: Circulating nucleic acids; plasma DNA; plasma RNA; acute medicine; stroke; cerebrovascular accident; nasopharyngeal cancer; molecular diagnostics.

Introduction

In recent years, there has been increasing interest in the development of molecular diagnostics based on the detection of extracellular DNA and RNA molecules that are found to circulate in the plasma/serum of human subjects [2, 6]. Numerous applications have been developed based on the realization that disease-associated nucleic acid signatures are found in the plasma of subjects suffering from certain pathologies. For example, tumor-associated microsatellite instability, point mutations and methylation changes have been detectable in plasma of subjects harboring tumors with such molecular changes [1, 4, 22, 23]. On the other hand, fetal-derived nucleic acid molecules have been shown to be released into maternal circulation during pregnancy and this phenomenon has formed the basis for the development of noninvasive prenatal diagnostic strategies [5, 17, 18]. Donor-derived DNA has also been shown to be detectable in the plasma of transplant recipients and thus offers the possibility of molecular monitoring of transplant rejection [21]. Hence, much potential has been shown for the development of blood-based tests for the diagnosis, monitoring and prognostication of diseases based on circulating nucleic acid analysis.

The potential clinical utility of circulating nucleic acids analysis is exemplified by the application of plasma Epstein-Barr virus (EBV) DNA detection in subjects with nasopharyngeal carcinoma (NPC) [16]. Plasma EBV DNA has been reported to be detectable in 96% of NPC patients but only 7% of healthy volunteers [16]. Though detectable in some normal individuals, the median plasma EBV DNA concentration in NPC patients is several orders of magnitude higher than that of controls [16] and correlates with tumor staging [14]. More importantly, plasma EBV DNA estimation after radiotherapy has been shown to be predictive of risk of tumor recurrence [15]. EBV DNA becomes undetectable in plasma of patients with complete tumor regression but remains detectable in patients who have clinical evidence of tumor recurrence [15]. Recently, a study has further demonstrated the potential value of serial monitoring of plasma EBV DNA concentration during neoadjuvant chemotherapy with radiotherapy for the treatment of NPC [3]. Thus, plasma EBV DNA measurement is a valuable tool for the diagnosis, prognostication and treatment monitoring of NPC [13]. It further serves as a prime example which demonstrates the clinical potential of molecular diagnostics based on circulating nucleic acid analysis.

When studying the rate of change of plasma EBV DNA levels in NPC patients undergoing radiotherapy, it is interesting to note the presence of an initial rise prior to the expected decline in EBV DNA concentrations [19]. It has been postulated that the observation is a result of EBV DNA liberation due to therapy-induced cancer cell death. In fact, other investigators have also suggested that cell death is a possible cause of circulating nucleic acid release [7–9]. Consequently, we hypothesize that conditions with significant cellular damage, such as blunt traumatic injuries, could also be associated with elevated levels of circulating nucleic acids.

Circulating DNA as a marker of cell death

We have pursued the investigation by recruiting patients who have sustained acute traumatic blunt injuries [20]. Blood specimens were collected on admission and the plasma DNA concentration as reflected by the plasma β-*globin* DNA concentration was compared between patients and control subjects. Plasma DNA concentration was indeed found to be significantly elevated in patients compared to controls [20]. Furthermore, the degree of elevation was seen to correlate positively with the injury severity score [20]. More specifically, significant correlation was observed between DNA concentrations in the plasma of trauma patients and the abbreviated injury score of the head and neck region, thorax and abdomen [20]. Remarkably, plasma DNA concentrations have also been found to be significantly higher among trauma patients who subsequently developed acute lung injury, acute respiratory distress syndrome or died than those who did not suffer those complications. Serial monitoring of trauma patients has revealed that plasma DNA was increased within 20 minutes of sustaining injury [11]. Plasma DNA has been found to be persistently elevated in patients who developed multiple organ dysfunction syndrome or required intensive care admission [11]. Consequent to these observations, we have further evaluated if an algorithm based on the admission plasma DNA concentration could be developed for the early prediction of post-traumatic organ failure and multiple organ dysfunction syndrome [28]. We have demonstrated that plasma DNA cut-off values allowing the sensitive and specific prediction of these two complications can be determined effectively [28].

Circulating DNA as a marker for acute stroke

Having demonstrated a potential relationship between tissue injury and elevated plasma DNA concentrations, we further investigated if similar relationships could be demonstrated for tissue damage of the central nervous system, namely acute stroke [29]. We hypothesized that disruption of the blood-brain barrier as a result of hemorrhagic or ischemic stroke would facilitate the release of nucleic acids into the peripheral circulation. Plasma DNA concentrations in blood taken at a median of 3 hours after symptom onset have been found to be five-fold higher in patients who subsequently died than those who survived at 6 months [29]. Furthermore, correlations are observed between plasma DNA concentrations and the Glasgow Coma Score, cerebral hemorrhage volume, post-stroke modified Rankin Score, or quality of life loss [29]. Hence, plasma DNA measurement is a potentially useful prognostic tool for the assessment of patients suffering acute stroke.

Future developments

Besides circulating DNA, elevated plasma RNA levels have also been documented in patients sustaining acute trauma [27]. In view of the inherent instability of RNA, the discovery of the presence of circulating RNA molecules in the plasma/serum of human subjects is rather surprising. Tsui *et al.* [31] has previously demonstrated that endogenous plasma RNA is surprisingly stable and contrasted markedly with free RNA molecules which have been shown to degrade immediately upon being spiked into plasma. The stability of endogenous circulating RNA could potentially be conferred by its association with subcellular particles. Ng *et al.* [26] demonstrated that a significant proportion of circulating RNA species in plasma is filterable. These data support the suggestion that plasma RNA exists in a particle-associated form in the circulation. Interestingly, by measuring plasma mRNA concentrations of a house-keeping gene, *glyceraldehyde-3-phosphate dehydrogenase (GAPDH)*, significantly higher levels were reported in the filtered plasma portions obtained from trauma patients as opposed to those from controls [27]. The effect was not observed in the unfiltered plasma portions, suggesting that traumatic injuries may lead to the predominant increase in the non-particle-associated, possibly free, fraction of plasma RNA. Furthermore, plasma RNA levels are found to be higher in patients sustaining severe injuries and those dying from their injuries [27].

In addition to *GAPDH*, investigators have reported the successful detection of tumor- and fetal-specific RNA species in the plasma of cancer patients [10, 12] and pregnant women [24], respectively. For example, placenta expressed genes are found to be readily detectable in the plasma of pregnant women [26, 30]. Furthermore, plasma RNA markers with relative disease-specificity have been developed by taking advantage of the aberrant RNA expression profiles of pathological tissues. One such example is the finding of elevated corticotropin releasing hormone mRNA concentrations in plasma of women whose pregnancy was complicated by preeclampsia [25]. Consequently, one could envisage that a similar approach could be adopted for the development of diagnostic tests that are not only sensitive to acute injuries but specific to particular organ systems based on the detection of tissue-specific RNA species.

Conclusions

Much potential has been shown for molecular diagnostics based on circulating nucleic acids analysis. Promising applications have been reported for the assessment of cancers and pregnancies. More recently, evidence has suggested that circulating nucleic acid measurement may be useful in the diagnosis or prognostication of acute pathologies, including blunt trauma and stroke. It is expected that disease-specific tests could potentially be developed by targeting RNA transcripts that are expressed specifically by certain tissues or pathological states. Therefore, one can be hopeful that new applications of circulating nucleic acids analysis, particularly in relation to acute pathologies, will be forthcoming.

Acknowledgments

THR and YMDL are supported by the Direct Grant Scheme (2040873) of the Chinese University of Hong Kong and an Earmarked Research Grant (CUHK4255/99M) from the Research Grants Council of the Hong Kong Special Administrative Region, China.

References

1. Anker P, Lefort F, Vasioukhin V, Lyautey J, Lederrey C, Chen XQ, Stroun M, Mulcahy HE, Farthing MJ (1997) K-ras mutations are found in DNA extracted from the plasma of patients with colorectal cancer. Gastroenterology 112: 1114–1120
2. Chan AKC, Chiu RWK, Lo YMD (2003) Cell-free nucleic acids in plasma, serum and urine: a new tool in molecular diagnosis. Ann Clin Biochem 40: 122–130
3. Chan AT, Ma BB, Lo YMD, Leung SF, Kwan WH, Hui EP, Mok TS, Kam M, Chan LS, Chiu SK, Yu KH, Cheung KY, Lai K, Lai M, Mo F, Yeo W, King A, Johnson PJ, Teo PM, Zee B (2004) Phase II study of neoadjuvant carboplatin and paclitaxel followed by radiotherapy and concurrent cisplatin in patients with locoregionally advanced nasopharyngeal carcinoma: therapeutic monitoring with plasma Epstein-Barr virus DNA. J Clin Oncol 22: 3053–3060
4. Chen XQ, Stroun M, Magnenat JL, Nicod LP, Kurt AM, Lyautey J, Lederrey C, Anker P (1996) Microsatellite alterations in plasma DNA of small cell lung cancer patients. Nat Med 2: 1033–1035
5. Chiu RWK, Lau TK, Leung TN, Chow KCK, Chui DHK, Lo YMD (2002) Prenatal exclusion of beta thalassaemia major by examination of maternal plasma. Lancet 360: 998–1000
6. Chiu RWK, Lo YMD (2004) The biology and diagnostic applications of fetal DNA and RNA in maternal plasma. Curr Top Dev Biol 61: 81–111
7. Fournie GJ, Courtin JP, Laval F, Chale JJ, Pourrat JP, Pujazon MC, Lauque D, Carles P (1995) Plasma DNA as a marker of cancerous cell death. Investigations in patients suffering from lung cancer and in nude mice bearing human tumours. Cancer Lett 91: 221–227
8. Giacona MB, Ruben GC, Iczkowski KA, Roos TB, Porter DM, Sorenson GD (1998) Cell-free DNA in human blood plasma: length measurements in patients with pancreatic cancer and healthy controls. Pancreas 17: 89–97
9. Jahr S, Hentze H, Englisch S, Hardt D, Fackelmayer FO, Hesch RD, Knippers R (2001) DNA fragments in the blood plasma of cancer patients: quantitations and evidence for their origin from apoptotic and necrotic cells. Cancer Res 61: 1659–1665
10. Kopreski MS, Benko FA, Kwak LW, Gocke CD (1999) Detection of tumor messenger RNA in the serum of patients with malignant melanoma. Clin Cancer Res 5: 1961–1965
11. Lam NY, Rainer TH, Chan LYS, Joynt GM, Lo YMD (2003) Time course of early and late changes in plasma DNA in trauma patients. Clin Chem 49: 1286–1291
12. Lo KW, Lo YMD, Leung SF, Tsang YS, Chan LYS, Johnson PJ, Hjelm NM, Lee JC, Huang DP (1999) Analysis of cell-free Epstein-Barr virus associated RNA in the plasma of patients with nasopharyngeal carcinoma. Clin Chem 45: 1292–1294
13. Lo YMD (2001) Quantitative analysis of Epstein-Barr virus DNA in plasma and serum: applications to tumor detection and monitoring. Ann N Y Acad Sci 945: 68–72
14. Lo YMD, Chan ATC, Chan LYS, Leung SF, Lam CW, Huang DP, Johnson PJ (2000) Molecular prognostication of nasopharyngeal carcinoma by quantitative analysis of circulating Epstein-Barr virus DNA. Cancer Res 60: 6878–6881
15. Lo YMD, Chan LYS, Chan ATC, Leung SF, Lo KW, Zhang J, Lee JC, Hjelm NM, Johnson PJ, Huang DP (1999) Quantitative and temporal correlation between circulating cell-free Epstein-Barr virus DNA and tumor recurrence in nasopharyngeal carcinoma. Cancer Res 59: 5452–5455
16. Lo YMD, Chan LYS, Lo KW, Leung SF, Zhang J, Chan ATC, Lee JC, Hjelm NM, Johnson PJ, Huang DP (1999) Quantitative analysis of cell-free Epstein-Barr virus DNA in plasma of patients with nasopharyngeal carcinoma. Cancer Res 59: 1188–1191
17. Lo YMD, Corbetta N, Chamberlain PF, Rai V, Sargent IL, Redman CW, Wainscoat JS (1997) Presence of fetal DNA in maternal plasma and serum. Lancet 350: 485–487
18. Lo YMD, Hjelm NM, Fidler C, Sargent IL, Murphy MF,

Chamberlain PF, Poon PM, Redman CW, Wainscoat JS (1998) Prenatal diagnosis of fetal RhD status by molecular analysis of maternal plasma. N Engl J Med 339: 1734–1738
19. Lo YMD, Leung SF, Chan LYS, Chan ATC, Lo KW, Johnson PJ, Huang DP (2000) Kinetics of plasma Epstein-Barr virus DNA during radiation therapy for nasopharyngeal carcinoma. Cancer Res 60: 2351–2355
20. Lo YMD, Rainer TH, Chan LYS, Hjelm NM, Cocks RA (2000) Plasma DNA as a prognostic marker in trauma patients. Clin Chem 46: 319–323
21. Lo YMD, Tein MS, Pang CC, Yeung CK, Tong KL, Hjelm NM (1998) Presence of donor-specific DNA in plasma of kidney and liver-transplant recipients [letter]. Lancet 351: 1329–1330
22. Lo YMD, Wong IH, Zhang J, Tein MS, Ng MH, Hjelm NM (1999) Quantitative analysis of aberrant p16 methylation using real-time quantitative methylation-specific polymerase chain reaction. Cancer Res 59: 3899–3903
23. Nawroz H, Koch W, Anker P, Stroun M, Sidransky D (1996) Microsatellite alterations in serum DNA of head and neck cancer patients. Nat Med 2: 1035–1037
24. Ng EKO, El-Sheikhah A, Chiu RWK, Chan KCA, Hogg M, Bindra R, Leung TN, Lau TK, Nicolaides KH, Lo YMD (2004) Evaluation of human chorionic gonadotropin beta-subunit mRNA concentrations in maternal serum in aneuploid pregnancies: a feasibility study. Clin Chem 50: 1055–1057
25. Ng EKO, Leung TN, Tsui NBY, Lau TK, Panesar NS, Chiu RWK, Lo YMD (2003) The concentration of circulating corticotropin-releasing hormone mRNA in maternal plasma is increased in preeclampsia. Clin Chem 49: 727–731
26. Ng EKO, Tsui NBY, Lam NY, Chiu RWK, Yu SC, Wong SC, Lo ES, Rainer TH, Johnson PJ, Lo YMD (2002) Presence of filterable and nonfilterable mRNA in the plasma of cancer patients and healthy individuals. Clin Chem 48: 1212–1217
27. Rainer TH, Lam NYL, Tsui NBY, Ng EKO, Chiu RWK, Joynt GM, Lo YMD (2004) Effects of filtration on glyceraldehyde-3-phosphate dehydrogenase mRNA in the plasma of trauma patients and healthy individuals. Clin Chem 50: 206–208
28. Rainer TH, Lo YMD, Chan LYS, Lam NYL, Lit LCW, Cocks RA (2001) Derivation of a prediction rule for posttraumatic organ failure using plasma DNA and other variables. Ann N Y Acad Sci 945: 211–220
29. Rainer TH, Wong LK, Lam W, Yuen E, Lam NYL, Metreweli C, Lo YM (2003) Prognostic use of circulating plasma nucleic acid concentrations in patients with acute stroke. Clin Chem 49: 562–569
30. Tsui NBY, Chim SSC, Chiu RWK, Lau TK, Ng EKO, Leung TN, Tong YK, Chan KCA, Lo YMD (2004) Systematic microarray based identification of placental mRNA in maternal plasma: towards non-invasive prenatal gene expression profiling. J Med Genet 41: 461–467
31. Tsui NBY, Ng EKO, Lo YMD (2002) Stability of endogenous and added RNA in blood specimens, serum, and plasma. Clin Chem 48: 1647–1653

Correspondence: Rossa W. K. Chiu, Department of Chemical Pathology, The Chinese University of Hong Kong, Prince of Wales Hospital, Shatin, New Territories, Hong Kong, China. e-mail: rossachiu@cuhk.edu.hk

Acta Neurochir (2005) [Suppl] 95: 475–480

The Lund concept: is this logical?

C.-H. Nordström

Department of Neurosurgery, Lund University Hospital, Lund, Sweden

Summary

The optimal therapy of sustained increase in intracranial pressure (ICP) is still controversial. The "Lund concept" is based on the physiological volume regulation of the intracranial compartments. In addition to its other functions the blood-brain barrier (BBB) is the most important regulator of brain volume. Water exchange across the intact BBB is counteracted by the low permeability to crystalloids (mainly Na^+ and Cl^-) combined with the high osmotic pressure (5,700 mmHg) on both sides of the BBB. If the BBB is disrupted transcapillary water transport will be determined by the differences in hydrostatic and colloid osmotic pressure between the intra- and extracapillary compartments. Under pathological conditions pressure autoregulation of cerebral blood flow is often impaired and intracapillary hydrostatic pressure will depend on variations in systemic blood pressure.

The "Lund concept" can be summarized in four paragraphs: I. Reduction of stress response and cerebral energy metabolism; II. Reduction of capillary hydrostatic pressure; III. Maintenance of colloid osmotic pressure and control of fluid balance; IV. Reduction of cerebral blood volume. The efficacy of the treatment protocol has been evaluated in experimental and clinical studies regarding the physiological and biochemical (utilizing intracerebral microdialysis) effects. The clinical experiences have been favourable.

Keywords: Brain oedema; blood-brain barrier; microdialysis.

Introduction

Except for decompressive craniectomy all treatments for increased intracranial pressure (ICP) are directed towards decreasing the volume of one or more of the intracranial components:

$$V_{intracran} = V_{blood} + V_{brain} + V_{CSF} + V_{mass\ lesion} \quad \text{(Eq. 1)}$$

Surgical treatments include evacuation of mass lesions ($V_{mass\ lesion}$) and, in some cases, drainage of CSF (V_{CSF}). The importance of early and adequate surgical evacuation of focal mass lesions is well documented and will not be discussed further in this presentation. The regulation of the remaining two volumes (V_{blood} and V_{brain}) is the focus for (non-surgical) intensive care. However, the surgical and non-surgical treatments are closely connected since all surgical therapies have one consequence in common: they all reduce tissue interstitial pressure. As discussed below, in a situation with increased permeability of the blood-brain barrier (BBB), a decrease in tissue pressure will lead to increased transcapillary water filtration. The consequences are well known to all neurosurgeons: the gradual increase in ICP almost always observed after evacuation of focal lesions; the collapse of the ventricles often induced by ventricular drainage; the bulging of cerebral tissue through the bone defect after craniectomy. The rapid change in volume obtained by surgical treatment should accordingly always be combined with a non-surgical therapy aiming primarily at a slow and lasting reduction in brain water content.

Blood-brain barrier permeability

Like other organs the volume regulation of the brain is primarily determined by mechanisms controlling the water exchange across the capillaries. Regarding these mechanisms the brain differs from other organs in its highly sophisticated capillary membrane function – the BBB. In addition to its other physiological functions the BBB is the most important regulator of cerebral volume [8].

The flux of water across a membrane or microvascular bed (J_v) is described by the equation

$$J_v = L_p \times A \times [\Delta P - \Sigma\sigma_s \times \Delta\Pi_s] \quad \text{(Eq. 2)}$$

where L_p is hydraulic conductivity, A is the surface area available for fluid exchange, ΔP is the transcapillary hydrostatic pressure difference, $\Delta\Pi_s$ the transcapillary osmotic pressure difference and σ_s the reflection coefficient of each solute (s) of the system. The transcapillary water transport is accordingly defined by the difference in hydrostatic pressure (ΔP), the osmotic pressure gradient ($\Delta \Pi_s$), the endothelial component which will determine which solutes are reflected and will contribute to the osmotic pressure gradient (σ_s), and the hydraulic conductance of the capillary wall ($L_p \times A$). The two major solutes of biological fluids (Na^+ and Cl^-) have a blood-brain barrier reflection coefficient of 1.0. Water passing the BBB in any direction will thus be virtually devoid of crystalloids and an opposing osmotic gradient, which counteracts further fluid transport, will immediately be created.

According to eq. 2 the magnitude of the hydraulic conductance ($L_p \times A$) describes the rate by which the fluid is transferred across the capillary membrane whenever there is a driving force ($\Delta P - \Sigma\sigma \times \Delta\Pi$). Accordingly, the only way of inducing transcapillary filtration or absorption is to disturb the balance between the hydrostatic and osmotic forces across the capillary membrane. For two reasons the magnitude of the cerebral perfusion pressure (CPP) is of less importance for brain volume under normal conditions. Firstly, intracapillary pressure and cerebral blood flow are physiologically tightly autoregulated and variations in systemic blood pressure are generally not transmitted to cerebral capillaries. Secondly, transcapillary fluid exchange is effectively counteracted by the low permeability to crystalloids combined with the high osmotic pressure of 5,700 mmHg on both sides of the BBB [8]. This contrasts to most other capillary regions where the osmotic pressure force is mainly derived from the difference between plasma and interstitial colloid osmotic pressure approximately balancing the transcapillary hydrostatic pressure (20–25 mmHg). Under pathophysiological conditions the regulation of brain volume may be completely altered. In these conditions pressure autoregulation is often impaired and the BBB may have an increased permeability for crystalloids or even for large molecules. The regulation of brain volume will then depend on the balance between the transcapillary hydrostatic pressure and the transcapillary colloidal osmotic pressure.

Cerebral blood flow and cerebral blood volume

The tight regulation of cerebral blood flow (CBF) primarily secures a continuous and sufficient supply of glucose and oxygen to cover the high cerebral metabolism. Two secondary effects of CBF regulation have important consequences for the regulation of ICP: the intracapillary hydrostatic pressure and the intracerebral blood volume (mainly within the venous compartment). Several important principles within intracranial dynamics can be described in relation to the conventional physiological regulatory mechanisms of the cerebral blood flow: pressure autoregulation, CO_2-regulation, and metabolic regulation.

Pressure autoregulation implies that blood flow remains relatively constant despite variations in CPP within defined limits. The phenomenon is present in many organs and appears to be most prominent in the brain and in the kidneys. In physiological literature autoregulation is usually described as a mechanism, which serves the purpose of keeping intracapillary hydrostatic pressure constant (c.f. above).

A rapid reduction of intracranial blood volume is conventionally obtained by controlled hyperventilation via the CO_2-regulation of CBF. The effect is attained by a pH dependent constriction of precapillary resistance vessels. Prolonged hyperventilation is probably of limited value or may even be harmful since pronounced hyperventilation carries the risk of inducing focal ischemia. Further, the reduction of CBF and CBV during hyperventilation is transient in spite of preserved hypocapnia. Finally, due to impaired cerebrovascular CO_2-reactivity hyperventilation is often not effective in reducing CBV in the most severely brain injured patients [10, 21].

A lasting vasoconstriction and reduction of CBV can be achieved through reduction of cerebral energy metabolism e.g. by barbiturates [15]. This mechanism appears to be effective only in patients with preserved cerebrovascular CO_2-reactivity [15]. In addition, prolonged high-dose barbiturate treatment is associated with pulmonary, cardiovascular and other serious complications [22].

Since in the brain, like in other organs, about 70% of blood volume is located within the venous compartment it might be possible to induce a lasting reduction of CBV with a minimal effect on CBF by inducing a pharmacological constriction of the capacitance vessels. Dihydroergotamine (DHE) is known to primarily constrict venous capacitance vessels within the peripheral circulation and has also been shown to decrease ICP in patients with severe head injuries [9]. This effect is obtained although CBF remains unaffected or increases [4, 13]. DHE has also been shown to decrease ICP in patients with impaired CO_2-reactivity [4] and in vitro studies have shown that DHE has a more pronounced constrictor effect on isolated human cortical veins than arteries [14].

Pathophysiological aspects of disturbed brain volume regulation

No techniques are available for quantitative measurements of cerebral intracapillary hydrostatic pressure or the effects of increased hydrostatic pressure on water transport across the BBB. Some aspects may however be studied in a model system utilizing a denervated cat skeletal muscle enclosed in a plethysmograph [1, 9]. This model simulates the pathophysiological situation in a damaged brain since muscle capillaries are permeable for water and most solutes but much less permeable for proteins ($\sigma \approx 0.8$–0.9). Further, pressure autoregulation of blood flow is impaired in this experimental model.

The clinical implications of the data obtained from these experiments may be summarized as follows. If the injured brain has an impaired pressure autoregulation and a BBB with an increased permeability to crystalloids then transcapillary fluid exchange will be highly dependent on variations in systemic arterial blood pressure. Under conditions with increased BBB permeability to crystalloids a decrease in tissue pressure will cause a net transport of fluid from the capillaries into the interstitial space and, as discussed above, all surgical treatments of increased ICP are associated with a decrease in tissue pressure.

The Lund concept

The "Lund concept" is based on the physiological considerations above and can be summarized in four paragraphs.

Reduction of stress response and cerebral energy metabolism

Stress response is reduced by liberal use of sedatives (benzodiazepines) and analgesics (opioids). A further reduction of the stress response and catecholamine release is obtained by a continuous infusion of low-dose thiopental (0.5–$3\ \mathrm{mg{\cdot}kg^{-1}{\cdot}h^{-1}}$) and fentanyl ($2$–$5\ \mu\mathrm{g{\cdot}kg^{-1}{\cdot}h^{-1}}$) but also by treatment with the β_1-antagonist metoprolol and the α_2-agonist clonidine (see below). The dose of thiopental is kept low to avoid cardiac inhibition, pulmonary complications and other side effects [22].

Reduction of capillary hydrostatic pressure

Mean arterial blood pressure is reduced to the physiological level for the age of the individual patient with a combination of the β_1-antagonist metoprolol (0.2–$0.3\ \mathrm{mg{\cdot}kg^{-1}{\cdot}24\ h^{-1}}$ *i.v.*) and the α_2-agonist clonidine (0.4–$0.8\ \mu\mathrm{g{\cdot}kg^{-1}} \times 4$–$6$ *i.v.*) [2, 3, 9]. The antihypertensive treatment is initiated after evacuation of focal mass lesions when the patients are clearly normovolemic as obtained by red cell and albumin/plasma transfusions to normal albumin and haemoglobin values and to a normal central venous pressure. A CPP of 60–70 mmHg is considered optimal but, if necessary to control ICP, a transient decrease in CPP to 50 mmHg for adults and 40 mmHg for children is accepted. Thiopental also has a precapillary vasoconstrictor effect, which will contribute to lowering the intracapillary hydrostatic pressure.

Maintenance of colloid osmotic pressure and control of fluid balance

Red cell transfusions and albumin are given to achieve normal values (Hb/s 125–140 g/l, alb/s ≈ 40 g/l) to ensure normovolemia and to optimise oxygen supply. The albumin/plasma/blood transfusions also serve the purpose of obtaining a normal colloid osmotic pressure favouring transcapillary absorption. A balanced or moderately negative fluid balance is a part of the treatment protocol and is achieved by diuretics (furosemide) and albumin infusion. All patients are given a low calorie enteral nutrition (max energy supply 15–$20\ \mathrm{kcal{\cdot}kg^{-1}{\cdot}24\ h^{-1}}$).

Reduction of cerebral blood volume

Intracranial blood volume may be reduced both on the arterial side with thiopental [15] and on the venous side with DHE [4, 9, 13, 14]. DHE is only given when other treatments are insufficient and always at the lowest dose necessary to reduce intracranial pressure. Since the therapy above (paragraph I–III) is usually successful DHE is rarely used. When given it should not be administered for more than five days in order to minimize the risks of compromising peripheral circulation, in particular in patients with fractures of the extremities or renal insufficiency. The maximum doses of DHE used are: $0.8\ \mu\mathrm{g{\cdot}kg^{-1}{\cdot}h^{-1}}$ day one, $0.6\ \mu\mathrm{g{\cdot}kg^{-1}{\cdot}h^{-1}}$ day two, $0.4\ \mu\mathrm{g{\cdot}kg^{-1}{\cdot}h^{-1}}$ day three, $0.2\ \mu\mathrm{g{\cdot}kg^{-1}{\cdot}h^{-1}}$ day four, and $0.1\ \mu\mathrm{g{\cdot}kg^{-1}{\cdot}h^{-1}}$ day five.

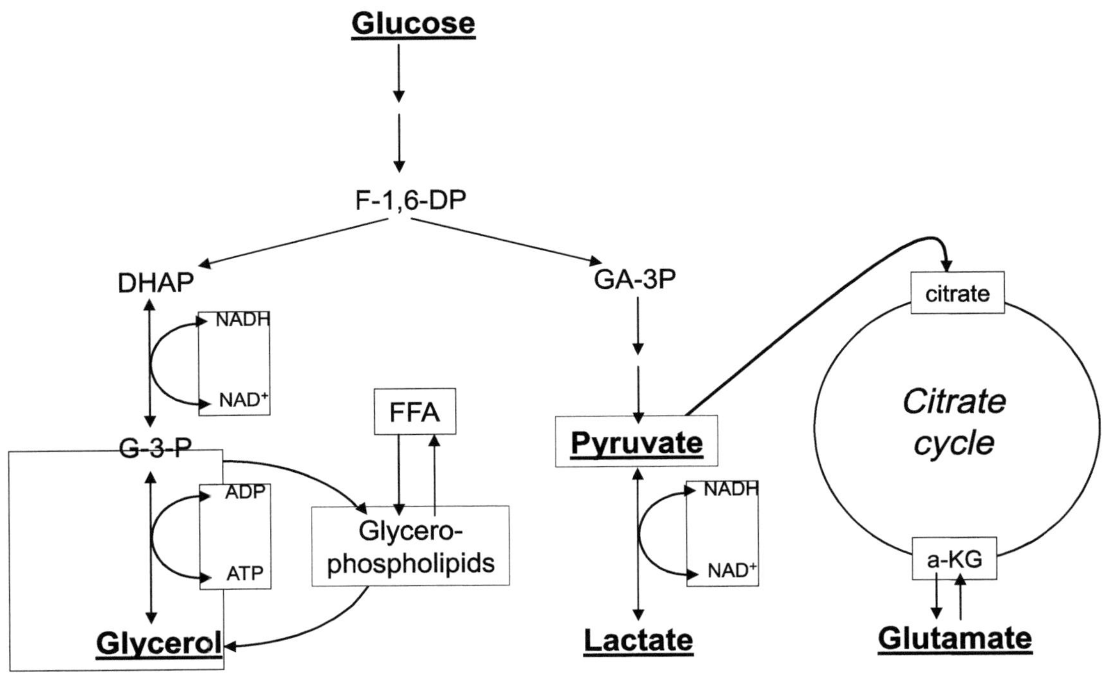

Normal human brain (perfusion rate 0.3 µl/min)	Glucose	Lactate	Pyruvate	La/py ratio	Glutamate	Glycerol
	1.7 ± 0.9 mmol/L	2.9 ± 0.9 mmol/L	166 ± 47 µmol/L	23 ± 4	16 ± 16 µmol/L	80 ± 40 µmol/L

Fig. 1. Simplified diagram of intermediary metabolism of the glycolytic chain and its relation to the formation of glycerol and glycerophospholipids and to the citric acid :ycle. Abbreviations: Fructose-1,6-diposphate (*F-1,6-DP*); Dihydroxyacetone-phosphate (*DHAP*); Glyceraldehyde-3-phosphate (*GA-3P*); Glycerol-3-phosphate (*G-3-P*); Free fatty acids (*FFA*); α-ketoglutarate (α-*KG*). Underlined metabolites are measured bedside with enzymatic techniques. References levels of the various metabolites for normal human brain obtained from [18]

Cerebral energy metabolism – relation to cerebral perfusion pressure

A reduction of CPP might jeopardize cerebral energy metabolism in patients with increased ICP especially in vulnerable regions such as the penumbra zones surrounding focal mass lesions [7]. The intracerebral microdialysis technique permits bedside analyses of compounds reflecting energy metabolism (glucose, pyruvate, lactate) and indicators of excessive concentration of excitatory transmitters (glutamate) as well as cell degradation (glycerol) (Fig. 1). The levels obtained during neuro intensive care can be compared to reference levels obtained in the normal human brain [18] and in patients with manifest cerebral ischemia [24]. By inserting multiple intracerebral catheters during open surgery it is possible to choose areas of interest *i.e.* the zones surrounding an evacuated focal lesion [7].

Figure 2 summarizes resent data, which support the view that in patients treated according to the "Lund concept" the lactate/pyruvate ratio (cytoplasmatic redox state) is usually not affected in the vulnerable penumbra zone until CPP below 50 mmHg [16]. Since cerebral biochemistry is monitored and displayed bedside therapeutic interventions are instituted if in an individual patient CPP is considered to be too low.

Clinical results

Ideally all clinical therapies should be based on randomised controlled studies. However, there are virtually no such studies to support any specific treatment for increased ICP [23]. The only randomised clinical trial that has compared the consequences of targeting different levels of CPP failed to show a long-term benefit of increasing CPP above 70 mmHg [19]. In a recent

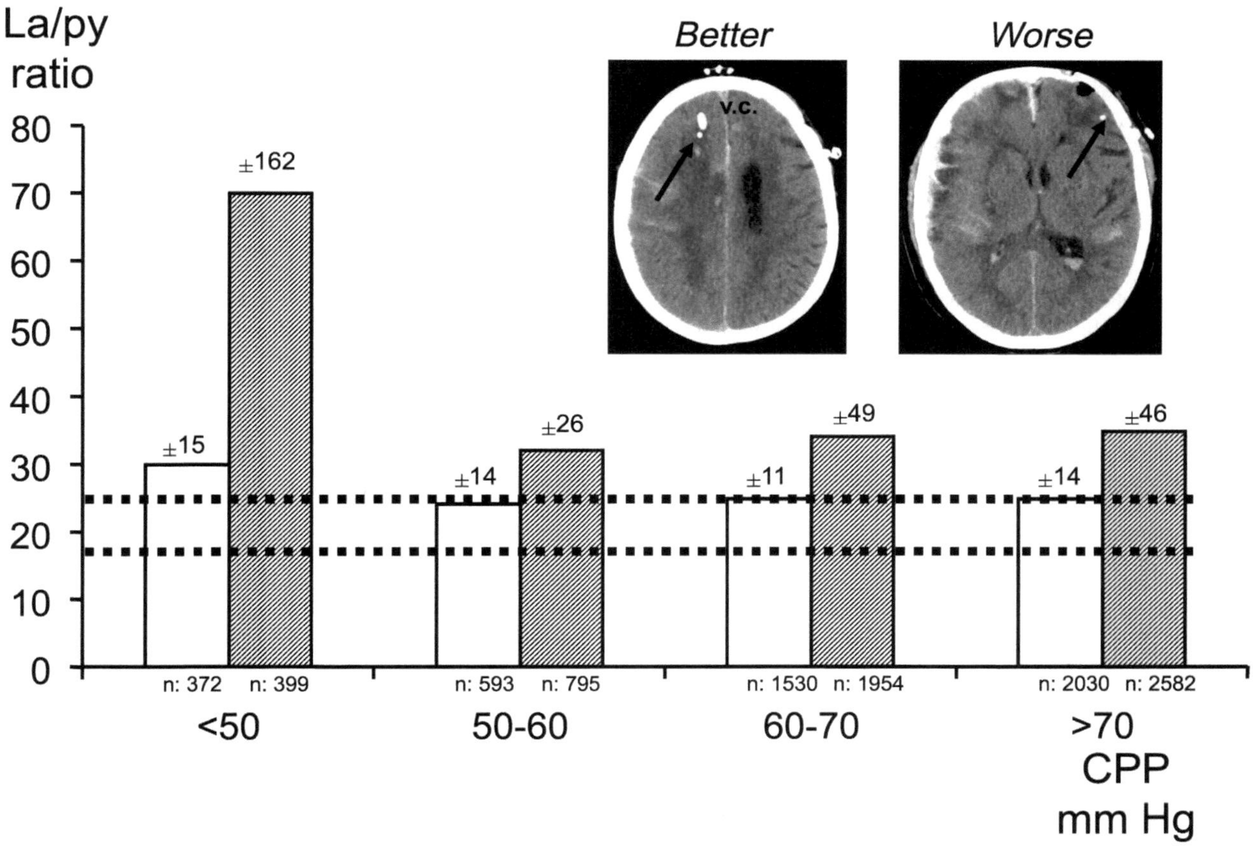

Fig. 2. Mean levels ± S.D. for the lactate/pyruvate (*la/py*) ratio (n: 7,704 measurements) in the "better" (white) and "worse" (striped) positions in relation to four ranges of CPP. The positioning of the microdialysis catheters in the "worse" (penumbra; surrounding the evacuated haemorrhagic contusion) and the "better" (close to the ventricular catheter – v.c.) areas are shown (arrows). The broken lines indicate the range (mean ± S.D.; 23 ± 4) in normal human brain during wakefulness. Data from Nordström *et al.* [16]

review on the management of cerebral perfusion pressure after head trauma it was concluded that it seemed likely that a CPP of 60 mmHg provided adequate perfusion for most patients with severe traumatic brain injuries [20]. Further, pharmacologically induced hypertension was recently shown to cause a further increase in ICP [17].

Clinical outcome of volume-targeted therapy according to the "Lund-concept" has been reported from four Swedish neurosurgical centres [5, 6, 11, 12]. All four studies have shown a remarkably low mortality. In the original study, which included a selected group of patients with very severe traumatic brain lesion and ICP above 25 mmHg in spite of conventional treatment, the decrease in mortality was from 47% to 8% [5]. The significant decrease in mortality was associated with significant increases in the groups of "good recovery" and "moderate disability" but did not increase the number of patients in the groups of "severe disability" or "vegetative state".

References

1. Asgeirsson B, Grände PO (1994) Effects of arterial and venous pressure alterations on transcapillary fluid exchange during raised tissue pressure. Intensive Care Med 20: 567–572
2. Asgeirsson B, Grände PO, Nordström CH (1994) A new therapy of post-trauma brain oedema based on haemodynamic principles for brain volume regulation. Intensive Care Med 20: 260–267
3. Asgeirsson B, Grände PO, Nordström CH, Berntman L, Messeter K, Ryding E (1995) Effects of hypotensive treatment with α_2-agonist and β_1-antagonist on cerebral hemodynamics in severe head injury. Acta Anasthesiol Scand 39: 347–351
4. Asgeirsson B, Grände PO, Nordström CH, Messeter K, Sjöholm A (1995) Cerebral hemodynamic effects of dihydroergotamine in patients with intracranial hypertension after severe head injury. Acta Anaesthesiol Scand 39: 922–930
5. Eker C, Asgeirsson B, Grände PO, Schalén W, Nordström C-H

(1998) Improved outcome after severe head injury with a new therapy based on principles for brain volume regulation and improved microcirculation. Critical Care Medicine 26: 1881–1886
6. Elf K, Nilsson P, Enblad P (2002) Outcome after traumatic brain injury improved by an organized secondary insult program and standardized neurointensive care. Crit Care Med 30: 2129–2134
7. Engström M, Polito A, Reinstrup P, Romner B, Ryding E, Ungerstedt U, Nordström CH (2005) Intracerebral microdialysis in clinical routine – the importance of catheter location. J Neurosurg In press
8. Fenstermacher JD (1984) Volume regulation of the central nervous system. In: Staub NC, Taylor AE (eds) Edema. Raven Press, New York, pp 383–404
9. Grände PO, Asgeirsson B, Nordström CH (2002) Volume targeted therapy of increased intracranial pressure: the Lund concept unifies surgical and non-surgical treatments. Acta Anaesthesiol Scand 46: 929–941
10. Messeter K, Nordström CH, Sundbärg G, Algotsson L, Ryding E (1986) Cerebral hemodynamics in patients with acute severe head trauma. J Neurosurg 64: 231–237
11. Naredi S, Edén E, Zäll S, Stephensen H, Rydenhag B (1998) A standardized neurosurgical/neurointensive therapy directed toward vasogenic edema after severe traumatic brain injury: clinical results. Intensive Care Med 24: 446–451
12. Naredi S, Olivecrona M, Lindgren A, Östlund AL, Grände PO, Koskinen LOD (2001) An outcome study of severe traumatic head injury using the "Lund therapy" with low-dose prostacyclin. Acta Anaesthesiol Scand 45: 401–405
13. Nilsson F, Messeter K, Grände PO, Rosén I, Ryding E, Nordström CH (1995) Effects of dihydroergotamine on cerebral circulation during experimental intracranial hypertension. Acta Anaesthesiol Scand 39: 916–921
14. Nilsson F, Nilsson T, Edvinsson L, Björkman S, Nordström CH (1997) Effects of dihydroergotamine and sumatriptan on isolated human cerebral and peripheral arteries and veins. Acta Anaesthesiol Scand 41: 1257–1262
15. Nordström CH, Messeter K, Sundbärg G, Schalén W, Werner M, Ryding E (1988) Cerebral blood flow, vasoreactivity, and oxygen consumption during barbiturate therapy in severe traumatic brain lesions. J Neurosurg 68: 424–431
16. Nordström CH, Reinstrup P, Xu W, Gärdenfors A, Ungerstedt U (2003) Assessment of the lower limit for cerebral perfusion pressure in severe head injuries by bedside monitoring of regional energy metabolism. Anesthesiology 98: 809–814
17. Oertel M, Kelly DF, Lee JH, McArthur DL, Glenn TC, Vespa P, Boscardin WJ, Hovda DA, Martin NA (2002) Efficacy of hyperventilation, blood pressure elevation, and metabolic suppression therapy in controlling intracranial pressure after head injury. J Neurosurg 97: 1045–1053
18. Reinstrup P, Ståhl N, Hallström Å, Mellergård P, Uski T, Ungerstedt U, Nordström CH (2000) Intracerebral microdialysis in clinical practice. Normal values and variations during anaesthesia and neurosurgical operations. Neurosurgery 47: 701–710
19. Robertson CS, Valadka AB, Hannay J, Contant CF, Gopinah SP, Cormio M, Uzura M, Grossman RG (1999) Prevention of secondary ischemic insults after severe head injury. Crit Care Med 27: 2086–2095
20. Robertson CS (2001) Management of cerebral perfusion pressure after traumatic brain injury. Anesthesiology 95: 1513–1517
21. Schalén W, Messeter K, Nordström CH (1991) Cerebral vasoreactivity and the prediction of outcome in severe traumatic brain lesions. Acta Anaesthesiol Scand 35: 113–122
22. Schalén W, Messeter K, Nordström CH (1992) Complications and side effects during thiopentone therapy in patients with severe head injuries. Acta Anaesthesiol Scand 36: 369–377
23. Slavic RS, Rhoney DH (2000) Pharmacological management of severe traumatic brain injury: an evidence-based review. J Inf Pharmacother, www.informedpharmacotherapy.com
24. Ståhl N, Mellergård P, Hallström Å, Ungerstedt U, Nordström CH (2001) Intracerebral microdialysis and bedside biochemical analysis in patients with fatal traumatic brain lesions. Acta Anaesthesiol Scand 45: 977–985

Correspondence: Carl-Henrik Nordström, Department of Neurosurgery, University Hospital, S-221 85 Lund, Sweden. e-mail: carl-henrik.nordstrom@neurokir.lu.se

Author Index

Index of Keywords

Springer and the Environment

MIX
Papier aus verantwortungsvollen Quellen
Paper from responsible sources
FSC® C105338

If you have any concerns about our products,
you can contact us on
ProductSafety@springernature.com

In case Publisher is established outside the EU,
the EU authorized representative is:
Springer Nature Customer Service Center GmbH
Europaplatz 3, 69115 Heidelberg, Germany

Printed by Libri Plureos GmbH
in Hamburg, Germany